Springer Series in Operations Research

Editors:
Peter W. Glynn   Stephen M. Robinson

Springer
New York
Berlin
Heidelberg
Barcelona
Hong Kong
London
Milan
Paris
Singapore
Tokyo

Springer Series in Operations Research

Altiok: Performance Analysis of Manufacturing Systems
Birge and Louveaux: Introduction to Stochastic Programming
Bonnans and Shapiro: Perturbation Analysis of Optimization Problems
Bramel and Simchi-Levi: The Logic of Logistics: Theory, Algorithms, and Applications for Logistics Management
Dantzig and Thapa: Linear Programming 1: Introduction
Drezner (Editor): Facility Location: A Survey of Applications and Methods
Fishman: Discrete-Event Simulation: Modeling, Programming, and Analysis
Fishman: Monte Carlo: Concepts, Algorithms, and Applications
Nocedal and Wright: Numerical Optimization
Olson: Decision Aids for Selection Problems
Whitt: Stochastic-Process Limits: An Introduction to Stochastic-Process Limits and Their Application to Queues
Yao (Editor): Stochastic Modeling and Analysis of Manufacturing Systems

Ward Whitt

Stochastic-Process Limits

An Introduction to
Stochastic-Process Limits and
Their Application to Queues

With 68 Illustrations

Ward Whitt
AT&T Labs–Research
Shannon Laboratory
180 Park Avenue
Florham Park, NJ 07932-0971
USA
wow@research.att.com

Series Editors:
Peter W. Glynn
Department of Management Science
   and Engineering
Terman Engineering Center
Stanford University
Stanford, CA 94305-4026
USA

Stephen M. Robinson
Department of Industrial Engineering
University of Wisconsin–Madison
1513 University Avenue
Madison, WI 53706-1572
USA

Library of Congress Cataloging-in-Publication Data
Whitt, Ward.
Stochastic-process limits: an introduction to stochastic-process limits and
their application to queues/Ward Whitt.
    p. cm. — (Springer series in operations research)
  Includes bibliographical references and index.
  ISBN 0-387-95358-2 (alk. paper)
  1. Queueing theory. 2. Stochastic processes. I. Title. II. Series.
  QA274.8. W45 2001
  519.8'2—dc21
                                                2001053053

Printed on acid-free paper.

Copyright © 2002 AT&T. All rights reserved.
This work may not be translated or copied in whole or in part without the written permission of the publisher (Springer-Verlag New York, Inc., 175 Fifth Avenue, New York, NY 10010, USA), except for brief excerpts in connection with reviews or scholarly analysis. Use in connection with any form of information storage and retrieval, electronic adaptation, computer software, or by similar or dissimilar methodology now known or hereafter developed is forbidden.
The use of general descriptive names, trade names, trademarks, etc., in this publication, even if the former are not especially identified, is not to be taken as a sign that such names, as understood by the Trade Marks and Merchandise Marks Act, may accordingly be used freely by anyone.

Production managed by Frank McGuckin; manufacturing supervised by Joe Quatela.
Camera-ready copy prepared from the author's LaTeX2e files using Springer's macros.
Printed and bound by Hamilton Printing Co., Rensselaer, NY.
Printed in the United States of America.

9 8 7 6 5 4 3 2 1

ISBN 0-387-95358-2   SPIN 10850716

Springer-Verlag New York Berlin Heidelberg
A member of BertelsmannSpringer Science+Business Media GmbH

I dedicate this book:

To my parents,
Millicent Whitt (1911–1996)
and Sidney Whitt (1908–   ),
who inspired my quest;

To the renowned probabilist,
Anatolii Skorohod (1930–   ),
who also inspired my quest;

To Albert Greenberg, Hamid Ahmadi
and the other good people of AT&T (1885–   ),
who supported my quest;

To my treasured mathematical typist,
Susan Pope,
who also supported my quest;

To Joe Abate, Arthur Berger, Gagan Choudhury,
Peter Glynn, Bill Massey and
my other treasured research collaborators,
who shared my quest;

To my beloved family members –
Joanna, Andrew, Daniel and Benjamin –
who, inevitably, also shared my quest;

And to any bold enough to venture here;
Godspeed on your quest.

> Ward Whitt
> The Red Rink
> Bridgewater, New Jersey
> April 8, 2001

Preface

What Is This Book About?

This book is about *stochastic-process limits*, i.e., limits in which a sequence of stochastic processes converges to another stochastic process. Since the converging stochastic processes are constructed from initial stochastic processes by appropriately scaling time and space, the stochastic-process limits provide a macroscopic view of uncertainty. The stochastic-process limits are interesting and important because they generate simple approximations for complicated stochastic processes and because they help explain the statistical regularity associated with a macroscopic view of uncertainty.

This book emphasizes the continuous-mapping approach to obtain new stochastic-process limits from previously established stochastic-process limits. The continuous-mapping approach is applied to obtain stochastic-process limits for *queues*, i.e., probability models of waiting lines or service systems. These limits for queues are called *heavy-traffic limits*, because they involve a sequence of models in which the offered loads are allowed to increase towards the critical value for stability. These heavy-traffic limits generate simple approximations for complicated queueing processes under normal loading and reveal the impact of variability upon queueing performance. By focusing on the application of stochastic-process limits to queues, this book also provides an introduction to heavy-traffic stochastic-process limits for queues.

In More Detail

More generally, this is a book about *probability theory* – a subject which has applications to every branch of science and engineering. Probability theory can help manage a portfolio and it can help engineer a communication network. As it should, probability theory tells us how to compute probabilities, but probability theory also has a more majestic goal: *Probability theory aims to explain the statistical regularity associated with a macroscopic view of uncertainty.*

In probability theory, there are many important ideas. But one idea might fairly lay claim to being the central idea: That idea is conveyed by the *central limit theorem*, which explains the ubiquitous bell-shaped curve: Following the giants – De Moivre, Laplace and Gauss – we have come to realize that, under regularity conditions, a sum of random variables will be approximately normally distributed if the number of terms is sufficiently large.

In the last half century, through the work of Erdös and Kac (1946, 1947), Doob (1949), Donsker (1951, 1952), Prohorov (1956), Skorohod (1956) and others, a broader view of the central limit theorem has emerged. We have discovered that there is not only statistical regularity in the nth sum as n gets large, but there also is statistical regularity in the first n sums. That statistical regularity is expressed via a stochastic-process limit, i.e., a limit in which a sequence of stochastic processes converges to another stochastic process: A sequence of continuous-time stochastic processes generated from the first n sums converges in distribution to Brownian motion as n increases. That generalization of the basic central limit theorem (CLT) is known as *Donsker's theorem*. It is also called a functional central limit theorem (FCLT), because it implies convergence in distribution for many functionals of interest, such as the maximum of the first n sums. The ordinary CLT becomes a simple consequence of Donsker's FCLT, obtained by applying a projection onto one coordinate, making the ordinary CLT look like a view from Abbott's (1952) *Flatland*.

As an extension of the CLT, Donsker's FCLT is important because it has many significant applications, beyond what we would imagine knowing the CLT alone. For example, there are many applications in Statistics: Donsker's FCLT enables us to determine asymptotically-exact approximate distributions for many test statistics. The classic example is the *Kolmogorov-Smirnov statistic*, which is used to test whether data from an unknown source can be regarded as an independent sample from a candidate distribution. The stochastic-process limit identifies a relatively simple approximate distribution for the test statistic, for any continuous candidate cumulative distribution function, that can be used when the sample size is large. Indeed, early work on the Kolmogorov-Smirnov statistic by Doob (1949) and Donsker (1952) provided a major impetus for the development of the general theory of stochastic-process limits. The evolution of that story can be seen in the books by Billingsley (1968, 1999), Csörgő and Horváth (1993), Pollard (1984), Shorack and Wellner (1986) and van der Waart and Wellner (1996).

Donsker's FCLT also has applications in other very different directions. The application that motivated this book is the application to queues: *Donsker's FCLT can be applied to establish heavy-traffic stochastic-process limits for queues.* A heavy-traffic limit for an open queueing model (with input from outside that eventually departs) is obtained by considering a sequence of queueing models, where the input load is allowed to increase toward the critical level for stability (where the input rate equals the maximum potential output rate). In such a heavy-traffic limit, the steady-state performance descriptions, such as the steady-state queue length, typically grow without bound. Nevertheless, with appropriate scaling of both time and space, there may be a nondegenerate stochastic-process limit for the entire queue-length process, which can yield useful approximations and can provide insight into system performance. The approximations can be useful even if the actual queueing system does not experience heavy traffic. *The stochastic-process limits strip away unessential details and reveal key features determining performance.*

Of special interest is the scaling of time and space that occurs in these heavy-traffic stochastic-process limits. It is natural to focus attention on the limit process, which serves as the approximation, but the scaling of time and space also provides important insights. For example, the scaling may reveal a *separation of time scales*, with different phenomena occurring at different time scales. In heavy-traffic limits for queues, the separation of time scales leads to unifying ideas, such as the *heavy-traffic averaging principle* (Section 2.4.2) and the *heavy-traffic snapshot principle* (Remark 5.9.1).

These many consequences of Donsker's FCLT can be obtained by applying the continuous-mapping approach: Continuous-mapping theorems imply that convergence in distribution is preserved under appropriate functions, with the simple case being a single function that is continuous. The continuous-mapping approach is much more effective with the FCLT than the CLT because many more random quantities of interest can be represented as functions of the first n partial sums than can be represented as functions of only the nth partial sum. Since the heavy-traffic stochastic-process limits for queues follow from Donsker's FCLT and the continuous-mapping approach, the statistical regularity revealed by the heavy-traffic limits for queues can be regarded as a consequence of the central limit theorem, when viewed appropriately.

In this book we tell the story about the expanded view of the central limit theorem in more detail. We focus on stochastic-process limits, Donsker's theorem and the continuous-mapping approach. We also put life into the general theory by providing a detailed discussion of one application — queues. We give an introductory account that should be widely accessible. To help visualize the statistical regularity associated with stochastic-process limits, we perform simulations and plot stochastic-process sample paths.

However, we hasten to point out that there already is a substantial literature on stochastic-process limits, Donsker's FCLT and the continuous-mapping approach, including two editions of the masterful book by Billingsley (1968, 1999). What distinguishes the present book from previous books on this topic is our focus on *stochastic process limits with nonstandard scaling and nonstandard limit processes.*

An important source of motivation for establishing such stochastic-process limits for queueing stochastic processes comes from evolving communication networks:

Beginning with the seminal work of Leland, Taqqu, Willinger and Wilson (1994), extensive *traffic measurements* have shown that the network traffic is remarkably bursty, exhibiting complex features such as *heavy-tailed probability distributions, strong (or long-range) dependence and self-similarity*. These features present difficult engineering challenges for network design and control; e.g., see Park and Willinger (2000). Accordingly, a goal in our work is to gain a better understanding of these complex features and the way they affect the performance of queueing models.

To a large extent, the complex features – the heavy-tailed probability distributions, strong dependence and self-similarity – can be *defined* through their impact on stochastic-process limits. Thus, a study of stochastic-process limits, in a sufficiently broad context, is directly a study of the complex features observed in network traffic. From that perspective, it should be clear that this book is intended as a response (but not nearly a solution) to the engineering challenge posed by the traffic measurements.

We are interested in the way complex traffic affects network performance. Since a major component of network performance is congestion (queueing effects), we abstract network performance and focus on the way the complex traffic affects the performance of queues. The heavy-traffic limits show that the complex traffic can have a dramatic impact on queueing performance! There are again heavy-traffic limits with these complex features, but both the scaling and the limit process may change. As in the standard case, the stochastic-process limits reveal key features determining performance.

Even though traffic measurements from evolving communication networks serve as an important source of motivation for our focus on stochastic-process limits with nonstandard scaling and nonstandard limit processes, they are not the only source of motivation. Once you are sensitized to heavy-tailed probability distributions and strong dependence, like Mandelbrot (1977, 1982), you start seeing these phenomena everywhere. There are natural applications of stochastic-process limits with nonstandard scaling and nonstandard limit processes to insurance, because insurance claim distributions often have heavy tails. There also are natural applications to finance, especially in the area of risk management; e.g., related to electricity derivatives. See Embrechts, Klüppelberg and Mikosch (1997), Adler, Feldman and Taqqu (1998) and Asmussen (2000).

The heavy-tailed distributions and strong dependence can lead to stochastic-process limits with *jumps in the limit process*, i.e., stochastic-process limits in which the limit process has discontinuous sample paths. The jumps have engineering significance, because they reveal sudden big changes, when viewed in a long time scale.

Much of the more technical material in the book is devoted to establishing stochastic-process limits with jumps in the limit process, but there already are books discussing stochastic-process limits with jumps in the limit process. Indeed, Jacod and Shiryaev (1987) establish many such stochastic-process limits. To be more precise, from the technical standpoint, what distinguishes this book from previous books on this topic is our focus on stochastic-process limits with *unmatched jumps* in the limit process; i.e., stochastic process limits in which the limit process has jumps unmatched in the converging processes.

For example, we may have a sequence of stochastic processes with continuous sample paths converging to a stochastic process with discontinuous sample paths. Alternatively, before scaling, we may have stochastic processes, such as queue-length stochastic processes, that move up and down by unit steps. Then, after introducing space scaling, the discontinuities are asymptotically negligible. Nevertheless, the sequence of scaled stochastic processes can converge in distribution to a limiting stochastic process with discontinuous sample paths.

Jumps are not part of Donsker's FCLT, because Brownian motion has continuous sample paths. But the classical CLT and Donsker's FCLT do not capture all possible forms of statistical regularity that can prevail! Other forms of statistical regularity emerge when the assumptions of the classical CLT no longer hold. For example, if the random variables being summed have heavy-tailed probability distributions (which here means having infinite variance), then the classical CLT for partial sums breaks down. Nevertheless, there still may be statistical regularity, but it assumes a new form. Then there is a different FCLT in which the limit process has jumps!

But the jumps in this new FCLT are *matched jumps*; each jump corresponds to an exceptionally large summand in the sums. At first glance, it is not so obvious that unmatched jumps can arise. Thus, we might regard stochastic-process limits with unmatched jumps in the limit process as pathological, and thus not worth serious attention. Part of the interest here lies in the fact that such limits, not only can occur, but routinely do occur in interesting applications. In particular, unmatched jumps in the limit process frequently occur in heavy-traffic limits for queues in the presence of heavy-tailed probability distributions. For example, in a single-server queue, the queue-length process usually moves up and down by unit steps. Hence, when space scaling is introduced, the jumps in the scaled queue-length process are asymptotically negligible. Nevertheless, occasional exceptionally long service times can cause a rapid buildup of customers, causing the sequence of scaled queue-length processes to converge to a limit process with discontinuous sample paths. We give several examples of stochastic-process limits with unmatched jumps in the limit process in Chapter 6.

Stochastic-process limits with unmatched jumps in the limit process present technical challenges: Stochastic-process limits are customarily established by exploiting the function space D of all right-continuous $\mathbb{R}^k$-valued functions with left limits, endowed with the Skorohod (1956) J_1 topology (notion of convergence), which is often called "the Skorohod topology." However, that topology does not permit stochastic-process limits with unmatched jumps in the limit process.

As a consequence, to establish stochastic-process limits with unmatched jumps in the limit process, we need to use a nonstandard topology on the underlying space D of stochastic-process sample paths. Instead of the standard J_1 topology on D, we use the M_1 topology on D, which also was introduced by Skorohod (1956). Even though the M_1 topology was introduced a long time ago, it has not received much attention. Thus, a major goal here is to provide a systematic development of the function space D with the M_1 topology and associated stochastic-process limits.

It turns out the standard J_1 topology is stronger (or finer) than the M_1 topology, so that previous stochastic-process limits established using the J_1 topology also hold with the M_1 topology. Thus, while the J_1 topology sometimes cannot be used, the M_1 topology can almost always be used. Moreover, the extra strength of the J_1 topology is rarely exploited. Thus, we would be so bold as to suggest that, *if only one topology on the function space D is to be considered, then it should be the M_1 topology.*

In some cases, the fluctuations in a stochastic process are so strong that no stochastic-process limit is possible with a limiting stochastic process having sample paths in the function space D. In order to establish stochastic-process limits involving such dramatic fluctuations, we introduce larger function spaces than D, which we call E and F. The names are chosen to suggest a natural progression starting from the space C of continuous functions and going beyond D. We define topologies on the spaces E and F analogous to the M_2 and M_1 topologies on D. Thus we exploit our study of the M topologies on D in this later work.

Even though the special focus here is on heavy-traffic stochastic-process limits for queues allowing unmatched jumps in the limit process, many heavy-traffic stochastic-process limits for queues have no jumps in the limit process. That is the case whenever we can directly apply the continuous-mapping approach with Donsker's FCLT. Then we deduce that reflected Brownian motion can serve as an asymptotically-exact approximation for several queueing processes in a heavy-traffic limit. In the queueing chapters we show how those classic heavy-traffic limits can be established and applied. Indeed, the book is also intended to serve as a general introduction to heavy-traffic stochastic-process limits for queues.

Organization of the Book

The book has fifteen chapters, which can be grouped roughly into four parts, ordered according to increasing difficulty. The level of difficulty is far from uniform: The first part is intended to be accessible with less background. It would be helpful (necessary?) to know something about probability and queues.

The *first part*, containing the first five chapters, provides an informal introduction to stochastic-process limits and their application to queues. The first part provides a broad overview, mostly without proofs, intending to complement and supplement other books, such as Billingsley (1968, 1999).

Chapter 1 uses simulation to help the reader directly experience the statistical regularity associated with stochastic-process limits. Chapter 2 discusses applications of the random walks simulated in Chapter 1. Chapter 3 introduces the mathematical framework for stochastic-process limits. Chapter 4 provides an overview of stochastic-process limits, presenting Donsker's theorem and some of its generalizations. Chapter 5 provides an introduction to heavy-traffic stochastic-process limits for queues.

The *second part*, containing Chapters 6 – 10, shows how the unmatched jumps can arise and expands the treatment of queueing models. The first chapter, Chapter 6 uses simulation to demonstrate that there should indeed be unmatched jumps in the limit process in several examples. Chapter 7 continues the overview of stochastic-process limits begun in Chapter 4. The remaining chapters in the second part apply the stochastic-process limits, with the continuous-mapping approach, to obtain more heavy-traffic limits for queues.

The *third part*, containing Chapters 11 – 14, is devoted to the technical foundations needed to establish stochastic-process limits with unmatched jumps in the limit process. The earlier queueing chapters draw on the third part to a large extent. The queueing chapters are presented first to provide motivation for the technical foundations.

The third part begins with Chapter 11, which provides more details on the mathematical framework for stochastic-process limits, expanding upon the brief introduction in Chapter 3. Chapter 12 focuses on the function space D of right-continuous $\mathbb{R}^k$-valued functions with left limits, endowed with one of the nonstandard Skorohod (1956) M topologies (M_1 or M_2). As a basis for applying the continuous-mapping approach to establish new stochastic-process limits in this context, Chapter 13 shows that commonly used functions from D or $D \times D$ to D preserve convergence with the M (and J_1) topologies. The third part concludes with Chapter 14, which establishes heavy-traffic limits for networks of queues.

The *fourth part*, containing Chapter 15, is more exploratory. It initiates new directions for research. Chapter 15 introduces the new spaces larger than D that can be used to express stochastic-process limits for scaled stochastic processes with even greater fluctuations. The fourth part is most difficult, not because what appears is so abstruse, but because so little appears of what originally was intended: Most of the theory is yet to be developed.

The organization of the book is described in more detail at the end of Chapter 3, in Section 3.6.

Even though the book never touches some of the topics originally intended, the book is quite long. To avoid excessive length, material was deleted from the book and placed in an *Internet Supplement*, which can be found via the Springer web site (*http://www.springer-ny.com/whitt*) or directly via:

http://www.research.att.com/~wow/supplement.html.

The first choice for cutting was the more technical material. Thus, the Internet Supplement contains many proofs for theorems in the book. The Internet Supplement also contains related supplementary material. Finally, the Internet Supplement provides a place to correct errors found after the book has been published. The initial contents of the Internet Supplement appear at the end of the book in Appendix B.

What is Missing?

Even though this book is long, it only provides introductions to stochastic-process limits and heavy-traffic stochastic-process limits for queues.

There are several different kinds of limits that can be considered for probability distributions and stochastic processes. Here we only consider central limit theorems and natural generalizations to the functions space D. We omit other kinds of limits such as large deviation principles. For large deviation principles, the continuous-mapping approach can be applied using contraction principles. Large deviations principles can be very useful for queues; see Shwartz and Weiss (1995). For a sample of other interesting probability limits (related to the Poisson clumping heuristic), see Aldous (1989).

Even though much of the book is devoted to queues, we only discuss heavy-traffic stochastic-process limits for queues. There is a large literature on queues. Nice general introductions to queues, at varying mathematical levels, are contained in the books by Asmussen (1987), Cooper (1982), Hall (1991), Kleinrock (1975, 1976) and Wolff (1989).

Queueing theory is intended to aid in the performance analysis of complex systems, such as computer, communication and manufacturing systems. We discuss performance implications of the heavy-traffic limits, but we do not discuss performance analysis in detail. Jain (1991) and Gunther (1998) discuss the performance analysis of computer systems; Bertsekas and Gallager (1987) discuss the performance analysis of communication networks; and Buzacott and Shanthikumar (1993) and Hopp and Spearman (1996) discuss the performance analysis of manufacturing systems.

Since we are motivated by evolving communication networks, we discuss queueing models that arise in that context, but we do not discuss the context itself. For background on evolving communication networks, see Keshav (1997), Kurose and Ross (2000) and Krishnamurthy and Rexford (2001). For research on communication network performance, see Park and Willinger (2000) and recent proceedings of *IEEE INFOCOM* and *ACM SIGCOMM*:

http://www.ieee-infocom.org/2000/
http://www.acm.org/pubs/contents/proceedings/series/comm/

Even within the relatively narrow domain of heavy-traffic stochastic-process limits for queues, we only provide an introduction. Harrison (1985) provided a previous introduction, focusing on Brownian motion and Brownian queues, the heavy-traffic limit processes rather than the heavy-traffic limits themselves. Harrison (1985) shows how martingales and the Ito stochastic calculus can be applied to calculate quantities of interest and solve control problems. Newell (1982) provides useful perspective as well with his focus on deterministic and diffusion approximations. Harrison and Newell show that the limit processes can be used directly as approximations without considering stochastic-process limits. In contrast, we emphasize insights that can be gained from the stochastic-process limits, e.g., from the scaling.

The subject of heavy-traffic stochastic-process limits remains a very active research topic. Most of the recent interest focuses on networks of queues with multiple classes of customers. A principal goal is to determine good polices for scheduling and routing. That focus places heavy-traffic stochastic-process limits in the mainstream of operations research.

Multiclass queueing networks are challenging because the obvious stability criterion – having the traffic intensity be less than one at each queue – can in fact fail to be sufficient for stability; see Bramson (1994a,b). Thus, for general multiclass queueing networks, the very definition of heavy traffic is in question. For some of the recent heavy-traffic stochastic-process limits in that setting, new methods beyond the continuous-mapping approach have been required; see Bramson (1998) and Williams (1998a,b).

Discussion of the heavy-traffic approach to multiclass queueing networks, including optimization issues, can be found in the recent books by Chen and Yao (2001) and Kushner (2001), in the collections of papers edited by Yao (1994), Kelly and Williams (1995), Kelly, Zachary and Ziedins (1996), Dai (1998), McDonald and Turner (2000) and Park and Willinger (2000), and in recent papers such as Bell and Williams (2001), Harrison (2000, 2001a,b), Kumar (2000) and Markowitz and Wein (2001). Hopefully, this book will help prepare readers to appreciate that important work and extend it in new directions.

Contents

Preface   vii

1 Experiencing Statistical Regularity   1
    1.1 A Simple Game of Chance 1
         1.1.1 Plotting Random Walks 2
         1.1.2 When the Game is Fair 3
         1.1.3 The Final Position 7
         1.1.4 Making an Interesting Game 12
    1.2 Stochastic-Process Limits 16
         1.2.1 A Probability Model 16
         1.2.2 Classical Probability Limits 20
         1.2.3 Identifying the Limit Process 22
         1.2.4 Limits for the Plots 25
    1.3 Invariance Principles 27
         1.3.1 The Range of Brownian Motion 28
         1.3.2 Relaxing the IID Conditions 30
         1.3.3 Different Step Distributions 31
    1.4 The Exception Makes the Rule 34
         1.4.1 Explaining the Irregularity 35
         1.4.2 The Centered Random Walk with $p = 3/2$ 37
         1.4.3 Back to the Uncentered Random Walk with $p = 1/2$ . . . 43
    1.5 Summary 45

2 Random Walks in Applications   49
  2.1 Stock Prices 49
  2.2 The Kolmogorov-Smirnov Statistic 51
  2.3 A Queueing Model for a Buffer in a Switch 55
       2.3.1 Deriving the Proper Scaling 56
       2.3.2 Simulation Examples 59
  2.4 Engineering Significance 63
       2.4.1 Buffer Sizing 64
       2.4.2 Scheduling Service for Multiple Sources 68

3 The Framework for Stochastic-Process Limits   75
  3.1 Introduction 75
  3.2 The Space $\mathcal{P}$ 76
  3.3 The Space D 78
  3.4 The Continuous-Mapping Approach 84
  3.5 Useful Functions 86
  3.6 Organization of the Book 89

4 A Panorama of Stochastic-Process Limits   95
  4.1 Introduction 95
  4.2 Self-Similar Processes 96
       4.2.1 General CLT's and FCLT's 96
       4.2.2 Self-Similarity 97
       4.2.3 The Noah and Joseph Effects 99
  4.3 Donsker's Theorem 101
       4.3.1 The Basic Theorems 101
       4.3.2 Multidimensional Versions 104
  4.4 Brownian Limits with Weak Dependence 106
  4.5 The Noah Effect: Heavy Tails 109
       4.5.1 Stable Laws 111
       4.5.2 Convergence to Stable Laws 114
       4.5.3 Convergence to Stable Lévy Motion 116
       4.5.4 Extreme-Value Limits 118
  4.6 The Joseph Effect: Strong Dependence 120
       4.6.1 Strong Positive Dependence 121
       4.6.2 Additional Structure 122
       4.6.3 Convergence to Fractional Brownian Motion 124
  4.7 Heavy Tails Plus Dependence 130
       4.7.1 Additional Structure 130
       4.7.2 Convergence to Stable Lévy Motion 131
       4.7.3 Linear Fractional Stable Motion 132
  4.8 Summary 136

5 Heavy-Traffic Limits for Fluid Queues — 137
- 5.1 Introduction — 137
- 5.2 A General Fluid-Queue Model — 139
  - 5.2.1 Input and Available-Processing Processes — 139
  - 5.2.2 Infinite Capacity — 140
  - 5.2.3 Finite Capacity — 143
- 5.3 Unstable Queues — 145
  - 5.3.1 Fluid Limits for Fluid Queues — 146
  - 5.3.2 Stochastic Refinements — 149
- 5.4 Heavy-Traffic Limits for Stable Queues — 153
- 5.5 Heavy-Traffic Scaling — 157
  - 5.5.1 The Impact of Scaling Upon Performance — 158
  - 5.5.2 Identifying Appropriate Scaling Functions — 160
- 5.6 Limits as the System Size Increases — 162
- 5.7 Brownian Approximations — 165
  - 5.7.1 The Brownian Limit — 166
  - 5.7.2 The Steady-State Distribution — 167
  - 5.7.3 The Overflow Process — 170
  - 5.7.4 One-Sided Reflection — 173
  - 5.7.5 First-Passage Times — 176
- 5.8 Planning Queueing Simulations — 178
  - 5.8.1 The Standard Statistical Procedure — 180
  - 5.8.2 Invoking the Brownian Approximation — 181
- 5.9 Heavy-Traffic Limits for Other Processes — 183
  - 5.9.1 The Departure Process — 183
  - 5.9.2 The Processing Time — 184
- 5.10 Priorities — 187
  - 5.10.1 A Heirarchical Approach — 189
  - 5.10.2 Processing Times — 190

6 Unmatched Jumps in the Limit Process — 193
- 6.1 Introduction — 193
- 6.2 Linearly Interpolated Random Walks — 195
  - 6.2.1 Asymptotic Equivalence with M_1 — 195
  - 6.2.2 Simulation Examples — 196
- 6.3 Heavy-Tailed Renewal Processes — 200
  - 6.3.1 Inverse Processes — 201
  - 6.3.2 The Special Case with $m = 1$ — 202
- 6.4 A Queue with Heavy-Tailed Distributions — 205
  - 6.4.1 The Standard Single-Server Queue — 206
  - 6.4.2 Heavy-Traffic Limits — 208
  - 6.4.3 Simulation Examples — 210
- 6.5 Rare Long Service Interruptions — 216
- 6.6 Time-Dependent Arrival Rates — 220

7 More Stochastic-Process Limits — 225
- 7.1 Introduction — 225
- 7.2 Central Limit Theorem for Processes — 226
  - 7.2.1 Hahn's Theorem — 226
  - 7.2.2 A Second Limit — 230
- 7.3 Counting Processes — 233
  - 7.3.1 CLT Equivalence — 234
  - 7.3.2 FCLT Equivalence — 235
- 7.4 Renewal-Reward Processes — 238

8 Fluid Queues with On-Off Sources — 243
- 8.1 Introduction — 243
- 8.2 A Fluid Queue Fed by On-Off Sources — 245
  - 8.2.1 The On-Off Source Model — 245
  - 8.2.2 Simulation Examples — 248
- 8.3 Heavy-Traffic Limits for the On-Off Sources — 250
  - 8.3.1 A Single Source — 251
  - 8.3.2 Multiple Sources — 255
  - 8.3.3 $M/G/\infty$ Sources — 259
- 8.4 Brownian Approximations — 260
  - 8.4.1 The Brownian Limit — 260
  - 8.4.2 Model Simplification — 263
- 8.5 Stable-Lévy Approximations — 264
  - 8.5.1 The RSLM Heavy-Traffic Limit — 265
  - 8.5.2 The Steady-State Distribution — 268
  - 8.5.3 Numerical Comparisons — 270
- 8.6 Second Stochastic-Process Limits — 272
  - 8.6.1 $M/G/1/K$ Approximations — 273
  - 8.6.2 Limits for Limit Processes — 277
- 8.7 Reflected Fractional Brownian Motion — 279
  - 8.7.1 An Increasing Number of Sources — 279
  - 8.7.2 Gaussian Input — 280
- 8.8 Reflected Gaussian Processes — 283

9 Single-Server Queues — 287
- 9.1 Introduction — 287
- 9.2 The Standard Single-Server Queue — 288
- 9.3 Heavy-Traffic Limits — 292
  - 9.3.1 The Scaled Processes — 292
  - 9.3.2 Discrete-Time Processes — 294
  - 9.3.3 Continuous-Time Processes — 297
- 9.4 Superposition Arrival Processes — 301
- 9.5 Split Processes — 305
- 9.6 Brownian Approximations — 306
  - 9.6.1 Variability Parameters — 307

		9.6.2	Models with More Structure	310
	9.7	Very Heavy Tails		313
		9.7.1	Heavy-Traffic Limits	314
		9.7.2	First Passage to High Levels	316
	9.8	An Increasing Number of Arrival Processes		318
		9.8.1	Iterated and Double Limits	318
		9.8.2	Separation of Time Scales	322
	9.9	Approximations for Queueing Networks		326
		9.9.1	Parametric-Decomposition Approximations	326
		9.9.2	Approximately Characterizing Arrival Processes	330
		9.9.3	A Network Calculus	331
		9.9.4	Exogenous Arrival Processes	337
		9.9.5	Concluding Remarks	338

10 Multiserver Queues   341

	10.1	Introduction		341
	10.2	Queues with Multiple Servers		342
		10.2.1	A Queue with Autonomous Service	342
		10.2.2	The Standard m-Server Model	345
	10.3	Infinitely Many Servers		348
		10.3.1	Heavy-Traffic Limits	349
		10.3.2	Gaussian Approximations	352
	10.4	An Increasing Number of Servers		355
		10.4.1	Infinite-Server Approximations	356
		10.4.2	Heavy-Traffic Limits for Delay Models	357
		10.4.3	Heavy-Traffic Limits for Loss Models	360
		10.4.4	Planning Simulations of Loss Models	361

11 More on the Mathematical Framework   367

	11.1	Introduction		367
	11.2	Topologies		368
		11.2.1	Definitions	368
		11.2.2	Separability and Completeness	371
	11.3	The Space $\mathcal{P}$		372
		11.3.1	Probability Spaces	372
		11.3.2	Characterizing Weak Convergence	373
		11.3.3	Random Elements	375
	11.4	Product Spaces		377
	11.5	The Space D		380
		11.5.1	J_2 and M_2 Metrics	381
		11.5.2	The Four Skorohod Topologies	382
		11.5.3	Measurability Issues	385
	11.6	The Compactness Approach		386

12 The Space D — 391

- 12.1 Introduction 391
- 12.2 Regularity Properties of D 392
- 12.3 Strong and Weak M_1 Topologies 394
  - 12.3.1 Definitions 394
  - 12.3.2 Metric Properties 396
  - 12.3.3 Properties of Parametric Representations 398
- 12.4 Local Uniform Convergence at Continuity Points 401
- 12.5 Alternative Characterizations of M_1 Convergence 403
  - 12.5.1 SM_1 Convergence 403
  - 12.5.2 WM_1 Convergence 408
- 12.6 Strengthening the Mode of Convergence 409
- 12.7 Characterizing Convergence with Mappings 410
- 12.8 Topological Completeness 413
- 12.9 Noncompact Domains 414
- 12.10 Strong and Weak M_2 Topologies 416
- 12.11 Alternative Characterizations of M_2 Convergence 418
  - 12.11.1 M_2 Parametric Representations 418
  - 12.11.2 SM_2 Convergence 419
  - 12.11.3 WM_2 Convergence 421
  - 12.11.4 Additional Properties of M_2 Convergence 422
- 12.12 Compactness 424

13 Useful Functions — 427

- 13.1 Introduction 427
- 13.2 Composition 428
- 13.3 Composition with Centering 431
- 13.4 Supremum 435
- 13.5 One-Dimensional Reflection 439
- 13.6 Inverse 441
  - 13.6.1 The Standard Topologies 442
  - 13.6.2 The M_1' Topology 444
  - 13.6.3 First Passage Times 446
- 13.7 Inverse with Centering 447
- 13.8 Counting Functions 453

14 Queueing Networks — 457

- 14.1 Introduction 457
- 14.2 The Multidimensional Reflection Map 460
  - 14.2.1 A Special Case 460
  - 14.2.2 Definition and Characterization 462
  - 14.2.3 Continuity and Lipschitz Properties 465
- 14.3 The Instantaneous Reflection Map 473
  - 14.3.1 Definition and Characterization 474
  - 14.3.2 Implications for the Reflection Map 480

14.4	Reflections of Parametric Representations	482
14.5	M_1 Continuity Results and Counterexamples	485
	14.5.1 M_1 Continuity Results	485
	14.5.2 Counterexamples	487
14.6	Limits for Stochastic Fluid Networks	490
	14.6.1 Model Continuity	492
	14.6.2 Heavy-Traffic Limits	493
14.7	Queueing Networks with Service Interruptions	495
	14.7.1 Model Definition	495
	14.7.2 Heavy-Traffic Limits	499
14.8	The Two-Sided Regulator	505
	14.8.1 Definition and Basic Properties	505
	14.8.2 With the M_1 Topologies	509
14.9	Related Literature	511

15 The Spaces E and F 515

15.1	Introduction	515
15.2	Three Time Scales	516
15.3	More Complicated Oscillations	519
15.4	The Space E	523
15.5	Characterizations of M_2 Convergence in E	527
15.6	Convergence to Extremal Processes	530
15.7	The Space F	533
15.8	Queueing Applications	535

References 541

Appendix A Regular Variation 569

Appendix B Contents of the Internet Supplement 573

Notation Index 577

Author Index 579

Subject Index 585

1
Experiencing Statistical Regularity

1.1. A Simple Game of Chance

A good way to experience statistical regularity is to repeatedly play a game of chance. So let us consider a simple game of chance using a spinner. To attract attention, it helps to have interesting outcomes, such as falling into an alligator pit or winning a dash for cash (e.g., you receive the opportunity to run into a bank vault and drag out as many money bags as you can within thirty seconds). However, to focus on statistical regularity, rather than fear or greed, we consider repeated plays with a simple outcome.

In our game, the payoff in each of several repeated plays is determined by spinning the spinner. We pay a fee for each play of the game and then receive the payoff indicated by the spinner. Let the payoff on the spinner be uniformly distributed around the circle; i.e., if the angle after the spin is θ, then we receive $\theta/2\pi$ dollars. Thus our payoff on one play is U dollars, where U is a uniform random number taking values in the interval $[0, 1]$.

We have yet to specify the fee to play the game, but first let us simulate the game to see what cumulative payoffs we might receive, not counting the fees, if we play the game repeatedly. We perform the simulation using our favorite random number generator, by generating n uniform random numbers $U_1, \ldots, U_n$, each taking values in the interval $[0, 1]$, and then forming associated partial sums by setting

$$S_k = U_1 + \cdots + U_k, \quad 1 \le k \le n,$$

and $S_0 \equiv 0$, where $\equiv$ denotes equality by definition. The nth partial sum S_n is the total payoff after n plays of the game (not counting the fees to play the game).

The successive partial sums form a *random walk*, with U_n being the nth step and S_n being the position after n steps.

1.1.1. *Plotting Random Walks*

Now, using our favorite plotting routine, let us plot the random walk, i.e., the $n+1$ partial sums S_k, $0 \leq k \leq n$, for a range of n values, e.g., for $n = 10^j$ for several values of j. This simulation experiment is very easy to perform. For example, it can be performed almost instantaneously with the statistical package S (or S-Plus), see Becker, Chambers and Wilks (1988) or Venables and Ripley (1994), using the function

```
walk <- function(j) {
uniforms <- runif(10^j)          # generate random numbers
firstsums <- cumsum(uniforms)    # form the partial sums
sums <- c(0, firstsums)          # include a 0^th sum
index <- order(sums) -1          # adjust the index
plot(index, sums) }              # do the plotting
```

Plots of the random walk with $n = 10^j$ for $j = 1, \ldots, 4$ are shown in Figure 1.1. For small n, e.g., for $n = 10$, we see irregularly spaced (vertically) points increasing to the right, but as n increases, the spacing between the points becomes blurred and regularity emerges: The plots approach a straight line with slope equal to $1/2$, the mean of a single step U_k. If we look at the pictures in successive plots, ignoring the units on the axes, we see that the plots become independent of n as n increases. Looking at the plot for large n produces a macroscopic view of uncertainty.

The plotter automatically plots the random walk $\{S_k : 0 \leq k \leq n\}$ in the available space. Ignoring the units on the axes is equivalent to regarding the plot as a display in the unit square. By "unit square" we do not mean that the rectangle containing the plot is necessarily a square, but that new units can range from 0 to 1 on both axes, independent of the original units. The plotter automatically plots the random walk in the available space by scaling time and space (the horizontal and vertical dimensions). Time is scaled by placing the $n+1$ points $1/n$ apart horizontally. Space is scaled by subtracting the minimum and dividing by the range (assuming that the range is not zero); i.e., we interpret the plot as

$$plot(\{S_k : 0 \leq k \leq n\}) \equiv plot(\{(S_k - min)/range : 0 \leq k \leq n\}),$$

where

$$min \equiv \min_{0 \leq k \leq n} S_k$$

and

$$range \equiv \max_{0 \leq k \leq n} S_k - \min_{0 \leq k \leq n} S_k .$$

Combining these two forms of scaling, the plotter displays the ordered pairs $(k/n, (S_k - min)/range)$ for $0 \leq k \leq n$. With that scaling, the ordered pairs

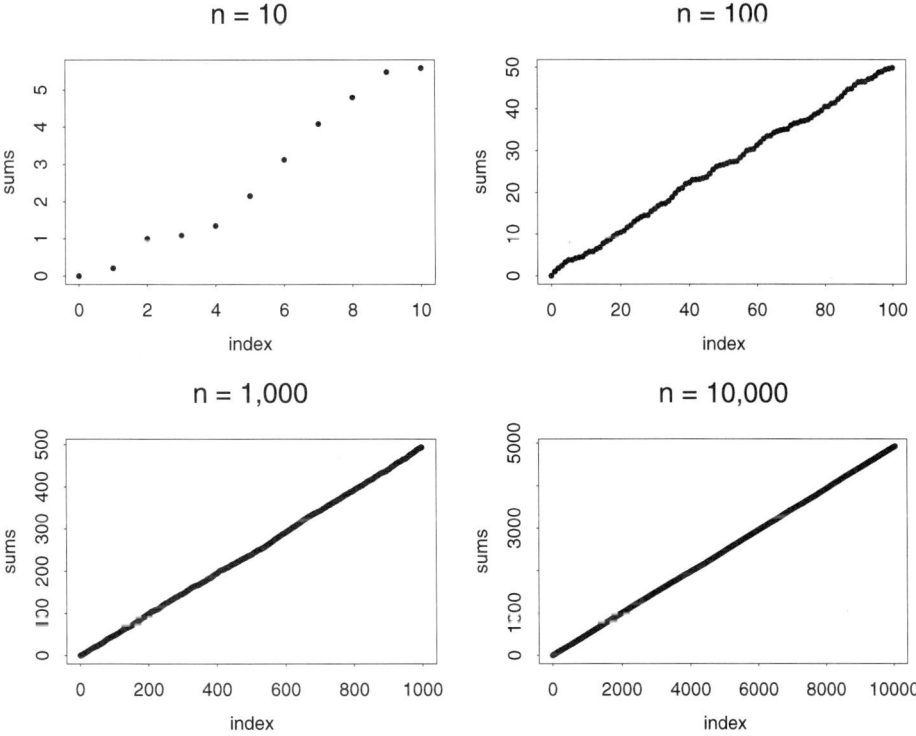

Figure 1.1. Possible realizations of the first 10^j steps of the random walk $\{S_k : k \geq 0\}$ with steps uniformly distributed in the interval $[0, 1]$ for $j = 1, \ldots, 4$.

do indeed fall in the unit square. Also note that $(S_k - min)/range$ must assume (approximately) the values 0 and 1 for at least one argument. That occurs because, without the rescaling, the plotting makes the units on the ordinate (y axis) range from the minimum value to the maximum value (approximately).

To confirm the regularity we see in Figure 1.1, we should repeat the experiment. When we repeat the experiment with different random number seeds (new uniform random numbers), the outcome for small n changes somewhat from experiment to experiment, but we always see essentially the same picture for large n. Thus the plots show regularity associated with both large n and repeated experiments.

1.1.2. When the Game is Fair

Now let us see what happens when the game is fair. Since the expected payoff is $1/2$ dollar each play of the game, the game is fair if the fee to play is $1/2$ dollar. To examine the consequences of making the game fair, we consider a minor modification of the simulation experiment above: We repeat the experiment after subtracting the mean $1/2$ from each step of the random walk; i.e., we plot the *centered random walk* (i.e., the centered partial sums $S_k - k/2$ for $0 \leq k \leq n$) for the same values of n as before.

If we consider the case $n = 10^4$, it is natural to expect to see a horizontal line instead of the line with slope $1/2$ in Figure 1.1. However, what we see is very different! Instead of a horizontal line, for $n = 10^4$ we see an irregular path, as shown in Figure 1.2.

We do not see the horizontal line because *the data have been automatically rescaled by the plotter*. The centering has let the plotter *blow up the picture* to show extra detail not apparent from Figure 1.1.

After centering, the range of values (the maximum minus the minimum) for the partial sums decreases dramatically. The first 10^4 uncentered partial sums assume values approximately in the interval $[0, 5000]$, whereas the first 10^4 centered partial sums all fall in the interval $[-60, 5]$. Thus, the range has decreased from $5,000$ to less than 100.

At first glance, it may not be evident that there is any regularity for large n in Figure 1.2. We would hope to be able to predict what we will see if we repeat the experiment with new uniform random numbers. However, when we repeat the simulation experiment with different random number seeds, we obtain different irregular paths. To illustrate, six independent plots for $n = 10^4$ are shown in Figure 1.3. The six path samples look somewhat similar, but each is different from the others.

In Figure 1.3, just as in Figures 1.1 and 1.2, we let the plotter automatically do the scaling. Thus, the units on vertical axis change from plot to plot. We plot in this manner throughout this chapter, by design. We will show that these "automatic plots" reveal statistical regularity if we ignore the units and think of the plot as being on the unit square. (Having the units on the axes barely legible reinforces this perspective.) But essentially the same conclusion can be drawn if we fix the units on the vertical axis. From Figure 1.3, after the fact, we can conclude that we could have fixed the units on the vertical axis, letting the values fall in the interval $[-100, 100]$. In either case, we are faced with the problem of understanding what we see.

We have arrived at a critical point, which may require us to adjust our thinking. To understand what we are seeing, we need to recognize that the irregular paths we see should be regarded as *random paths*. We then can understand that there actually is regularity underlying the six displayed paths in Figure 1.3, but it is *statistical regularity*.

We want to be able to predict what we will see when we increase n or perform additional experiments. For the uncentered random walks in Figure 1.1, we predict that the plot of $\{S_k : 0 \leq k \leq n\}$ will look like the diagonal line in the unit square for all n sufficiently large. However, for the centered random walks, the plots do not approach such a simple limit. What we should hope to predict when we repeat the experiment for the centered random walk (again ignoring the units on the axes) is the *probability distribution* of the random path. We should anticipate that the successive paths in repeated experiments will change from experiment to

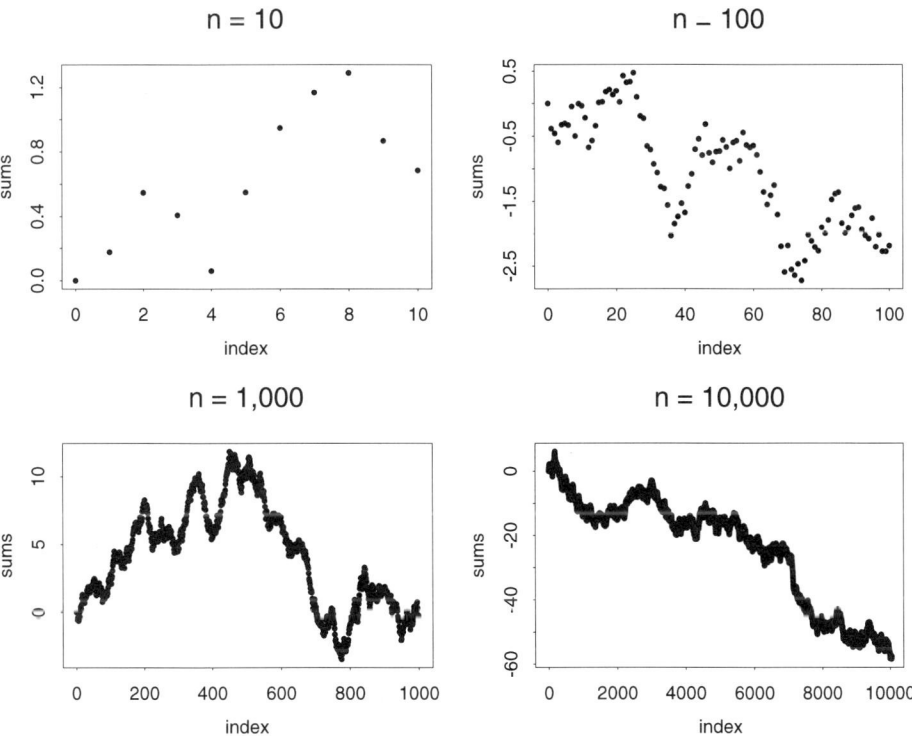

Figure 1.2. Possible realizations of the first 10^j steps of the centered random walk $\{S_k - k/2 : k \geq 0\}$ with steps uniformly distributed in the interval $[0, 1]$ for $j = 1, \ldots, 4$.

experiment, but we should look for a common probability distribution on the space of possible paths.

The simulation experiments suggest that, for all n sufficiently large, there tends to be a common probability distribution for the plotted random walk paths, where as before we ignore the units on the two axes or, equivalently, we regard the plot as being in the unit square. We can see part of the story when we generate new random walk paths for different values of n. For example, when we generate six centered random walk paths for $n = 10^5$ or $n = 10^6$, the plots look just like the plots in Figure 1.3. To make that clear, we plot six independent plots for the case $n = 10^6$ in Figure 1.4. As before, the units on the vertical axes change from plot to plot, but if we ignore the units on both axes, the plots in Figure 1.4 look just like the plots in Figure 1.3.

Looking at Figures 1.3 and 1.4, we should be confident about what we will see when $n = 10^8$ or $n = 10^{10}$. From Figure 1.4 and other similar plots, we see that, for n sufficiently large, the plots tend to be independent of n, provided that we ignore the units on the axes, and regard the plot as being in the unit square. Of course, as n increases, the units change on the two axes. And each new plot is a random path

6   1. Experiencing Statistical Regularity

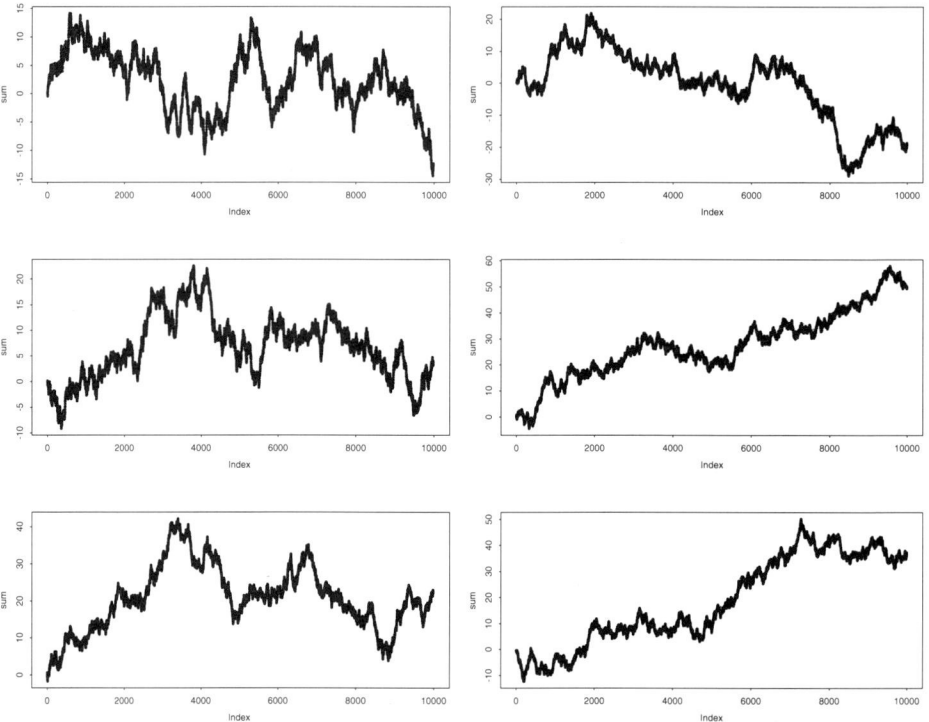

Figure 1.3. Six independent realizations of the first 10^4 steps of the centered random walk $\{S_k - k/2 : k \geq 0\}$ associated with steps uniformly distributed in the interval $[0, 1]$.

selected from the common probability distribution on the space of possible sample paths in the unit square.

As a consequence, we also see that the fluctuations in a smaller time scale are asymptotically negligible compared to the fluctuations in a larger time scale. Thus, for $j \geq 5$, the plots for 10^j are visually unchanged if we only keep the values at about 10^4 equally spaced indices. Indeed, such pruning of the data (reducing a data set of 10^j partial sums for $j \geq 5$ to 10^4 values) is useful to efficiently print the plots for large n.

The fact that the plots are independent of n for all n sufficiently large means that the plots tend to exhibit *self-similarity*. By self-similarity we mean that rescaled versions of the plot associated with increasing n tend to look like the original plot. More specifically, the probability distribution on the space of sample paths in the unit square tends to be unaffected by the scaling. Self-similarity will be a persistent theme; e.g., see Section 4.2.

When we consider rescaling, we can also decrease n. For instance, suppose that we consider the plot for $n = 10^7$ and select 10% of it from a subinterval of the plot. If we make a full plot of that 10% portion, then we obtain a plot for $n = 10^6$, which

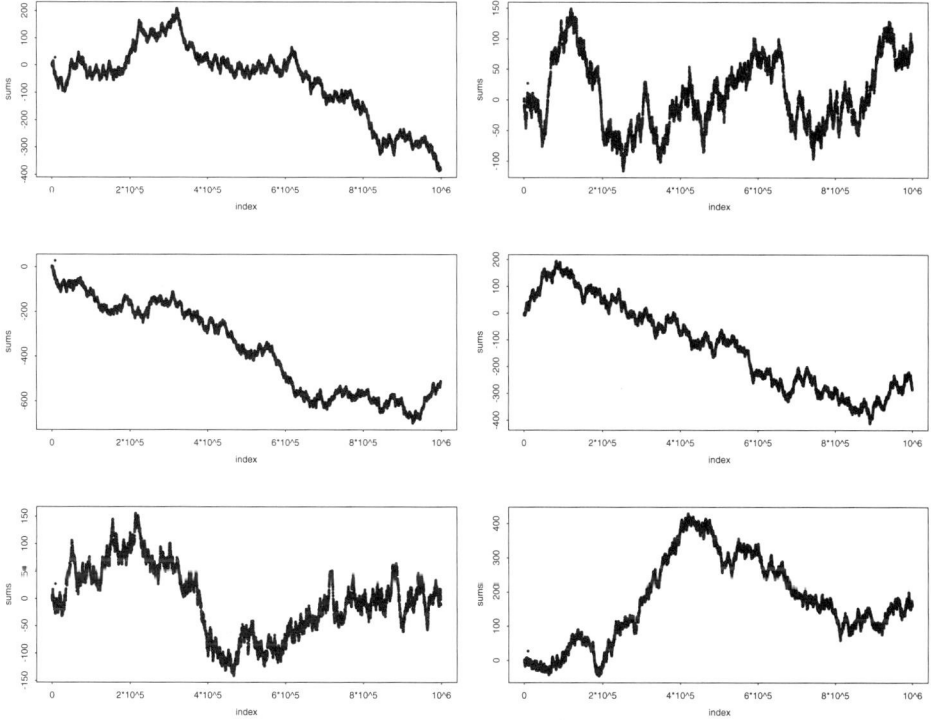

Figure 1.4. Six independent realizations of the first 10^6 steps of the centered random walk $\{S_k - k/2 : k \geq 0\}$ associated with steps uniformly distributed in the interval $[0, 1]$.

looks just like a random version of the original plot for $n = 10^7$. (By a "random version of the original plot" we mean that the probability distributions on the space of possible sample paths in the unit square tend to be the same.) Similarly, if we continue and select 10% of the new plot for $n = 10^6$ from any subinterval and plot it, then we obtain a plot for $n = 10^5$, which again looks like a random version of the original plot for 10^7. Of course, Figure 1.2 shows that the self-similarity for the random walks associated with decreasing n breaks down when n is too small. It is interesting to contemplate a limiting continuous-time random path that permits self-similarity without end!

1.1.3. The Final Position

It is difficult to actually see the probability distribution of the entire random path, because the path is multidimensional, but we can easily look at any one position of the random walk. For instance, suppose that we focus on the final position of the centered random walk, i.e., the single centered partial sum $S_n - n/2$ for one fixed (large) value of n.

8   1. Experiencing Statistical Regularity

It is evident that the final position of the centered random walk, $S_n - n/2$, changes from experiment to experiment. We find statistical regularity when we perform many independent replications of the experiment and look at the distribution of the final positions. So, let us do that.

Remark 1.1.1. *The final position and the relative final position.* For simplicity, we now want to look at the final position of the centered random walk, $S_n - n/2$, independent of the rest of the random walk. If instead we looked at the final position in the unit square, ignoring the original units, we would be looking at the *relative final position*, which must assume a value between 0 and 1. Letting $M_n \equiv \max_{1 \leq k \leq n}\{S_k - k/2\}$ and $m_n \equiv \min_{1 \leq k \leq n}\{S_k - k/2\}$, the relative final position is

$$R_n \equiv \frac{S_n - n/2 - m_n}{M_n - m_n}, \quad n \geq 1 \ . \tag{1.1}$$

It turns out that there is statistical regularity associated with the relative final position, just as there is statistical regularity associated with the entire plot, but the relative final position is more complicated than the final position. Hence, now we focus on the final position. We discuss the relative final position in Remark 1.2.2 at the end of Section 1.2.4. ∎

Suppose that we consider the final position of the centered random walk with uniform random steps for $n = 1000$, and suppose that we perform 1000 replications of the experiment. We thus obtain 1000 independent samples of the centered sum $S_{1000} - 500$. We can estimate the probability density of this distribution using the nonparametric probability density estimator *density* from S (with the default parameter settings). The estimated probability density of the final position $S_{1000} - 500$ is plotted in Figure 1.5.

Figure 1.5 shows that nonparametric density estimation does not achieve high resolution with only a modest amount of data, but it suggests that the final position of the random walk after 1000 steps is approximately normally distributed with zero mean. That conclusion is more strongly supported by the QQ plot in Figure 1.6. The QQ plot compares the empirical distribution of the data to the normal distribution; e.g., see p. 122 of Venables and Ripley (1994). Specifically, the QQ plot compares the sorted data to the quantiles of the normal distribution. If there are n data points, then we consider the $n - 1$ normal quantiles z_k, where

$$P(N(0,1) \leq z_k) = k/n, \quad 1 \leq k \leq n - 1 \ ,$$

with $N(m, \sigma^2)$ denoting a random variable with a normal (or Gaussian) distribution having mean m and variance σ^2. When $n = 1,000$, the normal quantiles range from -3.1 to $+3.1$, with there being more quantiles near 0 than at the extremes. (Since we focus on the shape of the QQ plot, the QQ plot compares the distributions independent of location and scale; e.g., the shape of the QQ plot is independent of the mean and variance of the reference normal distribution.)

The near-linear plot in Figure 1.6 is approximately the same as the QQ plot for 1000 independent samples from a normal distribution. To make that clear, a QQ

1.1. A Simple Game of Chance

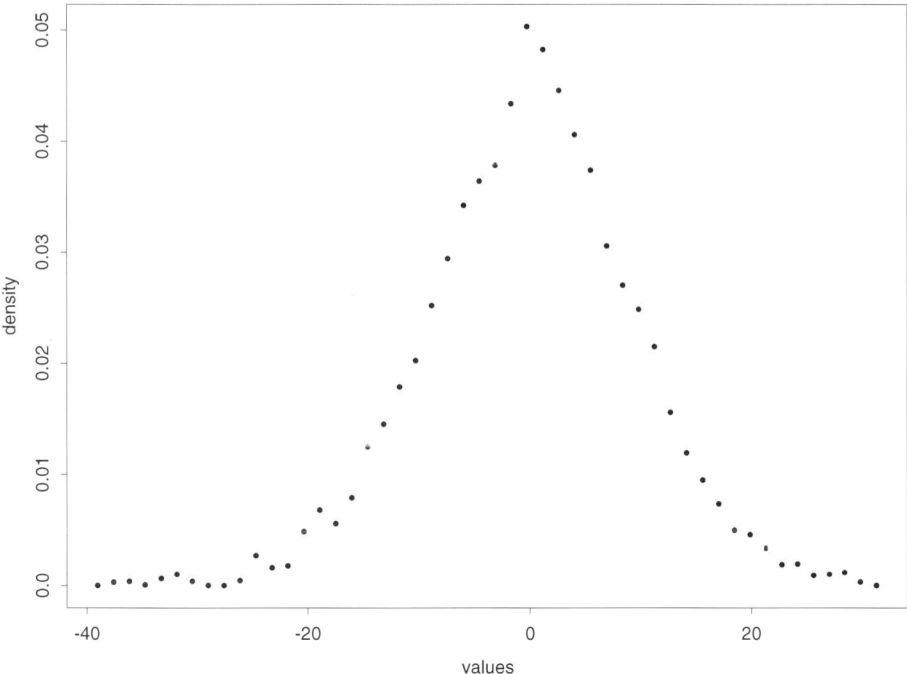

Figure 1.5. An estimate of the probability density of the final position of the random walk, obtained from 1000 independent samples of the centered partial sum $S_{1000} - 500$, where the steps U_k are uniformly distributed in the interval $[0, 1]$.

plot of a sample of 1000 observations from a normal distribution (with the same mean and variance) is also shown in Figure 1.6. Again the units are different in the two plots, because the range of values differs from sample to sample. The linearity that holds except for the tails strongly indicates that the final positions are indeed normally distributed.

But, in order to fairly draw that conclusion, we need more experience with QQ plots. We become more confident of the conclusion when we repeat these experiments a number of times; then we can observe the statistical variability in the QQ plots. We also gain confidence when we make QQ plots of various nonnormal distributions; then we can see how departures from normality are reflected in the plots. When you think hard about the figures, they become invitations to perform additional experiments. Our main point here is that analysis with the QQ plots indicates that the final position of the centered random walk is indeed approximately normally distributed.

That conclusion is also supported by density estimates based on more data. To illustrate how the density estimates perform as a function of sample size, we display the estimates of the probability density of the same final position $S_{1000} - 500$ based

10   1. Experiencing Statistical Regularity

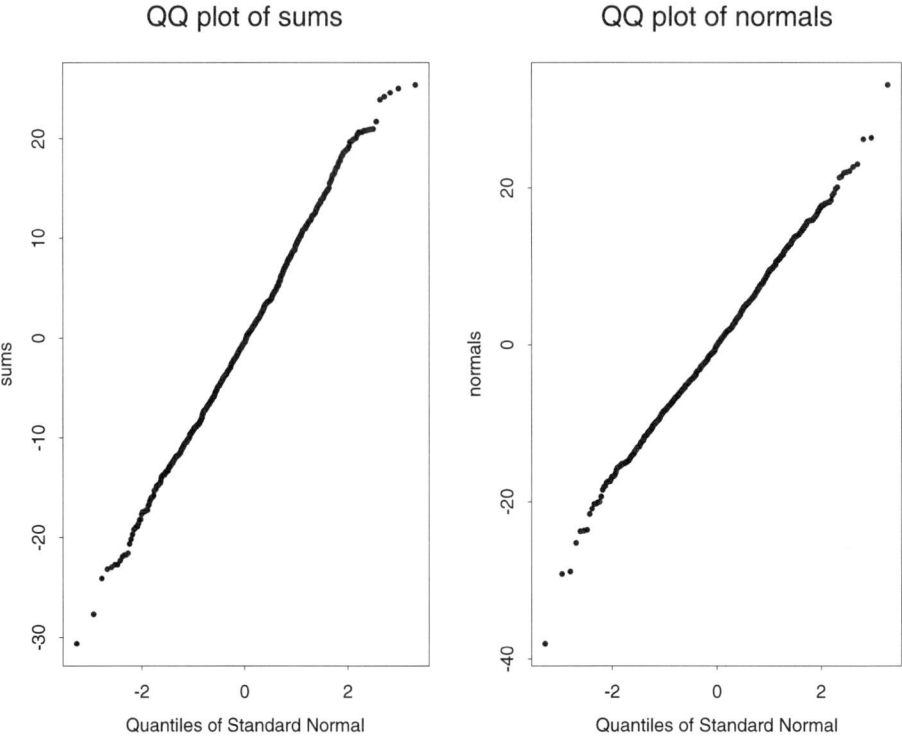

Figure 1.6. Two QQ plots of 1000 samples: the first for the sums, i.e., the final positions $S_{1000} - 500$ of the centered random walk, and the second for a normal distribution.

on 10^j samples for $j = 2, \ldots, 5$ in Figure 1.7 (again using the nonparametric density estimator *density* from S with the default parameter settings). Essentially the same plots are obtained for independent samples from normal distributions. From Figure 1.7, it is evident that the density estimates converge to a normal pdf as $n \to \infty$. For more on density estimation, see Devroye (1987).

It is not our purpose to delve deeply into statistical issues, but it is worth remarking that we obtain new interesting plots, like the random walk plots, when we do. Our brief examination of the distribution of the final position of the random walk suggests looking for a more precise statistical test to determine whether or not the final position of the random walk is indeed approximately normally distributed. To evaluate whether some data can be regarded as an independent sample any specified probability distribution, it is natural to carefully investigate how the empirical distribution of a sample from that probability distribution tends to differ from the underlying probability distribution itself.

Recall that the *cumulative distribution function* (cdf) F of a random variable X is the function

$$F(t) \equiv P(X \leq t) \quad \text{for} \quad t \in \mathbb{R} \ .$$

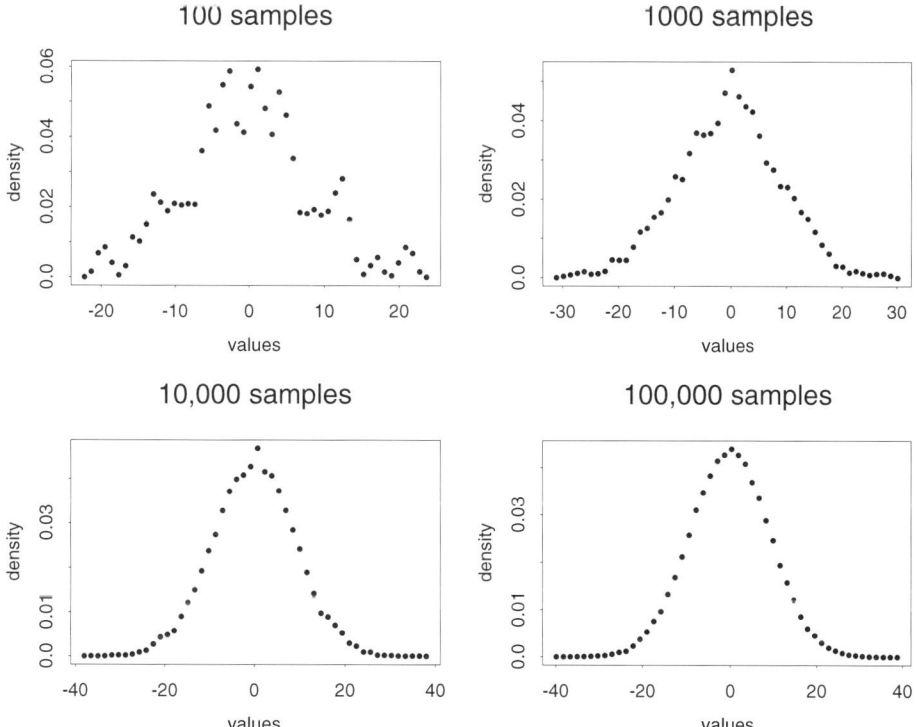

Figure 1.7. Estimates of the probability density of the final position of the random walk, obtained from 10^j independent samples of the centered partial sum $S_{1000} - 500$ for $j = 2, \ldots, 5$, for the case in which the steps U_k are uniformly distributed in the interval $[0, 1]$, based on the nonparametric density estimator *density* from S.

Similarly, the *empirical cdf* of a data set of size n is the proportion $F_n(t)$ of the n data points that are less than or equal to t, as a function of t.

The idea, then, is to look at the *difference* between a cdf and the empirical cdf obtained from an independent sample from that cdf. Moreover, it is natural to consider how that difference behaves as the sample size increases. Once we have made such a study, we can use the established behavior of samples from the specified probability distribution to *test* whether or not data from an unknown source can reasonably be regarded as a sample from the candidate probability distribution.

Example 1.1.1. *The empirical cdf of uniform random numbers.* To illustrate, we now consider the difference between the empirical cdf associated with n uniform random numbers on the interval $[0, 1]$ and the uniform cdf itself. Since the uniform cdf is $F(t) = t, 0 \le t \le 1$, we now want to plot $F_n(t) - t$ versus t for $0 \le t \le 1$. Since the function $F_n(t) - t, 0 \le t \le 1$, is a function of a continuous variable, the plotting is less routine than for the random walk. However, the empirical cdf F_n has special structure, making it possible to do the plotting quite easily. In particular, to do the plotting, let $U_k^{(n)}, 1 \le k \le n$, be the *order statistics* associated with the uniform random numbers $U_1, \ldots, U_n$, i.e., $U_k^{(n)}$ is the kth smallest of the uniform random numbers. Note that

$$F_n(U_k^{(n)}) = k/n \quad \text{and} \quad F_n(U_k^{(n)}-) = (k-1)/n \,,$$

$F_n(0) = 0$ and $F_n(1) = 1$, where $F_n(t-)$ is the left limit of the function F_n at t. Thus we can plot $F_n(t) - t$ versus t by plotting the points $(0,0)$, $(1,0)$, $(U_k^{(n)}, (k-1)/n - U_k^{(n)})$ and $(U_k^{(n)}, k/n - U_k^{(n)})$, $1 \le k \le n$, and connecting the points by lines (i.e., performing linear interpolation).

Plots for $n = 10^j$ for $j = 1, \ldots, 4$ are shown in Figure 1.8. The plots in Figure 1.8 look much like the plots of the uncentered random walks, but there is a subtle difference that can be confirmed by further replications of the experiment. Unlike before, here the final position is 0 just like the initial position. That makes sense as well, because both the empirical cdf and the actual cdf must assume the common value 1 at the right endpoint.

It turns out that there is statistical regularity in the empirical cdf's just like there is in the random walks. As before, the plots look the same for all sufficiently large n. Moreover, except for having the final position be 0, the plots look just like the random-walk plots. More generally, this example illustrates that statistical analysis is an important source of motivation for stochastic-process limits. We discuss this example further in Section 2.2. There we show how to develop a statistical test applicable to any continuous cdf, including the normal cdf that is of interest for the final position of the random walk. ∎

1.1.4. *Making an Interesting Game*

We have digressed from our original game of chance to consider the statistical regularity observed in the plots, which of course really is our main interest. But now let us return for a moment to the game of chance.

A gambling house cannot afford to make the game fair. The gambling house needs to charge a fee greater than the expected payoff in order to make a profit. What would be a good fee for the gambling house to charge?

From the perspective of the gambling house, one might think the larger the fee the better, but the players presumably have the choice of whether or not to play. If the gambling house charges too much, few players will want to play. The fee should be large enough for the gambling house to make money, but small enough so that potential players will want to play. We take that to mean that the individual players should have a good chance of winning.

One might think that those objectives are inconsistent, but they are not. The key to achieving those objectives is the realization that *the player and the gambling house experience the game in different time scales*. An individual player might contemplate playing the game 100 times on a single day, while the gambling house might offer the game to hundreds or thousands of players on each of many consecutive days.

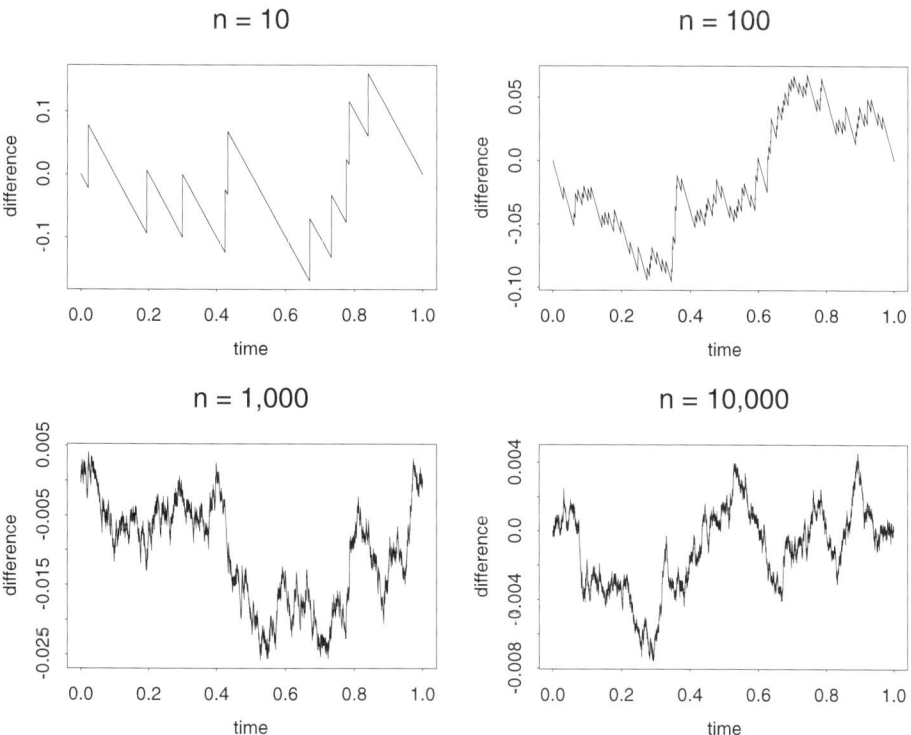

Figure 1.8. The difference between the empirical cdf and the actual cdf for samples of size 10^j from the uniform distribution over the interval $[0,1]$ for $j = 1, \ldots, 4$.

Thus, the player might evaluate his experience by the possible outcomes from about 100 plays of the game, while the gambling house might evaluate its experience by the possible outcomes from something like $10^4 - 10^6$ plays of the game. What we need, then, is a fee close enough to \$0.50 that the player has a good chance of winning in 100 plays, while the gambling house receives a good reliable return over $10^4 - 10^6$ games.

A reasonable fee might be \$0.51, giving the gambling house a 1 cent or 2% advantage on each play. (Gambling houses actually tend to take more, which shows the appeal of gambling despite the odds.) To see how the \$0.51 fee works, let us consider the possible experiences of the player and the gambling house. In Figure 1.9 we plot six independent realizations of a player's position during 100 plays of the game when there is a fee of \$0.51 for each play. The game looks pretty interesting for the player from Figure 1.9. The player has a reasonable chance of winning. Indeed, the player wins in plots 3 and 5, and finishes about even in plot 2. How do things look for the gambling house?

To see how the gambling house fares, we should look at the net payoffs over a much larger number of games. Hence, in Figures 1.10 and 1.11 we plot six independent

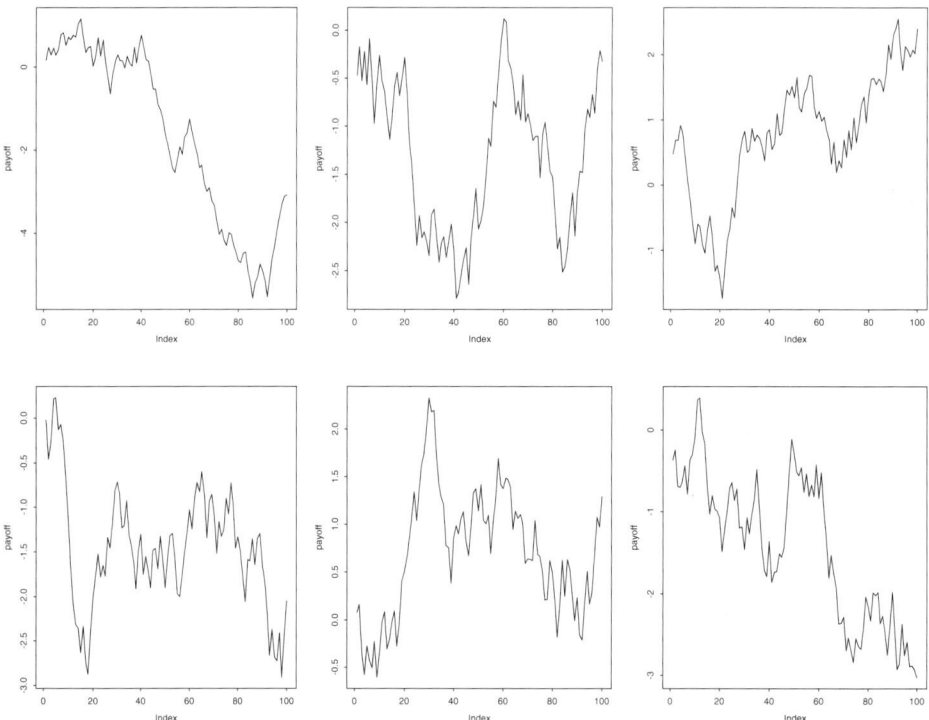

Figure 1.9. Six possible realizations of the first 100 net payoffs, positions of the random walk $\{S_k - 0.51k : k \geq 0\}$, with steps U_k uniformly distributed in the interval $[0,1]$ and a fee of \$0.51.

realizations of a player's position during 10^4 and 10^6 plays of the game. As before, we let the plotter automatically do the scaling, so that the units on the vertical axes change from plot to plot. But that does not alter the conclusions. In these larger time scales, we see that the player consistently loses money, so that a profit for the gambling house becomes essentially a sure thing. When we increase the number of plays to 10^6, there is little randomness left. That is shown in Figure 1.11. Further repetitions of the experiment confirm these observations. We again see the regularity associated with a macroscopic view of uncertainty.

Above we picked a candidate fee out of the air. We could instead be more systematic. For example, we might seek the largest fee such that the player satisfies some criteria indicating a good experience. Letting the fee for each game be f, we might want to constrain the probability p that a player wins at least a certain amount w, i.e., by requiring that

$$P(S_{100} - f(100) \geq w) \geq p \, .$$

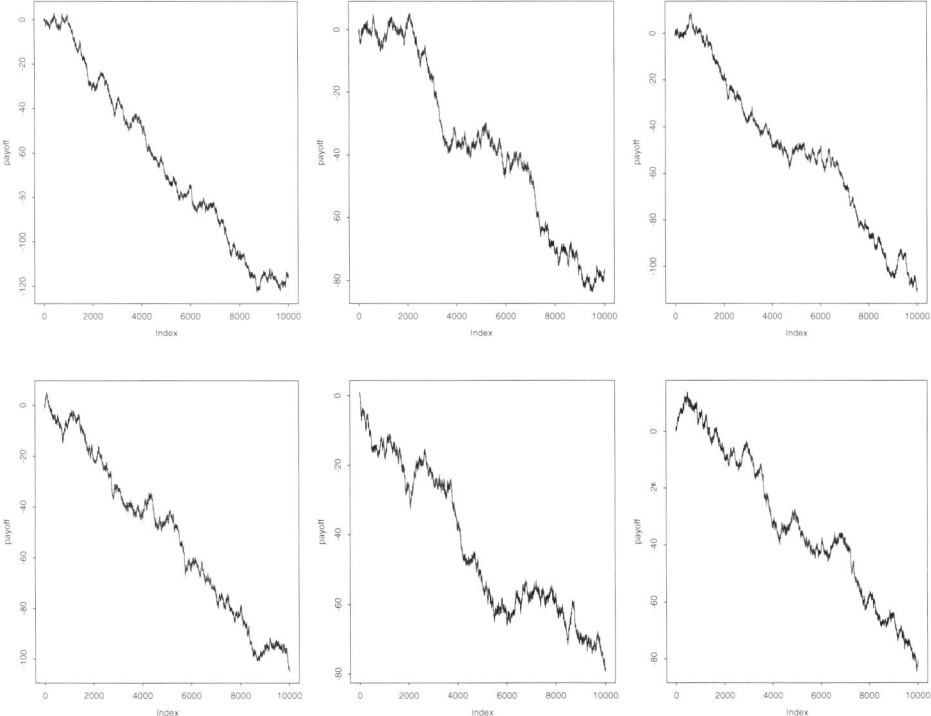

Figure 1.10. Possible realizations of the first 10^4 net payoffs (steps of the random walk $\{S_k - 0.51k : k \geq 0\}$ with steps U_k uniformly distributed in the interval $[0, 1]$.

Given such a formulation, we can determine the optimal fee f, i.e., the maximum fee f such that the constraint is satisfied, which is attained when the probability just equals p.

As noted at the outset, when we consider making the game interesting, we might well conclude that a uniform payoff distribution for each play is boring. We might want to have the possibility of much larger positive and/or negative payoffs on one play. It is easy to devise more interesting games with different payoff distributions, but the statistical regularity associated with large numbers observed above tends to be the same. Readers are invited to make their own games and look at the net payoffs for 10^j plays for various values of j.

An extreme case that is often attractive is to have, like a lottery, some small chance of a very large payoff. However, with independent trials, as determined by successive spins of the spinner, the gambling house faces the danger of having to make too many large payoffs. Such large losses are avoided in lotteries by not letting the game be based on independent trials. In a lottery only a few prizes are awarded (and possibly shared) so that the people running the lottery are guaranteed a positive return. However, an insurance company cannot control the outcomes so tightly,

16   1. Experiencing Statistical Regularity

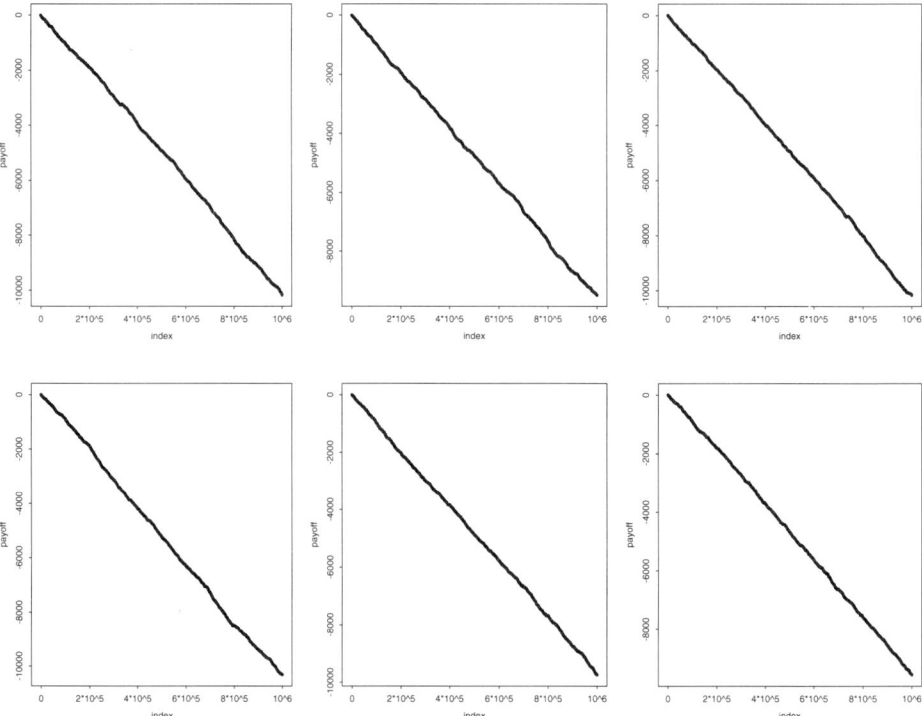

Figure 1.11. Possible realizations of the first 10^6 net payoffs (steps of the random walk $\{S_k - 0.51k : k \geq 0\}$ with steps U_k uniformly distributed in the interval $[0,1]$.

so that careful analysis of the possible outcomes is necessary; e.g., see Embrechts, Klüppelberg and Mikosch (1997). We too will be interested in the possibility of exceptionally large values in random events.

1.2. Stochastic-Process Limits

The plots we have looked at indicate that there is statistical regularity associated with large n, i.e., with large sample sizes. We now want to understand *why* we see what we see, and what we will see in other related situations. For that purpose, we turn to probability theory; see Ross (1993) and Feller (1968) for introductions.

1.2.1. A Probability Model

We can use probability theory to explain what we have seen in the random walk plots. The first step is to introduce an appropriate mathematical model: Assuming that our random number generator is working properly (an important issue, which we will not address, e.g., see p. 123 of Venables and Ripley (1994), L'Ecuyer(1998a,b) and references cited there), the observed values U_k, $1 \leq k \leq n$, should be distributed approximately as the first n values from a sequence of *independent and identically distributed* (IID) random variables uniformly distributed on $[0,1]$ (defined on an underlying probability space). Indeed, the model fit is usually so good that there is a tendency to identify the mathematical model with the physical experiment (a mistake), but since the model fit is so good, we need not doubt that the mathematical conclusions are applicable.

Remark 1.2.1. *Mathematics and the physical world.* It is important to realize that a physical phenomenon, a mathematical model of that physical phenomenon and a simulation of that mathematical model are three different things. But, if the mathematical model is well chosen, the three may be closely related. In particular, a mathematical model, whether simulated or analyzed, may provide useful desciptions of the physical phenomenon.

We are interested in mathematical queueing models because of their ability to explain queueing phenomena, but we should not expect a perfect match. For example, mathematical models often succeed by exploiting the infinite, even though the physical phenomenon is finite. Random numbers generated on a computer are inherently finite, and yet simulations based on random numbers can be well described by mathematical models exploiting the infinite.

Here, we perform stochastic simulations to reveal statistical regularity, and we introduce and analyze mathematical models to explain that statistical regularity. We expect to capture key features, but we do not expect a perfect fit. We want the the mathematics to explain key features observed in the simulations, and we want the simulations to confirm key features predicted by the mathematics. ∎

With that attitude, let us consider the probability model consisting of a sequence of IID uniform random numbers. Within the context of that probability model, we want to formulate stochastic process limits suggested by the plots. First, we see that as n increases the plotted random walk ceases to look discrete. For all sufficiently large n, the plotted random walk looks like a function of a continuous variable. Thus it is natural to seek a continuous-time representation of the original discrete-time random walk. We can do that by considering the associated continuous-time process $\{S_{\lfloor t \rfloor} : t \geq 0\}$, where $\lfloor \cdot \rfloor$ is the *floor function*, i.e., $\lfloor t \rfloor$ denotes the greatest integer less than or equal to t. If we also want to introduce centering, then we do the centering first, and instead consider the centered process $\{S_{\lfloor t \rfloor} - m\lfloor t \rfloor : t \geq 0\}$ for appropriate centering constant m, which here is $1/2$. Thus the continuous-time representation of the random walk is a step function, which coincides with the random walk at integer arguments.

However, the step function is not the only possible continuous-time representation of the random walk. We could instead form a process with continuous sample paths by connecting the points by lines, i.e., by performing a *linear interpolation*. Then, instead of $S_{\lfloor t \rfloor}$, we consider

$$\tilde{S}(t) \equiv (t - \lfloor t \rfloor)S_{\lfloor t \rfloor+1} + (1 + \lfloor t \rfloor - t)S_{\lfloor t \rfloor} \quad \text{for all} \quad t \geq 0 \;, \tag{2.1}$$

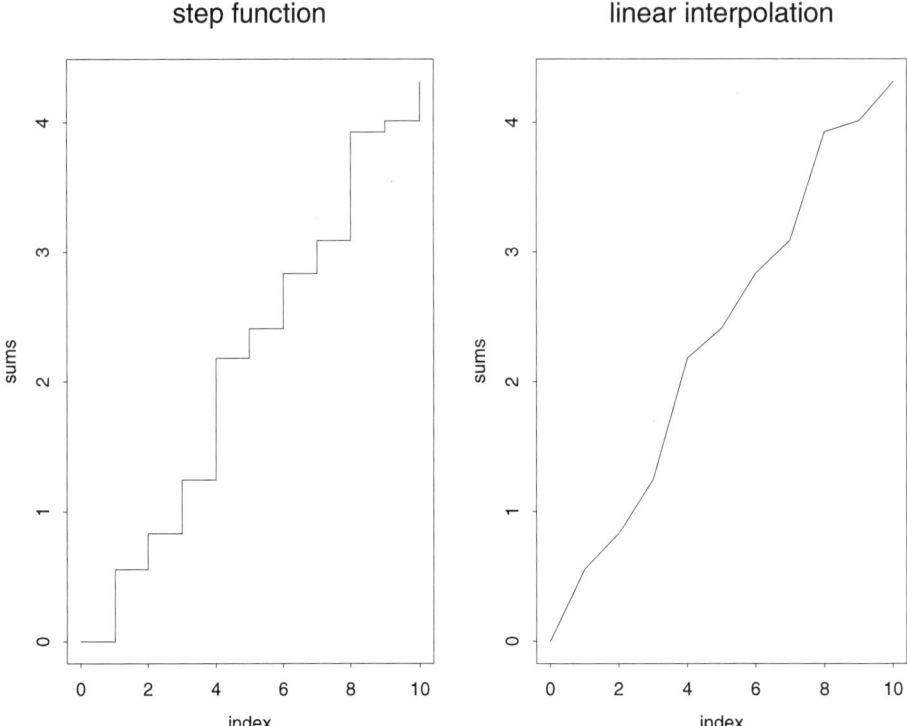

Figure 1.12. Possible initial segments of the two continuous-time stochastic processes constructed from one realization of an uncentered random walk with uniform steps for the case $n = 10$. The step-function representation appears on the left, while the linear-interpolation representation appears on the right.

and similarly if we do centering. (With centering, we do the centering before doing the linear interpolation.) Possible initial segments of the two continuous-time processes associated with the discrete-time (uncentered) random walk for the case $n = 10$ are shown in Figure 1.12. (The vertical lines in the plot are not really part of the step function.) Even though the 10 random walk steps are the same for both continuous-time representations, the two initial segments of the continuous-time stochastic processes look very different in Figure 1.12. However, for large n, plots of the two continuous-time representations of the discrete-time random walk look virtually identical. To make that important point clear, we plot the two continuous-time representations of the same discrete-time centered random walk (same sample paths) for $n = 10^j$ for $j = 1, \ldots, 4$ in Figure 1.13. Figure 1.13 shows that the two alternative representations indeed look the same for all n sufficiently large. Thus, when we focus on the random-walk plots for large n, we regard the two alternatives as equivalent. For our remaining discussion here, though, we will only discuss the step functions.

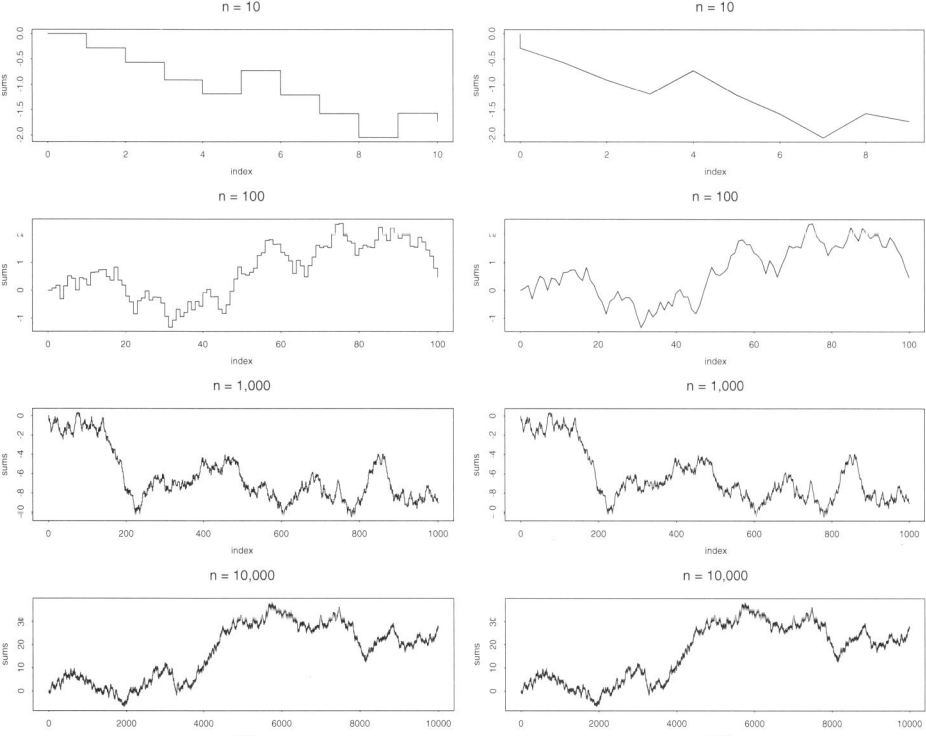

Figure 1.13. The two alternative continuous-time representations of a discrete-time process constructed from common realizations of a centered random walk with uniform steps for $n = 10^j$ with $j = 1, \ldots, 4$. The step-function representation appears on the left, while the linear-interpolation representation appears on the right.

We now want to scale time and space (the horizontal and vertical dimensions in the plots). Note that the plotter scales time by putting the $n+1$ random walk values in a region of fixed width. Thus, if we let 1 be the available width of the plot, then the $n+1$ random walk values are spaced $1/n$ apart. Equivalently, time is scaled automatically by the plotting routine by multiplying time t by n, i.e., by replacing t with nt. Then, for each n, we only look at the process for t in the closed interval $[0,1]$. The final position of the random walk for any n corresponds to $t = 1$.

We can also consider the space scaling in the same way. We can let 1 be the available height of the plot. Then the plotter automatically scales space by subtracting the minimum value and dividing by the range of the plotted values. Unfortunately, however, the range is random. Moreover, there is a complicated dependence between the path and its range. In formulating a stochastic-process limit, it is natural to try to perform the space scaling, like the time scaling, with a deterministic function of n. With such deterministic space scaling, we hope to achieve a nondegenerate limit as $n \to \infty$, but one for which the range is allowed to remain random. In the limit as $n \to \infty$, we will achieve essentially the same thing as the plots if the normalized range converges to a nondegenerate random limit.

What we do, then, is scale space by dividing by c_n, where $\{c_n : n \geq 1\}$ is a sequence of (deterministic) real numbers with $c_n \to \infty$ as $n \to \infty$. That is, for each n, we form the stochastic process

$$\mathbf{S}_n(t) \equiv c_n^{-1}(S_{\lfloor nt \rfloor} - m(\lfloor nt \rfloor)), \quad 0 \leq t \leq 1 \ . \tag{2.2}$$

We then want to find an appropriate sequence $\{c_n : n \geq 1\}$ so that

$$\{\mathbf{S}_n(t) : 0 \leq t \leq 1\} \to \{\mathbf{S}(t) : 0 \leq t \leq 1\} \quad \text{as} \quad n \to \infty \ , \tag{2.3}$$

where $\mathbf{S} \equiv \{\mathbf{S}(t) : 0 \leq t \leq 1\}$ is an appropriate limit process with t ranging over the interval $[0, 1]$ and $\to$ in (2.3) is an appropriate mode of convergence. When we have a limit as in (2.3), we have a *stochastic-process limit*.

1.2.2. Classical Probability Limits

Classical probability limits help explain the statistical regularity we have seen. First, referring to the asymptotically linear plots in Figures 1.1 and 1.11, the *strong law of large numbers* (SLLN) implies that the scaled partial sums $n^{-1}S_n$ approach the mean m as $n \to \infty$ with probability 1 (w.p.1); e.g., see Chapter X of Feller (1968), Chapter VII of Feller (1971) and Chapter 5 of Chung (1974). (In Figure 1.1 the mean is $1/2$; in Figure 1.11 the mean is -0.01.)

As an easy consequence of the SLLN, we can also conclude that

$$n^{-1}S_{\lfloor nt \rfloor} \to mt \quad \text{w.p.1} \quad \text{as} \quad n \to \infty$$

for each $t > 0$. Moreover, the pointwise convergence can actually be extended to uniform convergence over bounded intervals:

$$\{n^{-1}S_{\lfloor nt \rfloor} : 0 \leq t \leq 1\} \to \{mt : 0 \leq t \leq 1\} \quad \text{w.p.1} \quad \text{as} \quad n \to \infty \ ,$$

uniformly in t for t in the interval $[0,1]$. In other words,

$$\sup\{|n^{-1}S_{\lfloor nt \rfloor} - mt| : 0 \leq t \leq 1\} \to 0 \quad w.p.1 \quad \text{as} \quad n \to \infty \ .$$

Thus, in the setting of Figure 1.1, the limit (2.3) holds without centering (with $m = 0$) for $c_n = n$ with the limit process $\mathbf{S}$ being the line with slope $1/2$ (the one-step mean) defined over the interval $[0, 1]$. In this case, the mode of convergence in (2.3) is convergence w.p.1 on a space of functions with the uniform distance

$$\|x_1 - x_2\| \equiv \sup\{|x_1(t) - x_2(t)| : 0 \leq t \leq 1\} \ .$$

In this case, the stochastic-process limit is called a *functional strong law of large numbers* (FSLLN). Interestingly, the SLLN and the FSLLN are actually equivalent; see Theorem 3.2.1 in the Internet Supplement.

Next, turning to the plots of the centered random walks, with centering by the mean, in Figures 1.2, 1.3 and 1.4, we can appeal to the *central limit theorem* (CLT).

The CLT implies that

$$(\sigma^2 n)^{-1/2}(S_n - nm) \Rightarrow N(0,1) \quad \text{as} \quad n \to \infty, \qquad (2.4)$$

where $m \equiv EU_k = 1/2$ is the mean and $\sigma^2 \equiv Var\, U_k = 1/12$ is the variance of the uniform summand U_k, $\Rightarrow$ denotes convergence in distribution and the *standard* normal random variable $N(0,1)$ has cdf

$$\Phi(t) \equiv P(N(0,1) \le x) \equiv \int_{-\infty}^{x} (2\pi)^{-1/2} e^{-u^2/2}\, du ; \qquad (2.5)$$

e.g., see Section VIII.4 of Feller (1971) and Chapter 7 of Chung (1974).

It is useful to review what the limit (2.4) means: The convergence in distribution means that the cdf's converge, i.e.,

$$P(n^{-1/2}(S_n - mn) \le x) \to P(N(0,\sigma^2) \le x) \quad \text{as} \quad n \to \infty \qquad (2.6)$$

for all x. More generally, given real-valued random variables Z_n, $n \ge 1$, and Z, there is convergence in distribution, by the standard definition, denoted by $Z_n \Rightarrow Z$, if the associated cdf's converge, i.e., if

$$F_n(x) \equiv P(Z_n \le x) \to P(Z \le x) \equiv F(x) \quad \text{as} \quad n \to \infty \qquad (2.7)$$

for all x that are continuity points of the limiting cdf F, i.e., for which $P(Z = x) = 0$.

Since the normal distribution has a continuous cdf, the restriction to continuity points of the limiting cdf in (2.7) does not arise in (2.6). We need to allow non-convergence at discontinuity points in (2.7), because we want to say that we have convergence $Z_n \Rightarrow Z$ in situations such as the special case in which $P(Z = z) = 1$ and $P(Z_n = z_n) = 1$ for all n and $z_n \to z$ as $n \to \infty$. If $z_n \to z$ with $z_n > z$ for all n, then $F_n(z) \equiv P(Z_n \le z) = 0$ for all n, while $F(z) \equiv P(Z \le z) = 1$. Since $F_n(x) \to F(x)$ for all x except $x = z$, we obtain the desired convergence $Z_n \Rightarrow Z$ if we require pointwise convergence of the cdf's everywhere except at discontinuity points of the limiting cdf F.

There also are other convenient equivalent characterizations of convergence in distribution. In particular, (2.7) holds if and only if

$$E[h(Z_n)] \to E[h(Z)] \quad \text{as} \quad n \to \infty \qquad (2.8)$$

for every continuous bounded real-valued function h on $\mathbb{R}$, where E is the expectation operator. Moreover, (2.7) and (2.8) hold if and only if

$$g(Z_n) \Rightarrow g(Z) \quad \text{as} \quad n \to \infty \qquad (2.9)$$

for every continuous function g on $\mathbb{R}$. The alternative characterizations (2.8) and (2.9) are useful because they generalize to random elements of more general spaces.

The CLT in (2.4) explains the statistical regularity associated with the final positions of the centered random walks: In agreement with Figures 1.5–1.7, the CLT tells us that the centered partial sums $S_n - mn$ should be approximately normally distributed with mean 0 for all n sufficiently large.

We can also apply the CLT to obtain a corresponding limit for the scaled random walk $\mathbf{S}_n$ in (2.2) at an arbitrary time t in the interval $[0,1]$. More generally, we can consider an arbitrary $t \geq 0$. To do so, we set $c_n = \sqrt{n}$ and $m = 1/2$. In particular, it is an easy consequence of (2.4) that we must have

$$n^{-1/2}(S_{\lfloor nt \rfloor} - m\lfloor nt \rfloor) \Rightarrow \sigma N(0,t) \quad \text{in} \quad \mathbb{R} \quad \text{as} \quad n \to \infty \tag{2.10}$$

for each $t \geq 0$, where $m = 1/2$ and $\sigma^2 = 1/12$.

From (2.10) we clearly see that the space-scaling constants c_n in (2.2) must be asymptotically equivalent to $c\sqrt{n}$ for some constant c as $n \to \infty$. Moreover, the space scaling by $\sqrt{n}$ is consistent with the units on the axes in Figures 1.2–1.4. Indeed, if we instead scale by $c_n = n^p$ for $p > 1/2$, then the values converge to 0 as $n \to \infty$. Similarly, if we scale by $c_n = n^p$ for $p < 1/2$, then the values diverge as $n \to \infty$. (The absolute values diverge to infinity.) This property can be confirmed by further analysis of simulations, but we do not pursue it.

We now want to convert (2.10) into a stochastic-process limit of the form (2.3). Note that the left side of (2.10) coincides with $\mathbf{S}_n(t)$, but the right side of (2.10) is not a stochastic process evaluated at time t. What we need to do is identify the appropriate limit process $\mathbf{S}$ in (2.3).

1.2.3. Identifying the Limit Process

We should recognize that we have arrived at another critical point. Another important intellectual step is needed here. *We not only must identify the limit process; we need to realize that there indeed should be a limit process.*

The appropriate limit process turns out to be a *Brownian motion* (BM). Brownian motion stochastic processes can be characterized as the real-valued stochastic processes with stationary and independent increments having continuous sample paths. Brownian motion evaluated at time t turns out to be normally distributed with mean mt and variance $\sigma^2 t$ for some constants m and σ^2.

The special Brownian motion with parameters $m = 0$ and $\sigma^2 = 1$ is called *standard Brownian motion*; we shall refer to it by $\mathbf{B} \equiv \{\mathbf{B}(t) : t \geq 0\}$. It has marginal distributions

$$\mathbf{B}(t) \stackrel{\mathrm{d}}{=} N(0,t), \quad t \geq 0, \tag{2.11}$$

where $\stackrel{\mathrm{d}}{=}$ denotes equality in distribution.

An increment of Brownian motion is $\mathbf{B}(u) - \mathbf{B}(t)$ for $u > t$. By *stationary and independent increments*, we mean that the k-dimensional random vector

$$(\mathbf{B}(u_1 + h) - \mathbf{B}(t_1 + h), \ldots, \mathbf{B}(u_k + h) - \mathbf{B}(t_k + h))$$

has a distribution independent of h for all k, and that the k component random variables are independent, providing that $0 \leq t_1 \leq u_1 \leq t_2 \leq \cdots \leq u_k$.

Combining (2.10) and (2.11), we see that we can also express the limit (2.10) in terms of Brownian motion. In particular, after letting $c_n = \sqrt{n}$ in (2.2), we see that (2.10) is equivalent to

$$\mathbf{S}_n(t) \Rightarrow \sigma \mathbf{B}(t) \quad \text{in} \quad \mathbb{R} \quad \text{as} \quad n \to \infty \quad \text{for all} \quad t \geq 0, \tag{2.12}$$

where **B** is a standard Brownian motion,

$$\mathbf{S}_n(t) \equiv n^{-1/2}(S_{\lfloor nt \rfloor} - m(\lfloor nt \rfloor)), \quad t \geq 0, \qquad (2.13)$$

and $\sigma^2 = 1/12$ because the steps in the random walk are uniformly distributed over $[0,1]$. In equations (2.11), (2.12) and (2.13) we have let t range over the semi-infinite interval $[0, \infty)$, but we could also have restricted t to the closed interval $[0, 1]$ to be consistent with the plots.

We can apply the limit in (2.12) to generate approximations for the terms of the original random walk. To generate approximations, we replace the convergence in distribution by approximate equality in distribution. From (2.12), we obtain the approximation

$$S_{\lfloor nt \rfloor} \approx m\lfloor nt \rfloor + n^{1/2}\sigma \mathbf{B}(t) \qquad (2.14)$$

or

$$S_k \approx mk + n^{1/2}\sigma \mathbf{B}(k/n), \qquad (2.15)$$

where k is understood to be of order n and $\approx$ means approximately equal to in distribution. Note that the quality of the approximation for large n tends to depend more on the time scaling by n and the space scaling by $\sqrt{n}$ than the limit process $\sigma \mathbf{B}$.

The limit in (2.12) (with t ranging over the unit interval $[0,1]$) can be regarded as the explanation for what we have seen in the random-walk plots. The limit in (2.12) is a *stochastic-process limit*, because it establishes convergence of the sequence of stochastic processes $\{\{\mathbf{S}_n(t) : 0 \leq t \leq 1\} : n \geq 1\}$ in (2.13) to the limiting stochastic process $\{\sigma\mathbf{B}(t) : 0 \leq t \leq 1\}$. However, we want to go beyond the limit as expressed via (2.12). We want to strengthen the form of convergence in order to be able to deduce convergence of related quantities of interest; in particular, we want to show that plots of the centered random walk converge to plots of standard Brownian motion as $n \to \infty$.

The probability law or distribution of a stochastic process is usually specified by the family of its finite-dimensional distributions (f.d.d.'s). Hence, a natural first step is to go beyond convergence of the one-dimensional marginal distributions, which is provided by (2.12), to convergence of the f.d.d.'s, i.e., the k-dimensional marginal distributions for all k. From the assumed independence among the random walk steps, it is not difficult to see that (2.12) can be extended to obtain

$$(\mathbf{S}_n(t_1), \ldots, \mathbf{S}_n(t_k)) \Rightarrow (\sigma\mathbf{B}(t_1), \ldots, \sigma\mathbf{B}(t_k)) \quad \text{in} \quad \mathbb{R}^k \qquad (2.16)$$

as $n \to \infty$ for all positive integers k and all k time points $t_1, \ldots, t_k$ with $0 \leq t_1 < \cdots < t_k \leq 1$, where convergence in distribution of random elements of $\mathbb{R}^k$ is defined by the natural generalization of (2.7), (2.8) or (2.9). Because of the independence among the random walk steps in this example, there is little difference between (2.12) and (2.16), but in general (2.16) is a much stronger conclusion.

However, we want to go even further. We want to go beyond convergence of the f.d.d.'s in (2.16) to convergence of the plots. We want to establish limits for more general functions of the stochastic processes. To do so, *we regard* $\mathbf{S}_n$ *and* $\mathbf{B}$

as random elements of a function space containing all possible sample paths. (A function space is a space of functions.)

For **B**, we could consider the space $C \equiv C([0,1], \mathbb{R})$ of all continuous real-valued functions on the unit interval $[0,1]$, but to include $\mathbf{S}_n$, we need discontinuous functions. (We could work with the space C if we used linearly interpolated random walks, as in (2.1), but we are considering the step functions.) We could consider a space containing all continuous functions and the special step functions that capture the structure of $\mathbf{S}_n$, but with other applications in mind, we consider a larger set of functions. We let the function space be the set $D \equiv D([0,1], \mathbb{R})$ of all real-valued functions on $[0,1]$ that are right-continuous at all t in $[0,1)$ and have left limits everywhere in $(0,1]$, endowed with an appropriate topology (notion of convergence, see Chapter 3).

The desired generalization of (2.12) and (2.16) follows from *Donsker's theorem*. Donsker's theorem is a *functional central limit theorem* (FCLT), which implies here that

$$\mathbf{S}_n \Rightarrow \sigma \mathbf{B} \quad \text{in} \quad D, \tag{2.17}$$

where again $\mathbf{S}_n$ is the scaled random walk in (2.13), **B** is standard Brownian motion and the function space D is endowed with an appropriate topology. We discuss the topology on D and the precise meaning of (2.17) in Section 3.3.

Even though Brownian motion has a relatively simple characterization, it is a special stochastic process. For example, it has the self-similarity property observed in the plots (without limit). In particular, for all $c > 0$, the stochastic process $\{c^{-1/2}\mathbf{B}(ct) : 0 \le t \le 1\}$ has the same probability law on D; equivalently, it has the same finite-dimensional distributions, i.e., the random vector $(c^{-1/2}\mathbf{B}(ct_1), \ldots, c^{-1/2}\mathbf{B}(ct_k))$ has a distribution in $\mathbb{R}^k$ that is independent of c for any positive integer k and any k time points $t_i, 1 \le i \le k$, with $0 < t_1 < \cdots < t_k \le 1$.

Indeed, the self-similarity is a direct consequence of the stochastic-process limit in (2.17): First observe from (2.13) that, for any $c > 0$,

$$\mathbf{S}_{cn}(t) = c^{-1/2}\mathbf{S}_n(ct), \quad t \ge 0 \, . \tag{2.18}$$

By taking limits on both sides of (2.18), we obtain

$$\{\mathbf{B}(t) : 0 \le t \le 1\} \stackrel{\mathrm{d}}{=} \{c^{-1/2}\mathbf{B}(ct) : 0 \le t \le 1\} \, . \tag{2.19}$$

For further discussion, see Section 4.2.

Even though we are postponing a detailed discussion of the meaning of the convergence in (2.17), we can state a convenient characterization, which explains the applied value of (2.17) compared to (2.12) and (2.16). Just as in (2.8), the limit (2.17) means that

$$E[h(\mathbf{S}_n)] \to E[h(\sigma\mathbf{B})] \quad \text{as} \quad n \to \infty \tag{2.20}$$

for every continuous bounded real-valued function h on D. The topology on D enters in by determining which functions h are continuous. Just as with (2.9), (2.20) holds

if and only if
$$g(\mathbf{S}_n) \Rightarrow g(\sigma\mathbf{B}) \quad \text{in} \quad \mathbb{R} \qquad (2.21)$$
for every continuous real-valued function g on D. (It is easy to see that (2.20) implies (2.21) because the composition function $h \circ g$ is a bounded continuous real-valued function whenever g is continuous and h is a bounded continuous real-valued function.) Interestingly, (2.21) is the way that Donsker (1951) originally expressed his FCLT. The convergence of the functionals (real-valued functions) in (2.21) explains why the limit in (2.17) is called a FCLT.

It turns out that we also obtain (2.21) for every continuous function g, regardless of the range. For example, the function g could map D into D. Then we can obtain new stochastic-process limits from any given one. That is an example of the continuous-mapping approach for obtaining stochastic-process limits; see Section 3.4. The representation (2.21) is appealing because it exposes the applied value of (2.17) as an extension of (2.12) and (2.16). We obtain many associated limits from (2.21).

1.2.4. Limits for the Plots

We illustrate the continuous mapping approach by establishing a limit for the plotted random walks, where as before we regard the plot as being in the unit square $[0, 1] \times [0, 1]$.

To establish limits for the plotted random walks, we use the functions $sup : D \to \mathbb{R}$, $inf : D \to \mathbb{R}$, $range : D \to \mathbb{R}$ and $plot : D \to D$, defined for any $x \in D$ by

$$sup(x) \equiv \sup_{0 \leq t \leq 1} x(t),$$

$$inf(x) \equiv \inf_{0 \leq t \leq 1} x(t),$$

$$range(x) \equiv sup(x) - inf(x)$$

and

$$plot(x) \equiv (x - inf(x))/range(x) .$$

Note that $plot(x)$ is an element of D for each $x \in D$ such that $range(x) \neq 0$. Moreover, the function $plot$ is scale invariant, i.e., for each positive scalar c and $x \in D$ with $range(x) \neq 0$,

$$plot(cx) = plot(x) .$$

Fortunately, these functions turn out to preserve convergence in the topologies we consider. (The first three functions are continuous, while the final $plot$ function is continuous at all x for which $range(x) \neq 0$, which turns out to be sufficient.) Hence we obtain the initial limits

$$n^{-1/2} \max_{1 \leq k \leq n} \{S_k - mk\} = sup(\mathbf{S}_n) \Rightarrow sup(\sigma\mathbf{B}) \equiv \sup_{0 \leq t \leq 1} \{\sigma\mathbf{B}(t)\} ,$$

26   1. Experiencing Statistical Regularity

$$n^{-1/2} \min_{1 \leq k \leq n} \{S_k - mk\} = inf(\mathbf{S}_n) \Rightarrow inf(\sigma\mathbf{B}) \equiv \inf_{0 \leq t \leq 1} \{\sigma\mathbf{B}(t)\} ,$$

$$n^{-1/2} range(\{S_k - mk : 0 \leq k \leq n\}) \equiv range(\mathbf{S}_n) \Rightarrow range(\sigma\mathbf{B})$$

in $\mathbb{R}$ and the final desired limit

$$plot(\mathbf{S}_n) \Rightarrow plot(\sigma\mathbf{B}) = plot(\mathbf{B}) \quad \text{in} \quad D ,$$

where

$$plot(\{S_k - mk : 0 \leq k \leq n\}) = plot(\{c_n^{-1}(S_k - mk) : 0 \leq k \leq n\}) \equiv plot(\mathbf{S}_n) ,$$

from Donsker's theorem ((2.17) and (2.21)).

The limit $plot(\mathbf{S}_n) \Rightarrow plot(\mathbf{B})$ states that the plot of the scaled random walk converges to the plot of standard Brownian motion. Note that we use *plot*, not only as a function mapping D into D, but as a function mapping $\mathbb{R}^{n+1}$ into D taking the random walk segment into its plot.) Hence Donsker's theorem implies that the random walk plots can indeed be regarded as approximate plots of Brownian motion for all sufficiently large n. By using the FCLT refinement, we see that the stochastic-process limits do indeed explain the statistical regularity observed in the plots.

To highlight this important result, we state it formally as a theorem. Later chapters will provide a proof; specifically, we can apply Sections 3.4, 12.7 and 13.4.

Theorem 1.2.1. (convergence of plots to the plot of standard Brownian motion) *Consider an arbitrary stochastic sequence* $\{S_k : k \geq 0\}$. *Suppose that the limit in (2.3) holds in the space D with one of the Skorohod nonuniform topologies, where $c_n = \sqrt{n}$ and $\mathbf{S} = \sigma\mathbf{B}$ for some positive constant σ, with $\mathbf{B}$ being standard Brownian motion, as occurs in Donsker's theorem. Then*

$$plot(\{S_k - mk : 0 \leq k \leq n\}) \Rightarrow plot(\mathbf{B}) .$$

But an even more general result holds: *We have convergence of the plots for any space-scaling constants and almost any limit process.* We have the following more general theorem (proved in the same way as Theorem 1.2.1).

Theorem 1.2.2. (convergence of plots associated with any stochastic-process limit) *Consider an arbitrary stochastic sequence* $\{S_k : k \geq 0\}$. *Suppose that the limit in (2.3) holds in the space D with one of the Skorohod nonuniform topologies, where c_n and $\mathbf{S}$ are arbitrary. If*

$$P(range(\mathbf{S}) = 0) = 0 ,$$

then

$$plot(\{S_k - mk : 0 \leq k \leq n\}) \Rightarrow plot(\mathbf{S}) .$$

Note that the functions *sup, inf, range* and *plot* depend on more than one value $x(t)$ of the function x; they depend on the function over an initial segment. Thus, we exploit the strength of the limit in D in (2.17) as opposed to the limit in $\mathbb{R}$ in (2.12) or even the limit in $\mathbb{R}^k$ in (2.16). For the random walk we have considered (with IID uniform random steps), the three forms of convergence in (2.12), (2.16) and (2.17) all hold, but in general (2.16) is strictly stronger than (2.12) and (2.17) is strictly stronger than (2.16). Formulating the stochastic-process limits in D means that we can obtain many more limits for related quantities of interest, because many more quantitites of interest can be represented as images of continuous functions on the space of stochastic-process sample paths.

Remark 1.2.2. *Limits for the relative final position.* As noted in Remark 1.1.1, if we look at the final position of the centered random walk in the plots, ignoring the units on the axes, then we actually see the relative final position of the centered random walk, as defined in (1.1). Statistical regularity for the relative final position also follows directly from Theorems 1.2.1 and 1.2.2, because the relative final position is just the plot evaluated at time 1, i.e., $plot(x)(1)$. Provided that 1 is almost surely a continuity point of the limit process $\mathbf{S}$, under the conditions of Theorem 1.2.2 we have

$$R_n \Rightarrow plot(\mathbf{S})(1) \quad \text{in} \quad \mathbb{R} \quad \text{as} \quad n \to \infty,$$

as a consequence of the continuous-mapping approach, using the projection map that maps $x \in D$ into $x(1)$. ∎

To summarize, the random-walk plots *reveal* remarkable statistical regularity associated with large n because the plotter automatically does the required scaling. In turn, the stochastic-process limits *explain* the statistical regularity observed in the plots. In particular, Donsker's FCLT implies that the random-walk plots converge in distribution to the plots of standard Brownian motion as $n \to \infty$.

1.3. Invariance Principles

The random walks we have considered so far are very special: the steps are IID with a uniform distribution in the interval $[0, 1]$. However, the great power of the SLLN, FSLLN, CLT and FCLT is that they hold much more generally. Essentially the same limits hold in many situations in which the step distribution is changed or the IID condition is relaxed, or both. Moreover, the limits each depend on only a single parameter of the random walk. The limits in the SLLN and the FSLLN only involve the single parameter m, which is the mean step size in the IID case. Similarly, after centering is done, the limits in the CLT and FCLT only involve the single parameter σ^2, which is the variance of the step size in the IID case. Thus these limit theorems are *invariance principles*.

Moreover, the plots have an even stronger invariance property, because the limiting plots have no parameters at all! (We are thinking of the plot being in the unit square $[0, 1] \times [0, 1]$ in every case, ignoring the units on the axes.) Assuming only that the mean is positive, the plots of the uncentered random walk (with arbitrary step-size distribution) approach the identity function $e \equiv e(t) \equiv t, \quad 0 \leq t \leq 1$. If instead the mean is negative, then the limiting plot is $-e$ over the interval $[0, 1]$.

28   1. Experiencing Statistical Regularity

Similarly, the plots of the centered random walks approach the plot of standard Brownian motion over $[0, 1]$; i.e., the limiting plot does not depend on the variance σ^2. Thus, the random-walk plots reveal remarkable statistical regularity!

The power of the invariance principles is phenomenal. We will give some indication by giving a few examples and by indicating how they can be applied. We recommend further experimentation to become a true believer. For example, the plots of the partial sums – centered and uncentered – should be contrasted with corresponding plots for the random-walk steps. Even for large n, plots of uniform random numbers and exponential (exponentially distributed) random numbers look very different, whereas the plots of the corresponding partial sums look the same (for all n sufficiently large).

1.3.1. The Range of Brownian Motion

We can apply the invariance property to help determine limiting probability distributions. For example, we can apply the invariance property to help determine the distribution of the limiting random variables $sup(\mathbf{B})$ and $range(\mathbf{B})$.

We first consider the supremum $sup(\mathbf{B})$. We can use combinatorial methods to calculate the distribution of $\max_{1 \leq k \leq n}\{S_k - km\}$ for any given n for the special case of the *simple random walk*, with $P(X_1 = +1) = P(X_1 = -1) = 1/2$, as shown in Chapter III of Feller (1968) or Section 11 of Billingsley (1968). In that way, we obtain

$$P(sup(\mathbf{B}) > x) = 2P(N(0,1) > x) \equiv 2\Phi^c(x) ,\qquad(3.1)$$

where $\Phi^c(t) \equiv 1 - \Phi(t)$ for Φ in (2.5). Since $sup(\mathbf{B}) \stackrel{d}{=} |N(0,1)|$,

$$E[sup(\mathbf{B})] = \sqrt{2/\pi} \approx 0.8 \qquad(3.2)$$

and

$$E[sup(\mathbf{B})^2] = E[N(0,1)^2] = 1 .$$

These calculations are not entirely elementary; for details see 26.2.3, 26.2.41 and 26.2.46 in Abramowitz and Stegun (1972).

The limit $range(\sigma\mathbf{B})$ is more complicated, but it too can be characterized; see Section 11 of Billingsley (1968) and Borodin and Salminen (1996). There the combinatorial methods for the simple random walk are used again to determine the joint distribution of $inf(\mathbf{B})$ and $sup(\mathbf{B})$, yielding

$$P(a < inf(\mathbf{B}) < sup(\mathbf{B}) < b) = \sum_{k=-\infty}^{k=+\infty} (-1)^k [\Phi(b + k(b-a)) - \Phi(a + k(b-a))] ,$$

where Φ is again the standard normal cdf. From (3.2), we see that the mean of the range is

$$E[range(\mathbf{B})] = E[sup(\mathbf{B})] - E[inf(\mathbf{B})] = 2E[sup(\mathbf{B})] = 2\sqrt{2/\pi} \approx 1.6 .$$

We can perform multiple replications of random-walk simulations to estimate the distribution of $range(\mathbf{B})$ and associated summary characteristics such as the variance. We show the estimate of the probability density function of $range(\mathbf{B})$ based on $10{,}000$ samples of the random walk with $10{,}000$ steps, each uniformly distributed on $[0,1]$, in Figure 1.14 (again obtained using the nonparametric density estimator *density* from S). The range of the centered random walk should be approximately $\sigma\sqrt{n}$ times the range $range(\mathbf{B})$, so we divide the observed ranges in this experiment by $\sqrt{n/12} = 28.8675$. The estimated mean and standard deviation of $range(\mathbf{B})$ were 1.58 and 0.474, respectively. The estimated $0.1, 0.25, 0.5, 0.75$ and 0.9 quantiles were $1.05, 1.24, 1.50, 1.85$ and 2.23, respectively. This characterization of the distribution of $range(\mathbf{B})$ helps us interpret what we see in the random-walk plots.

From the analysis above, we know approximately what the mean and standard deviation of the range should be in the random-walk plots. Since $E[range(\mathbf{B})] \approx 1.6$, the mean of the random walk range should be about $1.6\sigma\sqrt{n} \approx 0.46\sqrt{n}$. Similarly, since the standard deviation of $range(\mathbf{B})$ is approximately 0.47, the standard deviation of the range in the random-walk plot should be approximately $0.47\sigma\sqrt{n} \approx 0.14\sqrt{n}$. Hence the (mean, standard deviation) pairs in Figures 1.3 and 1.4 with $n = 10^4$ and $n = 10^6$ are, respectively, $(46, 14)$ and $(460, 140)$. Note that the six observed values in each case are consistent with these pairs.

Historically, the development of the limiting behavior of $sup(\mathbf{S}_n)$ played a key role in the development of the general theory; e.g. see the papers by Erdös and Kac (1946), Donsker (1951), Prohorov (1956) and Skorohod (1956).

Remark 1.3.1. *Fixed space scaling.* In our plots, we have let the plotter automatically determine the units on the vertical axis. Theorems 1.2.1 and 1.2.2 show that there is striking statistical regularity associated with automatic plotting. However, for comparison, it is often desirable to have common units. Interestingly, Donsker's FCLT and the analysis of the range above shows how to determine appropriate units for the vertical axis for the centered random walk, before the simulations are run.

First, the CLT and FCLT tell us the range of values for the centered random walk should be of order $\sqrt{n}$ as the sample size n grows. The invariance principle tells us that, for suitably large n the scaling should depend on the random-walk-step distribution only through its variance σ^2.

The limit for the supremum $sup(\mathbf{S}_n) \equiv n^{-1/2}\max_{1 \leq k \leq n}\{S_k - mk\}$ tells us more precisely what fixed space scaling should be appropriate for the plots. Since $2P(N(0,1) \geq 4)$ may be judged suitably small, from (3.1) we conclude that it should usually be appropriate to let the values on the vertical axis for a centered random walk fall in the interval $[-4\sigma\sqrt{n}, 4\sigma\sqrt{n}]$ as a function of n and σ^2. For example, we could use this space scaling to replot the six random-walk plots in Figure 1.4. Since $n = 10^6$ and $\sigma^2 = 1/12$ there, we would let the values on the vertical axes in Figure 1.4 fall in the interval $[-1155, 1155]$. Notice that the values for the six plots all fall in the interval $[-700, 450]$, so that this fixed space scaling would work in Figure 1.4. ∎

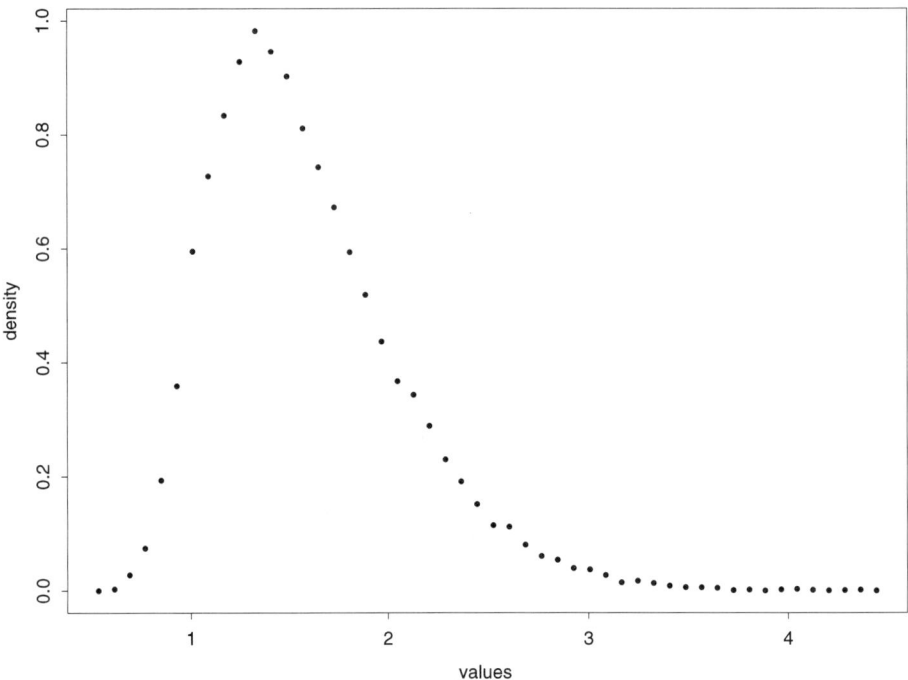

Figure 1.14. An estimate of the probability density of the range of Brownian motion over $[0, 1]$, obtained from $10,000$ independent samples of random walks with $10,000$ steps, each step being uniformly distributed in the interval $[0, 1]$.

To gain a better appreciation of the invariance property, we perform some more simulations. First, we want to see that the IID conditions are *not* necessary.

1.3.2. Relaxing the IID Conditions

To illustrate how the IID conditions can be relaxed, we consider *exponential smoothing*.

Example 1.3.1. *Exponential smoothing.* We now consider a simple example of a random walk in which the steps are neither independent nor identically distributed. We let the steps be constructed by exponential smoothing. Equivalently, the steps are an autoregressive moving-average (ARMA) process of order (1,0); see Section 4.6.

In particular, suppose that we generate uniform random numbers U_k on the interval $[0, 1]$, $k \geq 1$, as before, but we now let the kth step of the random walk be defined recursively by

$$X_k \equiv (1 - \gamma)X_{k-1} + \gamma U_k, \quad k \geq 1 , \tag{3.3}$$

where $X_0 = U_0$, where U_0 is another uniform random number on $[0, 1]$ and $0 < \gamma < 1$. Clearly, the new random variables X_k are neither independent nor identically distributed. Moreover, the distribution of X_k is no longer uniform. It is not difficult to see, though, that as k increases the distribution of X_k approaches a nondegenerate limit. More generally, the sequence $\{X_{n+k} : k \geq 0\}$ is asymptotically stationary as $n \to \infty$, but successive random variables remain dependent.

We now regard the random variables X_k as steps of a random walk; i.e., we let the successive positions of the random walk be

$$S_k \equiv X_1 + \cdots + X_k, \quad k \geq 1, \qquad (3.4)$$

where $S_0 \equiv 0$. Next we repeat the experiments done before. We display plots of the uncentered and centered random walks with $\gamma = 0.2$ for $n = 10^j$ with $j = 1, \ldots, 4$ in Figures 1.15 and 1.16. To determine the appropriate centering constant (the steady-state mean of X_k), we solve the equation

$$E[X] = (1 - \gamma)E[X] + \gamma E[U]$$

to obtain $m \equiv E[X] = E[U] = 1/2$. Even though the distribution of X_k changes with k, the mean remains unchanged because of our choice of the initial condition.

Figures 1.15 and 1.16 look much like Figures 1.1 and 1.2 for the IID case. However, there is some significant difference for small n because the successive steps are positively correlated, causing the initial steps to be alike. However, the plots look like the previous plots for larger n. For the centered random walks in Figure 1.16 with $n = 10^4$, what we see is again approximately a plot of Brownian motion. ∎

We can easily construct many other examples of random walks with dependent steps. For instance, we could consider a *random walk in a random environment*. A simple example has a two-state Markov-chain environment process with transition probabilities $P_{1,2} = 1 - P_{1,1} = p$ and $P_{2,1} = 1 - P_{2,2} = q$ for $0 < p < 1$ and $0 < q < 1$. We then let the kth step X_k have one distribution if the Markov chain is in state 1 at the kth step, and another distribution if the Markov chain is in state 2 then. We first run the Markov chain. Then, conditional on the realized states of the Markov chain, the random variables X_k are mutually independent with the appropriate distributions (depending upon the state of the Markov chain). If we consider a stationary version of the Markov chain, then the sequence $\{X_k : k \geq 1\}$ is stationary. Regardless of the initial conditions, we again see the same statistical regularity in the associated partial sums when n is sufficiently large. We invite the reader to consider such examples.

1.3.3. Different Step Distributions

Now let us return to random walks with IID steps and consider different possible step distributions. We now repeat the experiments above with various functions of the uniform random numbers, i.e., for $X_k \equiv f(U_k)$, $1 \leq k \leq n$, for different real-valued functions f. In particular, consider the following three cases:

$$\text{(i)} \quad X_k \equiv -m \log(1 - U_k) \quad \text{for} \quad m = 1, 10$$

32   1. Experiencing Statistical Regularity

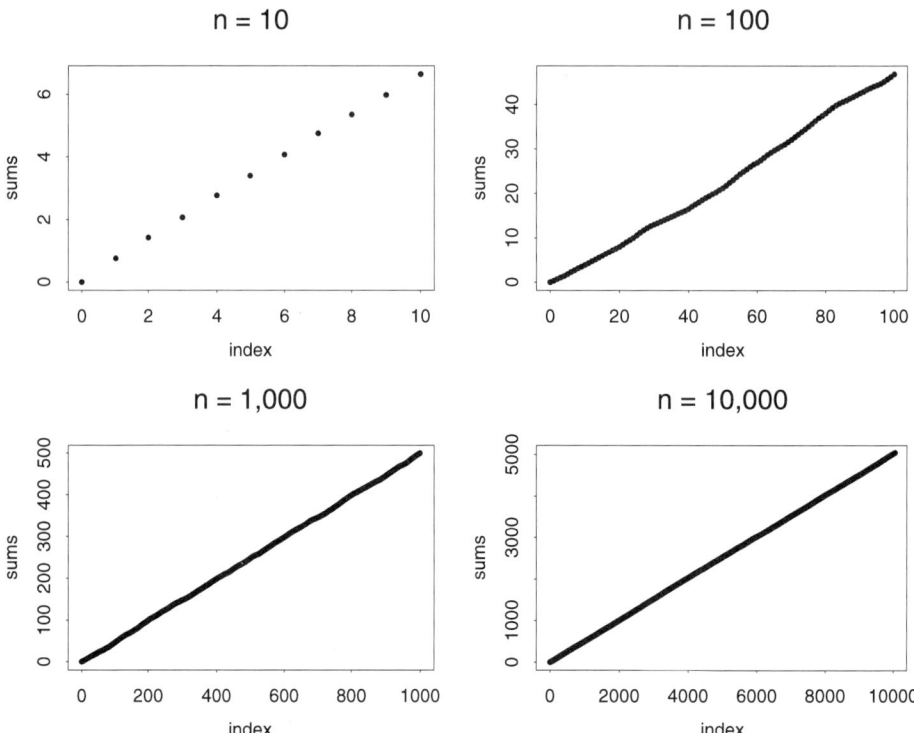

Figure 1.15. Possible realizations of the first 10^j steps of the uncentered random walk $\{S_k : k \geq 0\}$ with steps constructed by exponential smoothing, as in (3.3), for $j = 1, \ldots, 4$.

$$\begin{align} \text{(ii)} \quad & X_k \equiv U_k^p \quad \text{for} \quad p = 1/2, 3/2 \\ \text{(iii)} \quad & X_k \equiv U_k^{-1/p} \quad \text{for} \quad p = 1/2, 3/2 \,. \end{align} \tag{3.5}$$

As before, we form partial sums associated with the new summands X_k, just as in (3.4).

Before actually considering the plots, we observe that what we are doing covers the general IID case. Given the sequence of IID random variables $\{U_k : k \geq 1\}$, by the method above we can create an associated sequence of IID random variables $\{X_k : k \geq 1\}$ where X_k has an arbitrary cdf F. Letting the left-continuous inverse of F be

$$F^{\leftarrow}(t) \equiv \inf\{s : F(s) \geq t\}, \quad 0 < t < 1 \,,$$

we can obtain the desired random variables X_k with cdf F by letting

$$X_k \equiv F^{\leftarrow}(U_k), \quad k \geq 1 \,. \tag{3.6}$$

Since

$$F^{\leftarrow}(s) \leq t \quad \text{if and only if} \quad F(t) \geq s \,, \tag{3.7}$$

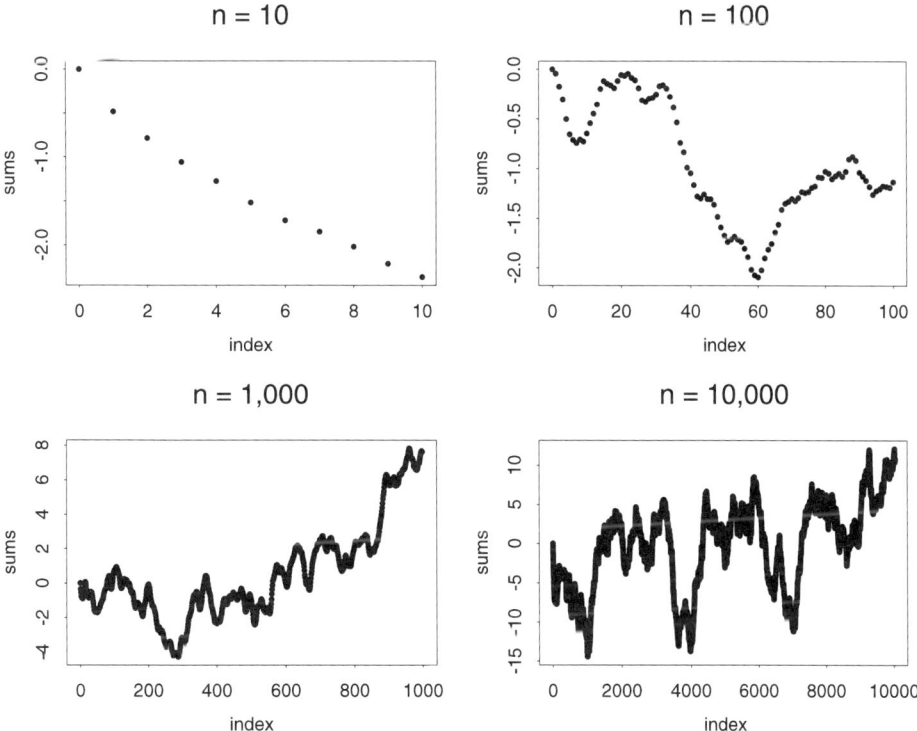

Figure 1.16. Possible realizations of the first 10^j steps of the centered random walk $\{S_k - k/2 : k \geq 0\}$ with steps constructed by exponential smoothing, as in (3.3), for $j = 1, \ldots, 4$.

we obtain
$$P(F^{\leftarrow}(U) \leq t) = P(U \leq F(t)) = F(t) ,$$
where U is a random variable uniformly distributed on $[0, 1]$, which implies that $F^{\leftarrow}(U)$ has cdf F for any cdf F when U is uniformly distributed on $[0, 1]$. For example, we see that X_k has an exponential distribution with mean m in case (i) of (3.5): If $F(t) = e^{-t/m}$, then $F^{\leftarrow}(t) = -m\log(1-t)$ and
$$P(X_k > t) = P(-m\log(1-U_k) > t) = P(1 - U_k < e^{-t/m}) = e^{-t/m} .$$
Incidentally, we could also work with the right-continuous inverse of F, defined by
$$F^{-1}(t) \equiv \inf\{s : F(s) > t\} = F^{\leftarrow}(t+), \quad 0 < t < 1 ,$$
where $F^{\leftarrow}(t+)$ is the right limit at t, because
$$P(F^{-1}(U) = F^{\leftarrow}(U)) = 1 ,$$
since $F^{\leftarrow}$ and F^{-1} differ at, at most, countably many points.

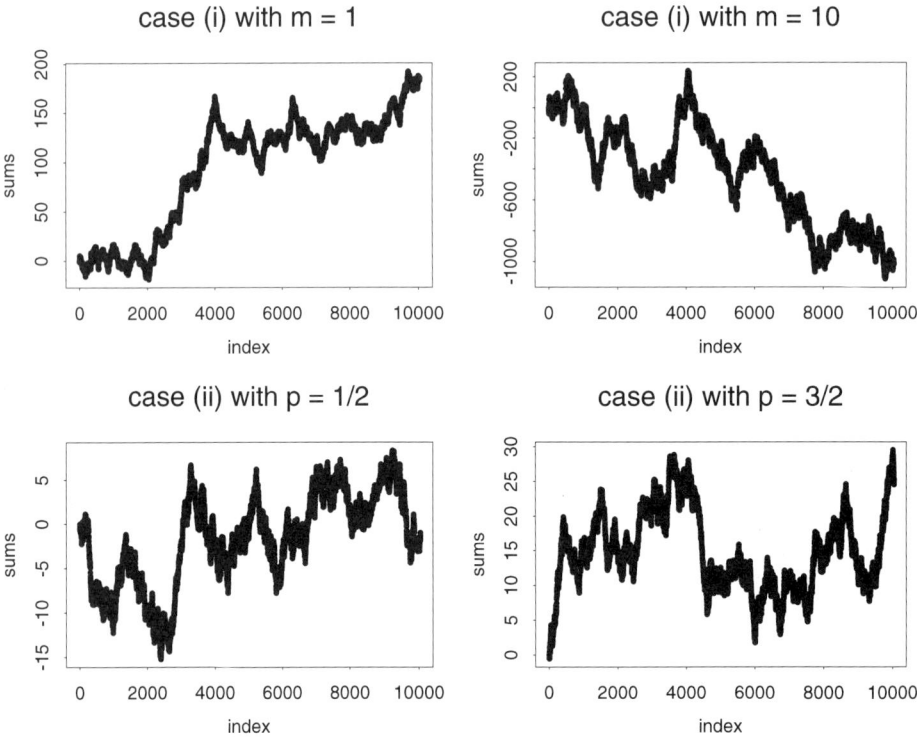

Figure 1.17. Possible realizations of the first 10^4 steps of the random walk $\{S_k - mk : k \geq 0\}$ with steps distributed as X_k in cases (i) and (ii) of (3.5).

Moreover, $F^{\leftarrow}(U_k)$, $k \geq 1$, are IID when U_k, $k \geq 1$, are IID. Of course, there also are other ways to generate IID random variables with specified distributions, but what we are doing is often a natural way.

So let us plot the uncentered and centered random walks with the step sizes in (3.5). When we do so for cases (i) and (ii), we see essentially the same pictures as before. For example, plots of the first 10^4 steps of the centered random walks in the four cases in (i) and (ii) of (3.5) are shown in Figure 1.17.

Again the plots look like plots of Brownian motion, indistinguishable from the plots for the uniform steps in Figure 1.3. Note that the units on the y axis change from plot to plot, but the plots themselves tend to have a common distribution.

1.4. The Exception Makes the Rule

Just when boredom has begun to set in, after seeing the same thing in cases (i) and (ii) in (3.5), we should be ready to appreciate the startlingly different large-n

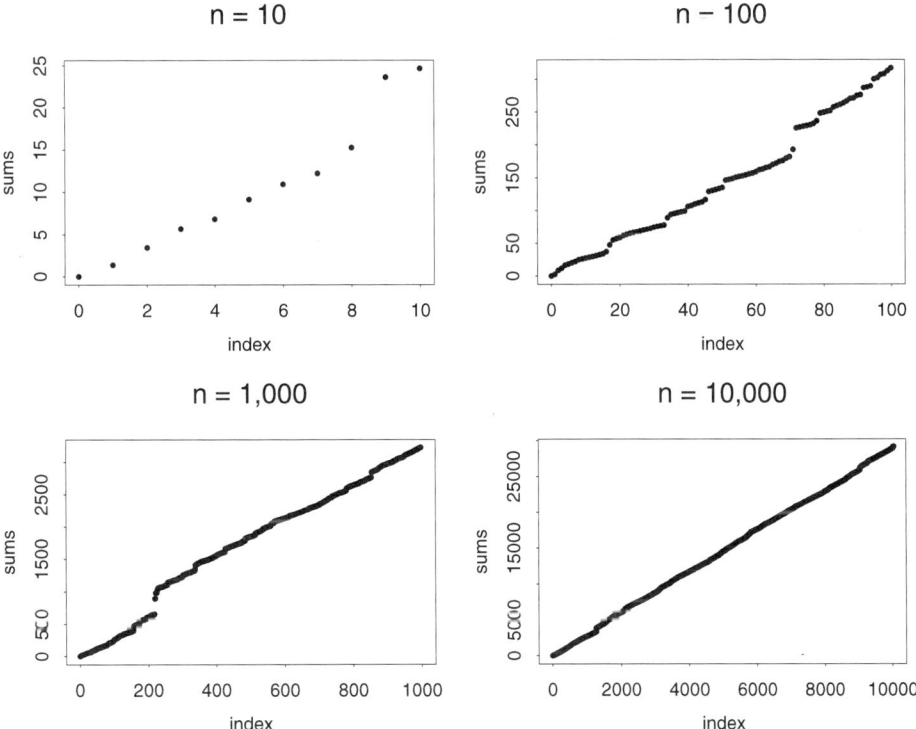

Figure 1.18. Possible realizations of the first 10^j steps of the uncentered random walk $\{S_k : k \geq 0\}$ with steps distributed as $U_k^{-1/p}$ in case (iii) of (3.5) for $p = 3/2$ and $j = 1, \ldots, 4$.

pictures in case (iii). Plots of the uncentered random walks are plotted in Figures 1.18 and 1.19.

In the case $p = 3/2$ in Figure 1.18, the plot of the uncentered random walk is again approaching a line as $n \to \infty$, but not as rapidly as before. (Again we ignore the units on the axes when we look at the plots.) However, in the case $p = 1/2$ in Figure 1.19 we something radically different: For large n, the plots have *jumps*!

1.4.1. Explaining the Irregularity

Fortunately, probability theory again provides an explanation for the *irregularity* that we now see: The SLLN states, under the prevailing IID assumptions, that scaled partial sums $n^{-1}S_n$ will approach the mean EX_1 w.p.1 as $n \to \infty$, regardless of other properties of the probability distribution of X_1, *provided that a finite mean exists*. Knowing the SLLN, we should expect to see lines when $n = 10^4$ in all experiments except possibly in case (iii).

36   1. Experiencing Statistical Regularity

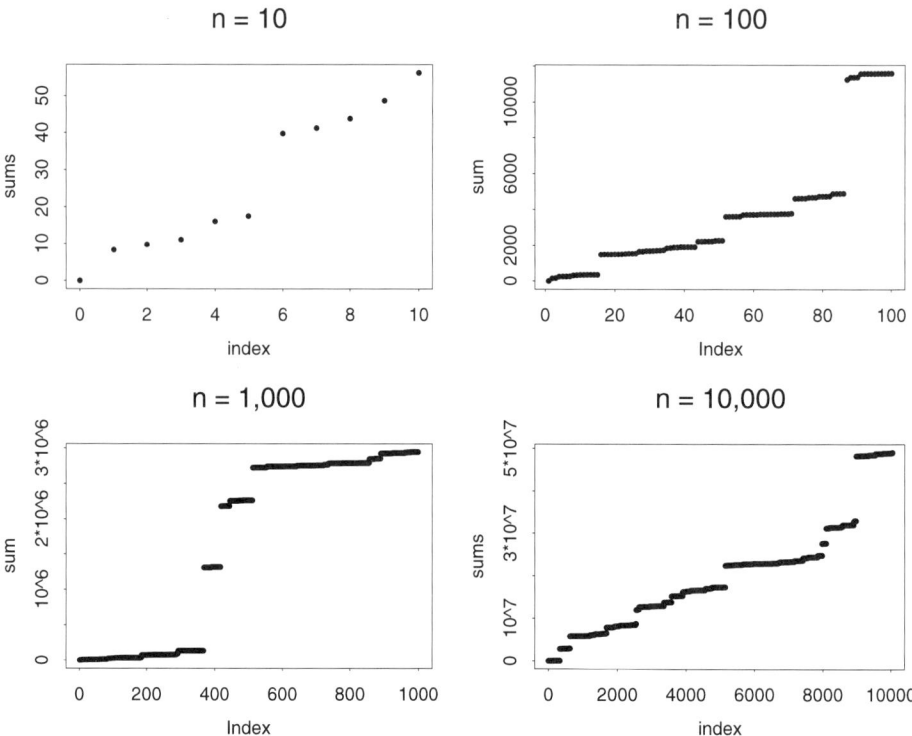

Figure 1.19. Possible realizations of the first 10^j steps of the uncentered random walk $\{S_k : k \geq 0\}$ with steps distributed as $U_k^{-1/p}$ in case (iii) of (3.5) for $p = 1/2$ and $j = 1, \ldots, 4$.

We might initially be fooled in case (iii), but we should anticipate occasional large steps because $U^{-1/p}$ involves *dividing* by very small values when U is small. Upon more careful examination, we see that $U^{-1/p}$ has a *Pareto distribution* with parameter p, which we refer to as Pareto(p), when U is uniformly distributed on $[0,1]$, i.e.,

$$P(U^{-1/p} > t) = P(U < t^{-p}) = t^{-p}, \quad t \geq 1, \tag{4.1}$$

with mean

$$E(U^{-1/p}) = \int_0^\infty P(U^{-1/p} > t)dt = 1 + \int_1^\infty t^{-p}dt, \tag{4.2}$$

which is finite, and equal to $1 + (p-1)^{-1}$, if and only if $p > 1$; see Chapter 19 of Johnson and Kotz (1970) for background on the Pareto distribution and Lemma 1 on p. 150 of Feller (1971) for the integral representation of the mean.

Thus the SLLN tells us not to expect the same behavior observed in the previous experiments in case (iii) when $p \leq 1$. Thus, unlike all previous random walks considered, the conditions of the SLLN are *not satisfied* in case (iii) with $p = 1/2$.

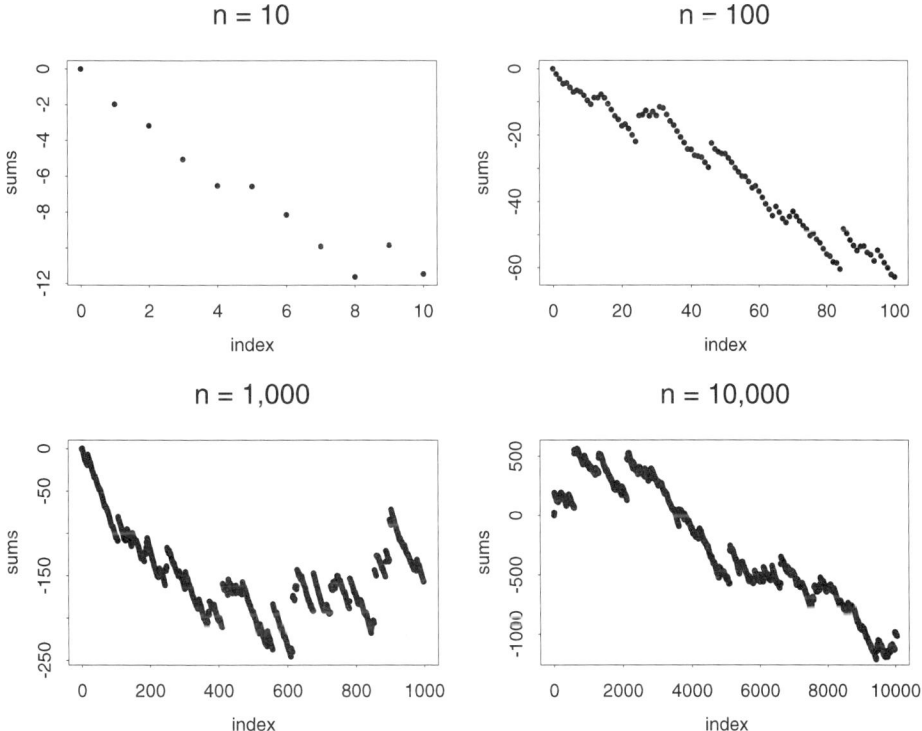

Figure 1.20. Possible realizations of the first 10^j steps of the centered random walk $\{S_k - 3k : k \geq 0\}$ associated with the Pareto steps $U_k^{-1/p}$ for $p = 3/2$, having mean 3 and infinite variance, for the cases $j = 1, \ldots, 4$.

Now let us consider the random walk with Pareto(p) steps for $p = 3/2$ in (3.5) (iii). Consistent with the SLLN, Figure 1.18 shows that the plots are approaching a straight line as $n \to \infty$ in this case. But what happens when we center?

1.4.2. The Centered Random Walk with $p = 3/2$

So now let us consider the centered random walk in case (iii) with $p = 3/2$. (Since the mean is infinite when $p = 1/2$, we cannot center when $p = 1/2$. We will return to the case $p = 1/2$ later.) We center by subtracting the mean, which in the case $p = 3/2$ is $1 + (p-1)^{-1} = 3$. Plots of the centered random walk with $p = 3/2$ for $n = 10^j$ with $j = 1, 2, 3, 4$ are shown in Figure 1.20. As before, the centering causes the plotter to automatically blow up the picture. However, now the slight departures from linearity for large n in Figure 1.18 are magnified. Now, just as in Figure 1.19, we see jumps in the plot!

Once again, probability theory offers an explanation. Just as the SLLN ceases to apply when the IID summands have infinite mean, so does the (classical) CLT cease to apply when the IID summands have finite mean but infinite variance. Such a case occurs with the Pareto(p) summands in case (iii) in (3.5) when $1 < p \le 2$. Thus, consistent with what we see in Figure 1.18, the SLLN holds, but the CLT does not, for the Pareto(p) random variable $U^{-1/p}$ in case (iii) when $p = 3/2$.

We have arrived at another critical point, where an important intellectual step is needed. We need to recognize that, *even though the sample paths are very different from the previous random-walk plots, which are approaching plots of Brownian motion, there may still be important statistical regularity in the new plots with jumps.*

To see the statistical regularity, we need to repeat the experiment and consider larger values of n. Even though the plots look quite different from the previous random-walk plots, we can see statistical regularity in the plots (again ignoring the units on the axes). To confirm that observation, six possible realizations for $p = 3/2$ in the cases $n = 10^4$ and $n = 10^6$ are shown in Figures 1.21 and 1.22. Figures 1.21 and 1.22 show more irregular paths, but with their own distinct character, much like handwriting. (We might contemplate the probability of the path writing a word. With a suitable font for the script, we might see "Null" but not "Set".) Again, Figures 1.21 and 1.22 show that there is statistical regularity associated with the irregularity we see. The plots are independent of n for all n sufficiently large. Again we see self-similarity in the plots.

Even though the irregular paths in Figures 1.19 – 1.22 have jumps, as before we can look for statistical regularity through the distribution of these random paths. Again, to be able to see something, we can focus on the final positions. Focusing first on the case with $p = 3/2$, we plot the estimated density of the centered sums $S_n - 3n$ for $n = 1,000$. Once again, we obtain the density estimate by performing independent replications of the experiment. To have more data this time, we use 10,000 independent replications. We display the resulting density estimate in Figure 1.23.

When we look at the estimated density of the final position, we see that it is radically different from the previous density plots in Figures 1.5 and 1.7. Clearly, *the final position is no longer normally distributed!*

Nevertheless, there is statistical regularity. As before, when we repeat the experiment with different random number seeds, we obtain essentially the same result for all sufficiently large n. Examination shows that there is statistical regularity, just as before, but the approximating distribution of the final position is now different. In Figure 1.23, the peak of the density looks like a spike because the range of values is now much greater. In turn, the range of values is greater because the distribution of $S_n - 3n$ has a heavy tail.

The heavier tails are more clearly revealed when we plot the tail of the empirical cdf of the observed values. (By the tail of a cdf F, we mean the *complementary cdf or ccdf*, defined by $F^c(t) \equiv 1 - F(t)$.)

To focus on the tail of the cdf F, we plot the tail of the empirical cdf in $log - log$ scale in Figure 1.18; i.e., we plot $\log F^c(t)$ versus $\log t$. To use $log - log$ scale, we consider only those values greater than 1, of which there were 3,121 when $n = 10^4$.

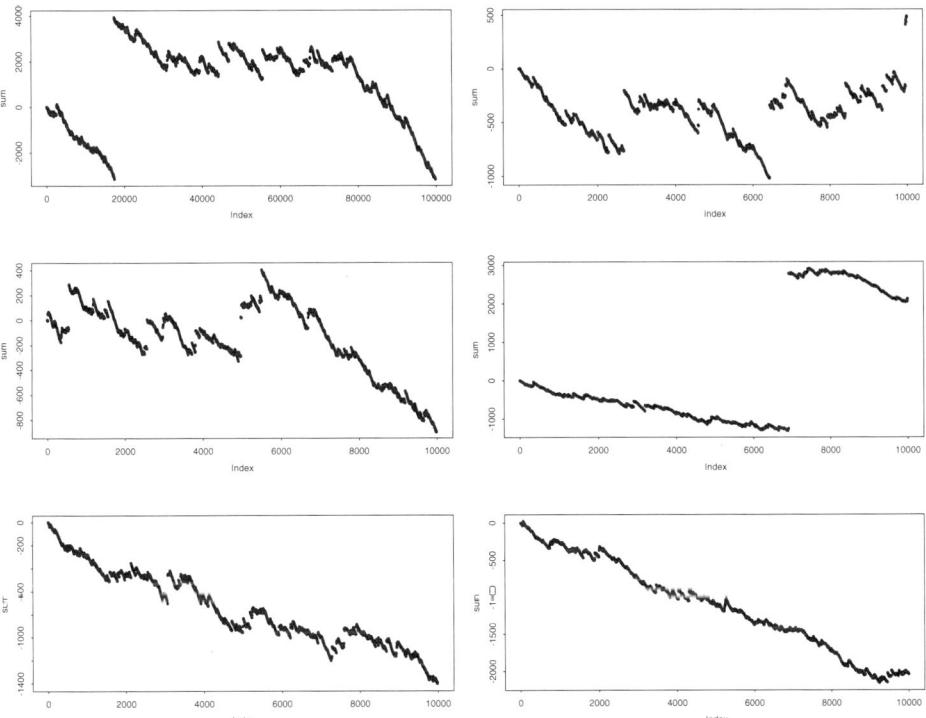

Figure 1.21. Six independent realizations of the first 10^4 steps of the centered random walk $\{S_k - 3k : k \geq 0\}$ associated with the Pareto steps $U_k^{-1/p}$ for $p = 3/2$, having mean 3 and infinite variance.

From Figure 1.24, we see that for larger values of the argument t, the empirical ccdf has a linear slope in $log-log$ scale. That indicates a *power tail*. Indeed, if the ccdf is of the form

$$F^c(t) = \alpha t^{-\beta} \quad \text{for} \quad t \geq t_0 > 1 , \qquad (4.3)$$

then

$$\log F^c(t) = -\beta \log t + \log \alpha \qquad (4.4)$$

for $t > t_0$. Then the parameters α and β in (4.3) can be seen as the intercept and slope in the $log-log$ plot.

Again there is supporting theory: A generalization of the CLT implies, under the IID assumptions and other regularity conditions (satisfied here), that properly scaled versions of the centered partial sums of Pareto(p) random steps converge in distribution, as in (2.7). In particular, when $1 < p < 2$,

$$n^{-1/p}(S_n - mn) \Rightarrow L \quad \text{in} \quad \mathbb{R} , \qquad (4.5)$$

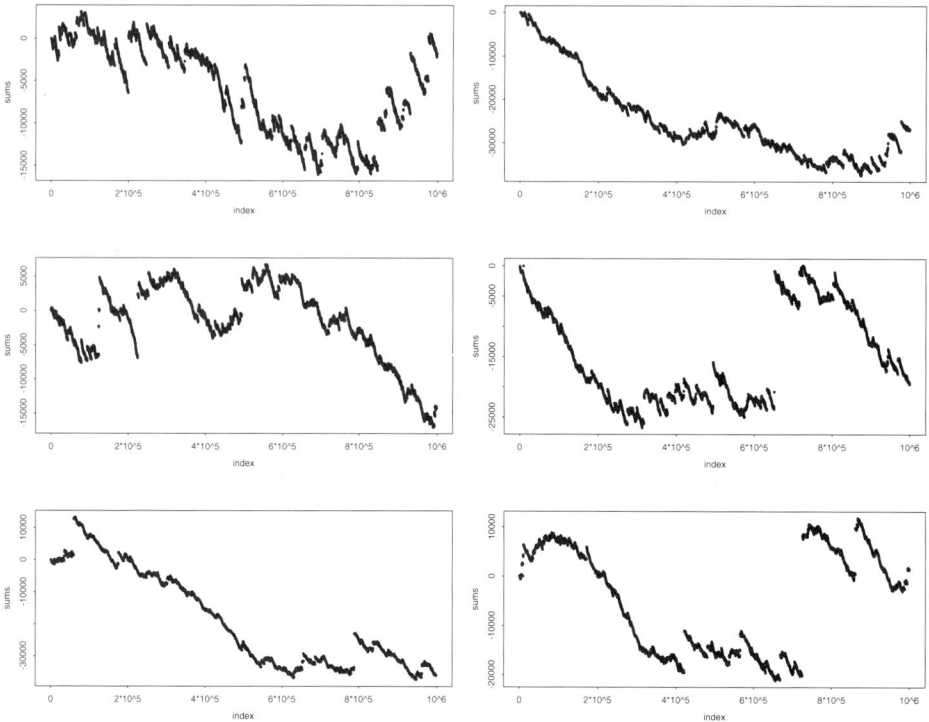

Figure 1.22. Six independent realizations of the first 10^6 steps of the centered random walk $\{S_k - 3k : k \geq 0\}$ associated with the Pareto steps $U_k^{-1/p}$ for $p = 3/2$, having mean 3 and infinite variance.

where $m = 1 + (p-1)^{-1}$ is the mean and the limiting random variable L has a *nonGaussian stable law* (depending upon p); e.g., see Chapter XVII of Feller (1971). In our specific case of $p = 3/2$, we have space scaling by $n^{2/3}$.

Unlike the Pareto distribution, the limiting stable law is not a pure power, but it has a power tail; i.e., it is asymptotically equivalent to a power: for $1 < p < 2$,

$$P(L > t) \sim ct^{-p} \quad \text{as} \quad t \to \infty \qquad (4.6)$$

for some positive constant c, where $f(t) \sim g(t)$ as $t \to \infty$ means that f is *asymptotically equivalent* to g, i.e., $f(t)/g(t) \to 1$ as $t \to \infty$. Thus the tail of the limiting stable law has the same asymptotic decay rate as the Pareto distribution of a single step.

Unlike the standard CLT in (2.4), the space scaling in (4.5) involves $c_n = n^{1/p}$ for $1 < p < 2$ instead of $c_n = n^{1/2}$. Nevertheless, the generalized CLT shows that there is again remarkable statistical regularity in the centered partial sums when the mean is finite and the variance is infinite. We again obtain essentially the same probability distribution for all n. We also obtain essentially the same probability distribution for other nonnegative step distributions, provided that they are centered by subtracting

1.4. The Exception Makes the Rule

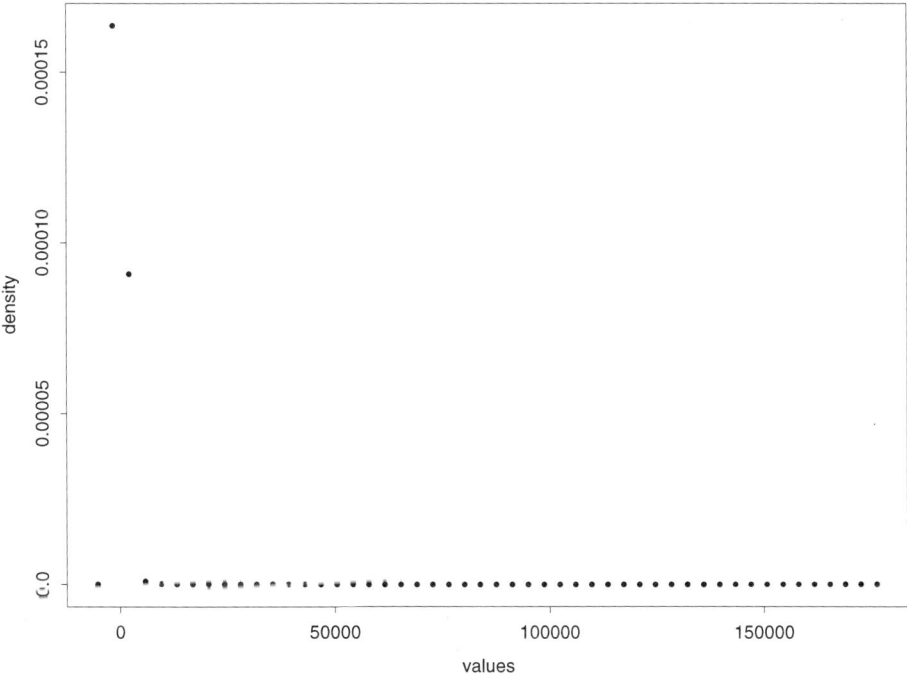

Figure 1.23. The density estimate obtained from $10,000$ independent samples of the final position of the centered random walk (i.e., the centered partial sum $S_{1000} - 3000$) associated with the Pareto steps $U_k^{-1/p}$ for $p = 3/2$.

the finite mean, and that the step-size ccdf $F^c(t)$ has the same asymptotic tail; i.e., we require that

$$F^c(t) \sim ct^{-p} \quad \text{as} \quad t \to \infty \tag{4.7}$$

for some positive constant c.

As before, there is also an associated stochastic-process limit. A generalization of Donsker's theorem (the FCLT) implies that the sequence of scaled random walks with Pareto(p) steps having $1 < p < 2$ converges in distribution to a *stable Lévy motion* as $n \to \infty$ in D. Now

$$\mathbf{S}_n \Rightarrow \mathbf{S} \quad \text{in} \quad D \,, \tag{4.8}$$

where

$$\mathbf{S}_n(t) \equiv n^{-1/p}(S_{\lfloor nt \rfloor} - m\lfloor nt \rfloor), \quad 0 \leq t \leq 1 \,, \tag{4.9}$$

for $n \geq 1$, m is the mean and $\mathbf{S}$ is a stable Lévy motion. That is, the stochastic-process limit (2.3) holds for $\mathbf{S}_n$ in (2.2), but now with $c_n = n^{1/p}$ and the limit

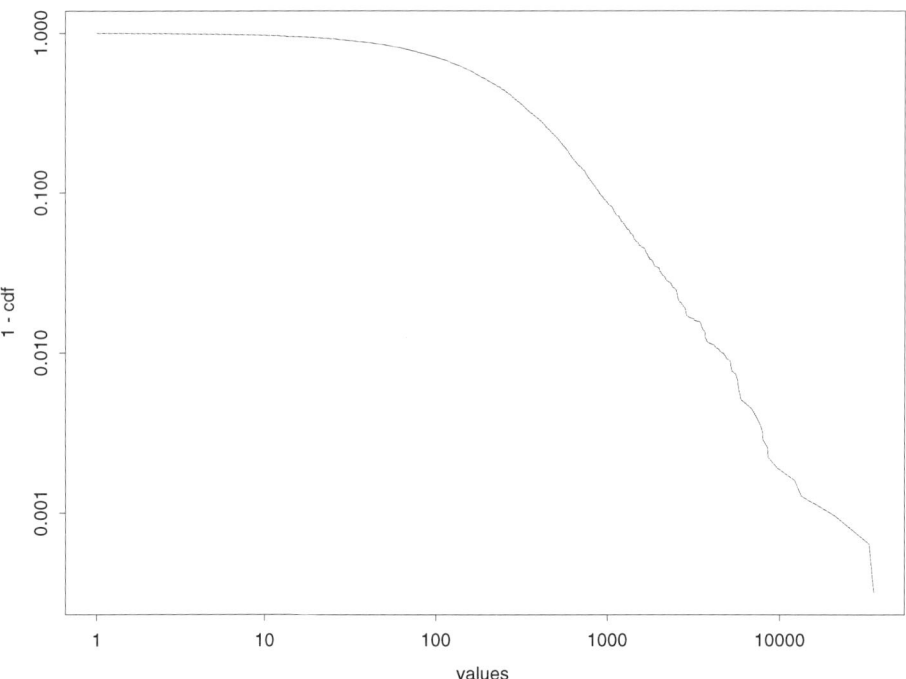

Figure 1.24. The tail of the empirical distribution function in $log-log$ scale obtained from $10,000$ independent samples of the final position of the centered random walk (i.e., the partial sum $S_{1000} - 3000$) associated with the Pareto steps $U_k^{-1/p}$ for $p = 3/2$ corresponding to the density in Figure 1.23. The results are based on the $3,121$ values greater than 1.

process **S** being stable Lévy motion instead of Brownian motion. Moreover, a variant of the previous invariance property holds here as well. For nonnegative random variables (the step sizes) satisfying (4.7), the limit process depends on its distribution only through the decay rate p and the single parameter c appearing in (4.7). We discuss this FCLT further in Chapter 4.

Since the random walk steps are IID, it is evident that the limiting stable Lévy motion must have stationary and independent increments, just like Brownian motion. However, the marginal distributions in $\mathbb{R}$ or $\mathbb{R}^k$ are nonnormal stable laws instead of the normal laws. Moreover, the stable Lévy motion has the self-similarity property, just like Brownian motion, but now with a different scaling. Now, for any $c > 0$, the stochastic process $\{c^{-1/p}\mathbf{S}(ct) : 0 \leq t \leq 1\}$ has a probability law on D, and thus finite-dimensional distributions, that are independent of c. Indeed, the proof is just like the proof for Brownian motion in (2.18).

It is significant that the space scaling to achieve statistical regularity is different now. In (4.9) above, we divide by $n^{1/p}$ for $1 < p < 2$ instead of by $n^{1/2}$. Similarly, in the self-similarity of the stable Lévy motion, we multiply by $c^{-1/p}$ instead of

$c^{-1/2}$. The new scaling can be confirmed by looking at the values on the y axis in the plots of Figures 1.20–1.22.

Figures 1.20–1.22 show that, unlike Brownian motion, stable Lévy motion must have *discontinuous sample paths*. Hence, *we have a stochastic-process limit in which the limit process has jumps*. The desire to consider such stochastic-process limits is a primary reason for this book.

1.4.3. Back to the Uncentered Random Walk with $p = 1/2$

Now let us return to the first Pareto(p) example with $p = 1/2$. The plots in Figure 1.19 are so irregular that we might not suspect that there is any statistical regularity there. However, after seeing the statistical regularity in the case $p = 3/2$, we might well think about reconsidering the case $p = 1/2$.

As before, we investigate by making some more plots. We have noted that we cannot center because the mean is infinite. So let us make more plots of the uncentered random walk with $p = 1/2$. Thus, in Figure 1.25 we plot six independent realizations of the uncentered random walk with 10^4 Pareto(0.5) steps. Now, even though these plots are highly irregular, with a single jump sometimes dominating the entire plot, we see remarkable statistical regularity.

Paralleling Figures 1.4 and 1.22, we confirm what we see in Figure 1.25 by plotting six independent samples of the uncentered random walk in case (iii) with $p = 1/2$ for $n = 10^6$ in Figure 1.26. Even though the plots of the uncentered random walks with Pareto(0.5) steps in Figures 1.19 – 1.26 are radically different from the previous plots of centered and uncentered random walks, we see remarkable statistical regularity in the new plots. As before, the plots tend to be independent of n for all n sufficiently large, provided we ignore the units on the axes. Thus we see self-similarity, just as in the plots of the centered random walks before. *From the random-walk plots, we see that statistical regularity can occur in many different forms.*

Given what we have just done, it is natural to again look for statistical regularity in the final positions. Thus we consider the final positions S_n (without centering) for $n = 1000$ and perform 10,000 independent replications. Paralleling Figures 1.23 and 1.24 above, an estimate of the probability density and the tail of the empirical cdf are plotted in Figures 1.27 and 1.28 below.

Figures 1.27 and 1.28 are quite similar to Figures 1.23 and 1.24, but now the distribution has an even heavier tail. Again there is supporting theory: A generalization of the CLT states, under the IID assumptions and other regularity conditions (satisfied here), that for $0 < p < 1$ there is convergence in distribution of the *uncentered partial sums* to a nonGaussian stable law if the partial sums are scaled appropriately, which requires that $c_n = n^{1/p}$. In particular, now with $p = 1/2$,

$$n^{-1/p} S_n \Rightarrow L \quad \text{in} \quad \mathbb{R}, \qquad (4.10)$$

where the limiting random variable again has a a nonGaussian stable law, which has an asymptotic power tail, i.e.,

$$P(L > t) \sim ct^{-p} \quad \text{as} \quad t \to \infty \qquad (4.11)$$

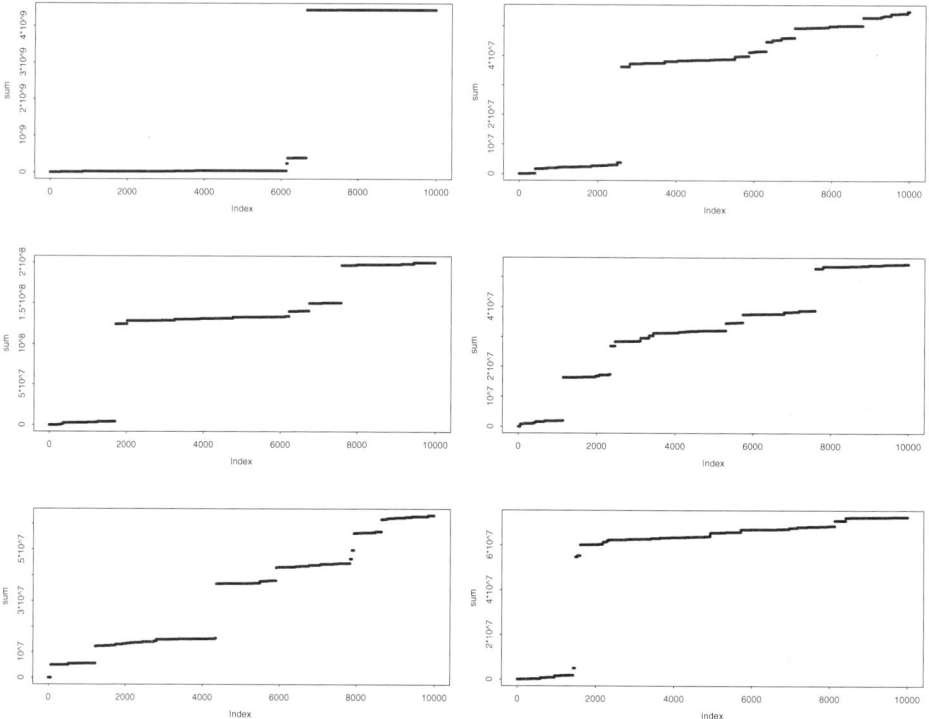

Figure 1.25. Six independent possible realizations of the first 10^4 steps of the uncentered random walk $\{S_k : k \geq 0\}$ with steps distributed as $U_k^{-1/p}$ in case (iii) of (3.5) for $p = 1/2$.

for $p = 1/2$ and some positive constant c; again see Chapter XVII of Feller (1971). As before, the tail of the stable law has the same asymptotic decay rate as a single step of the random walk.

Moreover, there again is an associated stochastic-process limit. Another generalization of Donsker's FCLT implies that there is the stochastic-process limit (4.8), where

$$\mathbf{S}_n(t) \equiv n^{-1/p} S_{\lfloor nt \rfloor}, \quad 0 \leq t \leq 1, \quad (4.12)$$

for $n \geq 1$, with the limit process $\mathbf{S}$ being another stable Lévy motion depending upon p.

Again there is an invariance property: Paralleling (4.7), we require that the random-walk step ccdf F^c satisfy

$$F^c(t) \sim ct^{-p} \quad \text{as} \quad t \to \infty, \quad (4.13)$$

where $p = 1/2$ and c is some positive constant. Any random walk with nonnegative (IID) steps having a ccdf satisfying (4.13) will satisfy the same FCLT, with the limit process depending on the step-size distribution only through the decay rate $p = 1/2$ and the constant c in (4.13).

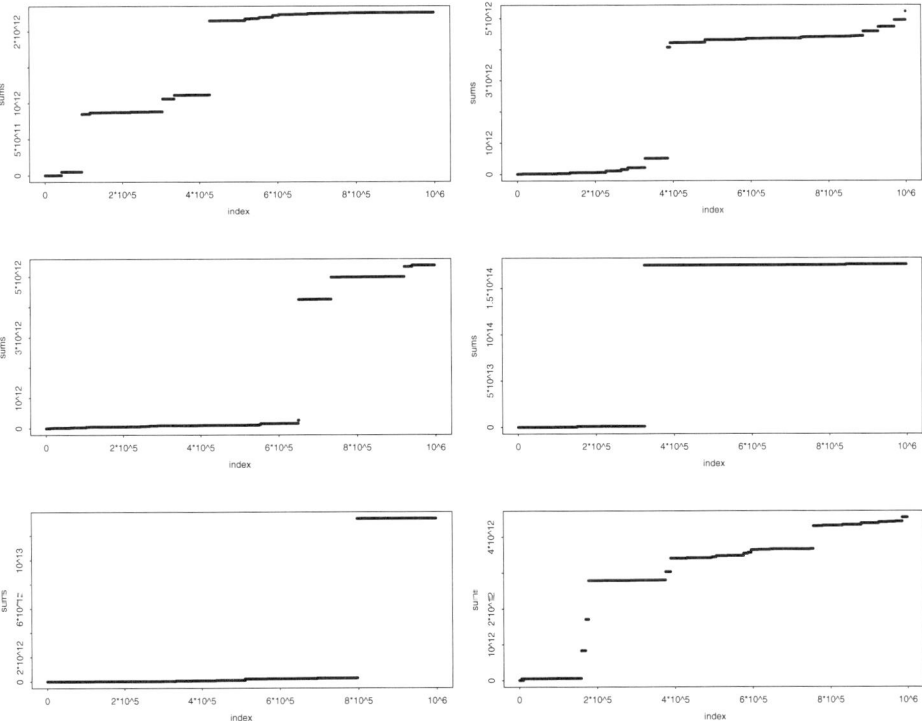

Figure 1.26. Six independent possible realizations of the first 10^6 steps of the uncentered random walk $\{S_k : k \geq 0\}$ with steps distributed as $U_k^{-1/p}$ in case (iii) of (3.5) for $p = 1/2$.

As before, the plotter automatically does the proper scaling. However, the space scaling is different from both the previous two cases, now requiring division by $n^{1/p}$ for $p = 1/2$. Again, we can verify that the space scaling by $n^{1/p}$ is appropriate by looking at the values in the plots in Figures 1.19–1.26. Just as before, the stochastic-process limit in D implies that the limit process must be self-similar. Now, for any $c > 0$, the stochastic processes $\{c^{-1/p}\mathbf{S}(ct) : 0 \leq t \leq 1\}$ have probability laws in D that are independent of c.

Figures 1.19 and 1.25 show that the limiting stable Lévy motion for the case $p = 1/2$ must also have discontinuous sample paths. So we have yet another stochastic-process limit in which the limit process has jumps.

1.5. Summary

To summarize, in this chapter we have seen that there is remarkable statistical regularity associated with random walks as the number n of steps increases. That statistical regularity is directly revealed when we plot the random walks. In great

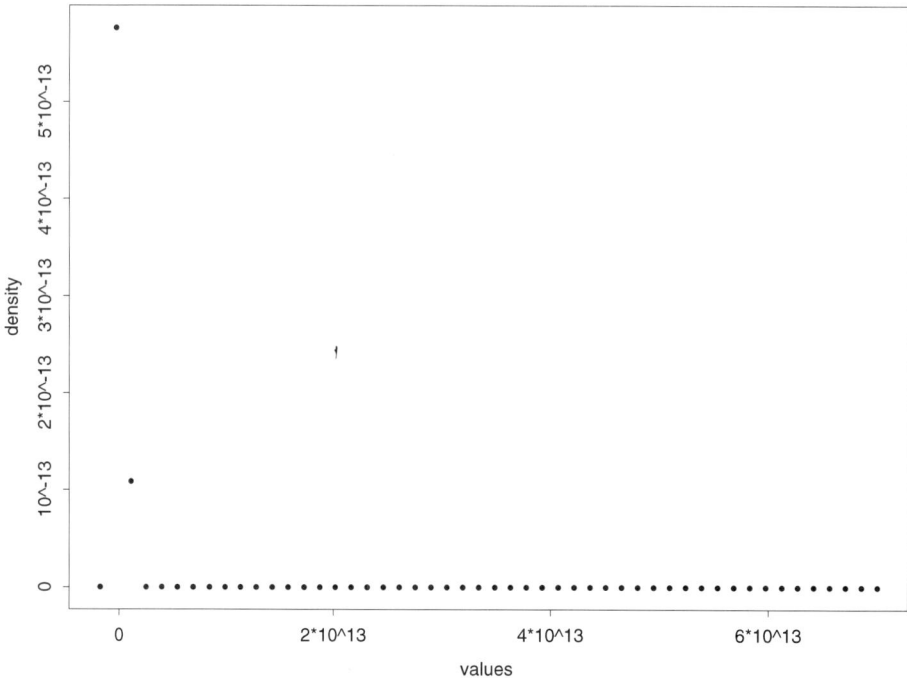

Figure 1.27. A density estimate obtained from 10,000 independent samples of the final position of the uncentered random walk (i.e., the partial sum S_{1000}) associated with the Pareto steps $U_k^{-1/p}$ in the case $p = 1/2$.

generality, as a consequence of Donsker's theorem, properly scaled versions of the centered random walks converge in distribution to Brownian motion as n increases. As a consequence, the random-walk plots converge to plots of standard Brownian motion.

The great generality of that result may make us forget that there are conditions for convergence to Brownian motion to hold. Through the exponential-smoothing example, we have seen that the conclusions of the classical limit theorems often still hold when the IID conditions are relaxed, but again there are limitations on the amount of dependence that can be allowed. That is easy to see by considering the extreme case in which *all the steps are identical*! Clearly, then the SLLN and the CLT break down. The classical limit theorems tend to remain valid when independence is replaced by *weak dependence*, but it is difficult to characterize the boundary exactly. We discuss FCLTs for weakly dependent sequences further in Chapter 4.

We also have seen for the case of IID steps that there are important situations in which the conditions of the FSLLN and Donsker's FCLT do not hold. We have seen

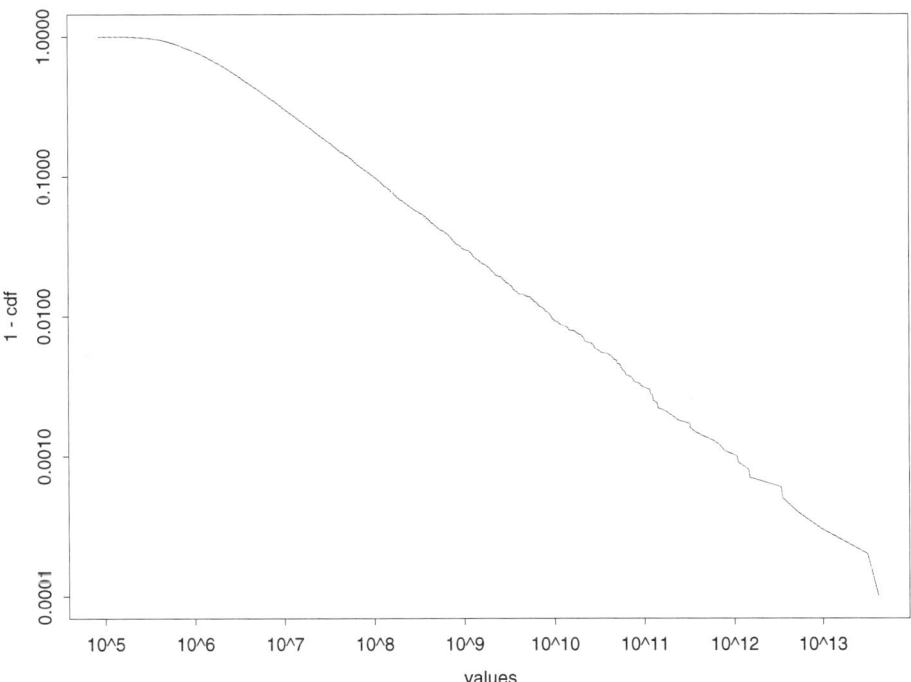

Figure 1.28. The tail of the empirical cumulative distribution function in $log - log$ scale obtained from $10,000$ independent samples of the final position of the uncentered random walk (i.e., the partial sum S_{1000}) associated with the Pareto steps $U_k^{-1/p}$ for $p = 1/2$ corresponding to the density in Figure 1.27.

that these fundamental theorems are not valid in the IID case when the step-size distribution has infinite mean (the FSLLN) or variance (the FCLT). Nevertheless, there often is remarkable statistical regularity associated with these heavy-tailed cases, but the limit process in the stochastic-process limit becomes a stable Lévy motion, which has jumps, i.e., it has discontinuous sample paths. We have thus seen examples of stochastic-process limits in which the limit process has jumps. We discuss such FCLTs further in Chapter 4.

If we allow greater dependence, which may well be appropriate in applications, then many more limit processes are possible, some of which will again have discontinuous sample paths. Again, see Chapter 4.

2
Random Walks in Applications

The random walks we have considered in Chapter 1 are easy to think about, because they have a relatively simple structure. However, the random walks are abstract, so that they may seem disconnected from reality. But that is not so!

Even though the random walks are abstract, they play a fundamental role in many applications. Many stochastic processes in applied probability models are very closely related to random walks. Indeed, we are able to obtain many stochastic-process limits for stochastic processes of interest in applied probability models directly from established probability limits for random walks, using the continuous-mapping approach.

To elaborate on this important point, we now give three examples of stochastic processes closely related to random walks. The examples involve stock prices, the Kolmogorov-Smirnov test statistic and a queueing model for a buffer in a switch. In the final section we discuss the engineering significance of the queueing model and the (heavy-traffic) stochastic-process limits.

2.1. Stock Prices

In some applications, random walks apply very directly. A good example is finance, which often can be regarded as yet another game of chance; see *A Random Walk Down Wall Street* by Malkiel (1996).

Indeed, we might model the price of a stock over time as a random walk; i.e., the position S_n can be the price in time period n. However, it is common to consider a refinement of the direct random-walk model, because the magnitude of any change is usually considered to be approximately proportional to the price.

A popular alternative model that captures that property is obtained by letting the price in period n be the *product* of the price in period $n-1$ and a *random multiplier* Y_n; i.e., if Z_n is the price in period n, then we have

$$Z_n = Z_{n-1} Y_n, \quad n \geq 1 . \tag{1.1}$$

That in turn implies that

$$Z_n = Z_0(Y_1 \times \cdots \times Y_n), \quad n \geq 1 . \tag{1.2}$$

Just as for random walks, for tractability we often assume that the successive random multipliers $Y_n : n \geq 1$, are IID. Hence, if we take logarithms, then we obtain

$$\log Z_n = \log Z_0 + S_n, \quad n \geq 0 ,$$

where $\{S_n : n \geq 0\}$ is a random walk, defined as in (3.4), with steps $X_n \equiv \log Y_n$, $n \geq 1$ that are IID. With this multiplicative framework, the *logarithms of successive prices constitute an initial position plus a random walk.* Approximations for random walks thus produce direct approximations for the logarithms of the prices.

It is natural to consider limits for the stock prices, in which the duration of the discrete time periods decreases in the limit, so that we can obtain convergence of the sequence of discrete-time price processes to a continuous-time limit, representing the evolution of the stock price in continuous time. To do so, we need to change the random multipliers as we change n. We thus define a sequence of price models indexed by n. We let Z_k^n and Y_k^n denote the price and multiplier, respectively, in period k in model n. For each n, we assume that the sequence of multipliers $\{Y_k^n : k \geq 1\}$ is IID. Since the periods are shrinking as $n \to \infty$, we want $Y_k^n \to 1$ as $n \to \infty$. The general idea is to have

$$E[\log Y_k^n] \approx m/n \quad \text{and} \quad Var[\log Y_k^n] \approx \sigma^2/n .$$

We let the initial price be independent of n; i.e., we let $Z_0^n \equiv Z_0$ for all n.

Thus, we incorporate the scaling within the partial sums for each n. We make further assumptions so that

$$\mathbf{S}_n(t) \equiv S_{\lfloor nt \rfloor}^n \Rightarrow \sigma \mathbf{B}(t) + mt \quad \text{as} \quad n \to \infty \tag{1.3}$$

for each $t > 0$, where $\mathbf{B}$ is standard Brownian motion. Given (1.3), we obtain

$$\log \mathbf{Z}_n(t) \equiv \log Z_{\lfloor nt \rfloor}^n = \log Z_0 + S_{\lfloor nt \rfloor}^n \Rightarrow \log Z_0 + \sigma \mathbf{B}(t) + mt ,$$

so that

$$\mathbf{Z}_n(t) \equiv Z_{\lfloor nt \rfloor}^n \Rightarrow \mathbf{Z}(t) \equiv Z_0 \exp\left(\sigma \mathbf{B}(t) + mt\right) ; \tag{1.4}$$

i.e., the price process converges in distribution as $n \to \infty$ to the stochastic process $\{\mathbf{Z}(t) : t \geq 0\}$, which is called *geometric Brownian motion*.

Geometric Brownian motion tends to inherit the tractability of Brownian motion. Since the moment generating function of a standard normal random variable is

$$\psi(\theta) \equiv E[\exp{(\theta N(0,1))}] = \exp{(\theta^2/2)},$$

the kth moment of geometric Brownian motion for any k can be expressed explicitly as

$$E[\mathbf{Z}(t)^k] = E[(Z_0)^k]\exp{(kmt + k^2t^2\sigma^2/2)}. \qquad (1.5)$$

See Section 10.4 of Ross (1993) for an introduction to the application of geometric Brownian motion to finance, including a derivation of the Black-Scholes option pricing formula. See Karatzas and Shreve (1988, 1998) for an extensive treatment.

The analysis so far is based on the assumption that the random-walk steps $X_k^n \equiv \log Y_k^n$ are IID with finite mean and variance. However, even though the steps must be finite, the volatility of the stock market has led people to consider alternative models. If we drop the finite-mean or finite-variance assumption, then we can still obtain a suitable continuous-time approximation, but it is likely to be a geometric stable Lévy motion (obtained by replacing the Brownian motion by a stable Lévy motion in the exponential representation in (1.4)). Even other limits are possible when the steps come from a double sequence $\{\{X_k^n : k \geq 1\} : n \geq 1\}$. When we consider models for volatile prices, we should be ready to see stochastic process limits with jumps. For further discussion, see Embrechts, Klüppelberg and Mikosch (1997), especially Section 7.6, and Rachev and Mittnik (2000).

In addition to illustrating how random walks can be applied, this example illustrates that we sometimes need to consider double sequences of random variables, such as $\{\{X_k^n : k \geq 1\} : n \geq 1\}$, in order to obtain the stochastic-process limit we want.

2.2. The Kolmogorov-Smirnov Statistic

For our second random-walk application, let us return to the empirical cdf's considered in Example 1.1.1 in Section 1.1.3. What we want to see now is a stochastic-process limit for the difference between the empirical cdf and the underlying cdf, explaining the statistical regularity we saw in Figure 1.8. The appropriate limit process is the *Brownian bridge* $\mathbf{B}_0$, which is just Brownian motion $\mathbf{B}$ over the interval $[0,1]$ conditioned to be 0 at the right endpoint $t = 1$.

Recall that the applied goal is to develop a statistical test to determine whether or not data from an unknown source can be regarded as an independent sample from a candidate cdf F. The idea is to base the test on the "difference" between the candidate cdf and the empirical cdf. We determine whether or not the observed difference is significantly greater than the difference for an independent sample from the candidate cdf F is likely to be. The problem, then, is to characterize the probability distribution of the difference between a cdf and the associated empirical cdf obtained from an independent sample. Interestingly, even here, random walks can play an important role.

Hence, let F be an arbitrary continuous candidate cdf and let F_n be the associated empirical cdf based on an independent sample of size n from F. A convenient test statistic, called the *Kolmogorov-Smirnov statistic*, can be based on the limit

$$D_n \equiv \sqrt{n} \sup_{t \in \mathbb{R}}\{|F_n(t) - F(t)|\} \Rightarrow sup(|\mathbf{B}_0|) \quad \text{as} \quad n \to \infty, \tag{2.1}$$

where $\mathbf{B}_0$ is the Brownian bridge, which can be represented as

$$\mathbf{B}_0(t) = \mathbf{B}(t) - t\mathbf{B}(1), \quad 0 \leq t \leq 1, \tag{2.2}$$

$$sup(|\mathbf{B}_0|) \equiv \sup_{0 \leq t \leq 1}\{|\mathbf{B}_0(t)|\}$$

and

$$P(sup(|\mathbf{B}_0|) > x) = 2\sum_{k=1}^{\infty}(-1)^{k+1}e^{-2k^2x^2}, \quad x > 0. \tag{2.3}$$

Notice that the limit in (2.1) is independent of the cdf F (assuming only that the cdf F is continuous). The candidate cdf F could be the uniform cdf in Example 1.1.1, a normal cdf, a Pareto cdf or a stable cdf. In particular, the limit process here is unaffected by the cdf F having a heavy tail.

In practice, we would compute the Kolmogorov-Smirnov statistic D_n in (2.1) for the empirical cdf associated with the data from the unknown source and the candidate cdf F. We then compute, using (2.3), the approximate probability of observing a value as large or larger than the observed value of the Kolmogorov-Smirnov statistic, under the assumption that the empirical cdf does in fact come from an independent sample from F. If that probability is very small, then we would reject the hypothesis that the data come from an independent sample from F.

As usual, good judgement is needed in the interpretation of the statistical analysis. When the sample size n is not large, we might be unable to reject the hypothesis that the data is an independent sample from a cdf F for more than one candidate cdf F. On the other hand, with genuine data (not a simulation directly from the cdf F), for any candidate cdf F, we are likely to be able to reject the hypothesis that the data is an independent sample from F for all n sufficiently large. Our concern here, though, is to justify the limit (2.1).

So, how do random walks enter in? Random walks appear in two ways. First, the empirical cdf $F_n(t)$ as a function of n itself is a minor modification of a random walk. In particular,

$$nF_n(t) = \sum_{k=1}^{n} I_{(-\infty, t]}(X_k),$$

where $I_A(x)$ is the *indicator function* of the set A, with $I_A(x) = 1$ if $x \in A$ and $I_A(x) = 0$ otherwise. Thus, for each t, $nF_n(t)$ is the sum of the n IID Bernoulli random variables $I_{(-\infty, t]}(X_k)$, $1 \leq k \leq n$, and is thus a random walk.

Note that the Bernoulli random variable $I_{(-\infty,t]}(X_k)$ has mean $F(t)$ and variance $F(t)F^c(t)$. Hence we can apply the SLLN and the CLT to deduce that

$$F_n(t) \to F(t) \quad \text{w.p.1} \quad \text{as} \quad n \to \infty$$

and

$$\sqrt{n}(F_n(t) - F(t)) \Rightarrow N(0, F(t)F^c(t)) \quad \text{in} \quad \mathbb{R} \quad \text{as} \quad n \to \infty \qquad (2.4)$$

for each $t \in \mathbb{R}$. Note that we have to multiply the difference by $\sqrt{n}$ in (2.4) in order to get a nondegenerate limit. That explains the multiplicative factor $\sqrt{n}$ in (2.1).

Paralleling the way we obtained stochastic-process limits for random walks in Section 1.2, we can go from the limit in (2.4) to the limit in (2.1) by extending the limit in (2.4) to a stochastic-process limit in the function space D. We can establish the desired stochastic-process limit in D in two steps: first, by reducing the case of a general continuous cdf F to the case of the uniform cdf (i.e., the cdf of the uniform distribution on $[0,1]$) and, second, by treating the case of the uniform cdf. Random walks can play a key role in the second step.

To carry out the first step, we show that the distribution of D_n in (2.1) is independent of the continuous cdf F. For that purpose, let $U_k, 1 \leq k \leq n$, be uniform random variables (on $[0,1]$) and let G_n be the associated empirical cdf. Recall from equation (3.7) in Section 1.3.3 that

$$F^{\leftarrow}(U_k) \leq t \quad \text{if and only if} \quad U_k \leq F(t) \,,$$

so that $F^{\leftarrow}(U_k) \stackrel{d}{=} X_k, 1 \leq k \leq n$, and

$$\{G_n(F(t)) : t \in \mathbb{R}\} \stackrel{d}{=} \{F_n(t) : t \in \mathbb{R}\} \,.$$

Hence,

$$D_n \equiv \sqrt{n} \sup_{t \in \mathbb{R}} \{|F_n(t) - F(t)|\} \stackrel{d}{=} \sqrt{n} \sup_{t \in \mathbb{R}} \{|G_n(F(t)) - F(t)|\} \,.$$

Moreover, since F is a continuous cdf, F maps $\mathbb{R}$ into the interval $(0,1)$ plus possibly $\{0\}$ and $\{1\}$. Since $P(U = 0) = P(U = 1) = 0$ for a uniform random variable U, we have

$$D_n \stackrel{d}{=} \sqrt{n} \sup_{0 \leq t \leq 1} \{|G_n(t) - t|\} \,, \qquad (2.5)$$

which of course is the special case for a uniform cdf.

Now we turn to the second step, carrying out the analysis for the special case of a uniform cdf, i.e., starting from (2.5). To make a connection to random walks, we exploit a well known property of Poisson processes. We start by focusing on the uniform order statistics: Let $U_k^{(n)}$ be the kth order statistic associated with n IID uniform random variables; i.e., $U_k^{(n)}$ is the kth smallest of the uniform random numbers. It is not difficult to see that the supremum in the expression for D_n in (2.5) must occur at one of the jumps in G_n (either the left or right limit) and these jumps occur at the random times $U_k^{(n)}$. Since each jump of D_n in (2.5) has magnitude $1/\sqrt{n}$,

$$|D_n - \sqrt{n}(\max_{1\leq k\leq n}\{|U_k^{(n)} - k/n|\})| \leq 1/\sqrt{n} \;. \tag{2.6}$$

Now we can make the desired connection to random walks: It turns out that

$$(U_1^{(n)},\ldots,U_n^{(n)}) \stackrel{d}{=} (S_1/S_{n+1},\ldots,S_n/S_{n+1}) \;, \tag{2.7}$$

where

$$S_k \equiv X_1 + \cdots + X_k, \quad 1 \leq k \leq n+1 \;,$$

with $X_k, 1 \leq k \leq n+1$, being IID exponential random variables with mean 1. To justify relation (2.7), consider a Poisson process and let the kth point be located at S_k (Which makes the intervals between points IID exponential random variables). It is well known, and easy to verify, that the first n points of the Poisson process are distributed in the interval $(0, S_{n+1})$ as the n uniform order statistics over the interval $(0, S_{n+1})$; e.g., see p. 223 of Ross (1993). When we divide by S_{n+1} we obtain the uniform order statistics over the interval $(0, 1)$, just as in the left side of (2.7).

With the connection to random walks established, we can apply Donsker's FCLT for the random walk $\{S_k : k \geq 0\}$ to establish the limit (2.1). In rough outline, here is the argument:

$$D_n \approx \sqrt{n}\max_{1\leq k\leq n}\{|(S_k/S_{n+1}) - (k/n)|\}$$

$$\approx (n/S_{n+1})\max_{1\leq k\leq n}\{|(S_k - k)/\sqrt{n} - (k/n)(S_{n+1} - n)/\sqrt{n}|\} \;. \tag{2.8}$$

Since $n/S_{n+1} \to 1$ as $n \to \infty$ and $(S_{n+1} - S_n)/\sqrt{n} \to 0$ as $n \to \infty$, we have

$$D_n \approx \sup_{0\leq t\leq 1}\{|(S_{\lfloor nt\rfloor} - \lfloor nt\rfloor)/\sqrt{n} - (\lfloor nt\rfloor/n)(S_n - n)/\sqrt{n}|\} \;. \tag{2.9}$$

To make the rough argument rigorous, and obtain (2.9), we repeatedly apply an important tool – the convergence-together theorem – which states that $X_n \Rightarrow X$ whenever $Y_n \Rightarrow X$ and $d(X_n, Y_n) \Rightarrow 0$, where d is an appropriate distance on the function space D; see Theorem 11.4.7.

Since the functions $\psi_1 : D \to D$ and $\psi_2 : D \to \mathbb{R}$, defined by

$$\psi_1(x)(t) \equiv x(t) - tx(1), \quad 0 \leq t \leq 1 \;, \tag{2.10}$$

and

$$\psi_2(x) \equiv \sup_{0\leq t\leq 1}\{|x(t)|\} \tag{2.11}$$

are continuous, from (2.9) we obtain the desired limit

$$D_n \Rightarrow \sup_{0\leq t\leq 1}\{|\mathbf{B}(t) - t\mathbf{B}(1)|\} \;. \tag{2.12}$$

Finally, it is possible to show that relations (2.2) and (2.3) hold.

The argument here follows Breiman (1968, pp. 283–290). Details can be found there, in Karlin and Taylor (1980, p. 343) or in Billingsley (1968, pp. 64, 83, 103,

141). See Pollard (1984) and Shorack and Wellner (1986) for further development. See Borodin and Salminen (1996) for more properties of Brownian motion.

Historically, the derivation of the limit in (2.1) is important because it provided a major impetus for the development of the general theory of stochastic-process limits; see the papers by Doob (1949) and Donsker (1951, 1952), and subsequent books such as Billingsley (1968).

2.3. A Queueing Model for a Buffer in a Switch

Another important application of random walks is to queueing models. We will be exploiting the connection between random walks and queueing models throughout the queueing chapters. We only try to convey the main idea now.

To illustrate the connection between random walks and queues, we consider a discrete-time queueing model of data in a buffer of a switch or router in a packet communication network.

Let W_k represent the workload (or buffer content, which may be measured in bits) at the end of period k. During period k there is a random input V_k and a deterministic constant output μ (corresponding to the available bandwidth) provided that there is content to process or transmit. We assume that the successive inputs V_k are IID, although that is not strictly necessary to obtain the stochastic-process limits.

More formally, we assume that the successive workloads can be defined recursively by

$$W_k \equiv \min\{K, \max\{0, W_{k-1} + V_k - \mu\}\}, \quad k \geq 1, \quad (3.1)$$

where the initial workload is W_0 and the buffer capacity is K. The *maximum* appears in (3.1) because the workload is never allowed to become negative; the output (up to μ) occurs only when there is content to emit. The *minimum* appears in (3.1) because the workload is not allowed to exceed the capacity K at the end of any period; we assume that input that would make the workload exceed K at the end of the period is lost.

The workload process $\{W_k : k \geq 1\}$ specified by the recursion (3.1) is quite elementary. Since the inputs V_k are assumed to be IID, the stochastic process $\{W_k\}$ is a discrete-time Markov process. If, in addition, we assume that the inputs V_k take values in a discrete set $\{ck : k \geq 0\}$ for some constant c (which is not a practical restriction), we can regard the stochastic process $\{W_k\}$ as a discrete-time Markov chain (DTMC). Since the state space of the DTMC $\{W_k\}$ is one-dimensional, the finite state space will usually not be prohibitively large. Thus, it is straightforward to exploit numerical methods for DTMC's, as in Kemeny and Snell (1960) and Stewart (1994), to describe the behavior of the workload process.

Nevertheless, we are interested in establishing stochastic-process limits for the workload process. In the present context, we are interested in seeing how the distribution of the inputs V_k affects the workload process. We can use heavy-traffic stochastic-process limit to produce simple formulas describing the performance. (We start giving the details in Chapter 5.) Those simple formulas provide insight that can be gained only with difficulty from a numerical algorithm for Markov chains.

We also are interested in the heavy-traffic stochastic-process limits to illustrate what can be done more generally. The heavy-traffic stochastic-process limits can be established for more complicated models, for which exact performance analysis is difficult, if not impossible. Since the heavy-traffic stochastic-process limits strip away unessential details, they reveal the key features determining the performance of the queueing system.

Now we want to see the statistical regularity associated with the workload process for large n. We could just plot the workload process for various candidate input processes $\{V_k : k \geq 1\}$ and parameters K and μ. However, the situation here is more complicated than for the the random walks we considered previously. We can simply plot the workload process and let the plotter automatically do the scaling for us, but it is not possible to automatically see the desired statistical regularity. For the queueing model, we need to do some analysis to determine how to do the proper scaling in order to achieve the desired statistical regularity. (That is worth verifying.)

2.3.1. Deriving the Proper Scaling

It turns out that stochastic-process limits for the workload process are intimately related to stochastic-process limits for the random walk $\{S_k : k \geq 0\}$ with steps

$$X_k \equiv V_k - \mu ,$$

but notice that in general this random walk is not centered. The random walk is only centered in the special case in which the input rate $E[V_k]$ exactly matches the potential output rate μ. However, to have a well-behaved system, we want the long-run potential output rate to exceed the long-run input rate.

In queueing applications we often characterize the system load by the *traffic intensity*, which is the rate in divided by the potential rate out. Here the traffic intensity is

$$\rho \equiv EV_1/\mu .$$

With an infinite-capacity buffer, we need $\rho < 1$ in order for the system to be stable (not blow up in the limit as $t \to \infty$).

We are able to obtain stochastic-process limits for the workload process by applying the continuous-mapping approach, starting from stochastic-process limits for the centered version of the random walk $\{S_k : k \geq 0\}$. However, to do so when $EX_k \neq 0$, we need to consider a sequence of models indexed by n to achieve the appropriate scaling. In the nth model, we let $X_{n,k}$ be the random-walk step X_k, and we let $EX_{n,k} \to 0$ as $n \to \infty$.

There is considerable freedom in the construction of a sequence of models, but from an applied perspective, it suffices to do something simple: We can keep a fixed input process $\{V_k : k \geq 1\}$, but we need to make the output rate μ and the buffer capacity K depend upon n. Let W_k^n denote the workload at the end of period k in

2.3. A Queueing Model for a Buffer in a Switch

model n. Following this plan, for model n the recursion (3.1) becomes

$$W_k^n \equiv \min\{K_n, \max\{0, W_{k-1}^n + V_k - \mu_n\}\}, \quad k \geq 1, \qquad (3.2)$$

where K_n and μ_n are the buffer capacity and constant potential one-period output in model n, respectively.

The problem now is to choose the sequences $\{K_n : n \geq 1\}$ and $\{\mu_n : n \geq 1\}$ so that we obtain a nondegenerate limit for an appropriately scaled version of the workload processes $\{W_k^n : k \geq 0\}$. If we choose these sequence of constants appropriately, then the plotter can do the scaling of the workload processes automatically.

Let $S_k^v \equiv V_1 + \cdots + V_k$ for $k \geq 1$ with $S_0^v \equiv 0$. The starting point is a FCLT for the random walk $\{S_k^v : k \geq 0\}$. Suppose that the mean $E[V_k]$ is finite, and let it equal m_v. Then the natural FCLT takes the form

$$\mathbf{S}_n^v \Rightarrow \mathbf{S}^v \quad \text{in} \quad D \quad \text{as} \quad n \to \infty, \qquad (3.3)$$

where

$$\mathbf{S}_n^v(t) \equiv n^{-H}(S_{\lfloor nt \rfloor}^v - m_v \lfloor nt \rfloor), \quad 0 \leq t \leq 1, \qquad (3.4)$$

the exponent H in the space scaling is a constant satisfying $0 < H < 1$ and $\mathbf{S}^v$ is the limit process. The common case has $H = 1/2$ and $\mathbf{S}^v = \sigma \mathbf{B}$, where $\mathbf{B}$ is standard Brownian motion. However, as seen for the random walks, if V_k has infinite variance, then we have $1/2 < H < 1$ and the limit process $\mathbf{S}^v$ is a stable Lévy motion (which has discontinuous sample paths). We elaborate on the case with $1/2 < H < 1$ in Section 4.5.

It turns out that a scaled version of the workload process $\{W_k^n : k \geq 0\}$ can be represented directly as the image of a two-sided reflection map applied to a scaled version of the uncentered random walk $\{S_k^n : k \geq 1\}$ with steps $V_k - \mu_n$. In particular,

$$\mathbf{W}_n = \phi_K(\mathbf{S}_n) \quad \text{for all} \quad n \geq 1, \qquad (3.5)$$

where

$$\mathbf{W}_n(t) \equiv n^{-H} W_{\lfloor nt \rfloor}^n, \quad 0 \leq t \leq 1, \qquad (3.6)$$

$$\mathbf{S}_n(t) \equiv n^{-H} S_{\lfloor nt \rfloor}^n, \quad 0 \leq t \leq 1, \qquad (3.7)$$

and $\phi_K : D \to D$ is the *two-sided reflection map*.

In fact, it is a challenge to even define the two-sided reflection map, which we may think of as serving as the continuous-time analog of (3.1) or (3.2); that is done in Sections 5.2 and 14.8; alternatively, see p. 22 of Harrison (1985). Consistent with intuition, it turns out that the two-sided reflection map ϕ_K is continuous on the function space D with appropriate definitions, so that we can apply the continuous-mapping approach with a limit for $\mathbf{S}_n$ in (3.7) to establish the desired limit for $\mathbf{W}_n$. But now we just want to determine how to do the plotting.

The next step is to relate the assumed limit for $\mathbf{S}_n^v$ to the required limit for $\mathbf{S}_n$. For that purpose, note from (3.4) and (3.7) that

$$\mathbf{S}_n(t) = \mathbf{S}_n^v(t) - n^{-H}(\mu_n - m_v)\lfloor nt \rfloor.$$

Hence we have the stochastic-process limit
$$\mathbf{S}_n \Rightarrow \mathbf{S} \quad \text{as} \quad n \to \infty , \tag{3.8}$$
where
$$\mathbf{S}(t) \equiv \mathbf{S}^v(t) - mt, \quad 0 \leq t \leq 1 , \tag{3.9}$$
if and only if
$$n^{-H}(\mu_n - m_v)\lfloor nt \rfloor \to mt \quad \text{as} \quad n \to \infty$$
for each $t > 0$ or, equivalently,
$$(\mu_n - m_v)n^{1-H} \to m \quad \text{as} \quad n \to \infty . \tag{3.10}$$

In addition, because of the space scaling by n^H in $\mathbf{S}_n$, we need to let
$$K_n \equiv n^H K . \tag{3.11}$$
Given the scaling in both (3.10) and (3.11), we are able to obtain the FCLT
$$\mathbf{W}_n \Rightarrow \mathbf{W} \equiv \phi_K(\mathbf{S}) , \tag{3.12}$$
where $\mathbf{W}_n$ is given in (3.6), $\mathbf{S}$ is given in (3.9) and ϕ_K is the two-sided reflection map.

The upshot is that we obtain the desired stochastic-process limit for the workload process, and the plotter can automatically do the appropriate scaling, if we let
$$\mu_n \equiv m_v + m/n^{1-H} \quad \text{and} \quad K_n \equiv n^H K \tag{3.13}$$
for any fixed m with $0 \leq m < \infty$ and K with $0 < K \leq \infty$, where H with $0 < H < 1$ is the scaling exponent appearing in (3.4).

At this point, it is appropriate to pause and reflect upon the significance of the scaling in (3.13). First note that time scaling by n (replacing t by nt) and space scaling by n^H (dividing by n^H) is determined by the FCLT in (3.3). Then the output rate and buffer size should satisfy (3.13). Note that the actual buffer capacity K_n in system n must increase, indeed go to infinity, as n increases. Also note that the output rate μ_n approaches m_v as n increases, so that the traffic intensity ρ_n approaches 1 as n increases. Specifically,
$$\rho_n \equiv \frac{E[V_1]}{\mu_n} = \frac{m_v}{m_v + mn^{-(1-H)}} = 1 - (m/m_v)n^{-(1-H)} + o(n^{-(1-H)})$$
as $n \to \infty$.

The obvious application of the stochastic-process limit in (3.12) is to generate approximations. The direct application of (3.12) is
$$\{n^{-H} W^n_{\lfloor nt \rfloor} : t \geq 0\} \approx \{\mathbf{W}(t) : t \geq 0\} , \tag{3.14}$$
where here $\approx$ means *approximately equal to in distribution*. Equivalently, by unscaling, we obtain the associated approximation (in distribution)
$$\{W^n_k : k \geq 0\} \approx \{n^{1/\alpha} \mathbf{W}(k/n) : k \geq 0\} . \tag{3.15}$$

Approximations such as (3.15), which are obtained directly from stochastic-process limits, may afterwards be refined by making modifications to meet other criteria, e.g., to match exact expressions known in special cases. Indeed, it is often possible to make refinements that remain asymptotically correct in the heavy-traffic limit, e.g., by including the traffic intensity ρ, which converges to 1 in the limit.

Often the initial goal in support of engineering applications is to develop a suitable approximation. Then heuristic approaches are perfectly acceptable, with convenience and accuracy being the criteria to judge the worth of alternative candidates. Even with such a pragmatic engineering approach, the stochastic-process limits are useful, because they generate initial candidate approximations, often capturing essential features, because the limit often is able to strip away unessential details. Moreover, the limits establish important theoretical reference points, demonstrating asymptotic correctness in certain limiting regimes.

2.3.2. Simulation Examples

Let us now look at two examples.

Example 2.3.1. *Workloads with exponential inputs.*

First let $\{V_k : k \geq 1\}$ be a sequence of IID exponential random variables with mean 1. Then the FCLT in (3.3) holds with $H = 1/2$ and $\mathbf{S}$ being standard Brownian motion $\mathbf{B}$. Thus, from (3.13), the appropriate scaling here is

$$\mu_n \equiv 1 + m/\sqrt{n} \quad \text{and} \quad K_n \equiv \sqrt{n}K \ . \tag{3.16}$$

To illustrate, we again perform simulations. Due to the recursive definition in (3.2), we can construct and plot the successive workloads just as easily as we constructed and plotted the random walks before. Paralleling our previous plots of random walks, we now plot the first n workloads, using the scaling in (3.16). In Figure 2.1 we plot the first n workloads for the case $H = 1/2$, $m = 1$ and $K = 0.5$ for $n = 10^j$ for $j = 1, \ldots, 4$. To supplement Figure 2.1, we show six independent replications for the case $n = 10^4$ in Figure 2.2.

What we see, as n becomes sufficiently large, is standard Brownian motion with drift $-m = -1$ modified by reflecting barriers at 0 and 0.5. Of course, just as for the random-walk plots before, the units on the axes are for the original queueing model. For example, for $n = 10^4$, the buffer capacity is $K_n = 0.5\sqrt{n} = 50$, so that the actual buffer content ranges from 0 to 50, even though the reflected Brownian motion ranges from 0 to 0.5. Similarly, for $n = 10^4$, the traffic intensity is $\rho_n = (1 + n^{-1/2})^{-1} = (1.01)^{-1} \approx 0.9901$ even though the Brownian motion has drift -1.

Unlike in the previous random-walk plots, the units on the vertical axes in Figure 2.2 are the same for all six plots. That happens because, in all six cases, the workload process takes values ranging from 0 to 50. The upper limit is 50 because for $n = 10^4$ the upper barrier in the queue is $0.5\sqrt{n} = 50$. The clipping at the upper barrier occurs because of overflows.

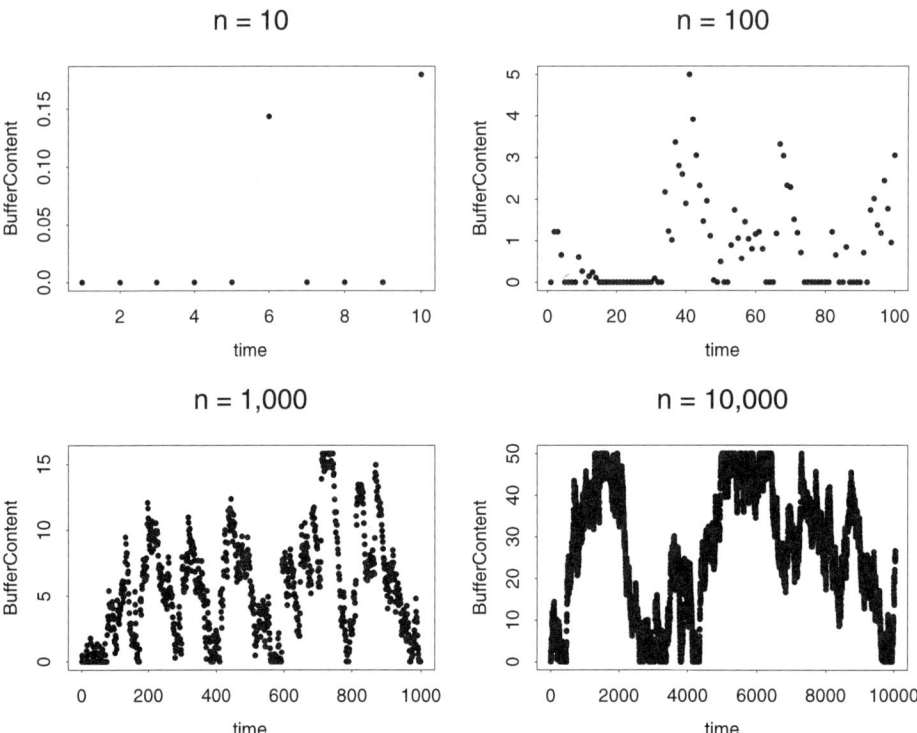

Figure 2.1. Possible realizations of the first n steps of the workload process $\{W_k^n : k \geq 0\}$ with IID exponential inputs having mean 1 for $n = 10^j$ with $j = 1, \ldots, 4$. The scaling is as in (3.16) with $m = 1$ and $K = 0.5$.

The traffic intensity 0.99 in Figure 2.2 is admittedly quite high. If we focus instead upon $n = 100$ or $n = 25$, then the traffic intensity is not so extreme, in particular, then $\rho_n = (1 + n^{-1/2})^{-1} = (1.1)^{-1} \approx 0.91$ or $(1.2)^{-1} \approx 0.83$.

In Figures 2.1 and 2.2 we see statistical regularity, just as in the early random-walk plots. Just as in the pairs of figures, (Figures 1.3 and 1.4) and (Figures 1.21 and 1.22), the plots for $n = 10^6$ look just like the plots for $n = 10^4$ when we ignore the units on the axes. The plots show that there should be a stochastic-process limit as $n \to \infty$. The plots demonstrate that a reflected Brownian motion approximation is appropriate with these parameters.

Moreover, our analysis of the stochastic processes to determine the appropriate scaling shows how we can obtain the stochastic-process limits. Indeed, we obtain the supporting stochastic-process limits for the workload process directly from the established stochastic-process limits for the random walks. In order to make the connection between the random walk and the workload process, we are constrained to use the scaling in (3.16). With that scaling, the plotter directly reveals the statistical regularity. ∎

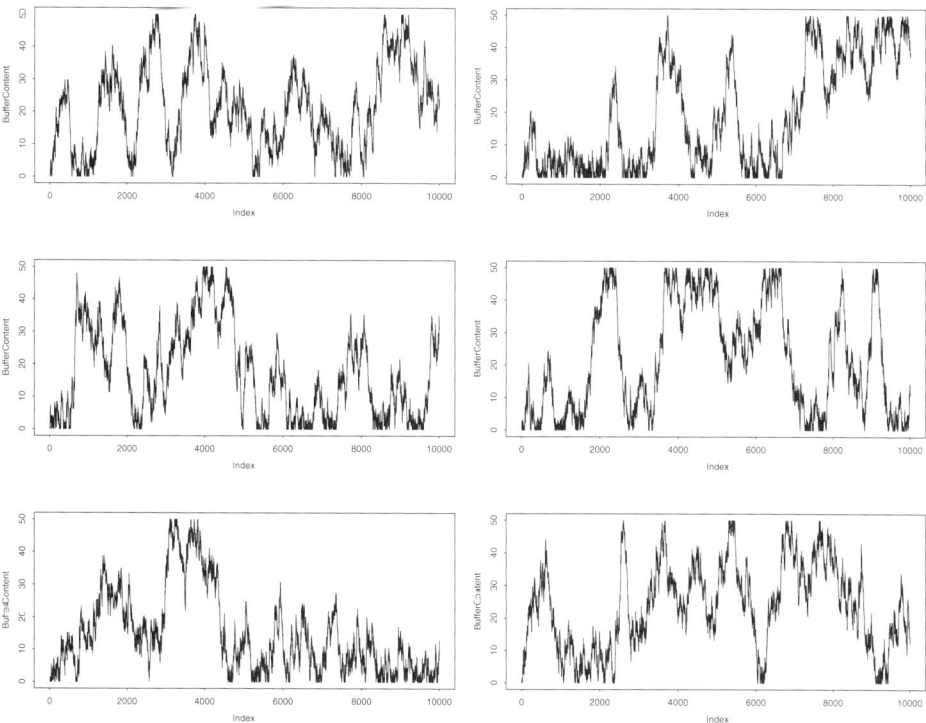

Figure 2.2. Six possible realizations of the first n steps of the workload process $\{W_k^n : k \geq 0\}$ with IID exponential inputs for $n = 10^4$. The scaling is as in (3.16) with $m = 1$ and $K = 0.5$.

Example 2.3.2. *Workloads with Pareto(3/2) inputs.*

For our second example, we assume that the inputs V_k have a Pareto(p) distribution with finite mean but infinite variance. In particular, we let

$$V_k \equiv U_k^{-1/p} \quad \text{for} \quad p = 3/2 \;, \tag{3.17}$$

just as in case (iii) of (3.5) in Section 1.3.3, which makes the distribution Pareto(p) for $p = 3/2$. Since $H = p^{-1}$ for $p = 3/2$, we need to use different scaling than we did in Example 2.3.1. In particular, instead of (3.16), we now use

$$\mu_n \equiv 1 + m/n^{1/3} \quad \text{and} \quad K_n \equiv n^{2/3} K \;, \tag{3.18}$$

with $m = 1$ and $K = 0.5$ just as before.

Since the scaling in (3.18) is different from the scaling in (3.16), for any given triple (m, K, n), the buffer size K_n is now larger, while the output rate differs more from the input rate. Assuming that $m > 0$, the traffic intensity in model n is now lower. That suggests that as H increases the heavy-traffic approximations may perform better at lower traffic intensities.

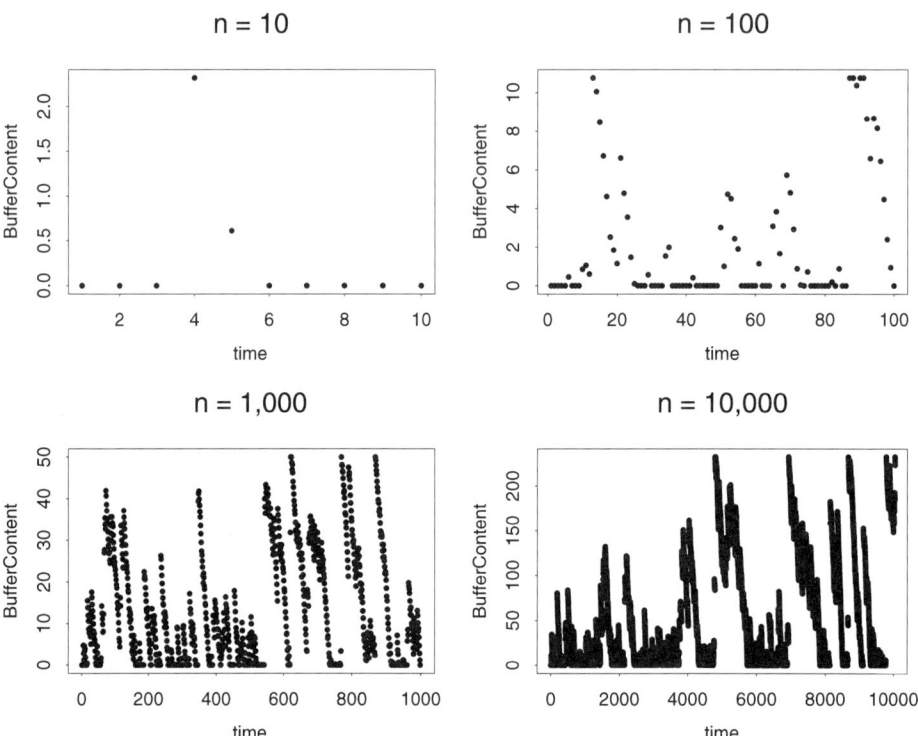

Figure 2.3. Possible realizations of the first n steps of the workload process $\{W_k^n : k \geq 0\}$ with IID Pareto(p) inputs having $p = 3/2$, mean 3 and infinite variance for $n = 10^j$ with $j = 1, \ldots, 4$. The scaling is as in (3.18) with $m = 1$ and $K = 0.5$.

We plot the first n workloads, using the scaling in (3.18), for $n = 10^j$ for $j = 1, \ldots, 4$ in Figure 2.3 for the case $m = 1$ and $K = 0.5$. What we see, as n becomes sufficiently large, is a stable Lévy motion with drift $-m = -1$ modified by reflecting barriers at 0 and 0.5. To supplement Figure 2.3, we show six independent replications for the case $n = 10^4$ in Figure 2.4. As before, the plots for $n = 10^6$ look just like the plots for $n = 10^4$ if we ignore the units on the axes. Just as in Figures 1.20–1.22 for the corresponding random walk, the plots here have jumps. ∎

In summary, the workload process $\{W_k\}$ in the queueing model is intimately related to the random walk $\{S_k\}$ with steps being the net inputs $V_k - \mu$ each period. With appropriate scaling, as in (3.13), which includes the queue being in heavy traffic, stochastic-process limits for a sequence of appropriately scaled workload processes can be obtained directly from associated stochastic-process limits for the underlying random walk.

Moreover, the limit process for the workload process is just the limit process for the random walk modified by having two reflecting barriers. Thus, the workload process in the queue exhibits the same statistical regularity for large sample sizes

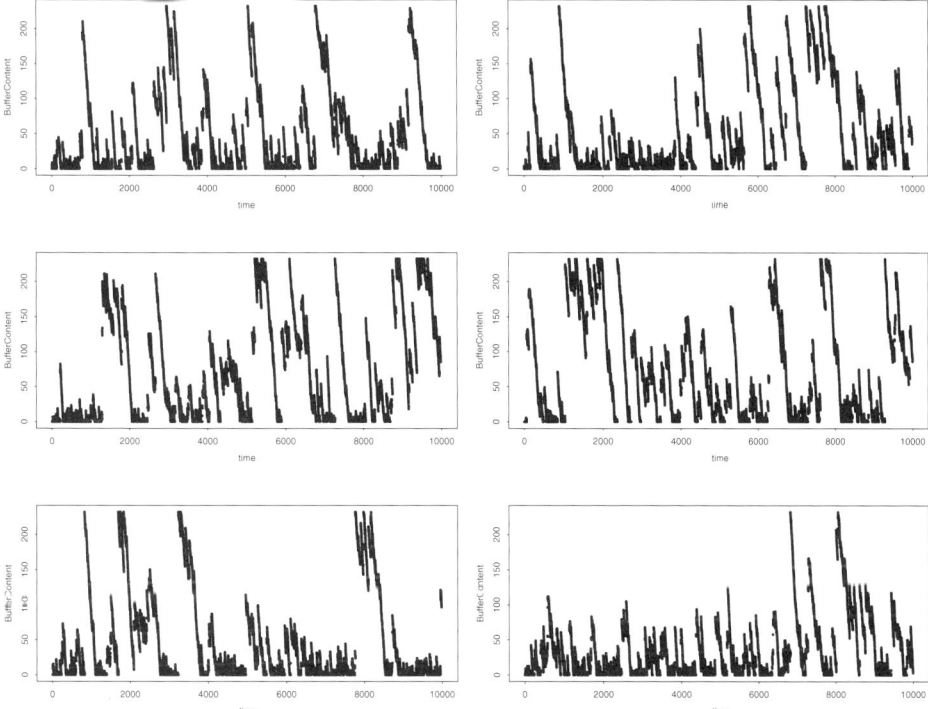

Figure 2.4. Six possible realizations of the first n steps of the workload process $\{W_k^n : k \geq 0\}$ with IID Pareto(p) inputs having $p = 3/2$, mean 3 and infinite variance for $n = 10^4$. The scaling is as in (3.18) with $m = 1$ and $K = 0.5$.

that we saw for the random walk. Indeed, the random walk is the source of that statistical regularity.

Just as for the random walks, the form of the statistical regularity may lead to the limit process for the workload process having discontinuous sample paths.

2.4. Engineering Significance

In the previous section, we saw that queueing models are closely related to random walks. With the proper (heavy-traffic) scaling, the same forms of statistical regularity that hold for random walks also hold for the workload process in the queueing model. But does it matter? Are there important engineering consequences?

To support an affirmative answer, in this final section we discuss the engineering significance of heavy-traffic stochastic-process limits for queues. First, in Section 2.4.1, we discuss buffer sizing in a switch or router in a communication network. Then, in Section 2.4.2, we discuss scheduling service with multiple sources, as occurs in manufacturing when scheduling production of multiple products on a single machine with setup costs or setup times for switching.

2.4.1. Buffer Sizing

The buffer (waiting space) in a network switch or router tends to be expensive to provide, so that economy dictates it be as small as possible. On the other hand, we want very few lost packets due to buffer overflow.

Queueing models are ideally suited to determine an appropriate buffer size. Let $L(K)$ be the long-run proportion of packets lost as a function of the buffer size K. We might specify a maximum allowable proportion of lost packets, ϵ. Given the function L, we then choose the buffer size K to satisfy the buffer-sizing equation

$$L(K) = \epsilon \, . \tag{4.1}$$

Classical queueing analysis, using standard models such as in Example 2.3.1, shows that $L(K)$ decays exponentially in K; specifically, L tends to have an *exponential tail*, satisfying

$$L(K) \sim \alpha e^{-\eta K} \quad \text{as} \quad K \to \infty \tag{4.2}$$

for asymptotic constants α and η depending upon the model details. (As in (4.6) of Section 1.4.2, $\sim$ means asymptotic equivalence. See Remark 5.4.1 for further discussion about asymptotics.)

It is natural to exploit the exponential tail asymptotics for L in (4.2) to generate the approximation

$$L(K) \approx \alpha e^{-\eta K} \tag{4.3}$$

for all K not too small. We then choose K to satisfy the *exponential buffer-sizing equation*

$$\alpha e^{-\eta K} = \epsilon \, , \tag{4.4}$$

from which we deduce that the target buffer size K^* should be

$$K^* = \eta^{-1} \log\left(\alpha/\epsilon\right) \, . \tag{4.5}$$

This analysis shows that the target buffer size should be directly proportional to η^{-1} and $\log \alpha$, and inversely proportional to $\log \epsilon$. It remains to determine appropriate values for the three constants η, α and ϵ, but the general relationships are clear. For example, if $\epsilon = 10^{-j}$, then K^* is proportional to the exponent j, which means that the cost of improving performance (as measured by the increase in buffer size K^* required to make ϵ significantly smaller) tends to be small.

So far, we have yet to exploit heavy-traffic limits. Heavy-traffic limits can play an important role because it actually is difficult to establish the exponential tail asymptotics in (4.2) directly for realistic models. As a first step toward analytic tractability, we may approximate the loss function $L(K)$ by the tail probability $P(W(\infty) > K)$, where $W(\infty)$ is the steady-state workload in the corresponding queue with unlimited waiting space. Experience indicates that the asymptotic form for $L(K)$ tends to be the same as the asymptotic form for the tail probability $P(W(\infty) > K)$ (sometimes with different asymptotic constants). From an applied point of view, we are not too concerned about great accuracy in this step, because the queueing model is crude (e.g., it ignores congestion controls) and the loss proportion $L(K)$ itself is only a rough performance indicator.

As a second step, we approximate $W(\infty)$ in the tail probability $P(W(\infty) > K)$ by the steady-state limit of the approximating process obtained from the heavy-traffic stochastic-process limit. For standard models, the approximating process is reflected Brownian motion, as in Example 2.3.1. Since the steady-state distribution of reflected Brownian motion with one-sided reflection is exponential (see Section 5.7), the heavy-traffic limit provides strong support for the approximations in (4.3)–(4.5) and helps identify approximate values for the asymptotic constants η and α. (The heavy-traffic limits also can generate approximations directly for the loss proportion $L(K)$; e.g., see Section 5.7.) The robustness of heavy-traffic limits (discussed in Chapters 4 and 5) suggests that the analysis should be insensitive to fine system details.

However, the story is not over! Traffic measurements from communication networks present a very different view of the world: These traffic measurements have shown that the traffic carried on these networks is remarkably bursty and complex, exhibiting features such as heavy-tailed probability distributions, strong positive dependence and self-similarity; e.g., see Leland et al. (1994), Garrett and Willinger (1994), Paxson and Floyd (1995), Willinger et al. (1995, 1997), Crovella and Bestavros (1996), Resnick (1997), Adler, Feldman and Taqqu (1998), Barford and Crovella (1998), Crovella, Bestavros and Taqqu (1998), Willinger and Paxson (1998), Park and Willinger (2000), Krishnamurthy and Rexford (2001) and references therein. These traffic studies suggest that different queueing models may be needed.

In particular, the presence of such traffic burstiness can significantly alter the behavior of the queue: *Alternative queueing analysis suggests alternative asymptotic forms for the function L*. Heavy-tailed probability distributions as in Example 2.3.2 lead to a different asymptotic form: When the inputs have power tails, like the Pareto inputs in Example 2.3.2, the function L tends to have a power tail as well: Instead of (4.2), we may have

$$L(K) \sim \alpha K^{-\eta} \quad \text{as} \quad K \to \infty , \qquad (4.6)$$

where again α and η are positive asymptotic constants; see Remark 5.4.1.

The change from the exponential tail in (4.2) to the power tail in (4.6) are contrary to the conclusions made above about the robustness of heavy-traffic approximations. Even though the standard heavy-traffic limits are remarkably robust, there is a limit to the robustness! The traffic burstiness can cause the robustness of the standard heavy-traffic limits to break down. Just as we saw in Example 2.3.2, the burstiness can have a major impact on the workload process.

However, we can still apply heavy-traffic limits: Just as before, we can approximate $L(K)$ by $P(W(\infty) > K)$, where $W(\infty)$ is the steady-state workload in the corresponding queue with unlimited waiting space. Then we can approximate $W(\infty)$

by the steady-state limit of the approximating process obtained from a heavy-traffic limit. However, when we properly take account of the traffic burstiness, the heavy-traffic limit process is no longer reflected Brownian motion. Instead, as in Example 2.3.2, it may be a reflected stable Lévy motion, for which $P(W(\infty) > K) \sim \alpha K^{-\eta}$. (For further discussion about the power tails, see Sections 4.5, 6.4 and 8.5.) Thus, different heavy-traffic limits support the power-tail asymptotics in (4.6) and yield approximations for the asymptotic constants.

Paralleling (4.3), we can use the approximation

$$L(K) \approx \alpha K^{-\eta} \tag{4.7}$$

for K not too small. Paralleling (4.4), we use the target equation (4.1) and (4.7) to obtain the *power buffer-sizing equation*

$$\alpha K^{-\eta} = \epsilon \tag{4.8}$$

from which we deduce that the *logarithm* of the target buffer size K^* should be

$$\log K^* = \eta^{-1} \log(\alpha/\epsilon) . \tag{4.9}$$

In this power-tail setting, we see that the required buffer size K^* is much more responsive to the parameters η, α and ϵ: Now the logarithm $\log K^*$ is related to the parameters η, α and ϵ the way K^* was before. For example, if $\epsilon = 10^{-j}$, then the logarithm of the target buffer size K^* is proportional to j, which means that the cost of improving performance (as measured by the increase in buffer size K^* required to make ϵ significantly smaller) tends to be large.

And that is not the end! The story is still not over. There are other possibilities: There are different forms of traffic burstiness. In Example 2.3.2 we focused on heavy-tailed distributions for IID inputs, but the traffic measurements also reveal strong dependence. The strong dependence observed in traffic measurements leads to considering fractional-Brownian-motion models of the input, which produce another asymptotic form for the function L; see Sections 4.6, 7.2 and 8.7. Unlike both the exponential tail in (4.2) and the power tail in (4.5), we may have a *Weibull tail*

$$L(K) \sim \alpha e^{-\eta K^\gamma} \quad \text{as} \quad K \to \infty \tag{4.10}$$

for positive constants α, η and γ, where $0 < \gamma < 1$; see (8.10) in Section 8.8. The available asymptotic results actually show that

$$P(W(\infty) > K) \sim \alpha K^{-\beta} e^{-\eta K^\gamma} \quad \text{as} \quad K \to \infty$$

for asymptotic constants η, α and β, where $W(\infty)$ is the steady-state of reflected fractional Brownian motion. Thus, the asymptotic results do not directly establish the asymptotic relation in (4.10), but they suggest the rough approximation

$$L(K) \approx \alpha e^{-\eta K^\gamma} \tag{4.11}$$

for all K not too small and the associated *Weibull buffer-sizing equation*

$$\alpha e^{-\eta K^\gamma} = \epsilon , \tag{4.12}$$

from which we deduce that the γth *power* of the target buffer size K^* should be

$$K^{*\gamma} = \eta^{-1} \log\left(\alpha/\epsilon\right) . \tag{4.13}$$

In (4.13) the γth power of K^* is related to the parameters α, η and ϵ the way K^* was in (4.5) and $\log K^*$ was in (4.9). Thus, consistent with the intermediate asymptotics in (4.10), since $0 < \gamma < 1$, we have the intermediate buffer requirements in (4.13).

Unfortunately, it is not yet clear which models are most appropriate. Evidence indicates that it depends on the context; e.g., see Heyman and Lakshman (1996, 2000), Ryu and Elwalid (1996), Grossglauser and Bolot (1999), Park and Willinger (2000), Guerin et al. (2000) and Mikosch et al. (2001). Consistent with observations by Sriram and Whitt (1986), long-term variability has relatively little impact on queueing performance when the buffers are small, but can be dramatic when the buffers are large.

Direct traffic measurements are difficult to interpret because they describe the carried traffic, not the offered traffic, and may be strongly influenced by congestion controls such as the Transmission Control Protocol (TCP); see Section 5.2 of Krishnamurthy and Rexford (2001) and Arvidsson and Karlsson (1999). Moreover, the networks and the dominant applications keep changing. For models of TCP, see Padhye et al. (2000), Bu and Towsley (2001), and references therein.

From an engineering perspective, it may be appropriate to ignore congestion controls when developing models for capacity planning. We may wish to provide sufficient capacity so that we usually meet the *offered load* (the original customer demand); e.g., see Duffield, Massey and Whitt (2001). When the system is heavily loaded, the controls slow down the stream of packets. From a careful analysis of traffic measurements, we may be able to reconstruct the intended flow. (For further discussion about offered-load models, see Remark 10.3.1.) However, heavy-traffic limits can also describe the performance with congestion-controlled sources, as shown by Das and Srikant (2000).

Our goal in this discussion, and more generally in the book, is not to draw engineering conclusions, but to describe an approach to engineering problems: Heavy-traffic limits yield simple approximations that can be used in engineering applications involving queues. Moreover, nonstandard heavy-traffic limits can capture the nonstandard features observed in network traffic. The simple analysis above shows that the consequences of the model choice can be dramatic, making order-of-magnitude differences in the predicted buffer requirements.

When the analysis indicates that very large buffers are required, instead of actually providing very large buffers, we may conclude that buffers are relatively ineffective for improving performance. Instead of providing very large buffers, we may choose to increase the available bandwidth (processing rate), introduce scheduling to reduce the impact of heavy users upon others, or regulate the source inputs (see Example 9.8.1). Indeed, all of these approaches are commonly used in practice. It is common to share the bandwidth among sources using a "fair queueing" discipline. Fair queueing disciplines are variants of the head-of-line processor-sharing discipline, which gives each of several active sources a guaranteed share of the available bandwidth. See Demers, Keshav and Shenker (1989), Greenberg and Madras

(1992), Parekh and Gallager (1993, 1994), Anantharam (1999) and Borst, Boxma and Jelenković (2000).

Many other issues remain to be considered: First, given any particular asymptotic form, it remains to estimate the asymptotic constants. Second, it remains to determine how the queueing system scales with increasing load. Third, it may be more appropriate to consider the transient or time-dependent performance measures instead of the customary steady-state performance measures. Fourth, it may be necessary to consider more than a single queue in order to capture network effects. Finally, it may be necessary to create appropriate controls, e.g., for scheduling and routing. Fortunately, for all these problems, and others, heavy-traffic stochastic-process limits can come to our aid.

2.4.2. Scheduling Service for Multiple Sources

In this final subsection we discuss *the engineering significance of the time-and-space scaling* that occurs in heavy-traffic limits for queues. The heavy-traffic scaling was already discussed in Section 2.3; now we want to point out its importance for system control.

We start by extending the queueing model in Section 2.3: Now we assume that there are inputs each time period from m separate sources. We let each source have its own infinite-capacity buffer, and assume that the work in each buffer is served in order of arrival, but otherwise we leave open the order of service provided to the different sources. As before, we can think of there being a single server, but now the server has to switch from queue to queue in order to perform the service, with there being a setup cost or a setup time to do the switching. (For background on the server scheduling problem, see Chapters 8 and 9 of Walrand (1988). For applications of heavy-traffic limits to study scheduling without setup costs or setup times, see van Mieghem (1995) and Ayhan and Olsen (2000).)

We initially assume that the server can switch from queue to queue instantaneously (within each discrete time period), but we assume that there are switchover costs for switching. To provide motivation for switching, we also assume that there are source-dependent holding costs for the workloads. To specify a concrete optimization problem, let W_k^i denote the source-i workload in its buffer at the end of period k and let $S_k^{i,j}$ be the number of switches from queue i to queue j in the first k periods. Let the total cost incurred in the first k periods be the sum of the total holding cost and the total switching cost, i.e.,

$$C_k \equiv H_k + S_k ,$$

where

$$H_k \equiv \sum_{i=1}^{m} \sum_{j=1}^{k} h_i W_j^i$$

and
$$S_k \equiv \sum_{i=1}^{m}\sum_{j=1}^{m} c_{i,j} S_k^{i,j} ,$$
where h_i is the source-i holding cost per period and $c_{i,j}$ is the switching cost per switch from source i to source j. Our goal then may be to choose a switching policy that minimizes the long-run average expected cost
$$\bar{C} \equiv \lim_{k \to \infty} k^{-1} E[C_k] .$$

This is a difficult control problem, even under the regularity condition that the inputs come from m independent sequences of IID random variables with finite means m_v^i. Under that regularity condition, the problem can be formulated as a *Markov sequential decision process*; e.g., see Puterman (1994): The state at the beginning of period $k+1$ is the workload vector $(W_k^1, \ldots, W_k^m)$ and the location of the server at the end of period k. An action is a specification of the sequence of queues visited and the allocation of the available processing per period, μ, during those visits. Both the state and action spaces are uncountably infinite, but we could make reasonable simplifying assumptions to make them finite.

To learn how we might approach the optimization problem, it is helpful to consider a simple scheduling policy: A *polling* policy serves the queues to exhaustion in a fixed cyclic order, with the server starting each period where it stopped the period before. We assume that the server keeps working until either its per-period capacity μ is exhausted or all the queues are empty.

There is a large literature on polling models; see Takagi (1986) and Boxma and Takagi (1992). For classical polling models, there are analytic solutions, which can be solved numerically. For those models, numerical transform inversion is remarkably effective; see Choudhury and Whitt (1996). However, analytical tractability is soon lost as model complexity increases, so there is a need for approximations.

The polling policy is said to be a *work-conserving service policy*, because the server continues serving as long as there is work in the system yet to be done (and service capacity yet to provide). An elementary, but important, observation is that the total workload process for any work-conserving policy is identical to the workload process with a single shared infinite-capacity buffer. Consequently, the heavy-traffic limit described in Section 2.3 in the special case of an infinite buffer ($K = \infty$) also holds for the total-workload process with polling; i.e., with the FCLT for the cumulative inputs in (3.3) and the heavy-traffic scaling in (3.10), we have the heavy-traffic limit for the scaled total-workload processes in (3.12), with the two-sided reflection map ϕ_K replaced by the one-sided reflection map. Given the space scaling by n^H and the time scaling by n, where $0 < H < 1$, the unscaled total workload at any time in the nth system is of order n^H and changes significantly over time intervals having length of order n.

The key observation is that the time scales are very different for the individual workloads at the source buffers. First, the individual workloads are bounded above by the total workload. Hence the unscaled individual workloads are also of order

n^H. Clearly, the mean inputs must satisfy the relation

$$m_v = m_{v,1} + \cdots + m_{v,m} .$$

Assuming that $0 < m_{v,i} < m_v$ for all i, we see that *each source by itself is not in heavy traffic when the server is dedicated to it*: With the heavy-traffic scaling in (3.10), the total traffic intensity approaches 1, i.e.,

$$\rho_n \equiv m_v/\mu_n \uparrow 1 \quad \text{as} \quad n \to \infty ,$$

but the instantaneous traffic intensity for source i when the server is devoted to it converges to a limit less than 1, i.e.,

$$\rho_{n,i} \equiv m_{v,i}/\mu_n \uparrow m_{v,i}/m_v \equiv \rho_i^* < 1 .$$

Since each source alone is not in heavy-traffic when the server is working on that source, the net output is at a constant positive rate when service is being provided, even in the heavy-traffic limit. Thus the server processes the order n^H unscaled work there in order n^H time, by the law of large numbers (see Section 5.3).

The upshot is that the unscaled individual workloads change significantly in order n^H time whenever the server is devoted to them, and the server cycles through the m queues in order n^H time, whereas the unscaled total workload changes significantly in order n time. Since $H < 1$, in the heavy-traffic limit the individual workloads change on a faster time scale. Thus, in the heavy-traffic limit we obtain a *separation of time scales*: When we consider the evolution of the individual workload processes in a short time scale, we can act as if the total workload is fixed.

Remark 2.4.1. *The classic setting: NCD Markov chains.* The separation of time scales in the polling model is somewhat surprising, because it occurs in the heavy-traffic limit. In other settings, a separation of time scales is more evident. With computers and communication networks, the relevant time scale for users is typically seconds, while the relevant time scale for system transactions is typically milliseconds. For those systems, engineers know that time scales are important.

There is a long tradition of treating different time scales in stochastic models using nearly-completely-decomposable (NCD) Markov chains; e.g., see Courtois (1977). With a NCD Markov chain, the state space can be decomposed into subsets such that most of the transitions occur between states in the same subset, and only rarely does the chain move from one subset to another. In a long time scale, the chain tends to move from one local steady-state regime to another, so that the long-run steady-state distribution is an appropriate average of the local steady-state distributions. In an intermediate time scale, the distribution of the chain is well approximated by the local steady-state distribution in the subset where the chain starts.

However, different behavior can occur if the chain does not approach steady-state locally within a subset. For example, that occurs in an infinite-capacity queue in a slowly changing environment when the queue is unstable in some environment states. Heavy-traffic limits for such queues were established by Choudhury, Mandelbaum, Reiman and Whitt (1997). Even though the queue content may ultimately approach a unique steady-state distribution, the local instability may cause significant fluctuations in an intermediate time scale. The transient behavior of the heavy-traffic limit process captures this behavior over the intermediate time scale. ∎

For the polling model, the separation of time scales suggests that in the heavy-traffic limit, given the fixed scaled total workload $\mathbf{W}_n(t) = w$, in the neighborhood of time t the vector of scaled individual workloads $(\mathbf{W}_n^1(t), \ldots, \mathbf{W}_n^m(t))$ rapidly traverses a deterministic piecewise-linear trajectory through points $(w^1, \ldots, w^m)$ in the hyperplane in $\mathbb{R}^m$ with $w^1 + \cdots + w^m = w$. For example, with three identical sources served in numerical cyclic order, the path is piecewise-linear, passing through the vertices $(2w/3, w/3, 0)$, $(0, 2w/3, w/3)$ and $(w/3, 0, 2w/3)$, corresponding to the instants the server is about to start service on sources 1, 2 and 3, respectively. In general, identifying the vertices is somewhat complicated, but the experience of each source is clear: it builds up to its peak workload at constant rate and then returns to emptiness at constant rate. And it does this many times before the total workload changes significantly. Hence at any given time its level can be regarded as uniformly distributed over its range.

As a consequence, we anticipate a *heavy-traffic averaging principle*: We should have a limit for the average of functions of the scaled individual workloads; i.e., for any $s, h > 0$ and any continuous real-valued function f,

$$h^{-1} \int_s^{s+h} f(\mathbf{W}_n^i(t)) dt \Rightarrow h^{-1} \int_s^{s+h} \left(\int_0^1 f(a_i u \mathbf{W}(t)) du \right) dt , \qquad (4.14)$$

where a_i is a constant satisfying $0 < a_i \leq 1$ for $1 \leq i \leq m$. In words, the time-average of the scaled individual-source workload process over the time interval $[s, s+h]$ approaches the corresponding time-average of a proportional space-average of the limit $\mathbf{W}$ for the scaled total workload process. (For other instances of the averaging principle, see Anisimov (1993), Freidlin and Wentzell (1993) and Sethi and Zhang (1994).)

This heavy-traffic averaging principle was rigorously established for the case of two queues by Coffman, Puhalskii and Reiman (1995) for a slightly different model in the Brownian case, with $H = 1/2$ and $\mathbf{W}$ reflected Brownian motion. They also determined the space-scaling constants a_i appearing in (4.14) for m sources: They showed that

$$a_i = \frac{\rho_i^*(1 - \rho_i^*)}{\sum_{1 \leq j < k \leq m} \rho_j^* \rho_k^*} , \qquad (4.15)$$

where ρ_i^* is the limiting source-i traffic intensity, i.e., $\rho_i^* \equiv m_{v,i}/m_v$ for our model. The upper limits a_i depend only on the means $m_{v,j}$, $1 \leq j \leq m$. For $m = 2$, $a_i = 1$; for m identical sources, $a_i = 2/m$. The variability affects the limit in (4.14) only through the scaling and the one-dimensional limit process $\mathbf{W}$.

Coffman, Puhalskii and Reiman (1998) also considered the two-queue polling model with unscaled switchover times. Even though the switchover times are asymptotically negligible in the heavy-traffic scaling, they have a significant impact because the relative amount of switching increases as the total workload decreases.

Coffman, Puhalskii and Reiman (1998) show that the heavy-traffic averaging principle is still valid with switchover times, with the scaled total workload processes converging to a Bessel diffusion process, which has state-dependent drift of the form $-a + b/x$ for positive constants a and b. (For additional heavy-traffic limits for polling models, see Olsen (2001), van der Mei and Levy (1997) and van der Mei (2000, 2001).)

Even though the polling models have yet to be analyzed for nonstandard scaling, with $H \neq 1/2$ and $\mathbf{W}$ not a diffusion process, it is evident that the heavy-traffic averaging principle still applies. We can anticipate that the other forms of variability (associated with heavy tails and strong dependence) affect the heavy-traffic limit only through the limit process $\mathbf{W}$.

The separation of time scales provides a way to attack complicated service control problems such as the one formulated at the beginning of this subsection. Even if all the desired supporting mathematics cannot be established, the heavy-traffic limits provide a useful perspective for approximately solving these problems. The heavy-traffic averaging principle reduces the dimension of the state-space in the control problem. It provides a form of *state-space collapse*; see Reiman (1984b), Harrison and van Mieghem (1997), Bramson (1998) and Williams (1998b). It lets us focus on the single process that is the heavy-traffic limit for the scaled total-workload process. For natural classes of service policies, we can express the local cost rate associated with a fixed total workload and then determine an expression for the long-run average total cost as a function of the controls that produces a tractable optimization problem. In the more challenging cases it may be necessary to apply numerical methods to solve the optimization problem, as in Kushner and Dupuis (2000).

By now, there has been substantial work on this heavy-traffic approach to scheduling, yielding excellent results. We do not try to tell the story here; instead we refer to Reiman and Wein (1998), Markowitz, Reiman and Wein (2000), Markowitz and Wein (2001) and Section 1.3 and Chapter 11 of Kushner (2001).

For these more complicated control problems, there are many open technical problems: It remains to establish the heavy-traffic averaging principle in more complicated settings and it remains to show that the derived policies are indeed asymptotically optimal in the heavy-traffic limit. Markowitz et al. (2000, 2001) restrict attention to dynamic cyclic policies in which each source is served once per cycle in the same fixed order. It is easy to construct examples in which larger classes of policies are needed: With three sources, it may be necessary to serve one source more frequently; e.g., the cycle $(1, 2, 1, 3)$ may be much better than either $(1, 2, 3)$ or $(1, 3, 2)$.

Nevertheless, the practical value of the heavy-traffic approach is well established: Numerical comparisons have shown that the policies generated from the heuristic heavy-traffic analysis perform well for systems under normal loading. Moreover, the heavy-traffic analysis produces important insight about the control problem, as illustrated by concluding remarks on p. 268 of Markowitz and Wein (2001) about the way model features – setups, due dates and product mix – affect the structure of policies. And there is opportunity for further work along these lines.

2.4. Engineering Significance

Heavy-traffic analysis has also been applied to other queueing control problems. We have discussed the scheduling of service for multiple sources by a single server. We may instead have to schedule and route input from multiple sources to several possible servers; see Bell and Williams (2001), Harrison and Lopez (1999) and references therein. More generally, we may have multiclass processing networks; see Harrison (1988, 2000, 2001a,b), Kumar (2000), Chapter 12 of Kushner (2001) and references therein.

In conclusion, the successful application of heavy-traffic analysis to these classic operations-research stochastic scheduling problems provides ample evidence that heavy-traffic stochastic-process limits for queues have engineering significance.

3
The Framework for Stochastic-Process Limits

3.1. Introduction

In Chapters 1 and 2 we saw that plots of stochastic-process sample paths can suggest stochastic-process limits. Now we want to define precisely what we mean by those stochastic-process limits.

The main idea is to think of a stochastic process as a random function. With that mindset, convergence of a sequence of stochastic processes naturally becomes convergence of a sequence of probability measures on a function space (space of functions). There then remain three problems: First, what should we mean by the convergence of a sequence of probability measures on an abstract space? Second, what should be the underlying function space containing the sample paths of the stochastic processes? And, third, what should be the topology (notion of convergence) in the underlying function space?

We start in Section 3.2 by defining the standard notion of convergence for a sequence of probability measures on a metric space. We also define the Prohorov metric on the space of all probability measures on the metric space, which induces that convergence.

In Section 3.3 we discuss the function space D that we will use to represent the space of possible sample paths of the stochastic processes. We define two different metrics on the functions space D: One is the standard J_1 metric, which induces the Skorohod (1956) J_1 topology. The other is the M_1 metric, which induces the Skorohod (1956) M_1 topology. The commonly used J_1 topology is often referred to as "the Skorohod topology." We use the M_1 topology in order to be able to establish stochastic-process limits with unmatched jumps in the limit process.

In Section 3.4 we state three versions of the continuous-mapping theorem that support the continuous-mapping approach for obtaining new stochastic-process limits from established stochastic-process limits. In Section 3.5 we introduce useful functions mapping D or the product space $D \times D$ into D that preserve convergence and thus facilitate the continuous-mapping approach. We conclude in Section 3.6 by describing the organization of the book.

This chapter is intended to be brief, providing background for the introductory chapters. We elaborate in Chapter 11 and refer to Billingsley (1968, 1999) for more details.

3.2. The Space $\mathcal{P}$

Our goal is to precisely define what we mean by a *stochastic-process limit*, i.e., the convergence of a sequence of stochastic processes. We use metrics for that purpose. We define a metric on a space of stochastic processes in two steps: First, we define a metric on the space of probability measures on a general metric space and, second, we define a metric on the underlying function space containing the sample paths of the stochastic processes.

A *metric* is a distance function satisfying certain axioms. In particular, a metric m on a set S is a nonnegative real-valued function on the product space $S \times S \equiv \{(s_1, s_2) : s_1 \in S, s_2 \in S\}$ such that $m(x,y) = 0$ if and only if $x = y$, satisfying the *symmetry property*

$$m(x,y) = m(y,x) \quad \text{for all} \quad x, y \in S$$

and the *triangle inequality*

$$m(x,z) \leq m(x,y) + m(y,z) \quad \text{for all} \quad x, y, z \in S\ .$$

A *sequence* in a set S is a function mapping the positive integers into S. A sequence $\{x_n : n \geq 1\}$ in a metric space (S, m) *converges* to a limit x in S if, for all $\epsilon > 0$, there exists an integer n_0 such that $m(x_n, x) < \epsilon$ for all $n \geq n_0$. If we use the metric only to specify which sequences converge, then we characterize the *topology* induced by the metric: In a metric space, the topology is a specification of which sequences converge. Topology is the more general concept, because different metrics can induce the same topology. For further discussion about topologies, see Section 11.2.

As a regularity condition, we assume that the metric space (S, m) is *separable*, which means that there is a countable dense subset; i.e., there is a countably infinite (or finite) subset S_0 of S such that, for all $x \in S$ and all $\epsilon > 0$, there exists $y \in S_0$ such that $m(x, y) < \epsilon$.

We first consider probability measures on a general separable metric space (S, m). In our applications, the underlying metric space S will be the function space D, but now S can be any nonempty set. To consider probability measures on (S, m), we make S a *measurable space* by endowing it with a σ-field of measurable sets (discussed further in Section 11.3). For the separable metric space (S, m), we always use the *Borel σ-field* $\mathcal{B}(S)$, which is the smallest σ-field containing the *open balls*

$$B_m(x, r) \equiv \{y \in S : m(x, y) < r\}.$$

The elements of $\mathcal{B}(S)$ are called measurable sets. We mention measurability and σ-fields because, in general, it is not possible to define a probability measure (satisfying the axioms of a probability measure) on all subsets; see p. 233 of Billingsley (1968).

We say that a sequence of probability measures $\{P_n : n \geq 1\}$ on (S, m) *converges weakly* or just *converges* to a probability measure P on (S, m), and we write $P_n \Rightarrow P$, if

$$\lim_{n \to \infty} \int_S f dP_n = \int_S f dP \qquad (2.1)$$

for all functions f in $C(S)$, the space of all continuous bounded real-valued functions on S. The metric m enters in by determining which functions f on S are continuous. It remains to show that this is a good definition; we discuss that point further in Section 11.3.

We now define the *Prohorov metric* on the space $\mathcal{P} \equiv \mathcal{P}(S)$ of all probability measures on the metric space (S, m); the metric was orginally defined by Prohorov (1956); see Dudley (1968) and Billingsley (1999). Let A^ϵ be the *open ϵ-neighborhood* of A, i.e.,

$$A^\epsilon \equiv \{y \in S : m(x, y) < \epsilon \quad \text{for some} \quad x \in A\}.$$

For $P_1, P_2 \in \mathcal{P}(S)$, the *Prohorov metric* is defined by

$$\pi(P_1, P_2) \equiv \inf\{\epsilon > 0 : P_1(A) \leq P_2(A^\epsilon) + \epsilon \quad \text{for all} \quad A \in \mathcal{B}(S)\}. \qquad (2.2)$$

At first glance, it may appear that π in (2.2) lacks the symmetry property, but it holds. We prove the following theorem in Section 1.2 of the Internet Supplement.

Theorem 3.2.1. (the Prohorov metric on $\mathcal{P}$) *For any separable metric space (S, m), the function π on $\mathcal{P}(S)$ in (2.2) is a separable metric. There is convergence $\pi(P_n, P) \to 0$ in $\mathcal{P}(S)$ if and only if $P_n \Rightarrow P$, as defined in (2.1).*

We primarily want to specify when weak convergence $P_n \Rightarrow P$ holds, thus we are primarily interested in the topology induced by the Prohorov metric. Indeed, there are other metrics inducing this topology; e.g., see Dudley (1968).

Instead of directly referring to probability measures, we often use random elements. A *random element* X of $(S, \mathcal{B}(S))$ is a (measurable; see Section 11.3) mapping from some underlying probability space $(\Omega, \mathcal{F}, P)$ to $(S, \mathcal{B}(S))$. (In the underlying probability space, Ω is a set, $\mathcal{F}$ is a σ-field and P is a probability measure.) The *probability law* of X or the *probability distribution* of X is the image probability measure PX^{-1} induced by X on $(S, \mathcal{B}(S))$; i.e.,

$$\begin{aligned} PX^{-1}(A) &\equiv P(X^{-1}(A)) \equiv P(\{\omega \in \Omega : X(\omega) \in A\}) \\ &\equiv P(X \in A) \quad \text{for} \quad A \in \mathcal{B}(S), \end{aligned}$$

where P is the probability measure in the underlying probability space $(\Omega, \mathcal{F}, P)$. We often use random elements, but when we do, we usually are primarily interested in their probability laws. Hence the underlying probability space $(\Omega, \mathcal{F}, P)$ is often left unspecified.

We say that a sequence of random elements $\{X_n : n \geq 1\}$ of a metric space (S, m) *converges in distribution* or *converges weakly* to a random element X of (S, m), and we write $X_n \Rightarrow X$, if the image probability measures converge weakly, i.e., if

$$P_n X_n^{-1} \Rightarrow P X^{-1} \quad \text{on} \quad (S, m) ,$$

using the definition in (2.1), where P_n and P are the underlying probability measures associated with X_n and X, respectively. It follows from (2.1) that $X_n \Rightarrow X$ if and only if

$$\lim_{n \to \infty} E f(X_n) = E f(X) \quad \text{for all} \quad f \in C(S) . \tag{2.3}$$

Thus convergence in distribution of random elements is just another way to talk about weak convergence of probability measures. When S is a function space, such as D, a random element of S becomes a *random function*, which we also call a *stochastic process*.

We can use the Skorohod representation theorem, also from Skorohod (1956), to help understand the topology of weak convergence in $\mathcal{P}(S)$. As before, $\stackrel{\mathrm{d}}{=}$ means equal in distribution.

Theorem 3.2.2. *(Skorohod representation theorem) If $X_n \Rightarrow X$ in a separable metric space (S, m), then there exist other random elements of (S, m), $\tilde{X}_n, n \geq 1$, and $\tilde{X}$, defined on a common underlying probability space, such that*

$$\tilde{X}_n \stackrel{\mathrm{d}}{=} X_n, n \geq 1, \quad \tilde{X} \stackrel{\mathrm{d}}{=} X$$

and

$$P(\lim_{n \to \infty} \tilde{X}_n = \tilde{X}) = 1 .$$

The Skorohod representation theorem is useful because it lets us relate the structure of the space of probability measures $(\mathcal{P}, \pi)$ to the structure of the underlying metric space (S, m). It also serves as a basis for the continuous-mapping approach; see Section 3.4 below. We prove the Skorohod representation theorem in Section 1.3 of the Internet Supplement.

3.3. The Space D

We now consider the underlying function space of possible sample paths for the stochastic processes. Since we want to consider stochastic processes with discontinuous, but not too irregular, sample paths, we consider the space D of all right-continuous $\mathbb{R}^k$-valued functions with left limits defined on a subinterval I of the real line, usually either $[0, 1]$ or $\mathbb{R}_+ \equiv [0, \infty)$; see Section 12.2 for additional details. We refer to the space as $D(I, \mathbb{R}^k)$, $D([0,1], \mathbb{R}^k)$ or $D([0,\infty), \mathbb{R}^k)$, depending upon the function domain, or just D when the function domain and range are clear from the context. The space D is also known as the space of *cadlag* or *càdlàg* functions – an acronym for the French *continu à droite, limites à gauche*.

The space D includes all continuous functions and the discontinuous functions of interest, but has useful regularity properties facilitating the development of a satisfactory theory. Let $C(I, \mathbb{R}^k)$, $C([0,1], \mathbb{R}^k)$ and $C([0,\infty), \mathbb{R}^k)$, or just C, denote the corresponding subsets of continuous functions.

We start by considering $D([0,1], \mathbb{R})$, i.e., by assuming that the domain is the unit interval $[0,1]$ and the range is $\mathbb{R}$. Recall that the space $D([0,1], \mathbb{R})$ was appropriate for the stochastic-process limits suggested by the plots in Chapter 1. The reference metric is the *uniform metric* $\|x_1 - x_2\|$, defined in terms of the *uniform norm*

$$\|x\| \equiv \sup_{0 \leq t \leq 1} \{|x(t)|\} . \tag{3.1}$$

On the subspace C the uniform metric works well, but it does *not* on D: When functions have discontinuities, we do not want to insist that corresponding jumps occur exactly at the same times in order for the functions to be close. Appropriate topologies were introduced by Skorohod (1956). For a celebration of Skorohod's impressive contributions to probability theory, see Korolyuk, Portenko and Syta (2000).

To define the first metric on D, let Λ be the set of strictly increasing functions λ mapping the domain $[0,1]$ onto itself, such that both λ and its inverse λ^{-1} are continuous. Let e be the *identity map* on $[0,1]$, i.e., $e(t) = t$, $0 \leq t \leq 1$. Then the standard J_1 metric on $D \equiv D([0,1], \mathbb{R})$ is

$$d_{J_1}(x_1, x_2) \equiv \inf_{\lambda \in \Lambda} \{\|x_1 \circ \lambda - x_2\| \vee \|\lambda - e\|\} , \tag{3.2}$$

where $a \vee b \equiv max\{a, b\}$.

The general idea in going from the uniform metric $\|\cdot\|$ *to the J_1 metric d_{J_1} is to say functions are close if they are uniformly close over* $[0,1]$ *after allowing small perturbations of time (the function argument).* For example, $d_{J_1}(x_n, x) \to 0$ as $n \to \infty$, while $\|x_n - x\| \geq 1$ for all n, in $D([0,1], \mathbb{R})$ when $x = I_{[2^{-1},1]}$ and $x_n = (1 + n^{-1}) I_{[2^{-1} + n^{-1}, 1]}$, $n \geq 3$.

In the example above, the limit function has a single jump of magnitude 1 at time 2^{-1}. The converging functions have jumps of size $1 + n^{-1}$ at time $2^{-1} + n^{-1}$; both the magnitudes and locations of the single jump in x_n converge to those of the limit function x. That is a characteristic property of the J_1 topology. Indeed, from definition (3.2) it follows that, if $d_{J_1}(x_n, x) \to 0$ in $D([0,1], \mathbb{R})$, then for any t with $0 < t \leq 1$ there necessarily exists a sequence $\{t_n : n \geq 1\}$ such that $t_n \to t$, $x_n(t_n) \to x(t)$, $x_n(t_n-) \to x(t-)$ and

$$x_n(t_n) - x_n(t_n-) \to x(t) - x(t-) \quad \text{as} \quad n \to \infty ;$$

i.e., the jumps converge. (It suffices to let $t_n = \lambda_n(t)$, where $\|\lambda_n - e\| \to 0$ and $\|x_n \circ \lambda_n - x\| \to 0$.) Thus, if x has a jump at t, i.e., if $x(t) \neq x(t-)$, and if $x_n \to x$, then for all n sufficiently large x_n must have a "matching jump" at some time t_n.

That is, for any $\epsilon > 0$, we can find n_0 such that, for all $n \geq n_0$, there is t_n with $|t_n - t| < \epsilon$ and
$$|(x_n(t_n) - x_n(t_n-)) - (x(t) - x(t-))| < \epsilon.$$

We need a different topology on D if we want the jump in a limit function to be unmatched in the converging functions. For example, we want to allow continuous functions to be arbitrarily close to a discontinuous function; e.g., we want to have $d(x_n, x) \to 0$ when $x = I_{[2^{-1},1]}$ and
$$x_n = n(t - 2^{-1} + n^{-1})I_{[2^{-1}-n^{-1}, 2^{-1})} + I_{[2^{-1},1]},$$
as shown in Figure 3.1. (We include dips in the axes because the points $2^{-1}-n^{-1}$ and 2^{-1} are not in scale. And similarly in later figures.) Notice that both $\|x_n - x\| = 1$

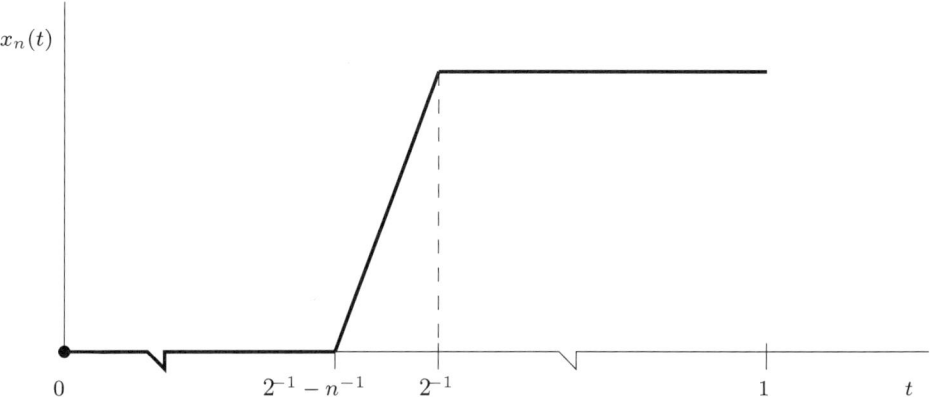

Figure 3.1. The continuous functions x_n that we want converging to the indicator function $x = I_{[2^{-1},1]}$ in D.

and $d_{J_1}(x_n, x) = 1$ for all n, so that both the uniform metric and the J_1 metric on D are too strong.

Another example has discontinuous converging functions, but converging functions in which a limiting jump is approached in more than one jump. With the same limit x above, let
$$x_n = 2^{-1} I_{[2^{-1}-n^{-1}, 2^{-1})} + I_{[2^{-1},1]},$$
as depicted in Figure 3.2. Again, $\|x_n - x\| \not\to 0$ and $d_{J_1}(x_n, x) \not\to 0$ in D as $n \to \infty$.

In order to establish limits with unmatched jumps in the limit function, we use the M_1 metric. We define the M_1 metric using the completed graphs of the functions. For $x \in D([0,1], \mathbb{R})$, the *completed graph* of x is the set
$$\Gamma_x \equiv \{(z,t) \in \mathbb{R} \times [0,1] :$$
$$z = \alpha x(t-) + (1-\alpha)x(t) \quad \text{for some} \quad \alpha, \quad 0 \leq \alpha \leq 1\}, \quad (3.3)$$

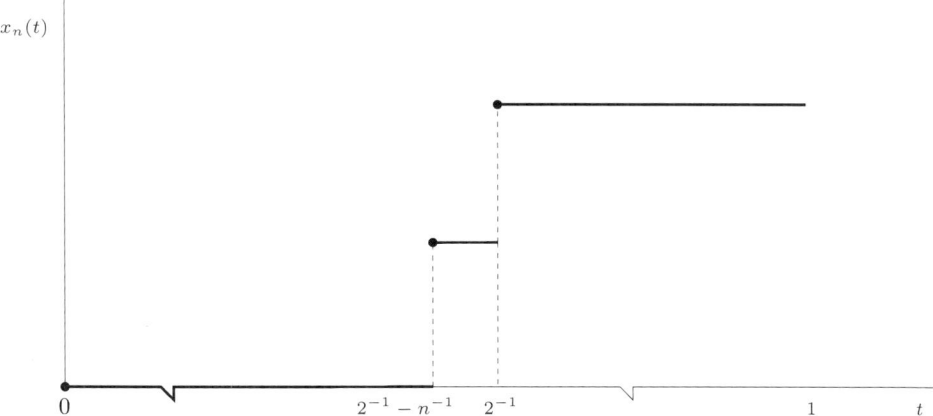

Figure 3.2. The two-jump discontinuous functions x_n that we want converging to the indicator function $x = I_{[2^{-1},1]}$.

where $x(t-)$ is the left limit of x at t with $x(0-) \equiv x(0)$. The completed graph is a connected subset of the plane $\mathbb{R}^2$ containing the line segment joining $(x(t), t)$ and $(x(t-), t)$ for all discontinuity points t. To illustrate, a function and its completed graph are displayed in Figure 3.3.

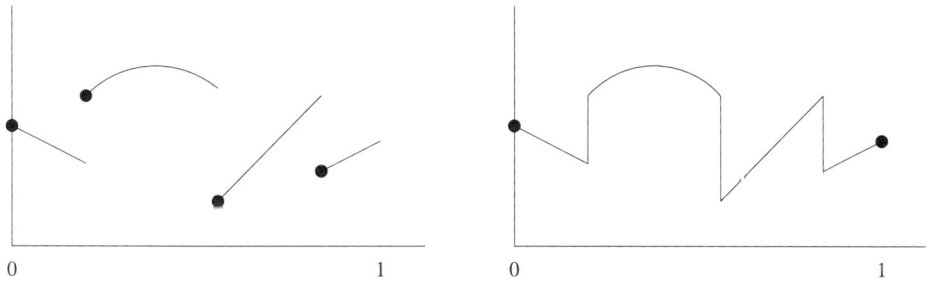

Figure 3.3. A function in $D([0,1], \mathbb{R})$ and its completed graph.

We define the M_1 metric using the uniform metric defined on parametric representations of the completed graphs of the functions. To define the parametric representations, we need an order on the completed graphs. We define an *order* on the graph Γ_x by saying that $(z_1, t_1) \leq (z_2, t_2)$ if either (i) $t_1 < t_2$ or (ii) $t_1 = t_2$ and $|x(t_1-) - z_1| \leq |x(t_2-) - z_2|$. Thus the order is a total order, starting from the "left end" of the completed graph and concluding on the "right end".

A *parametric representation* of the completed graph Γ_x (or of the function x) is a continuous nondecreasing function (u, r) mapping $[0, 1]$ onto Γ_x, with u being the spatial component and r being the time component. The parametric representation (u, r) is nondecreasing using the order just defined on the completed graph Γ_x.

Let $\Pi(x)$ be the set of parametric representations of x in $D \equiv D([0,1], \mathbb{R})$. For any $x_1, x_2 \in D$, the M_1 metric is

$$d_{M_1}(x_1, x_2) \equiv \inf_{\substack{(u_j, r_j) \in \Pi(x_j) \\ j=1,2}} \{\|u_1 - u_2\| \vee \|r_1 - r_2\|\}, \qquad (3.4)$$

where again $a \vee b \equiv max\{a, b\}$. It turns out that d_{M_1} in (3.4) is a bonafide metric on D. (The triangle inequality is not entirely obvious; see Theorem 12.3.1.)

It is easy to see that, if x is continuous, then $d_{M_1}(x_n, x) \to 0$ if and only if $\|x_n - x\| \to 0$. It is also easy to see that $d_{M_1}(x_n, x) \to 0$ as $n \to \infty$ for the examples in Figures 3.1 and 3.2. To illustrate, we display in Figure 3.4 specific parametric representations (u, r) and (u_n, r_n) of the completed graphs of x and x_n for the functions in Figure 3.1 that yield the distance $d_{M_1}(x_n, x) = n^{-1}$. The spatial

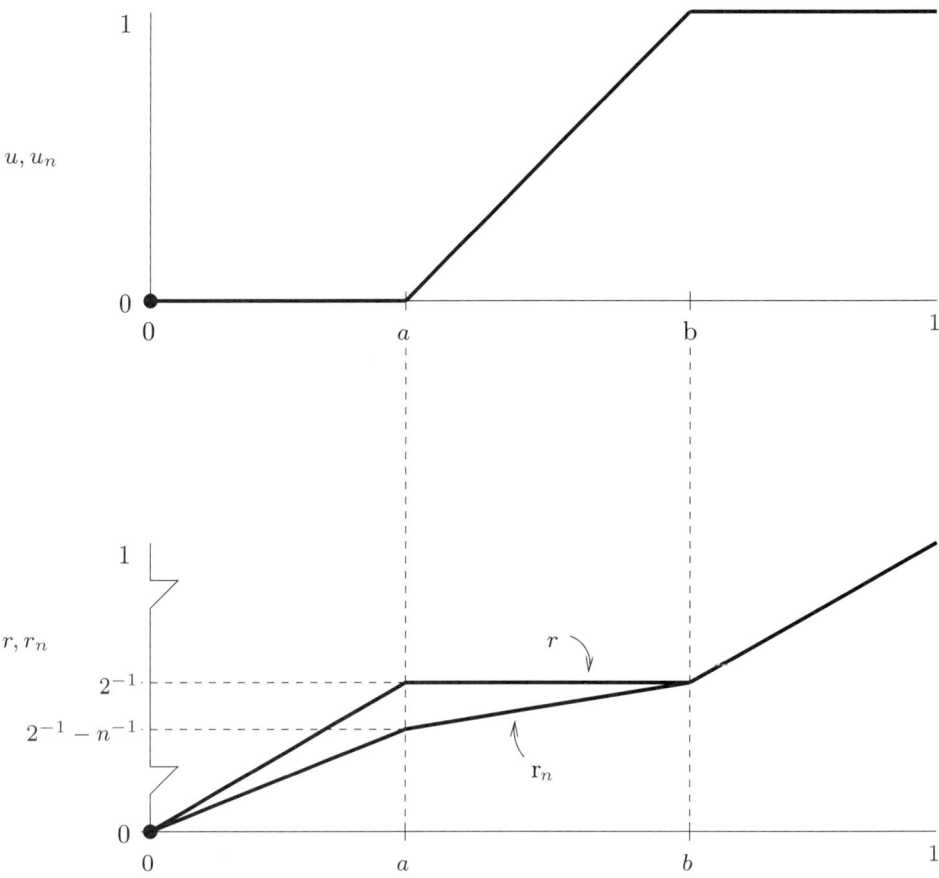

Figure 3.4. Plots of parametric representations (u, r) of Γ_x and (u_n, r_n) of Γ_{x_n} yielding $d_{M_1}(x_n, x) = n^{-1}$ for the functions in Figure 3.1. The points a and b are arbitrary, satisfying $0 < a < b < 1$.

components u and u_n are identical. The time components satisfy $\|r_n - r\| - n^{-1}$.

For applications, it is significant that previous limits for stochastic processes with the familiar J_1 topology on D will also hold when we use the M_1 topology instead, because the J_1 topology is stronger (or finer) than the M_1 topology; see Theorem 12.3.2.

We now want to modify the space $D([0,1],\mathbb{R})$ in two ways: We want to extend the range of the functions from $\mathbb{R}$ to $\mathbb{R}^k$ and we want to allow the domain of the functions be the semi-infinite interval $[0,\infty)$ instead of the unit interval $[0,1]$. First, the J_1 and M_1 metrics extend directly to $D^k \equiv D([0,1],\mathbb{R}^k)$ when the norm $|\cdot|$ on $\mathbb{R}$ in (3.1) is replaced by a corresponding norm on $\mathbb{R}^k$ such as the maximum norm

$$\|a\| \equiv \max_{1 \leq i \leq k} |a^i|,$$

for $a \equiv (a^1, \ldots, a^k) \in \mathbb{R}^k$. With the maximum norm on $\mathbb{R}^k$, we obtain the standard or strong J_1 and M_1 metrics on D^k. We call the topology induced by these metrics the standard or *strong topology*, and denote it by SJ_1 and SM_1, respectively.

We also use the *product topology* on D^k, regarding D^k as the product space $D \times \cdots \times D$, which has $x_n \to x$ as $n \to \infty$ for $x_n \equiv (x_n^1, \ldots, x_n^k)$ and $x \equiv (x^1, \ldots, x^k)$ in D^k if $x_n^i \to x^i$ as $n \to \infty$ in D for each i. The product topology on D^k is induced by the metric

$$d_p(x,y) \equiv \sum_{i=1}^{k} d(x^i, y^i), \tag{3.5}$$

where d is the metric on D^1. Since convergence in the strong topology implies convergence in the product topology, we also call the product topology the *weak topology*, and we denote it by WJ_1 and WM_1.

The definitions for $D([0,1],\mathbb{R}^k)$ extend directly to $D([0,t],\mathbb{R}^k)$ for any $t > 0$. It is natural to characterize convergence of a sequence $\{x_n : n \geq 1\}$ in $D([0,\infty),\mathbb{R}^k)$ in terms of associated convergence of the restrictions of x_n to the subintervals $[0,t]$ in the space $D([0,t],\mathbb{R}^k)$ for all $t > 0$. However, note that we encounter difficulties in $D([0,t],\mathbb{R}^k)$ if the right endpoint t is a discontinuity point of a prospective limit function x. For example, if $t_n \to t$ as $n \to \infty$, but $t_n > t$ for all n, then the restrictions of $I_{[t_n,\infty)}$ to the subinterval $[0,t]$ are the zero function, while $I_{[t,\infty)}$ is not, so we cannot get the desired convergence $I_{[t_n,\infty)} \to I_{[t,\infty)}$ we want. Thus we say that the sequence $\{x_n : n \geq 1\}$ converges to x as $n \to \infty$ in $D([0,\infty),\mathbb{R}^k)$ if the restrictions of x_n to $[0,t]$ converge to the restriction of x to $[0,t]$ in $D([0,t],\mathbb{R}^k)$ for all $t > 0$ that are continuity points of x.

The mode of convergence just defined can be achieved with metrics. Given a metric d_t on $D([0,t],\mathbb{R})$ applied to the restrictions of the functions to $[0,t]$, we define a metric d_∞ on $D([0,\infty),\mathbb{R})$ by letting

$$d_\infty(x_1, x_2) = \int_0^\infty e^{-t}[d_t(x_1,x_2) \wedge 1]dt, \tag{3.6}$$

where $a \wedge b \equiv \min\{a,b\}$ and $d_t(x_1,x_2)$ is understood to mean the distance d_t (either J_1 or M_1) applied to the restrictions of x_1 and x_2 to $[0,t]$. There are some technical complications in verifying that the integral in (3.6) is well defined (i.e., in showing that the integrand is integrable), but it is.

The function space D with the J_1 or M_1 topology is somewhat outside the mainstream of traditional functional analysis, because *addition is not a continuous map* from the product space $D \times D$ with the product topology to D.

Example 3.3.1. *Addition is not continuous.*

A simple example has $x = -y = I_{[2^{-1},1]}$ with

$$x_n = I_{[2^{-1}-n^{-1},1]} \quad \text{and} \quad y_n = -I_{[2^{-1}+n^{-1},1]} .$$

Then $(x+y)(t) = 0$ for all t, while

$$x_n + y_n = I_{[2^{-1}-n^{-1}, 2^{-1}+n^{-1}]} .$$

With the nonuniform Skorohod topologies, $x_n \to x$ and $y_n \to y$ as $n \to \infty$, but $x_n + y_n \not\to x + y$ as $n \to \infty$. ∎

Thus, even though D is a vector space (we can talk about the linear combinations $ax + by$ for functions x and y in D and numbers a and b in $\mathbb{R}$), D *is not a topological vector space* (and thus not a Banach space) with the J_1 and M_1 topologies (because those structures require addition to be continuous).

Nevertheless, in applications of the continuous-mapping approach to establish stochastic-process limits, we will often want to add or subtract two functions. Thus it is very important that addition can be made to preserve convergence. It turns out that addition on $D \times D$ is measurable and it is continuous at limits in a large subset of $D \times D$. For any of the nonuniform Skorohod topologies, it suffices to assume that the two limit functions x and y have no common discontinuity points; see Sections 12.6 and 12.7. With the M_1 topology (but not the J_1 topology), it suffices to assume that the two limit functions x and y have no common discontinuity points with jumps of opposite sign; see Theorem 12.7.3. (For instance, in Example 3.3.1, $x_n - y_n \to x - y$ in (D, M_1).) In many applications, we are able to show that the two-dimensional limiting stochastic process has sample paths in one of those subsets of pairs (x, y) w.p.1. Then we can apply the continuous-mapping theorem with addition.

3.4. The Continuous-Mapping Approach

The continuous-mapping approach to stochastic-process limits exploits previously established stochastic-process limits and the continuous-mapping theorem to obtain new stochastic-process limits of interest. Alternative approaches are the compactness approach described in Section 11.6 and various stochastic approaches (which usually exploit the compactness approach), which exploit special stochastic structure, such as Markov and martingale structure; e.g., see Billingsley (1968, 1999), Ethier and Kurtz (1986), Jacod and Shiryaev (1987) and Kushner (2001).

Here is a simple form of the continuous-mapping theorem:

Theorem 3.4.1. (simple continuous-mapping theorem). *If $X_n \Rightarrow X$ in (S, m) and $g : (S, m) \to (S', m')$ is continuous, then*
$$g(X_n) \Rightarrow g(X) \quad in \quad (S', m') .$$

Proof. Since g is continuous, $f \circ g$ is a continuous bounded real-valued function on (S, m) for each continuous bounded real-valued function f on (S', m'). Hence, under the conditions,
$$E[f \circ g(X_n)] \to E[f \circ g(X)]$$
for each continuous bounded real-valued function f on (S', m'), which implies the desired conclusion by (2.3). ■

Paralleling the simple continuous-mapping theorem above, we can use a *Lipschitz-mapping theorem* to show that distances, and thus rates of convergence with the Prohorov metric, are preserved under Lipschitz mappings: A function g mapping a metric space (S, m) into another metric space (S', m') is said to be *Lipschitz continuous*, or just *Lipschitz*, if there exists a constant K such that
$$m'(g(x), g(y)) \leq K m(x, y) \quad \text{for all} \quad x, y \in S . \tag{4.1}$$

The infimum of all constants K for which (4.1) holds is called the *Lipschitz constant*. As before, let $a \vee b \equiv \max\{a, b\}$. The following Lipschitz mapping theorem, taken from Whitt (1974a), is proved in Section 1.5 of the Internet Supplement. Applications to establish rates of convergence in stochastic-process limits are discussed in Section 2.2 of the Internet Supplement. We write $\pi(X, Y)$ for the distance between the probability laws of the random elements X and Y.

Theorem 3.4.2. (Lipschitz mapping theorem) *Suppose that $g : (S, m) \to (S', m')$ is Lipschitz as in (4.1) on a subset B of S. Then*
$$\pi(g(X), g(Y)) \leq (K \vee 1)\pi(X, Y)$$
for any random elements X and Y of (S, m) for which $P(Y \in B) = 1$.

We often need to go beyond the simple continuous-mapping theorem in Theorem 3.4.1. We often need to consider measurable functions that are only continuous almost everywhere or a sequence of such functions. Fortunately, the continuous-mapping theorem extends to such settings. We can work with a sequence of Borel measurable functions $\{g_n : n \geq 1\}$ all mapping one separable metric space (S, m) into another separable metric space (S', m'). It suffices to have $g_n(x_n) \to g(x)$ as $n \to \infty$ whenever $x_n \to x$ as $n \to \infty$ for a subset E of limits x in S such that $P(X \in E) = 1$. This generalization follows easily from the Skorohod representation theorem, Theorem 3.2.2: Starting with the convergence in distribution $X_n \Rightarrow X$, we apply the Skorohod representation theorem to obtain the special random elements $\tilde{X}_n$ and $\tilde{X}$ with the same distributions as X_n and X such that $\tilde{X}_n \to \tilde{X}$ w.p.1. Since $\tilde{X} \stackrel{d}{=} X$ and $P(X \in E) = 0$, we also have $P(\tilde{X} \in E) = 0$. We then apply the deterministic convergence preservation assumed for the functions g_n to get the limit
$$g(\tilde{X}_n) \to g(\tilde{X}) \quad \text{as} \quad n \to \infty \quad \text{in} \quad (S', m') \quad \text{w.p.1.}$$

Since convergence w.p.1 implies convergence in distribution, as a consequence we obtain

$$g(\tilde{X}_n) \Rightarrow g(\tilde{X}) \quad \text{in} \quad (S', m') \ .$$

Finally, since X_n and X are respectively equal in distribution to $\tilde{X}_n$ and $\tilde{X}$, also $g_n(X_n)$ and $g(X)$ are respectively equal in distribution to $g_n(\tilde{X}_n)$ and $g(\tilde{X})$. Thus, we obtain the desired generalization of the continuous-mapping theorem:

$$g_n(X_n) \Rightarrow g(X) \quad \text{in} \quad (S', m') \ .$$

It is also possible to establish such extensions of the simple continuous-mapping theorem in Theorem 3.4.1 directly, without resorting to the Skorohod representation theorem. We can use the continuous-mapping theorem or the generalized continuous-mapping theorem, proved in Section 1.5 of the Internet Supplement.

For $g : (S, m) \to (S', m')$, let $Disc(g)$ be the *set of discontinuity points* of g; i.e., $Disc(g)$ is the subset of x in S such that there exists a sequence $\{x_n : n \geq 1\}$ in S with $m(x_n, x) \to 0$ and $m'(g(x_n), g(x)) \not\to 0$.

Theorem 3.4.3. (continuous-mapping theorem) *If $X_n \Rightarrow X$ in (S, m) and $g : (S, m) \to (S', m')$ is measurable with $P(X \in Disc(g)) = 0$, then $g(X_n) \Rightarrow g(X)$.*

Theorem 3.4.4. (generalized continuous-mapping theorem) *Let g and g_n, $n \geq 1$, be measurable functions mapping (S, m) into (S', m'). Let the range (S', m') be separable. Let E be the set of x in S such that $g_n(x_n) \to g(x)$ fails for some sequence $\{x_n : n \geq 1\}$ with $x_n \to x$ in S. If $X_n \Rightarrow X$ in (S, m) and $P(X \in E) = 0$, then $g_n(X_n) \Rightarrow g(X)$ in (S', m').*

Note that $E = Disc(g)$ if $g_n = g$ for all n, so that Theorem 3.4.4 contains both Theorems 3.4.1 and 3.4.3 as special cases.

3.5. Useful Functions

In order to apply the continuous-mapping approach to establish stochastic-process limits, we need initial stochastic-process limits in D, the product space $D^k \equiv D \times \cdots \times D$ or some other space, and we need functions mapping D, D^k or the other space into D that preserve convergence. The initial limit is often Donsker's theorem or a generalization of it; see Chapters 4 and 7.

Since we are interested in obtaining stochastic-process limits, the functions preserving convergence must be D-valued rather than $\mathbb{R}$-valued or $\mathbb{R}^k$-valued. In this section we identify five basic functions from D or $D \times D$ to D that can be used to establish new stochastic-process limits from given ones: addition, composition, supremum, reflection and inverse. These functions will be carefully examined in Chapters 12 and 13. For the discussion here, let the function domain be $\mathbb{R}_+ \equiv [0, \infty)$.

The *addition map* takes $(x, y) \in D \times D$ into $x + y$, where

$$(x + y)(t) \equiv x(t) + y(t), \quad t \geq 0 \ . \tag{5.1}$$

The *composition map* takes $(x, y) \in D \times D$ into $x \circ y$, where
$$(x \circ y)(t) \equiv x(y(t)), \quad t \geq 0. \tag{5.2}$$

The *supremum map* takes $x \in D$ into $x^\uparrow$, where
$$x^\uparrow(t) \equiv \sup_{0 \leq s \leq t} x(s), \quad t \geq 0. \tag{5.3}$$

The (one-sided, one-dimensional) *reflection map* takes $x \in D$ into $\phi(x)$, where
$$\phi(x) \equiv x + (-x \vee 0)^\uparrow, \quad t \geq 0, \tag{5.4}$$

with $(x \vee 0)(t) \equiv x(t) \vee 0$. The *inverse* map takes x into x^{-1}, where
$$x^{-1}(t) \equiv \inf\{s \geq 0 : x(s) > t\}, \quad t \geq 0. \tag{5.5}$$

Regularity conditions are required in order for the composition $x \circ y$ in (5.2) and the inverse x^{-1} in (5.5) to belong to D; those conditions will be specified in Chapter 13.

The general idea is that, by some means, we have already established convergence in distribution
$$\mathbf{X}_n \Rightarrow \mathbf{X} \quad \text{in} \quad D,$$
and we wish to deduce that
$$\psi(\mathbf{X}_n) \Rightarrow \psi(\mathbf{X}) \quad \text{in} \quad D$$
for one of the functions ψ above. By virtue of the continuous-mapping theorem or the Skorohod representation theorem, it suffices to show that $\psi : D \to D$ is measurable and continuous at all $x \in A$, where $P(\mathbf{X} \in A) = 1$. Equivalently, in addition to the measurability, it suffices to show that ψ preserves convergence in D; i.e., that $\psi(x_n) \to \psi(x)$ whenever $x_n \to x$ for $x \in A$, where $P(\mathbf{X} \in A) = 1$.

There tends to be relatively little difficulty if A is a subset of continuous functions, but we are primarily interested in the case in which the limit has discontinuities. As illustrated by Example 3.3.1 for addition, when $x \notin C$, the basic functions often are not continuous in general. We must then identify an appropriate subset A in D, and work harder to demonstrate that convergence is indeed preserved.

Many applications of interest actually do not involve convergence preservation in such a simple direct form as above. Instead, the limits involve *centering*. In the deterministic framework (obtained after invoking the Skorohod representation theorem), we often start with
$$c_n(x_n - x) \to y \quad \text{in} \quad D, \tag{5.6}$$
where $c_n \to \infty$, from which we can deduce that
$$x_n \to x \quad \text{in} \quad D. \tag{5.7}$$

From (5.7) we can directly deduce that
$$\psi(x_n) \to \psi(x) \quad \text{in} \quad D$$

provided that ψ preserves convergence. However, we want more. We want to deduce that

$$c_n(\psi(x_n) - \psi(x)) \to z \quad \text{in} \quad D \tag{5.8}$$

and identify the limit z. We will want to show that (5.6) implies (5.8).

The common case is for x in (5.6)–(5.8) to be linear, i.e., for

$$x \equiv be, \quad \text{where } b \in \mathbb{R} \text{ and } e \in D \text{ with } e(t) \equiv t \quad \text{for all} \quad t\,.$$

We call that the case of *linear centering*. We will consider both linear and nonlinear centering.

The stochastic applications with centering are less straightforward. We might start with

$$\mathbf{X}_n \Rightarrow \mathbf{U} \quad \text{in} \quad D, \tag{5.9}$$

where

$$\mathbf{X}_n \equiv b_n^{-1}(X_n(nt) - \lambda nt), \quad t \geq 0,$$

for some stochastic processes $\{X_n(t) : t \geq 0\}$. Given (5.9), we wish to deduce that

$$\mathbf{Y}_n \Rightarrow \mathbf{V} \quad \text{in} \quad D \tag{5.10}$$

and identify the limit process $\mathbf{V}$ for

$$\mathbf{Y}_n \equiv b_n^{-1}(\psi(X_n)(nt) - \mu nt), \quad t \geq 0\,. \tag{5.11}$$

To apply the convergence-preservation results with centering, we can let

$$x_n(t) \equiv (n\lambda)^{-1} X_n(nt), \quad x(t) \equiv e(t) \equiv t, \quad c_n \equiv n\lambda/b_n$$

and assume that $|c_n| \to \infty$. The w.p.1 representation of the weak convergence in (5.9) yields

$$c_n(x_n - x) \to u \quad \text{w.p.1} \quad \text{in} \quad D,$$

where u is distributed as $\mathbf{U}$. The convergence-preservation result ((5.6) implies (5.8)) then yields

$$c_n[\psi(x_n) - \psi(x)] \to v \quad \text{w.p.1} \quad \text{in} \quad D\,. \tag{5.12}$$

We thus need to relate the established w.p.1 convergence in (5.12) to the desired convergence in distribution in (5.10). This last step depends upon the function ψ. To illustrate, suppose, as is the case for the supremum and reflection maps in (5.3) and (5.4), that $\psi(e) = e$ and ψ is homogeneous, i.e., that

$$\psi(ax) = a\psi(x) \quad \text{for} \quad x \in D \quad \text{and} \quad a > 0\,.$$

Then

$$c_n[\psi(x_n) - \psi(e)] = b_n^{-1}[\psi(X_n)(nt) - \lambda nt]\,.$$

Thus, under those conditions on ψ, we can deduce that (5.10) holds for $\mathbf{Y}_n$ in (5.11) with $\mu = \lambda$ and $\mathbf{V}$ distributed as v in (5.12).

In applications, our primary goal often is to obtain convergence in distribution for a sequence of real-valued random variables, for which we only need to consider the continuous mapping theorem with real-valued functions. However, it is often convenient to carry out the program in two steps: We start with a FCLT in D for a sequence of basic stochastic processes such as random walks. We then apply the continuous-mapping theorem with the kind of functions considered here to obtain new FCLT's for the basic stochastic processes in applied probability models, such as queue-length stochastic processes in a queueing model. Afterwards, we obtain desired limits for associated random variables of interest by applying the continuous-mapping theorem again with real-valued functions of interest. The final map may be the simple one-dimensional projection map π_t mapping $x \in D$ into $x(t) \in \mathbb{R}^k$ when $\mathbb{R}^k$ is the range of the functions in D, the average $t^{-1} \int_0^t x(s)ds$ or something more complicated.

3.6. Organization of the Book

We now expand upon the description of the organization of the book given at the end of the preface. As indicated there, the book has fifteen chapters, which can be roughly grouped into four parts, ordered according to increasing difficulty. The *first part*, containing the first five chapters, provides an informal introduction to stochastic-process limits and their application to queues.

Chapter 1 exposes the statistical regularity associated with a macroscopic view of uncertainty, with appropriate scaling, via plots of random walks, obtained from elementary stochastic simulations. Remarkably, the plotter automatically does the proper scaling when we plot the first n steps of the random walk for various values of n. The plots tend to look the same for all n sufficiently large, showing that there must be a stochastic-process limit. For random walks with IID steps having infinite variance, the plots show that the limit process must have jumps, i.e., discontinuous sample paths.

Chapter 2 shows that the abstract random walks considered in Chapter 1 have useful applications. Chapter 2 discusses applications to stock prices, the Kolmogorov-Smirnov statistic and queueing models. Chapter 2 also discusses the engineering significance of the queueing models and the heavy-traffic limits. The engineering significance is illustrated by applications to buffer sizing in network switches and service scheduling for multiple sources.

The present chapter, Chapter 3, introduces the mathematical framework for stochastic-process limits, involving the concept of weak convergence of a sequence of probability measures on a separable metric space and the function space D containing stochastic-process sample paths. Metrics inducing the Skorohod J_1 and M_1 topologies on D are defined. An overview of the continuous mapping approach to establish stochastic-process limits is also given.

Chapter 4 provides an overview of established stochastic-process limits. These stochastic-process limits are of interest in their own right, but they also serve as starting points in the continuous-mapping approach to establish new stochastic-process limits. The fundamental stochastic-process limit is provided by Donsker's theorem, which was already discussed in Chapter 1. The other stochastic-process limits are generalizations of Donsker's theorem. Of particular interest for the limits with jumps, is the generalization of Donsker's theorem in which the random-walk steps are IID with infinite variance. When the random-walk steps have such heavy-tailed distributions, the limit process is a stable Lévy motion in the case of a single sequence or a general Lévy process in the case of a triangular array or double sequence. When these limit processes are not Brownian motion, they have discontinuous sample paths. The stochastic-process limits with jumps in the limit process explain some of the jumps observed in the simulation plots in Chapter 1.

Lévy processes are very special because they have independent increments. Chapter 4 also discusses stochastic-process limits in which the limit process has dependent increments. The principal stochastic-process limits of this kind involve convergence to fractional Brownian motion and linear fractional stable motion. These limit processes with dependent increments arise when there is strong dependence in the converging stochastic processes. These particular limit processes have continuous sample paths, so the topology on D is not critical. Nevertheless, like heavy tails, strong dependence has a dramatic impact on the stochastic-process limit, changing both the scaling and the limit process.

Chapter 5 provides an introduction to heavy-traffic limits for queues. This first queueing chapter focuses on a general fluid queue model that captures the essence of many more-detailed queueing models. This fluid queue model is especially easy to analyze because the continuous-mapping approach with the reflection map can be applied directly. Section 5.5 derives scaling functions, expressed as functions of the traffic intensity in the queue, which provide insight into queueing performance. Proofs are provided in Chapter 5, but the emphasis is on the statement and applied value of the heavy-traffic limits rather than the technical details. This first queueing chapter emphasizes the classical Brownian approximation (involving a reflected Brownian motion limit process). The value of the Brownian approximation is illustrated in the Section 5.8, which discusses its application to plan queueing simulations: The heavy-traffic scaling produces a simple approximation for the simulation run length required to achieve desired statistical precision, as a function of model parameters.

The *second part*, containing Chapters 6 – 10, show how unmatched jumps can arise and expands the treatment of queueing models. Chapter 6 gives several examples of stochastic-process limits with unmatched jumps in the limit process. In all the examples it is obvious that either there are no jumps in the sample paths of the converging processes or the jumps in the converging processes are asymptotically negligible. What is not so obvious is that the limit process actually can have discontinuous sample paths. As in Chapter 1, simulations are used to provide convincing evidence.

Chapter 7 continues the overview of stochastic-process limits begun in Chapter 4. It first discusses process CLT's, which are central limit theorems for appropriately scaled sums of random elements of D. Process CLT's play an important

3.6. Organization of the Book

role in heavy-traffic stochastic-process limits for queues with superposition arrival processes, when the number of component arrival processes increases in the heavy-traffic limit.

Then Chapter 7 discusses CLT's and FCLT's for counting processes. They are shown to be equivalent to corresponding limits for partial sums. Chapter 7 concludes by applying the continuous-mapping approach with the composition and inverse maps, together with established stochastic-process limits in Chapter 4, to establish stochastic-process limits for renewal-reward stochastic processes. The M_1 topology plays an important role in Chapter 7.

The remaining chapters in the second part apply the stochastic-process limits, with the continuous-mapping approach, to obtain more heavy-traffic limits for queues. As in Chapter 5, Chapters 8 – 10 emphasize the applied value of the stochastic-process limits, but now more attention is given to technical details. Chapter 8 considers a more-detailed multisource on-off fluid-queue model that has been proposed to evaluate the performance of communication networks. That model illustrates how heavy-traffic limits can expose the essential features of complex models. This second queueing chapter also discusses nonclassical approximations involving reflected stable Lévy motion and reflected fractional Brownian motion, stemming from heavy-tailed probability distributions and strong dependence.

Chapter 9 focuses on standard single-server queues, while Chapter 10 focuses on standard multiserver queues. In addition to the standard heavy-traffic limits, we consider heavy-traffic limits in which the number of component arrival processes in a superposition arrival process or the number of servers in a multiserver queue increases in the heavy-traffic limit. Those limits tend to capture the behavior of systems with large numbers of sources or servers. Even with light-tailed probability distributions and weak dependence, these important limiting regimes produce nonBrownian (even nonMarkovian) limit processes.

The *third part*, containing Chapters 11 – 14, is devoted to the technical foundations needed to establish stochastic-process limits with unmatched jumps in the limit process. The third part begins with Chapter 11, which provides more details about the mathematical framework for stochastic-process limits, expanding upon the brief introduction in Chapter 3.

Chapter 12 presents the basic theory for the function space D. Four topologies are considered on D: strong and weak versions of the M_1 topology and strong and weak versions of the M_2 topology. The strong and weak topologies differ when the functions have range $\mathbb{R}^k$ for $k > 1$. The strong topologies agree with the standard topologies defined by Skorohod (1956), while the weak topologies agree with the product topology, regarding $D([0,T],\mathbb{R}^k)$ as the k-fold product of the space $D([0,T],\mathbb{R})$ with itself. The M topologies are defined and characterized in Chapter 12. The main ideas go back to Skorohod (1956), but more details are provided here. For example, several useful alternative characterizations of these topologies are given; e.g., see Theorem 12.5.1.

Chapter 13 focuses on the useful functions from D or $D \times D$ to D introduced in Section 3.5, which preserve convergence with the Skorohod topologies, and thus facilitate the continuous-mapping approach to establish new stochastic-process limits.

As illustrated in the queueing chapters, the functions in Chapter 13 can be combined with the FCLT's in Chapter 4 to obtain many new stochastic-process limits.

The third part concludes with a final chapter on queues: Chapter 14 establishes heavy-traffic limits for single-class networks of queues. The extension to networks of queues in Chapter 14 is more complicated than the single queues considered in Chapters 5, 8 and 9 because, unlike the one-dimensional reflection map used for the single queues, the multidimensional reflection map is not simply continuous in the M_1 topology. However, it is continuous, using the product M_1 topology, at all limit functions without simultaneous jumps of opposite sign in its coordinate functions.

The *fourth part*, containing only the final chapter, Chapter 15, introduces new function spaces larger than D. These spaces, called E and F, are intended to express limits for sequences of stochastic processes with oscillations in their sample paths so great that there is no limit in D. The names are chosen because of the ordering

$$C \subset D \subset E \subset F.$$

Example 3.6.1. *Motivation for the spaces E and F.* Suppose that the nth function in a sequence of continuous functions takes the value 4 in the interval $[0, 2^{-1} - n^{-1}]$, the value 5 in the interval $[2^{-1} + n^{-1}, 1]$ and has oscillations in the subinterval $[2^{-1} - n^{-1}, 2^{-1} + n^{-1}]$ for all $n \geq 3$. Specifically, within the subinterval $[2^{-1} - n^{-1}, 2^{-1} + n^{-1}]$, let this n^{th} function first increase from the value 4 at the left endpoint $2^{-1} - n^{-1}$ to 7, then decrease to 1, and then increase again to 5 at the right endpoint $2^{-1} + n^{-1}$, as shown in Figure 3.5.

That sequence of continuous functions converges pointwise to the limit function $x = 4I_{[0,2^{-1})} + 5I_{[2^{-1},1]}$ everywhere except possibly at $t = 1/2$, but it does not converge in D with any of the Skorohod topologies. Nevertheless, we might want to say that convergence does in fact occur, with the limit somehow revealing the oscillations of the functions in the neighborhood of $t = 1/2$. The spaces E and F allow for such limits

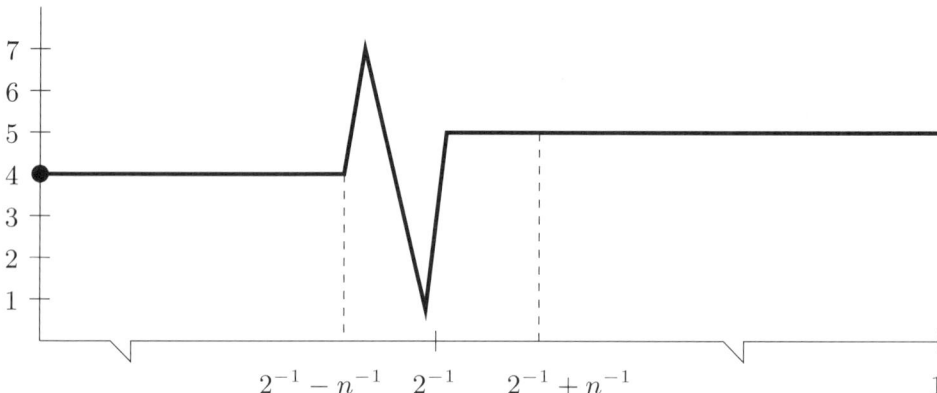

Figure 3.5. The nth function in $C[0,1]$ in a sequence of functions that converges to a proper limit in the space E but not in the space D.

In E, the limit corresponds to the set-valued function that is the one-point set $\{4\}$ for $t \in [0, 2^{-1})$, the one-point set $\{5\}$ for $t \in (2^{-1}, 1])$ and is the interval $[1, 7]$ at $t = 1/2$.

In E, the limit fails to capture the order in which the points are visited in the neighborhood of $t = 1/2$. The space F exploits parametric representations to also capture the order in which the points are visited. The larger spaces E and F are given topologies similar to the M_2 and M_1 topologies on D. Thus Chapter 15 draws heavily upon the development of the M topologies in Chapter 12. However, Chapter 15 only begins to develop the theory of E and F. Further development is a topic for future research. ∎

At the end of the book there are two appendices. Appendix A gives basic facts about regularly varying functions, while Appendix B gives the intial contents of the Internet Supplement. Also there are three indices: for notation, authors and subjects.

4
A Panorama of Stochastic-Process Limits

4.1. Introduction

In this chapter and Chapter 7 we give an overview of established stochastic-process limits for basic stochastic processes. These stochastic-process limits are of interest in their own right, but they also can serve as initial stochastic-process limits in the continuous-mapping approach to establish new stochastic-process limits. Indeed, the stochastic-process limits in this chapter all can be used to establish stochastic-process limits for queueing models. In fact, a queueing example was already given in Section 2.3. The FCLTs here, when applied to appropriate "cumulative-input" processes, translate into corresponding FCLTs for the workload process in that queueing model when we apply the continuous-mapping approach with the two-sided reflection map.

The fundamental stochastic-process limit is the convergence of a sequence of scaled random walks to Brownian motion in the function space D, provided by Donsker's (1951) theorem, which we already have discussed in Chapter 1. The other established stochastic-process limits mostly come from extensions of Donsker's theorem. In many cases, the limit process has continuous sample paths, in which case the topology on D can be the standard Skorohod J_1 topology. (The topologies on D were introduced in Section 3.3.) Even when the limit process has discontinuous sample paths, we often are able to use the standard J_1 topology, but there are cases in which the M_1 topology is needed. Even when the M_1 topology is not needed, it can be used. Thus, the M_1 topology can be used for all the stochastic-process limits here.

96   4. A Panorama

In this overview chapter we state results formally as theorems and give references, but we omit most proofs. We occasionally indicate how a secondary result follows from a primary result.

4.2. Self-Similar Processes

We start by looking at the lay of the land: In this section we consider the general form that functional central limit theorems can take, without imposing any stochastic assumptions such as independence, stationarity or the Markov property.

4.2.1. General CLT's and FCLT's

Consider a sequence $\{X_n : n \geq 1\}$ of $\mathbb{R}^k$-valued random vectors and form the associated *partial sums*

$$S_n \equiv X_1 + \cdots + X_n, \quad n \geq 1,$$

with $S_0 \equiv (0, \ldots, 0)$. We say that $\{X_n\}$ or $\{S_n\}$ obeys a *central limit theorem* (CLT) if there exist a sequence of constants $\{c_n : n \geq 1\}$, a vector m and a proper random vector S such that there is convergence in distribution (as $n \to \infty$)

$$c_n^{-1}(S_n - mn) \Rightarrow S \quad \text{in} \quad \mathbb{R}^k.$$

We call m the *translation scaling vector* (or constant if $\mathbb{R}^k = \mathbb{R}$) and $\{c_n\}$ the *space-scaling sequence*. (We might instead allow a sequence of translation vectors $\{m_n : n \geq 1\}$, but that is not the usual case and we do not consider that case here.)

Now form an associated sequence of *normalized partial-sum processes* in $D \equiv D([0, \infty), \mathbb{R}^k)$ by letting

$$\mathbf{S}_n(t) \equiv c_n^{-1}(S_{\lfloor nt \rfloor} - mnt), \quad t \geq 0, \qquad (2.1)$$

where $\lfloor t \rfloor$ is the greatest integer less than or equal to t. We say that $\{X_n\}$, $\{S_n\}$ or $\{\mathbf{S}_n\}$ obeys a *functional central limit theorem* (FCLT) if there exists a proper stochastic process $\mathbf{S} \equiv \{\mathbf{S}(t) : t \geq 0\}$ with sample paths in D such that

$$\mathbf{S}_n \Rightarrow \mathbf{S} \quad \text{in} \quad D, \qquad (2.2)$$

for $\mathbf{S}_n$ in (2.1) and some (unspecified here) topology on D.

The classical setting for the CLT and the FCLT occurs when the basic sequence $\{X_n : n \geq 1\}$ is a sequence of *independent and identically distributed* (IID) random vectors with finite second moments. However, we have not yet made those assumptions. Note that *any* sequence of $\mathbb{R}^k$-valued random vectors $\{S_n : n \geq 1\}$ can be viewed as a sequence of partial sums; just let

$$X_n \equiv S_n - S_{n-1}, \quad n \geq 1,$$

with $S_0 \equiv 0$. The partial sums of these new variables X_n obviously coincide with the given partial sums.

We also say a FCLT holds for a continuous-time $\mathbb{R}^k$-valued process $Y \equiv \{Y(t) : t \geq 0\}$ if (2.2) holds for $\mathbf{S}_n$ redefined as

$$\mathbf{S}_n(t) \equiv c_n^{-1}(Y(nt) - mnt), \quad t \geq 0. \tag{2.3}$$

Here $\{Y(t) : t \geq 0\}$ is the continuous-time analog of the partial-sum sequence $\{S_n : n \geq 1\}$ used in (2.1). Note that the discrete-time process $\{S_n\}$ is a special case, obtained by letting $Y(t) = S_{\lfloor t \rfloor}$, $t \geq 0$.

More generally, we can consider limits for continuous-time $\mathbb{R}^k$-valued stochastic processes indexed by a real variable s where $s \to \infty$. We then have

$$\mathbf{S}_s(t) \equiv c(s)^{-1}(Y(st) - mst), \quad t \geq 0, \tag{2.4}$$

for $s \geq s_0$. We say that a FCLT holds for $\mathbf{S}_s$ as $s \to \infty$ if $\mathbf{S}_s \Rightarrow \mathbf{S}$ in D as $s \to \infty$ for some (unspecified here) topology on D.

4.2.2. Self-Similarity

Before imposing any stochastic assumptions, it is natural to consider what can be said about the possible translation vectors and space-scaling functions $c(s)$ in (2.4) and the possible limit processes $\mathbf{S}$. Lamperti (1962) showed that convergence of all finite-dimensional distributions has strong structural implications. The possible limit processes are the self-similar processes, which were called semi-stable processes by Lamperti (1962) and then self-similar processes by Mandelbrot (1977); see Chapter 7 of Samorodnitsky and Taqqu (1994).

We say that an $\mathbb{R}^k$-valued stochastic process $\{Z(t) : t \geq 0\}$ is *self-similar with index $H > 0$* if, for all $a > 0$,

$$\{Z(at) : t \geq 0\} \stackrel{\mathrm{d}}{=} \{a^H Z(t) : t \geq 0\}, \tag{2.5}$$

where $\stackrel{\mathrm{d}}{=}$ denotes equality in distribution; i.e., if the stochastic process $\{Z(at) : t \geq 0\}$ has the same finite-dimensional distributions as the stochastic process $\{a^H Z(t) : t \geq 0\}$ for all $a > 0$. Necessarily $Z(0) = 0$ w.p.1. The classic example is Brownian motion, which is H-self-similar with $H = 1/2$. The scaling exponent H is often called the *Hurst parameter* in recognition of the early work by Hurst (1951, 1955).

Indeed, we already encountered self-similarity in Chapter 1. We saw self-similarity when we saw the plots looking the same for all sufficiently large n. As we can anticipate from the plots, the limit process should be self-similar: The limit process has identical plots for all time scalings. In the plots the appropriate space scaling is produced automatically by the plotter.

The class of self-similar processes is very large; e.g., if $\{Y(t) : -\infty < t < \infty\}$ is any stationary process, then

$$Z(t) \equiv t^H Y(\log t), \quad t > 0,$$

is H-self-similar. Conversely, if $\{Z(t) : t \geq 0\}$ is H-self-similar, then

$$Y(t) \equiv e^{-tH} Z(e^t), \quad -\infty < t < \infty,$$

is stationary; see p. 64 of Lamperti (1962) and p. 312 of Samorodnitsky and Taqqu (1994). In general, the sample paths of self-similar stochastic-processes can be very complicated, see Vervaat (1985), but Lamperti showed that self-similar processes must be *continuous in probability*, i.e.,

$$\|Z(t+h) - Z(t)\| \Rightarrow 0 \quad \text{in} \quad \mathbb{R}^k \quad \text{as} \quad h \to 0$$

for all $t \geq 0$. That does not imply that the stochastic process Z necessarily has a version with sample paths in D, however.

Convergence of the finite-dimensional distributions also implies that the space-scaling function has a special form. The space-scaling function $c(s)$ in (2.4) must be *regularly varying with index H*; see Appendix A. A regularly varying function is a generalization of a simple power; the canonical case is the simple power, i.e., $c(s) = cs^H$ for some constant c.

Theorem 4.2.1. (Lamperti's theorem) *If, for some $k \geq 1$,*

$$(\mathbf{S}_s(t_1), \ldots, \mathbf{S}_s(t_l)) \Rightarrow (\mathbf{S}(t_1), \ldots, \mathbf{S}(t_l)) \quad \text{in} \quad R^{kl} \quad \text{as} \quad s \to \infty$$

for all positive integers l and all l-tuples $(t_1, \ldots, t_l)$ with $0 \leq t_1 < \cdots < l_l$, where $\mathbf{S}_s$ is the scaled process in (2.4), then the limit process $\mathbf{S}$ is self-similar with index H for some $H > 0$ and continuous in probability. Then the space scaling function $c(s)$ must be regularly varying with the same index H.

Remark 4.2.1. *Self-similarity in network traffic.* Ever since the seminal work on network traffic measurements by Leland et al. (1994), there has been interest and controversy about the reported self-similarity observed in network traffic. From Lamperti's theorem, we see that, under minor regularity conditions, some form of self-similarity tends to be an inevitable consequence of any macroscopic view of uncertainty. Since network traffic data sets are very large, they naturally lead to a macroscopic view. *The engineering significance of the reported self-similarity lies in the self-similarity index H.* With centering about a finite mean (or without centering with an infinite mean), the observed index H with $H > 1/2$ indicates that there is extra variability beyond what is captured by the standard central limit theorem. As we saw in Section 2.3, and as we will show in later chapters, that extra traffic burstiness affects the performance in a queue to which it is offered. From the performance-analysis perspective, these are important new ideas, but the various forms of variability have been studied for a long time, as can be seen from Mandelbrot (1977, 1982), Taqqu (1986), Beran (1994), Samorodnitsky and Taqqu (1994) and Willinger, Taqqu and Erramilli (1996). ∎

From the point of view of generating simple parsimonious approximations, we are primarily interested in the special case in which the scaling function takes the relatively simple form $c(s) = cs^H$ for some constant c. (We will usually be considering the discrete case in which $c_n = cn^H$.) Then the approximation provided by the stochastic-process limit is characterized by the parameter triple (m, H, c) in addition to any parameters of the limit process. The parameter m is the centering constant, which usually is the mean; the parameter H is the self-similarity index and the space-scaling exponent; and the parameter c is the scale parameter, which appears as a constant multiplier in the space-scaling function. For example, there are no extra parameters beyond the triple (m, H, c) in the case of convergence to standard Brownian motion.

In most applications the underlying sequence $\{X_n\}$ of summands in the partial sums is stationary or asymptotically stationary, so that the limit process **S** in (2.2) must also have *stationary increments*, i.e.,

$$\{\mathbf{S}(t+u) - \mathbf{S}(u) : t \geq 0\} \stackrel{\mathrm{d}}{=} \{\mathbf{S}(t) - \mathbf{S}(0) : t \geq 0\}$$

for all $u > 0$. Thus, the prospective limit processes of primary interest are the H-*self-similar processes with stationary increments*, which we denote by H-sssi.

By far, the most frequently occurring H-sssi process is Brownian motion. Brownian motion also has independent increments, continuous sample paths and $H = 1/2$. Indeed, we saw plenty of Brownian motion sample paths in Chapter 1. The classical FCLT is covered by Donsker's theorem, which we review in Section 4.3. Then the basic sequence $\{X_n\}$ is IID with finite second moments. In Section 4.4 here and Section 2.3 of the Internet Supplement we show that essentially the same limit also occurs when the independence is replaced with weak dependence. The set of processes for which scaled versions converge to Brownian motion is very large. Thus these FCLTs describe remarkably persistent statistical regularity.

After Brownian motion, the most prominent H-sssi processes are the stable Lévy motion processes. The *stable Lévy motion processes* are the H-sssi processes with independent increments. The marginal probability distributions of stable Lévy motions are the stable laws. The Gaussian distribution is a special case of a stable law and Brownian motion is a special stable Lévy motion. The nonGaussian stable Lévy motion processes are the possible limit processes in (2.2) when $\{X_n\}$ is a sequence of IID random variables with infinite variance, as we will see in Section 4.5. For stable Lévy motions, the self-similarity index H can assume any value greater than or equal to $1/2$. Brownian motion is the special case of a stable Lévy motion with $H = 1/2$. For $H > 1/2$, the stable marginal distributions have power tails and infinite variance. NonGaussian stable Lévy motions have discontinuous sample paths, so jumps enter the picture, as we saw in Chapter 1.

4.2.3. The Noah and Joseph Effects

It is possible to have stochastic-process limits with self-similarity index H assuming any positive value. Values of H greater than $1/2$ tend to occur because of either exceptionally large values – the *Noah effect* – or exceptionally strong positive dependence – the *Joseph effect*. The Noah effect refers to the biblical figure Noah who experienced an extreme flood; the Joseph effect refers to the biblical figure Joseph who experienced long periods of plenty followed by long periods of famine; Genesis 41, 29-30: *Seven years of great abundance are coming throughout the land of Egypt, but seven years of famine will follow them*; see Mandelbrot and Wallis (1968) and Mandelbrot (1977, 1982).

The Joseph effect occurs when there is strong positive dependence. With the Joseph effect, but without heavy heavy tails, the canonical limit process is *fractional Brownian motion* (FBM). Like Brownian motion, FBM has normal marginal distributions and continuous sample paths. However, unlike Brownian motion, FBM has dependent increments. FBM is a natural H-sssi process exhibiting the Joseph effect, but unlike the stable Lévy motions arising with the Noah effect, FBM is by no means the only H-sssi limit process that can arise with strong dependence and finite second moments; see Vervaat (1985), O'Brien and Vervaat (1985) and Chapter 7 of Samorodnitsky and Taqqu (1994).

We will present FCLTs capturing the Noah effect in Sections 4.5 and 4.7, and the Joseph effect in Section 4.6. It is also possible to have FCLTs exhibiting *both* the Noah and Joseph effects, but unfortunately the theory is not yet so well developed in that area. However, many H-sssi stochastic processes exhibiting both the Noah and Joseph effects have been identified. A prominent example is the *linear fractional stable motion* (LFSM). Like FBM, but unlike stable Lévy motion, LFSM's with positive dependence have continuous sample paths.

The Noah and Joseph effects can be roughly quantified for the H-sssi processes that are also stable processes. A stochastic process $\{Z(t) : t \geq 0\}$ is said to be a *stable process* if all its finite-dimensional distributions are stable laws, all of which have a stable index α, $0 < \alpha \leq 2$; see Section 4.5 below and Chapters 1–3 of Samorodnitsky and Taqqu (1994). The normal or Gaussian distribution is the stable law with stable index $\alpha = 2$. A real-valued random variables X with a stable law with index α, $0 < \alpha < 2$, has a distribution with a power tail with exponent $-\alpha$, satisfying

$$P(|X| > t) \sim Ct^{-\alpha} \quad \text{as} \quad t \to \infty,$$

so that X necessarily has infinite variance. Thus, all the one-dimensional marginal distributions of a nonGaussian stable process necessarily have infinite variance. The class of H-sssi stable processes includes all the specific H-sssi processes mentioned so far: Brownian motion, stable Lévy motion, fractional Brownian motion and linear fractional stable motion.

For H-sssi stable processes with independent increments, $H = \alpha^{-1}$. We are thus led to say that we have *only the Noah effect* when $H = \alpha^{-1} > 2^{-1}$, and we quantify it by the difference $\alpha^{-1} - 2^{-1}$. Similarly, we say that we have *only the Joseph effect* when $H > 1/\alpha = 1/2$, and we quantify it by the difference $H - \alpha^{-1}$. We say that we have both the Noah and Joseph effects when $H > \alpha^{-1} > 2^{-1}$, and quantify the Noah and Joseph effects, respectively, by the differences $\alpha^{-1} - 2^{-1}$ and $H - \alpha^{-1}$, as before. Of course, we have neither the Noah effect nor the Joseph effect when $H = \alpha^{-1} = 2^{-1}$.

It is natural to ask which effect is more powerful. From the perspective of the indices, we see that the Noah effect can be more dramatic: The Noah effect $\alpha^{-1} - 2^{-1}$ can assume any positive value, whereas the Joseph effect $H - \alpha^{-1}$ can be at most $1/2$.

Obviously there should be no Joseph effect when the H-sssi stable process has independent increments. However, it is possible to have 0 Joseph effect without having independent increments; that occurs with the log-fractional stable motion in Section 7.6 of Samorodnitsky and Taqqu (1994). It is also possible to have convergence to nonstable H-sssi limit processes.

It is important to recognize that, even for H-sssi stable processes, we are not constrained to have $H \geq \alpha^{-1} \geq 2^{-1}$; the full range of possibilities is much greater. In this chapter we emphasize positive dependence, which makes the self-similarity index H larger than it would be with independent increments, but in general the possible forms of dependence are more complicated. For example, fractional Brownian motion can have any self-similarity index H with $0 < H \leq 1$; we need not have $H \geq 1/2$. The FBMs with $H < 1/2$ represent *negative dependence* instead of positive dependence. Such negative independence often arises when there is conscious human effort to smooth a stochastic process. For example, an arrival process to a queue may be smoothed by scheduling arrivals, as at a doctor's office. Then the actual arrival process may correspond to some random perturbation of the scheduled arrivals. Then the long-run variability tends to be substantially less than might be guessed from considering only the distribution of the interarrival times between successive customers.

In fact, for H-sssi stable processes with $\alpha < 1$ it is only possible to have negative dependence, because it is possible to have any H with $0 < H \leq \alpha^{-1}$, but *not* any H with $H > \alpha^{-1}$; see p. 316 of Samorodnitsky and Taqqu (1994). However, it is possible to have $1 < \alpha \leq 2$ and $\alpha^{-1} < H < 1$. Properties of H-sssi α-stable processes are described in Samorodnitsky and Taqqu (1994).

4.3. Donsker's Theorem

In this section we consider the classical case in which $\{X_n\}$ is a sequence of IID random variables with finite second moments. The FCLT is Donsker's theorem, which we now describe, expanding upon the discussion in Chapter 1.

4.3.1. The Basic Theorems

Since Donsker's theorem is a generalization of the classical CLT, we start by reviewing the classical CLT. For that purpose, let $N(m, \sigma^2)$ denote a random variable with a *normal or Gaussian distribution* with mean m and variance σ^2. We call the special case of the normal distribution with $m = 0$ and $\sigma^2 = 1$ the *standard normal distribution*. Let Φ be the *cumulative distribution function* (cdf) and n the *probability density function* (pdf) of the standard normal distribution, i.e.,

$$\Phi(x) = P(N(0,1) \leq x) \equiv \int_{-\infty}^{x} n(y) dy,$$

where

$$n(x) \equiv \frac{e^{-x^2}}{\sqrt{2\pi}}, \quad -\infty < x < \infty.$$

102   4. A Panorama

Recall that
$$N(m, \sigma^2) \stackrel{\mathrm{d}}{=} m + \sigma N(0,1)$$
for each $m \in \mathbb{R}$ and $\sigma^2 \in \mathbb{R}_+$.

Theorem 4.3.1. (classical central limit theorem) *Suppose that $\{X_n : n \geq 1\}$ is a sequence of IID random variables with mean $m \equiv EX_1$ and finite variance $\sigma^2 \equiv Var X_1$. Let $S_n \equiv X_1 + \cdots + X_n$, $n \geq 1$. Then (as $n \to \infty$)*
$$n^{-1/2}(S_n - mn) \Rightarrow \sigma N(0,1) \quad in \quad \mathbb{R} \; .$$

Donsker's theorem is a FCLT generalizing the CLT above. It is a limit for the entire sequence of partial sums, instead of just the nth partial sum. We express it via the normalized partial-sum process
$$\mathbf{S}_n(t) \equiv n^{-1/2}(S_{\lfloor nt \rfloor} - mnt), \quad t \geq 0 \; , \tag{3.1}$$
in $D \equiv D([0, \infty), \mathbb{R})$, i.e., as in (2.1) with $c_n = \sqrt{n}$.

Theorem 4.3.2. (Donsker's FCLT) *Under the conditions of the CLT in Theorem 4.3.1,*
$$\mathbf{S}_n \Rightarrow \sigma \mathbf{B} \quad in \quad (D, J_1) \; ,$$
where $\mathbf{S}_n$ is the normalized partial-sum process in (3.1) and $\mathbf{B} \equiv \{\mathbf{B}(t) : t \geq 0\}$ is standard Brownian motion.

The limiting Brownian motion in Donsker's FCLT is a Lévy process with continuous sample paths; a *Lévy process* is a stochastic process with stationary and independent increments; see Theorem 19.1 of Billingsley (1968). Those properties imply that an increment $B(s+t) - B(s)$ of Brownian motion $\{B(t) : t \geq 0\}$ is normally distributed with mean mt and variance $\sigma^2 t$ for some constants m and σ^2. *Standard Brownian motion* is Brownian motion with parameters $m = 0$ and $\sigma^2 = 1$.

The most important property of standard Brownian motion is that it exists. Existence is a consequence of Donsker's theorem; i.e., Brownian motion can be defined as the limit process once the limit for the normalized partial sums has been shown to exist.

In applications we often make use of the self-similarity scaling property
$$\{\mathbf{B}(ct) : t \geq 0\} \stackrel{\mathrm{d}}{=} \{\sqrt{c}\mathbf{B}(t) : t \geq 0\} \quad \text{for any} \quad c > 0 \; .$$
We can obtain Brownian motion with drift m, diffusion (or variance) coefficient σ^2 and initial position x for any $m, x \in \mathbb{R}$ and $\sigma^2 \in \mathbb{R}_+$, denoted by $\{\mathbf{B}(t; m, \sigma^2, x) : t \geq 0\}$, by simply scaling standard Brownian motion:
$$\mathbf{B}(t; m, \sigma^2, x) \equiv x + mt + \sigma \mathbf{B}(t), \quad t \geq 0 \; .$$

We have seen Donsker's theorem in action in Chapter 1. Plots of random walks with IID steps converging to Brownian motion are shown in Figures 1.2, 1.3, 1.4 and 1.17.

Donsker's FCLT is an *invariance principle* because the limit depends upon the distribution of X_1 only through its first two moments. By applying the continuous mapping theorem with various measurable real-valued functions on D that are continuous almost surely with respect to Brownian motion, we obtain many useful corollaries. For example, two useful functions are

$$f_1(x) \equiv \sup_{0 \leq t \leq 1} x(t) \qquad (3.2)$$

and

$$f_2(x) \equiv \lambda(\{t \in [0,1] : x(t) > 0\}) \qquad (3.3)$$

where λ is Lebesgue measure on $[0,1]$; see Section 11 of Billingsley (1968). The supremum function f_1 in (3.2) was discussed in Section 1.2 while establishing limits for the random-walk plots. The function f_2 in (3.3) is not continuous at all x, as can be seen by considering the constant functions $x_n(t) = n^{-1}$, $0 \leq t \leq 1$, $n \geq 1$, but f_2 is measurable and continuous almost surely with respect to Brownian motion. By applying the function f_2 in (3.3), we obtain the arc sine law. For general probability distributions, this result was first obtained directly by Erdös and Kac (1947). The distribution of $f_2(\mathbf{B})$ was found by Lévy (1939).

Corollary 4.3.1. (arc sine law) *Under the assumptions of the CLT in Theorem 4.3.1,*

$$n^{-1} Z_n \Rightarrow f_2(\mathbf{B}) \quad in \quad \mathbb{R},$$

where Z_n is the number of the first n partial sums $S_1, \ldots, S_n$ that are positive, $\mathbf{B}$ is standard Brownian motion and

$$P(f_2(\mathbf{B}) \leq x) = \frac{1}{\pi} \int_0^x \frac{dy}{\sqrt{y(1-y)}} = \frac{2}{\pi} arc\ sin(\sqrt{x}), \quad 0 < x < 1.$$

As indicated in Chapter 1, Donsker's FCLT can be used to derive the limiting distributions in Corollary 4.3.1. Since the limit depends on the distribution of X_1 only through its first two moments, we can work with the special case of a *simple random walk* in which

$$P(X_1 = 1) = P(X_1 = -1) = 1/2.$$

Combinatorial arguments can be used to calculate the limits for simple random walks; e.g., see Chapter 3 of Feller (1968).

It is interesting that the probability density function $f(y) = \pi^{-1}(y(1-y))^{-1/2}$ of $f_2(\mathbf{B})$ is U-shaped, having a minimum at $1/2$. For large n, having 99% of the partial sums positive is about 5 times more likely than having 50% of the partial sums positive.

4.3.2. *Multidimensional Versions*

It is significant that Theorems 4.3.1 and 4.3.2 extend easily to k dimensions. A key for establishing this extension is the Cramér-Wold device; see p. 49 of Billingsley (1968).

Theorem 4.3.3. (Cramér-Wold device) *For arbitrary random vectors $(X_{n,1}, \ldots, X_{n,k})$ in $\mathbb{R}^k$, there is convergence in distribution*

$$(X_{n,1}, \ldots, X_{n,k}) \Rightarrow (X_1, \ldots, X_k) \quad in \quad \mathbb{R}^k$$

if and only if

$$\sum_{i=1}^{k} a_i X_{n,i} \Rightarrow \sum_{i=1}^{k} a_i X_i \quad in \quad \mathbb{R}$$

for all $(a_1, \ldots, a_k) \in \mathbb{R}^k$.

The multivariate (k-dimensional) CLT involves convergence of normalized partial sums of random vectors to the multivariate normal distribution. We first describe the multivariate normal distribution. A pdf in $\mathbb{R}^k$ of the form

$$f(x_1, \ldots, x_k) = \gamma^{-1} \exp\left(-(1/2) \sum_{i=1}^{k} \sum_{j=1}^{k} x_i Q_{i,j} x_j\right), \tag{3.4}$$

where $Q \equiv (Q_{i,j})$ is a symmetric $k \times k$ matrix (necessarily with positive diagonal elements) and γ is a positive constant, is a *nondegenerate k-dimensional normal or Gaussian pdf centered at the origin*; see Section III.6 of Feller (1971). The pdf $f(x_1 - m_1, \ldots, x_k - m_k)$ for f in (3.4) is a nondegenerate k-dimensional normal pdf centered at $(m_1, \ldots, m_k)$. A random vector $(X_1, \ldots, X_k)$ with a nondegenerate k-dimensional normal pdf centered at $(m_1, \ldots, m_k)$ has means $EX_i = m_i$, $1 \leq i \leq k$. Let the *covariance matrix* of a random vector $(X_1, \ldots, X_k)$ in $\mathbb{R}^k$ with means $(m_1, \ldots, m_k)$ be $\Sigma \equiv (\sigma_{i,j}^2)$, where

$$\sigma_{i,j}^2 \equiv E(X_i - m_i)(X_j - m_j).$$

For a nondegenerate normal pdf, the matrices Q and Σ are nonsingular and related by

$$Q = \Sigma^{-1},$$

and the constant γ in (3.4) satisfies

$$\gamma^2 = (2\pi)^k |\Sigma|,$$

where $|\Sigma|$ is the determinant of Σ. Let $N(m, \Sigma)$ denote a random (row) vector with a nondegenerate normal pdf in $\mathbb{R}^k$ centered at $m \equiv (m_1, \ldots, m_k)$ and covariance matrix Σ. Note that

$$N(m, \Sigma) \stackrel{d}{=} m + N(0, \Sigma).$$

If Σ is the $k \times k$ covariance matrix of a nondegenerate k-dimensional normal pdf, then there exists a nonsingular $k \times k$ matrix C, which is not unique, such that

$$N(0, \Sigma) \stackrel{d}{=} N(0, I)C ,$$

where I is the identity matrix.

We can also allow degenerate k-dimensional normal distributions. We say that a $1 \times k$ row vector Y has a k-dimensional normal distribution with mean vector $m = (m_1, \ldots, m_k)$ and $k \times k$ covariance matrix Σ if

$$Y \stackrel{d}{=} m + XC ,$$

where X is a $1 \times j$ random vector for some $j \leq k$ with a nondegenerate j-dimensional normal pdf centered at the origin with covariance matrix I and C is a $j \times k$ matrix with

$$C^t C = \Sigma \tag{3.5}$$

where C^t is the transpose of C.

The following generalization of the CLT in Theorem 4.3.1 is obtained by applying the Cramér-Wold device in Theorem 4.3.3.

Theorem 4.3.4. *(k-dimensional CLT) Suppose that $\{X_n : n \geq 1\} = \{(X_{n,1}, \ldots, X_{n,k}) : n \geq 1\}$ is a sequence of IID random vectors in $\mathbb{R}^k$ with $EX_{1,i}^2 < \infty$ for $1 \leq i \leq k$. Let $m \equiv (m_1, \ldots, m_k)$ be the mean vector with $m_i \equiv EX_{1,i}$ and $\Sigma \equiv (\sigma_{i,j}^2)$ the covariance matrix with*

$$\sigma_{i,j}^2 = E(X_{1,i} - m_i)(X_{1,j} - m_j) \tag{3.6}$$

for all i, j with $1 \leq i \leq k$ and $1 \leq j \leq k$. Then

$$n^{-1/2}(S_n - mn) \Rightarrow N(0, \Sigma) \quad in \quad \mathbb{R}^k$$

where $S_n \equiv X_1 + \cdots + X_n$, $n \geq 1$.

A *standard k-dimensional Brownian motion* is a vector-valued stochastic process

$$\mathbf{B} \equiv (\mathbf{B}_1, \ldots, \mathbf{B}_k) \equiv \{\mathbf{B}(t) : t \geq 0\} \equiv \{(\mathbf{B}_1(t), \ldots, \mathbf{B}_k(t)) : t \geq 0\} ,$$

where $\mathbf{B}_1, \ldots, \mathbf{B}_k$ are k IID standard one-dimensional BMs. A general k-dimensional Brownian motion with drift vector $m \equiv (m_1, \ldots, m_k)$, $k \times k$ covariance vector Σ and initial vector $x \equiv (x_1, \ldots, x_k)$, denoted by $\{\mathbf{B}(t; m, \Sigma, x) : t \geq 0\}$ can be constructed by letting

$$\mathbf{B}(t; m, \Sigma, x) = x + mt + \mathbf{B}(t)C , \tag{3.7}$$

where $\mathbf{B}$ is a standard j-dimensional Brownian motion and C is a $j \times k$ matrix satisfying (3.5). In (3.7) we understand that

$$\{\mathbf{B}(t) : t \geq 0\} \stackrel{d}{=} \{\mathbf{B}(t; 0, I, 0) : t \geq 0\} ,$$

where I is the $j \times j$ identity matrix and 0 is the j-dimensional zero vector.

We now state the k-dimensional version of Donsker's theorem. The limit holds in the space $D^k \equiv D([0, \infty), \mathbb{R}^k)$ with the SJ_1 topology.

Theorem 4.3.5. (*k-dimensional Donsker FCLT*) *Under the conditions of the k-dimensional CLT in Theorem 4.3.4,*

$$\mathbf{S}_n \Rightarrow \mathbf{B}C \quad in \quad (D^2, SJ_1) \,,$$

where $\mathbf{S}_n$ *is the normalized partial-sum process in* (3.1), $\mathbf{B}$ *is a standard j-dimensional Brownian motion and C is a* $j \times k$ *matrix such that* (3.5) *holds, i.e.,*

$$\mathbf{B}C \stackrel{\mathrm{d}}{=} \{\mathbf{B}(t; 0, \Sigma) : t \geq 0\} \quad in \quad D^k \,,$$

where $\Sigma \equiv (\sigma_{i,j}^2)$ *is the covariance matrix of* $(X_{1,1}, \ldots, X_{1,k})$ *in* (3.6).

Proof. The one-dimensional marginals converge by Donsker's theorem, Theorem 4.3.2. That convergence implies that the marginal processes are tight by Prohorov's theorem, Theorem 11.6.1. Tightness of the marginal processes implies tightness of the overall processes by Theorem 11.6.7. Convergence of all the finite-dimensional distributions follows from the CLT in Theorem 4.3.4 and the Cramér-Wold device in Theorem 4.3.3. Finally, tightness plus convergence of the finite-dimensional distributions implies weak convergence in D by Corollary 11.6.1. ∎

It follows from either the k-dimensional Donsker FCLT or the one-dimensional Donsker FCLT that linear functions of the coordinate of the partial-sum process converge to a one-dimensional Brownian motion.

Corollary 4.3.2. *Under the conditions of Theorem 4.3.5,*

$$\sum_{i=1}^{k} a_i \mathbf{S}_{n,i} \Rightarrow \sigma \mathbf{B} \quad in \quad D$$

where $\mathbf{S}_n \equiv (\mathbf{S}_{n,1}, \ldots, \mathbf{S}_{n,k})$ *is the normalized partial-sum process in* (3.1), $\mathbf{B}$ *is a standard one-dimensional Brownian motion and*

$$\sigma^2 = \sum_{i=1}^{k} \sum_{j=1}^{k} a_i a_j \sigma_{i,j}^2 \,.$$

Donsker's FCLT was stated (as it was originally established) in the framework of a single sequence $\{X_n : n \geq 1\}$. There are extensions of Donsker's FCLT in the framework of a double sequence $\{X_{n,k} : n \geq 1, k \geq 1\}$, paralleling the extensions of the CLT. Indeed, a natural one is a special case of a martingale FCLT, Theorem 2.3.9 in the Internet Supplement.

It can be useful to go beyond the CLT and FCLT to establish bounds on the rate of convergence; see Section 2.2 of the Internet Supplement. For the FCLT, strong approximations can be exploited to produce bounds on the Prohorov distance.

4.4. Brownian Limits with Weak Dependence

For applications, it is significant that there are many generalizations of Donsker's theorem in which the IID assumption is relaxed. Many FCLTs establishing convergence to Brownian motion have been proved with independence replaced by weak dependence. In these theorems, only the space-scaling constant σ^2 in Donsker's theorem needs to be changed. Consequences of Donsker's theorem such as Corollary 4.3.1 thus still hold in these more general settings.

Suppose that we have a sequence of real-valued random variables $\{X_n : n \geq 1\}$. Let $S_n \equiv X_1 + \cdots + X_n$ be the nth partial sum and let $\mathbf{S}_n$ be the normalized partial-sum process

$$\mathbf{S}_n(t) = n^{-1/2}(S_{\lfloor nt \rfloor} - mnt), \quad t \geq 0 \ . \tag{4.1}$$

in D, just as in (3.1). We want to conclude that there is convergence in distribution

$$\mathbf{S}_n \Rightarrow \sigma \mathbf{B} \quad \text{in} \quad (D, J_1) \ , \tag{4.2}$$

where $\mathbf{B}$ is standard Brownian motion and identify the scaling parameters m and σ, without assuming that $\{X_n\}$ is necessarily a sequence of IID random variables.

In this section and in Section 2.3 of the Internet Supplement we review some of the sufficient conditions for (4.2) to hold with the IID condition relaxed. We give only a brief account, referring to Billingsley (1968, 1999), Jacod and Shiryaev (1987) and Philipp and Stout (1975) for more. First, assume that $\{X_n : -\infty < n < \infty\}$ is a two-sided *stationary sequence*, i.e., that $\{X_{k+n} : -\infty < n < \infty\}$ has a distribution (on $\mathbb{R}^\infty$) that is independent of k. (It is always possible to construct a two-sided stationary sequence starting from a one-sided stationary sequence $\{X_n : n \geq 1\}$, where the two sequences with positive indices have the same distribution; e.g., see p. 105 of Breiman (1968).) Moreover, assume that $EX_n^2 < \infty$. The obvious parameter values now are

$$m \equiv EX_n \quad \text{and} \quad \sigma^2 \equiv \lim_{n \to \infty} \frac{Var(S_n)}{n} \ ; \tag{4.3}$$

i.e., m should be the mean and σ^2 should be the *asymptotic variance*, where

$$\sigma^2 = Var\, X_n + 2 \sum_{k=1}^{\infty} Cov(X_1, X_{1+k}) \ . \tag{4.4}$$

Roughly speaking, we should anticipate that (4.2) holds with m and σ^2 in (4.3) whenever σ^2 in (4.4) is finite. However, additional conditions are actually required in the theorems. From a practical perspective, however, in applications it *is* usually reasonable to act as if the FCLT is valid if σ^2 in (4.4) is finite, and the main challenge is to find effective ways to calculate or estimate the asymptotic variance σ^2. There is a large literature on estimating the asymptotic variance σ^2 in (4.3), because the asymptotic variance is used to determine confidence intervals around the sample mean for estimates of the steady-state mean; for the sample mean $\bar{X}_n \equiv n^{-1} S_n$,

$$Var\, \bar{X}_n = n^{-2} Var\, S_n, \quad n \geq 1 \ .$$

Even for a mathematical model, statistical estimation is a viable way to compute the asymptotic variance. We can either estimate σ^2 from data collected from a system being modelled or from output of a computer simulation of the model. For more information, see Section 3.3 of Bratley, Fox and Schrage (1987), Damerdji (1994, 1995) and references therein.

In order for the FCLT in (4.2) to hold, the degree of dependence in the sequence $\{X_n\}$ needs to be controlled. One way to do this is via *uniform mixing conditions*. Here we follow Chapter 4 of Billingsley (1999); also see the papers in Section 2 of Eberlein and Taqqu (1986). To define uniform mixing conditions, let $\mathcal{F}_n \equiv \sigma[X_k : k \leq n]$ be the σ-field generated by $\{X_k : k \leq n\}$ and let $\mathcal{G}_n \equiv \sigma[X_k : k \geq n]$ be the σ-field generated by $\{X_k : k \geq n\}$. We write $X \in \mathcal{F}_k$ to indicate that X is $\mathcal{F}_k$-measurable. Let

$$\alpha_n \equiv \sup\{|P(A \cap B) - P(A)P(B)| : A \in \mathcal{F}_k,\ B \in \mathcal{G}_{k+n}\} \tag{4.5}$$

$$\rho_n \equiv \sup\{|E[XY]| : X \in \mathcal{F}_k,\ EX = 0,$$
$$EX^2 \leq 1,\ Y \in \mathcal{G}_{k+n},\ EY = 0,\ EY^2 \leq 1\} \tag{4.6}$$

$$\phi_n \equiv \sup\{|P(B|A) - P(B)| : A \in \mathcal{F}_k, P(A) > 0, B \in \mathcal{G}_{k+n}\}. \tag{4.7}$$

It turns out that these three measures of dependence are ordered by

$$\alpha_n \leq \rho_n \leq 2\sqrt{\phi_n}\ .$$

Theorem 4.4.1. (FCLT for stationary sequence with uniform mixing) *Assume that $\{X_n : -\infty < n < \infty\}$ is a two-sided stationary sequence with $Var\ X_n < \infty$ and*

$$\sum_{n=1}^{\infty} \rho_n < \infty\ . \tag{4.8}$$

for ρ_n in (4.6). Then the series in (4.4) converges absolutely and the FCLT (4.2) holds with $m = EX_1$ and σ^2 being the asymptotic variance in (4.4).

In many applications condition (4.8) will be hard to verify, but it does apply directly to finite-state discrete-time Markov chains (DTMC's), as shown on p. 201 of Billingsley (1999).

Theorem 4.4.2. (FCLT for stationary DTMCs) *Suppose that $\{Y_n : -\infty < n < \infty\}$ is the stationary version of an irreducible finite-state Markov chain and let $X_n = f(Y_n)$ for a real-valued function f on the state space. Then the conditions on $\{X_n\}$ in Theorem 4.4.1 and the conclusions there hold.*

It is also possible to replace the quantitative measure of dependence in (4.6) with a qualitative characterization of dependence. We say that the sequence $\{X_n : -\infty < n < \infty\}$ is *associated* if, for any k and any two (coordinatewise) nondecreasing real-valued functions f_1 and f_2 on $\mathbb{R}^k$ for which $E[f_i(X_1, \ldots, X_k)^2] < \infty$ for $i = 1, 2$,

$$Cov(f_1(X_1, \ldots, X_k),\ f_2(X_1, \ldots, X_k)) \geq 0\ .$$

(For further discussion of associated processes in queues and other discrete-event systems, see Glynn and Whitt (1989) and Chapter 8 of Glasserman and Yao (1994).)

The following FCLT is due to Newman and Wright (1981). See Cox and Grimmett (1984) and Dabrowski and Jakubowski (1994) for extensions.

Theorem 4.4.3. (FCLT for associated process) *If $\{X_n : -\infty < n < \infty\}$ is an associated stationary sequence with $EX_n^2 < \infty$ and $\sigma^2 < \infty$ for σ^2 in (4.4), then the FCLT (4.2) holds.*

Instead of uniform mixing conditions, we can use ergodicity and martingale properties; see p. 196 of Billingsley (1999). For a stationary process $\{X_n\}$, ergodicity essentially means that the SLLN holds: $n^{-1}S_n \to EX_1$ w.p.1 as $n \to \infty$, where $E|X_1| < \infty$; e.g., see Chapter 6 of Breiman (1968). The sequence of centered partial sums $\{S_n - mn : n \geq 1\}$ is a martingale if $E|X_1| < \infty$ and $E[X_n - m|\mathcal{F}_{n-1}] = 0$ for all $n \geq 1$, where as before $\mathcal{F}_n$ is the σ-field generated by $X_1, \ldots, X_n$.

Theorem 4.4.4. (stationary martingale FCLT) *Suppose that $\{X_n : -\infty < n < \infty\}$ is a two-sided stationary ergodic sequence with $Var\, X_n = \sigma^2$, $0 < \sigma^2 < \infty$, and $E[X_n - m|\mathcal{F}_{n-1}] = 0$ for all n for some constant m. Then the FCLT (4.2) holds with (m, σ^2) specified in the conditions here.*

There are two difficulties with the FCLT's stated so far. First, they require stationarity and, second, they do not contain tractable expressions for the asymptotic variance. In many applications, the stochastic process of interest does not start in steady state, but it is asymptotically stationary, and that should be enough. For those situations, it is convenient to exploit regenerative structure. Regenerative structure tends to encompass Markovian structure as a special case. The additional Markovian structure enables us to obtain formulas and algorithms for computing the asymptotic variance. We discuss FCLT's in Markov and regenerative settings in Section 2.3 of the Internet Supplement.

4.5. The Noah Effect: Heavy Tails

In the previous section we saw that the conclusion of Donsker's theorem still holds when the IID assumption is relaxed, with the finite-second-moment condition maintained; only the asymptotic-variance parameter σ^2 in (4.3) and (4.4) needs to be revised, with the key condition being that σ^2 be finite. We now see what happens when we keep the IID assumption but drop the finite-second-moment condition.

As we saw in Chapter 1, when the second moment is infinite, there is a dramatic change! When the second moments are infinite, there still may be limits, but the limits are very different. First, unlike in the finite-second-moment case, there may be no limit at all; the existence of a limit depends critically on regular behavior of the tails of the underlying probability distribution (of X_1). But that regular tail behavior is very natural to assume. When that regular tail behavior holds with infinite second moments, we obtain limits, but limits with different scaling and different limit processes.

Of particular importance to us, the new limit processes have discontinuous sample paths, so that the space D becomes truly important. In this setting we do not need the M_1 topology to establish the FCLT for partial sums of IID random variables, but we do often need the M_1 topology to successfully apply the continuous-mapping approach starting from the initial FCLTs to be described in this section. We illustrate the importance of the M_1 topology in Sections 6.3 and 7.3 below when we discuss FCLTs for counting processes.

The framework here will be a single sequence $\{X_n : n \geq 1\}$ of IID random variables, where $EX_n^2 = \infty$. As before, we will focus on the associated partial sums $S_n = X_1 + \cdots + X_n$, $n \geq 1$, with $S_0 = 0$. We form the normalized processes

$$\mathbf{S}_n(t) \equiv c_n^{-1}(S_{\lfloor nt \rfloor} - m_n nt), \quad t \geq 0 , \qquad (5.1)$$

in D where $\{m_n : n \geq 1\}$ and $\{c_n : n \geq 1\}$ are general deterministic sequences with $c_n \to \infty$ as $n \to \infty$. Usually we will have $m_n = m$ as in (2.1), but we need translation constants depending on n in one case (when the stable index is $\alpha = 1$). In Sections 4.3 and 4.4 we always had $c_n = \sqrt{n}$. Here will have $c_n/\sqrt{n} \to \infty$; a common case is $c_n = n^{1/\alpha}$ for $0 < \alpha < 2$, where α depends on the asymptotic behavior of the tail probability $P(|X_1| > t)$ as $t \to \infty$. Under regularity conditions, the normalized partial-sum process $\mathbf{S}_n$ in (5.1) will converge in (D, J_1) to a process called stable Lévy motion.

We consider the more general double-sequence (or triangular array) framework using $\{X_{n,k} : n \geq 1, k \geq 1\}$ in Section 2.4 of the Internet Supplement. Unlike in Sections 4.3 and 4.4 above, with heavy-tailed distributions, there is a big difference between a single sequence and a double sequence, because the class of possible limits is much larger in the double-sequence framework: With IID conditions, the possible limits in the framework of double sequences are all Lévy processes. Like the stable Lévy motion considered in this section, general Lévy processes have stationary and independent increments, but the marginal distributions need *not* be stable laws; the marginal distributions of Lévy processes are infinitely divisible distributions (a surprisingly large class). The smaller class of limits we obtain in the single-sequence framework has the advantage of producing more robust approximations; the larger class we obtain in the double-sequence framework has the advantage of producing more flexible approximations.

A *stable stochastic process* is a stochastic process all of whose finite-dimensional distributions are stable laws. The Gaussian distribution is a special case of a stable law, and a Gaussian process is a special case of a stable process, but the limits with infinite second moments will be nonGaussian stable processes, whose finite-dimensional distributions are nonGaussian stable laws. The nonGaussian stable distributions have heavy tails, so that exceptionally large increments are much more likely with a nonGaussian stable process than with a Gaussian process. We refer to Samorodnitsky and Taqqu (1994) for a thorough treatment of nonGaussian stable laws and nonGaussian stable processes. For additional background, see Bertoin (1996), Embrechts, Klüppelberg and Mikosch (1997), Feller (1971), Janicki and Weron (1993) and Zolotarev (1986).

4.5.1. Stable Laws

A random variable X is said to have a *stable law* if, for any positive numbers a_1 and a_2, there is a real number $b \equiv b(a_1, a_2)$ and a positive number $c \equiv c(a_1, a_2)$ such that

$$a_1 X_1 + a_2 X_2 \stackrel{\mathrm{d}}{=} b + cX, \qquad (5.2)$$

where X_1 and X_2 are independent copies of X and $\stackrel{\mathrm{d}}{=}$ denotes equality in distribution. A stable law is *strictly stable* if (5.2) holds with $b = 0$. Except in the pathological case $\alpha = 1$, a stable law always can be made strictly stable by appropriate centering. Note that a random variable concentrated at one point is always stable; that is a *degenerate* special case.

It turns out that the constant c in (5.2) must be related to the constants a_1 and a_2 there by

$$a_1^\alpha + a_2^\alpha = c^\alpha \qquad (5.3)$$

for some constant α, $0 < \alpha \leq 2$. Moreover, (5.2) implies that, for any $n \geq 2$, we must have

$$X_1 + \cdots + X_n \stackrel{\mathrm{d}}{=} n^{1/\alpha} X + b_n \qquad (5.4)$$

where $X_1, \ldots, X_n$ are independent copies of X and α is the same constant appearing in (5.3), which is called the *index* of the stable law.

The stable laws on $\mathbb{R}$ can be represented as a four-parameter family. Following Samorodnitsky and Taqqu (1994), let $S_\alpha(\sigma, \beta, \mu)$ denote a stable law (also called α-stable law) on the real line. Also let $S_\alpha(\sigma, \beta, \mu)$ denote a real-valued random variable with the associated stable law. The four parameters of the stable law are: the *index* α, $0 < \alpha \leq 2$; the *scale parameter* $\sigma > 0$; the *skewness parameter* β, $-1 \leq \beta \leq 1$, and the location or *shift parameter* μ, $-\infty < \mu < \infty$. When $1 < \alpha < 2$, the shift parameter is the mean. When $\alpha \leq 1$, the mean is infinite. The logarithm of the characteristic function of $S_\alpha(\sigma, \beta, \mu)$ is

$$\log E e^{i\theta S_\alpha(\sigma,\beta,\mu)}$$
$$= \begin{cases} -\sigma^\alpha |\theta|^\alpha (1 - i\beta (\mathrm{sign}\, \theta) \tan(\pi\alpha/2)) + i\mu\theta, & \alpha \neq 1 \\ -\sigma |\theta| (1 + i\beta(2/\pi)(\mathrm{sign}\, \theta) \log(|\theta|)) + i\mu\theta, & \alpha = 1, \end{cases} \qquad (5.5)$$

where $\mathrm{sign}(\theta) = 1$, 0 or -1 for $\theta > 0$, $\theta = 0$ and $\theta < 0$.

The cases $\alpha = 1$ and $\alpha = 2$ are singular cases, with special properties and special formulas. They are boundary cases, at which abrupt change of behavior occurs. The normal law is the special case with $\alpha = 2$; then μ is the mean, $2\sigma^2$ is the variance and β plays no role because $\tan(\pi) = 0$; i.e., $S_2(\sigma, 0, \mu) = N(\mu, 2\sigma^2)$. When $\beta = 1$ ($\beta = -1$), the stable distribution is said to be *totally skewed* to the right (left). For limits involving nonnegative summands, we will be interested in the centered totally-skewed stable laws $S_\alpha(\sigma, 1, 0)$.

With the notation $S_\alpha(\sigma, \beta, \mu)$ for stable laws, we can refine the stability property (5.4). If $X_1, \ldots, X_n$ are IID random variables distributed as $S_\alpha(\sigma, \beta, \mu)$, then

$$X_1 + \cdots + X_n \stackrel{d}{=} \begin{cases} n^{1/\alpha} X_1 + \mu(n - n^{1/\alpha}), & \alpha \neq 1 \\ nX_1 + \frac{2}{\pi}\sigma\beta n \log(n), & \alpha = 1. \end{cases} \quad (5.6)$$

From (5.6), we see that $S_\alpha(\sigma, \beta, 0)$ is strictly stable for all $\alpha \neq 1$ and that $S_1(\sigma, 0, \mu)$ is strictly stable.

All stable laws have continuous pdf's, but there are only three classes of these pdf's with convenient closed-form expressions: The first is the Gaussian distribution; as indicated above, $S_2(\sigma, 0, \mu) = N(\mu, 2\sigma^2)$. The second is the *Cauchy distribution* $S_1(\sigma, 0, \mu)$, whose pdf is

$$f(x) \equiv \frac{\sigma}{\pi((x-\mu)^2 + \sigma^2)}.$$

In the case $\mu = 0$,

$$P(S_1(\sigma, 0, 0) \leq x) \equiv \frac{1}{2} + \frac{1}{\pi} \text{Arctan}\left(\frac{x}{\sigma}\right).$$

The third is the *Lévy distribution* $S_{1/2}(\sigma, 1, \mu)$, whose pdf is

$$f(x) = \left(\frac{\sigma}{2\pi}\right)^{1/2} \frac{1}{(x-\mu)^{3/2}} \exp\left(\frac{-\sigma}{2(x-\mu)}\right), \quad x > \mu.$$

For the case $\mu = 0$, the cdf is

$$P(S_{1/2}(\sigma, 1, 0) \leq x) = 2(1 - \Phi(\sqrt{\sigma/x})), \quad x > 0,$$

where Φ is the standard normal cdf.

There are simple scaling relations among the nonGaussian stable laws: For any nonzero constant c,

$$S_\alpha(\sigma, \beta, \mu) + c \stackrel{d}{=} S_\alpha(\sigma, \beta, \mu + c), \quad (5.7)$$

$$cS_\alpha(\sigma, \beta, \mu) \stackrel{d}{=} \begin{cases} S_\alpha(|c|\sigma, \text{sign}(c)\beta, c\mu) & \text{if } \alpha \neq 1 \\ S_1(|c|\sigma, \text{sign}(c)\beta, c\mu - \frac{2c}{\pi}(\log(|c|))\sigma\beta) & \text{if } \alpha = 1, \end{cases} \quad (5.8)$$

$$-S_\alpha(\sigma, \beta, 0) \stackrel{d}{=} S_\alpha(\sigma, -\beta, 0). \quad (5.9)$$

If $S_\alpha(\sigma_i, \beta_i, \mu_i)$ are two independent α-stable random variables, then

$$S_\alpha(\sigma_1, \beta_1, \mu_1) + S_\alpha(\sigma_2, \beta_2, \mu_2) \stackrel{d}{=} S_\alpha(\sigma, \beta, \mu) \quad (5.10)$$

for

$$\sigma^\alpha = \sigma_1^\alpha + \sigma_2^\alpha, \quad \beta = \frac{\beta_1 \sigma_1^\alpha + \beta_2 \sigma_2^\alpha}{\sigma_1^\alpha + \sigma_2^\alpha}, \quad \mu = \mu_1 + \mu_2. \quad (5.11)$$

In general, the stable pdf's are continuous, positive and unimodal on their support. (Unimodality means that there is an argument t_0 such that the pdf is nondecreasing for $t < t_0$ and nonincreasing for $t > t_0$.) The stable laws $S_\alpha(\sigma, 1, \mu)$ with $0 < \alpha < 1$ have support (μ, ∞), while the stable laws $S_\alpha(\sigma, -1, \mu)$ with $0 < \alpha < 1$ have support $(-\infty, \mu)$. All other stable laws (if $\alpha \geq 1$ or if $\alpha < 1$ and $\beta \neq 1$) have support on the entire real line. See Samorodnitsky and Taqqu (1994) for plots of the pdf's.

It is significant that the nonGaussian stable laws have *power tails*. As in (4.6) in Section 1.4, we write $f(x) \sim g(x)$ as $x \to \infty$ if $f(x)/g(x) \to 1$ as $x \to \infty$. For $0 < \alpha < 2$,

$$P(S_\alpha(\sigma, \beta, \mu) > x) \sim x^{-\alpha} C_\alpha \frac{(1+\beta)}{2} \sigma^\alpha \qquad (5.12)$$

and

$$P(S_\alpha(\sigma, \beta, \mu) < -x) \sim x^{-\alpha} C_\alpha \frac{(1-\beta)}{2} \sigma^\alpha, \qquad (5.13)$$

where

$$C_\alpha = \left(\int_0^\infty x^{-\alpha} \sin x \, dx\right)^{-1} = \begin{cases} \frac{1-\alpha}{\Gamma(2-\alpha)\cos(\pi\alpha/2)} & \text{if } \alpha \neq 1 \\ 2/\pi & \text{if } \alpha = 1 \end{cases} \qquad (5.14)$$

with $\Gamma(x)$ being the gamma function.

Note that there is an abrupt change in tail behavior at the boundary $\alpha = 2$. For all $\alpha < 2$, the stable pdf has a power tail, but for $\alpha = 2$, the pdf is of order $e^{-x^2/2}$. There also is a discontinuity in the constant C_α in (5.14) at $\alpha = 1$; as $\alpha \to 1$, $C_\alpha \to 1$, but $C_1 = 2/\pi$.

When $\beta = 1$ ($\beta = -1$), the left (right) tail is asymptotically negligible. When also $\alpha < 1$, there is no other tail. When $1 < \alpha < 2$ and $\beta = 1$, the left tail decays faster than exponentially. Indeed, when $1 < \alpha < 2$,

$$P(S_\alpha(\sigma, 1, 0) < -x)$$
$$\sim A \left(\frac{x}{\alpha \hat\sigma_\alpha}\right)^{-\alpha/(2(\alpha-1))} \exp\left(-(\alpha-1)\left(\frac{x}{\alpha \hat\sigma_\alpha}\right)^{\alpha/(\alpha-1)}\right) \qquad (5.15)$$

where

$$A \equiv (2\pi\alpha(\alpha-1))^{-1/2} \quad \text{and} \quad \hat\sigma_\alpha \equiv \sigma(\cos((\pi/2)(2-\alpha)))^{-1/\alpha}.$$

When $\alpha = 1$ and $\beta = 1$,

$$P(S_1(\sigma, 1, 0) < -x) \sim \frac{1}{\sqrt{2\pi}} \exp\left(-\frac{(\pi/2\sigma)x - 1}{2} - e^{(\pi/2\sigma)x-1}\right). \qquad (5.16)$$

Consequently, the Laplace transform of $S_\alpha(\sigma, 1, 0)$ is well defined, even though the pdf has the entire real line for its support. In particular, the logarithm of the Laplace transform of $S_\alpha(\sigma, 1, 0)$ is

$$\psi_\alpha(s) \equiv \log E e^{-sS_\alpha(\sigma,1,0)} = \begin{cases} -\sigma^\alpha s^\alpha / \cos(\pi\alpha/2), & \alpha \neq 1 \\ 2\sigma s \log(s)/\pi, & \alpha = 1, \end{cases} \qquad (5.17)$$

for $\text{Re}(s) \geq 0$.

From the asymptotic form above, we can deduce properties of the moments. In particular, for $0 < \alpha < 2$,

$$E|S_\alpha(\sigma,\beta,\mu)|^p < \infty \quad \text{for} \quad 0 < p < \alpha \quad \text{and}$$
$$E|S_\alpha(\sigma,\beta,\mu)|^p = \infty \quad \text{for} \quad p \geq \alpha \ . \tag{5.18}$$

4.5.2. Convergence to Stable Laws

We now discuss convergence to stable laws. A cdf F on $\mathbb{R}$ is said to be in the *domain of attraction* of the stable law $S_\alpha(\sigma,\beta,\mu)$ if there exist constants m_n and c_n such that

$$c_n^{-1}(S_n - m_n) \Rightarrow S_\alpha(\sigma,\beta,\mu) \ , \tag{5.19}$$

where $S_n = X_1 + \cdots + X_n$, $n \geq 1$, $\{X_n : n \geq 1\}$ is a sequence of IID random variables and X_1 has cdf F. By (5.4), $S_\alpha(\sigma,\beta,\mu)$ is contained in the domain of attraction of $S_\alpha(\sigma,\beta,\mu)$ for all $(\alpha,\sigma,\beta,\mu)$. Clearly, by scaling, it suffices to let $\sigma = 1$ and $\mu = 0$. Hence, only the parameters α and β are unaltered by scaling. A cdf F is said to be in the *normal domain of attraction* of the stable law $S_\alpha(\sigma,\beta,\mu)$ if, in addition to being in the domain of attraction, the constants c_n in (5.19) can be chosen so that $\mu = 0$, $\sigma = 1$ and $c_n = cn^{1/\alpha}$ for some constant c.

This limit theory is classical; see Gnedenko and Kolmogorov (1968), Feller (1971) and p. 50 of Samorodnitsky and Taqqu (1994). Naturally, a key role is played by the cdf F of X_1. A big role is also played by the cdf of $|X_1|$; let G be its cdf and $G^c \equiv 1 - G$ its complementary cdf (ccdf), i.e.,

$$G^c(x) \equiv P(|X_1| > x) = 1 - F(x) + F(-x) \ . \tag{5.20}$$

The conditions make use of regularly varying functions; see Appendix A. We write $G^c \in \mathcal{R}(-\alpha)$ if the ccdf is regularly varying with index $-\alpha$. That holds if and only if $G^c(x) = x^{-\alpha}L(x)$ for some slowly varying function L.

Theorem 4.5.1. (stable-law CLT) *Let $\{X_n : n \geq 1\}$ be an IID sequence of real-valued random variables with cdf F. The cdf F belongs to the domain of attraction of $S_\alpha(1,\beta,0)$ for $0 < \alpha < 2$, i.e., (5.19) holds for $\sigma = 1$ and $\mu = 0$, if and only if both $G^c \in \mathcal{R}(-\alpha)$, i.e.,*

$$x^\alpha G^c(x) = L(x) \tag{5.21}$$

for G^c in (5.20), where L is slowly varying, and

$$F^c(x)/G^c(x) \to \frac{1+\beta}{2} \quad \text{as} \quad x \to \infty \ . \tag{5.22}$$

The space-scaling constants c_n in (5.19) then must satisfy

$$\lim_{n\to\infty} \frac{nL(c_n)}{c_n^\alpha} = C_\alpha \ , \tag{5.23}$$

for C_α in (5.14) and L in (5.21). The translation constants m_n in (5.19) may be chosen to satisfy

$$m_n = \begin{cases} 0 & \text{if } 0 < \alpha < 1 \\ nc_n \int_{-\infty}^{\infty} \sin(x/c_n) dF(x) & \text{if } \alpha = 1 \\ n \int_{-\infty}^{\infty} x dF(x) & \text{if } 1 < \alpha < 2 \end{cases} \quad (5.24)$$

If c_n satisfies (5.23), then $c_n = n^{1/\alpha} L_0(n)$, where L_0 is slowly varying (in general different from L in (5.21)).

At the expense of changing the scaling constants σ and μ in the limit, the normalization constants c_n in Theorem 4.5.1 can be chosen to be the $(1 - n^{-1})^{\text{th}}$ percentile of the cdf G instead of (5.23); i.e., we can let

$$c_n \equiv (1/G^c)^{\leftarrow}(n) \equiv \inf\{y : G(y) \geq n\} ; \quad (5.25)$$

see p. 3 of Resnick (1987) and p. 78 of Embrechts et. al. (1997).

Theorem 4.5.1 contains the result about normal domains of attraction as a special case. Note that the condition has the summand having a power law.

Theorem 4.5.2. (normal domain of attraction of a stable law) Let $\{X_n : n \geq 1\}$ be an IID sequence with cdf F. The cdf belongs to the normal domain of attraction of $S_\alpha(1, \beta, 0)$ for $0 < \alpha < 2$, i.e., (5.19) holds with $c_n = cn^{1/\alpha}$, $\sigma = 1$ and $\mu = 0$, if and only if both

$$G^c(x) \sim A x^{-\alpha} \quad \text{as} \quad x \to \infty \quad (5.26)$$

for G^c in (5.20) and positive constants A and α, and

$$\frac{F^c(x)}{G^c(x)} \to \frac{1 + \beta}{2} \quad \text{as} \quad x \to \infty . \quad (5.27)$$

The space-scaling constants can then be

$$c_n = (A/C_\alpha)^{1/\alpha} n^{1/\alpha} , \quad (5.28)$$

where the pair (A, α) is from (5.26) and C_α is the stable-law asymptote in (5.14). The translation constant m_n can then be as in (5.24).

Proof. Given Theorem 4.5.1, for $0 < \alpha < 2$, c_n can be chosen to be of the form $cn^{1/\alpha}$ for some constant c, while satisfying (5.23), if and only if the slowly varying function $L(t)$ approaches a constant as $t \to \infty$. Thus, a cdf belongs to the normal domain of attraction of a stable law of index α if and only if (5.21) and (5.22) hold with $L(t) \to A$ as $t \to \infty$ for some constant A. In other words, for the normal domain of attraction, (5.21) should be restated as (5.26). Then the left side of (5.23) becomes nA/c_n^α. If $nA/c_n^\alpha \to C_\alpha$ as $n \to \infty$, then $n^{1/\alpha} A^{1/\alpha}/c_n \to C_\alpha^{1/\alpha}$ as $n \to \infty$, so that it suffices to use (5.28). ∎

It is useful to have a sanity check to verify the form of the space-scaling constants in (5.28). That is provided by considering the special case in which

$$X_n \stackrel{\mathrm{d}}{=} (A/C_\alpha)^{1/\alpha} S_\alpha(1, \beta, 0) .$$

Note that this X_n satisfies (5.26) and (5.27); e.g., by (5.12) and (5.13),

$$P(A/C_\alpha)^{1/\alpha}|S_\alpha(1,\beta,0)| > x) = P(|S_\alpha(1,\beta,0)| > (C_\alpha/A)^{1/\alpha}x) \sim Ax^{-\alpha} \ .$$

However, by (5.6),

$$(C_\alpha/nA)^{1/\alpha}(X_1 + \cdots + X_n) \stackrel{d}{=} S_\alpha(1,\beta,0) \quad \text{for all} \quad n \geq 1 \ .$$

Hence we must have (5.28).

From a mathematical perspective, Theorem 4.5.1 is appealing because it fully characterizes when the limit exists and gives its value. However, from a practical perspective, the special case in Theorem 4.5.2 may be more useful because it yields a more parsimonious approximation as a function of n. For the case $0 < \alpha < 2$, Theorem 4.5.1 yields the approximation

$$S_n \stackrel{d}{\approx} ES_n + c_n S_\alpha(1,\beta,0) \ ,$$

with the approximation as a function of n being a function of α, β and the entire (in general complicated) sequence $\{c_n : n \geq 1\}$. On the other hand, for the same case, Theorem 4.5.2 yields the approximation

$$S_n \stackrel{d}{\approx} ES_n + cn^{1/\alpha} S_\alpha(1,\beta,0) \tag{5.29}$$

with the approximation as a function of n being a function only of the three parameters α, β and c.

In applications it is usually very difficult to distinguish between a power tail and a regularly-varying nonpower tail of the same index. Even estimating the stable index α itself can be a challenge; see Embrechts et al. (1997), Resnick (1997) and Adler, Feldman and Taqqu (1998).

4.5.3. Convergence to Stable Lévy Motion

We now want to obtain the FCLT generalization of the stable-law CLT in Theorem 4.5.1. The limit process is a stable Lévy motion, which is a special case of Lévy process. A *Lévy process* is a stochastic process $\mathbf{L} \equiv \{\mathbf{L}(t) : t \geq 0\}$ with sample paths in D such that $\mathbf{L}(0) = 0$ and $\mathbf{L}$ has stationary and independent increments; we discuss Lévy processes further in Section 2.4 of the Internet Supplement. A standard *stable (or α-stable) Lévy motion* is a Lévy process $\mathbf{S} \equiv \{\mathbf{S}(t) : t \geq 0\}$ such that the increments have stable laws, in particular,

$$\mathbf{S}(t+s) - \mathbf{S}(s) \stackrel{d}{=} S_\alpha(t^{1/\alpha},\beta,0) \stackrel{d}{=} t^{1/\alpha} S_\alpha(1,\beta,0) \tag{5.30}$$

for any $s \geq 0$ and $t > 0$, for some α and β with $0 < \alpha \leq 2$ and $-1 \leq \beta \leq 1$. The adjective "standard" is used because the shift and scale parameters of the stable law in (5.30) are $\mu = 0$ and $\sigma = t^{1/\alpha}$ (without an extra multiplicative constant). When we want to focus on the parameters, we call the process a standard (α,β)-stable Lévy motion. Formula (5.30) implies that a stable Lévy motion has stationary increments. When $\alpha = 2$, stable Lévy motion is Brownian motion. Except in the cases when $\alpha = 1$ and $\beta \neq 1$, a stable (or α-stable) Lévy motion is self-similar with self-similarity index $H = 1/\alpha$, i.e.,

$$\{\mathbf{S}(ct) : t \geq 0\} \stackrel{\mathrm{d}}{=} \{c^{1/\alpha}\mathbf{S}(t) : t \geq 0\} \;.$$

In many ways, nonBrownian ($\alpha < 2$) stable Lévy motion is like Brownian motion ($\alpha = 2$), but it is also strikingly different. For example, Brownian motion has continuous sample paths, whereas stable Lévy motion, except for its deterministic drift, is a *pure-jump process*. It has infinitely many discontinuities in any finite interval w.p.1. On the positive side, there is a version with sample paths in D, and we shall only consider that version. For $0 < \alpha < 1$ and $\beta = 1$, stable Lévy motion has nondecreasing sample paths, and is called a *stable subordinator*.

For $\alpha \geq 1$ and $\beta = 1$, stable Lévy motion has no negative jumps; it has positive jumps plus a negative drift. For $\alpha > 1$, stable Lévy motion (like Brownian motion) has sample paths of unbounded variation in each bounded interval. Like Brownian motion, stable Lévy motion has complicated structure from some points of view, but also admits many simple formulas.

In the case of IID summands (for both double and single sequences), Skorohod (1957) showed that all ordinary CLT's have FCLT counterparts in (D, J_1); see Jacod and Shiryaev (1987) for further discussion, in particular, see Theorems 2.52 and 3.4 on pages 368 and 373. Hence the FCLT generalization of Theorem 4.5.1 requires no new conditions.

Theorem 4.5.3. (stable FCLT) *Under the conditions of Theorem 4.5.1, in addition to the CLT*

$$c_n^{-1}(S_n - m_n) \Rightarrow S_\alpha(1, \beta, 0) \quad in \quad \mathbb{R} \;,$$

there is convergence in distribution

$$\mathbf{S}_n \Rightarrow \mathbf{S} \quad in \quad (D, J_1) \tag{5.31}$$

for the associated normalized process

$$\mathbf{S}_n(t) \equiv c_n^{-1}(S_{\lfloor nt \rfloor} - m_n t), \quad t \geq 0 \;, \tag{5.32}$$

where the limit $\mathbf{S}$ is a standard (α, β)-stable Lévy motion, with

$$\mathbf{S}(t) \stackrel{\mathrm{d}}{=} t^{1/\alpha} S_\alpha(1, \beta, 0) \stackrel{\mathrm{d}}{=} S_\alpha(t^{1/\alpha}, \beta, 0) \;.$$

We have seen the stable FCLT in action in Chapter 1. Plots of random walks with IID steps having Pareto(p) distributions converging to stable Lévy motion with $\alpha = p$ are shown in Figures 1.20, 1.21 and 1.22 for $p = 3/2$ and in Figures 1.19, 1.25 and 1.26 for $p = 1/2$. We have also seen how the stable FCLT can be applied with the continuous mapping approach to establish stochastic-process limits for queueing models. Plots of workload processes converging to reflected stable Lévy motion appear in Figures 2.3 and 2.4.

Of course, there is a corresponding FCLT generalization of Theorem 4.5.2. There also is a k-dimensional generalization of Theorem 4.5.3 paralleling Theorem 4.3.5 in Section 4.3. The proof is just like that for Theorem 4.3.5, again exploiting the

Cramér-Wold device in Theorem 4.3.3. To apply the Cramér-Wold device, we use the fact that a stochastic process is strictly stable (stable with index $\alpha \geq 1$) if and only if all linear combinations (over time points and coordinates) of the process are again strictly stable (stable with index $\alpha \geq 1$); combine Theorems 2.1.5 and 3.1.2 of Samorodnitsky and Taqqu (1994). (For $\alpha \neq 1$, we always work with the centered stable laws having $\mu = 0$, so that they are strictly stable.)

4.5.4. Extreme-Value Limits

We have observed that the sample paths of stable Lévy motion are discontinuous. For that to hold, the maximum jump X_n must be asymptotically of the same order as the centered partial sum $S_n - mn$ for $\alpha > 1$ and the uncentered sum S_n for $\alpha < 1$. That was illustrated by the random-walk sample paths in Section 1.4. Further insight into the sample-path structure, and to the limit more generally, can be obtained from extreme-value theory, for which we draw upon Resnick (1987) and Embrechts et al. (1997). We will focus on the successive maxima of the random variables $|X_n|$. Let

$$M_n \equiv \{|X_1|, |X_2|, \ldots, |X_n|\}, \quad n \geq 1 \ . \tag{5.33}$$

As in (5.20), $|X_1|$ has ccdf G^c. Extreme-value theory characterizes the possible limit behavior of the successive maxima M_n, with scaling. Of special concern to us is the case in which the limiting cdf is the *Fréchet cdf*

$$\Phi_\alpha(x) = \begin{cases} 0, & x \leq 0 \\ \exp(-x^{-\alpha}), & x > 0, \end{cases} \tag{5.34}$$

which is defined for all $\alpha > 0$. Let Φ_α also denote a random variable with cdf Φ_α. Here is the relevant extreme-value theorem (which uses the concept of regular variation; see Appendix A and Section 1.2 of Resnick (1987):

Theorem 4.5.4. (extreme-value limit) *Suppose that $\{|X_n| : n \geq 0\}$ is a sequence of IID random variables having cdf G with $EX_n^2 = \infty$. There exist constants c_n and b_n such that $c_n(M_n - b_n)$ converges in distribution to a nondegenerate limit for M_n in (5.33) if and only if $G^c \in \mathcal{R}(-\alpha)$, in which case*

$$c_n^{-1} M_n \Rightarrow \Phi_\alpha \quad in \quad \mathbb{R} \ , \tag{5.35}$$

where Φ_α has the Fréchet cdf in (5.34) and the scaling constants may be

$$c_n \equiv (1/G^c)^{\leftarrow}(n)$$

as in (5.25).

As noted after Theorem 4.5.1, we can also use the scaling constant c_n in (5.25) in the CLT and FCLT for partial sums; i.e., under the conditions of Theorem 4.5.1, we have

$$c_n^{-1} M_n \Rightarrow \Phi_\alpha \ ,$$

$$c_n^{-1}(S_n - nm_n) \Rightarrow S_\alpha(\sigma, \beta, 0)$$

and

$$\mathbf{S}_n \Rightarrow \mathbf{S} ,$$

where

$$\mathbf{S}_n(t) = c_n^{-1}(S_{\lfloor nt \rfloor} - m_n nt), \quad t \geq 0 ,$$

$$m_n = \begin{cases} 0 & \text{if } 0 < \alpha < 1 \\ EX_1 & \text{if } 1 < \alpha < 2 , \end{cases} \quad (5.36)$$

and $\mathbf{S}$ is a nondegenerate stable Lévy motion with

$$\mathbf{S}(1) \stackrel{\mathrm{d}}{=} S_\alpha(\sigma, \beta, 0)$$

for some σ, β and the scaling constants c_n throughout being as in (5.25).

It turns out that we can also obtain a limit for M_n by applying the continuous mapping theorem with the FCLT in (5.31). For that purpose, we exploit the *maximum-jump functional* $J : D \to \mathbb{R}$ defined by

$$J(x) = \sup_{0 \leq t \leq 1} \{|x(t) - x(t-)|\} . \quad (5.37)$$

In general, the maximum-jump function is not continuous on D, but it is almost surely with respect to stable Lévy motion; see p. 303 of Jacod and Shiryaev (1987). As before, let $Disc(x)$ be the set of discontinuities of x.

Theorem 4.5.5. (maximum jump function) *The maximum-jump function J in (5.37) is measurable and continuous on (D, J_1) at all $x \in D$ for which $1 \in Disc(x)^c$. Hence, J is continuous almost surely with respect to stable Lévy motion.*

Hence we can apply the continuous mapping theorem in Section 2.7 with Theorems 4.5.3–4.5.5 to obtain the following result. See Resnick (1986) for related results.

Theorem 4.5.6. (joint limit for normalized maximum and sum) *Under the conditions of Theorem 4.5.1, we have the FCLT (5.31) with (5.32) for c_n in (5.25) and*

$$c_n^{-1}(M_n, S_n - nm_n) \Rightarrow (J(\mathbf{S}), \mathbf{S}(1)) \quad \text{in} \quad \mathbb{R}^2 ,$$

where

$$J(\mathbf{S}) \stackrel{\mathrm{d}}{=} \Phi_\alpha$$

for J in (5.37), Φ_α in (5.34) and m_n in (5.36). Consequently, on any positive interval the stable process $\mathbf{S}$ has a jump w.p.1. and $(S_n - nm_n)/M_n$ has a nondegenerate limit as $n \to \infty$.

More generally, it is interesting to identify cases in which the largest single term M_n among $\{X_1, \ldots, X_n\}$, when $X_i \geq 0$, is (i) asymptotically negligible, (ii) asymptotically of the same order, or (iii) asymptotically dominant compared to the partial sum S_n or its centered version. Work on this problem is reviewed in Section 8.15 of Bingham et al. (1989); we summarize the main results below.

Theorem 4.5.7. (asymptotics for the ratio of the maximum to the sum) *Let $\{X_n : n \geq 1\}$ be a sequence of IID random variables with cdf F having support on $(0, \infty)$. Let S_n be the nth partial sum and M_n the nth maximum. Then*
 (a) $M_n/S_n \Rightarrow 0$ *if and only if $\int_0^x y dF(y)$ is slowly varying;*
 (b) $M_n/S_n \Rightarrow 1$ *if and only if F^c is slowly varying;*
 (c) M_n/S_n *converges in distribution to a nondegenerate limit if and only if F^c is regularly varying of index $-\alpha$ for some α, $0 < \alpha < 1$.*
 (d) *If, in addition F has finite mean μ, then $(S_n - n\mu)/M_n$ converges in distribution to a nondegenerate limit if and only if F^c is regularly varying of index $-\alpha$ for some α, $1 < \alpha < 2$.*

4.6. The Joseph Effect: Strong Dependence

In Section 4.4 we saw that the conclusion of Donsker's theorem still holds when the independence condition is replaced with weak dependence, provided that the finite-second-moment condition is maintained. The situation is very different when there is *strong dependence*, also called *long-range dependence*.

In fact, all hell breaks loose. The statistical regularity we have seen, both with light and heavy tails, depends critically on the independence. As we saw in Section 4.4, we can relax the independence considerably, but the results depend on the dependence being suitably controlled. By definition, strong dependence occurs when that control is lost.

When we allow too much dependence, many bizarre things can happen. A simple way to see the possible difficulties is to consider the extreme case in which the random variables X_n are all copies of a single random variable X, where X can have any distribution. Then the scaled partial sum $n^{-1}S_n$ has the law of X, so $n^{-1}S_n \Rightarrow X$. Obviously there is no unifying stochastic-process limit in this degenerate case.

Nevertheless, it is important to study strong dependence, because it can be present. With strong dependence, we need to find some appropriate way to introduce strong structure to replace the independence we are giving up. Fortunately, ways to do this have been discovered, but no doubt many more remain to be discovered. We refer to Beran (1994), Eberlein and Taqqu (1986) and Samorodnitsky and Taqqu (1994) for more discussion and references.

We will discuss two approaches to strong dependence in this section. One is to exploit Gaussian processes. Gaussian processes are highly structured because they are fully characterized by their first and second moments, i.e., the mean function and the covariance function. The other approach is to again exploit independence, but in a modified form.

When we introduce this additional structure, it often becomes possible to establish stochastic-process limits with strong dependence. Just as with the heavy tails considered in Section 4.5, the strong dependence has a dramatic impact on the form of the stochastic-process limits, changing both the scaling and the limit process.

4.6.1. Strong Positive Dependence

Consider a stationary sequence $\{X_n : n \geq 1\}$ with $EX_n = 0$ and $Var\, X_n < \infty$. Since the variance $Var\, X_n$ is assumed to be finite, we call this the light-tailed case; in the next section we consider the heavy-tailed case in which $Var\, X_n = \infty$. Strong dependence can be defined by saying that the natural mixing conditions characterizing weak dependence, as in Theorem 4.4.1, no longer hold. However, motivated by applications, we are interested in a particular form of strong dependence called *strong positive dependence*. Roughly speaking, with positive dependence, we have

$$Var(S_n) > nVar(X_1) \quad \text{for} \quad n > 1,$$

i.e., the variance of the nth partial sum is greater than it would be in the IID case. We are interested in the case in which this is true for all sufficiently large n (ignoring departures from the assumption in a short time scale).

Even though $Var\, X_n$ is finite, there may be so much dependence among the successive variables X_n that the variance of the partial sum $S_n \equiv X_1 + \cdots + X_n$ is not of order n. Unlike (4.3), we are now primarily interested in the case in which

$$\lim_{n \to \infty} \frac{Var(S_n)}{n} = \infty. \qquad (6.1)$$

In particular, we assume that $Var(S_n)$ is a regularly varying function with index $2H$ for some H with

$$1/2 < H < 1, \qquad (6.2)$$

i.e.,

$$Var(S_n) = n^{2H} L(n) \quad \text{as} \quad n \to \infty, \qquad (6.3)$$

where $L(t)$ is a slowly varying function; see Appendix A. The principal case of interest for applications is $L(t) \to c$ as $t \to \infty$ for some constant c. When (6.2) and (6.3) hold, we say that $\{X_n\}$ and $\{S_n\}$ exhibit *strong positive dependence*. Since $Var(S_n) \leq n^2 Var\, X_1$, (6.2) covers the natural range of possibilities when (6.1) holds. In fact, we allow $0 < H < 1$, which also includes negative dependence.

We primarily characterize and quantify the strong dependence through the asymptotic form of the variance of the partial sums, as in (6.3). However, it is important to realize that we still need to impose additional structure in order to allow us to focus only on these variances. We will impose appropriate structure below.

It is natural to deduce the asymptotic form of the variance $Var(S_n)$ in (6.3) directly, but we could instead start with a detailed characterization of the covariances between variables in the sequence $\{X_n\}$. We want to complement the weak-dependent case in (4.3) and (4.4), so we focus on the cases with $H \neq 1/2$. We state the result as a lemma; see p. 338 of Samorodnitsky and Taqqu (1994). We state the result for pure power asymptotics, but there is an extension to regularly varying functions.

Lemma 4.6.1. (from covariance asymptotics to variance asymptotics) *Suppose that the covariances have the asymptotic form*

$$r_n \equiv Cov(X_1, X_{1+n}) \equiv E[(X_1 - EX_1)(X_{1+n} - EX_{1+n})] \sim cn^{2H-2}$$

as $n \to \infty$. *If* $c > 0$ *and* $1/2 < H < 1$, *then*

$$Var(S_n) \sim c \frac{n^{2H}}{H(2H-1)} \quad as \quad n \to \infty.$$

If $c < 0$ *and* $0 < H < 1/2$, *then*

$$Var(S_n) \sim |c| \frac{n^{2H}}{H(1-2H)} \quad as \quad n \to \infty.$$

In this setting with $\text{Var}(X_n) < \infty$ and centering to zero mean, the natural scaled process is

$$\mathbf{S}_n(t) \equiv c_n^{-1} S_{\lfloor nt \rfloor}, \quad t \geq 0, \tag{6.4}$$

where

$$c_n \equiv (Var(S_n))^{1/2}. \tag{6.5}$$

With the scaling in (6.4), we have

$$E\mathbf{S}_n(t) = 0 \quad \text{and} \quad Var(\mathbf{S}_n(t)) = t, \quad t \geq 0.$$

Space scaling asymptotically equivalent to (6.5) is required to get convergence of the second moments to a proper limit. We will find conditions under which $\mathbf{S}_n \Rightarrow \mathbf{S}$ in D and identify the limit process $\mathbf{S}$. Note that the strong positive dependence causes $c_n/\sqrt{n} \to \infty$ as $n \to \infty$.

4.6.2. Additional Structure

We now impose the additional structure needed in order to obtain a FCLT. As indicated above, there are two cases that have been quite well studied. In the first case, $\{X_n\}$ is a zero-mean Gaussian sequence. Then the finite-dimensional distributions are determined by the covariance function. A generalization of this first case in which $X_n = g(Y_n)$, where $\{Y_n\}$ is Gaussian and $g : \mathbb{R} \to \mathbb{R}$ is a smooth nonlinear function, has also been studied, e.g., see Taqqu (1975, 1979) and Dobrushin and Major (1979), but we will not consider that case. It gives some idea of the complex forms possible for limit processes with strong dependence.

In the second case, the basic stationary sequence $\{X_n\}$ has the *linear-process representation*

$$X_n \equiv \sum_{j=0}^{\infty} a_j Y_{n-j}, \quad n \geq 1, \tag{6.6}$$

where $\{Y_n : -\infty < n < \infty\}$ is a two-sided sequence of IID random variables with $EY_n = 0$ and $EY_n^2 = 1$, and $\{a_j : j \geq 0\}$ is a sequence of (deterministic, finite) constants with

$$\sum_{j=0}^{\infty} a_j^2 < \infty. \tag{6.7}$$

With (6.6), the stochastic process $\{X_n\}$ is said to obtained from the underlying process $\{Y_n\}$ by applying a *linear filter*; e.g., see p. 8 of Box, Jenkins and Reinsel (1994).

In fact, the linear-process representation tends to include the Gaussian sequence as a special case, because if $\{X_n\}$ is a stationary Gaussian process, then under minor regularity conditions, $\{X_n\}$ can be represented as in (6.6), where $\{Y_n\}$ is a sequence of IID random variables distributed as $N(0,1)$; e.g., see Hida and Hitsuda (1976). Of course, in general the random variables in the linear-process representation need not be normally distributed. Thus, the linear-process representation includes the Gaussian sequence as a special case.

The second case can also be generalized by considering variables $g(X_n)$ for X_n in (6.6) and $g : \mathbb{R} \to \mathbb{R}$ a smooth nonlinear function, see Avram and Taqqu (1987) and references there, but we will not consider that generalization either. It provides a large class of stochastic-process limits in a setting where the strong dependence is still quite tightly controlled by the underlying linear-process representation.

It is elementary that $\{X_n\}$ in (6.6) is a stationary process and

$$Var(X_n) = \sum_{j=0}^{\infty} a_j^2,$$

so that condition (6.7) ensures that $Var(X_n) < \infty$, as assumed before. It is also easy to determine the covariance function for X_n:

$$r_n = \sum_{j=0}^{\infty} a_j a_{j+n}.$$

The nth partial sum can itself be represented as a weighted sum of the variables from the underlying sequence $\{Y_n\}$, namely,

$$S_n = \sum_{k=-\infty}^{n} Y_k a_{n,k},$$

where

$$a_{n,k} = \begin{cases} \sum_{j=0}^{n-k} a_j, & 1 \leq k \leq n, \\ \sum_{j=k}^{n+k} a_j, & k \leq 0. \end{cases}$$

Example 4.6.1. *Power weights.*

Suppose that the weights a_j in (6.6) have the relatively simple form

$$a_j = c j^{-\gamma}. \tag{6.8}$$

To get strong positive dependence with (6.7), we need to require that $1/2 < \gamma < 1$. The associated covariances are

$$r_n = c^2 \sum_{j=0}^{\infty} j^{-\gamma}(j+n)^{-\gamma} .$$

By applying the Euler-Maclaurin formula, Chapter 8 of Olver (1974), and the change of variables $x = nu$, we obtain the asymptotic form of r_n:

$$r_n \sim c^2 \int_0^{\infty} x^{-\gamma}(x+n)^{-\gamma}\, dx \sim n^{1-2\gamma} c^2 \int_0^{\infty} u^{-\gamma}(1+u)^{-\gamma}\, du$$

as $n \to \infty$, where

$$\int_0^{\infty} u^{-\gamma}(1+u)^{-\gamma}\, du = B(1-\gamma, 2\gamma-1) = \Gamma(1-\gamma)\Gamma(2\gamma-1)/\Gamma(\gamma)$$

with $B(z,w)$ and $\Gamma(z)$ the beta and gamma functions; see 6.1.1, 6.2.1 and 6.2.2 of Abramowitz and Stegun (1972). Hence

$$r_n \sim C_1 n^{1-2\gamma} \quad \text{as} \quad n \to \infty , \tag{6.9}$$

where

$$C_1 = c^2 \Gamma(1-\gamma)\Gamma(2\gamma-1)/\Gamma(\gamma) . \tag{6.10}$$

By Lemma 4.6.1, $H = (3-2\gamma)/2$ and

$$Var(S_n) \sim C_2 n^{3-2\gamma} \quad \text{as} \quad n \to \infty , \tag{6.11}$$

where

$$C_2 = 2c^2 \Gamma(1-\gamma)\Gamma(2\gamma-1)/\Gamma(\gamma)(3-2\gamma)^2 . \tag{6.12}$$

For instance, if $\gamma = 3/4$, then $H = 3/4$ and $C_2 = 41.95 c^2$ for c in (6.8). ∎

4.6.3. Convergence to Fractional Brownian Motion

We can deduce that the limit process $\mathbf{S}$ for $\mathbf{S}_n$ in (6.4), with (6.6) holding, must be a Gaussian process. First, if the basic sequence $\{X_n : n \geq 1\}$ is a Gaussian process, then the scaled partial-sum process $\{\mathbf{S}_n(t) : t \geq 0\}$ must also be a Gaussian process for each n, which implies that $\mathbf{S}$ must be Gaussian if $\mathbf{S}_n \Rightarrow \mathbf{S}$. Hence, if a limit holds more generally without the Gaussian condition, then the limit process must be as determined for the special case.

Alternatively, starting from the linear-process representation (6.6) with a general sequence $\{Y_n\}$ of IID random variables with $EY_n = 0$ and $EY_n^2 = 1$, we can apply the central limit theorem for nonidentically distributed summands, e.g., as on p. 262 of Feller (1971), and the Cramer-Wold device in Theorem 4.3.3 to deduce that

$$(\mathbf{S}_n(t_1), \ldots, \mathbf{S}_n(t_k)) \Rightarrow (\mathbf{S}(t_1), \ldots, \mathbf{S}(t_k)) \quad \text{in} \quad \mathbb{R}^k$$

for all positive integers k and all k-tuples $(t_1, \ldots, t_k)$ with $0 \le t_1 < \cdots < t_k$, where $(\mathbf{S}(t_1), \ldots, \mathbf{S}(t_k))$ must have a Gaussian distribution. Thus, weak convergence in D only requires in addition showing tightness.

The limit process in the FCLT is *fractional Brownian motion* (FBM). Standard FBM is the zero-mean Gaussian process $\mathbf{Z}_H \equiv \{\mathbf{Z}_H(t) : t \ge 0\}$ with covariance function

$$r_H(s,t) \equiv Cov(\mathbf{Z}_H(s), \mathbf{Z}_H(t)) \equiv \frac{1}{2}(t^{2H} + s^{2H} - (t-s)^{2H}), \qquad (6.13)$$

where any H with $0 < H < 1$ is allowed. For $H = 1/2$, standard FBM reduces to standard Brownian motion.

Standard FBM can also be expressed as a stochastic integral with respect to standard Brownian motion; in particular,

$$\mathbf{Z}_H(t) = \int_{-\infty}^{t} w_H(t,u) d\mathbf{B}(u), \qquad (6.14)$$

where

$$w_H(t,u) = \begin{cases} 0, & u \ge t, \\ (t-u)^{H-1/2}, & 0 \le u < t, \\ (t-u)^{H-1/2} - (-u)^{H-1/2} & u < 0. \end{cases} \qquad (6.15)$$

Of course, some care is needed in defining the stochastic integral with respect to Brownian motion, because the paths are of unbounded variation, but this problem has been addressed; e.g., see Karatzas and Shreve (1988), Protter (1992), Section 2.4 of Beran (1994) and Chapter 7 of Samorodnitsky and Taqqu (1994).

Note that (6.14) should be consistent with our expectations, given the initial weighted sum in (6.6). From (6.14) we can see how the dependence appears in FBM. We also see that FBM is a smoothed version of BM. For example, from (6.14) it is evident that FBM has continuous sample paths. The process FBM is also H-self-similar, which can be regarded as a consequence of being a weak-convergence limit, as discussed in Section 4.2.

We are now ready to state the FCLT, which is due to Davydov (1970); also see p. 288–289 of Taqqu (1975). Note that the theorem always holds for $1/2 \le H < 1$, but also holds for $0 < H < 1/2$ under extra moment conditions (in (6.17) below). These extra moment conditions are always satisfied in the Gaussian case. For refinements, see Avram and Taqqu (1987) and references therein.

Theorem 4.6.1. *(FCLT for strong dependence and light tails) Suppose that the basic stationary sequence $\{X_n : n \ge 1\}$ is either a zero-mean Gaussian process or a zero-mean linear process as in (6.6) and (6.7) with $E[X_n^2] < \infty$. If*

$$Var(S_n) = c_n^2 \equiv n^{2H} L(n), \quad n \ge 1, \qquad (6.16)$$

for $0 < H < 1$, where L is slowly varying, and, in the nonGaussian case,

$$E|S_n|^{2a} \le K(E[S_n^2])^a \quad \text{for some} \quad a > 1/H \qquad (6.17)$$

for some constant K, then

$$\mathbf{S}_n \Rightarrow \mathbf{Z}_H \quad in \quad (D, J_1) \qquad (6.18)$$

for $\mathbf{S}_n$ in (6.4) with c_n in (6.16) and $\mathbf{Z}_H$ standard FBM with self-similarity index H.

Remark 4.6.1. *Applying the continuous-mapping approach.* Considering the linear-process representations in (6.6) and (6.14), it is natural to view the limit in (6.18) as convergence of stochastic integrals

$$\int w_n dB_n \to \int w dB, \qquad (6.19)$$

where the integrands are deterministic, the limiting stochastic integral corresponds to (6.14) and

$$\mathbf{B}_n(t) = n^{-1/2} \sum_{i=1}^{\lfloor nt \rfloor} Y_i, \quad t \geq 0.$$

Donsker's theorem states that $\mathbf{B}_n \Rightarrow \mathbf{B}$ in D. It remains to show that $w_n \to w$ in a manner so that (6.19) holds. An approach to weak convergence of linear processes along this line is given by Kasahara and Maejima (1986). An earlier paper in this spirit for the special case of discounted processes is Whitt (1972). For more on convergence of stochastic integrals, see Kurtz and Protter (1991) and Jakubowski (1996). The point of this remark is that Theorem 4.6.1 should properly be viewed as a consequence of Donsker's FCLT and the continuous-mapping approach. ∎

The linear-process representation in (6.6) is convenient mathematically to impose structure, because we have constructed the stationary sequence $\{X_n\}$ from an underlying sequence of IID random variables with finite second moments, which we know how to analyze. What may not be evident, however, is that the linear-process representation can arise naturally from modelling. We show that it can arise naturally from time-series modeling in Section 2.5 of the Internet Supplement.

In Chapter 1, the random-walk simulations suggested stochastic-process limits. Having already proved convergence to FBM, we now can use the stochastic-process limits to provide a way to simulate FBM.

Example 4.6.2. *Simulating FBM.* We can simulate FBM, or more properly an approximation of FBM, by simulating a random walk $\{S_n\}$ with steps X_n satisfying the linear-process representation in (6.6), where $\{Y_n\}$ is IID with mean 0 and variance 1. We will let $Y_i \stackrel{d}{=} N(0,1)$. As part of the approximation, we truncate the series in (6.6). That can be done by assuming that $a_j = 0$ for $j \geq N$, where N is suitably large.

As in Chapter 1, the plotter does the appropriate space scaling automatically. In order to verify that what we see is consistent with the theory, we calculate the appropriate space-scaling constants. To be able to do so conveniently, we use the power weights in Example 4.6.1 with $c = 1$ and $\gamma = 3/4$. As indicated there, then

4.6. The Joseph Effect: Strong Dependence   127

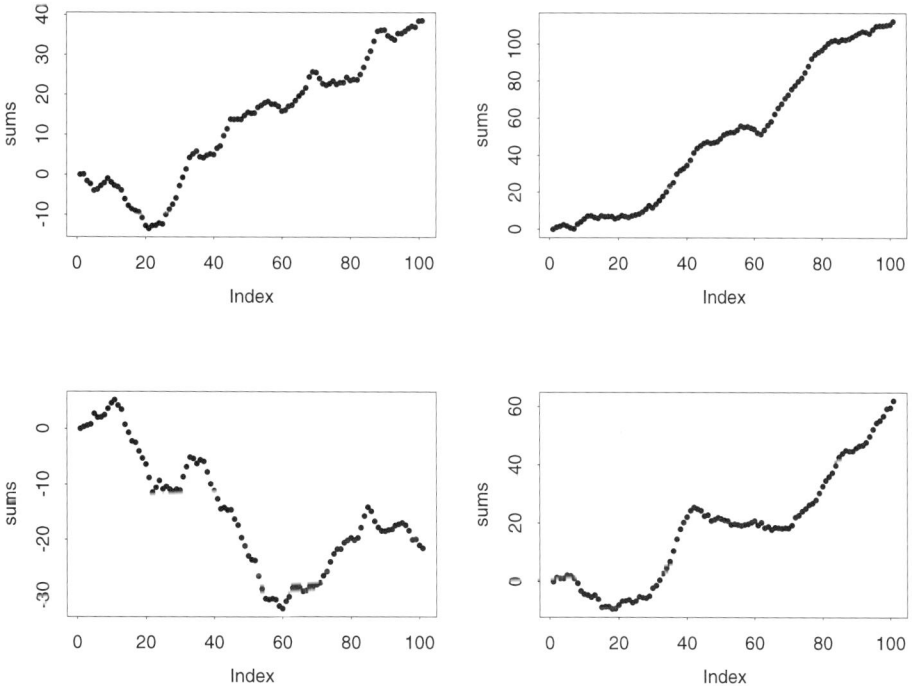

Figure 4.1. Four independent realizations of the first 10^2 steps of the unscaled random walk $\{S_k : 0 \leq k \leq n\}$ associated with the strongly dependent steps in Example 4.6.1.

the self-similarity index is $H = 3/4$, $\mathrm{Var}(S_n) \sim 41.95 n^{2H}$ and the space-scaling constants are

$$c_n = \sqrt{Var(S_n)} = 6.477 n^{3/4}\ .$$

We plot S_k for $0 \leq k \leq n$ for $n = 10^2$ and $n = 10^3$ in Figures 4.1 and 4.2. We plot four independent replications in each case. In these examples, we let $N = 10^4$. We use smaller n than in the IID case, because the computation is more complex, tending to require work of order nN. Comparing Figure 4.1 to Figures 1.3 and 1.4, we see that the sample paths of FBM are smoother than the paths of BM, as we should anticipate from (6.14).

As in Chapter 1, we can see emerging statistical regularity by considering successively larger values of n. The plots tend to look the same as n increases. However, as with the heavy-tailed distributions (the Noah effect), there is more variability from sample path to sample path than in the IID light-tailed case, as depicted in Figures 1.3 and 1.4. Even though the steps have mean 0, the strong dependence often make the plots look like the steps have nonzero mean. These sample paths show that it would be impossible to distinguish between strong dependence and

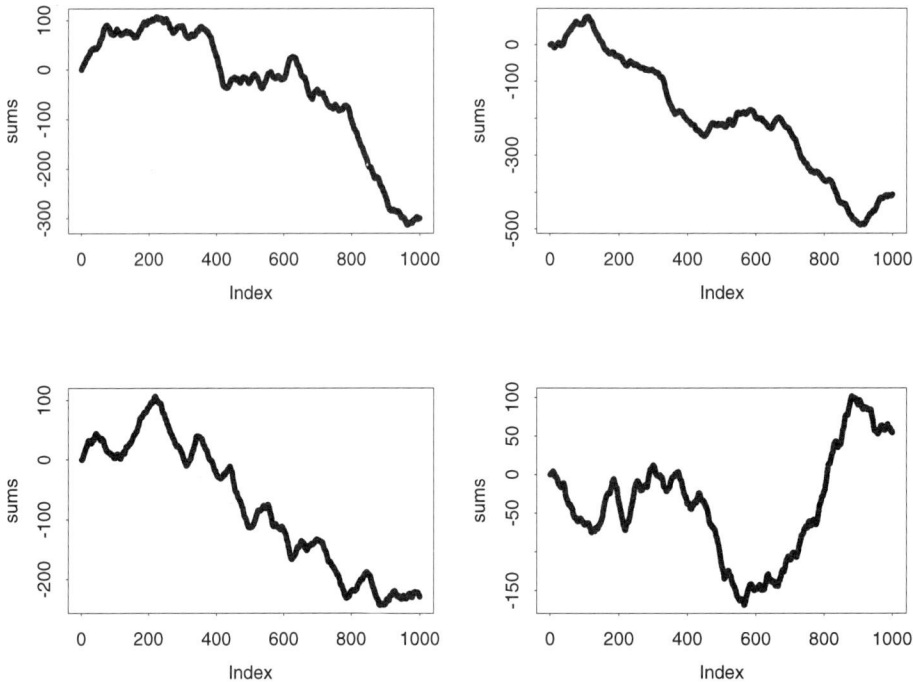

Figure 4.2. Four independent realizations of the first 10^3 steps of the unscaled random walk $\{S_k : 0 \leq k \leq n\}$ associated with the strongly dependent steps in Example 4.6.1.

nonstationarity from only a modest amount of data, e.g., from only a few sample paths like those in Figures 4.1 and 4.2.

The standard deviations of S_n for $n = 100$ and $n = 1,000$ are 205 and 1152, respectively. That is consistent with the final positions seen in Figures 4.1 and 4.2.

Since it is difficult to simulate the random walk S_n with dependent steps X_n, it is natural to seek more efficient methods to simulate FBM. For discussion of alternative methods, see pp. 370, 588 of Samorodnitsky and Taqqu (1994). ∎

The strong dependence poses a difficulty because of increased variability. The increased variability is indicated by the growth rate of $Var(S_n)$ as $n \to \infty$. However, the strong dependence also has a positive aspect, providing an opportunity for better *prediction*.

Remark 4.6.2. *Exploiting dependence for prediction.* The strong dependence helps to exploit observations of the past to predict process values in the not-too-distant future. To illustrate, suppose that we have a linear process as in (6.6), and that as time evolves we learn the values of the underlying sequence Y_n, so that after observing X_n and S_n we know the variables Y_j for $j \leq n$. From (6.6), the conditional means and variances are

$$E[X_{n+k}|Y_j, j \leq n] = \sum_{j=0}^{\infty} a_{k+j} Y_{n-j} , \qquad (6.20)$$

$$Var(X_{n+k}|Y_j, j \leq n) \equiv E[(X_{n+k} - E[X_{n+k}|Y_j, j \leq n])^2] = \sum_{j=0}^{k-1} a_j^2 , \qquad (6.21)$$

$$E[S_{n+k}|Y_j, j \leq n] = \sum_{j=0}^{\infty} (\sum_{i=j+1}^{j+k} a_i) Y_{n-j} \qquad (6.22)$$

and

$$Var(S_{n+k}|Y_j, j \leq n) \equiv E[(S_{n+k} - E[S_{n+k}|Y_j, j \leq n])^2] = \sum_{j=0}^{k-1} (\sum_{i=0}^{j} a_i)^2 .$$

If we use a criterion of mean-squared error, then the conditional mean is the best possible predictor of the true mean and the conditional variance is the resulting mean-squared error. A similar analysis applies to FBM, assuming that we learn the history of the underlying Brownian motion in the linear-process representation in (6.14). However, in many applications we can only directly observe the past of the sequence $\{X_n\}$, or the FBM $Z_H(t)$ in case of the limit process. Fortunately, prediction can still be done by exploiting time-series methods. We discuss prediction in queues in Remark 8.7.2. ∎

In some applications (e.g., at the end of Section 7.2 below) we will want continuous-time analogs of Theorem 4.6.1. With continuous-time processes, we need to work harder to establish tightness. We show how this can be done for Gaussian processes with continuous sample paths.

Theorem 4.6.2. (FCLT for Gaussian processes in C) *If $\{Y(t) : t \geq 0\}$ is a zero-mean Gaussian process with stationary increments, sample paths in C, $Y(0) = 0$,*

$$Var Y(t) \sim ct^{2H} \quad as \quad t \to \infty \qquad (6.23)$$

and

$$Var Y(t) \leq Kt^{2H} \quad for\ all \quad t \geq 0 \qquad (6.24)$$

for some constants c, K and H with $1/2 < H < 1$, then

$$\mathbf{Z}_n \Rightarrow c\mathbf{Z}_H \quad in \quad (C, U) ,$$

where $\mathbf{Z}_H$ is standard FBM and

$$\mathbf{Z}_n(t) \equiv n^{-H} Y(nt), \quad t \geq 0 .$$

Proof. For each n, $\mathbf{Z}_n$ is a Gaussian process. Given (6.23), it is elementary that $cov(\mathbf{Z}_n(s), \mathbf{Z}_n(t)) \to cov(\mathbf{Z}(s), \mathbf{Z}(t))$ as $n \to \infty$ for all s and t. That establishes convergence of the finite-dimensional distributions. By (6.24),

$$E[(\mathbf{Z}_n(t) - \mathbf{Z}_n(s))^2] = n^{-2H} Var Y(n(t-s)) \leq K(t-s)^{2H},$$

which implies tightness by Theorem 11.6.5. ∎

4.7. Heavy Tails Plus Dependence

The previous three sections described FCLTs with only heavy tails (Section 4.5) and with only dependence (Sections 4.4 and 4.6). The most complicated case involves both heavy tails and dependence. Unfortunately, there is not yet a well developed theory for stochastic-process limits in this case. Evidently, a significant part of the difficulty stems from the need to use nonstandard topologies on the function space D; e.g., see Avram and Taqqu (1992) and Jakubowski (1996). Hence, this interesting case provides additional motivation for the present book, but it remains to establish important new results.

We start by considering the natural analog of Section 4.4 to the case of heavy tails: stable limits with weak dependence. Since the random variables do not have finite variances, even describing dependence is complicated, because the covariance function is not well defined. However, alternatives to the covariance have been developed; see Samorodnitsky and Taqqu (1994). We understand weak dependence to hold when there is dependence but the stochastic-process limit is essentially the same as in the IID case.

We state one result for stable limits with weak dependence. It is a FCLT for linear processes with heavy tails. However, there is a significant complication caused by having dependence together with jumps in the limit process. To obtain a stochastic-process limit in D, it is necessary to use the M_1 topology on D. Moreover, even with the M_1 topology, it is necessary to impose additional conditions in order to establish the FCLT.

4.7.1. Additional Structure

Just as in the last section, in this section we assume that the basic sequence $\{X_n : n \geq 1\}$ is a stationary sequence with a linear-process representation

$$X_n \equiv \sum_{j=0}^{\infty} a_j Y_{n-j}, \tag{7.1}$$

where the *innovation process* $\{Y_n : -\infty < n < \infty\}$ is a sequence of IID random variables, but now we assume that Y_n has a heavy-tailed distribution. In particular, we assume that the distribution of Y_n is in the domain of attraction of a stable law $S_\alpha(1, \beta, 0)$ with $0 < \alpha < 2$; i.e., we assume that (5.21) and (5.22) hold. That in turn implies that $Var(Y_n) = \infty$.

Given the stable index α, we assume that

$$\sum_{j=0}^{\infty} |a_j|^{\alpha-\epsilon} < \infty \quad \text{for some} \quad \epsilon > 0. \tag{7.2}$$

Condition (7.2) ensures that the sum (7.1) converges in the L^p space for $p = \alpha - \epsilon$ and w.p.1; see Avram and Taqqu (1992). However, the variance $Var(X_n)$ is necessarily infinite.

We first remark that condition (7.2) permits quite strong dependence, because we can have

$$a_j \sim c j^{-\gamma} \quad \text{as} \quad j \to \infty \quad \text{for} \quad \text{any} \quad \gamma > \alpha^{-1}, \tag{7.3}$$

where c is a positive constant, so we might have $\sum_{j=1}^{\infty} |a_j| = \infty$.

For simplicity, we assume that $EY_n = 0$ if $1 < \alpha < 2$ and that the distribution of Y_n is symmetric if $\alpha = 1$. Then, under the assumptions above, Theorems 4.5.1 and 4.5.3 imply that

$$\mathbf{S}_n \Rightarrow \mathbf{S} \quad \text{in} \quad (D, J_1), \tag{7.4}$$

where

$$\mathbf{S}_n(t) \equiv c_n^{-1} \sum_{i=1}^{\lfloor nt \rfloor} Y_i, \quad t \geq 0, \tag{7.5}$$

$\mathbf{S}$ is a stable process with $\mathbf{S}(1) \stackrel{d}{=} S_\alpha(1, \beta, 0)$ and $c_n = n^{1/\alpha} L(n)$ for some slowly varying function L. We are interested in associated FCLTs for

$$\mathbf{Z}_n(t) \equiv c_n^{-1} \sum_{i=1}^{\lfloor nt \rfloor} X_i, \quad t \geq 0, \tag{7.6}$$

for $\{X_n\}$ in (7.1) and $\{c_n : n \geq 1\}$ in (7.5).

4.7.2. Convergence to Stable Lévy Motion

In considerable generality, $\mathbf{Z}_n$ in (7.6) satisfies essentially the same FCLT as $\mathbf{S}_n$ in (7.5), with the limit being a constant multiple of the previous limit $\mathbf{S}$. The following result is from Astrauskas (1983), Davis and Resnick (1985) and Avram and Taqqu (1992). Note that the M_1 topology is used. Let $Z_n \Rightarrow Z$ in $(D, f.d.d.)$ mean that there is convergence of all finite-dimensional distributions.

Theorem 4.7.1. (FCLT for a linear process with heavy tails) *Suppose that the sequence $\{X_n\}$ is the linear process in (7.1) satisfying the assumptions above, which imply (7.4). If, in addition,*

$$\sum_{j=0}^{\infty} |a_j| < \infty, \tag{7.7}$$

then

$$\mathbf{Z}_n \Rightarrow (\sum_{j=0}^{\infty} a_j)\mathbf{S} \quad in \quad (D, f.d.d.)$$

for **S** *in (7.4) and* $\mathbf{Z}_n$ *in (7.6). Suppose, in addition, that* $a_i \geq 0$ *for all i. If any one of the following conditions hold:*

(i) $0 < \alpha \leq 1$,

(ii) $a_i \neq 0$ *for only finitely many i,*

(iii) $\alpha > 1$, $\sum_{i=1}^{\infty} |a_i|^\nu < \infty$ *for some* $\nu < 1$ *and* $\{a_i\}$ *is a monotone sequence,*

then

$$\mathbf{Z}_n \Rightarrow (\sum_{j=0}^{\infty} a_j)\mathbf{S} \quad in \quad (D, M_1).$$

Avram and Taqqu (1992) actually established the M_1-convergence part of Theorem 4.7.1 under a somewhat weaker condition than stated above. Avram and Taqqu (1992) show that the M_1 topology is critical in Theorem 4.7.1; the result does not hold in (D, J_1) if there are at least two nonzero coefficients in (7.1). Indeed, that is evident because an exceptionally large value of Y_n will correspond to more than one exceptionally large value in the X_n; i.e., the jump in the limit process for $\mathbf{Z}_n$ will correspond to more than one jump in the converging processes. The linear-process structure is yet another setting leading to unmatched jumps in the limit process, requiring the M_1 topology instead of the familiar J_1 topology.

Note that the limit process in Theorem 4.7.1 has independent increments. Thus, just as in Section 4.4, the dependence in the original process is asymptotically negligible in the time scaling of the stochastic-process limit. Thus, the predicted value of $S_{\lfloor cn \rfloor}$ for $c > 1$ given $S_j, j \leq n$, is about S_n. At that time scale, there is not much opportunity to exploit past observations, beyond the present value, in order to predict future values.

Example 4.7.1. *Simulation to experience Theorem 4.7.1.* To illustrate Theorem 4.7.1, suppose that Y_1 has the Pareto(p) distribution with $p = 3/2$, just as in Section 1.4. Let the weights be $a_j = j^{-8}$ for $j \geq 0$. We simulate the random walk just as in Example 4.6.2. Since the weights decay faster here, it suffices to use a smaller truncation point N; we use $N = 100$. We plot four independent replications of the random walk $\{S_k : 0 \leq k \leq n\}$ for $n = 1,000$ in Figure 4.3. The plots look just like the plots of the random walk in the IID case in Figures 1.20, 1.21 and 1.22. Thus the simulation is consistent with Theorem 4.7.1. ∎

4.7.3. Linear Fractional Stable Motion

Note that the conditions of Theorem 4.7.1 do not cover the case in which

$$a_j \sim cj^{-\gamma} \quad as \quad j \to \infty \tag{7.8}$$

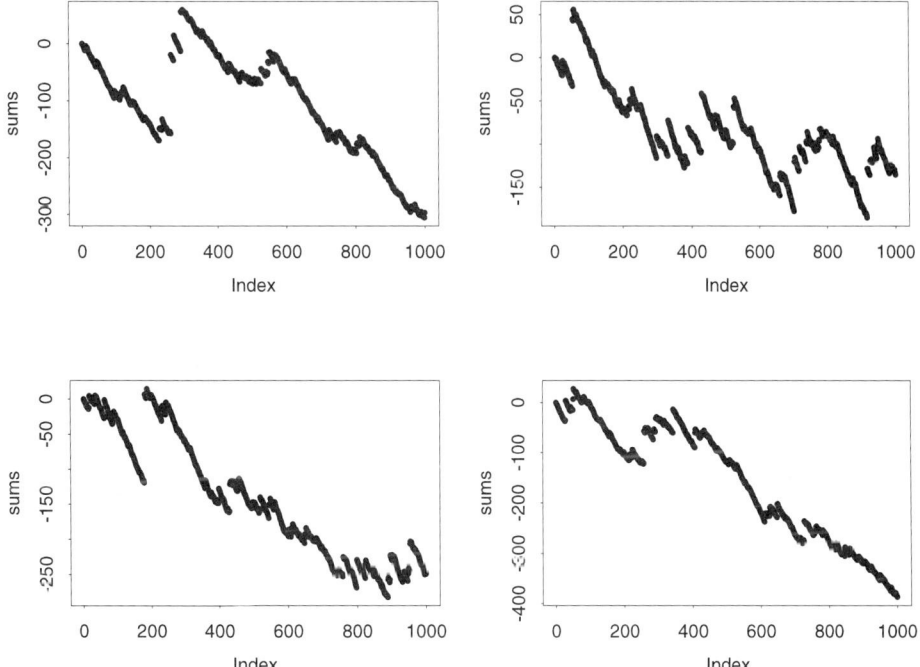

Figure 4.3. Four independent realizations of the first 10^3 steps of the unscaled random walk $\{S_k : 0 \le k \le n\}$ associated with the dependent heavy-tailed steps in Example 4.7.1.

for $c > 0$ and $\alpha^{-1} < \gamma \le 1$, where $1 < \alpha < 2$. We include $\gamma > \alpha^{-1}$ in (7.8) so that condition (7.2) is still satisfied, but condition (7.7) is violated. We refer to this case as strong positive dependence with heavy tails.

The limit process when (7.8) holds is *linear fractional stable motion* (LFSM), which is an H-sssi α-stable process with self-similarity index

$$H = \alpha^{-1} + 1 - \gamma > \alpha^{-1}, \qquad (7.9)$$

where $1 < \alpha < 2$, so that $2^{-1} < \alpha^{-1} < 1$, and $H < 1$; see Sections 7.3 and 7.4 of Samorodnitsky and Taqqu (1994).

Paralleling the representation of FBM as a stochastic integral with respect to standard Brownian motion in (6.14), we can represent LFSM as a stochastic integral with respect to stable Lévy motion; in particular, for $1 < \alpha < 2$ and $\alpha^{-1} < H < 1$,

$$\mathbf{Z}_{H,\alpha}(t) = \int_{-\infty}^{t} w_H(t,u) d\mathbf{S}_\alpha(u), \qquad (7.10)$$

where $\mathbf{S}_\alpha$ is an α-stable Lévy motion with $\mathbf{S}_\alpha(1) \stackrel{\mathrm{d}}{=} S_\alpha(1, \beta, 0)$ and

$$w_H(t, u) = \begin{cases} 0, & u \geq t, \\ (t-u)^{H-1/\alpha}, & 0 \leq u < t, \\ (t-u)^{H-1/\alpha} - (-u)^{H-1/\alpha} & u < 0; \end{cases} \quad (7.11)$$

The LFSM in (7.10) is natural because $\mathbf{Z}_{H,\alpha}(t)$ depends upon $\mathbf{S}_\alpha$ only over the interval $(-\infty, t]$ for any t, so that we can regard $\mathbf{S}_\alpha$ as an innovation process. For more general LFSMs, see Samorodnitsky and Taqqu (1994). It is significant that the LFSM above has continuous sample paths; see Theorem 12.4.1 of Samorodnitsky and Taqqu (1994).

Theorem 4.7.2. (FCLT with both the Noah and Joseph effects) *Suppose that the basic sequence $\{X_n\}$ has the linear-process representation (7.1), where $\{Y_n\}$ is a sequence of IID random variables with Y_1 in the normal domain of attraction of the stable law $S_\alpha(1, \beta, 0)$ i.e., such that (5.26) and (5.27) hold. If, in addition, (7.8) holds, then*

$$\mathbf{Z}_n \Rightarrow \mathbf{Z}_{H,\alpha} \quad in \quad (D, J_1) \, ,$$

where $\mathbf{Z}_{H,\alpha}$ is LFSM in (7.10) and $\mathbf{Z}_n$ is the scaled partial-sum process in (7.6) with space-scaling constants

$$c_n = n^H (A/C_\alpha)^{1/\alpha} (c/(1-\gamma)), \quad n \geq 1 \quad (7.12)$$

for A in (5.26), C_α in (5.14), (c, γ) in (7.8) and the self-similarity index H in (7.9).

By Theorem 4.5.2, under the assumptions in Theorem 4.7.2, the space-scaling constants for the partial sums of Y_n are $c_n = (nA/C_\alpha)^{1/\alpha}$. From (7.12), we see that the linear-process representation produces the extra multiplicative factor $n^{H-\alpha^{-1}} c(1-\gamma)^{-1}$.

We remark that Astrauskas (1983) actually proved a more general result, allowing both the tail probability $P(|Y_1| > x)$ and the weights a_j to be regularly varying at infinity instead of pure power tails. For extensions of Theorems 4.7.1 and 4.7.2, see Hsing (1999) and references therein.

Example 4.7.2. *Simulating LFSM.*

To illustrate Theorem 4.7.2, suppose that Y_1 has the Pareto(p) distribution with $p = 3/2$, just as in Example 4.7.1, but now let the weights be $a_j = j^{-\gamma}$ for $\gamma = 3/4$, just as in Example 4.6.2. Hence we have combined the heavy-tailed feature of Example 4.7.1 with the strong-dependence feature in Example 4.6.2. Since $\alpha = p$,

$$\gamma > \alpha^{-1} = 2/3$$

and (7.8) is satisfied. From (7.9), the self-similarity index in this example is

$$H = \alpha^{-1} + 1 - \gamma = 11/12 \, ,$$

so that H is much greater than $1/2$.

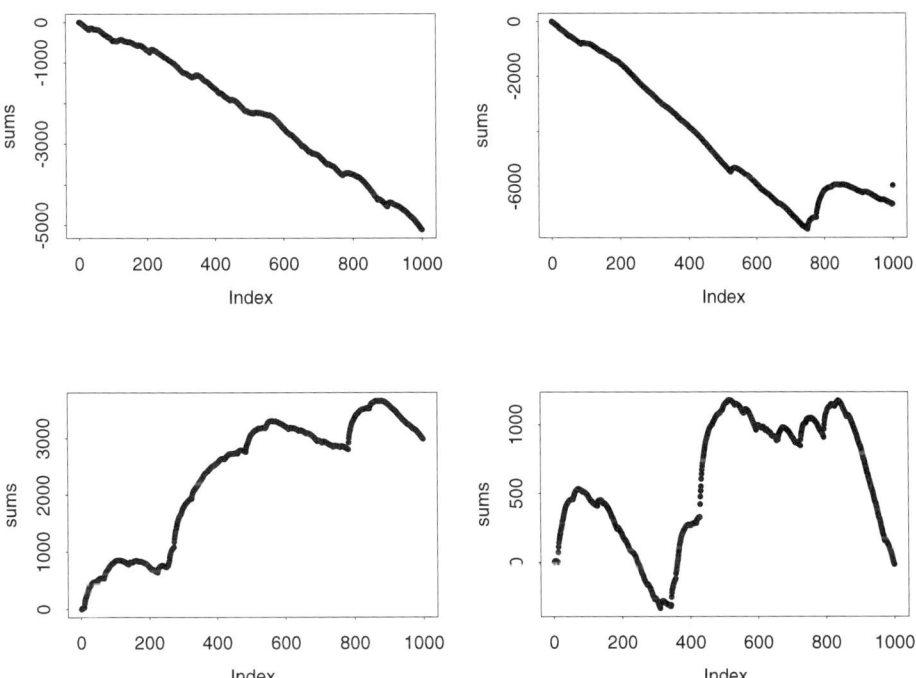

Figure 4.4. Four independent realizations of the first 10^3 steps of the unscaled random walk $\{S_k : 0 \leq k \leq n\}$ associated with the strongly dependent steps in Example 4.7.2.

We simulate the random walk just as in Example 4.7.1, except that we let the truncation point N be higher because of the more slowly decaying weights; in particular, now we let $N = 1,000$. We plot four independent replications of the random walk $\{S_k : 0 \leq k \leq n\}$ for $n = 1,000$ in Figure 4.4.

Unlike the plots in Figure 4.3, it is evident from Figure 4.4 that the sample paths are now continuous. However, the heavy tails plus strong dependence can induce strong surges up and down. The steady downward trend in the first plot occurs because there are relatively few larger values. The sudden steep upward surge at about $j = 420$ in the fourth plot in Figure 4.4 occurs because of a few exceptionally large values at that point. In particular, $Y_{416} = 17.0$, $Y_{426} = 88.1$ and $Y_{430} = 24.3$. In contrast, in the corresponding plot of FBM in Figure 4.2 with the same weights $a_j = j^{-3/4}$, all 2,000 of the normally distributed Y_j satisfy $|Y_j| \leq 3.2$. Finally, note that the large value of H is consistent with the large observed values of the range in the plots. ∎

From the dependent increments in the LFSM limit process, it is evident that there is again (as in Section 4.6) an opportunity to exploit the history of past observations in order to predict future values of the process. With strong dependence plus heavy-tailed distributions, the statistical techniques are more complicated, but there is a growing literature; see Samorodnitsky and Taqqu (1994), Kokoszka and Taqqu (1995, 1996a,b), Montanari, Rosso and Taqqu (1997), Embrechts, Klüppelberg and Mikosch (1997) and Adler, Feldman and Taqqu (1998).

4.8. Summary

We have now presented FCLTs for partial sums in each of the four cases – light or heavy tails with weak or strong dependence. We summarize the results in the table below.

		Dependence	
		Weak	Strong Joseph effect
Tails	light	Sections 4.3 and 4.4	Section 4.6
	heavy Noah effect	Section 4.5 Theorem 4.7.1	Theorem 4.7.2

Table 4.1. The four kinds of FCLTs discussed in Sections 4.3–4.7

In conlusion, we observe that the theory seems far from final form for the strong dependence discussed in Sections 4.6 and 4.7 and for heavy tails with any form of dependence. The results should be regarded as illustrative of what is possible. Careful study of specific applications is likely to unearth important new limit processes.

We next show how the continuous-mapping approach can be applied with established stochastic-process limits to establish heavy-traffic stochastic-process limits for queues. In Chapter 7 we present additional established stochastic-process limits.

5
Heavy-Traffic Limits for Fluid Queues

5.1. Introduction

In this chapter we see how the continuous-mapping approach can be applied to establish heavy-traffic stochastic-process limits for queueing models, and how those heavy-traffic stochastic-process limits, in turn, can be applied to obtain approximations for queueing processes and gain insight into queueing performance.

To establish the heavy-traffic stochastic-process limits, the general idea is to represent the queueing "content" process of interest as a reflection of a corresponding net-input process. For single queues with unlimited storage capacity, a one-sided one-dimensional reflection map is used; for single queues with finite storage capacity, a two-sided one-dimensional reflection map is used. These one-dimensional reflection maps are continuous as maps from D to D with all the principal topologies considered (but not M_2) by virtue of results in Sections 13.5 and 14.8. Hence, FCLT's for scaled net-input processes translate into corresponding FCLT's for scaled queueing processes.

Thus we see that the relatively tractable heavy-traffic approximations can be regarded as further instances of the statistical regularity stemming from the FCLT's in Chapter 4. The FCLT for the scaled net-input processes may be based on Donsker's theorem in Section 4.3 and involve convergence to Brownian motion; then the limit process for the scaled queueing processes is reflected Brownian motion (RBM). Alternatively, the FCLT for the scaled net input processes may be based on one of the other FCLT's in Sections 4.5 – 4.7 and involve convergence to a different limit process; then the limit process for the scaled queueing processes is the reflected version of that other limit process.

For example, when the net-input process can be constructed from partial sums of IID random variables with heavy-tailed distributions, Section 4.5 implies that the scaled net-input processes converge to a stable Lévy motion; then the limit process for the queueing processes is a reflected stable Lévy motion. The reflected stable Lévy motion heavy-traffic limit describes the effect of the extra burstiness due to the heavy-tailed distributions.

As indicated in Section 4.6, it is also possible to have more burstiness due to strong positive dependence or less burstiness due to strong negative dependence. When the net-input process has such strong dependence with light-tailed distributions, the scaled net-input processes may converge to fractional Brownian motion; then the limit process for the scaled queueing processes is reflected fractional Brownian motion.

In this chapter, attention will be focused on the "classical" Brownian approximation involving RBM and its application. For example, in Section 5.8 we show how the heavy-traffic stochastic-process limit with convergence to RBM can be used to help plan queueing simulations, i.e., to estimate the required run length to achieve desired statistical precision, as a function of model parameters. Reflected stable Lévy motion will be discussed in Sections 8.5 and 9.7, while reflected fractional Brownian motion will be discussed in Sections 8.7 and 8.8.

In simple cases, the continuous-mapping approach applies directly. In other cases, the required argument is somewhat more complicated. A specific simple case is the discrete-time queueing model in Section 2.3. In that case, the continuous-mapping argument applies directly: FCLT's for the partial sums of inputs V_k translate immediately into associated FCLT's for the workload (or buffer-content) process $\{W_k\}$, exploiting the continuity of the two-sided reflection map. The continuous-mapping approach applies directly because, as indicated in (3.5) in Chapter 1, the scaled workload process is exactly the reflection of the scaled net-input process, which itself is a scaled partial-sum process. Thus all the stochastic-process limits in Chapter 4 translate into corresponding heavy-traffic stochastic-process limits for the workload process in Section 2.3.

In this chapter we see how the continuous-mapping approach works with related continuous-time fluid-queue models. We start considering fluid queues, instead of standard queues (which we consider in Chapter 9), because fluid queues are easier to analyze and because fluid queues tend to serve as initial "rough-cut" models for a large class of queueing systems. The fluid-queue models have recently become popular because of applications to communication networks, but they have a long history. In the earlier literature they are usually called dams or stochastic storage models; see Moran (1959) and Prabhu (1998). In addition to queues, they have application to inventory and risk phenomena.

In this chapter we give proofs for the theorems, but the emphasis is on the statement and applied significance of the theorems. The proofs illustrate the continuous-mapping approach for establishing stochastic-process limits, exploiting the useful functions introduced in Section 3.5. Since the proofs draw on material from later chapters, upon first reading it should suffice to focus, first, on the theorem statements and their applied significance and, second, on the general flow of the argument in the proofs.

5.2. A General Fluid-Queue Model

In a fluid-queue model, a divisible commodity (fluid) arrives at a storage facility where it is stored in a buffer and gradually released. We consider an *open model* in which fluid arrives *exogenously* (from outside). For such open fluid-queue models, we describe the buffer content over time. In contrast, in a standard queueing model, which we consider in Chapter 9, individual customers (or jobs) arrive at a service facility, possibly wait, then receive service and depart. For such models, we count the number of customers in the system and describe the experience of individual customers. The fluid queue model can be used to represent the unfinished work in a standard queueing model. Then the input consists of the customer service requirements at their arrival epochs. And the unfinished work declines at unit rate as service is provided.

In considering fluid-queue models, we are motivated to a large extent by the need to analyze the performance of evolving communication networks. Since data carried by these networks are packaged in many small packets, it is natural to model the flow as fluid, i.e., to think of the flow coming continuously over time at a random rate. A congestion point in the network such as a switch or router can be regarded as a queue (dam or stochastic storage model), where input is processed at constant or variable rate (the available bandwidth). Thus, we are motivated to consider fluid queues. However, we should point out that other approaches besides queueing analysis are often required to engineer communication networks; to gain perspective, see Feldmann et al. (2000, 2001) and Krishnamurthy and Rexford (2001).

5.2.1. Input and Available-Processing Processes

In this section we consider a very general model: We consider a single fluid queue with general input and available-processing (or service) processes. For any $t > 0$, let $C(t)$ be the cumulative input of fluid over the interval $[0,t]$ and let $S(t)$ be the cumulative available processing over the interval $[0,t]$. If there is always fluid to process during the interval $[0,t]$, then the quantity processed during $[0,t]$ is $S(t)$. We assume that $\{C(t) : t \geq 0\}$ and $\{S(t) : t \geq 0\}$ are real-valued stochastic processes with nondecreasing nonnegative right-continuous sample paths. But at this point we make no further structural or stochastic assumptions.

A common case is processing at a constant rate μ whenever there is fluid to process; then

$$S(t) = \mu t, \quad t \geq 0 . \tag{2.1}$$

More generally, we could have input and output at random rates. Then

$$C(t) = \int_0^t R_i(s)ds \quad \text{and} \quad S(t) = \int_0^t R_o(s)ds, \quad t \geq 0, \quad (2.2)$$

where $\{R_i(t) : t \geq 0\}$ and $\{R_o(t) : t \geq 0\}$ are nonnegative real-valued stochastic processes with sample paths in D. For example, it is natural to have maximum possible input and processing rates ν_i and ν_o. Then, in addition to (2.2), we would assume that

$$0 \leq R_i(t) \leq \nu_i \quad \text{and} \quad 0 \leq R_o(t) \leq \nu_o \quad \text{for all} \quad t \quad \text{w.p.1} . \quad (2.3)$$

With (2.2), the stochastic processes C and S have continuous sample paths. We regard that as the standard case, but we allow C and S to be more general.

With the general framework, the discrete-time fluid-queue model in Section 2.3 is actually a special case of the continuous-time fluid-queue model considered here. The previous discrete-time fluid queue is put in the present framework by letting

$$C(t) \equiv \sum_{k=1}^{\lfloor t \rfloor} V_k \quad \text{and} \quad S(t) \equiv \mu \lfloor t \rfloor, \quad t \geq 0 ,$$

where $\lfloor t \rfloor$ is the greatest integer less than or equal to t.

5.2.2. Infinite Capacity

We will consider both the case of unlimited storage space and the case of finite storage space. First suppose that there is unlimited storage space. Let $W(t)$ represent the *workload* (or buffer content, i.e., the quantity of fluid waiting to be processed) at time t. Note that we can have significant fluid flow without ever having any workload. For example, if $W(0) = 0$, $C(t) = \lambda t$ and $S(t) = \mu t$ for all $t \geq 0$, where $\lambda < \mu$, then fluid is processed continuously at rate λ, but $W(t) = 0$ for all t. However, if C is a pure-jump process, then the processing occurs only when $W(t) > 0$. (The workload or virtual-waiting-time process in a standard queue is a pure-jump process.)

The workload $W(t)$ can be defined in terms of an *initial workload* $W(0)$ and a *net-input process* $C(t) - S(t)$, $t \geq 0$, via a *potential-workload process*

$$X(t) \equiv W(0) + C(t) - S(t), \quad t \geq 0, \quad (2.4)$$

by applying the *one-dimensional reflection map* to X, i.e., by letting

$$W(t) \equiv \phi(X)(t) \equiv X(t) - \inf_{0 \leq s \leq t}\{X(s) \wedge 0\}, \quad t \geq 0, \quad (2.5)$$

where $a \wedge b = \min\{a, b\}$.

We could incorporate the initial workload $W(0)$ into the cumulative-input process $\{C(t) : t \geq 0\}$ by letting $C(0) = W(0)$. Then X would simply be the net-input process. However, we elect not to do this, because it is convenient to treat the initial conditions separately in the limit theorems.

5.2. A General Fluid-Queue Model

The potential workload represents what the workload would be if we ignored the emptiness condition, and assumed that there is always output according to the available-processing process S. Then the workload at time t would be $X(t)$: the sum of the initial workload $W(0)$ plus the cumulative input $C(t)$ minus the cumulative output $S(t)$. Since emptiness may sometimes prevent output, we have definition (2.5).

Formula (2.5) is easy to understand by looking at a plot of the potential workload process $\{X(t) : t \geq 0\}$, as shown in Figure 5.1. Figure 5.1 shows a possible sample path of X when $S(t) = \mu t$ for $t \geq 0$ w.p.1 and there is only one on-off source that alternates between busy periods and idle periods, having input rate $r > \mu$ during busy periods and rate 0 during idle periods. Hence the queue alternates between net-input rates $r - \mu > 0$ and $-\mu < 0$. The plot of the potential workload process $\{X(t) : t \geq 0\}$ also can be interpreted as a plot of the actual workload process if we redefine what is meant by the origin. For the workload process, the origin is either 0, if X has not become negative, or the lowest point reached by X. The position of the origin for W is shown by the shaded dashed line in Figure 5.1.

An important observation is that the single value $W(t)$, for any $t > 0$, depends on the initial segment $\{X(s) : 0 \leq s \leq t\}$. To know $W(t)$, it is not enough to know the single value $X(t)$. However, by (2.5) it is evident that, for any $t > 0$, both $W(t)$ and the initial segment $\{W(s) : 0 \leq s \leq t\}$ are functions of the initial segment $\{X(s) : 0 \leq s \leq t\}$. With appropriate definitions, the reflection map in (2.5) taking the modified net-input process $\{X(t) : t \geq 0\}$ into the workload processes $\{W(t) : t \geq 0\}$ is a continuous function on the space of sample paths; see Section 13.5. Thus, by exploiting the continuous mapping theorem in a function space setting, a limit for a sequence of potential workload processes will translate into a corresponding limit for the associated sequence of workload processes.

Remark 5.2.1. *Model generality.* It may be hard to judge whether the fluid queue model we have introduced is exceptionally general or restrictive. It depends on the perspective: On the one hand, the model is very general because the basic stochastic processes C and S can be almost anything. We illustrate in Chapter 8 by allowing the input C to come from several on-off sources. We are able to treat that more complex model as a special case of the model studied here. On the other hand, the model is also quite restrictive because we assume that the workload stochastic process is directly a reflection of the potential-workload stochastic process. That makes the continuous-mapping approach especially easy to apply. In contrast, as we will see in Chapter 9, it is more difficult to treat the queue-length process in the standard single-server queue without special Markov assumptions. However, additional mathematical analysis shows that the model discrepancy is asymptotically negligible: In the heavy-traffic limit, the queue-length process in the standard single-server queue behaves as if it could be represented directly as a reflection of the associated net-input process. And similar stories hold for other models. The fluid model here is attractive, not only because it is easy to analyze, but also because it captures the essential nature of more complicated models. ∎

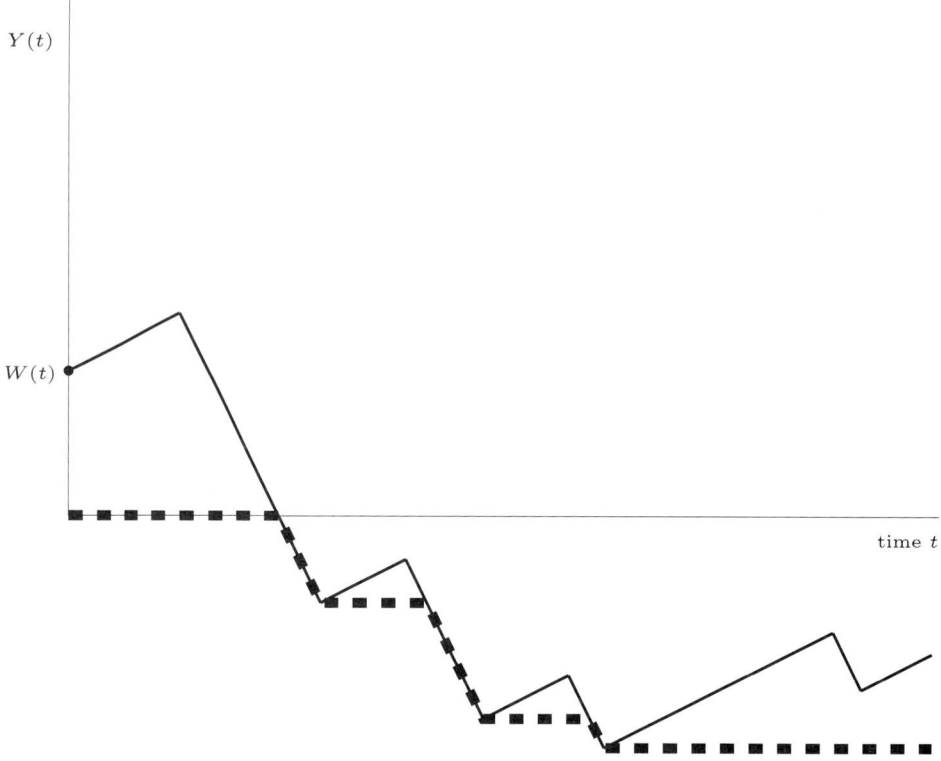

Figure 5.1. A possible realization of the potential workload process $\{X(t) : t \geq 0\}$ and the actual workload process $\{W(t) : t \geq 0\}$ with unlimited storage capacity: The actual workload process appears if the origin is the heavy shaded dashed line; i.e., solid line - dashed line = actual workload.

The general goal in studying this fluid-queue model is to understand how assumed behavior of the basic stochastic processes C and S affects the workload stochastic process W. For example, assuming that the net-input process $C - S$ has stationary increments and negative drift, under minor regularity conditions (see Chapter 1 of Borovkov (1976)), the workload $W(t)$ will have a limiting steady-state distribution. We want to understand how that steady-state distribution depends on the stochastic processes C and S. We also want to describe the transient (time-dependent) behavior of the workload process. Heavy-traffic limits can produce robust approximations that may be useful even when the queue is not in heavy traffic.

We now want to consider the case of a finite storage capacity, but before defining the finite-capacity workload process, we note that the one-sided reflection map in (2.5) can be expressed in an alternative way, which is convenient for treating generalizations such as the finite-capacity model and fluid networks; see Chapter 14 and Harrison (1985) for more discussion. Instead of (2.5), we can write

$$W(t) \equiv \phi(X)(t) \equiv X(t) + L(t), \qquad (2.6)$$

where X is the potential workload process in (2.4) and $\{L(t) : t \geq 0\}$ is a nondecreasing "regulator" process that increases only when $W(t) = 0$, i.e., such that

$$\int_0^t W(s)dL(s) = 0, \quad t \geq 0 . \tag{2.7}$$

From (2.5), we know that

$$L(t) = -\inf_{0 \leq s \leq t}\{X(s) \wedge 0\}, \quad t \geq 0 . \tag{2.8}$$

It can be shown that the characterization of the reflection map via (2.6) and (2.7) is equivalent to (2.5). For a detailed proof and further discussion, see Chapter 14, which focuses on the more complicated multidimensional generalization.

5.2.3. Finite Capacity

We now modify the definition in (2.6) and (2.7) to construct the finite-capacity workload process. Let the buffer capacity be K. Now we assume that any input that would make the workload process exceed K is lost. Let

$$W(t) \equiv \phi_K(X)(t) \equiv X(t) + L(t) - U(t), \quad t \geq 0 , \tag{2.9}$$

where again $X(t)$ is the potential workload process in (2.4), the initial condition is now assumed to satisfy $0 \leq W(0) \leq K$, and $L(t)$ and $U(t)$ are both nondecreasing processes. The *lower-boundary regulator process* $L \equiv \psi_L(X)$ increases only when $W(t) = 0$, while the *upper-boundary regulator process* $U \equiv \psi_U(X)$ increases only when $W(t) = K$; i.e., we require that

$$\int_0^t W(s)dL(s) = \int_0^t [K - W(s)]dU(s) = 0, \quad t \geq 0 . \tag{2.10}$$

The random variable $U(t)$ represents the quantity of fluid lost (the overflow) during the interval $[0,t]$. We are often interested in the overflow process $\{U(t) : t \geq 0\}$ as well as the workload process $\{W(t) : t \geq 0\}$.

Note that we can regard the infinite-capacity model as a special case of the finite-capacity model. When $K = \infty$, we can regard the second integral in (2.10) as implying that $U(t) = 0$ for all $t \geq 0$.

Closely paralleling Figure 5.1, for the finite-capacity model we can also depict possible realizations of the processes X and W together, as shown in Figure 5.2. As before, the potential workload process is plotted directly, but we also see the workload (buffer content) process W if we let the origin and upper barrier move according to the two heavily shaded dashed lines, which remain a distance K apart. Decreases in the dashed lines correspond to increases in the lower-barrier regulator process L, while increases in the shaded lines correspond to increases in the upper-barrier regulator process U. From the Figure 5.2, the validity of (2.9) and (2.10) is evident. Furthermore, it is evident that the two-sided reflection in (2.9) can be defined by successive applications of the one-sided reflection map in (2.5) and (2.6) corresponding to the lower and upper barriers separately. For further discussion, see Section 14.8.

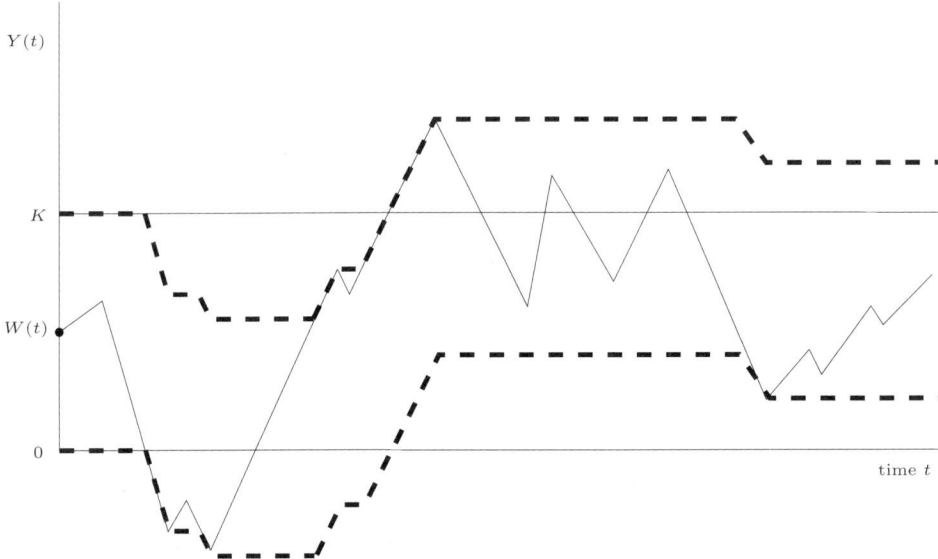

Figure 5.2. A possible realization of the potential workload process $\{X(t) : t \geq 0\}$ and the actual workload process $\{W(t) : t \geq 0\}$ with finite storage capacity K: The actual workload process appears if the origin and upper limit are the heavily shaded dashed lines always a distance K apart. As in Figure 5.1, solid line - lower dashed line = actual workload.

As in the infinite-capacity case, given K, the initial segment $\{W(s), L(s), U(s) : 0 \leq s \leq t\}$ depends on the potential-workload process X via the corresponding initial segment $\{X(s) : 0 \leq s \leq t\}$. Again, under regularity conditions, the reflection map in (2.9) taking $\{X(t) : t \geq 0\}$ into $\{(W(t), L(t), U(t) : t \geq 0\}$ is a continuous function on the space of sample paths (mapping initial segments into initial segments). Thus, stochastic-process limits for X translate into stochastic-process limits for (W, L, U), by exploiting the continuous-mapping approach with the full reflection map (ϕ_K, ψ_L, ψ_U) in a function space setting.

Let $D(t)$ represent the amount of fluid processed (not counting any overflow) during the time interval $[0, t]$. We call $\{D(t) : t \geq 0\}$ the *departure process*.

From (2.4) and (2.9),

$$\begin{aligned} D(t) &= W(0) + C(t) - W(t) - U(t) \\ &= S(t) - L(t), \quad t \geq 0 \,. \end{aligned} \quad (2.11)$$

Note that the departure process D in (2.11) is somewhat more complicated than the workload process W because, unlike the workload process, the departure process cannot be represented directly as a function of the potential workload process X or the net-input process $C - S$. Note that these processes do not capture the values of jumps in C and S that occur at the same time. Simultaneous jumps in C and S correspond to instants at which fluid arrives and some of it is instantaneously processed. The fluid that is instantaneously processed immediately upon arrival never affects the workload process. To obtain stochastic-process limits for the departure process, we will impose a condition to rule out such cancelling jumps in the limit processes associated with C and S. In particular, the departure process is considerably less complicated in the case of constant processing, as in (2.1).

We may also be interested in the *processing time* $T(t)$, i.e., the time required to process the work in the system at any time t, not counting any future input. For the processing time to correctly represent the actual processing time for the last particle of fluid in the queue, the fluid must be processed in the order of arrival. The processing time $T(t)$ is the first passage time to the level $W(t)$ by the future-available-processing process $\{S(t+u) - S(t) : u \geq 0\}$, i.e.,

$$T(t) \equiv inf\{u \geq 0 : S(t+u) - S(t) \geq W(t)\}, \quad t \geq 0 . \tag{2.12}$$

We can obtain an equivalent representation, involving a first passage time of the process S alone on the left in the infimum, if we use formula (2.9) for $W(t)$:

$$T(t) + t = t + inf\{u \geq 0 : S(t+u) - S(t) \geq X(t) + L(t) - U(t)\},$$
$$= inf\{u \geq 0 : S(u) \geq W(0) + C(t) + L(t) - U(t)\}, \quad t \geq 0 . \tag{2.13}$$

In general, the processing time is relatively complicated, but in the common case of constant processing in (2.1), $T(t)$ is a simple modification of $W(t)$, namely,

$$T(t) = W(t)/\mu, \quad t \geq 0 . \tag{2.14}$$

More generally, heavy-traffic limits also lead to such simplifications; see Section 5.9.2.

5.3. Unstable Queues

There are two main reasons queues experience congestion (which here means buildup of workload): First, the queue may be *unstable* (or overloaded); i.e., the input rate may exceed the output rate for an extended period of time, when there is ample storage capacity. Second, the queue may be stable, i.e., the long-run input rate may be less than the long-run output rate, but nevertheless short-run fluctuations produce temporary periods during which the input exceeds the output.

The unstable case tends to produce more severe congestion, but the stable case is more common, because systems are usually designed to be stable. Unstable queues typically arise in the presence of system failures. Since there is interest in system performance in the presence of failures, there is interest in the performance of unstable queues. For our discussion of unstable queues, we assume that there is unlimited storage capacity. We are interested in the buildup of congestion, which is described by the transient (or time-dependent) behavior of the queueing processes.

5.3.1. *Fluid Limits for Fluid Queues*

For unstable queues, useful insight can be gained from *fluid limits* associated with functional laws of large numbers (FLLN's). These stochastic-process limits are called fluid limits because the limit processes are deterministic functions of the form ct for some constant c. (More generally, with time-varying input and output rates, the limits could be deterministic functions of the form $\int_0^t r(s)ds$, $t \geq 0$, for some deterministic integrable function r.)

To express the FLLN's, we scale space and time both by n. As before, we use bold capitals to represent the scaled stochastic processes and associated limiting stochastic processes in the function space D. We use a hat to denote scaled stochastic processes with the fluid scaling (scaling space as well as time by n). Given the stochastic processes defined for the fluid-queue model in the previous section, form the associated scaled stochastic processes

$$
\begin{aligned}
\hat{\mathbf{C}}_n(t) &\equiv n^{-1}C(nt), \\
\hat{\mathbf{S}}_n(t) &\equiv n^{-1}S(nt), \\
\hat{\mathbf{X}}_n(t) &\equiv n^{-1}X(nt), \\
\hat{\mathbf{W}}_n(t) &\equiv n^{-1}W(nt), \\
\hat{\mathbf{L}}_n(t) &\equiv n^{-1}L(nt), \\
\hat{\mathbf{D}}_n(t) &\equiv n^{-1}D(nt), \\
\hat{\mathbf{T}}_n(t) &\equiv n^{-1}T(nt), \quad t \geq 0 \,.
\end{aligned} \quad (3.1)
$$

The continuous-mapping approach shows that FLLN's for C and S imply a joint FLLN for all the processes. As before, let $\mathbf{e}$ be the identity map, i.e., $\mathbf{e}(t) = t$, $t \geq 0$. Let $\mu \wedge \lambda \equiv min\{\mu, \lambda\}$ and $\lambda^+ \equiv max\{\lambda, 0\}$ for constants λ and μ.

We understand D to be the space $D([0, \infty), \mathbb{R})$, endowed with either the J_1 or the M_1 topology, as defined in Section 3.3. Since the limits are continuous deterministic functions, the J_1 and M_1 topologies here are equivalent to uniform convergence on compact subintervals. As in Section 3.3, we use D^k to denote the k-dimensional product space with the product topology; then $x_n \to x$, where $x_n \equiv (x_n^1, \ldots x_n^k)$ and $x \equiv (x^1, \ldots, x^k)$, if and only if $x_n^i \to x_i$ for each i.

We first establish a functional weak law of large numbers (FWLLN), involving convergence in probability or, equivalently (because of the deterministic limit), convergence in distribution (see p. 27 of Billingsley (1999)). As indicated above, we restrict attention to the infinite-capacity model. It is easy to extend the results to the finite-capacity model, provided that the capacity is allowed to increase with n, as in Section 2.3.

Theorem 5.3.1. (FWLLN for the fluid queue) *In the infinite-capacity fluid-queue model, if $\hat{\mathbf{C}}_n \Rightarrow \lambda \mathbf{e}$ and $\hat{\mathbf{S}}_n \Rightarrow \mu \mathbf{e}$ in (D, M_1), where $0 < \mu < \infty$ and $\hat{\mathbf{C}}_n$ and $\hat{\mathbf{S}}_n$ are given in (3.1), then*

5.3. Unstable Queues

$$(\hat{\mathbf{C}}_n, \hat{\mathbf{S}}_n, \hat{\mathbf{X}}_n, \hat{\mathbf{W}}_n, \hat{\mathbf{L}}_n, \hat{\mathbf{D}}_n, \hat{\mathbf{T}}_n) \Rightarrow$$
$$(\lambda \mathbf{e}, \mu \mathbf{e}, (\lambda - \mu)\mathbf{e}, (\lambda - \mu)^+\mathbf{e}, (\mu - \lambda)^+\mathbf{e}, (\lambda \wedge \mu)\mathbf{e}, (\rho - 1)^+\mathbf{e}) \quad (3.2)$$

in $(D, M_1)^7$ *for* $\rho \equiv \lambda/\mu$.

Proof. The single limits can be combined into joint limits because the limits are deterministic, by virtue of Theorem 11.4.5. So start with the joint convergence

$$(\hat{\mathbf{C}}_n, \hat{\mathbf{S}}_n, n^{-1}W(0)) \Rightarrow (\lambda \mathbf{e}, \mu \mathbf{e}, 0) \quad \text{in} \quad (D, M_1)^2 \times \mathbb{R} .$$

Since

$$\hat{\mathbf{X}}_n = \hat{\mathbf{C}}_n - \hat{\mathbf{S}}_n + n^{-1}W(0)$$

by (2.4), we can apply the continuous-mapping approach with addition, using the fact that addition on D^2 is measurable and continuous almost surely with respect to the limit process, to get the limit

$$\hat{\mathbf{X}}_n \Rightarrow \hat{\mathbf{X}} \equiv (\lambda - \mu)\mathbf{e} .$$

Specifically, we invoke Theorems 3.4.3 and 12.7.3 and Remark 12.7.1.

Then, because of (2.5) – (2.8), we can apply the simple continuous-mapping theorem, Theorem 3.4.1, with the reflection map to get

$$\hat{\mathbf{W}}_n \Rightarrow \hat{\mathbf{W}} \equiv \phi(\hat{\mathbf{X}}) = (\lambda - \mu)^+\mathbf{e}$$

and

$$\hat{\mathbf{L}}_n \Rightarrow \hat{\mathbf{L}} \equiv \psi_L(\hat{\mathbf{X}}) = (\mu - \lambda)^+\mathbf{e} ,$$

drawing on Theorems 13.5.1, 13.4.1 and 14.8.5. Then, by (2.11), we can apply the continuous-mapping approach with addition again to obtain $\hat{\mathbf{D}}_n \Rightarrow \hat{\mathbf{D}} = (\lambda \wedge \mu)\mathbf{e}$. Finally, by (2.13),

$$n^{-1}T(nt) + t = inf\{u \geq 0 : n^{-1}S(nu) \geq n^{-1}(C(nt) + L(nt) + W(0))\} \quad (3.3)$$

or, in more compact notation,

$$\hat{\mathbf{T}}_n + \mathbf{e} = \hat{\mathbf{S}}_n^{-1} \circ (\hat{\mathbf{C}}_n + \hat{\mathbf{L}}_n + n^{-1}W(0)) . \quad (3.4)$$

Hence, we can again apply the continuous-mapping approach, this time with the inverse and composition functions. As with addition used above, these functions as maps from D and $D \times D$ to D are measurable and continuous almost surely with respect to the deterministic, continuous, strictly increasing limits. Specifically, by Corollary 13.6.4 and Theorem 13.2.1, we obtain

$$\hat{\mathbf{T}}_n + \mathbf{e} \Rightarrow \mu^{-1}\mathbf{e} \circ (\lambda \mathbf{e} + (\mu - \lambda)^+\mathbf{e}) = (\rho \vee 1)\mathbf{e} ,$$

so that

$$\hat{\mathbf{T}}_n \Rightarrow (\rho - 1)^+\mathbf{e} ,$$

as claimed. By Theorem 11.4.5, all limits can be joint. ∎

From Theorem 5.3.1, we can characterize stable queues and unstable queues by the conditions $\lambda \leq \mu$ and $\lambda > \mu$, respectively, where λ and μ are the translation constants in the limits for the input process C and the available-processing process S. Equivalently, we can use the *traffic intensity* ρ, defined as

$$\rho \equiv \lambda/\mu . \tag{3.5}$$

From the relatively crude fluid-limit perspective, there is no congestion if $\rho \leq 1$; i.e., Theorem 5.3.1 implies that $\hat{\mathbf{W}}_n \Rightarrow 0\mathbf{e}$ if $\rho \leq 1$. On the other hand, if $\rho > 1$, then the workload tends to grow linearly at rate $\lambda - \mu$. Consistent with intuition, the fluid limits suggest using a simple deterministic analysis to describe congestion in unstable queues. When a queue is unstable for a significant time, the relatively simple deterministic analysis may capture the dominant congestion effect. The same reasoning applies to queues with time-dependent input and output rates that are unstable for substantial periods of time. See Oliver and Samuel (1962), Newell (1982) and Hall (1991) for discussions of direct deterministic analysis of the congestion in queues.

Ordinary weak laws of large numbers (WLLN's), such as

$$t^{-1}W(t) \Rightarrow (\lambda - \mu)^+ \quad \text{in} \quad \mathbb{R} \quad \text{as} \quad t \to \infty ,$$

follow immediately from the FWLLN's in Theorem 5.3.1 by applying the continuous-mapping approach with the projection map, which maps a function x into $x(1)$. We could not obtain these WLLN's or the stronger FWLLN's in Theorem 5.3.1 if we assumed only ordinary WLLN's for C and S, i.e., if we had started with limits such as

$$t^{-1}C(t) \Rightarrow \lambda \quad \text{in} \quad \mathbb{R} \quad \text{as} \quad t \to \infty ,$$

because we needed to exploit the continuous-mapping approach in the function space D. We cannot go directly from a WLLN to a FWLLN, because a FWLLN is strictly stronger than a WLLN.

However, we can obtain functional strong laws of large numbers (FSLLN's) starting from ordinary strong laws of large numbers (SLLN's), because a SLLN implies a corresponding FSLLN; see Theorem 3.2.1 and Corollary 3.2.1 in the Internet Supplement. To emphasize that point, we now state the SLLN version of Theorem 5.3.1. Once we go from the SLLN's for C and S to the FSLLN's, the proof is the same as for Theorem 5.3.1.

Theorem 5.3.2. (FSLLN for the fluid queue) *In the infinite-capacity fluid-queue model, if*

$$t^{-1}C(t) \to \lambda \quad \text{and} \quad t^{-1}S(t) \to \mu \quad \text{in} \quad \mathbb{R} \quad w.p.1 \quad \text{as} \quad t \to \infty ,$$

for $0 < \mu < \infty$, then

$$(\hat{\mathbf{C}}_n, \hat{\mathbf{S}}_n, \hat{\mathbf{X}}_n, \hat{\mathbf{W}}_n, \hat{\mathbf{L}}_n, \hat{\mathbf{D}}_n, \hat{\mathbf{T}}_n) \to$$
$$(\lambda\mathbf{e}, \mu\mathbf{e}, (\lambda - \mu)\mathbf{e}, (\lambda - \mu)^+\mathbf{e}, (\mu - \lambda)^+\mathbf{e}, (\lambda \wedge \mu)\mathbf{e}, (\rho - 1)^+\mathbf{e}) \tag{3.6}$$

w.p.1 in $(D, M_1)^7$ for ρ in (3.5).

5.3.2. Stochastic Refinements

We can also employ stochastic-process limits to obtain a more detailed description of congestion in unstable queues. These stochastic-process limits yield *stochastic refinements to the fluid limits* in Theorems 5.3.1 and 5.3.2 above. For the stochastic refinements, we introduce new scaled stochastic processes:

$$\begin{aligned}
\mathbf{C}_n(t) &\equiv c_n^{-1}(C(nt) - \lambda nt), \\
\mathbf{S}_n(t) &\equiv c_n^{-1}(S(nt) - \mu nt), \\
\mathbf{X}_n(t) &\equiv c_n^{-1}(X(nt) - (\lambda - \mu)nt), \\
\mathbf{W}_n(t) &\equiv c_n^{-1}(W(nt) - (\lambda - \mu)^+ nt), \\
\mathbf{L}_n(t) &\equiv c_n^{-1}(L(nt) - (\mu - \lambda)^+ nt), \\
\mathbf{D}_n(t) &\equiv c_n^{-1}(D(nt) - (\lambda \wedge \mu)nt), \\
\mathbf{T}_n(t) &\equiv c_n^{-1}(T(nt) - (\rho - 1)^+ nt), \quad t \geq 0 \;. \qquad(3.7)
\end{aligned}$$

As in the last chapter, the space scaling constants will be assumed to satisfy $c_n \to \infty$ and $n/c_n \to \infty$ as $n \to \infty$. The space-scaling constants will usually be a power, i.e., $c_n = n^H$ for $0 < H < 1$, but we allow other possibilities. In the following theorem we only discuss the cases $\rho < 1$ and $\rho > 1$. The more complex boundary case $\rho = 1$ is covered as a special case of results in the next section. Recall that D^k is the product space with the product topology; here we let the component space $D \equiv D^1$ have either the J_1 or the M_1 topology.

Since the limit processes $\mathbf{C}$ and $\mathbf{S}$ below may now have discontinuous sample paths, we need an extra condition to apply the continuous-mapping approach with addition. The extra condition depends on random sets of discontinuity points; e.g.,

$$Disc(\mathbf{S}) \equiv \{t : \mathbf{S}(t) \neq \mathbf{S}(t-)\} \;,$$

where $x(t-)$ is the left limit of the function x in D (see Section 12.2). The random *set of common discontinuity points* of $\mathbf{C}$ and $\mathbf{S}$ is $Disc(\mathbf{C}) \cap Disc(\mathbf{S})$. The *jump* in $\mathbf{S}$ associated with a discontinuity at t is $\mathbf{S}(t) - \mathbf{S}(t-)$. The required extra condition is somewhat weaker for the M_1 topology than for the J_1 topology.

Theorem 5.3.3. (FCLT's for the stable and unstable fluid queues) *In the infinite-capacity fluid queue, suppose that $c_n \to \infty$ and $c_n/n \to 0$ as $n \to \infty$. Suppose that*

$$(\mathbf{C}_n, \mathbf{S}_n) \Rightarrow (\mathbf{C}, \mathbf{S}) \quad in \quad D^2 \;, \qquad(3.8)$$

where D^2 has the product topology with the topology on D^1 being either J_1 or M_1, $\mathbf{C}_n$ and $\mathbf{S}_n$ are defined in (3.7) and

$$P(\mathbf{C}(0) = \mathbf{S}(0) = 0) = 1 \;. \qquad(3.9)$$

If the topology is J_1, assume that $\mathbf{C}$ and $\mathbf{S}$ almost surely have no common discontinuities. If the topology is M_1, assume that $\mathbf{C}$ and $\mathbf{S}$ almost surely have no common discontinuities with jumps of common sign.

(a) If $\rho < 1$ and $\mathbf{C} - \mathbf{S}$ has no positive jumps, then

$$(\mathbf{C}_n, \mathbf{S}_n, \mathbf{X}_n, \mathbf{W}_n, \mathbf{L}_n, \mathbf{D}_n) \Rightarrow$$
$$(\mathbf{C}, \mathbf{S}, \mathbf{C} - \mathbf{S}, 0\mathbf{e}, \mathbf{S} - \mathbf{C}, \mathbf{C}) \quad (3.10)$$

in D^6 with the same topology.

(b) If $\rho > 1$, then

$$(\mathbf{C}_n, \mathbf{S}_n, \mathbf{X}_n, \mathbf{W}_n, \mathbf{L}_n, \mathbf{D}_n) \Rightarrow$$
$$(\mathbf{C}, \mathbf{S}, \mathbf{C} - \mathbf{S}, \mathbf{C} - \mathbf{S}, 0\mathbf{e}, \mathbf{S}) \quad (3.11)$$

in D^6 with the same topology.

Proof. Paralleling the proof of Theorem 5.3.1 above, we start by applying condition (3.8) and Theorem 11.4.5 to obtain the joint convergence

$$(\mathbf{C}_n, \mathbf{S}_n, c_n^{-1}W(0)) \Rightarrow (\mathbf{C}, \mathbf{S}, 0) \quad \text{in} \quad D^2 \times \mathbb{R}.$$

Then, as before, we apply the continuous mapping approach with addition, now invoking the conditions on the discontinuities of $\mathbf{C}$ and $\mathbf{S}$, to get

$$(\mathbf{C}_n, \mathbf{S}_n, \mathbf{X}_n, c_n^{-1}W(0)) \Rightarrow (\mathbf{C}, \mathbf{S}, \mathbf{C} - \mathbf{S}, 0) \quad \text{in} \quad D^3 \times \mathbb{R}. \quad (3.12)$$

For the M_1 topology, we apply Theorems 3.4.3 and 12.7.3 and Remark 12.7.1. For J_1, we apply the J_1 analog of Corollary 12.7.1; see Remark 12.6.2.

The critical step is treating $\mathbf{W}_n$. For that purpose, we apply Theorem 13.5.2, for which we need to impose the extra condition that $C - S$ have no positive jumps in part (a). We also use condition (3.9), but it can be weakened. We can use the Skorohod representation theorem, Theorem 3.2.2, to carry out the argument for individual sample paths.

The limit for $\mathbf{L}_n$ in part (a) then follows from (2.6), again exploiting the continuous-mapping approach with addition. The limits for $\mathbf{L}_n$ in part (b) follows from Theorem 13.4.4, using (2.8) and condition (3.9). We can apply the convergence-together theorem, Theorem 11.4.7, to get limits for the scaled departure process $\mathbf{D}_n$. If $\lambda < \mu$, then

$$d_t(\mathbf{D}_n, \mathbf{C}_n) \le \|\mathbf{D}_n - \mathbf{C}_n\|_t \le \|c_n^{-1}W(0) - \mathbf{W}_n\|_t \Rightarrow 0$$

by (2.11), where d_t and $\|\cdot\|$ are the J_1 (or M_1) and uniform metrics for the time interval $[0,t]$, as in equations (3.2) and (3.1) of Section 3.3. If $\lambda > \mu$, then

$$d_t(\mathbf{D}_n, \mathbf{S}_n) \le \|\mathbf{D}_n - \mathbf{S}_n\|_t \le \|\mathbf{L}_n\|_t \Rightarrow 0$$

by (2.11). ∎

The obvious sufficient condition for the limit processes $\mathbf{C}$ and $\mathbf{S}$ to almost surely have no discontinuities with jumps of common sign is to have no common discontinuities at all. For that, it suffices for $\mathbf{C}$ and $\mathbf{S}$ to be independent processes without any fixed discontinuities; i.e., $\mathbf{C}$ has no fixed discontinuities if $P(t \in Disc(\mathbf{C})) = 0$ for all t.

With the J_1 topology, the conclusion can be strengthened to the strong SJ_1 topology instead of the product J_1 topology, but that is not true for M_1; see Remark 9.3.1 and Example 14.5.1.

When $\rho < 1$, we not only obtain the zero fluid limit $\hat{\mathbf{W}}_n \Rightarrow 0\mathbf{e}$ in Theorem 5.3.1, but we also obtain the zero limit $\mathbf{W}_n \Rightarrow 0\mathbf{e}$ in Theorem 5.3.3 (a) with the refined scaling in (3.7), provided that $C - S$ has no positive jumps. However, if $C - S$ has positive jumps, then the scaled workload process $\mathbf{W}_n$ fails to be uniformly negligible. That shows the impact of jumps in the limit process.

Under extra conditions, we get a limit for $\mathbf{T}_n$ jointly with the limit in Theorem 5.3.3.

Theorem 5.3.4. (FCLT for the processing time) *Let the conditions of Theorem 5.3.3 hold. If the topology is J_1, assume that S has no positive jumps.*

(a) If $\rho < 1$, then jointly with the limit in (3.10)

$$\mathbf{T}_n \Rightarrow 0\mathbf{e}$$

in D with the same topology.

(b) Suppose that $\rho > 1$. If the topology is J_1, assume that $\mathbf{C}$ and $\mathbf{S} \circ \rho\mathbf{e}$ almost surely have no common discontinuities. If the topology is M_1, assume that $\mathbf{C}$ and $\mathbf{S} \circ \rho\mathbf{e}$ almost surely have no common discontinuities with jumps of common sign. Then jointly with the limit in (3.11)

$$\mathbf{T}_n \Rightarrow \mu^{-1}(\mathbf{C} - \mathbf{S} \circ \rho\mathbf{e})$$

in D with the same topology.

Proof. We can apply Theorem 13.7.4 to treat $\mathbf{T}_n$, starting from (3.3) and (3.4). If $\lambda > \mu$, then

$$(n/c_n)(\hat{\mathbf{S}}_n - \mu\mathbf{e}, \hat{\mathbf{C}}_n + \hat{\mathbf{L}}_n + n^{-1}W(0) - \lambda\mathbf{e}) \Rightarrow (\mathbf{S}, \mathbf{C}) , \qquad (3.13)$$

because $\mathbf{L}_n \Rightarrow 0\mathbf{e}$ and $n^{-1}W(0) \to 0$. If $\lambda < \mu$, then

$$(n/c_n)(\hat{\mathbf{S}}_n - \mu\mathbf{e}, \hat{\mathbf{C}}_n + \hat{\mathbf{L}}_n + n^{-1}W(0) - \mu\mathbf{e}) \Rightarrow (\mathbf{S}, \mathbf{S}) , \qquad (3.14)$$

because, by (2.6),

$$d_t(\mathbf{C}_n + \mathbf{L}_n + c_n^{-1}W(0), \mathbf{S}_n) \le \|\mathbf{L}_n + \mathbf{X}_n\|_t = \|\mathbf{W}_n\|_t \Rightarrow 0 .$$

We can apply Theorem 13.7.4 to obtain limits for $\mathbf{T}_n$ jointly with the other limits because

$$\begin{aligned}\mathbf{T}_n &= (n/c_n)(\hat{\mathbf{T}}_n - (\rho - 1)^+\mathbf{e}) \\ &= (n/c_n)(\hat{\mathbf{S}}_n^{-1} \circ \hat{\mathbf{Z}}_n - (\rho \vee 1)\mathbf{e}) \\ &= (n/c_n)(\hat{\mathbf{S}}_n^{-1} \circ \hat{\mathbf{Z}}_n - \mu^{-1}\mathbf{e} \circ (\lambda \vee \mu)\mathbf{e})\end{aligned}$$

for appropriate $\mathbf{Z}_n$ (specified in (3.13) and (3.14) above), where $n/c_n \to \infty$ as $n \to \infty$. Theorem 13.7.4 requires condition (3.9) for $\mathbf{S}$. ∎

We regard the unstable case $\rho > 1$ as the case of primary interest for a single model. When $\rho > 1$, Theorem 5.3.3 (b) concludes that $W(t)$ obeys the same FCLT as $X(t)$. In a long time scale, the amount of reflection is negligible. Thus we obtain the approximation

$$W(t) \approx (\lambda - \mu)t + c_n \mathbf{X}(t/n) \tag{3.15}$$

for the workload, where $\mathbf{X} = \mathbf{C} - \mathbf{S}$. In the common setting of Donsker's theorem, $c_n = n^{1/2}$ and $\mathbf{X} = \sigma_X \mathbf{B}$, where $\mathbf{B}$ is standard Brownian motion. In that special case, (3.15) becomes

$$\begin{aligned} W(t) &\approx (\lambda - \mu)t + n^{1/2}\sigma_X \mathbf{B}(t/n) \\ &\approx N((\lambda - \mu)t, \sigma_X^2 t) \ . \end{aligned} \tag{3.16}$$

In this common special case, the stochastic refinement of the LLN shows that the workload obeys a CLT and, thus, the workload $W(t)$ should be approximately normally distributed with mean equal to the fluid limit $(\lambda - \mu)t$ and standard deviation proportional to $\sqrt{t}$, with the variability parameter given explicitly. With heavy tails or strong dependence (or both), but still with finite mean, the stochastic fluctuations about the mean will be greater, as is made precise by the stochastic-process limits.

Remark 5.3.1. *Implications for queues in series.* Part (a) of Theorem 5.3.3 has important implications for queues in series: If the first of two queues is stable with $\rho < 1$, then the departure process D at the first queue obeys the same FCLT as the input process C at that first queue. Thus, if we consider a heavy-traffic limit for the second queue (either because the second queue is unstable or because we consider a sequence of models for the second queue with the associated sequence of traffic intensities at the second queue approaching the critical level for stability, as in the next section), then the heavy-traffic limit at the second queue depends on the first queue only through the input stochastic process at that first queue. In other words, the heavy-traffic behavior of the second queue is the same as if the first queue were not even there. We obtain more general and more complicated heavy-traffic stochastic-process limits for the second queue only if we consider a sequence of models for both queues, and simultaneously let the sequences of traffic intensities at both queues approach the critical levels for stability, which puts us in the setting of Chapter 14. For further discussion, see Example 9.9.1, Chapter 14 and Karpelovich and Kreinin (1994). ∎

In this section we have seen how heavy-traffic stochastic-process limits can describe the congestion in an unstable queue. We have considered the relatively elementary case of constant input and output rates. Variations of the same approach apply to queues with time-varying input and output rates; see Massey and Whitt (1994a), Mandelbaum and Massey (1995), Mandelbaum, Massey and Reiman (1998) and Chapter 9 of the Internet Supplement.

5.4. Heavy-Traffic Limits for Stable Queues

We now want to establish nondegenerate heavy-traffic stochastic-process limits for stochastic processes in stable fluid queues (where the long-run input rate is less than the maximum potential output rate). (With a finite storage capacity, the workload will of course remain bounded even if the long-run input rate exceeds the output rate.)

The first heavy-traffic limits for queues were established by Kingman (1961, 1962, 1965). The treatment here is in the spirit of Iglehart and Whitt (1970a, b) and Whitt (1971a), although those papers focused on standard queueing models, as considered here in Chapters 9 and 10. An early heavy-traffic limit for finite-capacity queues was established by Kennedy (1973). See Whitt (1974b) and Borovkov (1976, 1984) for background on early heavy-traffic limits.

In order to establish the heavy-traffic stochastic-process limits for stable queues, we consider a sequence of models indexed by a subscript n, where the associated sequence of traffic intensities $\{\rho_n : n \geq 1\}$ converges to 1, the critical level for stability, as $n \to \infty$. We have in mind the case in which the traffic intensities approach 1 from below, denoted by $\rho_n \uparrow 1$, but that is not strictly required. For each n, there is a cumulative-input process C_n, an available processing process S_n, a storage capacity K_n with $0 < K_n \leq \infty$ and an initial workload $W_n(0)$ satisfying $0 \leq W_n(0) \leq K_n$. As before, we make no specific structural or stochastic assumptions about the stochastic processes C_n and S_n, so we have very general models. A more detailed model for the input is considered in Chapter 8.

To have the traffic intensity well defined in our setting, we assume that the limits

$$\lambda_n \equiv \lim_{t \to \infty} t^{-1} C_n(t) \tag{4.1}$$

and

$$\mu_n \equiv \lim_{t \to \infty} t^{-1} S_n(t) \tag{4.2}$$

exist w.p.1 for each n. We call λ_n the *input rate* and μ_n the *maximum potential output rate* for model n. (The actual output rate is the input rate minus the overflow rate.) Then the traffic intensity in model n is

$$\rho_n \equiv \lambda_n/\mu_n . \tag{4.3}$$

We will be letting $\rho_n \to 1$ as $n \to \infty$.

Given the basic model elements above, we can construct the potential-workload processes $\{X_n(t) : t \geq 0\}$, the workload processes $\{W_n(t) : t \geq 0\}$, the upper-barrier regulator (overflow) processes $\{U_n(t) : t \geq 0\}$, the lower-barrier regulator processes $\{L_n(t) : t \geq 0\}$ and the departure processes $\{D_n(t) : t \geq 0\}$ as described in Section 5.2.

We now form associated scaled processes. We could obtain fluid limits in this setting, paralleling Theorems 5.3.1 and 5.3.2, but they add little beyond the previous results. Hence we go directly to the generalizations of Theorem 5.3.3. We scale the processes as in (3.7), but now we have processes and translation constants for each

n. Let

$$\begin{align}
\mathbf{C}_n(t) &\equiv c_n^{-1}(C_n(nt) - \lambda_n nt) \,, \\
\mathbf{S}_n(t) &\equiv c_n^{-1}(S_n(nt) - \mu_n nt) \,, \\
\mathbf{X}_n(t) &\equiv c_n^{-1} X_n(nt) \,, \\
\mathbf{W}_n(t) &\equiv c_n^{-1} W_n(nt) \,, \\
\mathbf{U}_n(t) &\equiv c_n^{-1} U_n(nt) \,, \\
\mathbf{L}_n(t) &\equiv c_n^{-1} L_n(nt) \,, \quad t \geq 0 \,.
\end{align}$$
(4.4)

For the scaling constants, we have in mind $\lambda_n \to \lambda$ and $\mu_n \to \mu$ as $n \to \infty$, where $0 < \lambda < \infty$ and $0 < \mu < \infty$, with $c_n \to \infty$ and $n/c_n \to \infty$ as $n \to \infty$. As in Section 2.3, the upper barrier must grow as $n \to \infty$; specifically, we require that $K_n = c_n K$.

Our key assumption is a joint limit for $\mathbf{C}_n$ and $\mathbf{S}_n$ in (4.4). When there are limits for $\mathbf{C}_n$ and $\mathbf{S}_n$ with the translation terms involving λ_n and μ_n, the w.p.1 limits in (4.1) and (4.2) usually hold too, but (4.1) and (4.2) are actually not required. However, convergence in probability in (4.1) and (4.2) follows directly as a consequence of the convergence in distribution assumed below. Hence it is natural for the limits in (4.1) and (4.2) to hold as well.

Let (ϕ_K, ψ_U, ψ_L) be the reflection map mapping a potential-workload process X into the triple (W, U, L), as defined in Section 5.2. Here is the general heavy-traffic stochastic-process limit for stable fluid queues. It follows directly from the continuous-mapping approach using addition and reflection.

Theorem 5.4.1. (general heavy-traffic limit for stable fluid queues) *Consider a sequence of fluid queues indexed by n with capacities K_n, $0 < K_n \leq \infty$, general cumulative-input processes $\{C_n(t) : t \geq 0\}$ and general cumulative-available-processing processes $\{S_n(t) : t \geq 0\}$. Suppose that $K_n = c_n K$, $0 < K \leq \infty$, $0 \leq W_n(0) \leq K_n$,*

$$(c_n^{-1} W_n(0), \mathbf{C}_n, \mathbf{S}_n) \Rightarrow (W'(0), \mathbf{C}, \mathbf{S}) \quad in \quad \mathbb{R} \times D^2 \tag{4.5}$$

for $\mathbf{C}_n$ and $\mathbf{S}_n$ in (4.4), where the topology on D^2 is the product topology with the topology on D^1 being either J_1 or M_1, $c_n \to \infty$, $c_n/n \to 0$ and $\lambda_n - \mu_n \to 0$, so that

$$\eta_n \equiv n(\lambda_n - \mu_n)/c_n \to \eta \,, \tag{4.6}$$

where $-\infty < \eta < \infty$. If the topology is J_1, suppose that almost surely $\mathbf{C}$ and $\mathbf{S}$ have no common discontinuities. If the topology is M_1, suppose that almost surely $\mathbf{C}$ and $\mathbf{S}$ have no common discontinuities with jumps of common sign. Then, jointly with the limit in (4.5),

$$(\mathbf{X}_n, \mathbf{W}_n, \mathbf{U}_n, \mathbf{L}_n) \Rightarrow (\mathbf{X}, \mathbf{W}, \mathbf{U}, \mathbf{L}) \tag{4.7}$$

in D^4 with the same topology, where

$$\mathbf{X}(t) = W'(0) + \mathbf{C}(t) - \mathbf{S}(t) + \eta t, \quad t \geq 0 \,. \tag{4.8}$$

and
$$(\mathbf{W}, \mathbf{U}, \mathbf{L}) \equiv (\phi_K(\mathbf{X}), \psi_U(\mathbf{X}), \psi_L(\mathbf{X})) \qquad (4.9)$$
with (ϕ_K, ψ_U, ψ_L) being the reflection map associated with capacity K.

Proof. Note that
$$\mathbf{X}_n = c_n^{-1} W_n(0) + \mathbf{C}_n - \mathbf{S}_n + \eta_n \mathbf{e}, \qquad (4.10)$$
where $\mathbf{e}(t) \equiv t$ for $t \geq 0$. Thus, just as in Theorems 5.3.1 and 5.3.3 above, we can apply the continuous-mapping approach starting from the joint convergence
$$(c_n^{-1} W_n(0), \mathbf{C}_n, \mathbf{S}_n, \eta_n \mathbf{e}) \Rightarrow (W'(0), \mathbf{C}, \mathbf{S}, \eta \mathbf{e}) \qquad (4.11)$$
in $\mathbb{R} \times D^3$, which follows from (4.5), (4.6) and Theorem 11.4.5. We apply the continuous mapping theorem, Theorem 3.4.3, with addition to get $\mathbf{X}_n \Rightarrow \mathbf{X}$. (Alternatively, we could use the Skorohod representation theorem, Theorem 3.2.2.) We use the fact that addition is measurable and continuous almost surely with respect to the limit process, by virtue of the assumption about the discontinuities of $\mathbf{C}$ and $\mathbf{S}$. Specifically, for M_1 we apply Remark 12.7.1 and Theorem 12.7.3. For J_1 we apply the analog of Corollary 12.7.1; see Remark 12.6.2. Finally, we obtain the desired limit in (4.7) because
$$(\mathbf{W}_n, \mathbf{U}_n, \mathbf{L}_n) = (\phi_K(\mathbf{X}_n), \psi_U(\mathbf{X}_n), \psi_L(\mathbf{X}_n))$$
for all n. We apply the simple continuous-mapping theorem, Theorem 3.4.1, with the reflection maps, using the continuity established in Theorems 13.5.1 and 14.8.5. ∎

Just as in Theorem 5.3.3, with the J_1 topology the conclusion holds in the strong SJ_1 topology as well as the product J_1 topology. As before, the conditions on the common discontinuities of $\mathbf{C}$ and $\mathbf{S}$ hold if $\mathbf{C}$ and $\mathbf{S}$ are independent processes without fixed discontinuities.

In the standard heavy-traffic applications, in addition to (4.6), we have $\lambda_n < \mu_n$, $\mu_n \to \mu$ for $0 < \mu < \infty$, $\lambda_n - \mu_n \to 0$ and $\rho_n \equiv \lambda_n/\mu_n \uparrow 1$. However, we can have non-heavy-traffic limits by having $\lambda_n n/c_n \to a > 0$ and $\mu_n n/c_n \to b > 0$, so that $c = a - b$ and $\rho_n \equiv \lambda_n/\mu_n \to a/b$, where a/b can be any positive value. Nevertheless, the heavy-traffic limit with $\rho_n \uparrow 1$ is the principal case.

We discuss heavy-traffic stochastic-process limits for the departure process and the processing time in Section 5.9. Before discussing the implications of Theorem 5.4.1, we digress to put the heavy-traffic limits in perspective with other asymptotic methods.

Remark 5.4.1. *The long tradition of asymptotics.* Given interest in the distribution of the workload $W(t)$, we perform the heavy-traffic limit, allowing $\rho_n \uparrow 1$ as $n \to \infty$ in a sequence of models index by n, to obtain simplified expressions for the ccdf $P(W(t) > x)$ and the distribution of the entire process $\{W(t) : t \geq 0\}$. We describe the resulting approximation in the Brownian case in Section 5.7 below. To put the heavy-traffic limit in perspective, we should view it in the broader context of asymptotic methods: For general mathematical models, there is a long tradition of applying asymptotic methods to obtain tractable approximations; e.g., see Bender and Orszag (1978), Bleistein and Handelsman (1986) and Olver (1974). In this tradition are the heavy-traffic approximations and asymptotic expansions obtained by Knessl and Tier (1995, 1998) using singular perturbation methods.

For stochastic processes, it is customary to perform asymptotics. We usually simplify by letting $t \to \infty$: Under regularity conditions, we obtain $W(t) \Rightarrow W(\infty)$ as $t \to \infty$ and then we focus on the limiting steady-state ccdf $P(W(\infty) > x)$. (Or, similarly, we look for a stationary distribution of the process $\{W(t) : t \geq 0\}$.) This asymptotic step is so common that it is often done without thinking. See Asmussen (1987), Baccelli and Brémaud (1994) and Borovkov (1976) for supporting theory for basic queueing processes. See Bramson (1994a,b), Baccelli and Foss (1994), Dai (1994), Meyn and Down (1994) and Borovkov (1998) for related stability results for queueing networks and more general processes.

Given a steady-state ccdf $P(W(\infty) > x)$, we may go further and let $x \to \infty$ to find the steady-state tail-probability asymptotics. As noted in Section 2.4.1, a common case for a queue with unlimited waiting space is the exponential tail:

$$P(W(\infty) > x) \sim \alpha e^{-\eta x} \quad \text{as} \quad x \to \infty,$$

which yields the simple exponential approximation

$$P(W(\infty) > x) \approx \alpha e^{-\eta x}$$

for all x not too small; e.g., see Abate, Choudhury and Whitt (1994b, 1995).

With exponential tail-probability asymptotics, the key quantity is the asymptotic decay rate η. Since α is much less important than η, we may ignore α (i.e., let $\alpha = 1$), which corresponds to exploiting weaker large-deviation asymptotics of the form

$$\log P(W(\infty) > x) \sim -\eta x \quad \text{as} \quad x \to \infty;$$

e.g., see Glynn and Whitt (1994) and Shwartz and Weiss (1995).

The large deviations limit is associated with the concept of effective bandwidths used for admission control in communication networks; see Berger and Whitt (1998a,b), Chang and Thomas (1995), Choudhury, Lucantoni and Whitt (1996), de Veciana, Kesidis and Walrand (1995), Kelly (1996) and Whitt (1993b). The idea is to assign a deterministic quantity, called the effective bandwidth, to represent how much capacity a source will require. New sources are then admitted if the sum of the effective bandwidths does not exceed the available bandwidth.

We will also consider tail-probability asymptotics applied to the steady-state distribution of the heavy-traffic limit process. We could instead consider heavy-traffic limits after establishing tail-probability asymptotics. It is significant that the two iterated limits often agree: Often the heavy-traffic asymptotics for η as $\rho \uparrow 1$ matches the asymptotics as first $t \to \infty$ and then $x \to \infty$ in the heavy-traffic limit process; see Abate and Whitt (1994b) and Choudhury and Whitt (1994). More generally, Majewski (2000) has shown that large-deviation and heavy traffic limits for queues can be interchanged. The large-deviation and heavy-traffic views are directly linked by moderate-deviations limits, which involve a different scaling, including heavy traffic ($\rho_n \uparrow 1$); see Puhalskii (1999) and Wischik (2001b).

However, as noted in Section 2.4.1, other asymptotic forms are possible for queueing processes. We often have

$$P(W(\infty) > x) \sim \alpha x^{-\beta} e^{-\eta x} \quad \text{as} \quad x \to \infty , \tag{4.12}$$

for nonzero β; e.g., see Abate and Whitt (1997b), Choudhury and Whitt (1996) and Duffield (1997). Moreover, even other asymptotic forms are possible; e.g., see Flatto (1997).

With heavy-tailed distributions, we usually have a power tail, i.e., (4.12) holds with $\eta = 0$:

$$P(W(\infty) > x) \sim \alpha x^{-\beta} \quad \text{as} \quad x \to \infty .$$

When the steady-state distribution of the workload in a queue has a power tail, the heavy-traffic theory usually is consistent; i.e., the heavy-traffic limits usually capture the relevant tail asymptotics; see Section 8.5. For more on power-tail asymptotics, see Abate, Choudhury and Whitt (1994a), Duffield and O'Connell (1995), Boxma and Dumas (1998), Sigman (1999), Jelenković (1999, 2000), Likhanov and Mazumdar (2000), Whitt (2000c) and Zwart (2000, 2001).

With the asymptotic form in (4.12), numerical transform inversion can be used to calculate the asymptotic constants η, β and α from the Laplace transform, as shown in Abate, Choudhury, Lucantoni and Whitt (1995) and Choudhury and Whitt (1996). When $\eta = 0$, we can transform the distribution into one with $\eta > 0$ to perform the computation; see Section 5 of Abate, Choudhury and Whitt (1994a) and Section 3 of Abate and Whitt (1997b). See Abate and Whitt (1996, 1999a,b,c) for ways to construct heavy-tailed distributions with tractable Laplace transforms.

And there are many other kinds of asymptotics that can be considered. For example, with queueing networks, we can let the size of the network grow; e.g., see Whitt (1984e, 1985c), Kelly (1991), Vvedenskaya et al. (1996), Mitzenmacher (1996), and Turner (1998) ■

5.5. Heavy-Traffic Scaling

A primary reason for establishing the heavy-traffic stochastic-process limit for stable queues in the previous section is to generate approximations for the workload stochastic process in a stable fluid-queue model. However, it is not exactly clear how to do this, because in applications we have one given queueing system, not a sequence of queueing systems. The general idea is to regard our given queueing system as the nth queueing system in the sequence of queueing systems, but what should the value of n be?

The standard way to proceed is to choose n so that the traffic intensity ρ_n in the sequence of systems matches the actual traffic intensity in the given system. That procedure makes sense because the traffic intensity ρ is a robust first-order characterization of the system, not depending upon the stochastic fluctuations about long-term rates. As can be seen from (4.1) – (4.3) and Theorems 5.3.1 and 5.3.2, the traffic intensity appears in the fluid scaling. Thus, it is natural to think of the heavy-traffic stochastic-process limit as a way to capture the second-order variability effect beyond the traffic intensity ρ.

In controlled queueing systems, it may be necessary to solve an optimization problem to determine the relevant traffic intensity. Then the traffic intensity can not be regarded as given, but instead must be derived; see Harrison (2000, 2001a,b). After deriving the traffic intensity, we may proceed with further heavy-traffic analysis. Here we assume that the traffic intensity has been determined.

If we decide to choose n so that the traffic intensity ρ_n matches the given traffic intensity, then it is natural to index the models by the traffic intensity ρ from the outset, and then consider the limit as $\rho \uparrow 1$ (with $\uparrow$ indicating convergence upward from below). In this section we show how we can index the queueing models by the traffic intensity ρ instead of an arbitrary index n. We also discuss the applied significance of the scaling of space and time in heavy-traffic stochastic-process limits. We focus on the general fluid model considered in the last two sections, but the discussion applies to even more general models.

5.5.1. The Impact of Scaling Upon Performance

Let $W_\rho(t)$ denote the workload at time t in the infinite-capacity fluid-queue model with traffic intensity ρ. Let $c(\rho)$ and $b(\rho)$ denote the functions that scale space and time, to be identified in the next subsection. Then the scaled workload process is

$$\mathbf{W}_\rho(t) \equiv c(\rho)^{-1} W_\rho(b(\rho)t) \quad t \geq 0 \ . \tag{5.1}$$

The heavy-traffic stochastic-process limit can then be expressed as

$$\mathbf{W}_\rho \Rightarrow \mathbf{W} \quad \text{in} \quad (D, M_1) \quad \text{as} \quad \rho \uparrow 1 \ , \tag{5.2}$$

where $D \equiv D([0,\infty), \mathbb{R})$ and $\{\mathbf{W}(t) : t \geq 0\}$ is the limiting stochastic process. In the limits we consider, $c(\rho) \uparrow \infty$ and $b(\rho) \uparrow \infty$ as $\rho \uparrow 1$. Thus, the heavy-traffic stochastic-process limit provides a macroscopic view of uncertainty.

Given the heavy-traffic stochastic-process limit for the workload process in (5.2), the natural approximation is obtained by replacing the limit by approximate equality in distribution; i.e.,

$$c(\rho)^{-1} W_\rho(b(\rho)t) \approx \mathbf{W}(t), \quad t \geq 0 \ ,$$

or, equivalently, upon moving the scaling terms to the right side,

$$W_\rho(t) \approx c(\rho) \mathbf{W}(b(\rho)^{-1} t), \quad t \geq 0 \ , \tag{5.3}$$

where $\approx$ means approximately equal to in distribution (as stochastic processes).

We first discuss the applied significance of the two scaling functions $c(\rho)$ and $b(\rho)$ appearing in (5.1) and (5.3). Then, afterwards, we show how to identify these scaling functions for the fluid-queue model.

The scaling functions $c(\rho)$ and $b(\rho)$ provide important insight into queueing performance. The space-scaling factor $c(\rho)$ is relatively easy to interpret: The workload process (for times not too small) tends to be of order $c(\rho)$ as $\rho \uparrow 1$. The time-scaling factor $b(\rho)$ is somewhat more subtle: The workload process tends to make significant changes over time scales of order $b(\rho)$ as $\rho \uparrow 1$. Specifically, the change in the workload process, when adjusted for space scaling, from time $t_1 b(\rho)$ to time $t_2 b(\rho)$ is approximately characterized (for suitably high ρ) by the change in the limit process $\mathbf{W}$ from time t_1 to time t_2.

Consequently, over time intervals of length less than $b(\rho)$ the workload process tends to remain unchanged. Specifically, if we consider the change in the workload process W_ρ from time $t_1 b(\rho)$ to time $t_2(\rho)$, where $t_2(\rho) > t_1 b(\rho)$ but $t_2(\rho)/b(\rho) \to 0$ as $\rho \uparrow 1$, and if the limit process $\mathbf{W}$ is almost surely continuous at time t_1, then we conclude from the heavy-traffic limit in (5.2) that the relative change in the workload process over the time interval $[t_1 b(\rho), t_2(\rho)]$ is asymptotically negligible as ρ increases.

On the other hand, over time intervals of length greater than $b(\rho)$, the workload process W_ρ tends to approach its equilibrium steady-state distribution (assuming that both $\mathbf{W}(t)$ and $W_\rho(t)$ approach steady-state limits as $t \to \infty$). Specifically, when $t_2(\rho) > t_1 b(\rho)$ and $t_2(\rho)/b(\rho) \to \infty$ as $\rho \uparrow 1$, the workload process at time $t_2(\rho)$ tends to be in steady state, independent of its value at time $t_1 b(\rho)$. Thus, if we are considering the workload process over the time interval $[t_1 b(\rho), t_2(\rho)]$, we could use steady-state distributions to describe the distribution of $W_\rho(t_2(\rho))$, ignoring initial conditions at time $t_1 b(\rho)$ (In that step, we assume that $\mathbf{W}(t)$ approaches a steady-state distribution as $t \to \infty$, independent of initial conditions.) Thus, under regularity conditions, the time scaling in the heavy-traffic limit reveals the rate of convergence to steady state, as a function of the traffic intensity.

The use of steady-state distributions tends to be appropriate only over time intervals of length greater than $b(\rho)$. Since $b(\rho) \uparrow \infty$ as $\rho \uparrow 1$, transient (time-dependent) analysis becomes more important as ρ increases. Fortunately, the heavy-traffic stochastic-process limits provide a basis for analyzing the approximate transient behavior of the workload process as well as the approximate steady-state behavior. As indicated above, the change in the workload process (when adjusted for space scaling) between times $t_1 b(\rho)$ and $t_2 b(\rho)$ is approximately characterized by the change in the limit process $\mathbf{W}$ from time t_1 to time t_2. Fortunately, the limit processes often are sufficiently tractable that we can calculate such transient probabilities.

Remark 5.5.1. *Relaxation times.* The approximate time for a stochastic process to approach its steady-state distribution is called the *relaxation time*; e.g., see Section III.7.3 of Cohen (1982). The relaxation time can be defined in a variety of ways, but it invariably is based on the limiting behavior as $t \to \infty$ for fixed ρ. In the relatively nice light-tailed and weak-dependent case, it often can be shown, under regularity conditions, that

$$E[f(W_\rho(t))] - E[f(W_\rho(\infty))] \sim g(t,\rho) e^{-t/r(\rho)} \quad \text{as} \quad t \to \infty \,, \qquad (5.4)$$

for various real-valued functions f, with the functions g and r in general depending upon f. The standard asymptotic form for the second-order term g is $g(t,\rho) \sim c(\rho)$ or $g(t,\rho) \sim c(\rho) t^{\beta(\rho)}$ as $t \to \infty$. When (5.4) holds with such a g, $r(\rho)$ is called the relaxation time. Of course, a stochastic process that starts away from steady state usually does not reach steady state in finite time. Instead, it gradually approaches steady state in a manner such as described in (5.4). More properly, we should interpret $1/r(\rho)$ as the rate of approach to steady state.

With light tails and weak dependence, we usually have

$$r(\rho)/b(\rho) \to c \quad \text{as} \quad \rho \uparrow 1 ,$$

where c is a positive constant; i.e., the heavy-traffic time-scaling usually reveals the asymptotic form (as $\rho \uparrow 1$) of the relaxation time.

However, with heavy tails and strong dependence, the approach to steady state is usually much slower than in (5.4); see Asmussen and Teugels (1996) and Mikosch and Nagaev (2000). In these other settings, as well as in the light-tailed weak-dependent case, the time scaling in the heavy-traffic limit usually reveals the asymptotic form (as $\rho \uparrow 1$) of the approach to steady state. Thus, the heavy-traffic time scaling can provide important insight into the rate of approach to steady state. With heavy tails and strong dependence, the heavy-traffic limits show that transient analysis becomes more important. ∎

5.5.2. Identifying Appropriate Scaling Functions

We now consider how to identify appropriate scaling functions $b(\rho)$ and $c(\rho)$ in (5.1). We can apply the general stochastic-process limit in Theorem 5.4.1 to determine appropriate scaling functions. Specifically, the scaling functions $b(\rho)$ and $c(\rho)$ depend on the input rates λ_n, the output rates μ_n and the space-scaling factors c_n appearing in Theorem 5.4.1. The key limit is (4.6), which determines the drift η of the unreflected limit process **X**.

To cover most cases of practical interest, we make *three additional assumptions* about the scaling as a function of n in (4.4): First, we assume that the space scaling is by a simple power. Specifically, we assume that

$$c_n \equiv n^H \quad \text{for} \quad 0 < H < 1 . \tag{5.5}$$

(See Section 4.2 for discussion about the possible scaling functions.) We need the condition on the exponent H in (5.5) in order to have $c_n \to \infty$ and $c_n/n \to 0$ as $n \to \infty$, as assumed in Theorem 5.4.1.

Second, we assume that the translation terms λ_n and μ_n in (4.4) converge to finite positive limits as $n \to \infty$. In view of condition (4.6) in Theorem 5.4.1, it suffices to assume only that

$$\mu_n \to \mu \quad \text{as} \quad n \to \infty , \tag{5.6}$$

where $0 < \mu < \infty$.

Third, we assume that the basic limit in (4.6) holds with $\eta < 0$. That implies that the traffic intensities ρ_n are less than 1 for all n sufficiently large. Now, if we combine (4.6), (5.5) and (5.6) (and divide by μ_n in (4.6)), we obtain the condition

$$n^{1-H}(1 - \rho_n) \to \zeta \equiv -\eta/\mu > 0 \tag{5.7}$$

for $0 < \zeta < \infty$. From (5.7), we obtain the associated limit

$$n(1-\rho_n)^{1/(1-H)} \to \zeta^{1/(1-H)} \quad \text{as} \quad n \to \infty \tag{5.8}$$

or, equivalently,

$$n \sim \left(\frac{\zeta}{1-\rho_n}\right)^{\frac{1}{1-H}} \quad \text{as} \quad n \to \infty. \tag{5.9}$$

Thus the *canonical forms of the scaling functions* are

$$b(\rho) \equiv n \equiv \left(\frac{\zeta}{1-\rho}\right)^{\frac{1}{1-H}} \tag{5.10}$$

and

$$c(\rho) \equiv n^H \equiv \left(\frac{\zeta}{1-\rho}\right)^{\frac{H}{1-H}} \tag{5.11}$$

for $\zeta = -\eta/\mu$ as in (5.7).

To summarize, when the net-input process and potential-workload process satisfies a FCLT with time scaling by n and space scaling by n^H, the associated scaled workload processes, as functions of the traffic intensity ρ, have a heavy-traffic limit with the time-scaling function in (5.10) and space-scaling function in (5.11); i.e., as functions of ρ, the *time-scaling exponent* is $1/(1-H)$ and the *space-scaling exponent* is $H/(1-H)$.

The initial space-scaling exponent H (the Hurst parameter) depends on the burstiness; see Chapter 4. As the burstiness increases, H increases. Of course, the standard case, considered in most heavy-traffic limits for queues, is $H = 1/2$. The standard case with $H = 1/2$ occurs with Donsker's theorem and its variants with weak dependence and light tails, as discussed in Sections 4.3 and 4.4. Since $H = 1/2$ is the standard case, it is also the reference case. Values of H with $1/2 < H < 1$ indicate greater burstiness associated with heavy tails or strong positive dependence (or both). Values of H with $0 < H < 1/2$ are associated with strong negative dependence, as might occur with strong traffic shaping, e.g., scheduling.

From (5.10) and (5.11), we see that the scaling functions $b(\rho)$ and $c(\rho)$ increase rapidly as $H \uparrow 1$ for ρ near 1. Indeed, the scaling exponents increase as H increases from 0 toward 1. To make that important point clear, we display the two scaling exponents for a range of H values in Table 5.1.

Since H increases as the burstiness increases, we see that increased burstiness leads to greater scaling functions $c(\rho)$ and $b(\rho)$ for any given traffic intensity ρ. The larger value of $c(\rho)$ shows that the buffer content is likely to be larger (or that one needs larger buffers to avoid overflow). The larger values of $b(\rho)$ show that the time scales for statistical regularity are longer. When there is larger burstiness, transient analysis becomes more important in contrast to steady-state analysis.

From a practical engineering perspective, the analysis of the heavy-traffic scaling functions $b(\rho)$ and $c(\rho)$ indicates that, when exceptional variability is a possibility in a queueing setting, attention should be focused on the space-scaling exponent H

H	time-scaling exponent $1/(1-H)$	space-scaling exponent $H/(1-H)$
1/101	101/100	1/100
1/11	11/10	1/10
1/5	5/4	1/4
1/3	3/2	1/2
1/2	2	1
2/3	3	2
4/5	5	4
10/11	11	10
100/101	101	100

Table 5.1. The time-scaling and space-scaling exponents as a function of the Hurst parameter H.

for the net-input process as well as the traffic intensity ρ. Second-order refinements are provided by the constant ζ appearing in (5.7), (5.10) and (5.11) and the limit process **W** appearing in (5.2) and (5.3).

5.6. Limits as the System Size Increases

In this section we see how heavy-traffic stochastic-process limits for stable fluid queues change as the system size increases. The heavy-traffic limits thus show how performance scales as the system size increases. We will see that *the performance impact depends on the way that the system size increases*. We start with a base infinite-capacity fluid queue for which there is a heavy-traffic stochastic-process limit. We assume that there is a limit for the potential-workload processes of the form $\mathbf{X}_n \Rightarrow \mathbf{X}$, where

$$\mathbf{X}_n(t) \equiv n^{-H} X_n(nt), \quad t \geq 0, \qquad (6.1)$$

for $0 < H < 1$ and

$$\mathbf{X}(t) \equiv \eta t + \mathbf{Y}(t), \quad t \geq 0, \qquad (6.2)$$

with $\{\mathbf{Y}(t) : t \geq 0\}$ being H-self-similar, i.e.,

$$\{\mathbf{Y}(ct) : t \geq 0\} \stackrel{\mathrm{d}}{=} \{c^H \mathbf{Y}(t) : t \geq 0\} \qquad (6.3)$$

as in (2.5) in Section 4.2. Of course, there is a corresponding heavy-traffic stochastic-process limit for the workload process,

$$\mathbf{W}_n \Rightarrow \mathbf{W} \equiv \phi(\mathbf{X}),$$

where

$$\mathbf{W}_n \equiv \phi(\mathbf{X}_n).$$

It will be convenient to focus on the potential-workload processes $\mathbf{X}_n$ instead of the workload processes $\mathbf{W}_n$. We will focus on the scale factor σ when the limit process has the representation $\mathbf{X} \equiv \eta \mathbf{e} + \sigma \mathbf{Y}$. For fixed η and $\mathbf{Y}$, the associated reflection $\{\mathbf{W}(t) : t \geq 0\}$ tends to be increasing in σ (in a stochastic sense). For example, if $\mathbf{Y}$ is standard Brownian motion and $\eta < 0$, then the steady-state quantity $\mathbf{W}(\infty)$ has mean $\sigma^2/2|\eta|$; see (7.13) below. More generally, σ serves as a quantitative measure of the variability (for fixed $\mathbf{Y}$). The general principle is: *Increased variability in the potential workload process leads to larger workloads*, where "larger" is measured appropriately, e.g., by the mean or by a form of stochastic order.

We consider three ways to make the system larger: scaling space, scaling time and creating independent replicas. Let the *size-increase factor* be a positive integer m. We *scale space* (make it larger) by considering $m\mathbf{X}_n$; we *scale time* (make it faster) by considering $\mathbf{X}_n \circ m\mathbf{e}$; and we *create independent replicas* by considering $\mathbf{X}_{n,1} + \cdots + \mathbf{X}_{n,m}$, where $\mathbf{X}_{n,1}, \ldots, \mathbf{X}_{n,m}$ are m IID copies of the original stochastic processes $\mathbf{X}_n$.

For communication network applications, it is useful to think of constant deterministic processing, whose rate is being increased by a factor m. Scaling space then amounts to making the files or packets m times bigger to match the increased capacity. Scaling time amounts to sending the same input m times faster. Creating independent replicas means superposing (adding) m independent sources, each distributed as the original one. (We will be considering heavy-traffic limits for superposition input processes further in later chapters; see Sections 8.7.1, 9.4 and 9.8.)

In manufacturing, scaling space can also occur. Scaling space occurs in batching and unbatching; e.g., see Sections 8.5 and 9.3 of Hopp and Spearman (1996).

When we scale space, the limit process is

$$m\mathbf{X} = m\eta\mathbf{e} + m\mathbf{Y} \, . \tag{6.4}$$

When we scale time, the limit process is

$$\begin{aligned} \mathbf{X} \circ m\mathbf{e} &= m\eta\mathbf{e} + \mathbf{Y} \circ m\mathbf{e} \\ &\stackrel{\mathrm{d}}{=} m\eta\mathbf{e} + m^H \mathbf{Y} \, . \end{aligned} \tag{6.5}$$

When we create independent replicas, the limit process is

$$\sum_{i=1}^{m} \mathbf{X}_i = m\eta\mathbf{e} + \sum_{i=1}^{m} \mathbf{Y}_i \, . \tag{6.6}$$

The rate of the limit process increases by the same factor m in all three cases, but the impact on the stochastic component, characterized by the stochastic process $\mathbf{Y}$, is different for the three methods. Scaling time by m produces smaller stochastic fluctuations than scaling space by m, in the sense that the scale factors before $\mathbf{Y}$ in (6.4) and (6.5) are ordered: $m^H < m$. The advantage of time scaling over space scaling increases as H decreases (when the variability is smaller).

The impact of creating independent replicas depends on the properties of the stochastic process $\mathbf{Y}$. If $\mathbf{Y}$ is a Lévy process (has stationary and independent increments), then a concatenation of independent versions is equivalent to a longer version, i.e.,

$$\sum_{i=1}^{m} \mathbf{Y}_i \stackrel{d}{=} \mathbf{Y} \circ m\mathbf{e} . \tag{6.7}$$

Thus, if $\mathbf{Y}$ is a Lévy process, creating independent replicas is equivalent to scaling time, which we have seen produces better performance than scaling space.

On the other hand, suppose that $\mathbf{Y}$ is fractional Brownian motion (FBM), the principal example of a nonLévy limit process in Chapter 4. Since FBM is not a Lévy process, (6.7) does not hold. When $\mathbf{Y}$ is FBM, both $\mathbf{Y}$ and $\sum_{i=1}^{m} \mathbf{Y}_i$ are zero-mean Gaussian processes. For zero-mean Gaussian processes, it is natural to focus on the variances. With independent replicas, the variance is

$$Var \sum_{i=1}^{m} \mathbf{Y}_i(t) = m(Var\mathbf{Y}(t)), \quad t \geq 0 . \tag{6.8}$$

In contrast, with time scaling, because of the H-self-similarity, the variance is

$$Var\mathbf{Y}(mt) = Var(m^H \mathbf{Y}(t)) = m^{2H}(Var\mathbf{Y}(t)) . \tag{6.9}$$

Hence, the variance with independent replicas is less than, equal to or greater than the variance with time scaling, respectively, when $H > 1/2$, $H = 1/2$ or $H < 1/2$.

More generally, we can compare all three methods using the variance when $\mathbf{Y}(t)$ has finite variance. Using the H-self-similarity of $\mathbf{Y}$, we obtain

$$\begin{aligned} Var(m\mathbf{Y}(t)) &= m^2(Var\mathbf{Y}(t)), \\ Var\mathbf{Y}(mt) &= m^{2H}(Var\mathbf{Y}(t)), \\ Var \sum_{i=1}^{m} \mathbf{Y}_i(t) &= m(Var\mathbf{Y}(t)) . \end{aligned} \tag{6.10}$$

For $H < 1/2$, time scaling produces least variability; for $H > 1/2$, independent replicas produces least variability.

It is interesting to compare one large system (increased by factor m) to m separate independent systems, distributed as the original one. We say that there is *economy of scale* when the workload in the single large system tends to be smaller than the sum of the workloads in the separate systems. With finite variances, there is economy of scale when the ratio of the standard deviation to the mean is decreasing in m. From (6.10), we see that there is economy of scale with time scaling and independent replicas, but not with space scaling. For communication networks, the economy of scale associated with independent replicas is often called the *multiplexing gain*, i.e., the gain in efficiency from statistical multiplexing (combining independent sources). See Smith and Whitt (1981) for stochastic comparisons demonstrating the economy of scale in queueing systems. See Chapters 8 and 9 for more discussion.

Example 5.6.1. *Brownian motion.* Suppose that $\mathbf{X} = \eta \mathbf{e} + \sigma \mathbf{B}$, where $\eta < 0$, $\sigma > 0$ and $\mathbf{B}$ is standard Brownian motion. As noted above, the associated

RBM has steady-state mean $\sigma^2/2|\eta|$. With space scaling, time scaling and creating independent replicas, the steady-state mean of the RBM's become

$$m\sigma^2/2|\eta|, \quad \sigma^2/2|\eta| \quad \text{and} \quad \sigma^2/2|\eta|,$$

respectively. Thus, with space scaling, the steady-state mean is the same as the total steady-state mean in m separate systems. Otherwise, the steady-state mean is less by the factor m. ∎

In this section we have considered three different ways that the fluid queue can get larger. We have shown that the three different ways have different performance implications. It is important to realize, however, that in applications the situation may be more complicated. For example, a computer can be made larger by adding processors, but there invariably are limitations that prevent the maximum potential output rate from being proportional to the number of processors as the number of processors increases.

If the jobs are processed one at a time, then we must exploit *parallel processing*, i.e., the processors must share the processing of each job. However, usually a proportion of each job cannot be parallelized. Thus, with parallel processing, the capacity tends to increase nonlinearly with the number of processors; the marginal gain in capacity tends to be decreasing in m; e.g., see Amdahl (1967) and Chapters 5-7 and 14 of Gunther (1998). With deterministic processing, our analysis would still apply, provided that we interpret m as the actual increase in processing rate.

Even if we can accurately estimate the effective processing rate, there remain difficulties in applying the analysis in this section, because with parallel processing, it may not be appropriate to regard the processing as deterministic. It then becomes difficult to determine how the available-processing process S and its FCLT should change with m.

5.7. Brownian Approximations

In this section we apply the general heavy-traffic stochastic-process limits in Section 5.4 to establish Brownian heavy-traffic limits for fluid queues. In particular, under extra assumptions (corresponding to light tails and weak dependence), the limit for the normalized cumulative-input process will be a zero-drift Brownian motion (BM) and the limit for the normalized workload process will be a reflected Brownian motion (RBM), usually with negative drift.

The general heavy-traffic stochastic-process limits in Section 5.4 also generate nonBrownian approximations corresponding to the nonBrownian FCLT's in Chapter 4, but we do not discuss them here. We discuss approximations associated with stable Lévy motion and fractional Brownian approximations in Chapter 8.

Since Brownian motion has continuous sample paths and the reflection map maps continuous functions into continuous functions, RBM also has continuous sample paths. However, unlike Brownian motion, RBM does not have independent increments. But RBM is a Markov process. As a (well-behaved) Markov process with continuous sample paths, RBM is a diffusion process.

Harrison (1985) provides an excellent introduction to Brownian motion and "Brownian queues," showing how they can be analyzed using martingales and the Ito stochastic calculus. Other good introductions to Brownian motion and diffusion processes are Glynn (1990), Karatzas and Shreve (1988) and Chapter 15 of Karlin and Taylor (1981). Borodin and Salminen (1996) provide many Brownian formulas. Additional properties of RBM are contained in Abate and Whitt (1987a-b, 1988a-d).

5.7.1. The Brownian Limit

If $\mathbf{B}$ is a standard Brownian motion, then $\{y + \eta t + \sigma \mathbf{B}(t) : t \geq 0\}$ is a Brownian motion with *drift* η, *diffusion coefficient* (or variance coefficient) σ^2 and initial position y. We have the following elementary application of Section 5.4.

Theorem 5.7.1. (general RBM limit) *Suppose that the conditions of Theorem 5.4.1 are satisfied with* $W'(0) = y$, $c_n = \sqrt{n}$ *and* $(\mathbf{C}, \mathbf{S})$ *two-dimensional zero-drift Brownian motion with covariance matrix*

$$\Sigma = \begin{pmatrix} \sigma_C^2 & \sigma_{C,S}^2 \\ \sigma_{C,S}^2 & \sigma_S^2 \end{pmatrix} . \tag{7.1}$$

Then the conclusions of Theorems 5.4.1, 5.9.1 and 5.9.3 (b) hold with

$$(\mathbf{W}, \mathbf{U}, \mathbf{L}) \equiv (\phi_K(\mathbf{X}), \psi_U(\mathbf{X}), \psi_L(\mathbf{X}))$$

being reflected Brownian motion, i.e.,

$$\mathbf{X}(t) \stackrel{\mathrm{d}}{=} y + \eta t + \sigma_X \mathbf{B}(t) \tag{7.2}$$

for standard Brownian motion $\mathbf{B}$, *drift coefficient* η *in (4.6) and diffusion coefficient*

$$\sigma_X^2 = \sigma_C^2 + \sigma_S^2 - 2\sigma_{C,S}^2 . \tag{7.3}$$

Proof. Under the assumption on $(\mathbf{C}, \mathbf{S})$, $\mathbf{C} - \mathbf{S}$ is a zero-drift Brownian motion with diffusion coefficient σ_X^2 in (7.3). ∎

As indicated in Section 5.5, we can also index the queueing systems by the traffic intensity ρ and let $\rho \uparrow 1$. With $n = \zeta^2/(1-\rho)^2$ as in (5.10), the heavy-traffic limit becomes

$$\{\zeta^{-1}(1-\rho)W_\rho(t\zeta^2/(1-\rho)^2) : t \geq 0\} \Rightarrow \phi_K(\tilde{\mathbf{X}}) \quad \text{as} \quad \rho \uparrow 1 , \tag{7.4}$$

where W_ρ is the workload process in model ρ, which has output rate μ and traffic intensity ρ, and

$$\tilde{\mathbf{X}}(t) \stackrel{\mathrm{d}}{=} y - \zeta\mu t + \mathbf{B}(\sigma_X^2 t), \quad t \geq 0 , \tag{7.5}$$

with $\mathbf{B}$ being a standard Brownian motion. The capacity in model ρ is $K_\rho = \zeta K/(1-\rho)$.

We have freedom in the choice of the parameter ζ. If we let
$$\zeta = \sigma_X^2/\mu , \qquad (7.6)$$
and rescale time by replacing t by t/σ_X^2, then the limit in (7.4) can be expressed as
$$\{\sigma_X^{-2}\mu(1-\rho)W_\rho(t\sigma_X^2/\mu^2(1-\rho)^2) : t \geq 0\} \Rightarrow \phi_K(\mathbf{X}) \qquad (7.7)$$
where $\mathbf{X}$ is canonical Brownian motion with drift coefficient -1 and variance coefficient 1, plus initial position y, i.e.,
$$\{\mathbf{X}(t) : t \geq 0\} \stackrel{d}{=} \{y - t + \mathbf{B}(t) : t \geq 0\} .$$
That leads to the *Brownian approximation*
$$\{W_\rho(t) : t \geq 0\} \approx \{\sigma_X^2 \mu^{-1}(1-\rho)^{-1}\phi_K(\mathbf{X})(\mu^2(1-\rho)^2 t/\sigma_X^2) : t \geq 0\} , \qquad (7.8)$$
where $\mathbf{X}$ is again canonical Brownian motion.

Remark 5.7.1. *The impact of variability* The Brownian limit and the Brownian approximation provide insight into the way variability in the basic stochastic processes C and S affect queueing performance. In the heavy-traffic limit, the stochastic behavior of the processes C and S, beyond their rates λ and μ, affect the Brownian approximation solely via the single variance parameter σ_X^2 in (7.3), which can be identified from the CLT for $C - S$. For further discussion, see Section 9.6.1. ∎

We now show how the Brownian approximation applies to the steady-state workload.

5.7.2. The Steady-State Distribution.

The heavy-traffic limit in Theorem 5.7.1 does not directly imply that the steady-state distributions converge. Nevertheless, from (7.8), we obtain an approximation for the steady-state workload, namely,
$$W_\rho(\infty) \approx \frac{\sigma_X^2}{\mu(1-\rho)}\phi_K(\mathbf{X})(\infty) . \qquad (7.9)$$

Conditions for the convergence of steady-state distributions in heavy traffic have been established by Szczotka (1986, 1990, 1999).

We now give the steady-state distribution of RBM with two-sided reflection; see p. 90 of Harrison (1985). We are usually interested in the case of negative drift, but we allow positive drift as well when $K < \infty$.

Theorem 5.7.2. (steady-state distribution of RBM) *Let $\{\mathbf{W}(t) : t \geq 0\}$ be one-dimensional RBM with drift coefficient η, diffusion coefficient σ^2, initial value y and two-sided reflection at 0 and K. Then*
$$\mathbf{W}(t) \Rightarrow \mathbf{W}(\infty) \quad in \quad \mathbb{R} \quad as \quad t \to \infty ,$$

where $\mathbf{W}(\infty)$ has pdf

$$f(x) \equiv \begin{cases} 1/K & \text{if } \eta = 0 \\ \\ \frac{\theta e^{\theta x}}{e^{\theta K}-1} & \text{if } \eta \neq 0, \end{cases} \qquad (7.10)$$

with mean

$$E\mathbf{W}(\infty) = \begin{cases} K/2, & \text{if } \eta = 0 \\ \\ \frac{K}{1-e^{-\theta K}} - \frac{1}{\theta} & \text{if } \eta \neq 0 \end{cases} \qquad (7.11)$$

for

$$\theta \equiv 2\eta/\sigma^2 \qquad (7.12)$$

Note that the steady-state distribution of RBM in (7.10) depends only on the two parameters θ in (7.12) and K. The steady-state distribution is uniform in the zero-drift case; the steady-state distribution is an exponential distribution with mean $-\theta^{-1} = \sigma^2/2|\eta|$, conditional on being in the interval $[0, K]$, when $\eta < 0$ and $\theta < 0$; $K - \mathbf{W}(\infty)$ has an exponential distribution with mean $\theta^{-1} = \sigma^2/2\eta$, conditional on being in the interval $[0, K]$, when $\eta > 0$ and $\theta > 0$. Without the upper barrier at K, a steady-state distribution exists if and only if $\eta < 0$, in which case it is the exponential distribution with mean $-\theta^{-1}$ obtained by letting $K \to \infty$ in (7.10). As K gets large, the tails of the exponential distributions rapidly become negligible so that

$$E\mathbf{W}(\infty) \approx \begin{cases} |\theta|^{-1} & \text{if } \eta < 0 \\ \\ K - |\theta|^{-1} & \text{if } \eta > 0. \end{cases} \qquad (7.13)$$

Let us now consider the approximation indicated by the limit. Since $n^{-1/2}\mathbf{W}_n(nt) \Rightarrow \mathbf{W}(t)$, we use the approximations

$$\mathbf{W}_n(t) \approx \sqrt{n}\mathbf{W}(t/n) \qquad (7.14)$$

and

$$\mathbf{W}_n(\infty) \approx \sqrt{n}\mathbf{W}(\infty). \qquad (7.15)$$

Thus, when $K = \infty$, the Brownian approximation for $W_\rho(\infty)$ is an exponential random variable with mean

$$E[W_\rho(\infty)] \approx \frac{\sigma_X^2}{2\mu(1-\rho)}. \qquad (7.16)$$

The RBM's $\phi_K(\tilde{\mathbf{X}})$ in (7.4) and $\phi_K(\mathbf{X})$ in (7.7) and (7.8) are the Brownian queues, which serve as the approximating models. From the approximations in (7.8) – (7.16), we see the impact upon queueing performance of the processes C and S in the heavy-traffic limit. In the heavy-traffic limit, the processes C and S affect performance through their rates $\lambda = \rho\mu$ and μ and through the variance parameter σ_X^2, which depends on the elements of the covariance matrix Σ in (7.1) as indicated in (7.3).

Note in particular that the mean of RBM in (7.16) is directly proportional to the variability of $X = C - S$ through the variability parameter σ_X^2 in (7.3). The variability parameter σ_X^2 in turn is precisely the variance constant in the CLT for the net-input process $C - S$.

In (7.9)–(7.16) we have described the approximations for the steady-state workload distribution that follow directly from the heavy-traffic limit theorem in Theorem 5.7.1. It is also possible to modify or "refine" the approximations to satisfy other criteria. For example, extra terms that appear in known exact formulas for special cases, but which are negligible in the heavy-traffic limit, may be inserted. If the goal is to develop accurate numerical approximations, then it is natural to regard heavy-traffic limits as only one of the possible theoretical reference points. For the standard multiserver GI/G/s queue, for which the heavy-traffic limit is also RBM, heuristic refinements are discussed in Whitt (1982b, 1993a) and references therein.

For the fluid queue, an important reference case for which exact formulas are available is a single-source model with independent sequences of IID on times and off times (a special case of the model studied in Chapter 8). Kella and Whitt (1992b) show that the workload process and its steady-state distribution can be related to the virtual waiting time process in the standard GI/G/1 queue (studied here in Chapter 9). Relatively simple moment formulas are thus available in the M/G/1 special case. The steady-state workload distribution can be computed in the general GI/G/1 case using numerical transform inversion, following Abate, Choudhury and Whitt (1993, 1994a, 1999). Such computations were used to illustrate the performance of bounds for general fluid queues by Choudhury and Whitt (1997).

A specific way to generate refined approximations is to interpolate between light-traffic and heavy-traffic limits; see Burman and Smith (1983, 1986), Fendick and Whitt (1989), Reiman and Simon (1988, 1989), Reiman and Weiss (1989) and Whitt (1989b). Even though numerical accuracy can be improved by refinements, the direct heavy-traffic Brownian approximations remain appealing for their simplicity.

Example 5.7.1. *The M/G/1 steady state workload.* It is instructive to compare the approximations with exact values when we can determine them. For the standard M/G/1 queue with $K = \infty$, the mean steady-state workload has the simple exact formula

$$E[W_\rho(\infty)] = \frac{\rho \sigma_X^2}{2(1-\rho)}, \qquad (7.17)$$

which differs from (7.16) only by the factor ρ in the numerator of (7.17) and the factor μ in the denominator of (7.16). First, in the M/G/1 model the workload process has constant output rate 1, so $\mu = 1$. Hence, the only real difference between (7.16) and (7.17) is the factor ρ in the numerator of (7.17), which approaches 1 in the heavy-traffic limit.

To elaborate, in the M/G/1 queue, the cumulative input $C(t)$ equals the sum of the service times of all arrivals in the interval $[0, t]$, i.e., the cumulative input is

$$C(t) \equiv \sum_{k=1}^{A(t)} V_k, \quad t \geq 0 ,$$

where $\{A(t) : t \geq 0\}$ is a rate-ν Poisson arrival process independent of the sequence $\{V_k : k \geq 1\}$ of IID service times, with V_1 having a general distribution with mean EV_1. Thus, the traffic intensity is $\rho \equiv \nu EV_1$. The workload process is defined in terms of the net-input process $X(t) \equiv C(t) - t$ as described in Section 5.2.

The cumulative-input process is a special case of a renewal-reward process, considered in Section 7.4. Thus, by Theorem 7.4.1, if

$$\sigma_V^2 \equiv Var V_1 < \infty ,$$

then the cumulative-input process obeys a FCLT $\mathbf{C}_n \Rightarrow \mathbf{C}$ for $\mathbf{C}_n$ in (3.7) with translation constant $\lambda \equiv \rho$ and space-scaling function $c_n = n^{1/2}$. Then the limit process is $\sigma_C \mathbf{B}$, where $\mathbf{B}$ is standard Brownian motion and

$$\begin{aligned} \sigma_C^2 &= \nu \sigma_V^2 + \rho EV_1 \\ &= \rho EV_1 (c_V^2 + 1) , \end{aligned} \quad (7.18)$$

where c_V^2 is the squared coefficient of variation, defined by

$$c_V^2 \equiv \sigma_V^2 / (EV_1)^2 . \quad (7.19)$$

Therefore,

$$\sigma_X^2 = \sigma_C^2 = \rho EV_1 (c_V^2 + 1) . \quad (7.20)$$

With this notation, the exact formula for the mean steady-state workload in the M/G/1 queue is given in (7.17) above; e.g., see Chapter 5 of Kleinrock (1975). As indicated above, the approximation in (7.16) differs from the exact formula in (7.17) only by the factor ρ in the numerator of the exact formula, which of course disappears (becomes 1) in the heavy-traffic limit.

For the M/G/1 queue, it is known that

$$P(W_\rho(\infty) = 0) = 1 - \rho . \quad (7.21)$$

Thus, if we understand the approximation to be for the conditional mean $E[W_\rho(\infty)|W_\rho(\infty) > 0]$, then the approximation beomes exact. In general, however, the distribution of $W_\rho(\infty)$ is not exponential, so that the exponential distribution remains an approximation for the M/G/1 model, but the conditional distribution of $W(\infty)$ given that $W(\infty) > 0$ is exponential in the M/M/1 special case, in which the service-time distribution is exponential. ∎

5.7.3. The Overflow Process

In practice it is also of interest to describe the overflow process. In a communication network, the overflow process describes lost packets. An important design criterion is to keep the packet loss rate below a specified threshold. The *loss rate* in model n is

$$\beta_n \equiv \lim_{t \to \infty} t^{-1} U_n(t) \ . \tag{7.22}$$

The limits in Theorems 5.4.1 and 5.7.1 show that, with the heavy-traffic scaling, the loss rate should be asymptotically negligible as $n \to \infty$. Specifically, since $n^{-1/2} U_n(nt) \Rightarrow \mathbf{U}(t)$ as $n \to \infty$, where $\mathbf{U}$ is the upper-barrier regulator process of RBM, the cumulative loss in the interval $[0, n]$ is of order $\sqrt{n}$, so that the loss rate should be of order $1/\sqrt{n}$ as $n \to \infty$. (Of course, this asymptotic form depends on having the upper barriers grow as $K_n = \sqrt{n} K$ and $\rho_n \to 1$.) More precisely, we approximate the loss rate β_n by

$$\beta_n \approx \beta/\sqrt{n} \ , \tag{7.23}$$

where

$$\beta \equiv \lim_{t \to \infty} t^{-1} \mathbf{U}(t) \ . \tag{7.24}$$

Note that approximation (7.23) involves an unjustified interchange of limits, involving $n \to \infty$ and $t \to \infty$.

Berger and Whitt (1992b) make numerical comparisons (based on exact numerical algorithms) showing how the Brownian approximation in (7.23) performs for finite-capacity queues. For very small loss rates, such as 10^{-9}, it is not possible to achieve high accuracy. (Systems with the same heavy-traffic limit may have loss rates varying from 10^{-4} to 10^{-15}.) Such very small probabilities tend to be captured better by large-deviations limits. For a simple numerical comparison, see Srikant and Whitt (2001). Overall, the Brownian approximation provides important insight. That is illustrated by the sensitivity analysis in Section 9 of Berger and Whitt (1992b).

More generally, the heavy-traffic stochastic-process limits support the approximation

$$\mathbf{U}_n(t) \approx \sqrt{n} \mathbf{U}(t/n), \quad t \geq 0 \ , \tag{7.25}$$

where $\mathbf{U}$ is the upper-barrier regulator process of RBM. In order for the Brownian approximation for the overflow process in (7.25) to be useful, we need to obtain useful characterizations of the upper-barrier regulator process $\mathbf{U}$ associated with RBM. It suffices to describe one of the boundary regulation processes $\mathbf{U}$ and $\mathbf{L}$, because $\mathbf{L}$ has the same structure as $\mathbf{U}$ with a drift of the opposite sign. The rates of the process $\mathbf{L}$ and $\mathbf{U}$ are determined on p. 90 of Harrison (1985).

Theorem 5.7.3. (rates of boundary regulator processes) *The rates of the boundary regulator processes exist, satisfying*

$$\alpha \equiv \lim_{t \to \infty} \frac{\mathbf{L}(t)}{t} = \lim_{t \to \infty} \frac{E\mathbf{L}(t)}{t} = \begin{cases} \sigma^2/2K & \text{if } \eta = 0 \\ \\ \frac{\eta}{e^{\theta K} - 1} & \text{if } \eta \neq 0 \end{cases} \tag{7.26}$$

and

$$\beta \equiv \lim_{t \to \infty} \frac{\mathbf{U}(t)}{t} = \lim_{t \to \infty} \frac{E\mathbf{U}(t)}{t} = \begin{cases} \sigma^2/2K & \text{if } \eta = 0 \\ \frac{\eta}{1-e^{-\theta K}} & \text{if } \eta \neq 0 \ . \end{cases} \quad (7.27)$$

It is important to note that the loss rate β depends upon the variance σ^2, either directly (when $\eta = 0$) or via θ in (7.12). We can use regenerative analysis and martingales to further describe the Brownian boundary regulation processes $\mathbf{L}$ and $\mathbf{U}$; see Berger and Whitt (1992b) and Williams (1992). Let $T_{a,b}$ be the first passage time from level a to level b within $[0, K]$. Epochs at which RBM first hits 0 after first hitting K are regeneration points for the processes $\mathbf{L}$ and $\mathbf{U}$. Assuming that the RBM starts at 0, one regeneration cycle is completed at time $T_{0,K} + T_{K,0}$. Of course, $\mathbf{L}$ increases only during $[0, T_{0,K}]$, while $\mathbf{U}$ increases only during $[T_{0,K}, T_{0,K} + T_{K,0}]$. We can apply regenerative analysis and the central limit theorem for renewal processes to show that the following limits exist

$$\alpha \equiv \lim_{t \to \infty} \frac{\mathbf{L}(t)}{t} = \lim_{t \to \infty} \frac{E\mathbf{L}(t)}{t} = \frac{E\mathbf{L}(T_{0,K} + T_{K,0})}{E(T_{0,K} + T_{K,0})} \quad (7.28)$$

$$\beta \equiv \lim_{t \to \infty} \frac{\mathbf{U}(t)}{t} = \lim_{t \to \infty} \frac{E\mathbf{U}(t)}{t} = \frac{E\mathbf{U}(T_{0,K} + T_{K,0})}{E(T_{0,K} + T_{K,0})} \quad (7.29)$$

$$\sigma_L^2 \equiv \lim_{t \to \infty} \frac{Var\,\mathbf{L}(t)}{t} \quad \text{and} \quad \sigma_U^2 \equiv \lim \frac{Var\,\mathbf{U}(t)}{t} \ . \quad (7.30)$$

The parameters σ_L^2 and σ_U^2 in (7.30) are the *asymptotic variance parameters* of the processes $\mathbf{L}$ and $\mathbf{U}$. It is also natural to focus on the *normalized asymptotic variance parameters*

$$c_L^2 \equiv \sigma_L^2/\alpha \quad \text{and} \quad c_U^2 \equiv \sigma_U^2/\beta \ . \quad (7.31)$$

Theorem 5.7.4. (normalized asymptotic variance of boundary regulator processes) *The normalized asymptotic variance parameters in* (7.31) *satisfy*

$$c_U^2 = c_L^2 = E\left[\left(\mathbf{L}(T_{0,K}) - \frac{(T_{0,K} + T_{K,0})E\mathbf{L}(T_{0,K})}{E(T_{0,K} + T_{K,0})}\right)^2 \Big/ E\mathbf{L}(T_{0,K})\right]$$

$$= \begin{cases} 2K/3 & \text{if } \eta = 0 \\ \frac{2(1-e^{2\theta K}+4\theta K e^{\theta K})}{-\theta(1-e^{\theta K})^2} & \text{if } \eta \neq 0 \end{cases} \quad (7.32)$$

for $\theta \equiv 2\eta/\sigma^2$ *as in* (7.12).

In order to obtain the last line of (7.32) in Theorem 5.7.4, and for its own sake, we use an expression for the joint transform of $\mathbf{L}(T_{0,K})$ and $T_{0,K}$ from Williams (1992). Note that it suffices to let $\sigma^2 = 1$, because if $\sigma^2 > 0$ and $\mathbf{W}$ is a $(\eta/\sigma, 1)$ RBM on $[0, K/\sigma]$, then $\sigma\mathbf{W}$ is an (η, σ^2)-RBM on $[0, K]$.

Theorem 5.7.5. (joint distribution of key variables in the regenerative representation) *For $\sigma^2 = 1$ and all $s_1, s_2 \geq 0$,*

$$E[\exp(-s_1 \mathbf{L}(T_{0,K}) - s_2 T_{0,K})]$$

$$= \begin{cases} \frac{1}{1+s_1 K} & \text{if } \eta = 0, s_2 = 0 \\ \frac{1}{\cosh(\gamma K) + s_1 \gamma^{-1} \sinh(\gamma K)} & \text{if } \eta = 0, s_2 \neq 0 \\ \frac{e^{mK}}{\cos(\gamma K) + (s_1+m)\gamma^{-1} \sinh(\gamma K)} & \text{if } \eta \neq 0 , \end{cases} \quad (7.33)$$

where $\gamma = \sqrt{\eta^2 + 2s_2}$.

Since an explicit expression for the Laplace transform is available, we can exploit numerical transform inversion to calculate the joint probability distribution and the marginal probability distributions of $T_{0,K}$ and $\mathbf{L}(T_{0,K})$; see Abate and Whitt (1992a, 1995a), Choudhury, Lucantoni and Whitt (1994) and Abate, Choudhury and Whitt (1999).

Explicit expressions for the moments of $\mathbf{L}(T_{0,K})$ and $T_{0,K}$ can be obtained directly from Theorem 5.7.5.

Theorem 5.7.6. (associated moments of regenerative variables) *If $\eta = 0$ and $\sigma^2 = 1$, then*

$$\begin{aligned} ET_{0,K} &= K^2, \; ET_{0,K}^2 = 5K^4/3 , \\ E[\mathbf{L}(T_{0,K})] &= K, \; E[\mathbf{L}(T_{0,K})^2] = 2K^2 \\ E[T_{0,K}\mathbf{L}(T_{0,K})] &= 5K^3/3 . \end{aligned} \quad (7.34)$$

If $\eta \neq 0$ and $\sigma^2 = 1$, then

$$\begin{aligned} ET_{0,K} &= (e^{-2\eta K} - 1 + 2\eta K)/2\eta^2 , \\ E[T_{0,K}^2] &= (e^{-4\eta K} + e^{-2\eta K} + 6\eta K e^{-2\eta K} + 2\eta^2 K^2 - 2)/2\eta^4 , \\ E[\mathbf{L}(T_{0,K})] &= (1 - e^{-2\eta K})/2\eta , \\ E[\mathbf{L}(T_{0,K})^2] &= (1 - e^{-2\eta K})^2/2\eta^2 , \\ E[T_{0,K}\mathbf{L}(T_{0,K})] &= (e^{-2\eta K} - 3\eta K e^{-2\eta K} - e^{-4\eta K} + \eta K)/2\eta^3 . \end{aligned} \quad (7.35)$$

Fendick and Whitt (1998) show how a Brownian approximation can be used to help interpret loss measurements in a communication network.

5.7.4. One-Sided Reflection

Even nicer descriptions of RBM are possible when there is only one reflecting barrier at the origin (corresponding to an infinite buffer). Let $\mathbf{R} \equiv \{\mathbf{R}(t; \eta, \sigma^2, x) : t \geq 0\}$ denote RBM with one reflecting barrier at the origin, i.e., $\mathbf{R} = \phi(\mathbf{B})$ for $\mathbf{B} \equiv \{\mathbf{B}(t; \eta, \sigma^2, x) : t \geq 0\}$, where ϕ is the one-dimensional reflection map in (2.5) and **B** is Brownian motion. There is a relatively simple expression for the transient distribution of RBM when there is only a single barrier; see p. 49 of Harrison (1985).

Theorem 5.7.7. (transition probability of RBM with one reflecting barrier) *If* $\mathbf{R} \equiv \{\mathbf{R}(t; \eta, \sigma^2, x) : t \geq 0\}$ *is an* (η, σ^2)-*RBM then*

$$P(\mathbf{R}(t) \leq y | \mathbf{R}(0) = x) = 1 - \Phi\left(\frac{-y + x + \eta t}{\sigma\sqrt{t}}\right)$$
$$- \exp(2\eta y/\sigma^2)\Phi\left(\frac{-y - x - \eta t}{\sigma\sqrt{t}}\right),$$

where Φ *is the standard normal cdf.*

We now observe that we can express RBM with negative drift (and one reflecting barrier at the origin) in terms of *canonical RBM* with drift coefficient -1 and diffusion coefficient 1. We first state the result for Brownian motion and then for reflected Brownian motion.

Theorem 5.7.8. (scaling to canonical Brownian motion) *If* $m < 0$ *and* $\sigma^2 > 0$, *then*

$$\{a\mathbf{B}(bt; m, \sigma^2, x) : t \geq 0\} \stackrel{\mathrm{d}}{=} \{\mathbf{B}(t; -1, 1, ax) : t \geq 0\} \quad (7.36)$$

and

$$\{\mathbf{B}(t; m, \sigma^2, x) : t \geq 0\} \stackrel{\mathrm{d}}{=} \{a^{-1}\mathbf{B}(b^{-1}t; -1, 1, ax) : t \geq 0\} \quad (7.37)$$

for

$$a = \frac{|m|}{\sigma^2} > 0, \qquad b = \frac{\sigma^2}{m^2} > 0,$$
$$m = -\frac{1}{ab} < 0, \qquad \sigma^2 = \frac{1}{a^2 b} > 0. \quad (7.38)$$

Theorem 5.7.9. (scaling to canonical RBM). *If* $\eta < 0$ *and* $\sigma^2 > 0$, *then*

$$\{a\mathbf{R}(bt; \eta, \sigma^2, Y) : t \geq 0\} \stackrel{\mathrm{d}}{=} \{\mathbf{R}(t; -1, 1, aY) : t \geq 0\} \quad (7.39)$$

and

$$\{\mathbf{R}(t; \eta, \sigma^2, Y) : t \geq 0\} \stackrel{\mathrm{d}}{=} \{a^{-1}\mathbf{R}(b^{-1}t; -1, 1, aY) : t \geq 0\} \quad (7.40)$$

for

$$a \equiv \frac{|\eta|}{\sigma^2} > 0, \quad b \equiv \frac{\sigma^2}{\eta^2},$$
$$\eta = \frac{-1}{ab}, \quad \sigma^2 = \frac{1}{a^2 b}, \quad (7.41)$$

as in (7.38) *of Chapter* 4.

Theorem 5.7.9 is significant because it implies that we only need to do calculations for a single RBM — canonical RBM. Expressions for the moments of canonical

RBM are given Abate and Whitt (1987a,b) along with various approximations. There it is shown that the time-dependent moments can be characterized via cdf's. In particular, the time-dependent moments starting at 0, normalized by dividing by the steady-state moments are cdf's. Moreover the differences $E(\mathbf{R}(t)|\mathbf{R}(0) = x) - E[\mathbf{R}(t)|\mathbf{R}(0) = 0]$ divided by x are complementary cdf's (ccdf's), and all these cdf's have revealing structure. Here are explicit expressions for the first two moments.

Theorem 5.7.10. (moments of canonical RBM) *If $\mathbf{R}$ is canonical RBM, then*

$$E[\mathbf{R}(t)|\mathbf{R}(0) = x] = 2^{-1} + \sqrt{t}\phi\left(\frac{t-x}{\sqrt{t}}\right)$$

$$- (t - x + 2^{-1})\left[1 - \Phi\left(\frac{t-x}{\sqrt{t}}\right)\right]$$

$$- 2^{-1}e^{2x}\left[1 - \Phi\left(\frac{t+x}{\sqrt{t}}\right)\right]$$

and

$$E[\mathbf{R}(t)^2|\mathbf{R}(0) = x] = 2^{-1} + ((x-1)\sqrt{t} - \sqrt{t^3})\phi\left(\frac{t-x}{\sqrt{t}}\right)$$

$$+ ((t-x)^2 + t - 2^{-1})\left[1 - \Phi\left(\frac{t-x}{\sqrt{t}}\right)\right]$$

$$+ e^{2x}(t + x - 2^{-1})\left[1 - \Phi\left(\frac{t+x}{\sqrt{t}}\right)\right],$$

where Φ and ϕ are the standard normal cdf and pdf.

When thinking about RBM approximations for queues, it is sometimes useful to regard RBM as a special M/M/1 queue with $\rho = 1$. After doing appropriate scaling, the M/M/1 queue-length process approaches a nondegenerate limit as $\rho \to 1$. Thus structure of RBM can be deduced from structure for the M/M/1 queue; see Abate and Whitt (1988a-d). This is one way to characterize the covariance function of stationary RBM; see Abate and Whitt (1988c). Recall that a nonnegative-real-valued function f is completely monotone if it has derivatives of all orders that alternate in sign. Equivalently, f can be expressed as a mixture of exponential distributions; see p. 439 of Feller (1971).

Theorem 5.7.11. (covariance function of RBM) *Let $\mathbf{R}^*$ be canonical RBM initialized by giving $\mathbf{R}^*(0)$ an exponential distribution with mean $1/2$. The process $\mathbf{R}^*$ is a stationary process with completely monotone covariance function*

$$Cov(\mathbf{R}^*(0), \mathbf{R}^*(t)) = E[\mathbf{R}^*(t) - 2^{-1})(\mathbf{R}^*(0) - 2^{-1})]$$
$$= 2(1 - 2t - t^2)[1 - \Phi(\sqrt{t})] + 2\sqrt{t}(1+t)\phi(\sqrt{t})$$
$$= H_{1e}^c(t) = H_2^c(t), \quad t \geq 0,$$

where H_k is the kth-moment cdf and H^c_{1e} is the stationary-excess ccdf associated with the first-moment cdf, i.e.,

$$H_k(t) \equiv \frac{E[\mathbf{R}(t)^k|\mathbf{R}(0)=0]}{E\mathbf{R}(\infty)^k}, \quad t \geq 0 ,$$

and

$$H^c_{1e}(t) \equiv 1 - 2\int_0^t H^c_1(s)ds, \quad t \geq 0 .$$

Canonical RBM has asymptotic variance

$$\sigma^2_\mathbf{R} \equiv \lim_{t \to \infty} t^{-1} Var\left(\int_0^t \mathbf{R}(s)ds|\mathbf{R}(0)=x\right) = 1/2 .$$

5.7.5. First-Passage Times

We can also establish limits for first passage times. For a stochastic process $\{Z(t) : t \geq 0\}$, let $T_{a,b}(Z)$ denote the first passage time for Z to go from a to b. (We assume that $Z(0) = a$, and consider the first passage time to b.) In general, the first passage time functional is not continuous on D or even on the subset C, but the first passage time functional is continuous almost surely with respect to BM or RBM, because BM and RBM cross any level w.p.1 in a neighborhood of any time that they first hit a level. Hence we can invoke a version of the continuous mapping theorem to conclude that limits holds for the first passage times.

Theorem 5.7.12. (limits for first passage times) *Under the assumptions of Theorem 5.7.1,*

$$\frac{T_{a\sqrt{n},b\sqrt{n}}(W_n)}{n} \Rightarrow T_{a,b}(\mathbf{W})$$

for any positive a,b with $a \neq b$ and $0 \leq a, b \leq K$, where $\mathbf{W}$ is RBM and W_n is the unnormalized workload process in model n.

Now let $T_{a,b}(\mathbf{R})$ be the first-passage time from a to b for one-sided canonical RBM. The first passage time upward is the same as when there is a (higher) upper barrier (characterized in Theorems 5.7.5 and 5.7.6), but the first passage time down is new. Let $f(t; a, b)$ be the pdf of $T_{a,b}(\mathbf{R})$ and let $\hat{f}(s; a, b)$ be its Laplace transform, i.e.,

$$\hat{f}(s; a, b) \equiv \int_0^\infty e^{-st} f(t; a, b)dt ,$$

where s is a complex variable with positive real part. The Laplace transforms to and from the origin have a relatively simple form; see Abate and Whitt (1988a). Again, numerical transform inversion can be applied to compute the probability distributions themselves.

Theorem 5.7.13. (RBM first-passage-time transforms and moments) *For canonical RBM (with no upper barrier), the first-passage-time Laplace transforms to and from the origin are, respectively,*

$$\hat{f}(s;x,0) = e^{-xr_2}$$

and

$$\hat{f}(s;0,x) = \frac{r_1 + r_2}{r_1 e^{-xr_2} + r_2 e^{xr_1}}$$

for

$$r_1(s) = 1 + \sqrt{1+2s} \quad \text{and} \quad r_2(s) = \sqrt{1+2s} - 1 \,,$$

so that

$$\begin{aligned} ET_{x,0} &= x, \quad Var\, T_{x,0} = x \,, \\ ET_{0,x} &= 2^{-1}[e^{2x} - 1 - 2x] \quad \text{and} \\ Var\, T_{0,x} &= 4^{-1}[e^{4x} - 1 - 4x + 4e^{2x}(1-2x) - 4] \,. \end{aligned}$$

The first passage time down is closely related to the busy period of a queue, i.e., the time from when a buffer first becomes nonempty until it becomes empty again. This concept is somewhat more complicated for fluid queues than standard queues. In either case, the distribution of the busy period for small values tends to depend on the fine structure of the model, but the tail of the busy period often can be approximated robustly, and Brownian approximations can play a useful role; see Abate and Whitt (1988d, 1995b).

First-passage-time cdf's are closely related to extreme-value ccdf's because $T_{0,a}(W) \leq t$ if and only if $W^\uparrow(t) \equiv \sup_{0 \leq s \leq t} W(s) \geq a$. Extreme-value theory shows that there is statistical regularity associated with both first-passage times and extreme values as $t \to \infty$ and $u \to \infty$; see Resnick (1987). Heavy-traffic extreme-value approximations for queues are discussed by Berger and Whitt (1995a), Glynn and Whitt (1995) and Chang (1997). A key limit is

$$2R^\uparrow(t) - \log(2t) \Rightarrow Z \quad \text{as} \quad t \to \infty \,,$$

where R is canonical RBM and Z has the *Gumbel cdf*, i.e.,

$$P(Z \leq x) \equiv exp(-e^{-x}), \quad -\infty < x < \infty \,.$$

This limit can serve as a basis for extreme-value engineering.

To summarize, in this section we have displayed Brownian limits for a fluid queue, obtained by combining the general fluid-queue limits in Theorem 5.4.1 with the multidimensional version of Donsker's theorem in Theorem 4.3.5. We have also displayed various formulas for RBM that are helpful in applications of the Brownian limit. We discuss RBM limits and approximations further in the next section and in Sections 8.4 and 9.6.

5.8. Planning Queueing Simulations

In this section, following Whitt (1989a), we see how the Brownian approximation stemming from the Brownian heavy-traffic limit in Section 5.7 can be applied to plan simulations of queueing models. In particular, we show how the Brownian approximation can be used to estimate the required simulation run lengths needed to obtain desired statistical precision, before any data have been collected. These estimates can be used to help design the simulation experiment and even to determine whether or not a contemplated experiment should be conducted.

The queueing simulations considered are single replications (one long run) of a single queue conducted to estimate steady-state characteristics, such as long-run-average steady-state workload. For such simulations to be of genuine interest, the queueing model should be relatively complicated, so that exact numerical solution is difficult. On the other hand, the queueing model should be sufficiently tractable that we can determine an appropriate Brownian approximation.

We assume that both these criteria are met. Indeed, we specify the models that we consider by stipulating that scaled versions of the stochastic process of interest, with the standard normalization, converge to RBM as $\rho \uparrow 1$. For simplicity, we focus on the workload process in a fluid queue with infinite capacity, but the approach applies to other models as well.

Of course, such a Brownian approximation directly yields an approximation for the steady-state performance, but nevertheless we may be interested in the additional simulation in order to develop a more precise understanding of the steady-state behavior. Indeed, one use of such simulations is to evaluate how various candidate approximations perform. Then we often need to perform a large number of simulations in order to see how the approximations perform over a range of possible model parameters.

In order to exploit the Brownian approximation for a single queue, we focus on simulations of a single queue. However, the simulation actually might be for a network of queues. Then the analysis of a single queue is intended to apply to any one queue in that network. If we want to estimate the steady-state performance at all queues in the network, then the required simulation run length for the network would be the maximum required for any one queue in the network. Our analysis shows that it often suffices to focus on the bottleneck (most heavily loaded) queue in the network.

At first glance, the experimental design problem may not seem very difficult. To get a rough idea about how long the runs should be, one might do one "pilot" run to estimate the required simulation run lengths. However, such a preliminary experiment requires that you set up the entire simulation before you decide whether or not to conduct the experiment. Nevertheless, if such a sampling procedure could be employed, then the experimental design problem would indeed not be especially difficult. Interest stems from the fact that one sample run can be extremely misleading.

This queueing experimental design problem is interesting and important primarily because a uniform allocation of data over all cases (parameter values) is not nearly appropriate. Experience indicates that, for given statistical precision, the required amount of data increases as the traffic intensity increases and as the arrival-and-service variability (appropriately quantified) increases. Our goal is to quantify these phenomena.

To quantify these phenomena, we apply the space and time scaling functions. Our analysis indicates that to achieve a uniform relative error over all values of the traffic intensity ρ that the run length should be approximately proportional to the time-scaling factor $(1-\rho)^{-2}$ (for sufficiently high ρ). Relative error appears to be a good practical measure of statistical precision, except possibly when very small numbers are involved. Then absolute error might be preferred. It is interesting that the required run length depends strongly on the criterion used. With the absolute error criterion, the run length should be approximately proportional to $(1-\rho)^{-4}$. With either the relative or absolute error criteria, there obviously are great differences between the required run lengths for different values of ρ, e.g., for $\rho = 0.8, 0.9$ and 0.99.

We divide the simulation run-length problem into two components. First, there is the question: What should be the required run length given that the system starts in equilibrium (steady state)? Second, there is the question: What should we do in the customary situation in which it is not possible to start in equilibrium? We propose to delete an initial portion of each simulation run before collecting data in order to allow the system to (approximately) reach steady state. By that method, we reduce the bias (the systematic error that occurs when the expected value of the estimator differs from the quantity being estimated). The second question, then, can be restated as: How long should be the initial segment of the simulation run that is deleted?

Focusing on the first question first, we work with the workload stochastic process, assuming that we have a stationary version, denoted by W_ρ^*. First, however, note that specifying the run length has no meaning until we specify the time units. To fix the time units, we assume that the output rate in the queueing system is μ. (It usually suffices to let $\mu = 1$, but we keep general μ to show how it enters in.)

For the general fluid-queue model we have the RBM approximation in (7.8). However, since we are assuming that we start in equilibrium, instead of the Brownian approximation in (7.8), we assume that we have the associated *stationary Brownian approximation*

$$\{W_\rho^*(t) : t \geq 0\} \approx \{\sigma_X^2 \mu^{-1}(1-\rho)^{-1}\mathbf{R}^*(\sigma_X^{-2}\mu^2(1-\rho)^2 t; -1, 1) : t \geq 0\}, \quad (8.1)$$

where σ_X^2 is the variability parameter, just as in (7.8), and $\mathbf{R}^*$ is a stationary version of canonical RBM, with initial exponential distribution, i.e.,

$$\{\mathbf{R}^*(t; -1, 1) : t \geq 0\} \stackrel{d}{=} \{\mathbf{R}(t; -1, 1, Y) : t \geq 0\}, \quad (8.2)$$

where the initial position Y is an exponential random variable with mean $1/2$ independent of the standard Brownian motion being reflected; i.e., $\mathbf{R}^* = \phi(\mathbf{B} + Y)$ where ϕ is the reflection map and $\mathbf{B}$ is a standard Brownian motion independent of the exponential random variable Y.

180   5. Heavy-Traffic Limits

The obvious application is with $\{W_\rho^*(t) : t \geq 0\}$ being a stationary version of a workload process, as defined in Section 5.2. However, our analysis applies to any stationary process having the Brownian approximation in (8.1).

5.8.1. The Standard Statistical Procedure

To describe the standard statistical procedure, let $\{W(t) : t \geq 0\}$ be a stochastic process of interest and assume that is stationary with $EW(t)^2 < \infty$. (We use that notation because we are thinking of the workload process, but the statistical procedure is more general, not even depending upon the Brownian approximation.) Our object is to estimate the mean $E[W(0)]$ by the *sample mean*, i.e., by the time average

$$\bar{W}_t \equiv t^{-1} \int_0^t W(s)ds, \quad t \geq 0 . \tag{8.3}$$

The standard statistical procedure, assuming ample data, is based on a CLT for $\bar{W}_t$. We assume that

$$t^{1/2}(\bar{W}_t - E[W(0)]) \Rightarrow N(0, \sigma^2) \quad \text{as} \quad t \to \infty , \tag{8.4}$$

where σ^2 is the *asymptotic variance*, defined by

$$\sigma^2 \equiv \lim_{t \to \infty} t Var(\bar{W}_t) = 2 \int_0^\infty C(t)dt , \tag{8.5}$$

and $C(t)$ is the (auto) *covariance function*

$$C(t) \equiv E[W(t)W(0)] - (E[W(0)])^2, \quad t \geq 0 . \tag{8.6}$$

Of course, a key part of assumption (8.4) is the requirement that the asymptotic variance σ^2 be finite. The CLT in (8.4) is naturally associated with a Brownian approximation for the process $\{W(t) : t \geq 0\}$. Such CLTs for stationary processes with weak dependence were discussed in Section 4.4. Based on (8.4), we use the normal approximation

$$\bar{W}_t \approx N(E[W(0)], \sigma^2/t) \tag{8.7}$$

for the (large) t of interest, where σ^2 is the asymptotic variance in (8.5).

Based on (8.7), a $[(1 - \beta) \cdot 100]\%$ *confidence interval* for the mean $E[W(0)]$ is

$$[\bar{W}_t - z_{\beta/2}(\sigma^2/t)^{1/2}, \ \bar{W}_t + z_{\beta/2}(\sigma^2/t)^{1/2}] , \tag{8.8}$$

where

$$P(-z_{\beta/2} \leq N(0,1) \leq z_{\beta/2}) = 1 - \beta . \tag{8.9}$$

The width of the confidence interval in (8.8) provides a natural measure of the *statistical precision*. There are two natural criteria to consider: *absolute width* and *relative width*. Relative width looks at the ratio of the width to the quantity to be estimated, $E[W(0)]$.

For any given β, the absolute width and relative width of the $[(1 - \beta) \cdot 100]\%$ confidence intervals for the mean $E[W(0)]$ are, respectively,

$$w_a(\beta) = \frac{2\sigma z_{\beta/2}}{t^{1/2}} \quad \text{and} \quad w_r(\beta) = \frac{2\sigma z_{\beta/2}}{t^{1/2} E[W(0)]} \ . \tag{8.10}$$

For specified *absolute width* ϵ and specified *confidence level* $1 - \beta$, the required simulation run length, given (8.7), is

$$t_a(\epsilon, \beta) = \frac{4\sigma^2 z_{\beta/2}^2}{\epsilon^2} \ . \tag{8.11}$$

For specified *relative width* ϵ and specified *confidence level* $1 - \beta$, the required length of the estimation interval, given (8.7), is

$$t_r(\epsilon, \beta) = \frac{4\sigma^2 z_{\beta/2}^2}{\epsilon^2 (E[W(0)])^2} \ . \tag{8.12}$$

From (8.11) and (8.12) we draw the important and well-known conclusion that both $t_a(\epsilon, \beta)$ and $t_r(\epsilon, \beta)$ are inversely proportional to ϵ^2 and directly proportional to σ^2 and $z_{\beta/2}^2$.

Standard statistical theory describes how observations can be used to estimate the unknown quantities $E[W(0)]$ and σ^2. Instead, we apply additional information about the model to obtain rough preliminary estimates for $E[W(0)]$ and σ^2 without data.

5.8.2. Invoking the Brownian Approximation

At this point we invoke the Brownian approximation in (8.1). We assume that the process of interest is W_ρ^* and that it can be approximated by scaled stationary canonical RBM as in (8.1). The steady-state mean of canonical RBM and its asymptotic variance are both $1/2$; see Theorems 5.7.10 and 5.7.11. It thus remains to consider the scaling.

To consider the effect of scaling space and time in general, let W again be a general stationary process with covariance function C and let

$$W_{y,z}(t) \equiv yW(zt), \quad t \geq 0$$

for $y, z > 0$. Then the mean $E[W_{y,z}(t)]$, covariance function $C_{y,z}(t)$ and asymptotic variance of $W_{y,z}$ are, respectively,

$$E[W_{y,z}(t)] = yEW(zt) = yE[W(t)] \ ,$$
$$C_{y,z}(t) = y^2 C(zt) \quad \text{and} \quad \sigma_{y,z}^2 = y^2 \sigma^2/z \ . \tag{8.13}$$

Thus, from (8.1) and (8.13), we obtain the important approximations

$$E[W_\rho^*(0)] \approx \frac{\sigma_X^2}{2\mu(1-\rho)} \quad \text{and} \quad \sigma_{W_\rho^*}^2 \approx \frac{\sigma_X^6}{2\mu^4(1-\rho)^4} \ . \tag{8.14}$$

We have compared the approximation for the mean in (8.14) to the exact formula for the M/G/1 workload process in Example 5.7.1. Similarly, the exact formula for the asymptotic variance for the M/M/1 workload process, where $\mu = 1$, is

$$\sigma^2_{W_\rho} = \frac{2\rho(3-\rho)}{(1-\rho)^4} \; ; \qquad (8.15)$$

see (23) of Whitt (1989a). Formula (8.15) reveals limitations of the approximation in (8.14) in light traffic (as $\rho \downarrow 0$), but formula (8.15) agrees with the approximation in (8.14) in the limit as $\rho \to 1$, because $\sigma^2_X = 2\rho$ for the M/M/1 queue; let $EV_1 = 1$ and $c_V^2 = 1$ in (7.20). Numerical comparisons of the predictions with simulation estimates in more general models appear in Whitt (1989a). These formulas show that the approximations give good rough approximations for ρ not too small (e.g., for $\rho \geq 1/2$).

Combining (8.12) and (8.14), we see that the approximate required simulation run length for W_ρ^* given a specified *relative* width ϵ and confidence level $1-\beta$ for the confidence interval for $E[W_\rho^*(0)]$ is

$$t_r(\epsilon, \beta) \approx \frac{8\sigma^2_X z^2_{\beta/2}}{\epsilon^2 \mu^2 (1-\rho)^2} \; . \qquad (8.16)$$

Combining (8.11) and (8.14), we see that the approximate required simulation run length for W_ρ^* given a specified *absolute* width ϵ and confidence level $1-\beta$ for the confidence interval for $E[W_\rho(0)]$ is

$$t_a(\epsilon, \beta) \approx \frac{2\sigma^6_X z^2_{\beta/2}}{\epsilon^2 \mu^4 (1-\rho)^4} \; . \qquad (8.17)$$

In summary, the Brownian approximation in (8.1) dictates that, with a criterion based on the relative width of the confidence interval, the required run length should be directly proportional to both the the time-scaling term as a function of ρ alone, $(1-\rho)^{-2}$, and the heavy-traffic variability parameter σ^2_X. In contrast, with the absolute standard error criterion, the required run length should be directly proportional to $(1-\rho)^{-4}$, the *square* of the time-scaling term as a function of ρ alone, and σ^6_X, the *cube* of the heavy-traffic variability parameter σ^2_X.

The second question mentioned at the outset is: How to determine an initial transient portion of the simulation run to delete? To develop an approximate answer, we can again apply the Brownian approximation in (8.1). If the system starts empty, we can consider canonical RBM starting empty. By Theorem 5.7.10, the time-dependent mean of canonical RBM $E[\mathbf{R}(t)|\mathbf{R}(0) = 0]$ is within about 1% of its steady-state mean $1/2$ at $t = 4$. Hence, if we were simulating canonical RBM, then we might delete an initial portion of length 4. Thus, by (8.1), a rough rule of thumb for the queueing process W_ρ (with unit processing rate) is to delete an initial segment of length $4\sigma^2_X/\mu^2(1-\rho)^2$. When we compare this to formula (8.16), we see that the proportion of the total run that should be deleted should be about $\epsilon^2/2z^2_{\beta/2}$, which is small when ϵ is small.

We can also employ the Brownian approximation to estimate the bias due to starting away from steady-state. For example, the bias due to starting empty with canonical RBM is

$$E\bar{\mathbf{R}}_t - 1/2 = t^{-1} \int_0^t (E[\mathbf{R}(t; -1, 1, 0] - 1/2) ds$$
$$\approx t^{-1} \int_0^\infty (E[\mathbf{R}(s); -1, 1, 0] - 1/2) ds = 1/4t , \quad (8.18)$$

by Corollary 1.3.4 of Abate and Whitt (1987a). The approximate relative bias is thus $1/2t$. That same relative bias should apply approximately to the workload process in the queue. We can also estimate the reduced bias due to deleting an initial portion of the run, using Theorem 5.7.10 and the hyperexponential approximation

$$1/2 - E[\mathbf{R}(t; -1, 1, 0] \approx 0.36 e^{-5.23t} + 0.138 e^{-0.764t}, \quad t \geq 0 . \quad (8.19)$$

Our entire analysis depends on the normal approximation in (8.7), which in turn depends on the simulation run length t. Not only must t be sufficiently large so that the estimated statistical precision based on (8.7) is adequate, but t must be sufficiently large so that the normal approximation in (8.7) is itself reasonable. Consistent with intuition, experience indicates that the run length required for (8.7) to be a reasonable approximation also depends on the parameters ρ and σ_X^2, with t needing to increase as ρ and σ_X^2 increase. We can again apply the Brownian approximation to estimate the run length required. We can ask what run length is appropriate for a normal approximation to the distribution of the sample mean of canonical RBM. First, however, the time scaling alone tells us that the run length must be at least of order $\sigma_X^2/\mu^2(1-\rho)^2$. This rough analysis indicates that the requirement for (8.7) to be a reasonable approximation is approximately the same as the requirement to control the relative standard error. For further analysis supporting this conclusion, see Asmussen (1992).

5.9. Heavy-Traffic Limits for Other Processes

We now obtain heavy-traffic stochastic-process limits for other processes besides the workload process in the setting of Section 5.4. Specifically, we obtain limits for the departure process and the processing time.

5.9.1. The Departure Process

We first obtain limits for the departure process defined in (2.11), but in general we can have difficulties applying the continuous-mapping approach with addition starting from (2.11) because the limit processes $\mathbf{S}$ and $-\mathbf{L}$ can have common discontinuities of opposite sign. We can obtain positive results when we rule that out, again invoking Theorem 12.7.3.

Let the scaled departures processes be defined by

$$\mathbf{D}_n \equiv c_n^{-1}(D_n(nt) - \mu_n nt), \quad t \geq 0 . \tag{9.1}$$

Theorem 5.9.1. (limit for the departure process) *Let the conditions of Theorem 5.4.1 hold. If the topology on D is J_1, assume that $\mathbf{S}$ and $\mathbf{L}$ almost surely have no common discontinuities. If the topology on D is M_1, assume that $\mathbf{S}$ and $\mathbf{L}$ almost surely have no common discontinuities with jumps of common sign. Then, jointly with the limits in (4.5) and (4.7),*

$$\mathbf{D}_n \Rightarrow \mathbf{D} \equiv \mathbf{S} - \mathbf{L} \quad in \quad D \tag{9.2}$$

with the same topology, for $\mathbf{D}_n$ in (9.1), $\mathbf{S}$ in (4.5) and $\mathbf{L}$ in (4.9).

Proof. By (2.11),

$$\mathbf{D}_n = \mathbf{S}_n - \mathbf{L}_n .$$

By Theorem 5.4.1, $(\mathbf{S}_n, \mathbf{L}_n) \Rightarrow (\mathbf{S}, \mathbf{L})$ in D^2 jointly with the other limits. Just as in the proof of Theorem 5.4.1, we can apply the continuous mapping theorem, Theorem 3.4.3, with addition. Under the conditions on the discontinuities of $\mathbf{S}$ and $\mathbf{L}$, addition is measurable and almost surely continuous. Hence we obtain the desired limit in (9.2). ∎

The extra assumption in Theorem 5.9.1 is satisfied when $P(S_n(t) = \mu_n t, \ t \geq 0) = 1$ or when $\mathbf{X}$ has no negative jumps (which implies that $\mathbf{L} \equiv \psi_L(\mathbf{X})$ has continuous paths).

As an alternative to (9.1), we can use the input rate λ_n in the translation term of the normalized departure process; i.e., let

$$\mathbf{D}'_n \equiv c_n^{-1}(D_n(nt) - \lambda_n nt), \quad t \geq 0 . \tag{9.3}$$

When the input rate appears in the translation term, we can directly compare the departure processes D_n to the cumulative-input processes C_n.

Corollary 5.9.1. (limit for the departure process with input centering) *Under the assumptions of Theorem 5.9.1,*

$$\mathbf{D}'_n \Rightarrow \mathbf{D}' \equiv -\eta \mathbf{e} + \mathbf{S} - \mathbf{L} \quad in \quad (D, M_1) \tag{9.4}$$

for $\mathbf{D}'_n$ in (9.3), η in (4.6), $\mathbf{e}(t) = t$ for $t \geq 0$, $\mathbf{S}$ in (4.5) and $\mathbf{L}$ in (4.9).

Proof. Note that $\mathbf{D}'_n = \mathbf{D}_n - \eta_n \mathbf{e}$. Hence, as before, we can apply the continuous-mapping theorem, Theorem 3.4.3, with addition to the joint limit $(\mathbf{D}_n, \eta_n \mathbf{e}) \Rightarrow (\mathbf{D}, \eta \mathbf{e})$, which holds by virtue of Theorems 5.9.1 and 11.4.5. ∎

5.9.2. The Processing Time

We now establish heavy-traffic limits for the processing time $T(t)$ in (2.12). We first exploit (2.13) when $K = \infty$. Let the scaled processing-time processes be

$$\mathbf{T}_n(t) \equiv c_n^{-1} T_n(nt), \quad t \geq 0 . \tag{9.5}$$

5.9. Heavy-Traffic Limits for Other Processes 185

Theorem 5.9.2. (limit for the processing time when $K = \infty$) *Suppose that, in addition to the conditions of Theorem 5.4.1, $K = \infty$, $\mu_n \to \mu$ as $n \to \infty$, where $0 < \mu < \infty$,*

$$\eta_{C,n} \equiv n(\lambda_n - \mu)/c_n \to \eta_C \tag{9.6}$$

and

$$\eta_{S,n} \equiv n(\mu_n - \mu)/c_n \to \eta_S , \tag{9.7}$$

where $-\infty < \eta_C < \infty$ and $-\infty < \eta_S < \infty$, so that $\eta = \eta_C - \eta_S$. If the topology on D is J_1, suppose that almost surely no two of the limit processes $\mathbf{C}$, $\mathbf{S}$ and $\mathbf{L}$ have common discontinuities. If the topology on D is M_1, assume that $\mathbf{L}$ and $\mathbf{C}$ almost surely have no common discontinuities with jumps of opposite sign, and $\mathbf{S}$ and $\mathbf{L}$ almost surely have no common discontinuities with jumps of common sign. Suppose that

$$P(\mathbf{S}(0) = 0) = 1 . \tag{9.8}$$

Then

$$\mathbf{T}_n \Rightarrow \mu^{-1}\mathbf{W} \quad in \quad D \tag{9.9}$$

with the same topology on D, jointly with the limits in (4.5) and (4.7), for $\mathbf{T}_n$ in (9.5) and $\mathbf{W}$ in (4.9) with $K = \infty$.

Proof. We can apply the continuous-mapping approach with first passage times, using the inverse map with centering in Section 13.7. Specifically, we can apply Theorem 13.7.4 with the Skorohod representation theorem, Theorem 3.2.2. From (2.13),

$$n^{-1}T_n(nt) + nt = \inf\{u \geq 0 : n^{-1}S_n(nu) \geq n^{-1}(C_n(nt) + W'_n(0) + L_n(nt))\} .$$

By (4.5), (9.6) and (9.7),

$$(n/c_n)(\hat{\mathbf{S}}_n - \mu\mathbf{e}, \hat{\mathbf{Z}}_n - \mu\mathbf{e}) \Rightarrow (\mathbf{S} + \eta_S\mathbf{e}, \mathbf{Z} + \eta_C\mathbf{e}) , \tag{9.10}$$

where

$$\hat{\mathbf{S}}_n \equiv n^{-1}S_n(nt) \quad \text{and} \quad \hat{\mathbf{Z}}_n \equiv n^{-1}(C_n(nt) + W'_n(0) + L_n(nt)), \quad t \geq 0 .$$

We use the conditions on the discontinuities of $\mathbf{C}$ and $\mathbf{L}$ to obtain the limit

$$(n/c_n)(\hat{\mathbf{Z}}_n - \mu\mathbf{e}) \Rightarrow \mathbf{Z} + \eta_C\mathbf{e} ,$$

where

$$\mathbf{Z} = \mathbf{C} + W'(0) + \mathbf{L} ,$$

by virtue of Theorem 12.7.3. Since

$$\hat{\mathbf{T}}_n(t) \equiv n^{-1}T_n(nt) = (\hat{\mathbf{S}}_n^{-1} \circ \hat{\mathbf{Z}}_n)(t) - t, \quad t \geq 0 , \tag{9.11}$$

the desired limit for $\mathbf{T}_n$ follows from Theorem 13.7.4. In particular, (9.10), (9.11) and (9.8) imply the limit

$$(n/c_n)(\hat{\mathbf{S}}_n^{-1} \circ \hat{\mathbf{Z}}_n - \mu^{-1}\mathbf{e} \circ \mu\mathbf{e})$$
$$\Rightarrow \frac{(\mathbf{Z} + \eta_C \mathbf{e}) - (\mathbf{S} + \eta_S \mathbf{e}) \circ \mu^{-1}\mathbf{e} \circ \mu\mathbf{e}}{\mu} = \frac{\mathbf{W}}{\mu} \ . \quad \blacksquare$$

The continuity conditions in Theorem 5.9.2 are satisfied when $\mathbf{S}$ is almost surely continuous and $\mathbf{X}$ almost surely has no negative jumps (which makes $\mathbf{L}$ almost surely have continuous paths). That important case appears in the convergence to reflected stable Lévy motion in Theorem 8.5.1.

We can also obtain a FCLT for T_n when $K < \infty$ under stronger continuity conditions and pointwise convergence under weaker conditions. (It may be possible to establish analogs to part (b) below without such strong continuity conditions.)

Theorem 5.9.3. (limits for the processing time when $K \leq \infty$) *Suppose that the conditions of Theorem 5.4.1 hold with $0 < K \leq \infty$ and $\mu_n \to \mu$, where $0 < \mu < \infty$.*

(a) If

$$P(t \in Disc(\mathbf{S})) = P(t \in Disc(\mathbf{W})) = 0 \ , \quad (9.12)$$

then

$$\mathbf{T}_n(t) \Rightarrow \mu^{-1}\mathbf{W}(t) \quad in \quad \mathbb{R} \ . \quad (9.13)$$

(b) If

$$P(\mathbf{C} \in C) = P(\mathbf{S} \in C) = 1 \ , \quad (9.14)$$

then

$$\mathbf{T}_n \Rightarrow \mu^{-1}\mathbf{W} \quad in \quad (D, M_1) \ , \quad (9.15)$$

where $P(\mathbf{W} \in C) = 1$.

Proof. (a) By (2.12),

$$n^{-1}T_n(nt) = inf\{u \geq 0 : S_n(n(t+u)) - S_n(nt) \geq W_n(nt)\} \ ,$$

so that

$$\mathbf{T}_n(t) = inf\{u \geq 0 : \mu_n u + \mathbf{S}_n(t + u(c_n/n)) - \mathbf{S}_n(t) \geq \mathbf{W}_n(t)\} \ .$$

By the continuous-mapping approach, with condition (9.12),

$$\mathbf{T}_n(t) \Rightarrow inf\{u \geq 0 : \mu u \geq \mathbf{W}(t)\} \ ,$$

which implies the conclusion in (9.13).

(b) Under condition (9.14),

$$\sup_{0 \leq t \leq T} \{|\mathbf{S}_n(t + u(c_n/n)) - \mathbf{S}_n(t)|\} \to 0 \quad as \quad n \to \infty \quad w.p.1$$

for any T with $0 < T < \infty$; see Section 12.4. Hence the conclusion in part (a) holds uniformly over all bounded intervals. An alternative proof follows the proof of Theorem 5.9.2, including the process $\{U(t) : t \geq 0\}$ when $K < \infty$. $\blacksquare$

Remark 5.9.1. *The heavy-traffic snapshot principle.* With the previous heavy-traffic theorems in this section, Theorems 5.9.2 and 5.9.3 establish a version of the heavy-traffic snapshot principle, a term coined by Reiman (1982): *In the heavy-traffic limit, the processing time is asymptotically negligible compared to the time required for the workloads to change significantly.* Since time is scaled by n, the workloads can change significantly only over time intervals of length of order n. On the other hand, since the space scaling is by c_n, where $c_n \to \infty$ but $c_n/n \to 0$ as $n \to \infty$, the workload itself tends to be only of order c_n, which is asymptotically negligible compared to n. Correspondingly, Theorems 5.9.2 and 5.9.3 show that that processing times also are of order c_n. Thus, in the heavy-traffic limit, the workload when a particle of work departs is approximately the same as the workload when that particle of work arrived.

The heavy-traffic snapshot principle also holds in queueing networks. Thus the workload seen upon each visit to a queue in the network and upon departure from the network by a particle flowing through the network is the same, in the heavy-traffic limit, as seen by that particle upon initial arrival. The heavy-traffic snapshot principle implies that network status can be communicated effectively in a heavily loaded communication network. A special packet sent from source to destination may record the buffer content at each queue on its path. Then this information may be passed back to the source by a return packet. The snapshot principle implies that the buffer contents at the queues will tend to remain near their original levels (relative to heavy-loading levels), so that the information does not become stale. (A caveat: With the fluid-limit scaling in Section 5.3, the heavy-traffic snapshot principle is not valid. In practice, we need to check if the snapshot principle applies.) For more on the impact of old information on scheduling service in queues, see Mitzenmacher (1997).

5.10. Priorities

In this book we primarily consider the standard first-come first-served (FCFS) service discipline in which input is served in order of arrival, but it can be important to consider other service disciplines to meet performance goals. We now illustrate how we can apply heavy-traffic stochastic-process limits to analyze a queue with a non-FCFS service discipline. Specifically, we now consider the fluid-queue model with priority classes. We consider the relatively tractable *preemptive-resume priority discipline*; i.e., higher-priority work immediately preempts lower-priority work and lower-priority work resumes service where it stopped when it regains access to the server. Heavy-traffic limits for the standard single-server queue with the preemptive-resume priority discipline were established by Whitt (1971a).

In general, there may be any number m of priority classes, but it suffices to consider only two because, from the perspective of any given priority class, all lower priority work can be ignored, and all higher-priority work can be lumped together. Thus, the model we consider now is the same as in Section 5.2 except that there are two priority classes. Let class 1 have priority over class 2. For $i = 1, 2$, there is a class-i cumulative-input stochastic process $\{C_i(t) : t \geq 0\}$. As before, there is a single server, a buffer with capacity K and a single service process $\{S(t) : t \geq 0\}$. (There is only a single shared buffer, not a separate buffer for each class.)

Like the polling service discipline considered in Section 2.4.2, the preemptive-resume priority service discipline is a work-conserving service policy. Thus the total workload process is the same as for the FCFS discipline considered above. We analyze the priority model to determine the performance enhancement experienced by the high-priority class and the performance degradation experienced by the low-priority class.

We first define class-i available-processing processes by letting

$$\begin{aligned} S_1(t) &\equiv S(t), \\ S_2(t) &\equiv S_1(t) - D_1(t) \,, \end{aligned} \qquad (10.1)$$

where $D_1 \equiv \{D_1(t) : t \geq 0\}$ is the class-1 departure process, defined as in (2.11). We then can define the class-i potential-workload processes by

$$X_i(t) \equiv W_i(0) + C_i(t) - S_i(t) \,, \qquad (10.2)$$

just as in (2.4). Then the class-i workload, overflow and departure processes are $W_i \equiv \phi_K(X_i)$, $U_i \equiv \psi_U(X_i)$ and $D_i \equiv S_i - \psi_L(X_i)$, just as in Section 5.2.

We now want to consider heavy-traffic limits for the two-priority fluid-queue model. As in Section 5.4, we consider a sequence of queues indexed by n. Suppose that the per-class input rates $\lambda_{1,n}$ and $\lambda_{2,n}$ and a maximum-potential output rate μ_n are well defined for each n, with limits as in (4.1) and (4.2). Then the class-i traffic intensity in model n is

$$\rho_{i,n} \equiv \lambda_{i,n}/\mu_n \qquad (10.3)$$

and the overall traffic intensity in model n is

$$\rho_n \equiv \rho_{1,n} + \rho_{2,n} \,. \qquad (10.4)$$

As a regularity condition, we suppose that $\mu_n \to \mu$ as $n \to \infty$, where $0 < \mu < \infty$.

In this context, there is some difficulty in establishing a single stochastic-process limit that generates useful approximations for both classes. It is natural to let

$$\rho_{i,n} \to \rho_i \,, \qquad (10.5)$$

where $0 < \rho_i < \infty$. If we let $\rho \equiv \rho_1 + \rho_2 = 1$, then the full system is in heavy traffic, but the high-priority class is in light traffic: $\rho_{1,n} \to \rho_1 < 1$ as $n \to \infty$. That implies that the high-priority workload will be asymptotically negligible compared to the total workload in the heavy-traffic scaling. That observation is an important insight, but it does not produce useful approximations for the high-priority class.

On the other hand, if we let $\rho_1 = 1$, then the high-priority class is in heavy traffic, but $\rho \equiv \rho_1 + \rho_2 > 1$, so that the full system is unstable. Clearly, neither of these approaches is fully satisfactory. Yet another approach is to have *both* $\rho_n \to 1$ and $\rho_{1,n} \to 1$ as $n \to \infty$, but that forces $\rho_{2,n} \to 0$. Such a limit can be useful, but if the low-priority class does not contribute a small proportion of the load, then that approach will usually be unsatisfactory as well.

5.10.1. A Heirarchical Approach

What we suggest instead is a *heirarchical approach* based on considering the relevant scaling. From the scaling analysis in Section 5.5, including the time and space scaling in (5.10) and (5.11), we can see that the full system with higher traffic intensity has greater scaling than the high-priority class alone. Thus, we suggest *first* doing a heavy-traffic stochastic-process limit for the high-priority class alone, based on letting $\rho_{1,n} \uparrow 1$ and, *second*, afterwards doing a second heavy-traffic limit for both priority classes, based on fixing ρ_1 and letting $\rho_{2,n} \uparrow 1 - \rho_1$.

As a basis for these heavy-traffic limits, we assume that

$$(\mathbf{C}_{1,n}, \mathbf{C}_{2,n}, \mathbf{S}_n) \Rightarrow (\mathbf{C}_1, \mathbf{C}_2, \mathbf{S}) \tag{10.6}$$

where

$$\begin{aligned}
\mathbf{C}_{1,n}(t) &\equiv n^{-H_{C,1}}(C_{1,n}(nt) - \lambda_{1,n}nt), \\
\mathbf{C}_{2,n}(t) &\equiv n^{-H_{C,2}}(C_{2,n}(nt) - \lambda_{2,n}nt), \\
\mathbf{S}_n(t) &\equiv n^{-H_S}(S_n(nt) - \mu nt)
\end{aligned} \tag{10.7}$$

for $0 < H_{C,1} < 1$, $0 < H_{C,2} < 1$ and $0 < H_S < 1$. For simplicity, we let the processing rate μ be independent of n.

Note that a common case of considerable interest is the light-tailed weak-dependent case with space-scaling exponents

$$H_{C,1} = H_{C,2} = H_S = 1/2 \;, \tag{10.8}$$

but we allow other possibilities. We remark that in the light-tailed case with scaling exponents in (10.8) the heirarchical approach can be achieved directly using strong approximations; see Chen and Shen (2000). (See Section 2.2 of the Internet Supplement for a discussion of strong approximations.)

When (10.8) does not hold, then it is common for one of the three space-scaling exponents to dominate. That leads simplifications in the analysis that should be exploited. In the heavy-traffic limit, variability appears only for the processes with the largest scaling exponent.

Given a heavy-traffic stochastic-process limit as in Theorem 5.4.1 for the high-priority class alone with the space scaling factors in (10.7), we obtain the high-priority approximation

$$W_{1,\rho_1}(t) \approx \left(\frac{\zeta_1}{1-\rho_1}\right)^{\frac{H_1}{1-H_1}} \mathbf{W}_1\left(\left(\frac{1-\rho_1}{\zeta_1}\right)^{\frac{1}{1-H_1}} t\right), \quad t \geq 0 \;, \tag{10.9}$$

as in (5.3) with the scaling functions in (5.10) and (5.11) based on the traffic intensity ρ_1 and the space-scaling exponent

$$H_1 = max\{H_{C,1}, H_S\} \;. \tag{10.10}$$

The limit process $\mathbf{W}_1$ in (10.9) is $\phi_K(\mathbf{X}_1)$ as in (4.9), where

$$\mathbf{X}_1(t) = W_1'(0) + \mathbf{C}_1(t) - \mathbf{S}(t) + \eta_1 t, \quad t \geq 0,$$

as in (4.8). If $H_{C,1} > H_S$, then $\mathbf{S}(t) = 0$ in the limit; if $H_S > H_{C,1}$, then $\mathbf{C}_1(t) = 0$ in the limit. Instead of (4.6), here we have

$$\eta_{1,n} \equiv n(\lambda_{1,n} - \mu_n)/c_n \to \eta_1 .$$

Next we can treat the aggregate workload of both classes using traffic intensity $\rho = \rho_1 + \rho_2$. We can think of the high-priority traffic intensity ρ_1 as fixed with $\rho_1 < 1$ and let $\rho_{2,n} \uparrow 1 - \rho_1$. By the same argument leading to (10.9), we obtain a heavy-traffic stochastic-process limit supporting the approximation

$$W_\rho(t) \approx \left(\frac{\zeta}{1-\rho}\right)^{\frac{H}{1-H}} \mathbf{W}\left(\left(\frac{1-\rho}{\zeta}\right)^{\frac{1}{1-H}} t\right), \quad t \geq 0, \qquad (10.11)$$

where the space-scaling exponent now is

$$H = max\{H_{C,1}, H_{C,2}, H_S\} . \qquad (10.12)$$

The limit process $\mathbf{W}$ in (10.11) is $\phi_K(\mathbf{X})$ as in (4.9), where

$$\mathbf{X}(t) = W'(0) + \mathbf{C}_1(t) + \mathbf{C}_2(t) - \mathbf{S}(t) + \eta t, \quad t \geq 0,$$

as in (4.8). If $H_{C,i} < H$, then $\mathbf{C}_i(t) = 0$ in the limit; if $H_S < H$, then $\mathbf{S}(t) = 0$ in the limit. Instead of (4.6), here we have

$$\eta_n \equiv n(\lambda_{1,n} + \lambda_{2,n} - \mu_n)/c_n \to \eta .$$

Not only may the space-scaling exponent H in (10.11) differ from its counterpart H_1 in (10.9), but the parameters ρ and ζ in (10.11) routinely differ from their counterparts ρ_1 and ζ_1 in (10.9).

Of course, the low-priority workload is just the difference between the aggregate workload and the high-priority workload. If that difference is too complicated to work with, we can approximate the low-priority workload by the aggregate workload, since the high-priority workload should be relatively small, i.e.,

$$W_{2,\rho_2}(t) = W_\rho(t) - W_{1,\rho_1}(t) \approx W_\rho(t), \quad t \geq 0 . \qquad (10.13)$$

5.10.2. Processing Times

We now consider the *per-class processing times*, i.e., the times required to complete processing of all work of that class in the system. For the high-priority class, we can apply Theorems 5.9.2 and 5.9.3 to justify (only partially when $K < \infty$) the approximation

$$T_{1,\rho_1}(t) \approx W_{1,\rho_1}(t)/\mu . \qquad (10.14)$$

However, the low-priority processing time is more complicated because the last particle of low-priority work must wait, not only for the total aggregate workload to be processed, but also for the processing of all new high-priority work to arrive while that processing of the initial workload is going on. Nevertheless, the low-priority processing time is relatively tractable because it is the time required for the class-1 net input, starting from time t, to decrease far enough to remove the initial aggregate workload, i.e.,

$$T_2(t) \equiv inf\{u > 0 : X_1(t+u) - X_1(t) < -W(t)\} \ . \qquad (10.15)$$

Note that (10.15) is essentially of the same form as (2.12). Thus, we can apply (10.15) with the reasoning in Theorem 5.9.3 to establish an analog of Theorem 5.9.3, which partly justifies the heavy-traffic approximation

$$T_{2,\rho_1,\rho_2}(t) \approx \frac{W_\rho(t)}{\mu(1-\rho_1)} \ . \qquad (10.16)$$

In (10.16), $T_{2,\rho_1,\rho_2}(t)$ is the low-priority processing time as a function of the two traffic intensities and $W_\rho(t)$ is the aggregate workload at time t as a function of the total traffic intensity $\rho = \rho_1 + \rho_2$.

The heavy-traffic approximation in (10.16) should not be surprising because, as $\rho \uparrow 1$ with ρ_1 fixed, the stochastic fluctuations in X_1 should be negligible in the relatively short time required for the drift in X_1 to hit the target level; i.e., we have a separation of time scales just as in Section 2.4.2.

However, in applications, it may be important to account for the stochastic fluctuations in X_1. That is likely to be the case when ρ_1 is relatively high compared to ρ. Fortunately, the heavy-traffic limits also suggest a refined approximation. Appropriate heavy-traffic limits for X_1 alone suggest that the stochastic process $\{X_1(t) : t \geq 0\}$ can often be approximated by a Lévy process (a process with stationary and independent increments) without negative jumps. Moreover, the future net input $\{X_1(t+u) - X_1(t) : t \geq 0\}$ often can be regarded as approximately independent of $W(t)$. Under those approximating assumptions, the class-2 processing time in (10.15) becomes tractable. The Laplace transform of the conditional processing-time distribution given $W(t)$ is given on p.120 of Prabhu (1998). The conditional mean is the conditional mean in the heavy-traffic approximation in (10.16).

Remark 5.10.1. *Other service disciplines.* We conclude this section by referring to work establishing heavy-traffic limits for non-FCFS service disciplines. First, in addition to Chen and Shen (2000), Boxma, Cohen and Deng (1999) establish heavy-traffic limits for priority queues. As mentioned in Section 2.4.2, Coffman, Puhalskii and Reiman (1995, 1998), Olsen (2001), van der Mei and Levy (1997) and van der Mei (2000, 2001) establish heavy-traffic limits for polling service disciplines. Kingman (1982) showed how heavy-traffic limits can expose the behavior of a whole class of service disciplines related to random order of service. Yashkov (1993), Sengupta (1992), Grishechkin (1994), Zwart and Boxma (2000) and Boxma and Cohen (2000) establish heavy-traffic limits for the processor-sharing discipline. Fendick and Rodrigues (1991) develop a heavy-traffic approximation for the head-of-the-line generalized processor-sharing discipline. Abate and Whitt (1997a) and Limic

(1999) consider the last-in first-out service discipline. Doytchinov et al. (2001) and Kruk et al. (2000) consider "real-time" queues with due dates. These alternative service disciplines are important because they significantly affect queueing performance. As we saw for the high-priority class with two priority classes, the alternative service disciplines can effectively control congestion for some customers when the input of other customers is excessive. The derivations of the heavy-traffic limits with these alternative service disciplines are fascinating because they involve quite different arguments.

6
Unmatched Jumps in the Limit Process

6.1. Introduction

As illustrated by the random walks with Pareto steps in Section 1.4 and the workload process with Pareto inputs in Section 2.3, it can be important to consider stochastic-process limits in which the limit process has jumps, i.e., has discontinuous sample paths. The jumps observed in the plots in Chapters 1 and 2 correspond to exceptionally large increments in the plotted sequences, i.e., large steps in the simulated random walk and large inputs of required work in the simulated workload process of the queue. Thus, in the associated stochastic-process limit, the jumps in the limit process are *matched* by corresponding jumps in the converging processes. However, there are related situations in which the jumps in the limit process are not matched by jumps in the converging processes.

Indeed, a special focus of this book is on stochastic-process limits with unmatched jumps in the limit process. In the extreme case, the converging stochastic processes have continuous sample paths. Then the sample paths of the converging processes have portions with steep slope corresponding to the limiting jumps. In other cases, a single jump in the sample path of the limiting stochastic process corresponds to many small jumps in the sample path of one of the converging stochastic processes. In this chapter we give several examples showing how a stochastic-process limit with unmatched jumps in the limit process can arise. Most of these examples will be treated in detail later.

We give special attention to stochastic-process limits with unmatched jumps in the limit process because they represent an interesting phenomenon and because they require special treatment beyond the conventional theory. In particular, as discussed in Section 3.3, whenever there are unmatched jumps in the limit process, we cannot have a stochastic-process limit in the function space D with the conventional Skorohod (1956) J_1 topology. To establish the stochastic-process limit, we instead use the M_1 topology.

Just as in Chapter 1, we primarily draw our conclusions in this chapter by looking at pictures. By plotting initial segments of the stochastic processes for various sample sizes, we can see the stochastic-process limits emerging before our eyes. As before, the plots often do the proper scaling automatically, and thus reveal statistical regularity associated with a macroscopic view of uncertainty. The plots also show the relevance of stochastic-process limits with unmatched jumps in the limit process.

First, though, we should recognize that it is common for the limit process in a stochastic-process limit to have continuous sample paths. For example, that is true for Brownian motion, which is the canonical limiting stochastic process, occurring as the limit in Donsker's theorem, discussed in Chapters 1 and 4. In many books on stochastic-process limits, *all* the stochastic-process limits that are considered have limit processes with continuous sample paths, and there is much to consider.

Moreover, when a limit process in a stochastic-process limit does have discontinuous sample paths, the jumps in the limit process are often matched in the converging processes. We have already pointed out that only matched jumps appear in the examples in Chapters 1 and 2. Indeed, there is a substantial literature on stochastic-process limits where the limit process may have jumps and those jumps are matched in the converging processes. The extreme-value limits in Resnick (1987) and the many stochastic-process limits in Jacod and Shiryaev (1987) are all of this form.

However, even for the examples in Chapter 1 with limit processes having discontinuous sample paths, we would have stochastic-process limits with unmatched jumps in the limit process if we formed the continuous-time representation of the discrete-time process using linear interpolation, as in (2.1) in Chapter 1. We contend that the linearly interpolated processes should usually be regarded as asymptotically equivalent to the step-function versions used in Chapter 1; i.e., one sequence of scaled processes should converge if and only if the other does, and they should have the same limit process. That asymptotic eqivalence is suggested by Figure 1.13, which plots the two continuous-time representations of a random walk with uniform random steps. As the sample size n increases, both versions approach Brownian motion. Indeed, as n increases, the two alternative continuous-time representations become indistinguishable.

In Section 6.2 we look at more examples of random walks, comparing the linearly interpolated continuous-time representations (which always have continuous sample paths) to the standard step-function representation for the same random-walk sample paths. Now we make this comparison for random walks approaching a limit process with discontinuous sample paths. Just as in Chapter 1, we obtain jumps in the limit process by considering random walks with steps having a heavy-tailed distribution, in particular, a Pareto distribution. As before, the plots reveal statistical regularity. The plots also show that it is natural to regard the two continuous-time representations of scaled discrete-time processes as asymptotically equivalent.

However, the unmatched jumps in the limit process for the random walks in Section 6.2 can be avoided if we use the step-function representation instead of the linearly interpolated version. Since the step-function version seems more natural anyway, the case for considering unmatched jumps in the limit process is not yet very strong. In the rest of this chapter we give examples in which stochastic-process limits with unmatched jumps in the limit process cannot be avoided.

6.2. Linearly Interpolated Random Walks

All the stochastic-process limits with jumps in the limit process considered in Chapter 1 produce unmatched jumps when we form the continuous-time representation of the original discrete-time process by using linear interpolation. We now want to show, by example, that it is natural to regard the linearly interpolated continuous-time representation as asymptotically equivalent to the standard step-function representation in settings where the limit process has jumps.

Given a random walk or any discrete-time process $\{S_k : k \geq 0\}$, the scaled-and-centered step-function representations are defined for each $n \geq 1$ by

$$\mathbf{S}_n(t) \equiv c_n^{-1}(S_{\lfloor nt \rfloor} - m\lfloor nt \rfloor), \quad 0 \leq t \leq 1, \tag{2.1}$$

where $\lfloor x \rfloor$ is the greatest integer less than x and $c_n \to \infty$ as $n \to \infty$. The associated linearly interpolated versions are

$$\tilde{\mathbf{S}}_n(t) \equiv (nt - \lfloor nt \rfloor)\mathbf{S}_n((\lfloor nt \rfloor + 1)/n) + (1 + \lfloor nt \rfloor - nt)\mathbf{S}_n(\lfloor nt \rfloor/n), \tag{2.2}$$

for $0 \leq t \leq 1$. Clearly the sample paths of $\mathbf{S}_n$ in (2.1) are discontinuous for all n (except in the special case in which $S_k = S_0, 1 \leq k \leq n$), while the sample paths of $\tilde{\mathbf{S}}_n$ in (2.2) are continuous for all n.

6.2.1. Asymptotic Equivalence with M_1

We contend that the two sequences of processes $\{\mathbf{S}_n : n \geq 0\}$ and $\{\tilde{\mathbf{S}}_n : n \geq 0\}$ in the function space $D \equiv D([0,1], \mathbb{R})$ should be *asymptotically equivalent*, i.e., if either converges in distribution as $n \to \infty$, then so should the other, and they should have the same limit. It is easy to see that the desired asymptotic equivalence holds with the M_1 metric. In particular, we can show that $d_{M_1}(\mathbf{S}_n, \tilde{\mathbf{S}}_n) \Rightarrow 0$ as $n \to \infty$.

Theorem 6.2.1. (the M_1 distance between the continuous-time representations) *For any discrete-time process $\{S_k : k \geq 0\}$,*

$$d_{M_1}(\mathbf{S}_n, \tilde{\mathbf{S}}_n) \leq n^{-1} \quad \text{for all} \quad n \geq 1,$$

for $\mathbf{S}_n$ in (2.1) and $\tilde{\mathbf{S}}_n$ in (2.2).

196   6. Unmatched Jumps

Proof. For the M_1 metric, we can use an arbitrary parametric representation of the step-function representation $\mathbf{S}_n$. Then, for any $\epsilon > 0$, we can construct the associated parametric representation of $\tilde{\mathbf{S}}_n$ so that it agrees with the other parametric reprentation at the finitely many points in the domain $[0,1]$ mapping into the points $(k/n, \mathbf{S}_n(k/n))$ on the completed graph of $\mathbf{S}_n$ for $0 \le k \le n$, with the additional property that the spatial components of the two parametric representations differ by at most $n^{-1} + \epsilon$ anywhere. Since ϵ was arbitrary, we obtain the desired conclusion. ∎

We can apply Theorem 6.2.1 and the convergence-together theorem, Theorem 11.4.7, to establish the desired asymptotic equivalence with respect to convergence in distribution.

Corollary 6.2.1. (asymptotic equivalence of continuous-time representations) *If either $\mathbf{S}_n \Rightarrow \mathbf{S}$ in (D, M_1) or $\tilde{\mathbf{S}}_n \Rightarrow \mathbf{S}$ in (D, M_1), then both limits hold.*

Note that the conclusion of Theorem 6.2.1 is much stronger than the conclusion of Corollary 6.2.1. Corollary 6.2.1 concludes that $\mathbf{S}_n$, $\tilde{\mathbf{S}}_n$ and $\mathbf{S}$ all have approximately the same probability laws for all suitably large n, whereas Theorem 6.2.1 concludes that the individual sample paths of $\mathbf{S}_n$ and $\tilde{\mathbf{S}}_n$ are likely to be close for all suitably large n.

We used plots to illustrate the asymptotic equivalence of $\tilde{\mathbf{S}}_n$ and $\mathbf{S}_n$ for random walks with uniform steps, for which the limit process is Brownian motion, in Figure 1.13. That asymptotic equivalence is proved by Corollary 6.2.1. (Since the limit process has continuous sample paths, the various nonuniform Skorohod topologies are equivalent in this example.)

Now we use plots again to illustrate the asymptotic equivalence of $\tilde{\mathbf{S}}_n$ and $\mathbf{S}_n$ in random walks with jumps in the limit process. Since the asymptotic equivalence necessarily holds in the M_1 topology by virtue of Corollary 6.2.1, but not in the J_1 topology, we are presenting a case for using the M_1 topology.

6.2.2. Simulation Examples

We give three examples, all involving variants of the Pareto distribution.

Example 6.2.1. *Centered random walk with Pareto(p) steps.*

As in (3.5) (iii) in Section 1.3, we consider the random walk $\{S_k : k \ge 0\}$ with IID steps

$$X_k \equiv U_k^{-1/p} \tag{2.3}$$

for U_k uniformly distributed on the interval $[0, 1]$. The steps then have a Pareto(p) distribution with parameter p, having ccdf $F^c(t) = t^{-p}$ for $t \ge 1$. We first consider the case $1 < p \le 2$. In that case, the steps have a finite mean $m = 1 + (p-1)^{-1}$ but infinite variance. In Figures 1.20 – 1.22, we saw that the plots of the centered random walks give evidence of jumps. The supporting FCLT (in Section 4.5) states that the step-function representations converge in distribution to a stable Lévy motion, which indeed has discontinuous sample paths.

6.2. Linearly Interpolated Random Walks

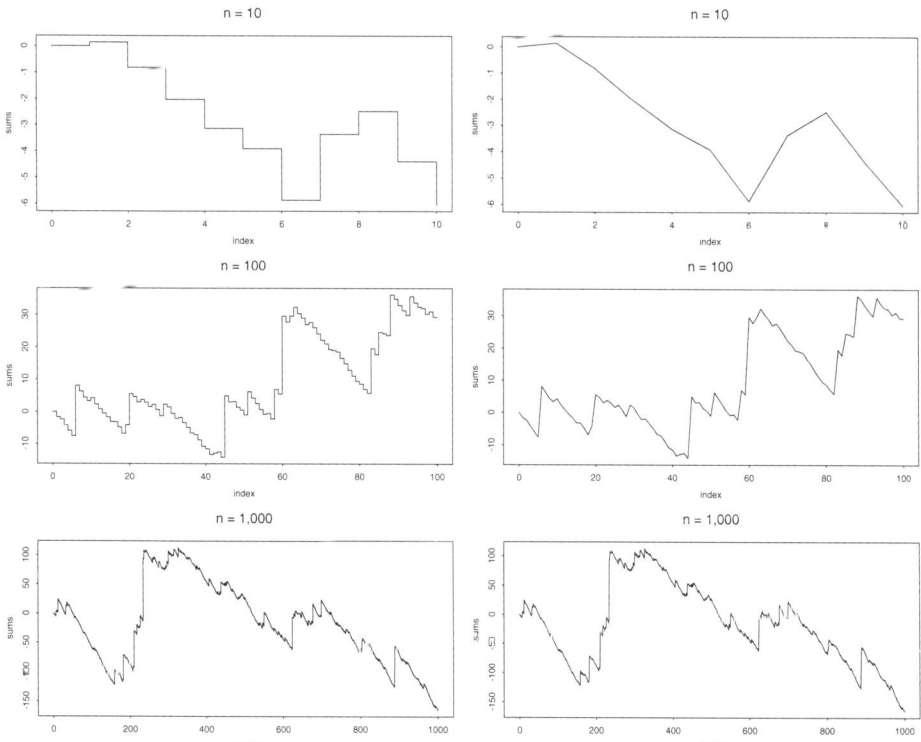

Figure 6.1. Plots of the two continuous-time representations of the centered random walk with Pareto(1.5) steps for $n = 10^j$ with $j = 1, 2, 3$. The step-function representation $\mathbf{S}_n$ in (2.1) appears on the left, while the linearly interpolated version $\tilde{\mathbf{S}}_n$ in (2.2) appears on the right.

Just as in Chapter 1, we use the statistical package S to simulate and plot the initial segments of the stochastic processes. Plots of the two continuous-time representations $\mathbf{S}_n$ and $\tilde{\mathbf{S}}_n$ for the same sample paths of the random walk are given for the case $p = 1.5$ and $n = 10^j$ with $j = 1, 2, 3$ in Figure 6.1. For $n = 10$, the two continuous-time representations look quite different. Indeed, at first it may seem that they cannot be corresponding continuous-time representations of the same realized segment of the random walk, but closer examination shows that the two continuous-time representations are correct. However, for $n = 100$ and beyond, the two continuous-time representations look very similar. For larger values of n such as $n = 10^4$ and beyond, the two continuous-time representations look virtually identical.

So far we have considered only $p = 1.5$. We now illustrate how the plots depend on p for $1 < p \le 2$. In Figure 6.2 we plot the two continuous-time representations of the random walk with Pareto(p) steps for three values of p, in particular for $p = 1.1, 1.5$ and 1.9. We do the plot for the case $n = 100$ using the *same uniform random numbers* (exploiting (2.3)). In each plot the largest steps stem from the

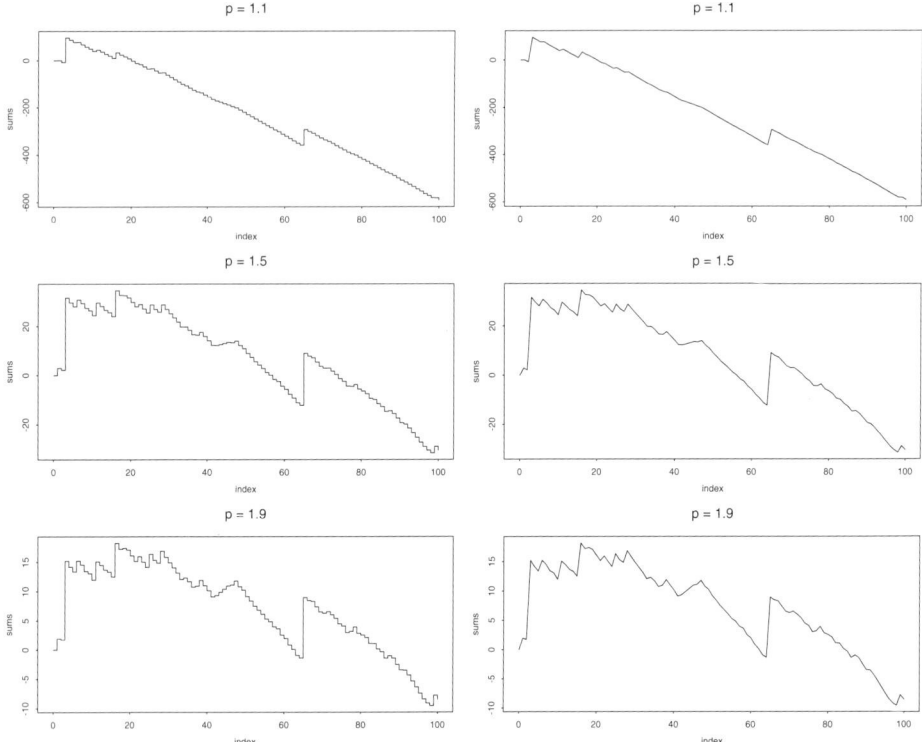

Figure 6.2. Plots of the two continuous-time representations of the centered random walk with Pareto(p) steps with $p = 1.1, 1.5$ and 1.9 for $n = 10^2$ based on the same uniform random numbers (using (2.3)). The step-function representation $\mathbf{S}_n$ in (2.1) appears on the left, while the linearly interpolated version $\tilde{\mathbf{S}}_n$ in (2.2) appears on the right.

smallest uniform random numbers. The three smallest uniform random numbers in this sample were $U_3 = 0.00542$, $U_{65} = 0.00836$ and $U_{16} = 0.0201$. The corresponding large steps can be seen in each case of Figure 6.2. Again, we see that the limiting stochastic process should have jumps (up). That conclusion is confirmed by considering larger and larger values of n. As in Figures 6.1 and 6.2, the two continuous-time representations look very similar. And the little difference we see for $n = 100$ deceases as n increases.

Example 6.2.2. *Uncentered random walk with Pareto(0.5) steps.* In Figures 1.19, 1.25 and 1.26 we saw that the *uncentered* random walk with Pareto(0.5) steps should have stochastic-process limits with jumps in the limit process. The supporting FCLT implies convergence to another stable Lévy motion as $n \to \infty$ (again see Section 4.5). Moreover, such a limit holds for IID Pareto(p) steps whenever $p \le 1$, because then the steps have infinite mean.

Now we look at the two continuous-time representations in this setting. We now plot the two continuous-time representations $\tilde{\mathbf{S}}_n$ and $\mathbf{S}_n$ associated with the un-

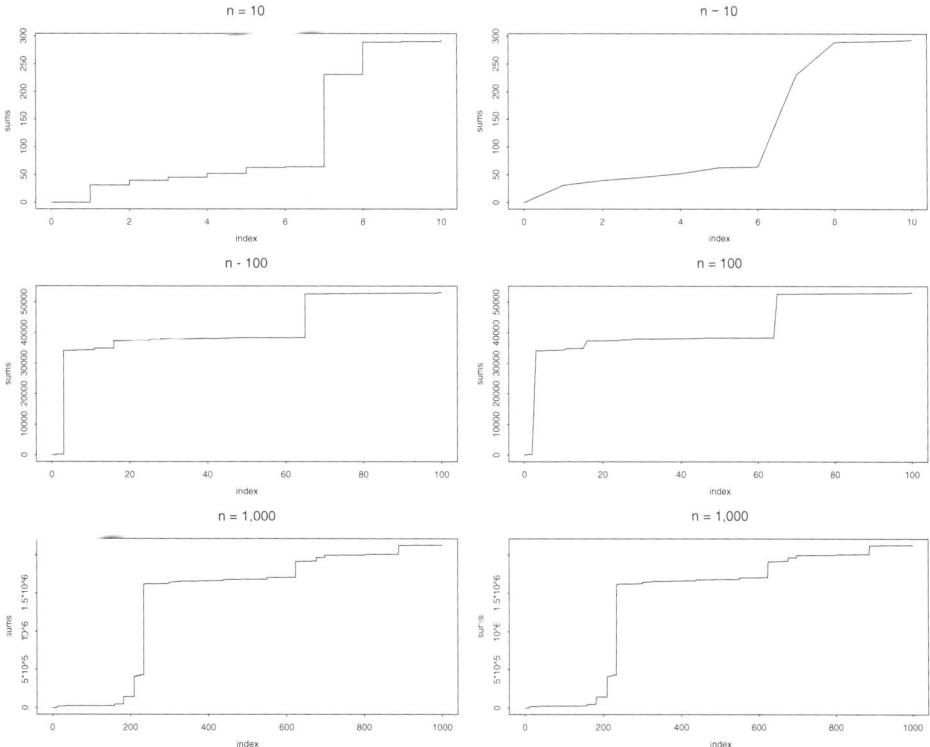

Plots of the uncentered random walk with Pareto(0.5) steps for $n = 10^j$ with $j = 1, 2, 3$. The step-function representation $\mathbf{S}_n$ in (2.1) appears on the left, while the linearly interpolated version $\tilde{\mathbf{S}}_n$ in (2.2) appears on the right.

centered random walk with Pareto(0.5) steps for $n = 10^j$ with $j = 1, 2, 3$ in Figure 6.2.2. Again, the two continuous-time representations initially (for small n) look quite different, but become indistinguishable as n increases. Just as in Chapter 1, even though there are jumps, we see statistical regularity associated with large n. Experiments with different n show the self-similarity discussed before.

Example 6.2.3. *Centered random walk with limiting jumps up and down.*

The Pareto distributions considered above have support on the interval $[1, \infty)$, so that, even with centering, the positive tail of the step-size distribution is heavy, but the negative tail of the step-size distribution is light. Consequently the limiting stochastic process in the stochastic-process limit for the random walks with Pareto steps can only have jumps up. (See Section 4.5)

We can obtain a limit process with both jumps up and jumps down if we again use (2.3) to define the steps, but we let U_k be uniformly distributed on the interval $[-1, 1]$ instead of in $[0, 1]$. Then we can have both arbitrarily large negative jumps and arbitrarily large positive jumps. We call the resulting distribution a *symmetric*

200   6. Unmatched Jumps

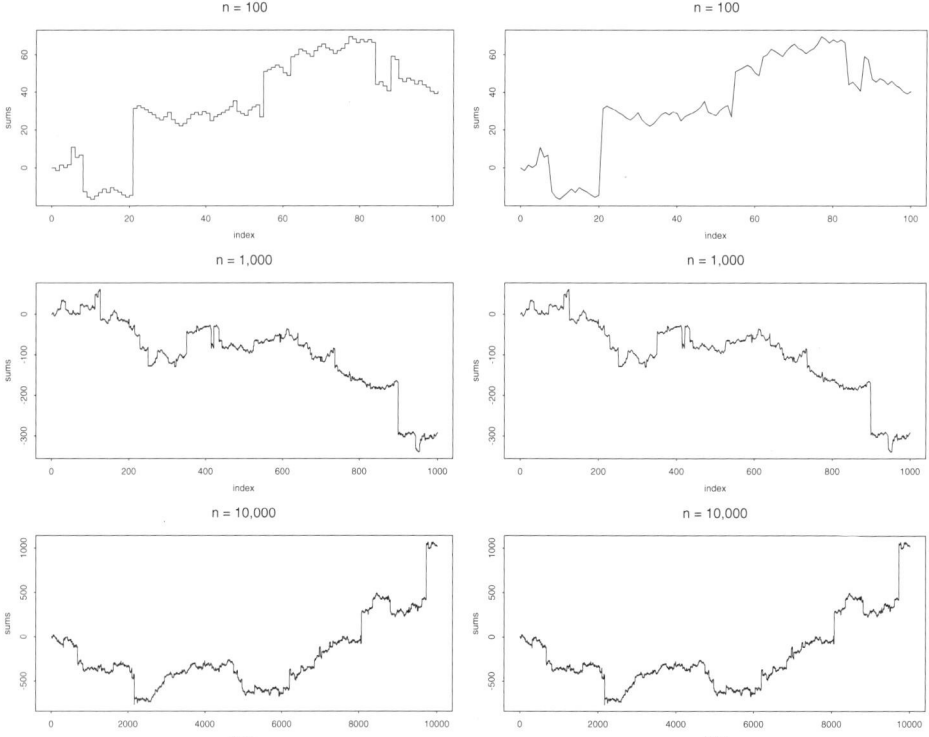

Figure 6.3. Plots of the two continuous-time representations of the random walk with symmetric Pareto(1.5) steps for $n = 10^j$ with $j = 2, 3, 4$. The step-function representation $\mathbf{S}_n$ in (2.1) appears on the left, while the linearly interpolated version $\tilde{\mathbf{S}}_n$ in (2.2) appears on the right.

Pareto distribution (with parameter p). Since the distribution is symmetric, no centering need be done for the plots or the stochastic-process limits.

To illustrate, we make additional comparisons between the linearly interpolated continuous-time representation and the step-function continuous-time representation of the random walk, now using the symmetric Pareto(p) steps for $p = 1.5$. The plots are shown in Figure 6.3. We plot the two continuous-time representations for $n = 10^j$ with $j = 2, 3, 4$. From the plots, it is evident that the limit process now should have jumps down as well as jumps up. Again, the two continuous-time representations look almost identical for large n.

6.3. Heavy-Tailed Renewal Processes

One common setting for stochastic-process limits with unmatched jumps in the limit process, which underlies many applications, is a heavy-tailed renewal process. Given partial sums $S_k \equiv X_1 + \cdots + X_k, k \geq 1$, from a sequence of nonnegative random variables $\{X_k : k \geq 1\}$ (without an IID assumption), the associated stochastic process $N \equiv \{N(t) : t \geq 0\}$ defined by

$$N(t) \equiv max\{k \geq 0 : S_k \leq t\}, \quad t \geq 0, \tag{3.1}$$

where $S_0 \equiv 0$, is called a *stochastic counting process*. When the random variables X_k are IID, the counting process is called a *renewal counting process* or just a *renewal process*.

6.3.1. Inverse Processes

Roughly speaking (we will be more precise in Chapter 13), the stochastic processes $\{S_k : k \geq 1\}$ and $N \equiv \{N(t) : t \geq 0\}$ can be regarded as *inverses* of each other, without imposing the IID condition, because

$$S_k \leq t \quad \text{if and only if} \quad N(t) \geq k. \tag{3.2}$$

The M_1 topology is convenient for relating limits for partial sums to associated limits for the counting processes, because the M_1-topology definition makes it easy to exploit the inverse relation in the continuous-mapping approach.

Moreover, it is not possible to use the standard J_1 topology to establish limits of scaled versions of the counting processes, because the J_1 topolgy requires all jumps in the limit process to be matched in the converging stochastic processes. The difficulty with the J_1 topology on D can easily be seen when the random variables X_k are strictly positive. Then the counting process N increases in unit jumps, and scaled versions of the counting process, such as

$$\mathbf{N}_n(t) \equiv c_n^{-1}(N(nt) - m^{-1}nt), \quad t \geq 0, \tag{3.3}$$

where $c_n \to \infty$, have jumps of magnitude $1/c_n$, which are asymptotically negligible as $n \to \infty$. Hence, if $\mathbf{N}_n$ in (3.3) is ever to converge as $n \to \infty$ to a limiting stochastic process with discontinuous sample paths, then we must have unmatched jumps in the limit process. Then we need the M_1 topology on D.

What is not so obvious, however, is that $\mathbf{N}_n$ will ever converge to a limiting stochastic process with discontinuous sample paths. However, such limits can indeed occur. Here is how: A long interrenewal time creates a long interval between jumps up in the renewal process. The long interrenewal time appears horizontally rather than vertically, not directly causing a jump. However, during such an interval, the scaled process in (3.3) will decrease linearly at rate n/mc_n, due to the translation term not being compensated for by any jumps up. When $n/c_n \to \infty$ (the usual case), the slope approaches $-\infty$. When the interrenewal times are long enough, these portions of the sample path with steep slope down can lead to jumps *down* in the limit process.

A good way to see how jumps can appear in the limit process for $\mathbf{N}_n$ is to see how limits for $\mathbf{N}_n$ in (3.3) are related to associated limits for $\mathbf{S}_n$ in (2.1) when both scaled processes are constructed from the same underlying process $\{S_k : k \geq 0\}$. A striking result from the continuous-mapping approach to stochastic-process limits (to be developed in Chapter 13) is an equivalence between stochastic-process limits for partial sums and associated counting processes, exploiting the M_1 topology (but not requiring any direct independence or common-distribution assumption). As a consequence of Corollary 13.8.1, we have the following result:

Theorem 6.3.1. (FCLT equivalence for counting processes and associated partial sums) *Suppose that* $0 < m < \infty$, $c_n \to \infty$, $n/c_n \to \infty$ *and* $\mathbf{S}(0) = 0$. *Then*

$$\mathbf{S}_n \Rightarrow \mathbf{S} \quad in \quad (D, M_1) \tag{3.4}$$

for $\mathbf{S}_n$ *in* (2.1) *if and only if*

$$\mathbf{N}_n \Rightarrow \mathbf{N} \quad in \quad (D, M_1) \tag{3.5}$$

for $\mathbf{N}_n$ *in* (3.3), *in which case*

$$(\mathbf{S}_n, \mathbf{N}_n) \Rightarrow (\mathbf{S}, \mathbf{N}) \quad in \quad (D^2, WM_1) , \tag{3.6}$$

where the limit processes are related by

$$\mathbf{N}(t) \equiv (m^{-1}\mathbf{S} \circ m^{-1}\mathbf{e})(t) \equiv m^{-1}\mathbf{S}(m^{-1}t), \quad t \geq 0 , \tag{3.7}$$

or, equivalently,

$$\mathbf{S}(t) = (m\mathbf{N} \circ m\mathbf{e})(t) \equiv m\mathbf{N}(mt), \quad t \geq 0 , \tag{3.8}$$

where $\mathbf{e}(t) = t$, $t \geq 0$.

Thus, *whenever the limit process* $\mathbf{S}$ *in* (3.4) *has discontinuous sample paths, the limit process* $\mathbf{N}$ *in* (3.5) *necessarily has discontinuous sample paths as well*. Moreover, $\mathbf{S}$ has only jumps up (down) if and only if $\mathbf{N}$ has only jumps down (up). Whenever $\mathbf{S}$ and $\mathbf{N}$ have discontinuous sample paths, the M_1 topology is needed to express the limit for $\mathbf{N}_n$ in (3.5). In contrast, the limit for $\mathbf{S}_n$ in (3.4) can hold in (D, J_1).

6.3.2. The Special Case with $m = 1$

The close relation between the limit processes $\mathbf{S}$ and $\mathbf{N}$ in (3.4) – (3.8) is easy to understand and visualize when we consider plots for the special case of strictly positive steps X_k with translation scaling constant $m = 1$. Note that the limit process $\mathbf{N}$ in (3.7) becomes simply $-\mathbf{S}$ when $m = 1$.

Also note that we can always scale so that $m = 1$ without loss of generality: For any given sequence $\{X_k : k \geq 0\}$, when we multiply X_k by m for all k, we replace $\mathbf{S}_n$ by $m\mathbf{S}_n$ and $\mathbf{N}_n$ by $\mathbf{N}_n \circ m^{-1}\mathbf{e}$. Hence, the limits $\mathbf{S}$ and $\mathbf{N}$ are replaced by $m\mathbf{S}$ and $\mathbf{N} \circ m^{-1}\mathbf{e}$, respectively.

Hence, suppose that $m = 1$. A useful observation, then, is that $N(S_k) = k$ for all k. (We use the assumption that the variables X_k are strictly positive.) With that in mind, note that we can plot $N(t) - t$ versus t, again using the statistical package S, by plotting the points $(0, 0)$, $(S_k, N(S_k) - 1 - S_k)$ and $(S_k, N(S_k) - S_k)$ in the plane $\mathbb{R}^2$ and then performing linear interpolation between successive points.

Roughly speaking, then, we can plot $N(t) - t$ versus t by plotting $N(S_k) - S_k$ versus S_k. On the other hand, when we plot the centered random walk $\{S_k - k :$

$k \geq 0\}$, we plot $(S_k - k)$ versus k. Since $N(S_k) = k$, we have

$$N(S_k) - S_k = k - S_k = -(S_k - k) \ .$$

Thus, the second component of the pair $(S_k, N(S_k) - S_k)$ is just minus 1 times the second component of the pair $(k, S_k - k)$. Thus, the plot of $N(t) - t$ versus t should be very close to the plot of $-(S_k - k)$ versus k. The major difference is in the first component: For the renewal process, the first component is S_k; for the random walk, the first component is k. However, since $n^{-1}S_n \to 1$ as $n \to \infty$ by the SLLN, that difference between these two first components disappears as $n \to \infty$.

Example 6.3.1. *Centered renewal processes with Pareto(p) steps for $1 < p < 2$.* By now, we are well acquainted with a situation in which the limit for $\mathbf{S}_n$ in (3.4) holds and the limit process $\mathbf{S}$ has discontinuous sample paths: That occurs when the underlying process $\{S_k : k \geq 0\}$ is a random walk with IID Pareto(p) steps for $1 < p < 2$. Then the limit (3.4) holds with $m = 1 + (p-1)^{-1}$ and $\mathbf{S}$ being a stable Lévy motion, which has discontinuous sample paths. The discontinuous sample paths are clearly revealed for the case $p = 1.5$ in Figures 1.20 – 1.22 and 6.1.

To make the relationship clear, we consider the case $m = 1$. We obtain $m = 1$ in our example with IID Pareto(1.5) steps by dividing the steps by 3; i.e., we let $X_k \equiv U_k^{-2/3}/3$. For this example with Pareto(1.5) steps having ccdf decay rate $p = 3/2$ and mean 1, we plot both the centered renewal process ($N(t) - t$ versus t) and minus 1 times the centered random walk ($-(S_k - k)$ versus k). We plot both sample paths, putting the centered renewal process on the left, for the cases $n = 10^j$ with $j = 1, 2, 3$ in Figure 6.4. We plot three possible representations of each for $n = 10^4$ in Figure 6.5. (We plot the centered random walk directly; i.e., we do not use either of the continuous-time represenations.)

For small n, the sample paths of the two centered processes look quite different, but as n increases, the sample paths begin to look alike. The jumps in the centered random walk plot are matched with portions of the centered-renewal-process plot with very steep slope. As n increases, the slopes in the portions of the centered-renewal-process plots corresponding to the random-walk jumps tend to get steeper and steeper, approaching the jump itself.

It is natural to wonder how the plots look as the decay rate p changes within the interval (1,2), which is the set of values yielding a finite mean but an infinite variance. We know that for smaller p the jumps are likely to be larger. To see what happens, we plot three realizations each of the centered renewal process and minus 1 times the centered random walk for Pareto steps having decay rates $p = 7/4$ and $p = 5/4$ (normalized as before to have mean 1) for $n = 10^4$ in Figures 6.6 and 6.7. From Figures 6.5 – 6.7, we see that the required space scaling decreases, the two irregular paths become closer, and the slopes in the renewal-process plot become steeper, as p increases from 5/4 to 3/2 to 7/4. For $p = 5/4$, we need larger n to see steeper slopes. However, in all cases we can see that there should be unmatched jumps in the limit process. ∎

204   6. Unmatched Jumps

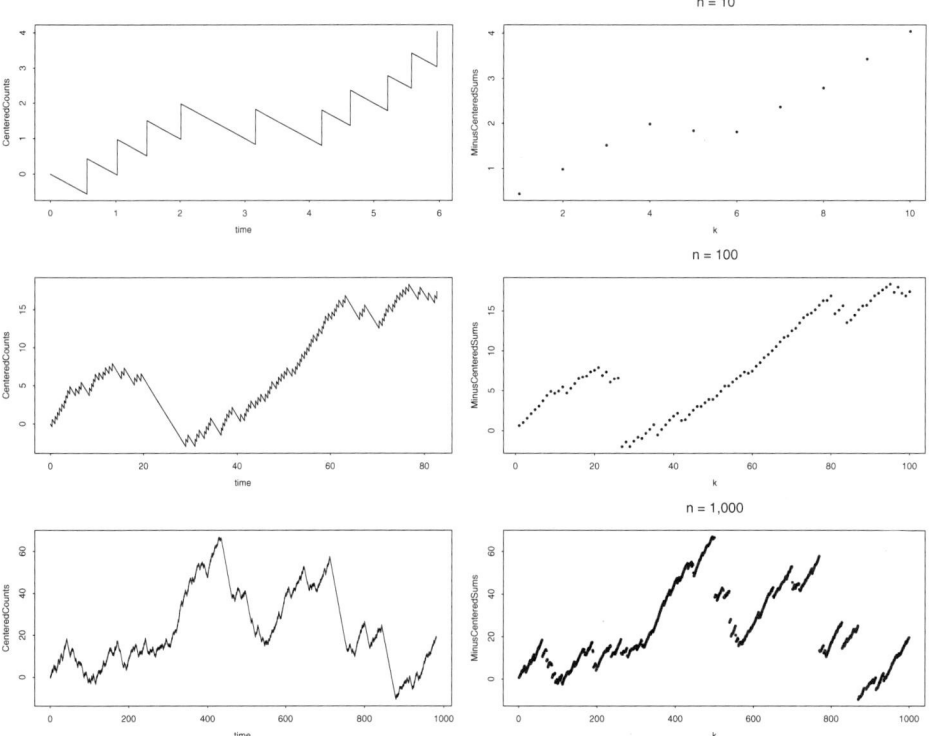

Figure 6.4. Plots of the centered renewal process (on the left) and minus 1 times the centered random walk (on the right) for Pareto(1.5) steps with mean $m = 1$ and $n = 10^j$ for $j = 1, 2, 3$.

For the Pareto-step random walk plots in Figures 6.4 – 6.7, we not only have $-\mathbf{S}_n \Rightarrow -\mathbf{S}$ and $\mathbf{N}_n \Rightarrow -\mathbf{S}$, but also the realizations of $\mathbf{N}_n$ and $-\mathbf{S}_n$ are becoming close to each other as $n \to \infty$. Such asymptotic equivalence follows from Theorem 6.3.1 by virtue of Theorem 11.4.8. Recall that we can start with any translation scaling constant m and rescale to $m = 1$.

Corollary 6.3.1. (asymptotic equivalence) *If, in addition to the assumptions of Theorem 6.3.1, the limit $\mathbf{S}_n \Rightarrow \mathbf{S}$ in (3.4) holds and $m = 1$, then*

$$d_{M_1}(\mathbf{N}_n, -\mathbf{S}_n) \Rightarrow 0 .$$

To summarize, properly scaled versions (with centering) of a renewal process (or, more generally, any counting process) are intimately connected with associated scaled versions (with centering) of random walks, so that FCLTs for random walks imply associated FCLTs for the scaled renewal process (and vice versa), provided that we use the M_1 topology. When the limit process for the random walk has discontinuous sample paths, so does the limit process for the renewal process,

6.4. A Queue with Heavy-Tailed Distributions

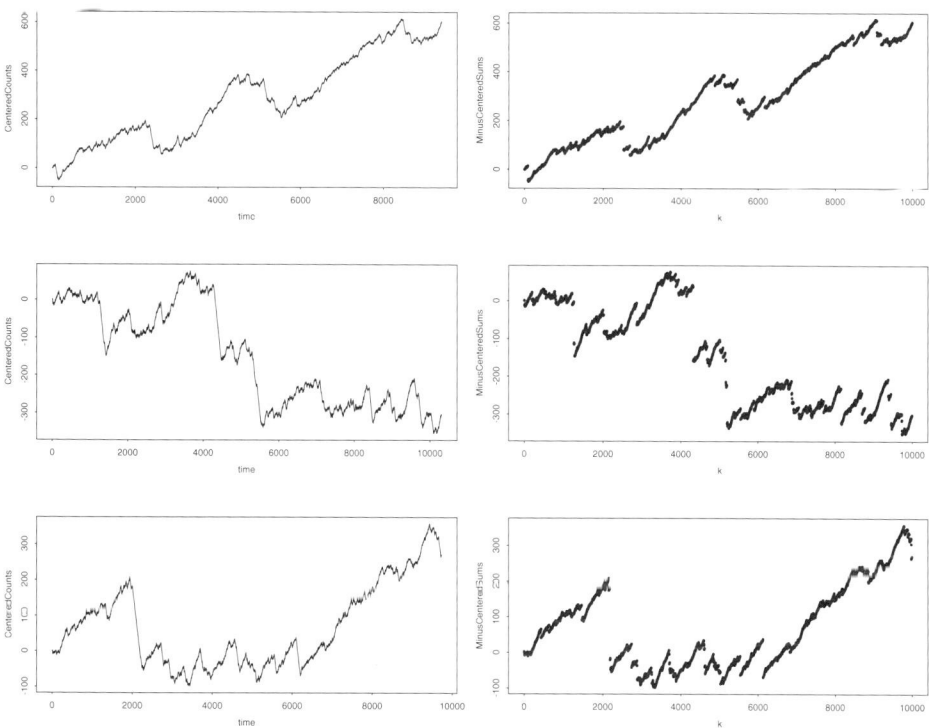

Figure 6.5. Plots of three independent realizations of the centered renewal process (on the left) and minus 1 times the centered random walk (on the right) for Pareto(1.5) steps with mean $m = 1$ and $n = 10^4$.

which necessarily produces unmatched jumps. We state specific FCLTs for renewal processes in Section 7.3.

6.4. A Queue with Heavy-Tailed Distributions

Closely paralleling the heavy-tailed renewal process just considered, heavy-traffic limits for the queue-length process in standard queueing models routinely produce stochastic-process limits with unmatched jumps in the limit process when the service times or interarrival times have heavy-tailed distributions (again meaning with infinite variance). In fact, renewal processes enter in directly, because the customer arrival process in the queueing model is a stochastic counting process, which is a renewal process when the interarrival times are IID.

We start by observing that jumps in the limit process associated with stochastic-process limits for the queue-length process almost always are unmatched jumps. That is easy to see when all the interarrival times and service times are strictly positive. (That is the case w.p.1 when the interarrival times and service times come

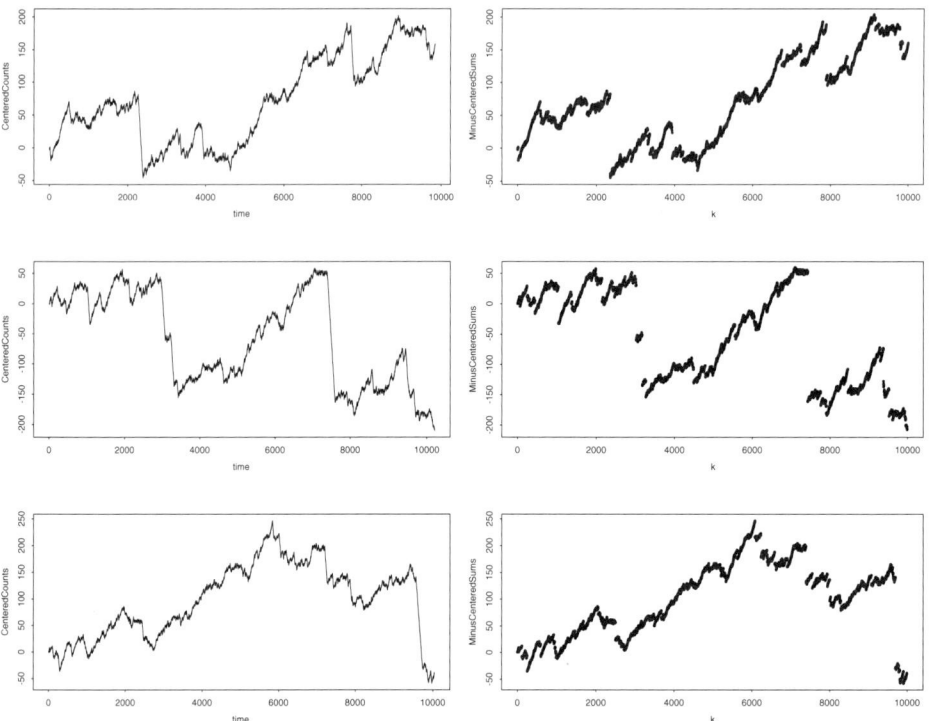

Figure 6.6. Plots of three independent realizations of the centered renewal process (on the left) and minus 1 times the centered random walk (on the right) associated with Pareto(p) steps in (2.3) with $p = 7/4$, $m = 1$ and $n = 10^4$.

from sequences of random variables with distributions assigning 0 probability to 0.) Then the queue length (i.e., the number of customers in the system) makes changes in unit steps. Thus, any jumps in the limit process associated with a stochastic-process limit for a sequence of queue-length processes with space scaling, where we divide by c_n with $c_n \to \infty$ as $n \to \infty$, must be unmatched jumps.

The real issue, then, is to show that jumps can appear in stochastic-process limits for the queue-length process. The stochastic-process limits we have in mind occur in a heavy-traffic setting, as in Section 2.3.

6.4.1. The Standard Single-Server Queue

To be specific, we consider a single-server queue with unlimited waiting room and the first-come first-served service discipline. (We will discuss this model further in Chapter 9. The model can be specified by a sequence of ordered pairs of nonnegative random variables $\{(U_k, V_k) : k \geq 1\}$. The variable U_k represents the *interarrival time* between customers k and $k - 1$, with U_1 being the arrival time of the first customer, while the variable V_k represents the *service time* of the customer k. The

6.4. A Queue with Heavy-Tailed Distributions

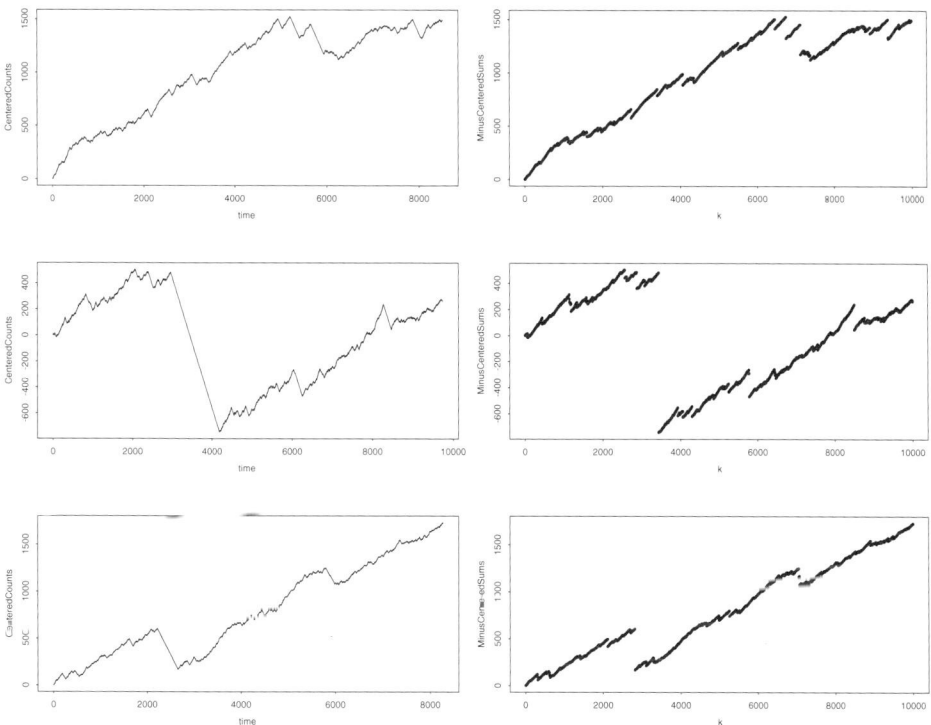

Figure 6.7. Plots of three independent realizations of the centered renewal process (on the left) and minus 1 times the centered random walk (on the right) associated with Pareto(p) steps in (2.3) with $p = 5/4$, $m = 1$ and $n = 10^4$.

arrival time of the customer k is thus

$$T_k \equiv U_1 + \cdots + U_k, \quad k \geq 1, \tag{4.1}$$

and the *departure time* of the customer k is

$$D_k \equiv T_k + W_k + V_k, \quad k \geq 1, \tag{4.2}$$

where W_k is the *waiting time* (before beginning service) of customer k. The waiting times can be defined recursively by

$$W_k \equiv [W_{k-1} + V_{k-1} - U_k]^+, \quad k \geq 2, \tag{4.3}$$

where $[x]^+ \equiv max\{x, 0\}$ and $W_1 \equiv 0$. (We have assumed that the system starts empty; that of course is not critical.)

We can now define associated continuous-time processes. The counting processes are defined just as in (3.1). The *arrival (counting) process* $\{A(t) : t \geq 0\}$ is defined by

$$A(t) \equiv max\{k \geq 0 : T_k \leq t\}, \quad t \geq 0, \tag{4.4}$$

the *departure (counting) process* $\{D(t) : t \geq 0\}$ is defined by

$$D(t) \equiv max\{k \geq 0 : D_k \leq t\}, \quad t \geq 0, \qquad (4.5)$$

and the *queue-length process* $\{Q(t) : t \geq 0\}$ is defined by

$$Q(t) \equiv A(t) - D(t), \quad t \geq 0. \qquad (4.6)$$

Here the queue length is the number in system, including the customer in service, if any.

The standard single-server queue that we consider now is closely related to the infinite-capacity version of the discrete-time fluid queue model considered in Section 2.3. Indeed, the recursive definition for the waiting times in (4.3) is essentially the same as the recursive definition for the workloads in (3.1) of Section 2.3 in the special case in which the waiting space is unlimited, i.e., when $K = \infty$. For the fluid queue model, we saw that the behavior of the workload process is intimately connected to the behavior of an associated random walk, and that heavy-tailed inputs lead directly to jumps in the limit process for appropriately scaled workload processes. The same is true for the waiting times here, as we will show in Section 9.2.

6.4.2. Heavy-Traffic Limits

Thus, just as in Section 2.3, we consider a sequence of models indexed by n in order to obtain interesting stochastic-process limits for stable queueing systems. We can achieve such a framework conveniently by scaling a single model. We use a superscript n to index the new quantities constructed in the nth model.

We start with a single sequence $\{(U_k, V_k) : k \geq 1\}$. Note that we have made no stochastic assumptions so far. The key assumption is a FCLT for the random walks, in particular,

$$(\mathbf{S}_n^u, \mathbf{S}_n^v) \Rightarrow (\mathbf{S}^u, \mathbf{S}^v) \quad \text{in} \quad (D^2, WM_1), \qquad (4.7)$$

where

$$\mathbf{S}_n^u \equiv c_n^{-1}(\sum_{i=1}^{\lfloor nt \rfloor} U_i - \lfloor nt \rfloor)$$

and

$$\mathbf{S}_n^v \equiv c_n^{-1}(\sum_{i=1}^{\lfloor nt \rfloor} V_i - \lfloor nt \rfloor).$$

The standard stochastic assumption to obtain (4.7) is for $\{U_k\}$ and $\{V_k\}$ to be independent sequences of IID random variables with

$$EV_k = EU_k = 1 \quad \text{for all} \quad k \geq 1. \qquad (4.8)$$

and other regularity conditions (finite variances to get convergence to Brownian motion or asymptotic power tails to get convergence to stable Lévy motions).

6.4. A Queue with Heavy-Tailed Distributions

Paralleling the scaling in (3.13) in Section 2.3, we form the nth model by letting
$$U_k^n \equiv b_n U_k \quad \text{and} \quad V_k^n \equiv V_k, \quad k \geq 1 , \tag{4.9}$$
where
$$b_n \equiv 1 + mc_n/n \quad \text{for} \quad n \geq 1 . \tag{4.10}$$
We assume that $c_n/n \downarrow 0$ as $n \to \infty$, so that $b_n \downarrow 1$ as $n \to \infty$. The scaling in (4.9) is a simple deterministic scaling of time in the arrival process; i.e., the arrival process in model n is
$$A^n(t) \equiv A(b_n^{-1}t), \quad t \geq 0 ,$$
for b_n in (4.10).

We now form scaled stochastic processes associated with the sequence of models by letting
$$\mathbf{W}_n(t) \equiv c_n^{-1} W_{\lfloor nt \rfloor}^n , \tag{4.11}$$
and
$$\mathbf{Q}_n(t) \equiv c_n^{-1} Q^n(nt), \quad t \geq 0 . \tag{4.12}$$

We now state the heavy-traffic stochastic-process limit, which follows from Theorems 9.3.3, 9.3.4 and 11.4.8. As before, for $x \in D$, let $Disc(x)$ be the set of discontinuities of x.

Theorem 6.4.1. (heavy-traffic limit for the waiting times and queue lengths) *Suppose that the stochastic-process limit in (4.7) holds and the scaling in (4.9) holds with $c_n \to \infty$ and $c_n/n \to 0$. Suppose that almost surely the sets $Disc(\mathbf{S}^u)$ and $Disc(\mathbf{S}^v)$ have empty intersection and*
$$P(\mathbf{S}^u(0) = 0) = P(\mathbf{S}^v(0) = 0) = 1 .$$
Then
$$\mathbf{W}_n \Rightarrow \mathbf{W} \equiv \phi(\mathbf{S}^v - \mathbf{S}^u - m\mathbf{e}) \quad \text{in} \quad (D, M_1) , \tag{4.13}$$
where ϕ is the one-sided reflection map in (5.4) in Section 3.5,
$$(\mathbf{W}_n, \mathbf{Q}_n) \Rightarrow (\mathbf{W}, \mathbf{W}) \quad \text{in} \quad (D^2, WM_1) \tag{4.14}$$
and
$$d_{M_1}(\mathbf{W}_n, \mathbf{Q}_n) \Rightarrow 0 . \tag{4.15}$$

We now explain why the limit process $\mathbf{Q}$ for the scaled queue-length processes can have jumps. Starting from (4.6), we have
$$\mathbf{Q}_n - \mathbf{A}_n - \mathbf{D}_n , \tag{4.16}$$
where
$$\mathbf{A}_n(t) \equiv c_n^{-1}(A^n(nt) - nt), \quad t \geq 0 \tag{4.17}$$

and
$$\mathbf{D}_n(t) \equiv c_n^{-1}(D^n(nt) - nt), \quad t \geq 0 . \tag{4.18}$$

Just as for the renewal processes in the previous section, an especially long service time (interarrival time) can cause a period of steep linear slope down in $\mathbf{D}_n$ ($\mathbf{A}_n$), which can correspond to jumps down in the associated limit process. The jump down from $\mathbf{D}_n$ ($\mathbf{A}_n$) corresponds to a jump up (down) in the limit process for $\mathbf{Q}_n$.

6.4.3. Simulation Examples

What we intend to do now is simulate and plot the waiting-time and queue-length processes under various assumptions on the interarrival-time and service-time distributions. Just as with the empirical cdf in Example 1.1.1 and the renewal process in Section 6.3, when we plot the queue-length process we need to plot a portion of a continuous-time process. Just as in the two previous cases, we can plot the queue-length process with the statistical package S, exploiting underlying random sequences. Here the relevant underlying random sequences are the arrival times $\{T_k\}$ and the departure times $\{D_k\}$, defined recursively above in (4.1) and (4.2).

Since the plotting procedure is less obvious now, we specify it in detail. We first form two dimensional vectors by appending a $+1$ to each arrival time and a -1 to each departure time. (Instead of the arrival time T_n, we have the vector $(T_n, 1)$; instead of the departure time D_n, we have the vector $(D_n, -1)$.) We then combine all the vectors (creating a matrix) and sort on the first component. The new first components are thus the successive times of any change in the queue length (arrival or departure). We then form the successive cumulative sums of the second components, which converts the second components into the queue lengths at the times of change. We could just plot the queue lengths at the successive times of change, but we go further to plot the full continuous-time queue-length process. We can plot by linear interpolation, if we include each queue length value twice, at the jump when the value is first attained and just before the next jump. (This method inserts a vertical line at each jump.)

We now give an S program to read in the first n interarrival times, service times and waiting times and plot the queue-length process over the time interval that these n customers are in the system (ignoring all subsequent arrivals). At the end of the time interval the system is necessarily empty. Our construction thus gives an odd end effect, but it can be truncated. Indeed, in our plots below we do truncate (at the expected time of the nth arrival).

Here is the S function:

```
QueueLength <- function(U, V, W) {
QueueLength <- vector("numeric", 2*length(U) + 1)
T <- cumsum(U)                    #construct arrival times
D <- T + W + V                    # departure times
TT <- cbind(T, +1)                #append +1 to arrivals
DD <- cbind(D, −1)                #append −1 to deps.
```

```
        m <- rbind(TT, DD)              #merge into one matrix
        msort <- m[sort.list(m[, 1]),]  #sort on first comp.
        time1 <- msort[, c(1)]          #extract change times
        QLchg <- msort[, c(2)]          #queue length changes
        QL1 <- cumsum(QLchg)            #successive q. lths.
        time2 <- c(0, time1, time1)     #times for lin.interp.
        time <- sort(time2)
        n <- length(time1)              #q. lths. for lin. int.
        QL <- c(0, QL1)
        for (k in seq(n)) {
        QueueLength[[2 * k − 1]] <- QL[[k]]
        QueueLength[[2 * k]] <- QL[[k]] }
        QueueLength[2 * n + 1] <- QL[n + 1]
        plot(time, QueueLength, type = "l")   #do the plotting
        }
```

We now consider a few examples. We use the *Kendall notation* to describe the model: $X/Y/c$ specifies a model with c servers, arrival process of type X and service process of type Y. For either X or Y, GI denotes an IID sequence with a general distribution, while M (for Markov) denotes (in addition) the exponential distribution. We use P_p for the Pareto distribution with parameter p.

Example 6.4.1. *The M/M/1 Queue.*

We first consider the standard M/M/1 queue. Thus, here we assume that the interarrival times and service times come from mutually independent sequences of IID exponentially distributed random variables. It suffices to specify the means of the interarrival time and the service time. Using the scaling in equations (4.9) and (4.10), we need to specify the constant m and the space-scaling sequence $\{c_n\}$.

At this point, we know what to do: There are no heavy tailed distributions, so we should let $c_n = \sqrt{n}$. We also let $m = 1$. Thus, we fully specify the sequence of M/M/1 models indexed by n by letting

$$EU_k^n = 1 + 1/\sqrt{n} \quad \text{and} \quad EV_k^n = 1 \quad \text{for all} \quad k \quad \text{and} \quad n \;. \tag{4.19}$$

With that choice, the plotter can do the appropriate scaling automatically.

We are primarily interested in the queue-length process, but we also plot the waiting times, because it is instructive to compare the plotted queue-length process to the plotted waiting times. Hence, we plot both the waiting times of the first n customers (linearly interpolated) and the queue-length process over the time interval $[0, nEU_1^n]$ for the cases $n = 10^j$ with $j = 1, 2, 3$ in Figure 6.8.

For small n, the queue-length process looks very different from the waiting time sequence, but as n increases, the sample path of the queue length process becomes very similar to the sample path of the waiting times, except possibly for the final portion, where the queue length experiences some of the end effect. To confirm what we see in Figure 6.8, we plot three possible realizations of the waiting times and the queue lengths for $n = 10^4$ in Figure 6.9.

212   6. Unmatched Jumps

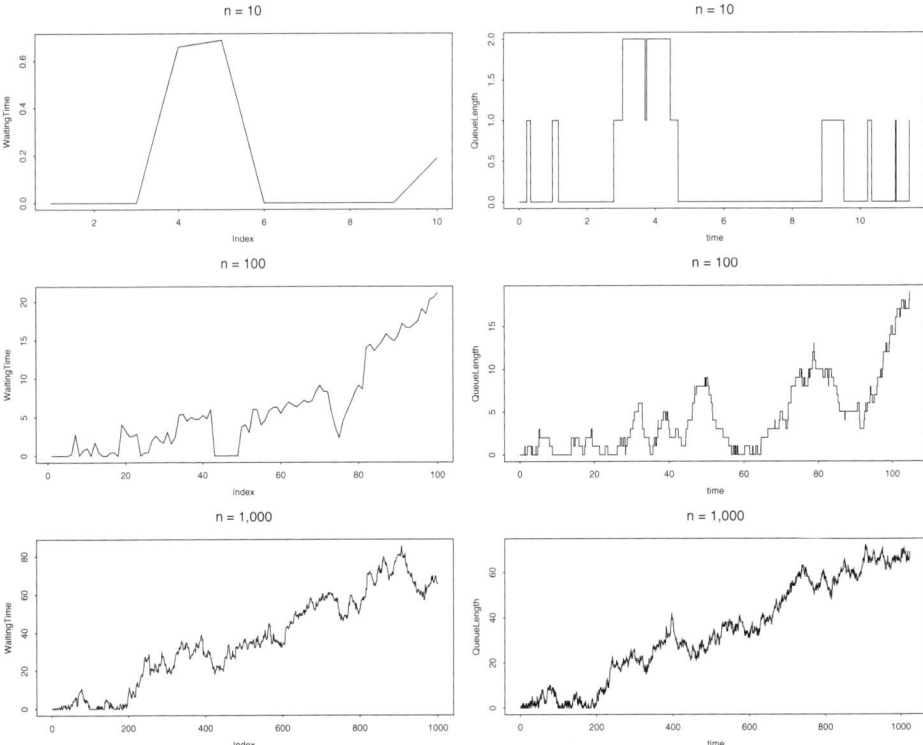

Figure 6.8. Plots of the waiting times of the first n arrivals (on the left) and the queue-length process over the interval $[0, nEU_1^n]$ (on the right) in the M/M/1 queue with scaling in (4.19) for $n = 10^j$ with $j = 1, 2, 3$.

From our experience so far, we should know what to expect: The plots are approaching plots of reflected Brownian motion with drift -1 (which does not have any jumps). Now the conditions and conclusions of Theorem 6.4.1 hold with $c_n = \sqrt{n}$ and $\mathbf{W} = \phi(\sigma \mathbf{B} - m\mathbf{e})$, where $\mathbf{B}$ is standard Brownian motion, $\mathbf{e}$ is the identity map, $\phi : D \to D$ is the one-sided reflection map and $\sigma^2 = Var(U_1) + Var(V_1) = 2$. We apply Donsker's theorem – Theorem 4.3.2.

Moreover, the plots show that the distance between the two scaled processes is indeed asymptotically negligible. Since the limit process here has continuous sample paths, we can express this asymptotic equivalence using the uniform norm over $[0, 1]$:

$$\| \mathbf{W}_n - \mathbf{Q}_n \| \Rightarrow 0 \quad \text{as} \quad n \to \infty . \qquad (4.20)$$

∎

Example 6.4.2. *The $M/P_{1.5}/1$ Queue.*

We now modify the previous example by letting the service-time distribution be Pareto(p) with $p = 1.5$ and mean 1. (In the framework of Section 1.3.3, we can use

6.4. A Queue with Heavy-Tailed Distributions 213

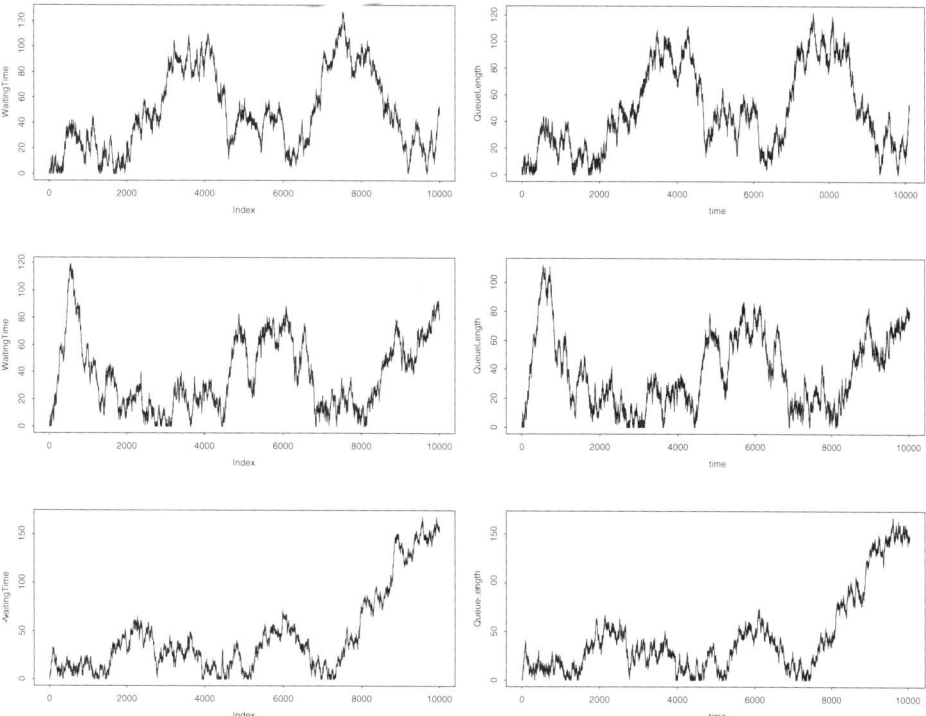

Figure 6.9. Three possible realizations of the waiting times of the first n arrivals (on the left) and the queue-length process over the interval $[0, nEU_1^n]$ (on the right) in the M/M/1 queue with scaling in (4.19) for $n = 10^4$.

$3^{-1}U^{-2/3}$, where U is uniform on the interval $[0, 1]$, which has ccdf $F^c(t) = (3t)^{-3/2}$ for $t \geq 1$.) With this heavy-tailed service-time distribtuion, we must scale space differently, because the space scaling in the FCLT for the random walk involves $c_n = n^{2/3}$ instead of $c_n = n^{1/2}$. Hence, instead of the scaling in (4.19), we now use

$$EU_k^n = 1 + n^{-1/3} \quad \text{and} \quad EV_k^n = 1 \quad \text{for all} \quad k \quad \text{and} \quad n. \tag{4.21}$$

The new scaling makes the traffic intensity ρ_n smaller than in Example 6.4.1 for any given n. For example, for $n = 10,000$, before we had $\rho_n = 1/1.01 \approx 0.990$, while now we have $\rho_n \approx 1/1.046 \approx 0.956$.

We plot three possible realizations of the waiting times of the first n customers (on the bottom or left) and queue-length process over the interval $[0, nEU_1^n]$ (on the top or right) for $n = 10^4$, in Figure 6.10. The first two plots look much like the M/M/1 plots in Figure 6.9 except now we can see upward jumps. But the third plot is very different!

There is now much more variability in the sample paths because of the possibility of the occasional very large jumps. The range of values is exceptionally small in case 2 and exceptionally large in case 3. The possibility of exceptionally large jumps produces large variations from plot to plot, as we saw for the random walks in Figure 1.21.

When we look at the third plots closely, it is not evident that the waiting-time and queue-length plots are for the same sample path. For instance, the second big jump in the waiting times occurs at about index 3100, whereas the corresponding second steep incline in the queue-length path begins at about time 4100. However, upon reflection, we see that these actually are consistent, because the waiting time of the customer having the second large service time is about 1000. Since the arrival rate is 1, that customer arrives at about time 3100. Hence that customer enters service, and begins occupying the server, at about time 4100. Thus the queue length should start building up at about time 4100, as it does.

The upward jumps are less sharp for the queue-length process, which we know actually increases by unit jumps, but the asymptotic behavior is evident from the plots. In this case, we are seeing a reflected stable Lévy motion with drift -1, which has discontinuous sample paths, instead of a reflected Brownian motion. Again we can explain the statistical regularity we see by Theorem 6.4.1. However, now the scaling involves $c_n = n^{2/3}$.

By Theorems 4.5.2 and 4.5.3, the limit process is $\mathbf{W} \equiv \phi(\sigma \mathbf{S}^v - \mathbf{e}) \equiv \sigma\phi(\mathbf{S}^v - \sigma^{-1}\mathbf{e})$, where $\sigma = 1/3 C_\alpha^{2/3}$ for C_α in (5.14) of Section 4.5.1, $\mathbf{S}^v$ is a centered α-stable Lévy motion with $\mathbf{S}^v(1) \stackrel{\mathrm{d}}{=} S_\alpha(1,1,0)$ and $\alpha = 3/2$. (Its steady-state distribution is given in Section 8.5.2.) Again, it is evident that the two scaled processes $\mathbf{W}_n$ and $\mathbf{Q}_n$ should now be asymptotically equivalent. ∎

Example 6.4.3. *The $P_{1.5}/M/1$ Queue.*

It is evident that a heavy-tailed service-time distribution should cause greater congestion, but it may not be evident that a heavy-tailed interarrival-time distribution can as well, because extra long interarrival times only serve to empty out the queue. However, heavy-tailed interarrival-time distributions can cause congestion as well. The reason is that, for given fixed mean, the occasionally exceptionally long interarrival times must be compensated for in the distribution by shorter interarrival times, and these shorter interarrival times lead to bursts of arrivals and thus increased queue lengths.

We illustrate by considering the $P_{1.5}/M/1$ queue, which has IID Pareto(1.5) interarrival times and IID exponential service times. This model is the *dual* of the model in Example 6.4.2, with the role of the interarrival times and service times switched (adjusted by scaling, so that the expected interarrival times are bigger than the expected service times in both cases).

In Figure 6.11 we plot three possible realizations of the waiting times of the first n arrivals (on the left) and the queue-length process over the interval $[0, nEU_1^n]$ (on the right) in the $P_{1.5}/M/1$ queue with the scaling in (4.21) for $n = 10^4$.

As in Figures 6.8 – 6.10, the queue-length plots are similar to the waiting-time plot, except possibly for the final portion of the queue-length plot, where the queue experiences its end effect. However, unlike in the previous figures, in Figure 6.11 we see evidence of jumps down.

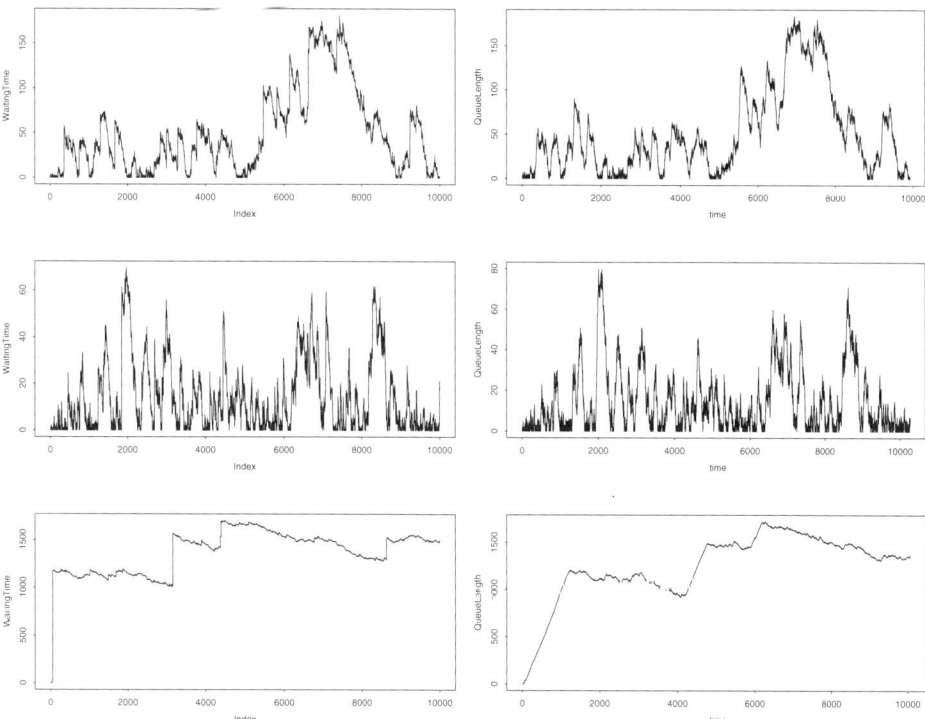

Figure 6.10. Three possible realizations of the waiting times of the first n arrivals (on the left) and the queue-length process over the interval $[0, nEU_1^n]$ (on the right) in the $M/P_{1.5}/1$ queue with the scaling in (4.21) for $n = 10^4$.

Just as for the $M/P_{1.5}/1$ model, the heavy-traffic FCLT in Theorem 6.4.1 applies to the $P_{1.5}/M/1$ and $P_{1.5}/P_{1.5}/1$ models. Indeed, we again have the same scaling, but now the limiting reflected stable Lévy motions are different, having jumps down only for the $P_{1.5}/M/1$ model and having jumps both up and down for the $P_{1.5}/P_{1.5}/1$ model, instead of having jumps up only for the $M/P_{1.5}/1$ model.

For the $P_{1.5}/M/1$ model, the heavy-traffic stochastic-process limit for the workload process is $\mathbf{W}_n \Rightarrow \mathbf{W}$, where again $c_n = n^{2/3}$, but now

$$\mathbf{W} = \phi(-\sigma \mathbf{S}^u - \mathbf{e}) \stackrel{d}{=} \sigma \phi(-\mathbf{S}^u - \sigma^{-1}\mathbf{e}),$$

where $\sigma = 1/3 C_\alpha^{2/3}$ for $\alpha = 3/2$, just as in Example 6.4.2. Here $-\mathbf{S}^u(1) \stackrel{d}{=} S_\alpha(1, -1, 0)$.

For the $P_{1.5}/P_{1.5}/1$ model, the limit process is

$$\mathbf{W} = \phi(\sigma \mathbf{S}^v - \sigma \mathbf{S}^u - \mathbf{e}) \stackrel{d}{=} \sigma \phi(\mathbf{S}^v - \mathbf{S}^u - \sigma^{-1}\mathbf{e}),$$

where $\mathbf{S}^v - \mathbf{S}^u \stackrel{d}{=} \mathbf{S}$ with $\mathbf{S}$ being a stable Lévy motion satisfying $\mathbf{S}(1) \stackrel{d}{=} 2^{2/3} S_\alpha(1, 0, 0)$; see (5.8) – (5.11) in Section 4.5.1.

216   6. Unmatched Jumps

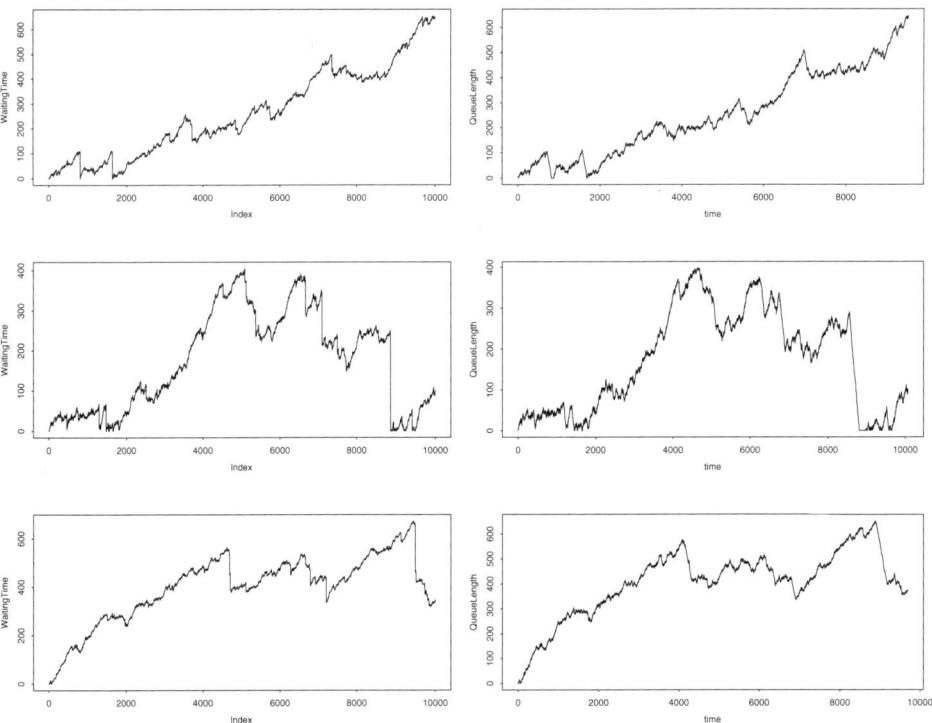

Figure 6.11. Three possible realizations of the waiting times of the first $n = 10^4$ arrivals (on the left) and the queue-length process over the interval $[0, nEU_1^n]$ (on the right) in the $P_{1.5}/M/1$ queue with the scaling in (4.21) for $n = 10^4$.

We should not be fooled by the jumps down for the $P_{1.5}/M/1$ model. Of course, the jumps down do constitute reductions in congestion, but elsewhere in the plot the sample path is rising, so that the range of values experienced can be substantial. Indeed, that is demonstrated by the heavy-traffic FCLT, which has space scaling by $n^{2/3}$, just as for the $M/P_{1.5}/1$ model in Example 6.4.2. ∎

6.5. Rare Long Service Interruptions

The queueing example just considered illustrates a common cause of congestion in queues: stochastic vartiability in the interarrival times and service times. However, congestion in queues can occur for other reasons: For example, the servers may be subject to breakdown and failure, causing service interruptions. In manufacturing systems, service interruptions due to machine failures or the unavailability of parts are often the dominant sources of congestion. With evolving communication networks, there is debate about whether the most important source of congestion is the

6.5. Rare Long Service Interruptions

uncertain burstiness of customer input or the uncertain failure of system elements. The biggest problems tend to occur when both happen together.

We can better understand the impact of service interruptions upon performance if we develop a probability model and establish appropriate stochastic-process limits. One such model, considered by Kella and Whitt (1990), is a queue with rare long service interruptions. The queue can be a standard single-server queue with unlimited waiting space, the first-come first-served service discipline and random arrivals and service times, as considered in the previous section. We can supplement that model by allowing random service interruptions. The interruptions can be triggered by queueing events; e.g., they could occur only when the queue becomes empty. Or they can occur exogenously. We will consider the case in which they occur exogenously.

Specifically, we will assume that the availability of the server is characterized by an alternating renewal process; i.e., there are alternating periods in which the server is available (up) or unavailable (down). For tractability, we assume that the up and down times come from mutually independent sequences of IID positive random variables with finite means and variances.

A revealing stochastic-process limit can be obtained by considering the queue in a heavy-traffic limit, in which the load is allowed to approach the critical value for stability. If the interruptions remain unchanged, then the service interruptions alter the conventional heavy-traffic limit with a reflected Brownian motion limit process only by increasing the traffic intensity and increasing the variance parameter of the Brownian motion, both of which cause increased congestion. However, we obtain a different nondegenerate limit, which is consistent with many applications, if we let the intervals between interruptions and the durations of the interruptions increase in the limit. If we let these quantities increase appropriately, with the duration of an interruption being asymptotically negligible compared to the time between interruptions, then we can obtain a revealing nondegenerate limit.

In particular, an interesting limiting regime has the random up times be of order n and the random down times be of order $\sqrt{n}$ as a function of the number n of customers being considered. Then, with the customary scaling of time by n and space by $\sqrt{n}$, the scaled up times become of order 1 and the scaled down times become of order $1/\sqrt{n}$. That makes the scaled down times asymptotically negligible. Thus, after scaling, the service interruptions occur in the limit according to a stochastic point process, with a finite positive expected number of interruptions in a finite time interval.

Since the scaled durations of the service interruptions are asymptotically negligible, the service interruptions occur instantaneously in the limit. Nevertheless, the service interruptions can have a significant spatial impact, because the number of arrivals during the order $\sqrt{n}$ down time is also of order $\sqrt{n}$. Thus, after scaling space by $\sqrt{n}$, the input during the down time causes a random jump of order 1 in the scaled queueing process at each interruption time.

The proposed scaling, with up times of order n and down times of order $\sqrt{n}$, thus produces random jumps of order-1 size, spaced at random order-1 intervals. In the limit, the proportion of time that the server is unavailable because of interruption is asymptotically negligible. Nevertheless, the asymptotic impact of the interruptions can be dramatic. With this limit, it is possible to compare the effects of the service interruptions (which appear in the limit process as jumps) to the customary stochastic fluctuations. Depending on the specific parameter settings, one or the other may dominate. In Section 14.7, following Kella and Whitt (1990) and Chen and Whitt (1993), we consider networks of queues with rare long service interruptions.

When we consider limits for sequences of queue-length stochastic processes affected by rare long interruptions of the kind just described, the jumps in the limit process are typically not matched in the converging scaled queue-length processes. In the queueing system, arrivals usually are coming one at a time. During a service interruption, service stops, but the arrivals keep coming. Thus the queue length process increases by many unit steps during such periods. After scaling time and space, the nth scaled queue-length process increases more rapidly (due to the time scaling) but by smaller asymptotically negligible amounts (due to the space scaling). Thus the resulting limit is a stochastic-process limit with unmatched jumps in the limit process.

In the rest of this subsection we illustrate the kind of limiting behavior provided by rare long service interruptions. To do so, we simplify the model: Even though service interruptions represent a different source of congestion than variability in customer demand, we often can represent service interruptions within the framework of a standard queueing model. We can simply include the interruption in the service time of one of the customers. Specifically, we can redefine the service-time distribution: The new service-time distribution becomes a mixture: With probability p, the new service time is the sum of an original service time and the interruption duration; with probability $1 - p$, the new service time reduces to an original service time. We then choose the probability p to match the probability that a customer is the first customer to experience a service interruption. If the timing of service interruptions needs to be modeled very precisely, then we can think of interruptions as special high-priority customers that preempt regular customers (in line or in service), but the simple model above often suffices

We have in mind rare long service interruptions occuring randomly, but to illustrate the interruption phenomenon, we let the interruptions occur in a fixed manner in our example below.

Example 6.5.1. *The $M/M/1$ queue with two fixed service interruptions.*

We construct a simple example to illustrate the kind of limit behavior associated with rare long service interruptions. Specifically, we consider the $M/M/1$ queue with the heavy-traffic scaling in (4.19), just as in Example 6.4.1, except that now we let customers number $n/4$ and $3n/4$ have service times of $2\sqrt{n}$ and $\sqrt{n}$, respectively, as a function of n. These special service times are introduced to represent interruptions that occur approximately at times $t/4$ and $3t/4$ in the scaled processes plotted over the interval $[0, 1]$. (By the SLLN, the scaled arrival time of customer number $n/4$ approaches $t/4$ as $n \to \infty$.) Note that the spacings between the interruptions is

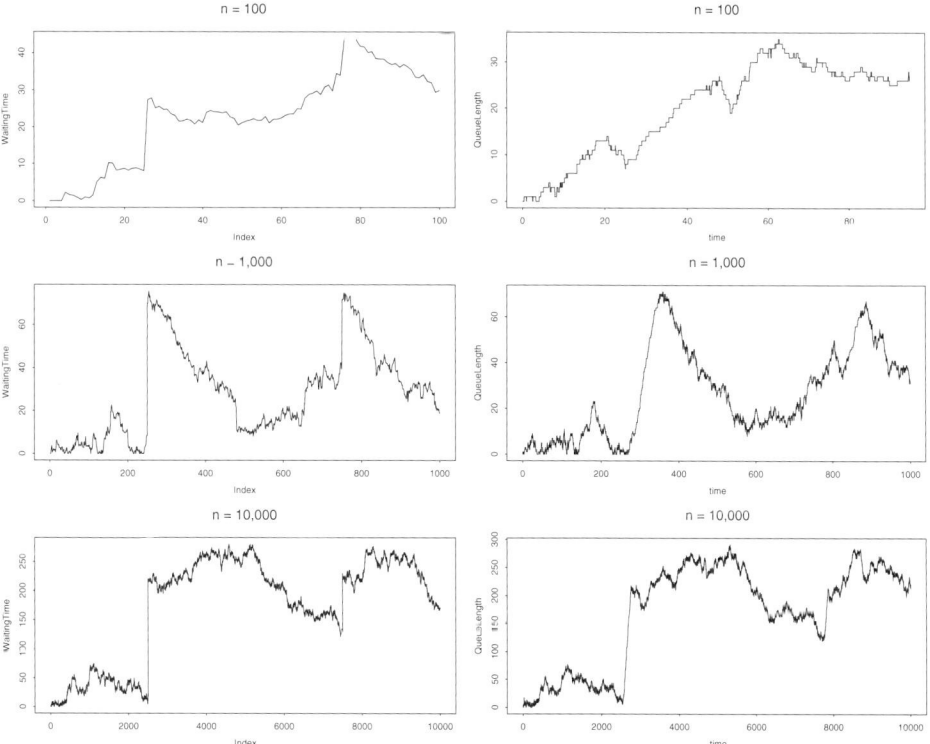

Figure 6.12. Plots of the waiting times of the first n arrivals (on the left) and the queue-length process over the interval $[0, nEU_1^n]$ (on the right) for in the M/M/1 queue with scaling in (4.19) and service interruptions of length $2\sqrt{n}$ and $\sqrt{n}$ associated with customers $n/4$ and $3n/4$ for $n = 10^j$ with $j = 2, 3, 4$.

indeed order n, while the durations of the interruptions (as captured by the special service times) are of order $\sqrt{n}$, as specified above.

We plot the waiting times of the first n customers and the queue-length process for the time interval $[0, nEU_1^n]$, the expected time for the n customers to arrive, for $n = 10^j$ with $j = 2, 3, 4$ in Figure 6.12. In Figure 6.12 the impact of the interruptions is clearer for the waiting times than for the queue lengths, especially for smaller n. For the queue-length process, the portion of the plot corresponding to the jump gets steeper as n increases. As before, we see that the queue-length and waiting-time plots coalesce as n increases. Now both scaled processes approach reflected Brownian motion with drift -1, modified by jumps of size 2 at time $t = 1/4$ and of size 1 at time $t = 3/4$. For the scaled queue-length process, the limit process must have unmatched jumps. ∎

Example 6.5.2. *The $P_{1.5}/M/1$ queue with two fixed service interruptions.*

Now, as in Example 6.4.3 we consider the $P_{1.5}/M/1$ queue with heavy-traffic scaling in (4.21), modified by having customers number $n/4$ and $3n/4$ experience interruptions. We choose the $P_{1.5}/M/1$ model instead of the $M/P_{1.5}/1$ model, because it naturally (without the interruptions) produces jumps down instead of up. Thus, it will be easier to recognize the new jumps up caused by the service interruptions.

In addition, the durations of the interruptions need to be scaled differently from the scaling in Example 6.5.1. In order to be consistent with the heavy-traffic limiting behavior in Example 6.4.3, we now need to scale the durations of the interruptions by $n^{2/3}$ instead of $n^{1/2}$. In particular, now we let the service times of customers number $n/4$ and $3n/4$ be $2n^{2/3}$ and $n^{2/3}$, respectively. We plot three possible realizations of the waiting times of the first n customers and the queue-length process over the time interval $[0, nEU_1^n]$, ignoring all arrivals after the first n, for the case $n = 10^4$ in Figure 6.13.

Just as we would expect from Figures 6.11 and 6.12, we see randomly occurring jumps down because of the $P_{1.5}$ arrival process and jumps up of magnitude 2 at time $t = 1/4$ and 1 at time $t = 3/4$. However, both kinds of jumps are much sharper for the waiting times than for the queue-length process. Hence, we evidently need larger n in this case to have the queue-length plots be visually similar to the waiting-time plots. The supporting FCLTs state that both scaled processes converge to a stable Lévy motion (with jumps down only) modified by the addition of two jumps up, a jump of size 2 at $t = 1/4$ and a jump of size 1 at $t = 3/4$; again, see Sections 4.5 and 14.7. Again, for the scaled queue-length process, that limit process must have unmatched jumps. ∎

The simple models of service interruptions considered in Examples 6.5.1 and 6.5.2 are of course quite artificial. However, from these examples, we can anticipate what we will see when we use the more realistic alternating renewal process model for up and down times.

6.6. Time-Dependent Arrival Rates

In many service systems, congestion occurs primarily because of systematic, deterministic variations in the input rate over time. Many service systems have arrival rates that vary systematically with time, so that there are known busy periods with higher loads than average. However, everything is not known. There remains uncertainty about the actual input; there are unanticipated fluctuations about the known time-varying deterministic rates.

To better understand the behavior of queues with time-varying arrival rates, we need to focus directly on queueing models with time-varying arrival rates. Just as for stationary queueing models, it can be helpful to consider heavy-traffic limits for queues with time-varying arrival rates. With time-varying arrival rates, we still scale time, but we think of expanding time immediately prior to the time of interest. We increase the overall arrival and service rate, which is tantamount to decreasing the rate of change in the arrival-rate and service-rate functions, so that temporary periods of overload or underload before the time of interest tend to persist longer and longer.

6.6. Time-Dependent Arrival Rates

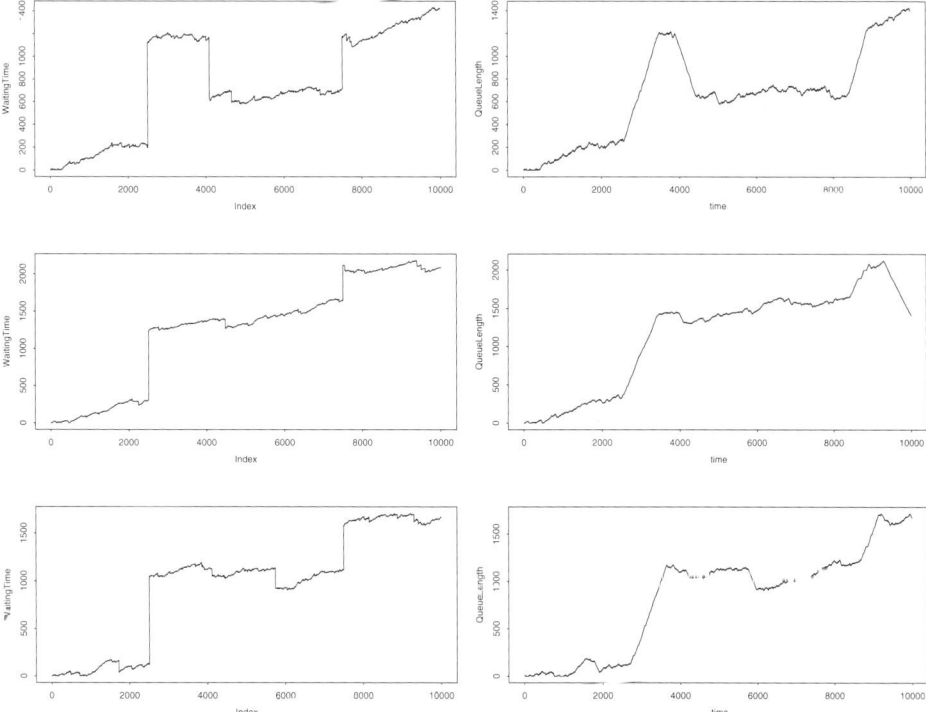

Figure 6.13. Three possible realizations of the waiting times of the first n arrivals (on the left) and the queue-length process over the interval $[0, nEU_1^n]$ (on the right) in the $P_{1.5}/M/1$ queue with scaling in (4.21) and service interruptions of length $2n^{2/3}$ and $n^{2/3}$ associated with customers $n/4$ and $3n/4$ for $n = 10^4$.

With such scaling, a law of large numbers can be established, in which the scaled queue-length process converges to a reflection of a deterministic net-input process, where the limiting deterministic net-input process satisfies an *ordinary differential equation* (ODE) driven by the original time-dependent arrival and service rates. That limit is identical to the direct deterministic ODE approximation we obtain if we ignore the stochastic aspects of the model. In the direct deterministic approximation, the net input becomes the solution an ODE driven by the time-dependent arrival and service rates; i.e., if λ is the arrival-rate function and μ is the service-rate function, then the deterministic approximation for the queue length is the function q satisfying

$$q(t) - \psi(x)(t) \equiv x(t) - \inf_{0 \leq s \leq t} x(s), \quad t \geq 0, \qquad (6.1)$$

where ϕ is again the one-sided reflection map, $q(0)$ is the initial queue length (assumed to satisfy $q(0) = 0$) and x is the deterministic net-input function, satifying

the ODE
$$\dot{x}(t) = \lambda(t) - \mu(t), \quad t \geq 0 \ . \tag{6.2}$$

When the deterministic fluctuations dominate the stochastic fluctuations, such a deterministic analysis can be very useful to describe system performance; e.g., see Oliver and Samuel (1962), Newell (1982) and Hall (1991).

However, in stochastic-process limits, we are primarily interested in going beyond the deterministic ODE limit described above. For example, Mandelbaum and Massey (1995) show that it is possible to establish a stochastic (FCLT) refinement to the deterministic ODE limit. It again can be obtained by applying the continuous-mapping approach to stochastic-process limits. In this setting, the continuous-mapping approach involves convergence preservation with nonlinear centering, and can be approached by identifying the directional derivative of the reflection map; see Chapter 6 of the Internet Supplement.

The behavior of the limit process in the stochastic-process limit depends on the deterministic function q. At any time, the deterministic function q must be in one of three states (based on the history of the build up prior to the time of interest): overloaded, critically loaded (when the cumulative input rate is in balance with the output rate) or underloaded. (Roughly speaking, these regimes correspond to the three cases $\rho > 1$, $\rho = 1$ and $\rho < 1$ in a stationary queueing model.)

With the usual stochastic assumptions (without any heavy-tailed distributions), the stochastic-process refinement is a diffusion process centered about the deterministic function q. The diffusion process corresponds to: ordinary Brownian motion when q is overloaded, reflected Brownian motion when q is critically loaded, and the zero function when q is underloaded.

Within each region, i.e., within any interval in which the determinisitic function q remains in one of its three basic states (overloaded, critically loaded or underloaded), the limiting stochastic process has continuous sample paths, but at the boundaries between different regions the limiting stochastic process can have jumps that are unmatched in the converging processes. Thus, the boundary points between different regions for the deterministic function q act as phase transitions for the queueing system. Relatively abrupt changes in the queueing process can occur at these transition times. And, once again, we have a stochastic-process limit with unmatched jumps.

Example 6.6.1. *A shift from critically loaded to underloaded.*

We now give a simple example. In the standard situation we have in mind, the arrival-rate function is changing continuously, so that we can obtain the deterministic net-input function by solving the ODE in (6.2). However, now we consider the more elementary situation in which there is a sudden shift down in the arrival rate at one time. As in the standard situation, we let the service rate be constant (although that is not required).

We let the queue initially be critically loaded, i.e., with $\rho = 1$, and then in the middle of the time period, we reduce the arrival rate, making the model underloaded. For simplicity, we again use the $M/M/1$ queue. We let the mean service

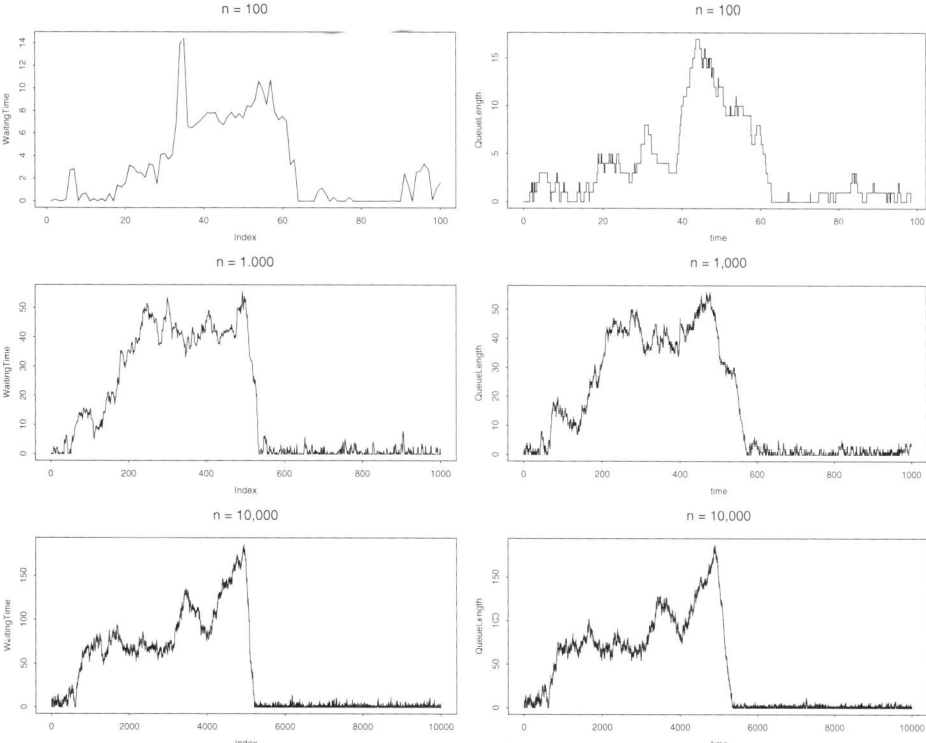

Figure 6.14. Plots of the waiting times of the first n arrivals (on the left) and the queue-length process over the interval $[0, n]$ (on the right) in the $M/M/1$ queue with $\rho = 1$ for the first $n/2$ arrivals and $\rho = 1/2$ for the last $n/2$ arrivals for $n = 10^j$ with $j = 2, 3, 4$.

time always be 1. We actually deviate slightly from the prescription for the arrival rate: We let the mean interarrival time for the first $n/2$ customers be 1 and the mean interarrival time of the next $n/2$ customers be 2. Hence, after $n/2$ arrivals, the instantaneous traffic intensity suddenly shifts from $\rho = 1$ to $\rho = 0.5$. Of course, with this definition, the shift in arrival rate occurs at a random time instead of a deterministic time, but after scaling time by n, that scaled random shift time converges to $t/2$ w.p.1. Thus, what we do is essentially the same as if we let the arrival-rate shift occur exactly at time $n/2$ when we consider n arrivals.

For the specified model, we plot the waiting times of the first n customers and the queue-length process over the time interval $[0, n]$ for $n = 10^j$ for $j = 2, 3, 4$ in Figure 6.14. As in previous plots, the situation is somewhat ambiguous for smaller n, but as n increases, we see statistical regularity. As before, the scaled waiting-time and queue-length plots coalesce as n increases. As n increases, a sharp jump down is visible when the traffic intensity shifts from $\rho = 1$ to $\rho = 1/2$. As we indicated before, asymptotically, this shift for the scaled processes occurs at time $t = 1/2$.

224   6. Unmatched Jumps

Again, we are able to establish supporting FCLTs. Both the scaled waiting-time process and the scaled queue-length process are approaching reflecting Brownian motion over the subinterval $[0, t/2)$ and the 0 function over the subinterval $[t/2, 1]$. As in the previous examples, the scaled queue-length and waiting-time processes are asymptotically equivalent.

Thus, the limit process for the scaled queue-length process has an unmatched jump at $t = 1/2$. In this example, the limit for the waiting-time process also has an unmatched jump at the same time.

7
More Stochastic-Process Limits

7.1. Introduction

This chapter is a sequel to Chapter 4: We continue providing an overview of established stochastic-process limits for basic stochastic processes. These stochastic-process limits are of interest in their own right, but they also can serve as initial stochastic-process limits in the continuous-mapping approach to establish new stochastic-process limits.

We start in Section 7.2 by considering a different kind of stochastic-process limit: We consider CLT's for sums of stochastic processes with sample paths in D. In Sections 8.7 and 9.8 we apply these CLT's for processes to treat queues with multiple sources, where the number of sources increases in the heavy-traffic limit.

In Section 7.3 we extend the discussion begun in Section 6.3 about stochastic-process limits for counting processes. As indicated before, stochastic-process limits for counting processes follow directly from stochastic-process limits for random walks, provided that the M_1 topology is used. In Chapters 9 and 10 we apply the stochastic-process limits for counting processes to obtain heavy-traffic limits for the standard queueing models.

In Section 7.4 we apply convergence-preservation results for the composition and inverse maps to establish stochastic-process limits for renewal-reward stochastic processes. Renewal-reward stochastic processes are random sums of IID random variables, where the random index is a renewal counting process. When the times between the renewals in the renewal counting process have a heavy-tailed distribution, we need the M_1 topology.

7.2. Central Limit Theorem for Processes

In this section we consider a different kind of stochastic-process limit. Instead of considering scaled partial sums of random vectors, we consider CLTs for partial sums of random elements of D, i.e., we consider the limiting behavior of the scaled partial sum

$$\mathbf{Z}_n(t) \equiv n^{-1/2} \sum_{i=1}^{n} [X_i(t) - EX_i(t)], \quad t \geq 0 , \qquad (2.1)$$

where $\{X_n : n \geq 1\} \equiv \{\{X_n(t) : t \geq 0\} : n \geq 1\}$ is a sequence of IID random elements of $D \equiv D([0,\infty),\mathbb{R})$.

In fact, we already discussed a special case of IID random elements of D when we considered the Kolmogorov-Smirnov statistic in Section 1.5. Recall that the emprical process associated with a sample of size n from a cdf F can be expressed as

$$F_n(t) \equiv \sum_{i=1}^{n} I_{(-\infty,t]}(Y_i), \quad t \in \mathbb{R} , \qquad (2.2)$$

where $\{Y_i\}$ is a sequence of IID real-valued random variables with cdf F.

For the normalized partial sums in (2.1), convergence of finite-dimensional distributions is elementary by the multidimensional CLT in Section 4.3, provided that $E[X_1(t)^2] < \infty$. As indicated in Section 11.6, to establish convergence in distribution in D with an appropriate topology, it only remains to establish tightness. Specifically, we can apply Theorem 11.6.6.

7.2.1. Hahn's Theorem

The following result is due to Hahn (1978). We have converted the original result from $D([0,1],\mathbb{R})$ to $D([0,\infty),\mathbb{R})$.

Theorem 7.2.1. (CLT for processes in D) *Let $\{X_n : n \geq 1\}$ be a sequence of IID random elements of $D \equiv D([0,\infty),\mathbb{R})$ with $EX_i(t) = 0$ and $E[X_i(t)^2] < \infty$ for all t. If, for all T, $0 < T < \infty$, there exist continuous nondeceasing real-valued* functions g and f on $[0,T]$ and numbers $\alpha > 1/2$ and $\beta > 1$ such that

$$E[(X(u) - X(s))^2] \leq (g(u) - g(s))^\alpha \tag{2.3}$$

and

$$E[(X(u) - X(t))^2(X(t) - X(s))^2] \leq (f(u) - f(t))^\beta \tag{2.4}$$

for all $0 \leq s \leq t \leq u \leq T$ with $u - s < 1$, then

$$\mathbf{Z}_n \Rightarrow \mathbf{Z} \quad in \quad (D, J_1) , \tag{2.5}$$

where $\mathbf{Z}_n$ is the normalized partial sum in (2.1), $\mathbf{Z}$ is a mean-zero Gaussian process with the covariance function of X_1 and $P(\mathbf{Z} \in C) = 1$.

Remark 7.2.1. *More elementary conditions.* The canonical sufficient conditions for (2.3) and (2.4) in applications of Theorem 7.2.1 have

$$f(t) = g(t) = Kt, \quad t \geq 0, \quad \alpha = 1 \quad \text{and} \quad \beta = 2 ,$$

for some constant K (depending upon T), yielding

$$E[(X(u) - X(s))^2] \leq K(u - s)) \tag{2.6}$$

and

$$E[(X(u) - X(t))^2(X(t) - X(s))^2] \leq K(u - t)^2 \tag{2.7}$$

for $0 \leq s \leq t \leq u \leq T$ with $u - s < 1$.

Note that conditions (2.6) and (2.7) apply to treat the empirical process in (2.2). There

$$E[(X(u) - X(s))^2] = P(s < Y_1 \leq u)$$

and

$$E[(X(u) - X(t))^2(X(t) - X(s))^2] = P(t < Y_1 \leq u \quad \text{and} \quad s < Y_1 \leq t) = 0 .$$

We see that condition (2.6) holds whenever the cdf F has a bounded density. However, the different approach in Section 1.5 shows that convergence in (D, J_1) actually holds whenever the cdf F is continuous. Plots of the scaled empirical process for the uniform cdf in Figure 1.8 illustrate Theorem 7.2.1. The limiting Gaussian process in that case is the Brownian bridge. ∎

Remark 7.2.2. *Extensions.* An analog of Theorem 7.2.1 for stable process limits in (D, J_1) when $E[X(t)^2] = \infty$ has been established by Bloznelis (1996). That opens the way for limits with jumps. It remains to develop conditions for the M topologies. Other extensions of Theorem 7.2.1 are contained in Bloznelis (1990), Bloznelis and Paulauskas (2000) and references therein; e.g., see Bass and Pyke (1987). See Araujo and Giné (1980) for CLTs for random elements of general Banach spaces. ∎

We now state some consequences of Theorem 7.2.1. We first apply Theorem 7.2.1 to establish a CLT for stochastic processes with smooth sample paths, such as cumulative-input stochastic processes to fluid queues. In that context, a standard model for the input from one source is an on-off model, in which there are alternating random on and off periods. During on periods, input arrives at a constant rate; during off periods there is no input. It is customary to assume that the successive on and off periods come from independent sequences of IID random variables, but we do not need to require that here. A generalization is to allow the source environment be governed by a k-state process instead of a two-state process. When the environment state is j, the input is transmitted at constant rate r_j. For example, the environment process might be a finite-state semi-Markov process; see Duffield and Whitt (1998, 2000). Again we do not require such specific assumptions.

Let $X(t)$ be the total input over the interval $[0,t]$. Since the input occurs at a random rate, with only finitely many possible rates, the sample paths are Lipschitz with probability one, i.e.,

$$|X(t) - X(s)| \leq K(t-s) \quad w.p.1 \tag{2.8}$$

for all $0 \leq s \leq t$, where K is the maximum possible rate. We are interested in the CLT (2.5) to describe the aggregate input from a large number of sources.

Corollary 7.2.1. (CLT for Lipschitz processes) *If $\{X_n\}$ is a sequence of IID random elements of C satisfying (2.8), then the CLT (2.5) holds.*

Proof. It is easy to see that conditions (2.6) and (2.7) hold. Indeed, (2.8) implies that

$$E[(X(u) - X(s))^2] \leq K(u-s)^2 \leq K(u-s)$$

and

$$E[(X(u) - X(t))^2(X(t) - X(s))^2] \leq K(u-t)^2(t-s)^2 \leq K(u-s)^4 \leq K(u-s)^2 .$$

■

Hahn (1978) applied Theorem 7.2.1 to establish the following CLTs for Markov processes. For any real-valued random variable Y, let the *essential supremum* be

$$ess\,sup\,(Y) \equiv \inf\{c : P(Y > c) = 0\} .$$

Theorem 7.2.2. (CLT for Markov processes) *Let $\{X_n : n \geq 1\}$ be a sequence of IID Markov processes with sample paths in D. If, for each T, $0 < T < \infty$, there exists a continuous nondecreasing real-valued function g on $[0,T]$ and a number $\alpha > 1/2$ such that either*

$$ess\,sup\,E[(X(t) - X(s))^2|X(s)] \leq (g(t) - g(s))^\alpha \tag{2.9}$$

or

$$ess\,sup\,E[(X(t) - X(s))^2|X(t)] \leq (g(t) - g(s))^\alpha , \tag{2.10}$$

for $0 \leq s \leq t \leq T$, with $t - s < 1$, then conditions (2.3) and (2.4) hold for $X(t) - EX(t)$, so that the conclusion of Theorem 7.2.1 holds.

Hahn also observed that Theorem 7.2.2 applies directly to finite-state CTMCs.

Corollary 7.2.2. (CLT for finite-state CTMCs) *If $\{X_n : n \geq 1\}$ be a sequence of IID finite-state continuous-time Markov chains determined by an infinitesimal generator matrix Q, Then the conditions of Theorem 7.2.2 hold, with $g(t) = t$ and $\alpha = 1$ in (2.9) and (2.10), so that the conclusion of Theorem 7.2.1 holds.*

Theorem 7.2.1 was also applied by Whitt (1985a) to obtain the following CLT for stationary renewal processes.

Theorem 7.2.3. (CLT for stationary renewal processes) *Let $\{X_n : n \geq 1\} \equiv \{\{X_n(t) : t \geq 0\} : n \geq 1\}$ be a sequence of IID stationary renewal counting processes with interrenewal-time cdf F. If*

$$\varlimsup_{t \to 0} t^{-1}(F(t) - F(0)) < \infty , \qquad (2.11)$$

then conditions (2.6) *and* (2.7) *hold, so that the conclusion of Theorem 7.2.1 holds.*

Remark 7.2.3. Whitt (1985a) showed that condition (2.11) is necessary for condition (2.7) to hold. Note that condition (2.11) allows an atom at 0 and is satisfied if the cdf F is otherwise absolutely continuous in a neighborhood of 0. Hence condition (2.11) is not very restrictive. ∎

From Theorems 7.2.1–7.2.3, we know that the limit process **Z** is a zero-mean Gaussian process with sample paths in C and the covariance function of one of the summands. Hence, to fully characterize the limit process it suffices to determine the covariance function of the original component process.

For example, let us consider the case of Theorem 7.2.3. There it suffices to determine the covariance function of a component stationary renewal process. For a stationary renewal process, now denoted by $\{A(t) : t \geq 0\}$, the covariance function can be computed from the interrenewal-time cdf F, exploiting numerical transform inversion to compute the renewal function; see Chapter 4 of Cox (1962) and Section 13 of Abate and Whitt (1992a). Recall that the renewal function $M(t)$ is the mean number of renewals in $[0, t]$ for the ordinary renewal process. We characterize the covariance function further in the following theorem.

Theorem 7.2.4. (covariance function for a stationary renewal process) *Suppose that A is a stationary renewal counting process (having stationary increments with $A(0) = 0$) with interrenewal-time cdf F having pdf f and mean λ^{-1}. Then, for $t < u$,*

$$Cov(A(t), A(u)) = Var\, A(t) + Cov(A(t), A(u) - A(t)) ,$$

where

$$V(t) = Var\, A(t) = 2\lambda \int_0^t [M(s) - \lambda s + 0.5]ds , \qquad (2.12)$$

$$E[A(t), A(u) - A(t)] = \lambda^3 \int_0^t da \int_0^u db\, f(a+b) M(u-b) M(t-a) \qquad (2.13)$$

and $M(t)$ is the renewal function of the associated ordinary renewal process, having Laplace transform

$$\hat{M}(s) = \frac{\hat{f}(s)}{s[1 - \hat{f}(s)]} ,$$

where

$$\hat{f}(s) \equiv \int_0^\infty e^{-st} f(t) dt \quad \text{and} \quad \hat{M}(s) \equiv \int_0^\infty e^{-st} M(t) dt .$$

Directly, the Laplace transform of $V(t)$ is

$$\hat{V}(s) \equiv \int_0^\infty e^{-st} V(t) dt = 2\lambda \left(\frac{\hat{M}(s)}{s} - \frac{\lambda}{s^3} + \frac{1}{2s^2} \right) .$$

Proof. For (2.12), see p. 57 of Cox (1962). For (2.13), consider the first point to the right of t. It falls at $t + b$ with the stationary-excess (or equilibrium lifetime) pdf $f_e(b) \equiv \lambda F^c(b)$. Conditional on that point being at $t + b$, the first point to the left of t falls at $t - a$ with pdf $f(a+b)/F^c(b)$. Conditional on the loction of these two points at $t - a$ and $t + b$, we can invoke the independence to conclude that the expected value of $A(t)[A(t + u) - A(t)]$ is $\lambda^2 M(t - a) M(u - b)$, where $M(t)$ is the ordinary renewal function (expected number of renewals in $[0, t]$). Integrating over all possible pairs (a, b) gives (2.13). ∎

7.2.2. A Second Limit

In many of the CLTs for processes, the component random elements of D have stationary increments. Then the limiting Gaussian process will also have stationary increments in addition to continuous sample paths, so that it is natural to consider an additional stochastic-process limit for the Gaussian process with time scaling; i.e., given $\mathbf{Z}_n \Rightarrow \mathbf{Z}$ as in (2.5), we can consider

$$\mathbf{Y}_n \Rightarrow \mathbf{Y} \quad \text{in} \quad (C, U) , \qquad (2.14)$$

where

$$\mathbf{Y}_n(t) \equiv c_n^{-1} \mathbf{Z}(nt), \quad t \geq 0 . \qquad (2.15)$$

Alternatively, if the component processes are stationary processes, then the limit process $\mathbf{Z}$ will be stationary, so that we have (2.14) with $\mathbf{Y}_n$ defined by

$$\mathbf{Y}_n(t) \equiv c_n^{-1} \int_0^{nt} \mathbf{Z}(s), \quad t \geq 0 . \qquad (2.16)$$

To obtain the second stochastic-process limit, we can often apply Theorem 4.6.2. The second limit allows us to replaces a Gaussian process with a general covariance function by a special Gaussian process – fractional Brownian motion – with the highly structured covariance function in (6.13). As with the first limit, we gain simplicity but lose structural detail by taking the limit. In considerable generality, we see that the large-time-scale behavior of the aggregate process should be like FBM. The second limit provides important new insight when that FBM is not Brownian motion.

We now illustrate by applying Corollary 7.2.1 and Theorem 4.6.2 to establish a double stochastic-process limit in (D, J_1) for the input from many on-off sources with heavy-tailed on-period or off-period distributions. Convergence of the finite-dimensional distributions was established by Taqqu, Willinger and Sherman (1997). As indicated there and in Willinger, Taqqu, Sherman and Wilson (1997), the stochastic-process limit is very helpful to understand the strong dependence and self-similarity observed in network traffic measurements, such as in Leland et al. (1994). The stochastic-process limit shows how high variability (the Noah effect) in the on and off periods can lead to strong positive dependence (the Joseph effect) in the cumulative input process. As indicated in Section 4.2, the very existence of the limit implies that the limit process must be self-similar.

The extension here to weak convergence from only convergence of the finite-dimensional distributions is important for establishing further stochastic-process limits, in particular, heavy traffic limits for queues with input from many on-off sources.

Now we assume that the on periods and off periods come from independent sequences of IID random variables. We let the input rate be 1 during each on period and 0 during each off period. Let the on periods have cdf F_1, ccdf $F_1^c \equiv 1 - F_1$ and finite mean m_1; let the off periods have cdf F_2, ccdf $F_2^c \equiv 1 - F_2$ and finite mean m_2. We assume that the cdf's have probability density functions, although that can be generalized; e.g., see Section VI.1 of Asmussen (1987).

The critical assumtion is that the ccdf's F_1 and F_2 can have heavy tails. Specifically, we assume that the cdf F_i either has finite variance σ_i^2 or that

$$F_i^c(t) \sim c_i t^{-\alpha_i} \quad \text{as} \quad t \to \infty$$

for $1 < \alpha < 2$. (That can be generalized to regularly varying tails; see Taqqu, Willinger and Sherman (1997).)

We now specify the limiting scaling constant. When $\sigma^2 < \infty$, let $\alpha_i = 2$ and $a_i = c_i \Gamma(2 - \alpha_i)/(\alpha_i - 1)$, where Γ is the gamma function. When $\alpha_1 = \alpha_2$, let $\alpha_{min} = \alpha_1$ and

$$\sigma_{lim}^2 \equiv \frac{2(m_1^2 + m_2^2)}{(m_1 + m_2)^3 \Gamma(4 - \alpha_{min})} \ . \qquad (2.17)$$

When $\alpha_1 \neq \alpha_2$, let $\alpha_{min} \equiv \alpha_1 \wedge \alpha_2 \equiv \min \alpha_1, \alpha_2$ and

$$\sigma_{lim}^2 \equiv \frac{2 m_{max}^2 a_{min}}{(m_1 + m_2)^3 \Gamma(4 - \alpha_{min})} \ , \qquad (2.18)$$

where (min, max) is the pair of indices (1, 2) if $\alpha_1 < \alpha_2$ and (2, 1) if $\alpha_2 < \alpha_1$.

Let $X(t)$ be the cumulative input from one on-off source over the interval $[0, t]$. We assume that the process X has been initialized so that X has stationary increments. Then

$$EX(t) = m_1/(m_1 + m_2) \quad \text{for all} \quad t \geq 0 \ .$$

(That can be generalized as well.) We are interested in the limiting behavior of scaled sum of IID versions of this cumulative-input stochastic process, in particular,

$$\mathbf{Z}_{n,\tau} \equiv \tau^{-H} n^{-1/2} \sum_{i=1}^{n} [X_i(\tau t) - m_1 \tau t/(m_1 + m_2)] . \qquad (2.19)$$

Theorem 7.2.5. (iterated limit for time-scaled sum of on-off cumulative-input processes) *If $\{X_i : i \geq 1\}$ is a sequence of IID cumulative-input stochastic processes satisfying the assumptions above, then*

$$\mathbf{Z}_{n,\tau} \Rightarrow \sigma_{lim} \mathbf{Z}_H \quad in \quad (D, J_1)$$

as first $n \to \infty$ and then $\tau \to \infty$ for

$$H = (3 - \alpha_{min})/2 , \qquad (2.20)$$

$\mathbf{Z}_{n,\tau}$ *in (2.19), σ^2 in (2.17) or (2.18) and $\mathbf{Z}_H$ standard FBM.*

Proof. First the CLT in D as $n \to \infty$ follows from Corollary 7.2.1. As a consequence, the limit process, say Y, has paths in C. We establish weak convergence as $\tau \to \infty$ in (C, U) by applying Theorem 4.6.2. Convergence of the finite-dimensional distributions was established by Taqqu, Willinger and Sherman (1997). As an important part of that step, they established (6.23) for H in (2.20). They claim to establish weak convergence in the second limit as $\tau \to \infty$ by applying Theorem 11.6.5 using only (6.23), but their argument at the end of Section 3 has a gap, because it does not control the small-time behavior. However, that gap can be filled quite easily by establishing (6.24). First, by (6.23), there exists t_0 such that

$$VarY(t) \leq 2ct^{2H} \quad \text{for all} \quad t > t_0 .$$

However, given t_0 (where, without loss of generality we may assume that t_0 is a positive integer),

$$VarY(t) \leq t_0^2 VarY(t/t_0) \quad \text{for all} \quad t, \quad 0 \leq t \leq t_0 ,$$

by writing $Y(t)$ as the sum of t_0 random variables $Y(kt/t_0) - Y((k-1)t/t_0)$, $1 \leq k \leq t_0$. Hence, it suffices to consider $t \leq 1$. However, by the Lipschitz sample-path structure of X_1,

$$|X_1(t)| \leq t \quad w.p.1 ,$$

so that

$$VarY(t) = VarX_1(t) \leq t^2 ,$$

which is less than t^{2H} for $t \leq 1$. ∎

Theorem 7.2.5 involves an iterated limit, in which first $n \to \infty$ and then afterward $\tau \to \infty$. Mikosch et al. (2001) consider the double limit with $\tau_n \to \infty$ as $n \to \infty$. They show that convergence to FBM still holds when $\tau_n \to \infty$ slowly enough. See Remark 8.7.1 for further discussion.

Theorem 7.2.5 extends to more general source traffic models, such as when the rate process is a finite state semi-Markov process (SMP), as in Duffield and Whitt (1998, 2000). In that setting we call the SMP environment states *levels*. The self-similarity index H is again determined by the level-holding-time cdf's $F_{i,j}$, giving the distribution of the time spent at level i given that the next level will be j. Assuming that the DTMC governing the state transitions is irreducible, If any of these cdf's has a heavy tail, then we get convergence to FBM with $1/2 < H < 1$. In particular, H again is given by (2.20), where α_{min} is the minimum among the stable indices of the cdf's $F_{i,j}$, assumed to satisfy $1 < \alpha_{min} < 2$.

A related stochastic-process limit for the aggregate input from many sources is in Kurtz (1996); it also leads to FBM under appropriate conditions.

We remark that we encounter difficulties when we try to establish the second limit for the stationary renewal processes treated in Theorem 7.2.3 when $H > 1/2$, because Var $(X(t)) = O(t)$ as $t \to 0$, so that we cannot establish (6.24) for $H > 1/2$. We do get the second limit when $H = 1/2$ though. Convergence of the finite-dimensional distributions for a more general model was established by Taqqu and Levy (1986).

In this section we have only discussed CLTs for processes that converge to Gaussian processes with continuous sample paths. It is also of interest to establish convergence to stable processes. Such stochastic-process limits (only convergence of finite-dimensional distributions) have been established by Levy and Taqqu (1987, 2000). As indicated before, criteria for weak convergence in D to a stable process have been determined by Bloznelis (1996). More work needs to be done in that area.

7.3. Counting Processes

With queueing models and many other applications, the basic random variables X_n are often nonnegative. For example, in a queueing model X_n may represent a service time, interarrival time, busy period or idle period. With nonnegative random variables, in addition to the partial sums $S_n = X_1 + \cdots + X_n, n \geq 1$, with $S_0 = 0$, we are often interested in the associated *counting process* $N \equiv \{N(t) : t \geq 0\}$, defined in (3.1) of Section 6.3. For example, the arrival and service counting processes are used to establish heavy-traffic limits for the standard single-server queue.

When the random variables X_n are nonnegative, we can think of the partial sums S_n as points on the positive halfline $\mathbb{R}_+$. Then $N(t)$ counts the number of points in the interval $[0, t]$. The two processes $\{S_n : n \geq 0\}$ and $\{N(t) : t \geq 0\}$ thus serve as equivalent representations of a *stochastic point process*. When $\{X_n : n \geq 1\}$ is a sequence of IID random variables, the counting process N is also called a *renewal process*.

As indicated in Section 6.3, the partial sums $\{S_n : n \geq 0\}$ and the counting process $\{N(t) : t \geq 0\}$ are inverse processes. Fortunately, we are able to exploit to inverse relation to great advantage for establishing limit theorems. Under minimal regularity conditions, we are able to show that CLTs and FCLTs hold for $\{N(t)\}$ if and only if they hold for $\{S_n\}$, without making any direct probability assumptions about the sequence $\{X_n\}$. These equivalence results are applications of the continuous-mapping approach, which we carefully develop in Chapter 13. Since these limits are also frequently applied in the continuous-mapping approach, we state the key results here.

Before proceeding, however, we point out that, even though the nonnegativity condition on the summands X_j is often natural, it is actually not required to obtain limits for the counting processes from associated limits for the partial sums. Without the nonnegativity, we can go from a CLT or FCLT for partial sums to an associated CLT or FCLT for the associated sequence of successive maxima M_n, where

$$M_n \equiv \max\{S_1, \ldots, S_n\}, \quad n \geq 1 .$$

It turns out that limits for S_n imply corresponding limits for M_n. Then M_n itself can be regarded as a partial-sum process with nonnegative steps, so that we can apply the results in this section to M_n. We then obtain limits for the associated counting process, defined as

$$N(t) \equiv \max\{k : M_k \leq t\}, \quad t \geq 0 .$$

The details are in Chapter 13.

7.3.1. CLT Equivalence

We now return to the case of nonnegative random variables. We first state an equivalence result for CLTs; it is Theorem 3.5.1 in the Internet Supplement, which extends Theorem 6 of Glynn and Whitt (1988) and Theorem 4.2 of Massey and Whitt (1994a). Note that there are no direct probability assumptions on the basic sequence $\{X_k\}$ and the limit is arbitrary. Also note that space scaling is done by a regularly varying function with index p, $0 < p < 1$, which covers the standard scaling by $\sqrt{n}$ and is consistent with the CLT for IID random variables in the domain of attraction of a stable law with index α, $1 < \alpha \leq 2$, in Section 4.5. (See Appendix A for more on regularly varying functions.)

Theorem 7.3.1. (CLT equivalence for partial sums and counting processes) *Suppose that $\{X_n : n \geq 1\}$ is a sequence of nonnegative random variables, $m > 0$ and ψ is a regularly varying real-valued function on $(0, \infty)$ with index p, $0 < p < 1$. Then*

$$\psi(n)^{-1}(S_n - mn) \Rightarrow L \quad in \quad \mathbb{R} \quad as \quad n \to \infty ,$$

where $S_n \equiv X_1 + \cdots + X_n$, $n \geq 1$, if and only if

$$\psi(t)^{-1}(N(t) - m^{-1}t) \Rightarrow -m^{-(1+p)}L \quad in \quad \mathbb{R} \quad as \quad t \to \infty ,$$

where $N(t) \equiv \max\{k \geq 0 : S_k \leq t\}$, $t \geq 0$.

7.3.2. FCLT Equivalence

Next we present an equivalence result for FCLTs allowing double sequences $\{X_{n,k}\}$. Thus, there is a sequence of partial sums $\{S_{n,k} : k \geq 1\}$ and an associated counting process $\{N_n(t) : t \geq 0\}$ for each n, so that the result here also apply to the convergence of sequences of random walks to general Lévy processes in Section 2.4 of the Internet Supplement. (Theorem 7.3.1 above does not extend to double sequences.)

Again no direct probability assumptions are made on the basic sequences $\{X_{n,k} : k \geq 1\}$ and the limit process can be anything. Finally, note that we use the M_1 topology. As we have seen in previous sections of this chapter, it is usually possible to establish the FCLT for partial sums in (3.1) below in the stronger J_1 topology, but nevertheless as discussed in Section 6.3, the FCLT for the counting process only holds in the M_1 topology when the limit process **S** for the normalized partial sums has sample paths containing positive jumps, as occurs with the stable Lévy motion limit in Section 4.5.

To state the result we use the composition function, mapping x, y into $x \circ y \equiv x(y(t))$, $t \geq 0$.

Theorem 7.3.2. (FCLT equivalence for partial sums and counting processes) *Suppose that $\{X_{n,k} : k \geq 1\}$ is a sequence of nonnegative random variables for each $n \geq 1$, $c_n \to \infty$, $n/c_n \to \infty$, $m_n \to m$, $0 < m < \infty$ and $\mathbf{S}(0) = 0$. Then*

$$\mathbf{S}_n \Rightarrow \mathbf{S} \quad in \quad D([0,\infty), \mathbb{R}, M_1) , \qquad (3.1)$$

where $S_{n,k} \equiv X_{n,1} + \cdots + X_{n,k}$, $k \geq 1$, $S_{n,0} \equiv 0$ and

$$\mathbf{S}_n(t) \equiv c_n^{-1}(S_{n,\lfloor nt \rfloor} - m_n nt), \quad t \geq 0 , \qquad (3.2)$$

if and only if

$$\mathbf{N}_n \to -m^{-1}\mathbf{S} \circ m^{-1}\mathbf{e} \quad in \quad D([0,\infty), \mathbb{R}, M_1) , \qquad (3.3)$$

where $N_n(t) \equiv \max\{k \geq 0 : S_{n,k} \leq t\}$, $t \geq 0$, and

$$\mathbf{N}_n(t) \equiv c_n^{-1}(N_n(nt) - m_n^{-1}nt), \quad t \geq 0 . \qquad (3.4)$$

If the limits in (3.1) and (3.3) hold, then

$$(\mathbf{S}_n, \mathbf{N}_n) \Rightarrow (\mathbf{S}, \mathbf{N}) \quad in \quad (D, M_1)^2 . \qquad (3.5)$$

Theorem 7.3.2 comes from Section 7 of Whitt (1980), which extends related results by Iglehart and Whitt (1971) and Vervaat (1972). Theorem 7.3.2 is proved by applying the continuous-mapping approach with the the inverse map x^{-1} defined in (5.5) in Section 3.5, using linear centering; see Section 13.8. Specifically, the result is implied by Corollary 13.8.1.

We now show how the FCLT equivalence in Theorem 7.3.2 can be applied with previous FCLTs for partial sums to obtain FCLTs for counting processes. We start with Brownian motion limits.

Corollary 7.3.1. (Brownian FCLT for counting processes) *Suppose that the conditions of Theorem 7.3.2 hold with $c_n = \sqrt{n}$. If the FCLT for partial sums in (3.1) holds with $\mathbf{S} = \sigma \mathbf{B}$, where $\mathbf{B}$ is standard Brownian motion, then*

$$\mathbf{N}_n \Rightarrow m^{-3/2} \sigma \mathbf{B} \quad in \quad (D, J_1)$$

for $\mathbf{N}_n$ in (3.4) with $c_n = \sqrt{n}$.

Proof. Apply Theorem 7.3.2, noting that the J_1 and M_1 topologies are equivalent when the limit has continuous sample paths and

$$-m^{-1} \sigma \mathbf{B} \circ m^{-1} \mathbf{e} \stackrel{d}{=} m^{-3/2} \sigma \mathbf{B} \, . \quad \blacksquare$$

Now we consider FCLTs with stable Lévy motion limits. Recall that $S_\alpha(\sigma, \beta, \mu)$ denotes a stable law with index α, scale parameter σ, skewness parameter β and shift parameter μ; see Section 4.5. We first consider the case $1 < \alpha < 2$. When the summands are IID nonnegative random variables, the skewness parameter will be $\beta = 1$.

Corollary 7.3.2. (stable Lévy FCLT for counting processes when $\alpha > 1$) *Suppose that the conditions of Theorem 7.3.2 hold with $c_n = n^{1/\alpha}$, $1 < \alpha < 2$. If the FCLT for the partial sums in (3.1) holds with $\mathbf{S}$ a stable Lévy motion with $\mathbf{S}(1) \stackrel{d}{=} \sigma S_\alpha(1, \beta, 0)$, then*

$$\mathbf{N}_n \Rightarrow -m^{-(1+\alpha^{-1})} \mathbf{S} \quad in \quad (D, M_1)$$

for $\mathbf{N}_n$ in (3.4) with $c_n = n^{1/\alpha}$ and

$$-m^{-(1+\alpha^{-1})} \mathbf{S}(1) \stackrel{d}{=} m^{-(1+\alpha^{-1})} \sigma S_\alpha(1, -\beta, 0) \, .$$

Proof. Apply Theorem 7.3.2, noting that

$$-m^{-1} \mathbf{S} \circ m^{-1} \mathbf{e} \stackrel{d}{=} -m^{-(1+\alpha^{-1})} \mathbf{S}$$

and

$$-S_\alpha(\sigma, \beta, 0) \stackrel{d}{=} S_\alpha(\sigma, -\beta, 0) \, . \quad \blacksquare$$

We now present FCLTs for counting processes that capture the Joseph effect.

Corollary 7.3.3. (FBM FCLT for counting processes) *Suppose that $\{X_n\}$ is a stationary sequence of nonnegative random variables with mean $m = EX_n$ and $Var(X_n) < \infty$. If $\{X_n - m\}$ satisfies the conditions of Theorem 4.6.1, which requires that $Y_n \geq 0$ and $a_j \geq 0$ in (6.6), then*

$$\mathbf{N}_n \Rightarrow -m^{-1} \mathbf{S} \circ m^{-1} \mathbf{e} \quad in \quad (D, M_1) \, , \tag{3.6}$$

where

$$\mathbf{N}_n(t) \equiv c_n^{-1}(N(nt) - m^{-1} nt), \quad t \geq 0 \, , \tag{3.7}$$

for c_n in (6.16) and $\mathbf{S}$ is standard FBM.

Proof. Apply Theorems 7.3.2 and 4.6.1, noting that $-\mathbf{S} \stackrel{\mathrm{d}}{=} \mathbf{S}$. ∎

Corollary 7.3.4. (LFSM FCLT for counting processes) *Suppose that $\{X_n\}$ is a stationary sequence of nonnegative random variables with finite mean $m = EX_n$. If $\{X_n - m\}$ also satisfies the conditions of Theorem 4.7.2, which with the nonnegativity requires that $Y_j \geq 0$ and $a_j \geq 0$ in (6.6), then the stochastic-process limit in (3.6) holds with $\mathbf{S}$ being LSFM in (7.10) with $\beta = 1$ and $\mathbf{N}_n$ in (3.7) with c_n in (7.12).*

We have yet to state general equivalence theorems that cover FCLTs with stable Lévy motion limits having index $\alpha \leq 1$. When $\alpha = 1$, we have space scaling by n, but a translation term that grows faster than n. The following covers the special case of $\alpha = 1$.

Theorem 7.3.3. (FCLT equivalence to cover stable limits with index $\alpha = 1$) *Suppose that $\{X_{n,k} : k \geq 1\}$ is a sequence of nonnegative random variables for each n, $c_n \to \infty$, $m_n \to \infty$, $nm_n/c_n \to \infty$ and $\mathbf{S}(0) = 0$. Then*

$$\mathbf{S}_n \Rightarrow \mathbf{S} \quad in \quad D([0,\infty), \mathbb{R}, M_1)$$

where $S_{n,k} \equiv X_{n,1} + \cdots + X_{n,k}$, $k \geq 1$, $S_{n,0} \equiv 0$ and

$$\mathbf{S}_n(t) \equiv c_n^{-1}[S_{n,\lfloor nt \rfloor} - m_n nt], \quad t \geq 0, \qquad (3.8)$$

if and only if

$$\mathbf{N}_n \Rightarrow -\mathbf{S} \quad in \quad D([0,\infty), \mathbb{R}, M_1), \qquad (3.9)$$

where

$$\mathbf{N}_n(t) \equiv c_n^{-1}[m_n N_n(nm_n t) - nm_n t], \quad t \geq 0.$$

Proof. We can apply Theorem 7.3.2 after we express (3.8) in the appropriate form: Letting $x_n(t) \equiv S_{n,\lfloor nt \rfloor}/nm_n$ and scaling space by nm_n/c_n, we obtain

$$\mathbf{S}_n(t) = \frac{nm_n}{c_n}\left[\frac{S_{n,\lfloor nt \rfloor}}{nm_n} - t\right], \quad t \geq 0.$$

Then $x_n^{-1}(t) \approx N_n(nm_n t)/n$ and (3.9) essentially follows from Theorem 7.3.2. ∎

We now state a FCLT equivalence theorem to cover the case of stable Lévy motion limits with $\alpha < 1$. In the standard framework with IID random variables, the random variables have infinite mean. In this case, there is no translation term. We now use the inverse map without centering to characterize the limit process. Specifically, we apply Theorem 13.6.1.

Theorem 7.3.4. (FCLT equivalence to cover stable limits with $\alpha < 1$) *Suppose that $\{X_{n,k} : k \geq 1\}$ is a sequence of nonnegative random variables such that $S_{n,k} \equiv X_{n,1} + \cdots + X_{n,k} \to \infty$ w.p.1 as $k \to \infty$ and $N_n(t) \equiv \max\{k \geq 0 : S_{n,k} \leq t\} \to \infty$ w.p.1 as $t \to \infty$ for each n. Also suppose that $c_n \to \infty$ as $n \to \infty$. Then*

$$\mathbf{S}_n \Rightarrow \mathbf{S} \quad in \quad (D, M_1)$$

with $\mathbf{S}^{-1}(0) = 0$ for
$$\mathbf{S}_n(t) \equiv c_n^{-1} S_{n,\lfloor nt \rfloor}, \quad t \geq 0,$$
if and only if
$$\mathbf{N}_n \Rightarrow \mathbf{S}^{-1} \quad in \quad (D, M_1)$$
with $\mathbf{S}(0) = 0$ for
$$\mathbf{N}_n(t) \equiv n^{-1} N_n(c_n t), \quad t \geq 0.$$

Since the process sample paths are nondecreasing, the M_1 convergence in Theorem 7.3.4 is equivalent to convergence of the finite-dimensional distributions. Theorem 7.3.4 is easy to apply because
$$\{\mathbf{S}^{-1}(s) \geq t\} = \{\mathbf{S}(t) \leq s\},$$
so that
$$P(\mathbf{S}^{-1}(s) \geq t) = P(\mathbf{S}(t) \leq s)$$
for all positive s and t; see Lemma 13.6.3 in Section 13.6. When $\mathbf{S}(t)$ has a stable law with index α, $0 < \alpha < 1$, the marginal distributions are easy to compute by numerical transform inversion.

7.4. Renewal-Reward Processes

We now apply the convergence-preservation results for the composition and inverse maps established in Chapter 13 to obtain FCLTs for renewal-reward processes. Renewal-reward processes are random sums of IID random variables, where the random index is an independent renewal counting process. The results for the renewal process alone follow from the previous section.

Let $\{X_n : n \geq 1\}$ and $\{Y_n : n \geq 1\}$ be independent sequences of IID random variables, where Y_n is nonnegative. Let $S_n^x \equiv X_1 + \cdots + X_n$ and $S_n^y \equiv Y_1 + \cdots + Y_n$ be the associated partial sums, with $S_0^x \equiv S_0^y \equiv 0$. Let N be the renewal counting process associated with $\{Y_n\}$, i.e.
$$N(t) \equiv \max\{k \geq 0 : S_k^y \leq t\}, \quad t \geq 0.$$
The renewal-reward process is the random sum
$$Z(t) \equiv \sum_{i=1}^{N(t)} X_i, \quad t \geq 0. \tag{4.1}$$

The random variable $Z(t)$ represents the cumulative input of required work in a service system during the interval $[0, t]$ when customers with random service requirements X_n arrive at random times S_n^y.

We assume that X_1 and Y_1 have finite means $m \equiv EX_1$ and $\lambda^{-1} \equiv EY_1 > 0$. Thus we have the SLLNs: $n^{-1} S_n^x \to m$, $n^{-1} S_n^y \to \lambda^{-1}$ w.p.1 as $n \to \infty$ and $t^{-1} N(t) \to \lambda$, $t^{-1} Z(t) \to \lambda m$ w.p.1 as $t \to \infty$.

We are interested in the FCLT refinements. We assume that X_1 and Y_1 either have finite variances σ_x^2 and σ_y^2 or are in the normal domain of attraction of stable laws. In the case of infinite variances, let

$$P(|X_1| > t) \sim \gamma_x t^{-\alpha_x} \tag{4.2}$$

$$P(X_1 > t) \sim \frac{(1+\beta_x)}{2} P(|X_1| > t) \tag{4.3}$$

$$P(Y_1 > t) \sim \gamma_y t^{-\alpha_y} \tag{4.4}$$

as $t \to \infty$, where $1 < \alpha_x < 2$ and $1 < \alpha_y < 2$. (Recall that $Y_n \geq 0$ w.p.1.)

We form the normalized process

$$\mathbf{Z}_n(t) \equiv n^{-1/\alpha}(Z(nt) - \lambda m n t), \quad t \geq 0, \tag{4.5}$$

where $\alpha = \min\{\alpha_x, \alpha_y\}$ with $\alpha_x \equiv 2$ if $\sigma_x^2 < \infty$ and $\alpha_y \equiv 2$ if $\sigma_y^2 < \infty$.

The starting point for obtaining FCLTs for $Z(t)$ are FCLTs for S_n^x and S_n^y, which involve the scaled stochastic processes

$$\mathbf{S}_n^x(t) \equiv n^{-1/\alpha_x}(S_{\lfloor nt \rfloor}^x - mnt), \quad t > 0, \tag{4.6}$$

and

$$\mathbf{S}_n^y(t) \equiv n^{-1/\alpha_y}(S_{\lfloor nt \rfloor}^y - \lambda^{-1} n t), \quad t > 0. \tag{4.7}$$

The associated scaled process for the renewal counting process $N(t)$ is

$$\mathbf{Y}_n(t) \equiv n^{-1/\alpha_y}(N(nt) - \lambda n t), \quad t \geq 0. \tag{4.8}$$

Connections between limits for $\mathbf{S}_n^y$ and $\mathbf{Y}_n$ follow from Section Section 13.8; an overview was given in Section 7.3.

Theorem 7.4.1. (renewal-reward FCLT with finite variances) *If the random variables X_1 and Y_1 have finite variances σ_x^2 and σ_y^2, then*

$$\mathbf{Z}_n \Rightarrow \sigma \mathbf{B} \quad in \quad (D, J_1),$$

where $\mathbf{Z}_n$ is in (4.5) with $\alpha = 2$, $\mathbf{B}$ is standard Brownian motion and

$$\sigma^2 \equiv \lambda \sigma_x^2 + m^2 \lambda^3 \sigma_y^2. \tag{4.9}$$

Proof. We apply Corollary 13.3.2. For that purpose, let $(\mathbf{X}_n, \mathbf{Y}_n)$ in (3.8) be defined by $\mathbf{X}_n = \mathbf{S}_n^x$ in (4.6) and $\mathbf{Y}_n$ in (4.8). The convergence $\mathbf{X}_n \Rightarrow \mathbf{U}$, where $\mathbf{U} \stackrel{d}{=} \sigma_x \mathbf{B}$ follows from Donsker's theorem, Theorem 4.3.2. The convergence $\mathbf{Y}_n \Rightarrow \mathbf{V}$, where $\mathbf{V} \stackrel{d}{=} \lambda^{3/2} \sigma_y \mathbf{B}$ follows from Donsker's Theorem and Corollary 13.8.1, as indicated in Corollary 7.3.1. By independence and Theorem 11.4.4, we obtain the joint convergence $(\mathbf{X}_n, \mathbf{Y}_n) \Rightarrow (\mathbf{U}, \mathbf{V})$ in (3.8) from the two marginal limits. Since the sample paths of $(\mathbf{U}, \mathbf{V})$ are continuous, condition (3.9) automatically holds. Finally, the limit in (3.10) simplifies, because

$$\sigma_x \mathbf{B}_1 \circ \lambda \mathbf{e} + m \lambda^{3/2} \sigma_y \mathbf{B}_2 \stackrel{d}{=} (\lambda \sigma_x^2 + m^2 \lambda^3 \sigma_y^2)^{1/2} \mathbf{B}$$

where $\mathbf{B}_1$ and $\mathbf{B}_2$ are independent standard Brownian motions. ∎

It is often insightful to replace variances by dimensionless parameters describing the variability independent of the scale. Thus, let c_x^2 and c_y^2 be the *squared coefficients of variation* or SCVs, defined by $c_x^2 \equiv \sigma_x^2/m^2$ and $c_y^2 \equiv \lambda^2 \sigma_y^2$. We can alternatively express σ^2 in (4.9) as

$$\sigma^2 = \rho m(c_x^2 + c_y^2) , \qquad (4.10)$$

where $\rho \equiv \lambda m$.

The limit process $\sigma \mathbf{B}$ in Theorem 7.4.1 has continuous sample paths. In contrast, when either X_1 or Y_1 has a heavy-tailed distribution, the limit process has discontinuous sample paths.

Theorem 7.4.2. (renewal-reward FCLT for heavy-tailed summands) *Suppose that (4.2) and (4.3) hold with $1 < \alpha_x < 2$, and that either $\sigma_y^2 < \infty$ or $\sigma_y^2 = \infty$ and (4.4) holds with $\alpha_x < \alpha_y < 2$. Then*

$$\mathbf{Z}_n \Rightarrow \sigma \mathbf{S}_\alpha \quad \text{in} \quad (D, J_1) ,$$

where $\mathbf{Z}_n$ is in (4.5), $\alpha = \alpha_x$, $\mathbf{S}_\alpha$ is a stable Lévy motion with $\mathbf{S}_\alpha(1) \stackrel{d}{=} S_\alpha(1, \beta_x, 0)$ for β_x in (4.3) and

$$\sigma^\alpha \equiv (\gamma_x \lambda / C_\alpha) \qquad (4.11)$$

for γ_x in (4.2) above and C_α in (5.14) of Section 4.5.

Proof. We again apply Corollary 13.3.2 and Theorem 11.4.4, with $\mathbf{X}_n = \mathbf{S}_n^x$ in (4.6) and $\mathbf{Y}_n$ in (4.8). From Theorems 4.5.2 and 4.5.3, we get $\mathbf{X}_n \Rightarrow \mathbf{U}$, as needed in the condition of Corollary 13.3.2, with $\mathbf{U} \equiv \delta \mathbf{S}_\alpha$, where $\mathbf{S}_\alpha(1) \stackrel{d}{=} S_\alpha(1, \beta_x, 0)$ for β_x in (4.3) and

$$\delta = (\gamma_x / C_\alpha)^{1/\alpha}$$

for γ_x in (4.2) and C_α in (5.14). With the space scaling by $n^{-\alpha}$, we get $\mathbf{Y}_n \Rightarrow \mathbf{V}$ for $\mathbf{V} = 0\mathbf{e}$. Since $\mathbf{V} = 0\mathbf{e}$, condition (3.9) holds trivially. Hence we get $\mathbf{Z}_n \Rightarrow \mathbf{U} \circ \lambda \mathbf{e} \stackrel{d}{=} \lambda^{1/\alpha} \mathbf{U}$ from Corollary 13.3.2. Hence σ is as in (4.11). ■

Even though the limit process $\sigma \mathbf{S}_\alpha$ in Theorem 7.4.2 has discontinuous sample paths, we can use the J_1 topology on D. That is no longer true for the next two theorems.

Theorem 7.4.3. (renewal-reward FCLT for a heavy-tailed renewal process) *Suppose that (4.4) holds for $1 < \alpha_y < 2$ and that either $\sigma_x^2 < \infty$ or $\sigma_x^2 = \infty$ and (4.2) holds with $\alpha_y < \alpha_x < 2$. Then*

$$\mathbf{Z}_n \Rightarrow \sigma \mathbf{S}_\alpha \quad \text{in} \quad (D, M_1)$$

where $\mathbf{Z}_n$ is in (4.5), $\alpha = \alpha_y$, $\mathbf{S}_\alpha$ is stable Lévy motion with $\mathbf{S}_\alpha(1) \stackrel{d}{=} S_\alpha(1, -1, 0)$ and

$$\sigma^\alpha \equiv m^\alpha \lambda^\alpha \gamma_y \lambda / C_\alpha$$

for γ_y in (4.4) above and C_α in (5.14) of Section 4.5.

Proof. We again apply Corollary 13.3.2 and Theorem 11.4.4 with $\mathbf{X}_n = \mathbf{S}_n^x$ in (4.6) and $\mathbf{Y}_n$ in (4.8). Using the scaling function $n^{-1/\alpha}$, we obtain $\mathbf{U} = 0\mathbf{e}$. We obtain the FCLT for the renewal counting process from Theorems 4.5.2, 4.5.3, 7.3.2 and Corollary 7.3.2, making use of the M_1 topology. The limit process $\mathbf{V}$ then is

$$\mathbf{V} \stackrel{\mathrm{d}}{=} -\delta\lambda \mathbf{S}'_\alpha \circ \lambda e$$

where $\mathbf{S}'_\alpha(1) \stackrel{\mathrm{d}}{=} S_\alpha(1, 1, 0)$ and $\delta = (\gamma_y/C_\alpha)^{1/\alpha}$. Thus

$$m\mathbf{V} \stackrel{\mathrm{d}}{=} m\lambda^{1+\alpha^{-1}}(\gamma_y/C_\alpha)^{1/\alpha}\mathbf{S}_\alpha$$

where $\mathbf{S}_\alpha(1) \stackrel{\mathrm{d}}{=} S_\alpha(1, -1, 0)$, by virtue of (5.9) in Section 4.5. ■

We now treat the case in which the two processes have heavy tails with the same index.

Theorem 7.4.4. (renewal-reward FCLT for heavy-tailed summands and renewal process) *Suppose that (4.2)–(4.4) hold with $1 < \alpha_x < 2$ and $\alpha_y = \alpha_x$. Then*

$$\mathbf{Z}_n \Rightarrow \sigma\mathbf{S}_\alpha \quad in \quad (D, M_1),$$

where $\mathbf{Z}_n$ is in (4.5), $\alpha = \alpha_x = \alpha_y$, $\mathbf{S}_\alpha$ is a stable Lévy motion with $\mathbf{S}_\alpha(1) \stackrel{\mathrm{d}}{=} S_\alpha(1, \beta', 0)$ for

$$\beta' \equiv \frac{\gamma_x \beta_x - m^\alpha \lambda^\alpha \gamma_y}{\gamma_x + m^\alpha \lambda^\alpha \gamma_y},$$

$$\sigma^\alpha \equiv \frac{\lambda}{C_\alpha}(\gamma_x + m^\alpha \lambda^\alpha \gamma_y),$$

γ_x *in (4.2), γ_y in (4.4) and C_α in (5.14) of Section 4.5.*

Proof. We again apply Corollary 13.3.2 and Theorem 11.4.4 with $(\mathbf{X}_n, \mathbf{Y}_n)$ defined as in the previous theorems. As in Theorem 7.4.2, we get $\mathbf{X}_n \Rightarrow \mathbf{U}$ for $\mathbf{U} \stackrel{\mathrm{d}}{=} \delta\mathbf{S}_\alpha$ with $\delta = (\gamma_x/C_\alpha)^{1/\alpha}$. As in Theorem 7.4.3, we get $\mathbf{Y}_n \Rightarrow \mathbf{V}$ for $\mathbf{V} \stackrel{\mathrm{d}}{=} \eta S_\alpha$ with $\mathbf{S}_\alpha(1) \stackrel{\mathrm{d}}{=} S_\alpha(1, -1, 0)$ and $\eta = \lambda^{1+\alpha^{-1}}(\gamma_y/C_\alpha)^{1/\alpha}$. Since $\mathbf{U}$ and $\mathbf{V}$ are independent processes without any fixed discontinuities, condition (3.9) holds. Hence we get $\mathbf{Z}_n \Rightarrow \mathbf{Z}$, where

$$\mathbf{Z} \stackrel{\mathrm{d}}{=} \delta\mathbf{S}_\alpha^1 \circ \lambda e + m\eta\mathbf{S}_\alpha^2$$

where $\mathbf{S}_\alpha^1$ and $\mathbf{S}_\alpha^2$ are two independent α-stable Lévy motions with $\mathbf{S}_\alpha^1(1) \stackrel{\mathrm{d}}{=} S_\alpha(1, \beta_x, 0)$ and $\mathbf{S}_\alpha^2(1) \stackrel{\mathrm{d}}{=} S_\alpha(1, -1, 0)$. Thus we obtain

$$\begin{aligned}\sigma^\alpha &= \delta^\alpha \lambda + m^\alpha \eta^\alpha \\ &= \frac{\gamma_x}{C_\alpha}\lambda + m^\alpha \frac{\gamma_y}{C_\alpha}\lambda^\alpha \lambda \\ &= (\lambda/C_\alpha)(\gamma_x + m^\alpha \lambda^\alpha \gamma_y) \\ \beta' &= \frac{\delta\lambda^{1/\alpha}\beta_x + m\eta(-1)}{\delta\lambda^{1/\alpha} + m\eta}\end{aligned}$$

using scaling relations (5.9) and (5.10) in Section 4.5. ■

We used the strong (mostly independence) assumptions to get

$$(\mathbf{X}_n, \mathbf{Y}_n) \Rightarrow (\mathbf{U}, \mathbf{V}) \quad \text{in} \quad D^2 \tag{4.12}$$

for $\mathbf{X}_n = \mathbf{S}_n^x$ in (4.6) and $\mathbf{Y}_n$ in (4.8). Given (4.12) without the other specific assumptions we would still get limits for $\mathbf{Z}_n$ in (4.5) by applying the convergence-preservation results for composition and inverse. Chapter 4 contains alternative FCLTs that can be employed.

8
Fluid Queues with On-Off Sources

8.1. Introduction

In this chapter we consider a queueing model introduced to help understand the performance of evolving communication networks. As indicated in Section 2.4.1, traffic measurements have shown that the traffic carried on these networks is remarkably bursty and complex, exhibiting features such as heavy-tailed probability distributions, strong positive dependence and self-similarity. These traffic studies have generated strong interest in the impact of heavy-tailed probability distributions and other forms of traffic burstiness upon queueing performance.

A useful model for studying such phenomena is a fluid queue having input from multiple on-off sources; e.g., see Anick et al. (1982), Roberts (1992), Willinger et al. (1997), Taqqu et al. (1997), Choudhury and Whitt (1997), Boxma and Dumas (1998) and Zwart (2001). The queue represents a switch or router in the communication network, where data must be temporarily stored and then forwarded to its destination. The queue may have constant or random release rate, representing the available bandwidth. The actual flow of data in many small packets is modelled as fluid. Each of the many sources alternates between periods when it is busy (active or on) and periods when it is idle (inactive or off). During busy periods, the source transmits data at a constant or random rate; during idle periods, the source is idle, not transmitting anything. The total input to the queue is the superposition (sum) of the inputs from the separate sources, which usually are assumed to be stochastically independent. Given such a model, with stochastic elements specified in more detail, the object is to describe the distributions of quantities such as the buffer content, data loss and end-to-end delay experienced by users.

In this chapter we establish heavy-traffic stochastic-process limits for fluid-queue models with multiple on-off sources. Much of the literature on the fluid queue with multiple on-off sources focuses on the relatively tractable special case of homogeneous (IID) sources in which the busy periods and idle periods come from independent sequences of IID exponentially distributed random variables; e.g., see Anick et al. (1982). We consider more general models, aiming to capture the performance impact of features such as heavy-tailed distributions and strong dependence. We show how the heavy-traffic limits can identify key features determining performance in more complicated queueing models.

The on-off source traffic model represents stochastic fluctuations at two different time scales. The pattern of busy periods and idle periods produces stochastic fluctuations in a longer time scale. The stochastic process depicting the fluid flow during busy periods represents stochastic fluctuations in a shorter time scale. If we let the flow during a busy period be at a deterministic constant rate, then we are deciding in advance that the stochastic fluctuations in the shorter time scale are neglibible compared to the stochastic fluctuations in the longer time scale. More generally, the model gives us the opportunity to compare the impact of stochastic fluctuations in the two different time scales.

We could consider fluctuations in an even longer time scale by letting the number of sources itself evolve as a stochastic process, but here we consider a fixed number of sources. The heavy-traffic stochastic-process limits can be extended to the more general setting by treating the number of sources as a random environment, as in Example 9.6.2.

It will be obvious that in the principal cases the stochastic processes of interest have continuous sample paths. Thus, if there is convergence of a sequence of fluid-queue stochastic processes to a limiting stochastic process with discontinuous sample paths, as we establish in Section 8.5, then the Skorohod M_1 topology must be used. Such limits with discontinuous sample paths will arise under heavy-traffic conditions and heavy-traffic scaling when the busy-period or idle-period distributions of some sources have heavy tails (infinite variance).

Here is how this chapter is organized: In Section 8.2 we introduce the more-detailed multisource on-off model for the input to a fluid queue, Then in Section 8.3 we apply the continuous-mapping approach again to establish heavy-traffic stochastic-process limits for the more-detailed fluid-queue model.

We consider the special cases of Brownian-motion and stable-Lévy-motion heavy-traffic stochastic-process limits for fluid-queue models in Sections 8.4 and 8.5. In these sections we discuss properties of the reflected limit processes to demonstrate that the stochastic-process limits lead to tractable approximations. In some cases, probability distributions of random quantities associated with the limit process can be given explicitly in closed form. In other cases, the probability distributions can be conveniently characterized via transforms. Then numerical transform inversion can be exploited to calculate the probability distributions. There is a great potential for combining asymptotic and numerical methods.

In some cases, such as with convergence to reflected stable Lévy motion, the limit process is relatively complicated. Then, for applications, there may be interest in developing approximations for the limit process. In Section 8.6 we show how a second stochastic-process limit can be used for that purpose.

We consider strongly-dependent net-input processes in Section 8.7. When the input comes from many independent sources, the central limit theorem for processes in Section 7.2 implies that the net-input process can be approximated by a Gaussian process. With strong dependence, the scaled net-input processes converge to fractional Brownian motion (FBM). Then the associated sequence of scaled workload processes converges to a reflected FBM (RFBM). We develop approximations for the steady-state distribution of RFBM and more general reflected stationary Gaussian processes in Section 8.8.

As in Chapter 5, we give proofs for the theorems in this chapter, but we are primarily interested in the result statements and their applied significance. The proofs draw on material in later chapters.

Remark 8.1.1. *Literature on nonBrownian heavy-traffic limits.* Our discussion of heavy-traffic stochastic-process limits for fluid queues, emphasizing nonBrownian limit processes, follows Whitt (2000a, b). NonBrownian heavy-traffic limits for queues and related models have been established by Brichet et al. (1996, 2000), Boxma and Cohen (1998, 1999, 2000), Cohen (1998), Furrer, Michna and Weron (1997), Konstantopoulos and Lin (1996, 1998), Kurtz (1996), Resnick and Rootzén (2000), Resnick and Samorodnitsky (2000), Resnick and van den Berg (2000) and Tsoukatos and Makowski (1997, 2000).

8.2. A Fluid Queue Fed by On-Off Sources

In this section we add extra detail to the fluid-queue model introduced in Section 5.2. In particular, we consider multiple on-off sources.

In the fluid queue model, there can be infinite (unlimited) or finite storage space, as discussed in Section 5.2. We assume that there is a single shared buffer receiving the input from all the sources. (We discussed the case of separate source queues in Section 2.4.2.) Fluid can be processed according to the general stochastic process S, as described in Section 5.2. As indicated there, an important special case is $S(t) = \mu t$ for all $t \geq 0$ w.p.1; i.e., the fluid can be processed continuously at constant rate μ whenever there is work to process. However, we consider the general case. For example, it allows us to model service interruptions; for further discussion about service interruptions, see Remark 8.3.2.

8.2.1. The On-Off Source Model

Input arrives from each of m sources. Each source is alternatively busy (active or on) and idle (inactive or off) for random *busy periods* B_i and *idle periods* I_i, $i \geq 1$. Without loss of generality, let the first busy period begin at time 0. (That is without loss of generality, because we can redefine B_1 and I_1 to represent alternative initial conditions; e.g., to start idle, let $B_1 = 0$. To have the busy and idle periods well defined, we assume that $I_i > 0$ for all i and $B_i > 0$ for all $i \geq 2$.) For mathematical tractability, it is natural to assume that the successive pairs (B_i, I_i) after the first are IID, but we do not make that assumption. Our key assumption will be a FCLT for the associated partial sums, which from Chapter 4 we know can hold without that IID assumption.

A *busy cycle* is a busy period plus the following idle period. Thus the *termination time* of the jth busy cycle is

$$T_j \equiv \sum_{i=1}^{j} (B_i + I_i) \, . \tag{2.1}$$

(As before, we use $\equiv$ instead of $=$ to designate equality by a definition.) As a regularity condition, we assume that $T_j \to \infty$ with probability one (w.p.1) as $j \to \infty$. Let $N \equiv \{N(t) : t \geq 0\}$ be the *busy-cycle counting process*, defined by

$$N(t) = \min\{j \geq 0 : T_j \leq t\}, \quad t \geq 0, \tag{2.2}$$

where $T_0 \equiv 0$. Let A_j be the set of times when the jth busy period occurs, i.e.,

$$A_j \equiv \{t : T_{j-1} \leq t < T_{j-1} + B_j) \, , \tag{2.3}$$

where $T_0 \equiv 0$. Let A be the source *activity period* — the set of times when the source is busy; i.e.,

$$A \equiv \bigcup_{n=1}^{\infty} A_n \, . \tag{2.4}$$

For any set S, let I_S be the *indicator function* of the set S; i.e., $I_S(x) = 1$ if $x \in S$ and $I_S(x) = 0$ otherwise. Thus, for A in (2.4), $\{I_A(t) : t \geq 0\}$ is the *activity process*; $I_A(t) = 1$ if the source is active at time t and $I_A(t) = 0$ otherwise. Let $B(t)$ represent the *cumulative busy time* in $[0, t]$; i.e., let

$$B(t) \equiv \int_0^t 1_A(s) ds, \quad t \geq 0 \, . \tag{2.5}$$

Possible realizations for $\{I_A(t) : t \geq 0\}$ and $\{B(t) : t \geq 0\}$ are shown in Figure 8.1.

Let input come from the source when it is active according to the stochastic process $\{\Lambda(t) : t \geq 0\}$; i.e., let the *cumulative input* during the interval $[0, t]$ be

$$C(t) \equiv (\Lambda \circ B)(t) \equiv \Lambda(B(t)), \quad t \geq 0 \tag{2.6}$$

where $\circ$ is the composition map. (This definition ignores complicated end effects at the beginning and end of busy periods and idle periods.)

We assume that the sample paths of Λ are nondecreasing. A principal case is

$$P(\Lambda(t) = \hat{\lambda} t, \ t \geq 0) = 1$$

for a positive deterministic scalar $\hat{\lambda}$, in which case $C(t) = \hat{\lambda} B(t)$, but we allow other possibilities. (We use the notation $\hat{\lambda}$ here because we have already used λ as the overall input rate, i.e., the rate of $C(t)$.) To be consistent with the busy-period concept, one might require that the sample paths of Λ be strictly increasing, but we do not require it. To be consistent with the fluid concept, the sample paths of Λ should be continuous, in which case the sample paths of C will be continuous. That is the intended case, but we do not require it either.

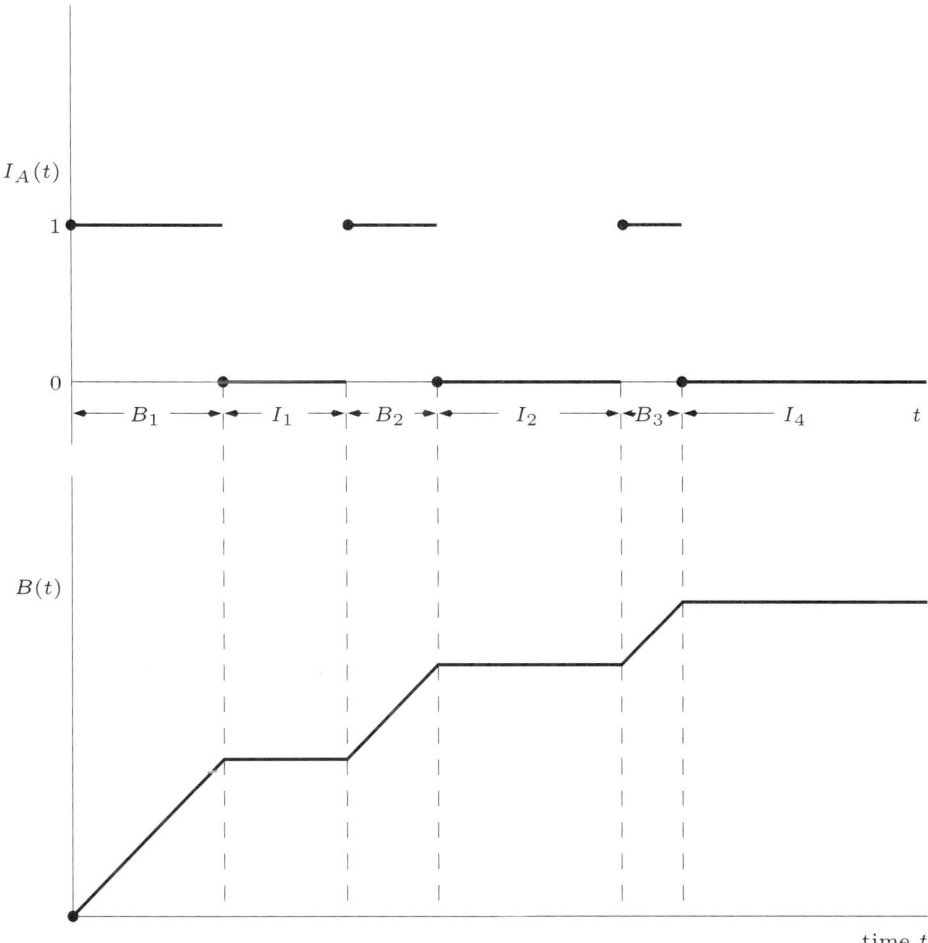

Figure 8.1. Possible realizations for the initial segments of a source-activity process $\{I_A(t) : t \geq 0\}$ and cumulative-busy-time process $\{B(t) : t \geq 0\}$.

Notice that the random quantities $\{T_j : j > 1\}$ in (2.1), $\{A_j : j \geq 1\}$ in (2.3), A in (2.4), $\{I_A(t) : t \geq 0\}$ and $\{B(t) : t \geq 0\}$ in (2.5) and $\{C(t) : t \geq 0\}$ in (2.6) are all defined in terms of the *basic model elements* — the stochastic processes $\{(B_j, I_j) : j \geq 1\}$ and $\{\Lambda(t) : t \geq 0\}$. Many measures of system performance will depend only on these source characteristics only via the cumulative-input process $\{C(t) : t \geq 0\}$. Also notice that we have imposed no stochastic assumptions yet. By using the continuous-mapping approach, we are able to show that desired heavy-traffic stochastic-process limits hold whenever corresponding stochastic-process limits hold for stochastic processes associated with the basic model data.

Now suppose that we have m sources of the kind defined above. We add an extra subscript l, $1 \leq l \leq m$, to index the source. Thus the basic model elements are $\{\{(B_{l,j}, I_{l,j}) : j \geq 1\}, 1 \leq l \leq m\}$ and $\{\{\Lambda_l(t) : t \geq 0\}, 1 \leq l \leq m\}$. The cumulative input from source l over $[0, t]$ is $C_l(t)$. The cumulative input from all m sources over $[0, t]$ is

$$C(t) \equiv \sum_{l=1}^{m} C_l(t), \quad t \geq 0 . \tag{2.7}$$

As indicated above, we suppose that the input from these m sources is fed to a single queue where work is processed according to the available-processing stochastic process S. At this point we can apply Section 5.2 to map the cumulative-input process C and the available-processing stochastic process S into a net-input process, a potential-workload process and the workload process, exploiting a reflection map. We use the one-sided reflection map if there is unlimited storage capacity and the two-sided reflection map if there is limited storage capacity.

Motivated by the fluid notion, it is natural to assume that the sample paths of the single-source cumulative-input processes are continuous. Then the aggregate (for all sources) cumulative-input stochastic process and the buffer-content stochastic process in the fluid queue model also have continuous sample paths. However, as a consequence of the heavy-tailed busy-period and idle-period distributions, the limiting stochastic processes for appropriately scaled versions of these stochastic processes have discontinuous sample paths. Thus, stochastic-process limits with unmatched jumps in the limit process arise naturally in this setting.

Just as for the renewal processes in Section 6.3, it is obvious here that any jumps in the limit process must be unmatched. What is not so obvious, again, is that there can indeed be jumps in the limit process. Moreover, the setting here is substantially more complicated, so that it is more difficult to explain where the jumps come from. To show that there can indeed be jumps in the limit process, we once again resort to simulations and plots. Specifically, we plot the buffer-content process for several specific cases.

8.2.2. Simulation Examples

To have a concrete example to focus on, suppose that we consider a fluid queue with two IID sources. Let the mean busy period and mean idle period both be 1, so that each source is busy half of the time. Let the sources transmit at constant rate during their busy periods. Let the input rate for each source during its busy period be 1. Thus the long-run input rate for each source is $1/2$ and the overall input rate is 1. The instantaneous input rate then must be one of $0, 1$ or 2.

In the queueing example in Section 2.3 we saw that it was necessary to do some careful analysis to obtain the appropriate scaling. In particular, in that discrete-time model, we had to let the finite capacity K and the output rate μ depend on the sample size n in an appropriate way. We avoid that complication here by assuming that the capacity is infinite (by letting $K = \infty$) and by letting the output rate exactly equal the input rate (by letting $\mu = 1$, where μ here is the output rate).

With those parameter choices, the instantaneous net-input rate (input rate minus output rate) at any time must be one of $-1, 0$ or $+1$. To understand the plots, it is good to think about the consequence of having exceptionally long busy periods and idle periods. If both sources are simultaneouly in long busy periods, then there will be a long interval over which the net-input rate is $+1$. Similarly, if both sources are simultaneously in long idle periods, then there will be a long interval over which the net-input rate is -1. If only one source is in a long busy period, then the other source will oscillate between busy and idle periods, so that the net-input rate will oscillate between 0 and $+1$, yielding an average rate of about $+1/2$. Similarly, if only one source is in a long idle period, then again the other source will oscillate between busy and idle periods, so that the net-input rate will oscillate between 0 and -1, yielding an average rate of about $-1/2$.

The likelihood of exceptionally long busy periods (idle periods) depends on the busy-period (idle-period) probability distribution, and these probability distributions have yet to be specified. To illustrate light-tailed and heavy-tailed alternatives, we consider the exponential and Pareto(1.5) distributions We consider four cases: We consider every combination of the two possible distributions assigned to the busy-period distribution and the idle-period distribution. We call the model Pareto/exponential if the busy-period distribution is Pareto and the idle-period distribution is exponential, and so on.

We plot four possible realizations of the workload sample path for each of the four combinations of busy-period and idle-period distributions in Figures 8.2 – 8.5. We generate $1,000$ busy cycles (idle period plus following busy period) for each source, starting at the beginning of an idle period. We plot the workload process in the interval $[0, T]$, where $T = \min(T_1, T_2)$ with T_i being the time that the $1,000^{th}$ busy cycle ends for source i. Note that $E[T_i] = 2000$.

The differences among the plots are less obvious than before. Places in the plots corresponding to jumps in the limit process have steep slopes. Jumps up only are clearly discernable in the plot of the Pareto/exponential workload in Figure 8.3, while jumps down only are clearly discernable in the plot of the exponential/Pareto workload in Figure 8.4. In contrast, both jumps up and down are discernable in the plots of the Pareto/Pareto workload in Figure 8.5. However, these jumps are not always apparent in every realization. The plots of the exponential/exponential workloads in Figure 8.2 are approaching plots of reflected Brownian motion. Just as for plots of random walks, sometimes plots that are approaching a relected stable Lévy motion appear quite similar to other plots that are approaching a reflected Brownian motion. However, additional replications and statistical tests can confirm the differences.

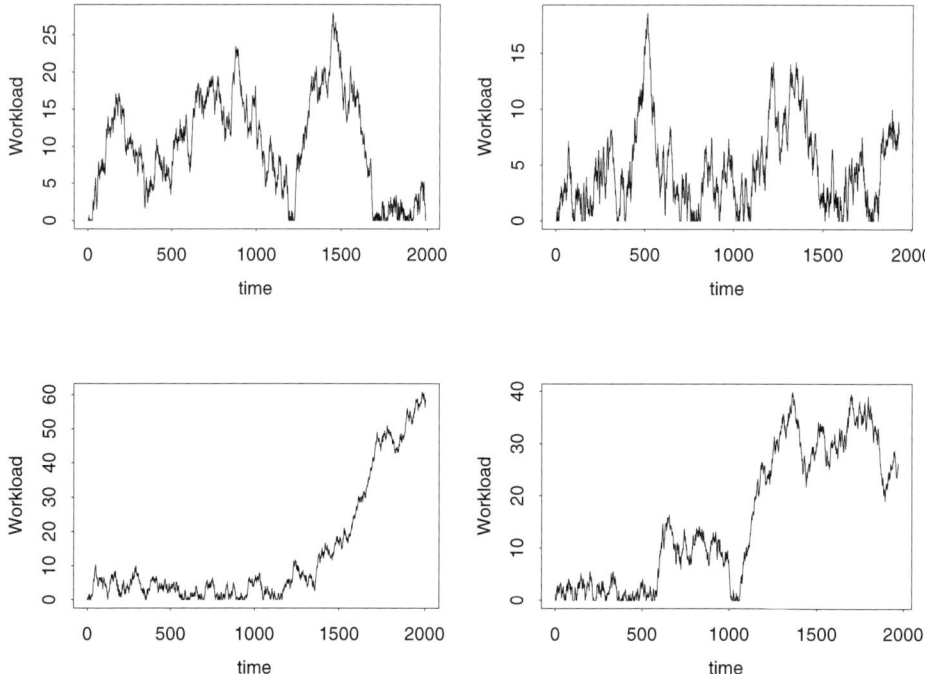

Figure 8.2. Plots of four possible realizations of the workload process in the two-source infinite-capacity exponential/exponential fluid queue having exponential busy-period and idle-period distributions with mean 1, where the input rate equals the output rate. Each source has up to 10^3 busy cycles.

8.3. Heavy-Traffic Limits for the On-Off Sources

In this section, following Whitt (2000b), we establish stochastic-process limits for the more-detailed multisource on-off fluid-queue model in Section 8.2. From Theorems 5.4.1 and 5.9.1, we see that it suffices to establish a stochastic-process limit for the cumulative-input processes, so that is our goal.

Let model n have m_n on-off sources. Thus the basic model random elements become $\{\{\{(B_{n,l,i}, I_{n,l,i}) : i \geq 1\}, 1 \leq l \leq m_n\}, n \geq 1\}$ and $\{\{\{\Lambda_{n,l}(t) : t \geq 0\}, 1 \leq l \leq m_n\}, n \geq 1\}$. Let the source cumulative-input processes $C_{n,l}$ be defined as in Section 8.2. Let the aggregate cumulative-input process be the sum as before, i.e.,

$$C_n(t) \equiv C_{n,1}(t) + \cdots + C_{n,m_n}(t), \quad t \geq 0 . \tag{3.1}$$

We establish general limit theorems for the stochastic processes $\{N_{n,l}(t) : t \geq 0\}$, $\{B_{n,l}(t) : t \geq 0\}$ and $\{C_{n,l}(t) : t \geq 0\}$. Recall that $N_{n,l}(t)$ represents the number of completed busy cycles in the interval $[0,t]$, while $B_{n,l}(t)$ represents the cumulative

8.3. Heavy-Traffic Limits for the On-Off Sources   251

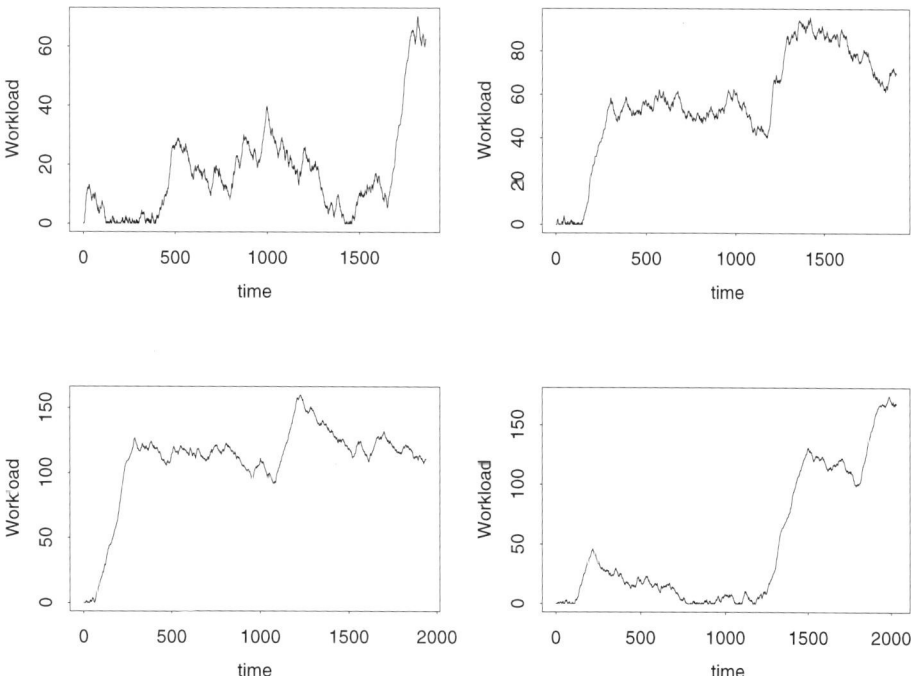

Figure 8.3. Plots of four possible realizations of the workload process in the two-source infinite-capacity Pareto/exponential fluid queue having Pareto(1.5) busy-period distributions and exponential idle-period distributions with mean 1, where the input rate equals the output rate. Each source has up to 10^3 busy cycles.

busy time in the interval $[0, t]$, both for source l in model n. We first consider the stochastic processes $\{N_{n,l}(t) : t \geq 0\}$ and $\{B_{n,l}(t) : t \geq 0\}$.

8.3.1. A Single Source

We first focus on a single source, so we omit the subscript l here. We now define the scaled stochastic processes in (D, M_1). As before, we use bold capitals to represent the scaled stochastic processes and associated limiting stochastic processes in D. We use the same scaling as in Section 5.4; i.e., we scale time by n and space by c_n, where $c_n/n \to 0$ as $n \to \infty$. Let

$$\mathbf{B}_n(t) \equiv c_n^{-1} \sum_{i=1}^{\lfloor nt \rfloor} (B_{n,i} - m_{B,n})$$

$$\mathbf{I}_n(t) \equiv c_n^{-1} \sum_{i=1}^{\lfloor nt \rfloor} (I_{n,i} - m_{I,n})$$

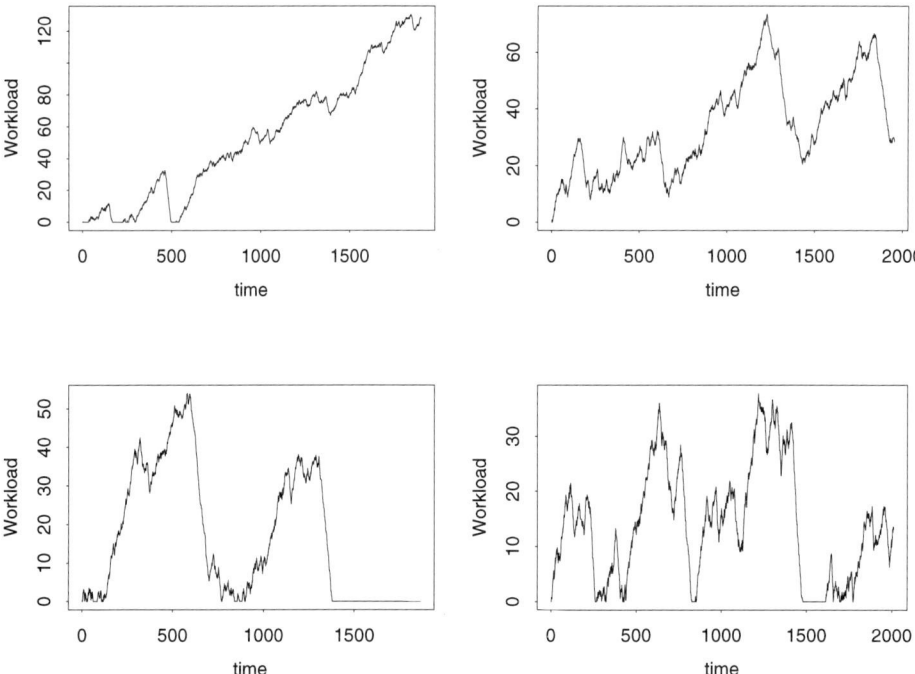

Figure 8.4. Plots of four possible realizations of the workload process in the two-source infinite-capacity exponential/Pareto fluid queue having exponential busy-period distributions and Pareto(1.5) idle-period distributions with mean 1, where the input rate equals the output rate. Each source has up to 10^3 busy cycles.

$$\begin{aligned} \mathbf{N}_n(t) &\equiv c_n^{-1}[N_n(nt) - \gamma_n nt] \\ \mathbf{B}'_n(t) &\equiv c_n^{-1}[B_n(nt) - \xi_n nt], \quad t \geq 0 \end{aligned} \quad (3.2)$$

where again $\lfloor nt \rfloor$ is the integer part of nt,

$$\xi_n \equiv \frac{m_{B,n}}{m_{B,n} + m_{I,n}} \quad \text{and} \quad \gamma_n \equiv \frac{1}{m_{B,n} + m_{I,n}} . \quad (3.3)$$

We think of $m_{B,n}$ in (3.2) as the mean busy period, $EB_{n,i}$, and $m_{I,n}$ as the mean idle period, $EI_{n,i}$, in the case $\{(B_{n,i}, I_{n,i}) : i \geq 1\}$ is a stationary sequence for each n, but in general that is not required. Similarly, we think of ξ_n in (3.3) as the *source on rate* and γ_n^{-1} in (3.3) as the *mean source cycle time*.

When we consider a sequence of models, as we have done, scaling constants can be incorporated in the random variables. However, in (3.2) and later, we include space and time scaling consistent with what occurs with a single model in heavy traffic.

8.3. Heavy-Traffic Limits for the On-Off Sources   253

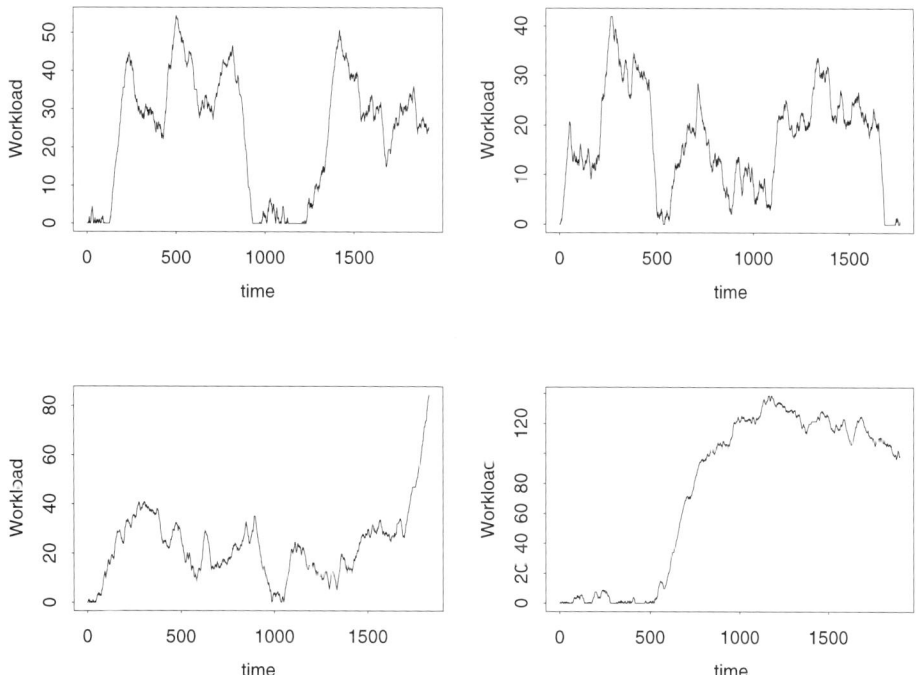

Figure 8.5. Plots of four possible realizations of the workload process in the two-source infinite-capacity Pareto/Pareto fluid queue having Pareto(1.5) busy-period and idle-period distributions with mean 1, where the input rate equals the output rate. Each source has up to 10^3 busy cycles.

We first show that a limit for $(\mathbf{B}_n, \mathbf{I}_n)$ implies a limit for $(\mathbf{N}_n, \mathbf{B}'_n)$ jointly with $(\mathbf{B}_n, \mathbf{I}_n)$. That follows by applying the continuous-mapping approach with the addition, inverse and composition functions. Here is a quick sketch of the argument: Since N_n is the counting process associated with the partial sums of $B_{n,i} + I_{n,i}$, we first apply the addition function to treat $\mathbf{B}_n + \mathbf{I}_n$ and then the inverse function to treat $\mathbf{N}_n$. The reasoning for the counting process is just as in Section 7.3. Then the cumulative busy-time $B_n(t)$ is approximately a random sum, i.e.,

$$B_n(t) \approx \sum_{i=1}^{N_n(t)} B_{n,i},$$

so that we can apply composition plus addition, see Chapter 13. For the case of limit processes with discontinuous sample paths, the last step of the argument is somewhat complicated. Thus, the proof has been put in the Internet Supplement; see Section 5.3 there.

254   8. On-Off Sources

As before, let $Disc(X)$ be the random set of discontinuities in $[0, \infty)$ of X. Let ϕ be the empty set. (We will refer to the reflection map by ϕ_K to avoid any possible confusion.)

Theorem 8.3.1. (FCLT for the cumulative busy time) *If*
$$(\mathbf{B}_n, \mathbf{I}_n) \Rightarrow (\mathbf{B}, \mathbf{I}) \quad in \quad (D, M_1)^2 \tag{3.4}$$
for $\mathbf{B}_n$ and $\mathbf{I}_n$ in (3.2), $c_n \to \infty$, $c_n/n \to 0$, $m_{B,n} \to m_B$, $m_{I,n} \to m_I$, with $0 < m_B + m_I < \infty$, so that $\xi_n \to \xi$ with $0 \le \xi \le 1$ and $\gamma_n \to \gamma > 0$ for ξ_n and γ_n in (3.3), and
$$P(Disc(\mathbf{B}) \cap Disc(\mathbf{I}) = \phi) = 1 , \tag{3.5}$$
then
$$(\mathbf{B}_n, \mathbf{I}_n, \mathbf{N}_n, \mathbf{B}'_n) \Rightarrow (\mathbf{B}, \mathbf{I}, \mathbf{N}, \mathbf{B}') \quad in \quad (D, M_1)^4 , \tag{3.6}$$
for $\mathbf{N}_n$, $\mathbf{B}'_n$ in (3.2) and
$$\begin{aligned} \mathbf{N}(t) &\equiv -\gamma[\mathbf{B}(\gamma t) + \mathbf{I}(\gamma t)] \\ \mathbf{B}'(t) &\equiv (1-\xi)\mathbf{B}(\gamma t) - \xi \mathbf{I}(\gamma t) . \end{aligned} \tag{3.7}$$

Next, given that the single-source cumulative-input process is defined as a composition in (2.6), the continuous-mapping approach can be applied with the convergence preservation of the composition map established in Chapter 13 to show that a joint limit for $B_n(t)$ and $\Lambda_n(t)$ implies a limit for $C_n(t)$, all for one source. Note that this limit does not depend on the way that the cumulative-busy-time processes $B_n(t)$ are defined. Also note that the process $\{\Lambda_n(t) : t \ge 0\}$ representing the input when the source is active can have general sample paths in D. The fluid idea suggests continuous sample paths, but that is not required.

We again omit the source subscript l. Let $\mathbf{e}$ be the identity map on $\mathbb{R}_+$, i.e., $\mathbf{e}(t) = t$, $t \ge 0$. Let
$$\begin{aligned} \mathbf{\Lambda}_n(t) &\equiv c_n^{-1}[\Lambda_n(nt) - \hat{\lambda}_n nt] \\ \mathbf{C}_n(t) &\equiv c_n^{-1}[C_n(nt) - \lambda_n nt], \quad t \ge 0 . \end{aligned} \tag{3.8}$$

Theorem 8.3.2. (FCLT for the cumulative input) *If*
$$(\mathbf{\Lambda}_n, \mathbf{B}'_n) \Rightarrow (\mathbf{\Lambda}, \mathbf{B}') \quad in \quad (D, M_1)^2 \tag{3.9}$$
for $\mathbf{\Lambda}_n$ in (3.8) and $\mathbf{B}'_n$ in (3.2), where $c_n \to \infty$, $c_n/n \to 0$, $\xi_n \to \xi$, $\hat{\lambda}_n \to \hat{\lambda}$ for $0 < \hat{\lambda} < \infty$ and $\mathbf{\Lambda} \circ \xi \mathbf{e}$ and $\mathbf{B}'$ have no common discontinuities of opposite sign, then
$$(\mathbf{\Lambda}_n, \mathbf{B}'_n, \mathbf{C}_n) \Rightarrow (\mathbf{\Lambda}, \mathbf{B}', \mathbf{C}) \quad in \quad (D, M_1)^3 . \tag{3.10}$$
for $\mathbf{C}_n$ in (3.8) with $\lambda_n = \xi_n \hat{\lambda}_n$ and
$$\mathbf{C}(t) \equiv \mathbf{\Lambda}(\xi t) + \hat{\lambda} \mathbf{B}'(t) . \tag{3.11}$$

Proof. Given that $\mathbf{B}'_n \Rightarrow \mathbf{B}'$, we have $\mathbf{B}''_n \Rightarrow \xi e$ for $\mathbf{B}''_n(t) = n^{-1}B(nt)$. Then we can apply the continuous-mapping approach with composition and addition to treat $\mathbf{C}_n$, i.e., by Corollary 13.3.2,

$$\mathbf{C}_n = \mathbf{\Lambda}_n \circ \mathbf{B}''_n + \hat{\lambda}_n \mathbf{B}'_n \Rightarrow \mathbf{\Lambda} \circ \xi e + \hat{\lambda}\mathbf{B}' \; . \quad \blacksquare$$

From (3.11), we see that the limit processes $\mathbf{\Lambda}$ and $\mathbf{B}'$ appear in $\mathbf{C}$ as a simple sum with deterministic scalar modification. The stochastic fluctuations in the cumulative-input process over a longer (shorter) time scale are captured by the component $\hat{\lambda}\mathbf{B}'(t)$ ($\mathbf{\Lambda}(\xi t)$). Thus, the contribution of each component to the limit process $\mathbf{C}$ can easily be identified and quantified.

As shown in Section 5.4, we can apply the continuous mapping theorem with the reflection map ϕ_K in (2.9) to convert a limit for the cumulative-input processes C_n into a limit for the workload processes W_n and the associated processes L_n, U_n and D_n.

8.3.2. Multiple Sources

Now we are ready to combine Theorems 5.4.1, 8.3.1 and 8.3.2 to obtain simultaneous joint limits for all processes with m sources. To treat the fluid queues, we introduce the sequence of available-processing processes $\{\{S_n(t) : t \geq 0\} : n \geq 1\}$ and the sequence of storage capacities $\{K_n : n \geq 1\}$.

We define the following random elements of (D, M_1) associated with source l, $1 \leq l \leq m$:

$$\mathbf{B}_{n,l}(t) \equiv c_n^{-1} \sum_{i=1}^{\lfloor nt \rfloor} (B_{n,l,i} - m_{B,n,l})$$

$$\mathbf{I}_{n,l}(t) \equiv c_n^{-1} \sum_{i=1}^{\lfloor nt \rfloor} (I_{n,l,i} - m_{I,n,l})$$

$$\mathbf{\Lambda}_{n,l}(t) \equiv c_n^{-1}[\Lambda_{n,l}(nt) - \hat{\lambda}_{n,l}nt]$$

$$\mathbf{N}_{n,l}(t) \equiv c_n^{-1}[N_{n,l}(nt) - \gamma_{n,l}nt]$$

$$\mathbf{B}'_{n,l}(t) \equiv c_n^{-1}[B_{n,l}(nt) - \xi_{n,l}nt]$$

$$\mathbf{C}_{n,l}(t) \equiv c_n^{-1}[C_{n,l}(nt) - \hat{\lambda}_{n,l}\xi_{n,l}nt] \tag{3.12}$$

Theorem 8.3.3. (heavy-traffic limit for the fluid-queue model with m sources) *Consider a sequence of fluid queues indexed by n with $m_n = m$ sources, capacities K_n, $0 < K_n \leq \infty$, and cumulative-available-processing processes $\{S_n(t) : t \geq 0\}$. Suppose that $K_n = c_n K$, $0 < K \leq \infty$, $W_{n,l}(0) \geq 0$ and $\sum_{l=1}^{m} W_{n,l}(0) \leq K_n$,*

$$(\mathbf{B}_{n,l}, \mathbf{I}_{n,l}, \mathbf{\Lambda}_{n,l}, c_n^{-1}W_{n,l}(0), 1 \leq l < m, \mathbf{S}_n)$$
$$\Rightarrow (\mathbf{B}_l, \mathbf{I}_l, \mathbf{\Lambda}_l, W'_l(0), 1 \leq l \leq m, \mathbf{S}) \tag{3.13}$$

in $(D, M_1)^{3m+1} \times \mathbb{R}^m$ for $\mathbf{B}_{n,l}$, $\mathbf{I}_{n,l}$, $\mathbf{\Lambda}_{n,l}$ in (3.12) and $\mathbf{S}_n$ in (4.4) of Chapter 5, where $c_n \to \infty$, $c_n/n \to 0$, $\hat{\lambda}_{n,l} \to \hat{\lambda}_l$ for $0 < \hat{\lambda}_l < \infty$, $m_{B,n,l} \to m_{B,l}$ and

$m_{I,n,l} \to m_{I,l}$ with $0 < m_{B,l} + m_{I,l} < \infty$, so that

$$\xi_{n,l} \equiv \frac{m_{B,n,l}}{m_{B,n,l} + m_{I,n,l}} \to \xi_l \qquad (3.14)$$

and

$$\gamma_{n,l} \equiv \frac{1}{m_{B,n,l} + m_{I,n,l}} \to \gamma_l > 0 . \qquad (3.15)$$

If, in addition, $Disc(\mathbf{B}_l \circ \gamma_l \mathbf{e})$, $Disc(\mathbf{I}_l \circ \gamma_l \mathbf{e})$ and $Disc(\mathbf{\Lambda}_l \circ \xi_l \mathbf{e})$, $1 \le l \le m$, are pairwise disjoint w.p.1, and $\lambda_n - \mu_n \to 0$ so that

$$\eta_n \equiv n\left(\lambda_n - \mu_n\right)\Big/c_n \to \eta \quad as \quad n \to \infty \qquad (3.16)$$

for $-\infty < \eta < \infty$, where

$$\lambda_n \equiv \sum_{l=1}^{m} \lambda_{n,l} \quad and \quad \lambda_{n,l} \equiv \xi_{n,l} \hat{\lambda}_{n,l} \qquad (3.17)$$

for each l, $1 \le l \le m$, then

$$(\mathbf{B}_{n,l}, \mathbf{I}_{n,l}, \mathbf{\Lambda}_{n,l}, \mathbf{N}_{n,l}, \mathbf{B}'_{n,l}, \mathbf{C}_{n,l}, 1 \le l \le m, \mathbf{C}_n, \mathbf{S}_n, \mathbf{X}_n, \mathbf{W}_n, \mathbf{U}_n, \mathbf{L}_n)$$
$$\Rightarrow (\mathbf{B}_l, \mathbf{I}_l, \mathbf{\Lambda}_l, \mathbf{N}_l, \mathbf{B}'_l, \mathbf{C}_l, 1 \le l \le m, \mathbf{C}, \mathbf{S}, \mathbf{X}, \mathbf{W}, \mathbf{U}, \mathbf{L})$$

in $(D, M_1)^{6m+6}$ for $\mathbf{N}_{n,l}, \mathbf{B}'_{n,l}, \mathbf{C}_{n,l}$ in (3.12), $\mathbf{C}_n$ in (3.8) with $(\mathbf{X}_n, \mathbf{W}_n, \mathbf{U}_n, \mathbf{L}_n)$ in (4.4) of Chapter 5 and

$$\begin{aligned}
\mathbf{N}_l(t) &\equiv -\gamma_l[\mathbf{B}_l(\gamma_l t) + \mathbf{I}_l(\gamma_l t)] \\
\mathbf{B}'_l(t) &\equiv (1 - \xi_l)\mathbf{B}_l(\gamma_l t) - \xi_l \mathbf{I}_l(\gamma_l t) \\
\mathbf{C}_l(t) &\equiv \mathbf{\Lambda}_l(\xi_l t) + \hat{\lambda}_l[(1 - \xi_l)\mathbf{B}_l(\gamma_l t) - \xi_l \mathbf{I}_l(\gamma_l t)] \\
\mathbf{C}(t) &\equiv \mathbf{C}_1(t) + \cdots + \mathbf{C}_m(t) \\
\mathbf{X}(t) &\equiv \sum_{l=1}^{m} W'_l(0) + \mathbf{C}(t) - \mathbf{S}(t) + \eta t \\
\mathbf{W}(t) &\equiv \phi_K(\mathbf{X})(t) \\
(\mathbf{U}(t), \mathbf{L}(t)) &\equiv (\psi_U(\mathbf{X})(t), \psi_L(\mathbf{X})(t)), \quad t \ge 0,
\end{aligned} \qquad (3.18)$$

for ϕ_K and (ψ_U, ψ_L) in (2.9) or (2.10) and η in (3.16).

In many heavy-tailed applications, the fluctuations of the idle periods $I_{n,l,i}$ and the process $\Lambda_{n,l}(t)$ will be asymptotically negligible compared to the fluctuations of the busy periods $B_{n,l,i}$. In that case, condition (3.13) will hold with $\mathbf{I}_l(t) = \mathbf{\Lambda}_l(t) = 0$, $1 \le l \le m$. Then $\mathbf{C}_l$ in (3.18) simplifies to a simple scaling of the limit process $\mathbf{B}_l$, i.e., then

$$\mathbf{C}_l(t) = \hat{\lambda}_l(1 - \xi_l)\mathbf{B}_l(\gamma_l t), \quad t \ge 0 . \qquad (3.19)$$

Moreover, if some sources have more bursty busy periods than others, then only the ones with highest burstiness will impact the limit. In the extreme case, $\mathbf{B}_l(t)$

will be the zero function for all but one l, say l^*, and

$$\mathbf{C}(t) = \mathbf{C}_{l^*}(t) = \hat{\lambda}_{l^*}(1 - \xi_{l^*})\mathbf{B}_{l^*}(\gamma_{l^*}t), \quad t \geq 0, \tag{3.20}$$

and the stochastic nature of the limit for the workload process will be determined, asymptotically, by the single limit process $\{\mathbf{B}_{l^*}(t) : t \geq 0\}$ in (3.13).

In summary, we have shown that a heavy-traffic stochastic-process limit holds for the scaled workload process, with appropriate scaling (including (3.16)), whenever associated stochastic-process limits hold for the basic model elements. In the common case in which we have no initial workloads, deterministic processing according to the rates μ_n and the source input during busy periods is deterministic (i.e., when $\Lambda_{n,l}(t) = \hat{\lambda}_{n,l}t$ for deterministic constants $\hat{\lambda}_{n,l}$), there is a heavy-traffic stochastic-process limit for the workload process whenever the partial sums of the busy periods and idle periods satisfy a joint FCLT. Donsker's theorem and its extensions in Chapter 4 thus yield the required initial FCLTs.

Applying Chapter 4 and Section 2.4 of the Internet Supplement, we obtain convergence of the appropriate scaled cumulative-input processes to Brownian motion, stable Lévy motion and more general Lévy processes, all of which have stationary and independent increments. We describe consequences of those stochastic-process limits in Sections 8.4 and 8.5 below and in Section 5.2 in the Internet Supplement.

Remark 8.3.1. *The effect of dependence.* At first glance, the independent-increments property of the limit process may seem inconsistent with the dependence observed in network traffic measurements in the communications network context, but recall that the limits are obtained under time scaling. Even if successive busy and idle periods for each source are nearly independent, the cumulative inputs

$$C(t_1 + h) - C(t_1) \quad \text{and} \quad C(t_2 + h) - C(t_2)$$

in disjoint intervals $(t_1, t_1 + h]$ and $(t_2, t + h]$ with $t_1 + h < t_2$ are likely to be dependent because a single busy cycle for one source can fall within both intervals.

However, when we introduce time scaling by n, as in (4.4), the associated two scaled intervals $(nt_1, n(t_1 + h)]$ and $(nt_2, n(t_2 + h)]$ become far apart, so that it is natural that the dependence should disappear in the limit. It is also intuitively clear that the dependence in the original cumulative-input process (before scaling space and time) should have an impact upon performance, causing greater congestion. And that is confirmed by measurements. It is thus important that the limit can capture that performance impact.

Even though the dependence present in the cumulative-input process disappears in the limit, that dependence has a significant impact upon the limit process: *After scaling time, the large busy periods which cause strong dependence over time tend to cause large fluctuations in space.* This effect of time scaling is shown pictorially in Figure 8.6. Thus, even though the limit processes have independent increments, the burstiness in the original cumulative-input process (e.g., caused by a heavy-tailed busy-period distribution) leads to approximating workload distributions that reflect the burstiness. For example, with heavy-tailed busy-period distributions, the limiting workload process will have marginal and steady-state distributions with heavy tails (with related decay rates), which is consistent with what is seen when a single-server queue is simulated with the trace of measured cumulative-input processes. (However, as noted in Section 2.4.1, the presence of flow controls such

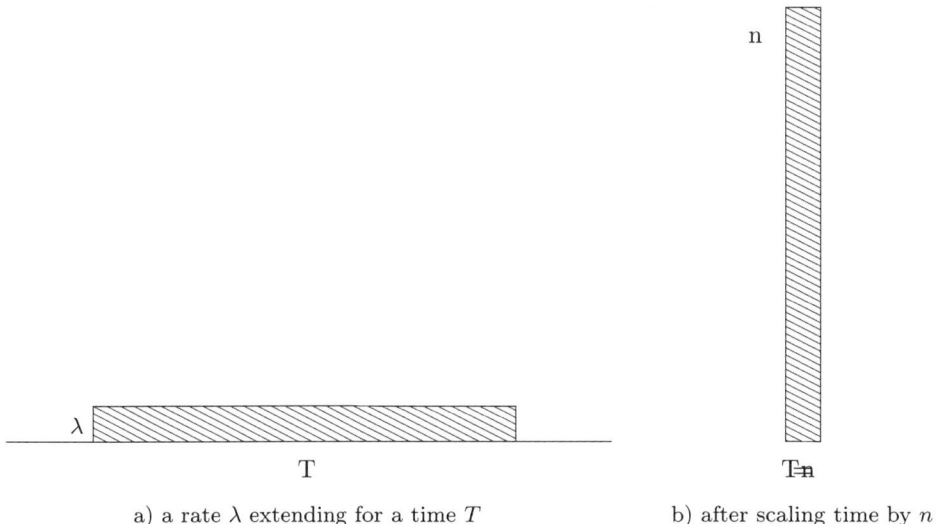

a) a rate λ extending for a time T b) after scaling time by n

Figure 8.6. The effect of time scaling: transforming dependence over time into jumps in space.

as that used by the Transmission Control Protocol (TCP) significantly complicates interpretations of simulations based on actual network traces.)

Remark 8.3.2. *Structure for the available-processing processes.* Theorem 8.3.3 goes beyond Theorem 5.4.1 by giving the cumulative-input process additional structure. However, in both cases, the available-processing processes $\{S_n(t) : t \geq 0\}$ are kept abstract; we simply assumed that the available-processing processes satisfy a FCLT. However, structure also can be given to the available-processing processes, just as we gave structure to the cumulative-input processes. Indeed, the *same structure* is meaningful. In particular, we can have multiple output channels or servers, each of which is subject to service interruption. We can stipulate that the total available processing in the interval $[0, t]$ is the superposition of the component available-processing processes, i.e., with q output channels,

$$S(t) = S_1(t) + \cdots + S_q(t), \quad t \geq 0.$$

If server i is subject to service interruptions, then it is natural to model the available-processing process S_i by an on-off model. The busy periods then represent intervals when processing can be done, and the idle periods represent intervals when processing cannot be done. Thus, Theorems 8.3.1 and 8.3.2 can be applied directly to obtain stochastic-process limits for structured available-processing processes. And we immediately obtain generalizations of Theorem 8.3.3 and the Brownian limit in

Theorem 8.4.1. For further discussion about queues with service interruptions, see Section 14.7. ∎

8.3.3. $M/G/\infty$ Sources

So far we have assumed that the number of sources is fixed, i.e., that $m_n = m$ as $n \to \infty$. An alternative limiting regime is to have $m_n \to \infty$ as $n \to \infty$. We can let $m_n \to \infty$ and still keep the total input rate unchanged by letting the rate from each source decrease. One way to do this is by letting the off periods in each source grow appropriately; e.g., we can let $I_{n,l,i} = m_n I_{l,i}$ and $B_{n,l,i} = B_{l,i}$. Let $N_n(t)$ be the number of new busy cycles started in $[0, t]$ by all sources.

Under regularity conditions, with this scaling of the off periods, $N_n \Rightarrow N$ as $n \to \infty$, where N is a Poisson process. Indeed that stochastic-process limit is the classical stochastic-process limit in which a superposition of independent renewal processes with fixed total intensity (in which each component process is asymptotically negligible) converges to a Poisson process; see Theorem 9.8.1. Moreover, by a simple continuity argument, the associated cumulative busy-time for all sources, say $\{B_n(t) : t \geq 0\}$ converges to the integral of the number of busy servers in an $M/G/\infty$ queue (a system with infinitely many servers, IID service times and a Poisson arrival process). We state the result as another theorem. Note that there is no additional scaling of space and time here.

Theorem 8.3.4. (an increasing number of sources) *Consider a sequence of fluid models indexed by n with m_n IID fluid sources in model n, where $m_n \to \infty$. Let $I_{n,l,i} = m_n I_{l,i}$ and $B_{n,l,i} = B_{l,i}$ where $\{(B_{l,i}, I_{l,i}) : i \geq 1\}$ is an IID sequence. Then*

$$(N_n, B_n) \Rightarrow (N, B) \quad in \quad (D, M_1)^2 , \qquad (3.21)$$

where N is a Poisson process with arrival rate $1/EI_{1,1}$,

$$B(t) = \int_0^t Q(s)ds, \quad t \geq 0 , \qquad (3.22)$$

and $Q(t)$ is the number of busy servers at time t in an $M/G/\infty$ queueing model with arrival process N and IID service times distributed as $B_{1,1}$.

Heavy-traffic stochastic-process limits can be obtained by considering a sequence of models with changing $M/G/\infty$ inputs, as has been done by Tsoukatos and Makowski (1997, 2000) and Resnick and van der Berg (2000).

Let $\{Q_n(t) : t \geq 0\}$ be a sequence processes counting the number of busy servers in $M/G/\infty$ systems. Let $B_n(t)$ be defined in terms of Q_n by (3.22). Let

$$\mathbf{Q}_n(t) \equiv c_n^{-1}[Q_n(t) - \eta_n], \quad t \geq 0 , \qquad (3.23)$$

and

$$\mathbf{B}'_n(t) \equiv c_n^{-1}[B_n(t) - \eta_n t], \quad t \geq 0 . \qquad (3.24)$$

(We use the prime to be consistent with (3.2).) We think of $\{Q_n(t) : t \geq 0\}$ as being a stationary process, or at least an asymptotically stationary process, for each n.

Hence its translation term in (3.23) is η_n independent of t. In contrast, $B_n(t)$ grows with t, so the translation term of $\mathbf{B}'_n(t)$ in (3.24) is $\eta_n t$.

We can apply the continuous mapping theorem with the integral function in (3.22), using Theorems 3.4.1 and 11.5.1, to show that limits for $\mathbf{Q}_n$ imply corresponding limits for $\mathbf{B}'_n$.

Theorem 8.3.5. (changing M/G/∞ inputs) *Consider a sequence of fluid queue models indexed by n with M/G/∞ inputs characterized by the processes $\{Q_n(t) : t \geq 0\}$. If $\mathbf{Q}_n \Rightarrow \mathbf{Q}$ in (D, M_1) for $\mathbf{Q}_n$ in (3.23), then*

$$\mathbf{B}'_n \Rightarrow \mathbf{B}' \quad in \quad (D, M_1) ,$$

for $\mathbf{B}'_n$ in (3.24) and

$$\mathbf{B}'(t) = \int_0^t \mathbf{Q}(s)ds, \quad t \geq 0 . \tag{3.25}$$

From heavy-traffic limit theorems for the $M/G/\infty$ queue due to Borovkov (1967) [also see Section 10.3], we know that the condition of Theorem 8.3.5 holds for proper initial conditions when the Poisson arrival rate is allowed to approach infinity. Then $\mathbf{Q}$ is a Gaussian process, which can be fully characterized. In the case of exponential service times, the limit process $\mathbf{Q}$ is the relative tractable Ornstein-Uhlenbeck diffusion process.

We discuss infinite-server models further in Section 10.3. As noted in Remark 10.3.1, the $M/G/\infty$ queue is remarkably tractable, even with a Poisson arrival process having a time-dependent rate. We discuss limits with an increasing number of sources further in Section 8.7.

8.4. Brownian Approximations

We now supplement the discussion of Brownian approximations in Section 5.7 by establishing a Brownian limit for the more detailed multisource on-off fluid-queue model discussed in Sections 8.2 and 8.3. We discuss Brownian approximations further in Section 9.6.

8.4.1. The Brownian Limit

For simplicity, we let heavy traffic be achieved by only suitably changing the processing rate μ_n, so that the output rate μ_n approaches a fixed input rate λ. Moreover, we can construct the available-processing processes indexed by n by scaling time in a fixed available-processing process S. Let the cumulative available processing in the interval $[0,t]$ in model n be $S_n(t)$. We assume that the available-processing processes S_n satisfy a FCLT, in particular,

$$\mathbf{S}_n \Rightarrow \mathbf{S} \quad in \quad (D, M_1) , \tag{4.1}$$

where $\mathbf{S}$ is a zero-drift Brownian motion with variance parameter σ_S^2 and

$$\mathbf{S}_n(t) \equiv n^{-1/2}(S_n(nt - \mu_n nt), \quad t \geq 0 . \tag{4.2}$$

We use the M_1 topology in (4.1) and later, but it does not play a crucial role in this section because the limit processes here have continuous sample paths.

Since we only change the processing rate μ_n, we can have a single model for the cumulative input process. In the setting of Theorem 8.3.3, we thus assume that the random variables $B_{n,l,i}$ and $I_{n,l,i}$ are independent of n. Moreover, we assume that the processes $\{\Lambda_{n,l}(t) : t \geq 0\}$ are independent of n. Hence we drop the subscript n from these random quantities. We call this a *single fluid model*.

In order to be able to invoke Donsker's FCLT (specifically, the multidimensional version in Theorem 4.3.5), we make several independence assumptions. First, we assume that the m sources are mutually independent. By that we mean that the stochastic processes $\{(B_{l,i}, I_{l,i}) : i \geq 1\}$ and $\{\Lambda_l(t) : t \geq 0\}$ for different l, $1 \leq l \leq m$ are mutually independent. Second, for each source l, we assume that the rate process $\{\Lambda_l(t) : t \geq 0\}$ is independent of the sequence $\{(B_{l,i}, I_{l,i}) : i \geq 1\}$. We also assume that the sources are independent of the available-processing process.

To invoke Donsker's FCLT for $\{(B_{l,i}, I_{l,i}) : i \geq 1\}$, we assume that $\{(B_{l,i}, I_{l,i}) : i \geq 1\}$ is a sequence of IID random vectors in $\mathbb{R}^2$ with finite second moments. In particular, let

$$m_{B,l} \equiv EB_{l,1}, \quad m_{I,l} \equiv EI_{l,1},$$
$$\sigma_{B,l}^2 \equiv Var\, B_{l,1}, \quad \sigma_{I,l}^2 \equiv Var\, I_{l,1},$$
$$\sigma_{B,I,l}^2 \equiv Cov(B_{l,1}, I_{l,1}) . \tag{4.3}$$

The second-moment assumption implies that all the quantities in (4.3) are finite. The positivity assumption on $B_{l,i}$ and $I_{l,i}$ implies that $m_{B,l} > 0$ and $m_{I,l} > 0$.

Instead of describing the rate processes $\{\Lambda_l(t) : t \geq 0\}$ in detail, we simply assume that they satisfy FCLT's. In particular, we assume that

$$\mathbf{\Lambda}_{n,l} \Rightarrow \mathbf{\Lambda}_l \quad \text{in} \quad (D, M_1) \tag{4.4}$$

for each l, where $\mathbf{\Lambda}_l$ is a zero-drift Brownian motion with variance coefficient $\sigma_{\Lambda,l}^2$ (possibly zero) and

$$\mathbf{\Lambda}_{n,l}(t) \equiv n^{-1/2}[\Lambda_l(nt) - \hat{\lambda}_l nt], \quad t \geq 0 . \tag{4.5}$$

Given $\hat{\lambda}_l$ in (4.5), we can define the overall "input rate" λ by

$$\lambda \equiv \sum_{l=1}^{m} \lambda_l , \tag{4.6}$$

where

$$\lambda_l \equiv \xi_l \hat{\lambda}_l \quad \text{and} \quad \xi_l \equiv \frac{m_{B,l}}{m_{B,l} + m_{I,l}} . \tag{4.7}$$

The Brownian heavy-traffic stochastic-process limit follows directly from Theorems 8.3.3, 4.3.5 and 11.4.4. We obtain convergence to the same Brownian limit (with different variance parameter) if we replace the assumed independence by associated weak dependence, as discussed in Section 4.4.

Theorem 8.4.1. (Brownian limit for the fluid queue with m on-off sources) *Consider a single fluid-queue model with m on-off sources satisfying the independence assumptions above, the moment assumptions in (4.3), the scaling assumption in (4.6) and the convergence assumptions in (4.4) and (4.5). Let a sequence of systems indexed by n be formed by letting the capacity in system n be $K_n = \sqrt{n}K$ for some K, $0 < K \leq \infty$, and introducing available-processing processes satisfying the limits in (4.1) and (4.2). Let the initial random workloads from the m sources in system n satisfy*

$$W_{n,l}(0) \geq 0, \quad 1 \leq l \leq m, \quad \sum_{l=1}^{m} W_{n,l}(0) \leq K_n \tag{4.8}$$

and

$$n^{-1/2} W_{n,l}(0) \Rightarrow y_l \quad in \quad \mathbb{R}, \tag{4.9}$$

where y_l is a deterministic scalar. Assume that

$$\eta_n \equiv \sqrt{n}(\lambda - \mu_n) \to \eta \quad as \quad n \to \infty \tag{4.10}$$

for $-\infty < \eta < \infty$, λ in (4.6) and μ_n in (4.2). Then the conditions and conclusions of Theorems 5.4.1, 5.9.1, 8.3.1, 8.3.2 and 8.3.3 hold with $c_n \equiv \sqrt{n}$, $W'_l(0) \equiv y_l$, $\gamma_l \equiv (m_{B,l} + m_{I,l})^{-1}$, ξ_l in (4.7) and $(\mathbf{B}_l, \mathbf{I}_l, \mathbf{\Lambda}_l)$, $1 \leq l \leq m$, being mutually independent three-dimensional zero-drift Brownian motions, independent of the standard Brownian motion $\mathbf{S}$ in (4.1). For each l, the limit processes $(\mathbf{B}_l, \mathbf{I}_l)$ and $\mathbf{\Lambda}_l$, are mutually independent zero-drift Brownian motions, with $(\mathbf{B}_l, \mathbf{I}_l)$ having covariance matrix

$$\Sigma_l = \begin{pmatrix} \sigma_{B,l}^2 & \sigma_{B,I,l}^2 \\ \sigma_{B,I,l}^2 & \sigma_{I,l}^2 \end{pmatrix}. \tag{4.11}$$

The limit processes $\mathbf{N}_l$, $\mathbf{B}'_l$, $\mathbf{C}_l$ and $\mathbf{C}$ are all one-dimensional zero-drift Brownian motions. In particular,

$$\mathbf{N}_l \stackrel{d}{=} \sigma_{N,l} \mathbf{B}, \qquad \mathbf{B}'_l \stackrel{d}{=} \sigma_{B,l} \mathbf{B}$$
$$\mathbf{C}_l \stackrel{d}{=} \sigma_{C,l} \mathbf{B}, \qquad \mathbf{C} \stackrel{d}{=} \sigma_C \mathbf{B}, \tag{4.12}$$

where $\mathbf{B}$ is a standard Brownian motion and

$$\begin{aligned}
\sigma_{N,l}^2 &\equiv \gamma_l^3 (\sigma_{B,l}^2 + 2\sigma_{B,I,l}^2 + \sigma_{I,l}^2), \\
\sigma_{B',l}^2 &\equiv \gamma_l ((1-\xi_l)^2 \sigma_{B,l}^2 - 2(1-\xi_l)\xi_l \sigma_{B,I,l}^2 + \xi_l^2 \sigma_{I,l}^2), \\
\sigma_{C,l}^2 &\equiv \xi_l \sigma_{\Lambda,l}^2 + \hat{\lambda}_l^2 \gamma_l ((1-\xi_l)^2 \sigma_{B,l}^2 - 2(1-\xi_l)\xi_l \sigma_{B,I,l}^2 + \xi_l^2 \sigma_{I,l}^2), \\
\sigma_C^2 &\equiv \sum_{l=1}^{m} \sigma_{C,l}^2, \\
\sigma_X^2 &\equiv \sigma_C^2 + \sigma_S^2.
\end{aligned} \tag{4.13}$$

The limit process **X** *is distributed as*

$$\mathbf{X} \stackrel{\mathrm{d}}{=} \{\mathbf{B}(t; \eta, \sigma_X^2, y) : t \geq 0\} \stackrel{\mathrm{d}}{=} \{y + \eta t + \sigma_X \mathbf{B}(t) : t \geq 0\}, \quad (4.14)$$

for η in (4.10), σ_X^2 in (4.13) and $y = y_1 + \cdots + y_m$ for y_l in (4.9).

8.4.2. Model Simplification

We now observe that the heavy-traffic stochastic-process limit produces a significant model simplification. As a consequence of Theorem 8.4.1, the scaled workload processes converge to a one-dimension reflected Brownian motion (RBM). In particular, the limit **W** is $\phi_K(\mathbf{X})$, where **X** is the one-dimensional Brownian motion in (4.14) and ϕ_K is the two-sided reflection map in (2.9) and Section 14.8. The process **W** depends on only four parameters: the initial position y specified in (4.9), the drift η specified in (4.10), the diffusion coefficient σ_X^2 specified in (4.13) and the upper barrier K for the two-sided reflection. If we are only interested in the steady-state distribution, then we can ignore the initial position y, which leaves only three parameters. From the expression for σ_X^2 in (4.13), we can determine the impact of various component sources of variability.

In contrast, the original fluid model has much more structure. The structure was reduced substantially by our independence assumptions, but still there are many model data. First, we need to know the m probability distributions in $\mathbb{R}^2$ of the busy periods $B_{l,1}$ and idle periods $I_{l,1}$, $1 \leq l \leq m$. These m distributions only affect the limit in Theorem 8.4.1 through their first two moments and covariances. Second, we have the m rate processes $\{\Lambda_l(t) : t \geq 0\}$ and the available-processing process $\{S(t) : t \geq 0\}$. The m rate processes affect the limit only through the scaling parameters $\hat{\lambda}_l$ and $\sigma_{\Lambda,l}^2$ in the assumed FCLT in (4.4) and (4.5). The available-processing process affects the limit only through the scaling parameter σ_S^2 in (4.1). We also have the m initial random workloads $W_{n,l}(0)$. The random vector $(W_{n,1}(0), \ldots, W_{n,m}(0))$ can have a very general m-dimensional distribution. That distribution affects the limit only via the deterministic limits y_l in (4.8). Since we have convergence in distribution to a deterministic limit, we automatically get convergence in the product space; i.e.,

$$n^{-1}(W_{n,1}(0), \ldots, W_{n,m}(0)) \Rightarrow (y_1, \ldots, y_m) \quad \text{in} \quad \mathbb{R}^m$$

as $n \to \infty$ by Theorem 11.4.5. Finally, we have the capacity K_n and the processing rate μ_n. Thus, for the purpose of determining the limit, there are $8m + 3$ relevant parameters.

It is significant that the stochastic-process limit clearly shows how the $8m + 3$ parameters in the original fluid model should be combined to produce the final four parameters characterizing the limiting RBM. First, the final drift η depends only upon the parameters μ_n, ξ_l and λ_l for $1 \leq l \leq m$ and n as indicated in (4.10). Second, the final variance parameter σ_X^2 depends on the basic model parameters as indicated in (4.13). We can thus quickly evaluate the consequence of altering the rate or variability of the stochastic processes characterizing the model.

We now consider the Brownian approximation for the distribution of the steady-state workload in the multisource on-off model. The steady-state distribution of RBM is given in Theorem 5.7.2. By Theorem 8.4.1, the key parameter is

$$\theta \equiv \frac{2\eta}{\sigma^2} \approx \frac{2\sqrt{n}\left(\left(\sum_{l=1}^{m} \lambda_l\right) - \mu_n\right)}{\sigma_X^2} \qquad (4.15)$$

where $\lambda_l \equiv \xi_l \hat{\lambda}_l$ and $\mathbf{W}(\infty)$ has pdf in (7.10). As $K \to \infty$, the mean of $\mathbf{W}(\infty)$ approaches $-\theta^{-1}$ when $\eta < 0$. Then $W_n(\infty)$ has approximately an exponential distribution with mean

$$
\begin{aligned}
E\mathbf{W}_n(\infty) &\approx \sqrt{n} E\mathbf{W}(\infty) \\
&\approx \frac{\sqrt{n}\sigma_X^2}{2\sqrt{n}\left(\mu_n - \sum_{l=1}^{m} \lambda_l\right)} = \frac{\sigma_X^2}{2\left(\mu_n - \sum_{l=1}^{m} \lambda_l\right)} \, .
\end{aligned} \qquad (4.16)
$$

Notice that the $\sqrt{n}$ factors in (7.14) and (4.15) cancel out in (4.16).

It is instructive to consider the contribution of each source to the overall mean steady-state workload. Expanding upon formula (4.16) in the case $K = \infty$, we obtain the approximation

$$EW(\infty) \approx \frac{(\sigma_S^2 + \sum_{l=1}^{m} \sigma_{C,l}^2)}{2(\mu - \sum_{l=1}^{m} \lambda_l)} \, . \qquad (4.17)$$

Source l contributes to the approximate mean steady-state workload via both its rate λ_l and its variance parameter $\sigma_{C,l}^2$.

8.5. Stable-Lévy Approximations

Paralleling Theorem 8.4.1, we now combine the FCLT obtaining convergence of normalized partial-sum processes to stable Lévy motion (SLM) in Section 4.5 and the general limits for multisource on-off fluid queues in Section 8.3 in order to obtain a reflected-stable-Lévy-motion (RSLM) stochastic-process limit for the multisource on-off fluid queue. In particular, we assume that the busy-period distribution has a heavy tail, which makes the limit for the normalized cumulative-input process be a centered totally-skewed stable Lévy motion (having $\mu = 0$ and $\beta = 1$). The limit for the normalized workload process will be the reflection of a constant drift plus this centered totally-skewed Lévy stable motion. See Section 6.4.3 for RSLM limits for a more conventional queueing model.

It is possible to obtain more general reflected Lévy process limits for general sequences of models, which also have remarkably tractable steady-state distributions when the underlying Lévy process has no negative jumps. We discuss these more general limits in Sections 2.4 and 5.2 of the Internet Supplement.

8.5.1. *The RSLM Heavy-Traffic Limit*

As in Section 5.7, the limit will be achieved under heavy-traffic conditions. As before, we can let heavy-traffic be achieved by suitably changing the deterministic processing rate μ_n by scaling a fixed available-processing stochastic process S, so that the output rate approaches the constant input rate. As in Section 5.7, we thus have a single model for the cumulative-input process in the fluid queue, so that we are in the setting of the single-sequence limit in Section 4.5. In the setting of Theorem 8.3.3, the variables $B_{n,l,i}$ and $I_{n,l,i}$ are independent of n. As in Section 5.7, we assume that the processes $\{\Lambda_{n,l}(t) : t \geq 0\}$ also are independent of n. Hence we drop the subscript n from these quantities. We again call this a single fluid model.

When the random variables have infinite second moments and appropriately scaled versions of the random walk converge to a stable process, the scaling depends critically on the tail probability decay rate or, equivalently, the stable index α. Hence it is natural for one component in the model to dominate in the sense that it has a heavier tail than the other components. We will assume that the busy-period distributions have the heaviest tail, so that the stochastic fluctuations in the idle periods I_l, the rate processes $\{\Lambda_t(t) : t \geq 0\}$ and the available-processing process $\{S(t) : t \geq 0\}$ become asymptotically negligible. (That is conveyed by assumption (5.5) in Theorem 8.5.1 below.) It is straightforward to obtain the corresponding limits in the other cases, but we regard this case as the common case. It has the advantage of not requiring conditions involving joint convergence.

We need fewer independence assumptions than we needed in Section 5.7. In particular, now we assume that the m busy-period sequences $\{B_{l,i} : i \geq 1\}$ are mutually independent sequences of IID random variables. We also assume that the idle-periods come from sequences $\{I_{l,i} : i \geq 1\}$ of IID random variables for each l, but that could easily be weakened. As in Section 5.7, we make a stochastic-process-limit assumption on the rate processes $\{\Lambda_l(t) : t \geq 0\}$ and the available-processing process $\{S(t) : t \geq 0\}$ instead of specifying the structure in detail.

Theorem 8.5.1. (RSLM limit for the multisource fluid queue) *Consider a single fluid model with m sources satisfying the independence assumptions above and the scaling assumption in (4.6). Let a sequence of systems indexed by n be formed by having the capacity in system n be $K_n = n^{1/\alpha} K$ for some K, $0 < K \leq \infty$. . Suppose that $1 < \alpha < 2$. Let the initial random workloads from the m sources in system n satisfy*

$$W_{n,l}(0) \geq 0, \quad 1 \leq l \leq m, \quad \sum_{l=1}^{m} W_n(0) \leq K_n, \tag{5.1}$$

and

$$n^{-1/\alpha} W_{n,l}(0) \Rightarrow y_l \quad in \quad \mathbb{R}, \quad 1 \leq l \leq m. \tag{5.2}$$

Assume that

$$x^\alpha P(B_{l,1} > x) \to A_l \quad as \quad x \to \infty; \tag{5.3}$$

8. On-Off Sources

where $0 \leq A_l < \infty$, and

$$x^\alpha P(I_{l,1} > x) \to 0 \quad as \quad x \to \infty \tag{5.4}$$

for $1 \leq l \leq m$. Assume that

$$\mathbf{\Lambda}_{n,l} \Rightarrow \mathbf{0} \quad and \quad \mathbf{S}_n \Rightarrow \mathbf{0} \quad in \quad (D, M_1), \tag{5.5}$$

where

$$\mathbf{\Lambda}_{n,l}(t) \equiv n^{-1/\alpha}(\Lambda_l(nt) - \hat{\lambda}_l nt), \quad t \geq 0 \tag{5.6}$$

for each l, $1 \leq l \leq m$ and

$$\mathbf{S}_{n,l}(t) \equiv n^{-1/\alpha}(S(nt) - \mu_n nt), \quad t \geq 0. \tag{5.7}$$

Assume that

$$\eta_n \equiv n^{(1-\alpha^{-1})}(\lambda - \mu_n) \to \eta \quad as \quad n \to \infty \tag{5.8}$$

for λ in (4.6) and μ_n in (5.7). Then the conditions and conclusions of Theorems 8.3.3, 5.9.1 and 5.9.2 (when $K = \infty$ in the last case) hold with $c_n = n^{1/\alpha}$, $m_{B,n,l} = m_{B,l} = EB_{l,1}$, $m_{I,n,l} = m_{I,l} = EI_{l,1}$, $\xi_{n,l} = \xi_l$, $\gamma_{n,l} = \gamma_l > 0$, $\mathbf{I}_l = \mathbf{\Lambda}_l = \mathbf{S} = \mathbf{0}$, $W'_l(0) = y_l$, $1 \leq l \leq m$, and $\mathbf{B}_1, \ldots, \mathbf{B}_m$ being m mutually independent stable Lévy motions with

$$\mathbf{B}_l(t) \stackrel{\mathrm{d}}{=} S_\alpha(\sigma_l t^{1/\alpha}, 1, 0), \quad t \geq 0, \tag{5.9}$$

where

$$\sigma_l \equiv (A_l/C_\alpha)^{1/\alpha} \tag{5.10}$$

for A_l in (5.3) and C_α in (5.14) in Section 4.5. Moreover,

$$\begin{aligned}(\mathbf{N}_l, \mathbf{B}'_l, \mathbf{C}_l, \mathbf{C})(t) \\ = (-\gamma_l \mathbf{B}_l(\gamma_l t), (1-\xi_l)\mathbf{B}_l(\gamma_l t), \\ \hat{\lambda}_l(1-\xi_l)\mathbf{B}_l(\gamma_l t), \sum_{l=1}^{m} \hat{\lambda}_l(1-\xi_l)\mathbf{B}_l(\gamma_l t)) \\ \stackrel{\mathrm{d}}{=} \Big(-\gamma_l^{1+\alpha^{-1}}\mathbf{B}_l(t), (1-\xi_l)\gamma_p^{\alpha^{-1}}\mathbf{B}_l(t), \\ \hat{\lambda}_l(1-\xi_l)\gamma_l^{\alpha^{-1}}\mathbf{B}_l(t), \sum_{l=1}^{m} \hat{\lambda}_l(1-\xi_l)\gamma_l^{\alpha^{-1}}\mathbf{B}_l(t)\Big),\end{aligned} \tag{5.11}$$

so that $\mathbf{C}$ is a centered stable Lévy motion with

$$\mathbf{C}(t) \stackrel{\mathrm{d}}{=} \left(\sum_{l=1}^{m} \hat{\lambda}_l^\alpha (1-\xi_l)^\alpha \gamma_l A_l/C_l\right)^{1/\alpha} S_\alpha(t^{\alpha^{-1}}, 1, 0) \tag{5.12}$$

and

$$\mathbf{X}(t) = y + \eta t + \mathbf{C}(t), \quad t \geq 0, \tag{5.13}$$

for $y = y_1 + \cdots + y_m$. Hence the limit $\mathbf{W} = \phi_K(\mathbf{X})$ is the reflection of the linear drift $\eta \mathbf{e}$ plus standard stable Lévy motion $\mathbf{C}$ having parameter four-tuple $(\alpha, \sigma, \beta, \mu) = (\alpha, \sigma_C, 1, 0)$ with

$$\sigma_C = \left(\sum_{l=1}^{m} \hat{\lambda}_l^{\alpha} (1 - \xi_l)^{\alpha} \gamma_l \sigma_l^{\alpha} \right)^{1/\alpha}, \quad (5.14)$$

starting at initial position y, $0 \leq y \leq K$, with reflection having upper barrier at K, $0 < K \leq \infty$.

Proof. We apply Theorem 8.3.3. First note that (5.3) and (5.4) imply that

$$0 < EB_{l,1} < \infty \quad \text{and} \quad 0 < EI_{l,1} < \infty . \quad (5.15)$$

To apply Theorem 8.3.3, we need to verify condition (3.13). The limits for $\mathbf{B}_{n,l}$ individually follow from Theorem 4.5.3, using the normal-domain of attraction result in Theorem 4.5.2. The joint FCLT for $(\mathbf{B}_{n,1}, \ldots, \mathbf{B}_{n,m})$ then follows from Theorem 11.4.4. Since the limits for $\mathbf{I}_{n,l}$, $\mathbf{\Lambda}_{n,l}$ and $\mathbf{S}_n$ are deterministic, we obtain an overall joint FCLT from the individual FCLTs, using Theorem 11.4.5. That yields the required stochastic-process limit in (3.13). See (5.7) – (5.11) in Section 4.5 for the scaling yielding (5.12). For Theorem 5.9.2 with $K = \infty$, we use the fact that $\mathbf{S} = 0\mathbf{e}$ and $\mathbf{C}$ has nonnegative jumps to conclude that $P(\mathbf{L} \in C) = 1$ and that all the conditions on the discontinuities of the limit processes in Theorem 5.9.2 are satisfied in this setting. ∎

Theorem 8.5.1 tells us what are the key parameters governing system performance. First, the nonstandard space scaling by $c_n = n^{\alpha}$ shows that the scaling exponent α in (5.3) is critical. But clearly the values of the means, which necessarily are finite with $1 < \alpha < 2$, are also critical. The values of the means appear via the asymptotic drift η in (5.8).

Since the stable laws with $1 < \alpha < 2$ have infinite variance, there are no variance parameters describing the impact of variability, as there were in Section 5.7. However, essentially the same variability parameters appear; they just cannot be interpreted as variances. In both cases, the variability parameters appear as *scale factors* multiplying canonical limit processes. In Theorem 8.4.1, the variability parameter is the parameter σ_X appearing as a multiplicative factor before the standard Brownian motion $\mathbf{B}$ in (4.14). Correspondingly, in Theorem 8.5.1, the variability parameter is the stable scale parameter σ_C in (5.14) and (5.12). That scale parameter depends critically on the parameters A_l, which are the second order parameters appearing in the tail-probability asymptotics for the busy-period distributions in (5.3).

The stable Lévy motions and reflected stable Lévy motions appearing in Theorem 8.5.1 are less familiar than the corresponding Brownian motions and reflected Brownian motions appearing in Section 5.7, but they have been studied quite extensively too; e.g., see Samorodnitsky and Taqqu (1994) and Bertoin (1996).

8.5.2. The Steady-State Distribution

It is significant that the limiting stable Lévy motion for the cumulative-input process is *totally skewed*, i.e., has skewness parameter $\beta = 1$, so that the stable Lévy motion has *sample paths without negative jumps*. That important property implies that the reflected stable Lévy motion with negative drift has a relatively simple steady-state distribution, both with and without an upper barrier. That is also true for more general reflected Lévy processes; see Section 5.2 in the Internet Supplement, Theorem 4.2 of Kella and Whitt (1991) and Section 4 (a) of Kella and Whitt (1992c). With a finite upper barrier, the steady-state distribution is the steady-state distribution with infinite capacity, truncated and renormalized. Equivalently, the finite-capacity distribution is the conditional distribution of the infinite-capacity content, given that the infinite-capacity content is less than the upper barrier.

Theorem 8.5.2. (steady-state distribution of RSLM) *Let $\mathbf{R} \equiv \phi_K(\eta \mathbf{e} + \mathbf{S})$ be the reflected stable Lévy motion with negative drift ($\eta < 0$) and SLM $\mathbf{S}$ having stable index α, $1 < \alpha < 2$, and scaling parameter σ in (5.12) (with $\beta = 1$ and $\mu = 0$) arising as the limit in Theorem 8.5.1 (where $\mathbf{S}(1) \stackrel{d}{=} S_\alpha(\sigma,1,0) \stackrel{d}{=} \sigma S_\alpha(1,1,0)$).*
(a) *If $K = \infty$, then*

$$\lim_{t \to \infty} P(\mathbf{R}(t) \leq x) = H(x) \quad \text{for all} \quad x, \tag{5.16}$$

where H is a proper cdf with pdf h on $(0,\infty)$ with Laplace transform

$$\hat{h}(s) \equiv \int_0^\infty e^{-sx} h(x) dx = \frac{s\hat{\phi}'(0)}{\hat{\phi}(s)} = \frac{1}{1 + (\nu s)^{\alpha-1}}, \tag{5.17}$$

with

$$\hat{\phi}(s) \equiv \log E e^{-s[\eta + S_\alpha(\sigma,1,0)]} = -\eta s(\sigma^\alpha s^\alpha / \cos(\pi\alpha/2)) \tag{5.18}$$

and the scale factor ν ($H_\nu(x) = H_1(x/\nu)$) being

$$\nu^{\alpha-1} \equiv \frac{\sigma^\alpha}{\eta \cos(\pi\alpha/2)} > 0. \tag{5.19}$$

The associated ccdf $H^c \equiv 1 - H$ has Laplace transform

$$\hat{H}^c(s) \equiv \int_0^\infty e^{-sx} H^c(x) dx = \frac{1 - \hat{h}(s)}{s}$$

$$= \frac{\nu}{(\nu s)^{2-\alpha}(1 + (\nu s)^{\alpha-1})}. \tag{5.20}$$

(b) *If $c < 0$ and $K < \infty$, then*

$$\lim_{t \to \infty} P(\mathbf{R}(t) \leq x) = \frac{H(x)}{H(K)}, \quad 0 \leq x \leq K, \tag{5.21}$$

where H is the cdf in (5.16).

We have observed that the heavy-traffic limit for the workload processes depends on the parameter five-tuple $(\alpha, \eta, \sigma, K, y)$. When we consider the steady-state distribution, the initial value y obviously plays no role, but we can reduce the number of relevant parameters even further: From Theorem 8.5.2, we see that the steady-state distribution depends on the parameter four-tuple $(\alpha, \eta, \sigma, K)$ only via the parameter triple (α, ν, K) for ν in (5.19).

As should be anticipated, the steady-state distribution in Theorem 8.5.2 has a heavy tail. Indeed it has infinite mean. We can apply Heaviside's theorem, p. 254 of Doetsch (1974) to deduce the asymptotic form of the ccdf H^c and pdf h in Theorem 8.5.2. We display the first two terms of the asymptotic expansions below. (The second term for $\alpha = 3/2$ seems inconsistent with the second term for $\alpha \neq 3/2$; that occurs because the second term for $\alpha = 3/2$ actually corresponds to the third term for $\alpha \neq 3/2$, while the second term is zero.)

Theorem 8.5.3. (tail asymptotics for steady-state distribution) *For $\nu = 1$, the steady-state ccdf and pdf in Theorem 8.5.2 satisfy*

$$H^c(x) \sim \begin{cases} \frac{1}{\Gamma(2-\alpha)x^{\alpha-1}} - \frac{1}{\Gamma(3-2\alpha)x^{2(\alpha-1)}}, & \alpha \neq 3/2 \\ \\ \frac{1}{\sqrt{\pi}x^{1/2}} - \frac{1}{2\sqrt{\pi}x^{3/2}}, & \alpha = 3/2 \end{cases} \quad (5.22)$$

and

$$h(x) \sim \begin{cases} \frac{-1}{\Gamma(1-\alpha)x^{\alpha}} + \frac{1}{\Gamma(2-2\alpha)x^{2\alpha-1}}, & \alpha \neq 3/2 \\ \\ \frac{1}{2\sqrt{\pi}x^{3/2}} - \frac{3}{4\sqrt{\pi}x^{5/2}}, & \alpha = 3/2 \end{cases} \quad (5.23)$$

as $x \to \infty$, where $\Gamma(x)$ is the gamma function.

For the special case $\alpha = 3/2$, the limiting pdf h and cdf H can be expressed in convenient closed form. We can apply 29.3.37 and 29.3.43 of Abramowitz and Stegun (1972) to invert the Laplace transforms analytically in terms of the error function, which is closely related to the standard normal cdf. A similar explicit expression is possible for a class of $M/G/1$ steady-state workload distributions; see Abate and Whitt (1998). That can be used to make numerical comparisons.

Theorem 8.5.4. (explicit expressions for $\alpha = 3/2$) *For $\alpha = 3/2$ and $\nu = 1$, the limiting pdf and ccdf in Theorem 8.5.2 are*

$$h(x) = \frac{1}{\sqrt{\pi x}} - 2e^x \Phi^c(\sqrt{2x}), \quad x \geq 0, \quad (5.24)$$

and

$$H^c(x) = 2e^x \Phi^c(\sqrt{2x}), \quad x \geq 0, \quad (5.25)$$

where Φ^c is again the standard normal ccdf.

Theorems 8.5.2 – 8.5.4 provide important practical engineering insight. Note the the ccdf $H^c(x)$ of the steady-state distribution of RSLM with no upper barrier decays slowly. Specifically, from (5.22), we see that it decays as the power $x^{-(\alpha-1)}$. In fact, the exponent $-(\alpha-1)$, implies a slower decay than the decay rate $-\alpha$ of the heavy-tailed busy-period ccdf $B_{l,1}^c(x)$ in (5.3). Indeed, the density $h(x)$ of the steady-state RSLM decays at exactly the same rate as the busy-period ccdf.

In contrast, for the RBM limit (with negative drift) in Section 5.7, the steady-state distribution decays exponentially. In both cases, the steady-state distribution with a finite buffer is the steady-state distribution with an infinite buffer truncated and renormalized. Equivalently, the steady state distribution with a finite buffer is the steady-state distribution with an infinite buffer condioned to be less than the buffer capacity K. That property supports corresponding approximations for finite-capacity queueing systems; see Whitt (1984a). The fact the ccdf $H^c(x)$ of the steady-state distribution of RSLM decays as a power implies that a buffer will be less effective in reducing losses than it would be for a workload process that can be approximated by RBM.

Also note that the asymptote in (5.22) provides a useful "back-of-the envelope" approximation for the ccdf of the steady-state distribution:

$$P(W_\rho(\infty) > x) \approx P((\zeta/(1-\rho))^{1/(\alpha-1)} Z > x) , \qquad (5.26)$$

where

$$P(Z > x) \approx C x^{-(\alpha-1)} \qquad (5.27)$$

for constants ζ and C determined by (5.11) and (5.22).

The asymptotic tail in (5.26) and (5.27) is consistent with known asymptotics for the exact steady-state workload ccdf in special cases; e.g., see the power-tail references cited in Remark 5.4.1.

By comparing the second term to the first term of the asymptotic expansion of the ccdf $H^c(x)$ in Theorem 8.5.3, we can see that the one-term asymptote should tend to be an upper bound for $\alpha < 1.5$ and a lower bound for $\alpha > 1.5$. We also should anticipate that the one-term asymptote should be more accurate for α near $3/2$ than for other values for α. We draw this conclusion for two reasons: first, at $\alpha = 3/2$ a potential second term in the expansion does not appear; so that the relative error (ratio of appearing second term to first term) is of order $x^{-2(\alpha-1)}$ instead of $x^{-(\alpha-1)}$ for $\alpha \neq 3/2$. Second, for $\alpha \neq 3/2$ but α near $3/2$, the constant $\Gamma(3-2\alpha)$ in the denominator of the second term tends to be large, i.e., $\Gamma(x) \to \infty$ as $x \to 0$.

8.5.3. Numerical Comparisons

We now show that the anticipated structure deduced from examining Theorem 8.5.3 actually holds by making numerical comparisons with exact values computed by numerically inverting the Laplace transform in (5.20). To do the inversion, we use the Fourier series method in Abate and Whitt (1995a); see Abate, Choudhury and Whitt (1999) for an overview.

We display results for $\alpha = 1.5$, $\alpha = 1.9$ and $\alpha = 1.1$ in Tables 8.1–8.3. For $\alpha = 1.5$, Table 8.1 shows that the one-term asymptote is a remarkably accurate

8.5. Stable-Lévy Approximations

approximation for x such that $H^c(x) < 0.20$. In Table 8.1 we also demonstrate a strong sensitivity to the value of α by showing the exact values for $\alpha = 1.49$ and $\alpha = 1.40$. For $x = 10^4$ when $H^c(x) = 0.056$ for $\alpha = 1.50$, the corresponding values of $H^c(x)$ for $\alpha = 1.49$ and $\alpha = 1.40$ differ by about 12% and 200%, respectively.

	$H^c(x)$ for $\alpha = 1.5$ $(\nu = 1)$			
x	exact $\alpha = 1.5$	one-term asymptote	exact $\alpha = 1.49$	exact $\alpha = 1.40$
10^{-1}	0.7236	1.78	0.7190	0.6778
10^0	0.4276	0.5642	0.4290	0.4421
10^1	0.1706	0.1784	0.1760	0.2278
10^2	$0.5614\,e-1$	$0.5642\,e-1$	$0.5970\,e-1$	0.1004
10^3	$0.1783\,e-1$	$0.1784\,e-1$	$0.1946\,e-1$	$0.4146\,e-1$
10^4	$0.5641\,e-2$	$0.5642\,e-2$	$0.6304\,e-2$	$0.1673\,e-1$
10^5	$0.1784\,e-2$	$0.1784\,e-2$	$0.2041\,e-2$	$0.6693\,e-2$
10^6	$0.5642\,e-3$	$0.5642\,e-3$	$0.6604\,e-3$	$0.2670\,e-2$
10^7	$0.1784\,e-3$	$0.1784\,e-3$	$0.2137\,e-3$	$0.1064\,e-2$
10^8	$0.5642\,e-4$	$0.5642\,e-4$	$0.6916\,e-4$	$0.4236\,e-3$
10^{16}	$0.5642\,e-8$	$0.5642\,e-8$	$0.8315\,e-8$	$0.2673\,e-6$

Table 8.1. A comparison of the limiting ccdf $H^c(x)$ in Theorem 8.5.2 for $\alpha = 1.5$ and $\nu = 1$ with the one-term asymptote and the alternative exact values for $\alpha = 1.49$ and $\alpha = 1.40$.

Tables 8.2 and 8.3 show that the one-term asymptote is a much less accurate approximation for α away from 1.5. In the case $\alpha = 1.9$ ($\alpha = 1.1$), the one-term asymptote is a lower (upper) bound for the exact value, as anticipated. For $\alpha = 1.9$, we also compare the ccdf values $H^c(x)$ to the corresponding ccdf values for a mean-1 exponential variable (the case $\alpha = 2$). The ccdf values differ drastically in the tail, but are quite close for small x. A reasonable rough approximation for $H^c(x)$ for all x when α is near (but less than) 2 is the maximum of the one-term asymptote and the exponential ccdf e^{-x}. It is certainly far superior to either approximation alone.

The exponent $\alpha = 2$ is a critical boundary point for the ccdf tail behavior: Suppose that the random variable $A_1(1)$ has a power tail decaying as $x^{-\alpha}$. If $\alpha > 2$, then the limiting ccdf $H^c(x)$ is exponential, i.e., $H^c(x) = e^{-x}$, but for $\alpha < 2$ the ccdf decays as $x^{-(\alpha-1)}$. This drastic change can be seen at the large x values in Table 8.2.

Table 8.3 also illustrates how we can use the asymptotics to numerically determine its accuracy. We can conclude that the one-term asymptote is accurate at those x for which the one-term and two-term asymptotes are very close. Similarly, we can conclude that the two-term asymptote is accurate at those x for which the two-term and three-term asymptotes are close, and so on.

Remark 8.5.1. *First passage times.* We have applied numerical transform inversion to calculate steady-state tail probabilities of RSLM. To describe the transient behavior, we might want to compute first-passage-time probabilities.

272   8. On-Off Sources

	$H^c(x)$ for $\alpha = 1.9$ ($\nu = 1$)		
		one-term	exponential
x	exact	asymptote	$\alpha = 2$
$0.1 \times 2^0 = 0.1$	0.878	0.835	0.905
$0.1 \times 2^1 = 0.2$	0.786	0.447	0.819
$0.1 \times 2^2 = 0.4$	0.641	0.240	0.670
$0.1 \times 2^4 = 1.6$	0.238	0.069	0.202
$0.1 \times 2^6 = 6.4$	$0.312\ e{-1}$	$0.198\ e{-1}$	$0.166\ e{-2}$
$0.1 \times 2^8 = 25.6$	$0.626\ e{-2}$	$0.568\ e{-2}$	$0.76\ e{-11}$
$0.1 \times 2^{12} = 409.6$	$0.472\ e{-3}$	$0.468\ e{-3}$	≈ 0
$0.1 \times 2^{16} = 6553.6$	$0.386\ e{-4}$	$0.386\ e{-4}$	≈ 0

Table 8.2. A comparison of the reflected stable ccdf $H^c(x)$ in Theorem 8.5.2 for $\alpha = 1.9$ and $\nu = 1$ with the one-term asymptote and the mean-1 exponential ccdf corresponding to $\alpha = 2$.

	$H^c(x)$ for $\alpha = 1.1$ ($\nu = 1$)		
		one-term	two-term
x	exact	asymptote	asymptote
10^{-1}	0.543	1.18	-0.19
10^0	0.486	0.94	0.08
10^1	0.428	0.74	0.20
10^2	0.373	0.59	0.25
10^4	0.272	0.373	0.237
10^6	0.191	0.235	0.181
10^8	0.129	0.148	0.126
10^{12}	$0.558\ e{-1}$	$0.590\ e{-1}$	$0.555\ e{-1}$
10^{16}	$0.230\ e{-1}$	$0.235\ e{-1}$	$0.230\ e{-1}$
10^{24}	$0.371\ e{-2}$	$0.373\ e{-2}$	$0.371\ e{-2}$
10^{32}	$0.590\ e{-3}$	$0.590\ e{-3}$	$0.590\ e{-3}$

Table 8.3. A comparison of the reflected stable ccdf $H^c(x)$ in Theorem 8.5.2 for $\alpha = 1.1$ and $\nu = 1$ with the one-term and two-term asymptotes from Theorem 8.5.3.

Rogers (2000) has developed a numerical transform inversion algorithm for calculating first-passage-time probabilities in Lévy processes with jumps in at most one direction.

8.6. Second Stochastic-Process Limits

The usual approximation for a stochastic process $\mathbf{X}_n$ based on a stochastic-process limit $\mathbf{X}_n \Rightarrow \mathbf{X}$ is $\mathbf{X}_n \approx \mathbf{X}$. However, if the limit process $\mathbf{X}$ is not convenient, then we may want to consider developing approximations for the limit process $\mathbf{X}$. We can obtain such additional approximations by considering yet another stochastic-process limit. We describe two approaches in this section.

The first approach is based on having another stochastic-process limit with the same limit process $\mathbf{X}$: We may be able to establish both $\mathbf{X}_n \Rightarrow \mathbf{X}$ and $\mathbf{Y}_n \Rightarrow \mathbf{X}$, where the process $\mathbf{X}_n$ is the scaled version of the relatively complicated process of interest and $\mathbf{Y}_n$ is the scaled version of another more tractable process. Then we can use the double approximation

$$\mathbf{X}_n \approx \mathbf{X} \approx \mathbf{Y}_m ,$$

where n and m are suitably large. We discuss two possible approximations for processes associated with a reflected-stable-Lévy-motion (RSLM) in the first subsection below.

The second approach is to approximate the limit process $\mathbf{X}$ by establishing a further stochastic-process limit for scaled versions of the limit process $\mathbf{X}$ itself. That produces an approximation that should be relevant in an even longer time scale, since time is now scaled twice. We discuss this second approach in the second subsection below.

8.6.1. M/G/1/K Approximations

We now show how the two-limit approach can be used to generate $M/G/1/K$ approximations for complicated queueing models such as the multisource on-off fluid-queue model with heavy tailed busy-period distributions. We first apply Theorem 8.5.1 to obtain a reflected-stable-Lévy-motion (RSLM) approximation for the workload process in the fluid queue. We then construct a sequence of $M/G/1/K$ models with scaled workload processes converging to the RSLM.

Consider a stable Lévy motion $\sigma\mathbf{S} + \eta\mathbf{e}$, where $\mathbf{S}(1) \stackrel{d}{=} S_\alpha(1,1,0)$ with $1 < \alpha < 2$ and $\sigma > 0$, modified by two-sided reflection at 0 and K. We now show that, for any choice of the four parameters α, σ, η and K, we can construct a sequence of M/G/1/K fluid queues such that the scaled M/G/1/K net-input processes, workload processes, overflow processes and departure processes converge to those associated with the given RSLM. Since we can apply the continuous-mapping approach with the two-sided reflection map, it suffices to show that the scaled net-input processes converge to $\sigma\mathbf{S} + \eta\mathbf{e}$. We can apply Theorems 5.4.1 and 5.9.1.

The specific M/G/1/K fluid-queue model corresponds to the standard M/G/1/K queue with bounded virtual waiting time, which is often called the *finite dam*; see Section III.5 of Cohen (1982). For the workload process, there is a constant output rate of 1. As in Example 5.7.1, The cumulative input over the interval $[0,t]$ is the sum of the service times of all arrivals in the interval $[0,t]$, i.e., the cumulative input is

$$C(t) \equiv \sum_{k=1}^{A(t)} V_k, \quad t \geq 0 ,$$

where $\{A(t) : t \geq 0\}$ is a rate-λ Poisson arrival process independent of a sequence $\{V_k : k \geq 1\}$ of IID service times. The workload process is defined in terms of the net-input process $X(t) \equiv C(t) - t$ as described in Section 5.2. Any input that would take the workload above the storage capacity K overflows and is lost.

The $M/G/1/K$ model is appealing because $C \equiv \{C(t) : t \geq 0\}$ is a compound Poisson process, which is a special case of a renewal-reward process; see Section 7.4. Like the SLM, the processes C and X are Lévy processes, but unlike the SLM, the processes C and X almost surely have only finitely many jumps in a bounded interval.

The key to achieving the asymptotic parameters α and $\beta = 1$ is to use a heavy-tailed service-time distribution with power tail, where the ccdf decays as $x^{-\alpha}$ as $x \to \infty$. Thus, we let the service times be mV_k, where $\{V_k : k \geq 1\}$ is a sequence of IID nonnegative random variables with mean $EV_1 = 1$ and an appropriate power tail:

$$P(V_1 > x) \sim \gamma_1 x^{-\alpha} \quad \text{as} \quad x \to \infty .$$

Let $A \equiv \{A(t) : t \geq 0\}$ be a rate-1 Poisson process. Let the net-input process in the nth $M/G/1/K$ model be

$$X_n(t) \equiv \sum_{k=1}^{A(\lambda_n t)} mV_k - t, \quad t \geq 0 ;$$

i.e., we let the service times be distributed as mV_1 and we let the arrival process be a Poisson process with rate λ_n. The associated *scaled net-input processes* are

$$\mathbf{X}_n(t) \equiv n^{-1/\alpha} X_n(nt), \quad t \geq 0 .$$

We will choose the parameters m and λ_n in order to obtain the desired convergence

$$\mathbf{X}_n \Rightarrow \sigma \mathbf{S} + \eta \mathbf{e} .$$

We must also make appropriate definitions for the queue. Since we scale space by dividing by $n^{1/\alpha}$ in the stochastic-process limit, we let the upper barrier in the nth $M/G/1/K$ queue be $K_n = n^{1/\alpha} K$. We also must match the initial conditions appropriately. If the RSLM starts at the origin, then the $M/G/1/K$ queue starts empty.

To determine the parameters m and λ_n, observe that

$$\mathbf{X}_n = \mathbf{S}_n + \eta_n \mathbf{e} ,$$

where

$$\mathbf{S}_n(t) \equiv n^{-1/\alpha} \Big(\sum_{k=1}^{A(\lambda_n nt)} mV_k - \lambda_n mnt \Big), \quad t \geq 0 ,$$

and

$$\eta_n \equiv n^{1-\alpha^{-1}} (\lambda_n m - 1), \quad n \geq 1 .$$

8.6. Second Stochastic-Process Limits

Hence, we can obtain $\eta_n = \eta$ by letting

$$\lambda_n = m^{-1}(1 + \eta n^{-(1-\alpha^{-1})}), \quad n \geq 1. \tag{6.1}$$

Since $\lambda_n \to m^{-1}$ as $n \to \infty$, $\mathbf{S}_n$ has the same limit as

$$\tilde{\mathbf{S}}_n(t) \equiv n^{-1/\alpha}\left(\sum_{k=1}^{A(m^{-1}nt)} mV_k - nt\right), \quad t \geq 0.$$

Note that

$$P(mV_k > x) \sim \gamma_1 m^\alpha x^{-\alpha} \quad \text{as} \quad x \to \infty.$$

Hence, we can apply Theorem 7.4.2 for renewal-reward processes to deduce that

$$\tilde{\mathbf{S}}_n \Rightarrow \tilde{\sigma}\mathbf{S} \quad \text{in} \quad (D, J_1),$$

where

$$\tilde{\sigma}^\alpha = \gamma_1 m^\alpha m^{-1}/C_\alpha$$

for C_α in (5.14) of Section 4.5.1. To achieve our goal of $\tilde{\sigma} = \sigma$, we must have

$$m = (\sigma C_\alpha/\gamma_1)^{1/(\alpha-1)}. \tag{6.2}$$

With that choice of m, we have $\tilde{\mathbf{S}}_n \Rightarrow \sigma\mathbf{S}$. For λ_n in (6.1) and m in (6.2), we obtain $\mathbf{S}_n \Rightarrow \sigma\mathbf{S}$ and $\mathbf{X}_n \Rightarrow \sigma\mathbf{S} + \eta\mathbf{e}$ as desired.

For example, as the service-time distribution in the M/G/1/K fluid queue, we can use the Pareto distribution used previously in Chapter 1. Recall that a Pareto(p) random variable Z_p with $p > 1$ has ccdf

$$F_p^c(x) \equiv P(Z_p > x) \equiv x^{-p}, \quad x \geq 1, \tag{6.3}$$

and mean

$$m_p = 1 + (p-1)^{-1}. \tag{6.4}$$

To achieve the specified power tail, we must let $p = \alpha$ for $1 < \alpha < 2$.

To put the Pareto distribution in the framework above, we need to rescale to obtain mean 1. Clearly, $m_\alpha^{-1}Z_\alpha$ has mean 1 and

$$P(m_\alpha^{-1}Z_\alpha > x) = P(Z_\alpha > m_\alpha x) = (m_\alpha x)^{-\alpha}, \quad x \geq 1,$$

so that we let $V_1 \stackrel{d}{=} m_\alpha^{-1}Z_\alpha$ and have

$$\gamma_1 = m_\alpha^{-\alpha} = (1 + (\alpha-1)^{-1})^{-\alpha}.$$

Given those model specifications, we approximate the limiting RSLM, say $\mathbf{W}$, by the nth scaled M/G/1/K workload process, i.e.,

$$\{\mathbf{W}(t) : t \geq 0\} \approx \{n^{-1/\alpha}(W_n(nt) : t \geq 0\} \tag{6.5}$$

for suitably large n, where $\{W_n(t) : t \geq 0\}$ is the workload (or virtual waiting time) process in the nth M/G/1/K queueing system with upper barrier at $K_n = n^{1/\alpha}K$ specified above.

This approximation can be very useful because the M/G/1/K fluid-queue model or finite dam has been quite thoroughly studied and is known to be tractable; e.g., see Takács (1967), Cohen (1982) and Chapter 1 of Neuts (1989).

For the remaining discussion, consider an M/G/1/K model with sevice times V_k and arrival rate λ. For $K = \infty$ and $\rho = \lambda E V_1 < 1$, the M/G/1 workload process is especially tractable; e.g., see Abate and Whitt (1994a). Then the steady-state distribution is characterized by the Pollaczek-Khintchine transform. For $K < \infty$, the steady-state distribution is the infinite-capacity steady-state truncated and renormalized, just as in Theorems 8.5.2 here and Theorem 5.2.1 in the Internet Supplement.

Given the steady-state workload $W(\infty)$ with $K < \infty$, the overflow rate is

$$\beta = \lambda E[W(\infty) + V_1 - K]^+ ,$$

where V_1 is a service time independent of $W(\infty)$. The rate of overflows of at least size x is

$$\beta_x = \lambda P(W(\infty) + V_1 - K > x) .$$

(For asymptotics, see Zwart (2000).) The ccdf $P(W(\infty) + V_1 > x + K)$ is easily computed by numerical transform inversion. For that purpose, we first remove the known atom (positive probability mass) at zero in the distribution of $W(\infty)$, as discussed in Abate and Whitt (1992a). Since $P(W(\infty) = 0) = 1 - \rho$, we can write

$$\begin{aligned} P(W(\infty) + V_1 > x + K) &= (1-\rho)P(V_1 > x + K) \\ &\quad + \rho P(W(\infty) + V_1 > x + K | W(\infty) > 0) . \end{aligned}$$

We then calculate the conditional ccdf $P(W(\infty) + V_1 > x + K | W(\infty) > 0)$ by numerically inverting its Laplace transform $(1 - Ee^{-s(W\infty)|W(\infty)>0})Ee^{-sV_1})/s$.

For any K with $0 < K \leq \infty$, the M/G/1/K fluid-queue departure process is an on-off cumulative-input process with mutually independent sequences of IID busy periods and IID idle periods, with the idle periods being exponentially distributed with mean λ^{-1}. The departure rate during busy periods is 1. For $K = \infty$, the busy-period distribution is characterized by its Laplace transform, which satisfies the Kendall functional equation; e.g., as in equation (28) of Abate and Whitt (1994a). The numerical values of the Laplace transform for complex arguments needed for numerical transform inversion can be computed by iterating the Kendall functional equation; see Abate and Whitt (1992b).

In the heavy-tailed case under consideration, $P(V_1 > x) \sim Ax^{-\alpha}$, where $A = C_\alpha \sigma^\alpha$. By de Meyer and Teugels (1980), the busy-period distribution in this M/G/1/∞ model inherits the power tail; i.e.,

$$P(B_1 > x) \sim A(1-\rho)^{-(\alpha+1)} x^{-\alpha} \quad \text{as} \quad x \to \infty .$$

Hence, we can apply Theorems 8.3.1 and 8.3.2 to establish a stochastic-process limit for the M/G/1 departure process with ρ assumed fixed. However, as noted at the end of Section 5.3.2, for fixed ρ with $\rho < 1$, the departure process obeys the same FCLT as the input process, given in Theorem 7.4.2. The two approaches to this FCLT can be seen to be consistent.

We have indicated that many M/G/1/K fluid-queue random quantities can be conveniently expressed in terms of the Laplace transform of the service-time distribution. Unfortunately, the Laplace transform of the Pareto distribution does not have a convenient simple explicit form, but because of its connection to the gamma integral, the transform values can easily be computed by exploiting efficient algorithms based on continued fractions, as shown by Abate and Whitt (1999a). That algorithm to compute Pareto distributions by numerical transform inversion is applied in Ward and Whitt (2000).

For some applications it might be desirable to have even more tractable "Markovian" service-time distributions. One approach is to approximate the Pareto distribution by a mixture of exponentials. A specific procedure is described in Feldmann and Whitt (1998). Since the Pareto distribution is completely monotone, it is directly a continuous mixture of exponentials. Thus, finite mixtures can provide excellent approximations for the Pareto distribution, but since it is impossible to match the entire tail, it is often necessary to use ten or more component exponentials to obtain a good fit for applications. Having only two component exponentials usually produces a poor match.

Alternatively, the M/G/1/K approximations can be produced with different heavy-tailed service-time distributions. Other heavy-tailed distributions that can be used instead of the scaled Pareto distribution or the approximating hyperexponential distribution are described in Abate, Choudhury and Whitt (1994) and Abate and Whitt (1996, 1999b, c).

We can also consider other approximating processes converging to the RSLM. Attractive alternatives to the M/G/1/K models just considered are discrete-time random walks on a discrete lattice. Such random-walk approximations are appealing because we can then employ well-known numerical methods for finite-state Markov chains, as in Kemeny and Snell (1960) and Stewart (1994). For the RSLM based on the totally skewed ($\beta = 1$) α-stable Lévy motion that arises as the stochastic-process limit in Theorem 8.5.1, it is natural to construct the random walk by appropriately modifying an initial random walk with IID steps distributed as the random variable Z with the zeta distribution, which has probability mass function

$$p(k) \equiv P(Z = k) \equiv 1/\zeta(\alpha + 1)k^{\alpha+1}, \quad k \geq 1, \qquad (6.6)$$

where $\zeta(s)$ is the Riemann zeta function; see p. 240 of Johnson and Kotz (1969) and Chapter 23 of Abramowitz and Stegun (1972). The zeta distribution has mean $\zeta(\alpha)/\zeta(\alpha + 1)$ and the appropriate tail asymptotics, i.e.,

$$F^c(x) \equiv P(Z > x) \sim 1/\alpha\zeta(\alpha + 1)x^\alpha \quad \text{as} \quad x \to \infty. \qquad (6.7)$$

We can apply Section 4.5 to construct scaled random walks converging to the totally skewed α-stable Lévy motion.

8.6.2. Limits for Limit Processes

We now consider the second approach for approximating the limit process $\mathbf{X}$ associated with a stochastic process-limit $\mathbf{X}_n \Rightarrow \mathbf{X}$. Now we assume that $\mathbf{X}_n$ is the

scaled process
$$\mathbf{X}_n(t) \equiv n^{-H}(X_n(nt) - \nu_n nt), \quad t \geq 0, \tag{6.8}$$

where $0 < H < 1$. If $\mathbf{X}$ is not sufficiently tractable, we may try to establish to establish a further stochastic-process limit for scaled versions of the limit process $\mathbf{X}$.

Suppose that we can construct the new scaled process
$$\tilde{\mathbf{X}}_n(t) \equiv n^{-\tilde{H}}(\mathbf{X}(nt) - \eta nt), \quad t \geq 0 \tag{6.9}$$

and show that $\tilde{\mathbf{X}}_n \Rightarrow \tilde{\mathbf{X}}$ as $n \to \infty$. Then we can approximate the first limit process $\mathbf{X}$ by
$$\{\mathbf{X}(t) : t \geq 0\} \approx \{\eta t + n^{\tilde{H}} \tilde{\mathbf{X}}(t/n) : t \geq 0\}. \tag{6.10}$$

Given the two stochastic-process limits, we can combine the two approximations with the scaling in (6.8) and (6.9) to obtain the overall approximation
$$\begin{aligned}\{X_n(t) : t \geq 0\} &\approx \{\nu_n t + n^H \mathbf{X}(t/n) : t \geq 0\} \\ &\approx \{\nu_n t + n^H \eta t/n + n^H m^{\tilde{H}} \tilde{\mathbf{X}}(t/nm) : t \geq 0\}\end{aligned} \tag{6.11}$$

for appropriate n and m. In the queueing context, we can fix n by letting n be such that the traffic intensity ρ_n coincides with the traffic intensity of the queue being approximated.

Since the limit processes for scaled overflow and departure processes in the fluid-queue model are somewhat complicated, even in the light-tailed weakly-dependent case considered in Section 5.7, it is natural to apply this two-limit approach to overflow and departure processes. We can apply Theorem 5.7.4 to obtain FCLT's for the boundary regulator processes $\mathbf{U}$ and $\mathbf{L}$ in the Brownian case. For that purpose, introduce the normalized processes
$$\begin{aligned}\tilde{\mathbf{U}}_n(t) &\equiv n^{-1/2}(\mathbf{U}(nt) - \beta nt), \\ \tilde{\mathbf{L}}_n(t) &\equiv n^{-1/2}(\mathbf{L}(nt) - \alpha nt), \quad t \geq 0,\end{aligned} \tag{6.12}$$

where α and β are the rates determined in Theorem 5.7.3.

By using the regenerative structure associated with Theorem 5.7.4 (see Theorem 2.3.8 in the Internet Supplement), we obtain the following stochastic-process limit.

Theorem 8.6.1. (FCLT for RBM boundary regulator processes) *The normalized boundary regulation processes for RBM in (6.12) satisfy*
$$\begin{aligned}\tilde{\mathbf{U}}_n &\Rightarrow \sigma_U \mathbf{B}, \\ \tilde{\mathbf{L}}_n &\Rightarrow \sigma_L \mathbf{B} \quad in \quad D,\end{aligned} \tag{6.13}$$

where $\mathbf{B}$ is standard BM and σ_U^2 and σ_L^2 are given in (7.26), (7.27) and (7.30)–(7.32).

In general, a corresponding limit for the departure process is more complicated, but one follows directly from Theorems 8.6.1 and 5.9.1 when the processing is deterministic, so that the limit process $\mathbf{S}$ in (9.2) is the zero function.

Corollary 8.6.1. (second limit for the departure process) *Let* $\mathbf{D} = \mathbf{S} - \mathbf{L}$ *be the heavy-traffic limit for the departure process in Theorem 5.9.1 under the conditions of the general Brownian limit in Theorem 5.7.1. If* $\mathbf{S} = 0\mathbf{e}$, *then*

$$\tilde{\mathbf{D}}_n = -\tilde{\mathbf{L}}_n \Rightarrow \sigma_L \mathbf{B} \quad in \quad (D, M_1)$$

for $\tilde{\mathbf{L}}_n$ *in* (6.12), $\mathbf{B}$ *standard Brownian motion and* σ_L *as in* (7.26), (7.30) *and* (7.32).

The condition $\mathbf{S} = 0\mathbf{e}$ in Corollary 8.6.1 is satisfied under the common assumption of deterministic processing. We can apply Corollary 8.6.1 to obtain the approximation

$$\begin{aligned}\{D_n(t) : t \geq 0\} &\approx \mu_n t - n^{1/2} \mathbf{L}(t/n) \\ &\approx \mu_n t - n^{1/2} \alpha t/n - n^{1/2} m^{1/2} \sigma_L \mathbf{B}(t/nm) \:. \end{aligned} \quad (6.14)$$

where

$$\{m^{1/2} \mathbf{B}(t/m) : t \geq 0\} \stackrel{\mathrm{d}}{=} \{\mathbf{B}(t) : t \geq 0\} \:.$$

As above, we may fix n by choing ρ_n to match the traffic intensity in the given queueing system.

8.7. Reflected Fractional Brownian Motion

In this section we discuss heavy-traffic stochastic-process limits in which the limit process for the scaled net-input process does *not* have independent increments. Specifically, we consider heavy-traffic limits in which the scaled workload processes converge to reflected fractional Brownian motion (RFBM). With time scaling, such limits arise because of strong dependence. As in the previous section, the RFBM limit occurs in a second stochastic-process limit.

8.7.1. An Increasing Number of Sources

We consider the same on-off model introduced in Section 8.2, but now we let the number of sources become large before we do the heavy-traffic scaling. (We also discuss limits in which the number of sources grows in Sections 8.3.3 and 9.8.) When we let the number of sources become large, we can apply the central limit theorem for processes in Section 7.2 to obtain convergence to a Gaussian process. Then, again following Taqqu, Willinger and Sherman (1997), we can let the busy-period and idle-period distributions have heavy tails, and scale time, to obtain a stochastic-process limit for the Gaussian process in which the limit process is FBM. The continuous-mapping approach with the reflection map then yields convergence of the scaled workload processes to RFBM. This heavy-traffic limit supplements and supports direct modeling using FBM, as in Norros (1994, 2000).

With the on-off model, heavy-tailed busy-period and idle-period distributions cause the scaling exponent H to exceed $1/2$. Specifically, as indicated in (2.20) in

Section 7.2,
$$H = (3 - \alpha_{min})/2, \tag{7.1}$$
where α_{min} is the minimum of the idle-period and busy-period ccdf decay rates if both have power tails with decay rates $\alpha_i < 2$. Otherwise, it is the decay rate of the one ccdf with a power tail if only one has a power tail. If both the busy-period and the idle-period have finite variance, then $\alpha_{min} = 2$ and $H = 1/2$. The following result is an immediate consequence of Theorem 7.2.5 and the continuous-mapping approach.

Theorem 8.7.1. (RFBM limit for the on-off model with many sources) *Consider a family of fluid-queue models indexed by the parameter pair (n, τ). Let all the models be based on a single on-off source model as specified before Theorem 7.2.5. In model (n, τ) there are n IID on-off sources and time scaling by τ. Let the processing in model (n, τ) be at a constant deterministic rate*
$$\mu_{n,\tau} \equiv \lambda n\tau + cn^{1/2}\tau^H, \tag{7.2}$$
where $\lambda \equiv m_1/(m_1 + m_2)$ is the single-source input rate. Let the capacity in model (n, τ) be $K_{n,\tau} \equiv n^{1/2}\tau^H K$. Let $W_{n,\tau}(0) = 0$. Then
$$(\mathbf{X}_{n,\tau}, \mathbf{W}_{n,\tau}) \Rightarrow (\sigma_{lim}\mathbf{Z}_H - c\mathbf{e}, \phi_K(\sigma_{lim}\mathbf{Z}_H - c\mathbf{e})) \tag{7.3}$$
in $(C, U)^2$ as first $n \to \infty$ and then $\tau \to \infty$, where H is in (7.1), σ_{lim}^2 is in (2.17) or (2.18) in Chapter 4, $\mathbf{Z}_H$ is standard FBM, $\mathbf{X}_{n,\tau}$ is the scaled net-input process
$$\mathbf{X}_{n,\tau} \equiv \tau^{-H} n^{-1/2} (\sum_{i=1}^{n} C_i(\tau t) - \mu_{n,\tau} t), \quad t \geq 0, \tag{7.4}$$
and $\mathbf{W}_{n,\tau}$ is the scaled workload process, i.e.,
$$\mathbf{W}_{n,\tau}(t) \equiv \phi_K(\mathbf{X}_{n,\tau}). \tag{7.5}$$

Remark 8.7.1. *Double limits.* Theorems 8.5.1 and 8.7.1 establish heavy-traffic limits with *different* limit processes for the *same* fluid queue model with multiple on-off sources having heavy-tailed busy and idle periods. Theorem 8.5.1 establishes convergence to RSLM with a fixed number of sources, whereas Theorem 8.7.1 establishes convergence to RFBM when the number of sources is sent to infinity before the time (and space) scaling is performed. As indicated in Section 7.2.2, Mikosch et al. (2001) establish more general double limits that provide additional insight. Then RFBM (RSLM) is obtained if the number of sources increases sufficiently quickly (slowly) in the double limit. ∎

8.7.2. Gaussian Input

When looking at traffic data from communication networks, we do not directly see the source busy and idle periods of the on-off model. Instead, we see an irregular stream of packets. Given such a packet stream, we must estimate source busy and idle periods if we are to use the on-off source traffic model. When we do fit busy and idle periods to traffic data, we may find that the busy and idle periods do not nearly have the independence assumed in Sections 5.7 and 8.5. Thus we might elect not to use the on-off model.

Instead, we might directly analyze the aggregate cumulative-input process. Measurements of the aggregate cumulative-input process $\{C(t) : t \geq 0\}$ may reveal strong positive dependence. From the data, we may fairly conclude that the variance of $C(t)$ is finite, but that it grows rapidly with t. In particular, we may see asymptotic growth of the variance as a function of t according to a power as

$$Var(C(t)) \sim t^{2H} \quad \text{as} \quad t \to \infty \tag{7.6}$$

for

$$1/2 < H < 1 , \tag{7.7}$$

which indicates strong positive dependence, as we saw in Section 4.6.

Given the asymptotic growth rate of the variance in (7.6), it is natural to look for a FCLT capturing strong positive dependence with light tails. From Section 4.6, it is natural to anticipate that properly scaled cumulative-input processes should converge to fractional Brownian motion (FBM). In particular, letting the input rate be 1 as before, it is natural to anticipate that

$$\mathbf{C}_n \Rightarrow \sigma \mathbf{Z}_H \quad \text{in} \quad (C, U) \quad \text{as} \quad n \to \infty , \tag{7.8}$$

where

$$\mathbf{C}_n(t) \equiv n^{-H}(C(nt) - nt), \quad t \geq 0 , \tag{7.9}$$

the process $\mathbf{Z}_H$ is standard FBM, as characterized by (6.13) or (6.14) in Section 4.6, and σ is a positive scaling constant. Since both $\mathbf{C}_n$ and $\mathbf{Z}_H$ have continuous sample paths, we can work in the function space C.

However, the principal theorems in Section 4.6, Theorems 4.6.1 and 4.6.2, do *not* directly imply the stochastic-process limit in (7.8), because there are extra structural conditions beyond the variance asymptotics in (7.6). In order to apply the theorems in Section 4.6, we need additional structure – either linear structure or Gaussian structure.

Fortunately, it is often reasonable to assume additional Gaussian structure in order to obtain convergence to FBM. With communication networks, it is often natural to regard the aggregate cumulative-input stochastic process as a Gaussian process, because the aggregate cumulative-input process is usually the superposition of a large number of component cumulative-input processes associated with different sources, which may be regarded as approximately independent. The intended activities of different users can usually be regarded as approximately independent. However, network controls in response to lost packets such as contained in TCP can induce dependence among sources. Hence approximate independence needs to be checked.

Not only are sources approximately independent, but the individual source input over bounded intervals is usually bounded because of a limited access rate. Thus we may apply Theorem 7.2.1 to justify approximating the aggregate cumulative-input process as a Gaussian process, just as we did for Theorem 8.7.1. Then there is a strong theoretical basis for approximating the workload process by reflected FBM without assuming that we have on-off sources.

We summarize by stating the theorem. Define additional random elements of C by

$$\mathbf{X}_n(t) \equiv n^{-H}(C(nt) - \mu_n nt) = \mathbf{C}_n(t) + n^{1-H}(1 - \mu_n)t \; ,$$
$$\mathbf{W}_n(t) \equiv \phi_K(\mathbf{X}_n)(t), \quad t \geq 0 \; . \tag{7.10}$$

We apply Theorems 5.4.1 and 4.6.2 to obtain the following result. We apply Theorem 7.2.1 to justify the Gaussian assumption.

Theorem 8.7.2. (RFBM limit with strongly-dependent Gaussian input) *Consider a sequence of fluid queues indexed by n with capacities K_n, $0 < K_n \leq \infty$, and output rates μ_n, $n \geq 1$. Suppose that $\{C(t) - t : t \geq 0\}$ is a zero-mean Gaussian process with*

$$Var(C(t)) \sim \sigma t^{2H} \quad as \quad t \to \infty \tag{7.11}$$

and

$$Var(C(t)) \leq Mt^{2H} \quad for\ all \quad t > 0 \tag{7.12}$$

for some positive constants H, σ and M with $1/2 \leq H < 1$. If, in addition, $K_n = n^H K$, $0 < K \leq \infty$, $0 \leq W_n(0) \leq K_n$ for all n,

$$n^{-H} W_n(0) \Rightarrow y \quad in \quad \mathbb{R} \quad as \quad n \to \infty \tag{7.13}$$

and

$$n^{1-H}(1 - \mu_n) \to \eta \quad as \quad n \to \infty \tag{7.14}$$

for $-\infty < \eta < \infty$, then

$$(\mathbf{C}_n, \mathbf{X}_n, \mathbf{W}_n) \Rightarrow (\sigma \mathbf{Z}_H, y + \sigma \mathbf{Z}_H + \eta \mathbf{e}, \phi_K(y + \sigma \mathbf{Z}_H + \eta \mathbf{e})) \tag{7.15}$$

in $(C, U)^3$, where $\mathbf{Z}_H$ is standard FBM with parameter H and $(\mathbf{C}_n, \mathbf{X}_n, \mathbf{W}_n)$ is in (7.9) and (7.10).

Unfortunately, the limit process RFBM is relatively intractable; see Norros (2000) for a discussion. The asymptotic behavior of first passage times to high levels has been characterized by Zeevi and Glynn (2000). We discuss approximations for the steady-state distribution in the next section.

Of course, it may happen that traffic measurements indicate, not only that the aggregate cumulative-input process fails to have independent increments, but also that the aggregate cumulative-input process is not nearly Gaussian. Nevertheless, the FBM approximation might be reasonable after scaling. The FBM approximation would be supported by the theory in Section 4.6 if the cumulative-input process had the linear structure described in Section 4.6. With different structure, there might be a relevant stochastic-process limit to a different limit process. An invariance principle like that associated with Donsker's theorem evidently does not hold in the strongly-dependent case. Thus, careful analysis in specific settings may lead to different approximating processes.

Remark 8.7.2. *Bad news and good news.* From an applied perspective, the FBM heavy-traffic stochastic-process limit provides both bad news and good news. The main bad news is the greater congestion associated with greater space scaling as H increases. Part of the bad news also is the fact that the limit process is relatively difficult to analyze. Moreover, as discussed in Section 5.5, the greater time scaling as H increases means that significant relative changes in the process take longer to occur. That is part of the bad news if we are concerned about recovery from a large congestion event; see Duffield and Whitt (1997).

The good news is the greater possibility of *real-time prediction* of future behavior based on observations of the system up to the present time. As discussed in Section 4.6, the dependence of the increments in FBM provides a basis for exploiting the history, beyond the present state, to predict the future. For further discussion about predicting congestion in queues, see Duffield and Whitt (1997, 1998, 2000), Srikant and Whitt (2001), Ward and Whitt (2000) and Whitt (1999a,b). ∎

8.8. Reflected Gaussian Processes

In the last section we established convergence to reflected fractional Brownian motion (RFBM) for workload processes in fluid-queue models. In this section we consider approximations that can be obtained for the steady-state distributions of RFBM and more general stationary Gaussian processes when we have one-sided reflection. To do so, we exploit a lower bound due to Norros (1994), which has been found to be often an excellent approximation; e.g. see Addie and Zuckerman (1994) and Choe and Shroff (1998, 1999).

First let X be a general stochastic process with stationary increments defined on the entire real line $(-\infty, \infty)$ with $X(0) = 0$. Then the (one-sided) reflection of X is

$$\phi(X)(t) \equiv X(t) - \inf_{0 \leq s \leq t} X(s)$$
$$= \sup_{0 \leq s \leq t} \{X(t) - X(s)\} \stackrel{d}{=} \sup_{0 \leq s \leq t} \{-X(-s)\}. \quad (8.1)$$

Assuming that

$$\sup_{0 \leq s \leq t} \{-X(s)\} \to \sup_{s \geq 0} \{-X(-s)\} < \infty \quad \text{w.p.1} \quad \text{as} \quad t \to \infty, \quad (8.2)$$

$$\phi(X)(t) \Rightarrow \phi(X)(\infty) \stackrel{d}{=} \sup_{s \geq 0} \{-X(-s)\} \quad \text{as} \quad t \to \infty. \quad (8.3)$$

A finite limit in (8.2) will hold when the process X has negative drift, i.e., when $EX(t) = -mt$ for some $m > 0$ and X is ergodic.

We propose approximating the steady-state tail probability $P(\phi(X)(\infty) > x)$ by a lower bound obtained by interchanging the probability and the supremum, i.e.,

$$P(\phi(X)(\infty) > x) = P\left(\sup_{t \geq 0}\{-X(-t)\} > x\right)$$
$$\geq \sup_{t \geq 0} P(-X(-t) > x) . \qquad (8.4)$$

Assuming that $X(t) \to -\infty$ as $t \to \infty$, we have $-X(-t) \to -\infty$ as $t \to \infty$. Hence $P(-X(-t) > x)$ will be small for both small t and large t, so it is natural to anticipate that there is an intermediate value yielding the maximum in (8.4). Moreover, large deviation arguments can be developed to show that the lower bound is asymptotically correct, in a logarithmic sense, as $x \to \infty$ under regularity conditions; see Duffield and O'Connell (1995), Botvich and Duffield (1995) Choe and Shroff (1998, 1999), Norros (2000) and Wischik (2001b). The moderate-deviations limit by Wischik (2001b) is especially insightful because it applies, not just to the Gaussian process, but also to superposition processes converging to Gaussing processes (in a CLT for processes).

In general, the lower bound may not get us very far, because it tends to be intractable. However, if we assume that X is also a Gaussian process, then we can conveniently evaluate the lower bound. For a Gaussian process X, we can calculate the lower bound in (8.4) Once we find the optimum t, t^*, the probability $P(-X(-t^*) > x)$ is Gaussian.

We can find t^* by transforming the variables to variables with zero means. In particular, let

$$Z(t) = \frac{-X(t) - E[-X(-t)]}{x - E[-X(-t)]}, \quad t \geq 0 , \qquad (8.5)$$

and note that $Z(t)$ has mean 0 (assuming that $E[-X(-t)] \leq 0$) and

$$Z(t) \geq 1 \quad \text{if and only if} \quad -X(t) > x . \qquad (8.6)$$

Hence, the optimum t^* for the lower bound in (8.4) is the t^* maximizing the variance of $Z(t)$, where

$$Var\, Z(t) = \frac{Var\, X(-t)}{(x - E[-X(-t)])^2} . \qquad (8.7)$$

Given the mean and covariance function of the Gaussian process X, the variance $Var\, Z(t)$ in (8.7) is computable. The final approximation is thus

$$P(\phi(X)(\infty) > x) \approx P(-X(-t^*) > x)$$
$$= \Phi^c([x - E[-X(-t^*)]]/\sqrt{Var\, X(-t^*)})$$
$$= \Phi^c(1/\sqrt{Var\, Z(t^*)}) \qquad (8.8)$$

where $\Phi^c(x) \equiv 1 - \Phi(x)$ is the standard normal ccdf.

The approximation can be applied to any stationary Gaussian process with negative drift. From the last section, we are especially interested in the case in which

X is FBM. So suppose that
$$X(t) \equiv \sigma \mathbf{Z}_H(t) + \eta t, \quad t \geq 0 ,$$
where $\eta < 0$ and $\mathbf{Z}_H$ is standard FBM. Since $Var(X(t)) = \sigma^2 t^{2H}$, the variance of $Z(t)$ in (8.5) is mazimized for
$$t^* = xH/|\eta|(1-H).$$
Consequently, the desired lower bound is
$$P(\phi(X)(\infty) > x) \geq \Phi^c(\sigma^{-1}(|\eta|/H)^H (x/(1-H))^{1-H}) , \qquad (8.9)$$
where Φ^c is the standard normal ccdf. Using the approximation
$$\Phi^c(x) \approx e^{-x^2/2} ,$$
we obtain the approximation
$$P(\phi(X)(\infty) > x) \approx e^{-\gamma x^{2(1-H)}} , \qquad (8.10)$$
where
$$\gamma \equiv \frac{1}{2\sigma^2}\Big(\frac{|\eta|}{H}\Big)^{2H}\Big(\frac{1}{1-H}\Big)^{2(1-H)} . \qquad (8.11)$$

For $H > 1/2$, approximation (8.10) is a Weibull distribution with relatively heavy tail. Thus the lower-bound distribution has a Weibull tail. Note that for $H = 1/2$ approximation (8.10) agrees with the exact value for RBM in Theorem 5.7.2. Additional theoretical support for approximation (8.10) and asymptotic refinements are contained in Narayan (1998), Hüsler and Piterbarg (1999) and Massoulie and Simonian (1999); see Section 4.5 of Norros (2000). They show that there is an additional prefactor $Kx^{-\gamma}$ in (8.10) as $x \to \infty$ for $\gamma = (1-H)(2H-1)/H$. The exponent in approximation (8.10) was shown to be asymptotically correct as $x \to \infty$ by Duffield and O'Connell in their large-deviations limit.

9
Single-Server Queues

9.1. Introduction

In this chapter we continue applying the continuous-mapping approach to establish heavy-traffic stochastic-process limits for stochastic processes of interest in queueing models, but now we consider the standard single-server queue, which has unlimited waiting space and the first-come first-served service discipline. That model is closely related to the infinite-capacity fluid queue considered in Chapter 5, but instead of continuous divisible fluid, customers with random service requirements arrive at random arrival times. Thus, attention here is naturally focused on integer-valued stochastic processes counting the number of arrivals, the number of departures and the number of customers in the queue.

Here is how this chapter is organized: We start in Section 9.2 by defining the basic stochastic processes in the single-server queue. Then in Section 9.3 we establish general heavy-traffic stochastic-process limits for a sequence of single-server queue models. Included among the general heavy-traffic limits for queues in Section 9.3 are limits for departure processes, so the displayed heavy-traffic limits for one queue immediately imply associated heavy-traffic limits for single-server queues in series. However, in general the limit processes for departure processes are complicated, so that the heavy-traffic limits are more useful for single queues.

In Section 9.4 and 9.5 we obtain FCLT's for arrival processes that are superpositions or splittings of other arrival processes. Along with the limits for departure processes, these results make the heavy-traffic FCLT's in Section 9.3 applicable to general acyclic open networks of single-server queues. Corresponding (more complicated) heavy-traffic limits for general single-class open networks of single-server queues (allowing feedback) are obtained in Chapter 14 by applying the multidimensional reflection map, again with variants of the M_1 topology to treat limit processes with discontinuous sample paths.

In Section 9.6, we amplify the discussion in Section 5.7 by discussing the reflected-Brownian-motion (RBM) limit that commonly occurs in the light-tailed weak-dependent case and the associated RBM approximation that stems from it. We show how the useful functions in Chapter 13 can be applied again to yield the initial FCLTs for the arrival and service processes (required for the heavy-traffic limits) in more detailed models, e.g., where the input is from a multiclass batch-renewal process with class-dependent service times or a Markov-modulated point process.

We discuss the special case of very heavy tails in Section 9.7. When the service-time ccdf decays like $x^{-\alpha}$ as $x \to \infty$ for $0 < \alpha < 1$, the service-time mean is infinite. The queueing processes then fail to have proper steady-state distributions. The heavy-traffic stochastic-process limits are useful to describe how the queueing processes grow. Very large values then tend to be reached by jumps. The heavy-traffic limits yield useful approximations for the distributions of the time a high level is first crossed and the positions before and after that high level are crossed.

In Section 9.8 we extend the discussion in Section 8.7 by establishing heavy-traffic stochastic-process limits for single-server queues with superposition arrival processes, when the number of component arrival processes in the superposition increases in the limit. When the number of component arrival processes increases in the limit with the total rate held fixed, burstiness greater than that of a Poisson process in the component arrival processes tends to be dissipated because the superposition process approaches a Poisson process. For example, even with heavy-tailed interarrival times, the superposition process may satisfy a FCLT with a limit process having continuous sample paths. On the other hand, the limit process tends to be more complicated because it fails to have independent increments.

Finally, in Section 9.9 we discuss heuristic parametric-decomposition approximations for open queueing networks. In these approximations, arrival and service processes are each partially characterized by two parameters, one describing the rate and the other describing the variability. We show how the heavy-traffic stochastic-process limits can be used to help determine appropriate variability parameters.

9.2. The Standard Single-Server Queue

In this section we define the basic stochastic processes in the standard single-server queue, going beyond the introduction in Section 6.4.1. In this model, there is a single server and unlimited waiting space. Successive customers with random service requirements arrive at random arrival times. Upon arrival, customers wait in queue until it is their turn to receive service. Service is provided to the customers on a first-come first-served basis. After the customers enter service, they receive service without interruption and then depart.

The model can be specified by a sequence $\{(U_k, V_k) : k \geq 1\}$ of ordered pairs of nonnegative random variables. The variable U_k represents the interarrival time between customers k and $k-1$, while the variable V_k represents the service time of customer k. To fully specify the system, we also need to specify the initial conditions. For simplicity, we will assume that the first customer arrives at time $U_1 \geq 0$ to find an empty system. It is easy to extend the heavy-traffic limits to cover other initial conditions, as was done in Chapter 5.

The model here is similar to the fluid-queue model in Chapter 5, but the differences lead us to consider different processes, for which we use different notation. The main descriptive quantities of interest here are: W_k, the waiting time until beginning service for customer k; $L(t)$, the workload (in unfinished service time facing the server) at time t; $Q(t)$, the queue length (number in system, including the one in service, if any) at time t; Q_k^A, the queue length just before the kth arrival; and Q_k^D, the queue length just after the kth departure. The workload $L(t)$ is also the waiting time of a potential or "virtual" arrival at time t; thus the workload $L(t)$ is also called the virtual waiting time.

The waiting time of the kth customer, W_k, can be expressed in terms of the waiting time of the previous customer, W_{k-1}, the service time of the previous customer, V_{k-1}, and the interarrival time between customers $k-1$ and k, U_k, by the classical *Lindley recursion*; i.e., we can make the definition

$$W_k \equiv [W_{k-1} + V_{k-1} - U_k]^+, \quad k \geq 2 , \tag{2.1}$$

where $[x]^+ = \max\{x, 0\}$ and $W_1 = 0$. We can apply mathematical induction to show that W_k can be expressed in terms of appropriate partial sums by a discrete-analog of the reflection map in (5.4) in Chapter 13 and in (2.5) below.

Theorem 9.2.1. *The waiting times satisfy*

$$W_k = S_k - \min\{S_j : 0 \leq j \leq k\}, \quad k \geq 0 , \tag{2.2}$$

where

$$S_k \equiv S_{k-1}^v - S_k^u, \quad S_k^v \equiv V_1 + \cdots + V_k \quad \text{and} \quad S_k^u \equiv U_1 + \cdots + U_k, \quad k \geq 1 ,$$

with $S_0 \equiv S_0^v \equiv V_0 \equiv S_0^u \equiv U_0 \equiv 0$.

Note that the indices in S_k^v in the definition of S_k are offset by 1. Nevertheless, S_k is the kth partial sum

$$S_k = X_1 + \cdots + X_k, \quad k \geq 1 ,$$

where

$$X_i = V_{i-1} - U_i, \quad i \geq 1 ,$$

with $V_0 = 0$. Note that $W_1 = W_0 = 0$ with our definition, because $S_0 \equiv 0$ and $S_1 \leq 0$ since $V_0 \equiv 0$.

We define the arrival counting process by letting

$$A(t) \equiv \max\{k \geq 0 : S_k^u \leq t\}, \quad t \geq 0 . \tag{2.3}$$

Since S_k^u is the arrival time of customer k, $A(t)$ counts the number of arrivals in the interval $[0, t]$. We use the arrival process A to define the *cumulative-input process*. The cumulative input of work in the interval in $[0, t]$ is the sum of the service times of all arrivals in $[0, t]$, i.e., so the cumulative input can be defined as the random sum

$$C(t) \equiv S_{A(t)}^v \equiv \sum_{i=1}^{A(t)} V_i, \quad t \geq 0 . \tag{2.4}$$

The associated *net-input process* is

$$X(t) \equiv C(t) - t, \quad t \geq 0 .$$

As in the fluid queue, the workload is the one-sided reflection of X, i.e.,

$$L(t) \equiv \phi(X)(t) \equiv X(t) - \inf_{0 \leq s \leq t}\{X(s) \wedge 0\}, \quad t \geq 0 ; \tag{2.5}$$

i.e., ϕ is the reflection map in (2.5) and (2.6) in Chapter 8 and in (5.4) of Chapter 13. The workload process in the single-server queue coincides with the workload process in the fluid queue with cumulative-input process C and deterministic processing rate 1.

Since the cumulative-input process here is a pure-jump process, the server is working if and only if the workload is positive. Thus, the *cumulative busy time* of the server is easy to express in terms of C and L, in particular,

$$B(t) \equiv C(t) - L(t), \quad t \geq 0 . \tag{2.6}$$

Equivalently, the cumulative idle time in $[0, t]$ is the lower-boundary regulator function associated with the reflection map, i.e.,

$$I(t) \equiv \psi_L(X)(t) \equiv -\inf_{0 \leq s \leq t}\{X(s) \wedge 0\}, \quad t \geq 0 , \tag{2.7}$$

and the cumulative busy time is

$$B(t) = t - I(t), \quad t \geq 0 . \tag{2.8}$$

Paralleling the definition of the arrival counting process A in (2.3), define a counting process associated with the service times by letting

$$N(t) \equiv \max\{k \geq 0 : S_k^v \leq t\}, \quad t \geq 0 . \tag{2.9}$$

Following our treatment of the waiting times and workload, we would like to think of the queue-length process $\{Q(t) : t \geq 0\}$ as the reflection of an appropriate "net-input" process. However, that is not possible in general. When the service times come from a sequence of IID exponential random variables, independent of the arrival process, we can exploit the lack of memory property of the exponential distribution to conclude that the queue-length process is distributed the same as the reflection of the process $\{A(t) - N'(t) : t \geq 0\}$, where $\{N'(t) : t \geq 0\}$ is a

Poisson process counting "potential" service times. However, more generally, we do not have such a direct reflection representation, so we will have to work harder.

Let $D(t)$ count the number of departures in $[0,t]$. The *departure process* can be defined by

$$D(t) \equiv N(B(t)), \quad t \geq 0 . \tag{2.10}$$

We then can define the *queue-length process* by

$$Q(t) \equiv A(t) - D(t), \quad t \geq 0 , \tag{2.11}$$

because we have stipulated that the first arrival finds an empty system.

Let D_k^A be the time of the kth departure (the departure time of the kth arrival). We can use the inverse relation for counting processes and associated partial sums to define the *departure-time sequence* $\{D_k^A : k \geq 1\}$ in terms of the departure process $\{D(t) : t \geq 0\}$, i.e.,

$$D_k^A \equiv \inf\{t \geq 0 : D(t) \geq k\}, \quad k \geq 1 . \tag{2.12}$$

However, it is convenient to start with another expression for D_k^A. First, let T_k^A be the *service-start time* of customer k, with $T_0^A \equiv 0$. Since a customer must arrive and wait before starting service,

$$T_k^A \equiv S_k^u + W_k, \quad k \geq 1 . \tag{2.13}$$

Since service is not interrupted,

$$D_k^A = T_k^A + V_k, \quad k \geq 1 . \tag{2.14}$$

From Theorem 9.2.1 and (2.14), we obtain the following.

Corollary 9.2.1. (departure time representation) *The departure times satisfy*

$$D_k^A = S_k^v - \min_{0 \leq j \leq k}\{S_{j-1}^v - S_j^u\}, \quad k \geq 1 . \tag{2.15}$$

We next define the continuous-time *service start-time process* $\{T(t) : t \geq 0\}$ by letting

$$T(t) \equiv \max\{k \geq 0 : T_k^A \leq t\}, \quad t \geq 0 . \tag{2.16}$$

Given the continuous-time process $\{Q(t) : t \geq 0\}$, we can define the sequences $\{Q_k^A : k \geq 1\}$ and $\{Q_k^D : k \geq 1\}$ as *embedded sequences*, i.e.,

$$\begin{aligned} Q_k^A &\equiv Q(S_k^u-) \\ Q_k^D &\equiv Q(D_k^A) . \end{aligned} \tag{2.17}$$

Note that the definitions in (2.17) make Q_k^A the queue length before *all arrivals* at arrival epoch S_k^u, and Q_k^D is the queue length *after all departures* at departure epoch D_k^A. (Other definitions are possible, for Q_k^A if there are 0 interarrival times, and for Q_k^D if there are 0 service times.)

So far, we have not introduced any probabilistic assumptions. The standard assumption is that $\{U_k : k \geq 1\}$ and $\{V_k : k \geq 1\}$ are independent sequences of IID random variables with general distributions, in which case the counting processes A

and N are called renewal processes. Then, with the Kendall notation, the queueing model is called GI/GI/1, with GI denoting independence (I) with general distributions (G). The first GI refers to the interarrival times, while the second refers to the service times. The final "1" indicates a single server. Unlimited waiting space and the FCFS service discipline are understood.

If in addition one of the distributions is exponential, deterministic, Erlang of order k (convolution of k IID exponentials) or hyperexponential of order k (mixture of k exponentials), then GI is replaced by M, D, E_k and H_k, respectively. Thus the $M/E_k/1$ model has a Poisson arrival process (associated with exponential interarrival times) with Erlang service times, while the $H_k/M/1$ model has a renewal arrival process with hyperexponential interarrival times and exponential service times. An attractive feature of the heavy-traffic limits is that they do not depend critically on the distributional assumptions or even the IID assumptions associated with the GI/GI/1 model.

9.3. Heavy-Traffic Limits

We now establish heavy-traffic stochastic-process limits for the stochastic processes in the stable single-server queue. We can obtain fluid limits (FLLN's) just as in Section 5.3, but we omit them. We go directly to the heavy-traffic limits for stable queues, as in Section 5.4.

9.3.1. The Scaled Processes

As in Section 5.4, we introduce a sequence of queueing models indexed by n. In model n, $U_{n,k}$ is the interarrival time between customers k and $k-1$, and $V_{n,k}$ is the service time of customer k. The partial sums for model n are $S_{n,k}$, $S_{n,k}^v$ and $S_{n,k}^u$, defined just as in Theorem 9.2.1 with $S_{n,0} \equiv S_{n,0}^u \equiv S_{n,0}^v \equiv 0$ for all n. The other stochastic processes are defined just as in Section 9.2, with an extra subscript n to indicate the model number.

We convert the initial model data as represented via the partial sums $S_{n,k}^u$ and $S_{n,k}^v$ into two sequences of random elements of $D \equiv D([0,\infty), \mathbb{R})$ by introducing translation and scaling, i.e., by letting

$$\begin{aligned} \mathbf{S}_n^u(t) &\equiv c_n^{-1}[S_{n,\lfloor nt \rfloor}^u - \lambda_n^{-1} nt], \\ \mathbf{S}_n^v(t) &\equiv c_n^{-1}[S_{n,\lfloor nt \rfloor}^v - \mu_n^{-1} nt], \quad t \geq 0, \end{aligned} \quad (3.1)$$

where λ_n, μ_n and c_n are positive constants and $\lfloor x \rfloor$ is the greatest integer less than or equal to x. We think of λ_n and μ_n in (3.1) as the arrival rate and service rate.

Since the indices of S_k^v are shifted by one, we also form the associated modification of $\mathbf{S}_n^v$ above by setting

$$\bar{\mathbf{S}}_n^v(t) \equiv c_n^{-1}[S_{n,\lfloor nt \rfloor -1}^v - \mu_n^{-1} nt], \quad t \geq 0, \quad (3.2)$$

where $S_{n,-1}^v \equiv 0$. (Recall that $S_{n,0}^v \equiv 0$ too.)

9.3. Heavy-Traffic Limits

We then define associated random elements of D induced by the partial sums $S_{n,k}$, waiting times $W_{n,k}$, service-start times $T^A_{n,k}$ and departure times $D^A_{n,k}$ by letting

$$\begin{align}
\mathbf{S}_n(t) &\equiv c_n^{-1} S_{n,\lfloor nt \rfloor} = (\bar{\mathbf{S}}^v_n - \mathbf{S}^u_n)(t) , \\
\mathbf{W}_n(t) &\equiv c_n^{-1} W_{n,\lfloor nt \rfloor} , \\
\mathbf{T}^A_n(t) &\equiv c_n^{-1} [T^A_{n,\lfloor nt \rfloor} - \lambda_n^{-1} nt] , \\
\mathbf{D}^A_n(t) &\equiv c_n^{-1} [D^A_{n,\lfloor nt \rfloor} - \lambda_n^{-1} nt], \quad t \geq 0 ,
\end{align} \quad (3.3)$$

where

$$S^v_{n,-1} \equiv S^v_{n,0} \equiv S^u_{n,0} \equiv W_{n,0} \equiv T^A_{n,0} \equiv D^A_{n,0} \equiv 0 .$$

We next define normalized random elements of D induced by the associated continuous-time processes by letting

$$\begin{align}
\mathbf{A}_n(t) &\equiv c_n^{-1}[A_n(nt) - \lambda_n nt] , \\
\mathbf{N}_n(t) &\equiv c_n^{-1}[N_n(nt) - \mu_n nt] , \\
\mathbf{C}_n(t) &\equiv c_n^{-1}[C_n(nt) - \lambda_n \mu_n^{-1} nt] , \\
\mathbf{X}_n(t) &\equiv c_n^{-1} X_n(nt) , \\
\mathbf{L}_n(t) &\equiv c_n^{-1} L_n(nt) , \\
\mathbf{B}_n(t) &\equiv c_n^{-1}[B_n(nt) - nt] , \\
\mathbf{T}_n(t) &\equiv c_n^{-1}[T_n(nt) - \lambda_n nt] , \\
\mathbf{D}_n(t) &\equiv c_n^{-1}[D_n(nt) - \lambda_n nt] , \\
\mathbf{Q}_n(t) &\equiv c_n^{-1} Q_n(nt), \quad t \geq 0 .
\end{align} \quad (3.4)$$

Finally, we define two sequences of random functions induced by the queue lengths at arrival epochs and departure epochs by letting

$$\begin{align}
\mathbf{Q}^A_n(t) &\equiv c_n^{-1} Q^A_{n,\lfloor nt \rfloor} \\
\mathbf{Q}^D_n(t) &\equiv c_n^{-1} Q^D_{n,\lfloor nt \rfloor}, \quad t \geq 0 ,
\end{align} \quad (3.5)$$

where $Q^A_{n,0} \equiv Q^D_{n,0} \equiv 0$.

Notice that there are no translation terms in $\mathbf{S}_n$ and $\mathbf{W}_n$ in (3.3) or in $\mathbf{X}_n$ and $\mathbf{L}_n$ in (3.4). Thus we can apply the continuous-mapping theorem with the reflection map in (2.5) to directly obtain some initial results.

Theorem 9.3.1. (single-server-queue heavy-traffic limits directly from the reflection map) *Consider the sequence of single-server queues with the random elements in (3.3) and (3.4).*

(a) if

$$\mathbf{S}_n \Rightarrow \mathbf{S} \quad in \quad D$$

with the topology J_1 or M_1, then

$$\mathbf{W}_n \Rightarrow \phi(\mathbf{S}) \quad in \quad D$$

with the same topology, where ϕ is the reflection map in (2.5).

(b) If
$$\mathbf{X}_n \Rightarrow \mathbf{X} \quad in \quad D$$
with the topology J_1 or M_1, then
$$\mathbf{L}_n \Rightarrow \phi(\mathbf{X}) \quad in \quad D$$
with the same topology.

Proof. It follows from Theorem 9.2.1 that
$$\mathbf{W}_n = \phi(\mathbf{S}_n), \quad n \geq 1 \, ,$$
for the reflection map $\phi : D \to D$ in (2.5) and $\mathbf{S}_n$ and $\mathbf{W}_n$ in (3.3). Similarly, it follow from (2.5) that
$$\mathbf{L}_n = \phi(\mathbf{X}_n), \quad n \geq 1$$
for $\mathbf{X}_n$ and $\mathbf{L}_n$ in (3.4). Hence the stated results follow directly from the simple continuous mapping theorem, Theorem 3.4.1, because the reflection map is continuous by Theorem 13.5.1. ∎

Remark 9.3.1. *Strong and weak topologies on D^2.* Let SJ_1 and SM_1 denote the strong or standard J_1 and M_1 topologies on the product space D^k, and let WJ_1 and WM_1 denote the associated weak or product topologies on D^k. Given the limit $\mathbf{S}_n \Rightarrow \mathbf{S}$ in (D, J_1) assumed in Theorem 9.3.1 (a), we obtain the joint convergence
$$(\mathbf{S}_n, \mathbf{W}_n) \Rightarrow (\mathbf{S}, \phi(\mathbf{S})) \quad in \quad D([0, \infty), \mathbb{R}^2, SJ_1) \, . \tag{3.6}$$
However, given the same limit in (D, M_1), we do not obtain the analog of (3.6) in (D^2, SM_1). Example (14.5.1) shows that the map taking x into $(x, \phi(x))$ is *not* continuous when the range has the SM_1 topology. Hence, we use the WM_1 topology on the product space D^k. ∎

We now want to obtain limits for the random elements in (3.3), starting from convergence of the pair $(\mathbf{S}_n^u, \mathbf{S}_n^v)$ in (3.1). We start by establishing limits for the discrete-time processes.

9.3.2. Discrete-Time Processes

Before stating limits for the discrete-time processes, we establish conditions under which the two scaled service processes $\mathbf{S}_n^v$ and $\bar{\mathbf{S}}_n^v$ are asymptotically equivalent.

Theorem 9.3.2. (asymptotic equivalence of the scaled service processes) *If either $\mathbf{S}_n^v \Rightarrow \mathbf{S}^v$ or $\bar{\mathbf{S}}_n^v \Rightarrow \mathbf{S}$ in $D([0, \infty), \mathbb{R}, \mathcal{T})$, where $\mathcal{T}$ is the topology J_1, M_1 or M_2, then*
$$d_{J_1}(\mathbf{S}_n^v, \bar{\mathbf{S}}_n^v) \Rightarrow 0 \quad in \quad D([0, \infty), \mathbb{R}) \tag{3.7}$$
for d_{J_1} in (3.2) in Section 3.3 and
$$(\mathbf{S}_n^v, \bar{\mathbf{S}}_n^v) \Rightarrow (\mathbf{S}^v, \mathbf{S}^v) \quad in \quad D^2 \tag{3.8}$$

with the product-$\mathcal{T}$ topology.

Proof. Assume that $\mathbf{S}_n^v \Rightarrow \mathbf{S}^v$. (The argument is essentially the same starting with $\bar{\mathbf{S}}_n^v \Rightarrow \mathbf{S}^v$.) Use the Skorohod representation theorem to replace convergence in distribution by convergence w.p.1. By Section 12.4, the assumed convergence implies local uniform convergence at continuity points. Let t be such that $P(t \in Disc(\mathbf{S}^v)) = 0$, which holds for all but at most countably many t. By the right continuity at 0 for functions in D, the local uniform convergence holds at 0 and t. We now define homeomorphisms of $[0,t]$ needed for J_1 convergence in $D([0,t],\mathbb{R})$: Let $\nu_n : [0,t] \to [0,t]$ be defined by $\nu_n(0) = 0$, $\nu_n(t) = t$, $\nu_n(n^{-1}) = 2n^{-1}$ and $\nu_n(t - 2n^{-1}) = t - n^{-1}$ with ν_n defined by linear interpolation elsewhere. Let $\|\cdot\|_t$ be the uniform norm on $D([0,t],\mathbb{R})$. Since $\|\nu_n - \mathbf{e}\|_t = n^{-1}$ and

$$\begin{aligned}\|\bar{\mathbf{S}}_n^v \circ \nu_n - \mathbf{S}_n^v\|_t &\leq 2\sup\{|\mathbf{S}_n^v(s)| : 0 \leq s \leq 2n^{-1}\} \\ &\quad + 2\sup\{|\mathbf{S}_n^v(s)| : t - n^{-1} \leq s \leq t + n^{-1}\} \\ &\to 0 \quad \text{as} \quad n \to \infty ,\end{aligned}$$

the limit in (3.7) holds in $D([0,t],\mathbb{R})$. Since such limits hold for a sequence $\{t_n\}$ with $t_n \to \infty$, we have (3.7), which implies (3.8) by Theorem 11.4.7. ∎

We apply Theorem 9.3.2 to establish heavy-traffic limits for the discrete time processes.

Theorem 9.3.3. (heavy-traffic limits starting from arrival times and service times) *Suppose that*

$$(\mathbf{S}_n^u, \mathbf{S}_n^v) \Rightarrow (\mathbf{S}^u, \mathbf{S}^v) \quad in \quad D^2 , \tag{3.9}$$

where the topology is either WJ_1 or WM_1, $\mathbf{S}_n^u$ and $\mathbf{S}_n^v$ are defined in (3.1),

$$P(Disc(\mathbf{S}^u) \cap Disc(\mathbf{S}^v) = \phi) = 1 , \tag{3.10}$$

$c_n \to \infty$, $n/c_n \to \infty$, $\lambda_n^{-1} \to \lambda^{-1}$, $0 < \lambda^{-1} < \infty$, *and*

$$\eta_n \equiv n(\mu_n^{-1} - \lambda_n^{-1})/c_n \to \eta \quad as \quad n \to \infty . \tag{3.11}$$

(a) If the topology in (3.9) is WJ_1, then

$$(\mathbf{S}_n^u, \bar{\mathbf{S}}_n^v, \mathbf{S}_n, \mathbf{W}_n, \mathbf{T}_n^A) \Rightarrow (\mathbf{S}^u, \mathbf{S}^v, \mathbf{S}, \mathbf{W}, \mathbf{T}^A) \tag{3.12}$$

in $D([0,\infty), \mathbb{R}^5, SJ_1)$, where

$$\mathbf{S} \equiv \mathbf{S}^v - \mathbf{S}^u + \eta \mathbf{e}, \quad \mathbf{W} = \phi(\mathbf{S}) \quad and \quad \mathbf{T}^A = \delta(\mathbf{S}^u, \mathbf{S}^v + \eta \mathbf{e}) , \tag{3.13}$$

with ϕ being the reflection map and $\delta : D \times D \to D$ defined by

$$\delta(x_1, x_2) = x_2 + (x_1 - x_2)^\uparrow , \tag{3.14}$$

where $x^\uparrow$ is the supremum of x defined by

$$x^\uparrow(t) \equiv \sup_{0 \leq s \leq t} x(s), \quad t \geq 0 . \tag{3.15}$$

Then the limit processes $\mathbf{S}^u$, $\mathbf{S}^v$ and $\mathbf{T}^A$ have no negative jumps.

(b) If the topology in (3.9) above is WM_1, then the limit in (3.12) holds in (D^5, WM_1), with the limit processes being as in (3.13).

Proof. We start by invoking the Skorohod representation theorem, Theorem 3.2.2, to replace the convergence in distribution in (3.9) by convergence w.p.1 for special versions. For simplicity, we do not introduce extra notation to refer to the special versions. By Theorem 9.3.2, we obtain convergence

$$(\mathbf{S}_n^u, \bar{\mathbf{S}}_n^v) \Rightarrow (\mathbf{S}^u, \mathbf{S}^v) \qquad (3.16)$$

from the initial limit in (3.9), with the same topology on D^2. We then apply condition (3.10) to strengthen the convergence to be in $D([0,\infty), \mathbb{R}^2, S\mathcal{T}_1)$, drawing upon Section 12.6. Let t be any time point in $(0, \infty)$ for which

$$P(t \in Disc(\mathbf{S}^u) \cup Disc(\mathbf{S}^v)) = 0 \ .$$

There necessarily exists infinitely many such t in any bounded interval. We thus have convergence of the restrictions of $(\mathbf{S}_n^u, \bar{\mathbf{S}}_n^v)$ in $D([0,t,], \mathbb{R}^2, S\mathcal{T}_1)$, for which we again use the same notation.

(a) First, suppose that $S\mathcal{T}_1 = SJ_1$. By the definition of SJ_1 convergence, we can find increasing homeomorphisms ν_n of $[0, t]$ such that

$$\|(\mathbf{S}_n^u, \bar{\mathbf{S}}_n^v) - (\mathbf{S}^u, \mathbf{S}^v) \circ \nu_n\|_t \to 0 \quad w.p.1 \ ,$$

where $\|\cdot\|_t$ is the uniform norm on $[0,t]$. Since

$$\begin{aligned}
\mathbf{S}_n &= \bar{\mathbf{S}}_n^v - \mathbf{S}_n^u + \eta_n \mathbf{e} \ , \\
\mathbf{W}_n &= \phi(\mathbf{S}_n) \ , \\
\mathbf{T}_n^A &= \delta(\mathbf{S}_n^u, \bar{\mathbf{S}}_n^v + \eta_n \mathbf{e})
\end{aligned}$$

where ϕ is the reflection map and δ is the map in (3.14), both regarded as maps from $D([0,t], \mathbb{R})$ or $D([0,t], \mathbb{R})^2$ to $D([0,t], \mathbb{R})$, which are easily seen to be Lipschitz continuous, first with respect to the uniform metric and then for d_{J_1}, it follows that

$$\|(\mathbf{S}_n^u, \bar{\mathbf{S}}_n^v, \mathbf{S}_n, \mathbf{W}_n, \mathbf{T}_n^A) - (\mathbf{S}^u, \mathbf{S}^v, \mathbf{S}, \mathbf{W}, \mathbf{T}) \circ \nu_n\|_t \to 0 \text{ w.p.1} \ ,$$

for $\mathbf{S}$, $\mathbf{W}$ and $\mathbf{T}^A$ in (3.13), so that

$$(\mathbf{S}_n^u, \bar{\mathbf{S}}_n^v, \mathbf{S}_n, \mathbf{W}_n, \mathbf{T}_n^A) \to (\mathbf{S}^u, \mathbf{S}^v, \mathbf{S}, \mathbf{W}, \mathbf{T}^A) \quad \text{in} \quad D([0, \infty), \mathbb{R}^5, SJ_1)$$

w.p.1 as claimed. Next, let J_t^+ and J_t^- be the maximum-positive-jump and maximum-negative-jump functions over $[0,t]$, i.e.,

$$J_t^+(x) \equiv \sup_{0 \leq s \leq t} \{x(s) - x(s-)\} \qquad (3.17)$$

and

$$J_t^-(x) \equiv -\inf_{0 \leq s \leq t} \{x(s) - x(s-)\} \ . \qquad (3.18)$$

If the topology is J_1, then the functions J_t^+ and J_t^- are continuous at all $x \in D$ for which $t \notin Disc(x)$. Since $\mathbf{S}_n^u$ and $\bar{\mathbf{S}}_n^v$ have no negative jumps, then neither do $\mathbf{S}^u$

and $\mathbf{S}^v$ if the topology limit is J_1. For any $x \in D$, $x^\uparrow \equiv \sup_{0 \le s \le t}\{x(s)\}$, $t \ge 0$, has no negative jumps. Thus, since

$$\mathbf{T}^A = \mathbf{S}^v + \eta e + (\mathbf{S}^u - \mathbf{S}^v - \eta \mathbf{e})^\uparrow \; ,$$

$\mathbf{T}^A$ also has no negative jumps when the topology is J_1.

(b) Suppose that the topology on D^2 for the convergence in (3.16) is $ST_1 = SM_1$, after applying condition (3.10) to strengthen the mode of convergence from WM_1. Then we can apply the continuous maps to get the limit (3.12) with the WM_1 topology. We need the SM_1 topology on the domain in order for δ in (3.14) to be continuous. As indicated in Remark 9.3.1, unlike with the J_1 topology, we need the weaker WM_1 topology on the range product space D^k. ■

Remark 9.3.2. *Alternative conditions.* In Theorem 9.3.3 we only use condition (3.10) to strengthen the mode of convergence in (3.9) and (3.16) to the strong topology from the weak product topology. Thus, instead of condition (3.10), we could assume that (3.9) holds with the strong topology. That could hold without (3.10) holding.

Moreover, to obtain limits for $\mathbf{S}_n$ and $\mathbf{W}_n$ with the M_1 topology, instead of (3.10), we could assume that the two limit processes $\mathbf{S}^u$ and $\mathbf{S}^v$ have no common discontinuities of common sign. Then addition is continuous by virtue of Theorem 12.7.3. However, then extra conditions would be needed to establish limits for $\mathbf{T}_n$ in (3.12) and $\mathbf{D}_n$ in Theorem 9.3.4 below. ■

9.3.3. Continuous-Time Processes

We now establish limits for the normalized continuous-time processes in (3.4) and the embedded queue-length processes in (3.5). Now we need the M_1 topology to treat stochastic-process limits with discontinuous sample paths, because we must go from partial sums to counting processes. The limits for departure processes imply limits for queues in series and contribute to establishing limits for acyclic networks of queues.

Theorem 9.3.4. *(heavy-traffic limits for continuous-time processes) Suppose that, in addition to the conditions of Theorem 9.3.3 (with the topology in (3.9) being either WJ_1 or WM_1),*

$$P(\mathbf{S}^u(0) = 0) = P(\mathbf{S}^v(0) = 0) = 1 \; . \tag{3.19}$$

Then

$$(\mathbf{A}_n, \mathbf{N}_n, \mathbf{C}_n, \mathbf{X}_n, \mathbf{L}_n, \mathbf{B}_n, \mathbf{Q}_n, \mathbf{Q}_n^A, \mathbf{Q}_n^D, \mathbf{T}_n, \mathbf{D}_n, \mathbf{D}_n^A)$$
$$\Rightarrow (\mathbf{A}, \mathbf{N}, \mathbf{C}, \mathbf{X}, \mathbf{L}, \mathbf{B}, \mathbf{Q}, \mathbf{Q}^A, \mathbf{Q}^A, \mathbf{T}, \mathbf{D}, \mathbf{D}^A) \quad in \; (D^{12}, WM_1)$$

jointly with the limits in (3.12), where

$$\mathbf{A} \equiv -\lambda \mathbf{S}^u \circ \lambda \mathbf{e}, \quad \mathbf{N} \equiv -\lambda \mathbf{S}^v \circ \lambda \mathbf{e} \; ,$$
$$\mathbf{C} \equiv (\mathbf{S}^v - \mathbf{S}^u) \circ \lambda \mathbf{e}, \quad \mathbf{X} \equiv \mathbf{S} \circ \lambda \mathbf{e} \; ,$$
$$\mathbf{L} \equiv \phi(\mathbf{X}) = \mathbf{W} \circ \lambda \mathbf{e}, \quad \mathbf{B} \equiv \mathbf{X}^\downarrow = \mathbf{S}^\downarrow \circ \lambda \mathbf{e} \; ,$$

298   9. Single-Server Queues

$$\mathbf{Q} \equiv \lambda \mathbf{L}, \quad \mathbf{Q}^A \equiv \mathbf{Q} \circ \lambda^{-1}\mathbf{e} = \lambda \mathbf{W}, \quad \mathbf{T} \equiv -\lambda \mathbf{T}^A \circ \lambda \mathbf{e}$$
$$\mathbf{D} \equiv \hat{\delta}(\mathbf{A}, \mathbf{N} - \lambda^2 c\mathbf{e}), \quad \mathbf{D}^A = -\lambda^{-1}\mathbf{D} \circ \lambda^{-1}\mathbf{e} \quad (3.20)$$

where $\hat{\delta} : D \times D \to D$ is defined by

$$\hat{\delta}(x_1, x_2) \equiv x_2 + (x_1 - x_2)^{\downarrow} . \quad (3.21)$$

Proof. Again we start with the Skorohod representation theorem, Theorem 3.2.2, to replace convergence in distribution with convergence w.p.1 for the associated special versions, without introducing new notation for the special versions. By exploiting the convergence preservation of the inverse map with centering in Theorem 13.7.1 as applied to counting functions in Section 13.8, we obtain $(\mathbf{A}_n, \mathbf{N}_n) \to (\mathbf{A}, \mathbf{N})$ in (D^2, WM_1) for $(\mathbf{A}_n, \mathbf{N}_n)$ in (3.1) and $(\mathbf{A}, \mathbf{N})$ in (3.20). (We use condition (3.19) at this point.) Since $\mathbf{C}_n$ involves a random sum, we apply composition with translation as in Corollary 13.3.2 to get its limit. In particular, note that

$$\mathbf{C}_n(t) = \mathbf{S}_n^v \circ \hat{\mathbf{A}}_n(t) + \mu_n^{-1}\mathbf{A}_n(t)$$

for $\mathbf{C}_n$ and $\mathbf{A}_n$ in (3.4), where

$$\hat{\mathbf{A}}_n(t) \equiv n^{-1}A_n(nt), \quad t \geq 0 .$$

Since $\mathbf{A}_n \to \mathbf{A}$, $\hat{\mathbf{A}}_n \to \lambda \mathbf{e}$. Condition (3.10) and the form of $\mathbf{A}$ in (3.20) implies that

$$P(Disc(\mathbf{A}) \cap Disc(\mathbf{S}^v \circ \lambda \mathbf{e}) = \phi) = 1 . \quad (3.22)$$

Thus we can apply Corollary 13.3.2 to get

$$\mathbf{C}_n \to \mathbf{S}^v \circ \lambda \mathbf{e} + \mu^{-1}(-\lambda \mathbf{S}^u \circ \lambda \mathbf{e}) = (\mathbf{S}^v - \mathbf{S}^u) \circ \lambda \mathbf{e} .$$

Since

$$\mathbf{X}_n(t) = \mathbf{C}_n(t) + nt(\lambda_n \mu_n^{-1} - 1)/c_n$$

and condition (3.11) holds,

$$\mathbf{X}_n \to \mathbf{C} + \lambda \eta \mathbf{e} = \mathbf{S} \circ \lambda \mathbf{e} \quad \text{in} \quad (D, M_1) .$$

Since $\mathbf{L}_n = \phi(\mathbf{X}_n)$, we can apply the reflection map again to treat $\mathbf{L}_n$. By (2.7) and (2.8), $\mathbf{B}_n = \mathbf{X}_n^{\downarrow}$, where $x^{\downarrow} \equiv -(-x)^{\uparrow}$ and $x^{\uparrow}$ is the supremum map. Hence we can apply the supremum map to establish the convergence of $\mathbf{B}_n$.

The argument for the queue-length process is somewhat more complicated. The idea is to represent the random function $\mathbf{Q}_n$ as the image of the reflection map applied to an appropriate function. It turns out that we can write

$$\mathbf{Q}_n = \phi(\mathbf{A}_n - \mathbf{N}_n \circ \hat{\mathbf{B}}_n + \lambda_n \mu_n \eta_n \mathbf{e}) , \quad (3.23)$$

where

$$\hat{\mathbf{B}}_n(t) \equiv n^{-1}B(nt), \quad t \geq 0 , \quad (3.24)$$

and
$$\hat{\mathbf{B}}_n \to \mathbf{e} \quad \text{in} \quad D \quad \text{w.p.1} \tag{3.25}$$
since $\mathbf{B}_n \Rightarrow \mathbf{B}$. We can write (3.23) because
$$\begin{aligned} Q_n(t) &= A_n(t) - N_n(B_n(t)) \\ &= A_n(t) - N_n(B_n(t)) - \mu_n[t - B_n(t)] + \mu_n[t - B_n(t)] \\ &= \phi(A_n - N_n \circ B_n - \mu_n[e - B_n])(t) \ . \end{aligned} \tag{3.26}$$

The second line of (3.26) is obtained by adding and subtracting $\mu_n[t - B_n(t)]$. The third line holds because the resulting expression is equivalent to the reflection representation since $\mu_n[t - B_n(t)]$, being μ_n times the cumulative idle time in $[0,t]$, is necessarily nondecreasing and increases only when $Q_n(t) = 0$. (See Theorem 14.2.3 for more on this point.) When we introduce the scaling in the random functions, the third line of (3.26) becomes (3.23), because

$$\begin{aligned} &(\mathbf{A}_n - \mathbf{N}_n \circ \hat{\mathbf{B}}_n + \lambda_n \mu_n \eta_n \mathbf{e})(t) \\ &= c_n^{-1}(A_n(nt) - \lambda_n nt - N_n(B_n(nt)) + \mu_n B_n(nt) + (\lambda_n - \mu_n)nt) \\ &= c_n^{-1}(A_n - N_n \circ B_n - \mu_n[e - B_n])(nt), \quad t \geq 0 \ , \end{aligned} \tag{3.27}$$

and $\phi(cx \circ be) = c\phi(x) \circ be$ for all $b > 0$ and $c > 0$. We have already noted that $(\mathbf{A}_n, \mathbf{N}_n) \to (\mathbf{A}, \mathbf{N})$ in (D^2, WM_1). By (3.25) and Theorem 11.4.5, we have
$$(\mathbf{A}_n, \mathbf{N}_n, \hat{\mathbf{B}}_n) \to (\mathbf{A}, \mathbf{N}, \mathbf{e}) \quad \text{in} \quad (D^3, WM_1) \ . \tag{3.28}$$

Given (3.28), we can apply composition with Theorem 13.2.3 to get
$$(\mathbf{A}_n, \mathbf{N}_n \circ \hat{\mathbf{B}}_n) \to (\mathbf{A}, \mathbf{N}) \quad \text{in} \quad (D^2, WM_1) \ . \tag{3.29}$$

By condition (3.10) and the form of $\mathbf{A}$ and $\mathbf{N}$ in (3.20),
$$P(Disc(\mathbf{A}) \cap Disc(\mathbf{N}) = \phi) = 1 \ . \tag{3.30}$$

Hence the mode of convergence in (3.28) and (3.29) can be strengthened to SM_1. Thus, we can apply the subtraction map to get
$$\mathbf{A}_n - \mathbf{N}_n \circ \hat{\mathbf{B}}_n \to \mathbf{A} - \mathbf{N} \quad \text{in} \quad (D, M_1) \ . \tag{3.31}$$

Combining (3.23) and (3.31), we obtain
$$\mathbf{Q}_n \to \phi(\mathbf{A} - \mathbf{N} + \lambda^2 \eta \mathbf{e}) = \mathbf{Q} \quad \text{in} \quad (D, M_1) \ . \tag{3.32}$$

Next, to treat the departure processes, note that $\mathbf{D}_n = \mathbf{A}_n - \mathbf{Q}_n$. By (3.23),
$$\begin{aligned} \mathbf{D}_n &= \mathbf{N}_n \circ \hat{\mathbf{B}}_n - \lambda_n \mu_n \eta_n \mathbf{e} + (\mathbf{A}_n - \mathbf{N}_n \circ \hat{\mathbf{B}}_n + \lambda_n \mu_n \eta_n \mathbf{e})^{\downarrow} \\ &= \hat{\delta}(\mathbf{A}_n, \mathbf{N}_n \circ \hat{\mathbf{B}}_n - \lambda_n \mu_n \eta_n \mathbf{e}) \end{aligned} \tag{3.33}$$

for $\hat{\delta}$ in (3.21). Since $\hat{\delta}$ is continuous as a map from (D^2, SM_1) to (D, M_1),
$$\mathbf{D}_n \to \hat{\delta}(\mathbf{A}, \mathbf{N} - \lambda^2 \eta \mathbf{e}) \quad \text{in} \quad (D, M_1) \ .$$

We then apply the convergence-preservation property of the inverse map with centering in the context of counting functions to obtain the limit for $\mathbf{D}_n^A$ from (3.33).

We can apply the composition map to treat $\mathbf{Q}_n^A$ and $\mathbf{Q}_n^D$. First, as a consequence of (3.9) and (3.11), since $n/c_n \to \infty$, we have

$$\hat{\mathbf{S}}_n^A \to \lambda^{-1}\mathbf{e} \quad \text{and} \quad \hat{\mathbf{D}}_n^A \to \lambda^{-1}\mathbf{e} \tag{3.34}$$

for

$$\hat{\mathbf{S}}_n^A(t) \equiv n^{-1} S_{n,\lfloor nt \rfloor}^u \quad \text{and} \quad \hat{\mathbf{D}}_n^A(t) \equiv n^{-1} D_{n,\lfloor nt \rfloor}^A . \tag{3.35}$$

Applying Theorem 11.4.5 with (3.32) and (3.34), we obtain

$$(\mathbf{Q}_n, \hat{\mathbf{S}}_n^A, \hat{\mathbf{D}}_n^D) \to (\mathbf{Q}, \lambda^{-1}\mathbf{e}, \lambda^{-1}\mathbf{e}) \quad \text{in} \quad (D^3, WM_1)$$

and, then applying Theorem 13.2.3, we obtain

$$\mathbf{Q}_n^A = \mathbf{Q}_n \circ \hat{\mathbf{S}}_n^A \to \mathbf{Q} \circ \lambda^{-1}\mathbf{e} \quad \text{and} \quad \mathbf{Q}_n^D = \mathbf{Q}_n \circ \hat{\mathbf{D}}_n^A \to \mathbf{Q} \circ \lambda^{-1}\mathbf{e}$$

in (D, M_1). Finally, limits for the normalized continuous-time service-start-time processes $\mathbf{T}_n$ follow by applying the inverse map with centering as applied to counting functions to the previous limits for $\mathbf{T}_n^A$, just as we obtained limits for $\mathbf{A}_n$ and $\mathbf{N}_n$ starting from $\mathbf{S}_n^u$ and $\mathbf{S}_n^v$. ∎

Remark 9.3.3. *Impossibility of improving from M_1 to J_1.* When the limit processes have discontinuous sample paths, the M_1 mode of convergence in Theorem 9.3.4 cannot be improved to J_1. First, it is known that $\mathbf{S}_n^u \Rightarrow \mathbf{S}^u$ and $\mathbf{A}_n \Rightarrow -\lambda \mathbf{S}^u \circ \lambda e$ both hold in (D, J_1) if and only if $P(\mathbf{S}^u \in C) = 1$; see Lemma 13.7.1. In the special case of identical deterministic service times, the processes $\mathbf{C}_n$, $\mathbf{X}_n$, $\mathbf{L}_n$ and $\mathbf{Q}_n$ are simple functions of $\mathbf{A}_n$, so their convergence also cannot be strengthened to J_1. Similarly, the limits $\mathbf{T}_n^A \Rightarrow \mathbf{T}^A$ and $\mathbf{T}_n \to -\lambda \mathbf{T} \circ \lambda e$ cannot both hold in J_1.

Similarly, we cannot have convergence

$$(\mathbf{T}_n^A, \mathbf{D}_n^A) \to (\mathbf{T}^A, \mathbf{T}^A)$$

in $D([0,\infty), \mathbb{R}^2, SJ_1)$ if $\mathbf{S}^v$ has discontinuities, because that would imply that

$$\mathbf{D}_n^A - \mathbf{T}_n^A \to \mathbf{0} \quad \text{in} \quad (D, J_1) . \tag{3.36}$$

The limit (3.36) is a contradiction because $\mathbf{S}_n^v \to \mathbf{S}^v$ in (D, J_1) implies that

$$J_t(\mathbf{S}_n^v) \equiv \sup_{0 \leq s \leq t} \{|\mathbf{S}_n^v(s) - \mathbf{S}_n^v(s-)|\} = J_t(\mathbf{D}_n^A - \mathbf{T}_n^A)$$

$$= \sup_{0 \leq s \leq t} \{c_n^{-1} V_{n,\lfloor ns \rfloor}\} \to J_t(\mathbf{S}^v)$$

for any t such that $P(t \in Disc(\mathbf{S}^v)) = 0$, and $P(J_t(\mathbf{S}^v) = 0) < 1$ for all sufficiently large t if $\mathbf{S}^v$ fails to have continuous sample paths. ∎

It is natural to choose the measuring units so that the mean service time is 1. Then $\lambda = \mu = \mu_n = 1$ for all n. When $\lambda = 1$, the limit processes $\mathbf{W}$, $\mathbf{L}$, $\mathbf{Q}$ and $\mathbf{Q}^A$ all coincide; they all become $\phi(\mathbf{S})$, the reflection of $\mathbf{S}$. We observed the coincidence of $\mathbf{W}$ and $\mathbf{Q}$ when $\lambda = 1$ in Chapter 6.

The discussion about heavy-traffic scaling in Section 5.5 applies here as well. There are slight differences because (3.11) differs from (4.6) in Section 5.4. If $c_n = n^H$ for $0 < H < 1$ and $\eta < 0$, then just as in (5.10) in Section 5.5, we obtain

$$n = (\zeta/(1-\rho))^{1/(1-H)} , \tag{3.37}$$

but now

$$\zeta = -\eta\lambda > 0 . \tag{3.38}$$

(Now λ^{-1} plays the role of μ before.)

Remark 9.3.4. *From queue lengths to waiting times.* We conclude this section by mentioning some supplementary material in the Internet Supplement. In Section 5.4 of the Internet Supplement, drawing upon Puhalskii (1994), we show how heavy-traffic limits for workload and waiting-time processes can be obtained directly from associated heavy-traffic limits for arrival, departure and queue-length processes. These results apply the FCLT for inverse processes with nonlinear centering in Section 13.7. Following Puhalskii (1994), we apply the result to establish a limit for a single-server queue in a central-server model (i.e., a special closed queueing network) as the number of customers in the network increases. In that setting it is not easy to verify the conditions in the earlier limit theorems for waiting times and the workload because the arrival and service processes are state-dependent. That result has also been applied by Mandelbaum, Massey, Reiman and Stolyar (1999). ∎

9.4. Superposition Arrival Processes

In Section 8.3 we established heavy-traffic stochastic-process limits for a fluid queue with input from multiple sources. We now establish analogous heavy-traffic stochastic-process limits for the standard single-server queue with arrivals from multiple sources. We use the inverse map with centering (Sections 13.7 and 13.8) to relate the arrival-time sequences to the arrival counting processes.

With multiple sources, the arrival process in the single-server queue is the superposition of m component arrival processes, i.e.,

$$A(t) \equiv A_1(t) + \cdots + A_m(t), \quad t \geq 0 , \tag{4.1}$$

where $\{A_i(t) : t \geq 0\}$ is the ith component arrival counting process with associated arrival times (partial sums)

$$S_{i,k}^u \equiv \inf\{t \geq 0 : A_i(t) > k\}, \quad k \geq 0 , \tag{4.2}$$

and interarrival times

$$U_{i,k} \equiv S_{i,k}^u - S_{i,k-1}^u, \quad k \geq 1 . \tag{4.3}$$

We extend the previous limit theorems in Section 9.3 to this setting by showing how limits for the m partial-sum sequences $\{S_{i,k}^u : k \geq 1\}$, $1 \leq i \leq m$, imply limits

302   9. Single-Server Queues

for the overall partial-sum sequence $\{S_k^u : k \geq 1\}$, where

$$S_k^u \equiv \inf\{t \geq 0 : A(t) \geq k\}, \quad k \geq 0 . \tag{4.4}$$

As in Section 9.3, we consider a sequence of models indexed by n; e.g., let $S_{n,i,k}^u$ be the kth partial sum (arrival time of customer k) in the ith component arrival process of model n. Let the random elements of D be defined by

$$\begin{aligned}
\mathbf{S}_{n,i}^u(t) &\equiv c_n^{-1}[S_{n,i,\lfloor nt \rfloor}^u - \lambda_{n,i}^{-1} nt] \\
\mathbf{A}_{n,i}(t) &\equiv c_n^{-1}[A_{n,i}(nt) - \lambda_{n,i} nt] \\
\mathbf{S}_n^u(t) &\equiv c_n^{-1}[S_{n,\lfloor nt \rfloor}^u - \lambda_n^{-1} nt] \\
\mathbf{A}_n(t) &\equiv c_n^{-1}[A_n(nt) - \lambda_n nt], \quad t \geq 0 ,
\end{aligned} \tag{4.5}$$

for $n \geq 1$. The M_1 topology plays an important role when the limit processes can have discontinuous sample paths.

Theorem 9.4.1. (FCLT for superposition arrival processes) *Suppose that*

$$(\mathbf{S}_{n,1}^u, \ldots, \mathbf{S}_{n,m}^u) \Rightarrow (\mathbf{S}_1^u, \ldots, \mathbf{S}_m^u) \quad \text{in} \quad (D^m, WM_1) , \tag{4.6}$$

where $\mathbf{S}_{n,i}^u$ is defined in (4.5),

$$P(Disc(\mathbf{S}_i^u \circ \lambda_i \mathbf{e}) \cap Disc(\mathbf{S}_j^u \circ \lambda_j \mathbf{e}) = \phi) = 1 \tag{4.7}$$

for all i, j with $1 \leq i, j \leq m$ and $i \neq j$, and

$$P(\mathbf{S}_i^u(0) = 0) = 1, \quad 1 \leq i \leq m . \tag{4.8}$$

If, in addition $c_n \to \infty$, $n/c_n \to \infty$ and $\lambda_{n,i} \to \lambda_i$, $0 < \lambda_i < \infty$, for $1 \leq i \leq m$, then

$$\begin{aligned}
(\mathbf{S}_{n,1}^u, \ldots, \mathbf{S}_{n,m}^u, \mathbf{A}_{n,1}, \ldots, \mathbf{A}_{n,m}, \mathbf{A}_n, \mathbf{S}_n^u) \\
\Rightarrow (\mathbf{S}_1^u, \ldots, \mathbf{S}_m^u, \mathbf{A}_1, \ldots, \mathbf{A}_m, \mathbf{A}, \mathbf{S}^u)
\end{aligned} \tag{4.9}$$

in (D^{2m+2}, WM_1), where

$$\lambda_n \equiv \lambda_{n,1} + \cdots + \lambda_{n,m} ,$$

$$\begin{aligned}
\mathbf{A}_i &\equiv -\lambda_i \mathbf{S}_i^u \circ \lambda_i \mathbf{e}, \quad \mathbf{A} \equiv \mathbf{A}_1 + \cdots + \mathbf{A}_m \\
\mathbf{S}^u &\equiv -\lambda^{-1} \mathbf{A} \circ \lambda^{-1} \mathbf{e} = \sum_{i=1}^m \gamma_i \mathbf{S}_i^u \circ \gamma_i \mathbf{e}
\end{aligned} \tag{4.10}$$

for

$$\lambda \equiv \lambda_1 + \cdots + \lambda_m \quad \text{and} \quad \gamma_i \equiv \lambda_i/\lambda, \quad 1 \leq i \leq m . \tag{4.11}$$

Proof. We apply the Skorohod representation theorem, Theorem 3.2.2, to replace the convergence in distribution in (4.6) by convergence w.p.1 for special versions. We then apply the convergence-preservation results for the inverse map with centering, as applied to counting functions, in Corollary 13.8.1 to obtain, first, the limits for $\mathbf{A}_{n,i}$ from the limits for $\mathbf{S}_{n,i}^u$ and, second, the limit for $\mathbf{S}_n^u$ from the limit for $\mathbf{A}_n$.

We use addition with condition (4.7) to obtain the convergence of $\mathbf{A}_n$ from the convergence of $(\mathbf{A}_{n,1}, \ldots, \mathbf{A}_{n,m})$. ∎

Remark 9.4.1. *The case of IID Lévy processes.* Suppose that the limit processes $\mathbf{S}_1^u, \ldots, \mathbf{S}_m^u$ in Theorem 9.4.1 are IID Lévy processes. Then $\gamma_i = m^{-1}$,

$$\mathbf{A} \stackrel{\mathrm{d}}{=} \mathbf{A}_1 \circ m\mathbf{e} \quad \text{and} \quad \sum_{i=1}^{m} \mathbf{S}_i^u \stackrel{\mathrm{d}}{=} \mathbf{S}_1^u \circ m\mathbf{e} ,$$

so that

$$\mathbf{S}^u \stackrel{\mathrm{d}}{=} m^{-1} \mathbf{S}_1^u .$$

In this case, $\mathbf{A}$ and $\mathbf{S}$ differ from $\mathbf{A}_1$ and $\mathbf{S}_1^u$ only by the deterministic scale factor m.

We can remove the deterministic scale factor by rescaling to make the overall arrival rate independent of m. We can do that for any given m by replacing $A_i(t)$ by $A_i(t/m)$ for $t \geq 0$ or, equivalently, by replacing $S_{i,k}^u$ by $m S_{i,k}^u$ for $k \geq 0$. If we make that scale change at the outset, then the limit processes $\mathbf{A}$ and $\mathbf{S}^u$ become independent of m. However, we cannot draw that conclusion if the limit processes $\mathbf{S}_1^u, \ldots, \mathbf{S}_m^u$ are not Lévy processes. For further discussion, see Section 5.6 and Remarks 10.2.2 and 10.2.4. ∎

We can combine Theorems 9.3.3, 9.3.4 and 9.4.1 to obtain a heavy-traffic limit for queues with superposition arrival processes.

Theorem 9.4.2. (heavy-traffic limit for a queue with a superposition arrival process) *Suppose that*

$$(\mathbf{S}_{n,1}^u, \ldots, \mathbf{S}_{n,m}^u, \mathbf{S}_n^v) \Rightarrow (\mathbf{S}_1^u, \ldots, \mathbf{S}_m^u, \mathbf{S}^v) \quad \text{in} \quad (D^{m+1}, WM_1) \tag{4.12}$$

for $\mathbf{S}_{n,i}^u$ in (4.5) and $\mathbf{S}_n^v$ in (3.1), where

$$P(Disc(\mathbf{S}_i^u \circ \gamma_i \mathbf{e}) \cap Disc(\mathbf{S}_j^u \circ \gamma_j \mathbf{e})) = \phi) = 1 \tag{4.13}$$

and

$$P(Disc(\mathbf{S}_i^u \circ \gamma_i \mathbf{e}) \cap Disc(\mathbf{S}^v) = \phi) = 1 \tag{4.14}$$

for all i, j with $1 \leq i, j \leq m$, $i \neq j$ and γ_i in (4.11). Suppose that, for $1 \leq i \leq m$,

$$P(\mathbf{S}_i^u(0) = 0) = P(\mathbf{S}^v(0) = 0) = 1 , \tag{4.15}$$

$c_n \to \infty$, $n/c_n \to \infty$, $\lambda_{n,i}^{-1} \to \lambda_i^{-1}$, $0 < \lambda_i^{-1} < \infty$, *and*

$$\eta_n \equiv n(\mu_n^{-1} - \lambda_n^{-1})c_n \to \eta \quad as \quad n \to \infty \tag{4.16}$$

for λ_n in (1.10). Then the conditions and conclusions of Theorems 9.3.3 and 9.3.4 hold with $\mathbf{S}^u$ and $\mathbf{A}$ in (4.10) and λ in (4.11).

Proof. As usual, start by applying the Skorohod representation theorem to replace convergence in distribution by convergence w.p.1, without introducing special notation for the special versions. Then conditions (4.12)–(4.15) plus Theorem 9.4.1

imply that conditions (3.9) and (3.10) in Theorem 9.3.3 and condition (3.19) in Theorem 9.3.4 hold. Thus the conditions of Theorems 9.3.3 and 9.3.4 hold with $\mathbf{S}^u$ and $\mathbf{A}$ in (4.10) and λ in (4.11). ∎

We now show what Theorem 9.4.2 yields in the standard light-tailed weak-dependent case. The following results closely parallels Theorem 8.4.1.

Corollary 9.4.1. (the Brownian case) *Suppose that the conditions of Theorem 9.4.2 hold with $c_n = \sqrt{n}$, $\mathbf{S}_i^u = \sigma_{u,i} \mathbf{B}_i^u$, $1 \le i \le m$, and $\mathbf{S}^v = \sigma_v \mathbf{B}^v$, where $\mathbf{B}_1^u, \dots, \mathbf{B}_m^u, \mathbf{B}^v$ are $m+1$ IID standard Brownian motions. Then the conclusions of Theorem 9.4.2 hold with*

$$\mathbf{S}^u \stackrel{\mathrm{d}}{=} \sigma_u \mathbf{B} \tag{4.17}$$

for

$$\sigma_u^2 = \sum_{i=1}^{m} \gamma_i^3 \sigma_{u,i}^2 \tag{4.18}$$

and

$$\mathbf{S} \stackrel{\mathrm{d}}{=} \sigma_S \mathbf{B} + \eta \mathbf{e} \tag{4.19}$$

for η in (4.16),

$$\sigma_S^2 = \sigma_u^2 + \sigma_v^2 , \tag{4.20}$$

and $\mathbf{B}$ being a standard Brownian motion.

A corresponding corollary is easy to establish in the heavy-tailed case, when the limits are scaled versions of independent stable Lévy motions. For the IID case, using essentially a single model, we apply Theorem 4.5.3. Since the random variables $U_{n,i,k}$ and $V_{n,k}$ are nonnegative, we get totally skewed stable Lévy motion limits (with $\beta = 1$) for $\mathbf{S}_i^u$ and $\mathbf{S}^v$.

Corollary 9.4.2. (the stable-Lévy-motion case) *Suppose that the conditions of Theorem 9.4.2 hold with the limit processes $\mathbf{S}_i^u$, $1 \le i \le m$, and $\mathbf{S}^v$ being mutually independent stable Lévy motions with index α, $1 < \alpha < 2$, where*

$$\mathbf{S}_i^u(1) \stackrel{\mathrm{d}}{=} \sigma_{u,i} S_\alpha(1,1,0), \quad 1 \le i \le m , \tag{4.21}$$

and

$$\mathbf{S}^v(1) \stackrel{\mathrm{d}}{=} \sigma_v S_\alpha(1,1,0) . \tag{4.22}$$

Then the conclusions of Theorem 9.4.2 hold with $\mathbf{S}^u$ and $\mathbf{S}$ being stable Lévy motions with index α, where

$$\mathbf{S}^u(1) \stackrel{\mathrm{d}}{=} \sigma_u S_\alpha(1,1,0) \tag{4.23}$$

for

$$\sigma_u = \left(\sum_{i=1}^{m} \gamma_i^{\alpha+1} \right)^{1/\alpha}$$

and
$$\mathbf{S}(1) \stackrel{\mathrm{d}}{=} S_\alpha(\sigma, \beta, 0) , \qquad (4.24)$$
where
$$\sigma = (\sigma_v^\alpha + \sigma_u^\alpha)^{1/\alpha}$$
and
$$\beta = \frac{\sigma_v^\alpha - \sigma_u^\alpha}{\sigma_v^\alpha + \sigma_u^\alpha} .$$

Proof. Again we apply Theorem 9.4.2. We obtain (4.23) and (4.24) by applying the basic scaling relations in (5.7)–(5.11) of Section 4.5. ∎

9.5. Split Processes

In this section we obtain a FCLT for counting processes that are split from other counting processes. For example, the original counting process might be a departure process, and each of these departures may be routed to one of several other queues. We then want to consider the arrival counting processes at these other queues. We also allow new points to be created in these split arrival processes. (Events in the original process may trigger or cause one or more events of different kinds. In manufacturing there may be batching and unbatching. In communication networks there may be multicasting; the same packet received may be simultaneously sent out on several outgoing links.)

Let $\tilde{A}(t)$ count the number of points in the original process in the time interval $[0, t]$. Let $X_{i,j}$ be the number of points assigned to split process i at the epoch of the jth point in the original arrival process $\tilde{A}$. With the standard splitting, for each j, $X_{i,j} = 1$ for some i and $X_{i,j} = 0$ for all other i, but we allow other possibilities.

Under the assumptions above, the number of points in the ith split counting process in the time interval $[0, t]$ is

$$A_i(t) \equiv \sum_{j=1}^{\tilde{A}(t)} X_{i,j}, \quad t \geq 0 . \qquad (5.1)$$

Now we assume that we have processes as above for each n, i.e., $\{\tilde{A}_n(t) : t \geq 0\}$, $\{X_{n,i,j} : i \geq 1\}$ and $\{A_{n,i}(t) : t \geq 0\}$. We form associated random elements of $D \equiv D([0, \infty), \mathbb{R})$ by setting

$$\tilde{\mathbf{A}}_n(t) \equiv c_n^{-1}[A_n(nt) - \lambda_n nt]$$

$$\mathbf{S}_{n,i}(t) \equiv c_n^{-1}\left[\sum_{j=1}^{\lfloor nt \rfloor} X_{n,i,j} - p_{n,i} nt\right]$$

$$\mathbf{A}_{n,i}(t) \equiv c_n^{-1}[A_{n,i}(nt) - \lambda_n p_{n,i} nt], \quad t \geq 0 , \qquad (5.2)$$

where λ_n is a positive scalar and $p_n \equiv (p_{n,1}, \ldots, p_{n,m})$ is an element of $\mathbb{R}^m$ with nonnegative components.

We can apply Corollary 13.3.2 to establish a FCLT for the vector-valued split processes $\mathbf{A}_n \equiv (\mathbf{A}_{n,1}, \ldots, \mathbf{A}_{n,m})$ in D^m. Let $\mathbf{S}_n \equiv (\mathbf{S}_{n,1}, \ldots, \mathbf{S}_{n,m})$.

Theorem 9.5.1. *(FCLT for split processes) Suppose that*

$$(\tilde{\mathbf{A}}_n, \mathbf{S}_n) \Rightarrow (\tilde{\mathbf{A}}, \mathbf{S}) \quad in \quad D^{1+m} \tag{5.3}$$

with the topology WJ_1 or WM_1. Also suppose that $c_n \to \infty$, $n/c_n \to \infty$, $\lambda_n \to \lambda$, $p_n \to p$, where $p_i > 0$ for each i, and almost surely $\tilde{\mathbf{A}}$ and $\mathbf{S}_i \circ \lambda \mathbf{e}$ have no common discontinuities of opposite sign for $1 \leq i \leq m$. Then

$$\mathbf{A}_n \Rightarrow \mathbf{A} \quad in \quad D^m \tag{5.4}$$

with the same topology, where

$$\mathbf{A}_i \equiv p_i \tilde{\mathbf{A}} + \mathbf{S}_i \circ \lambda \mathbf{e} \ . \tag{5.5}$$

Proof. Since $A_i(t)$ in (5.1) is a random sum, we can apply the continuous mapping theorem, 3.4.3, with composition and addition. Specifically, we apply Corollary 13.3.2 after noting that

$$\mathbf{A}_{n,i} = \mathbf{S}_{n,i} \circ \hat{\mathbf{A}}_{n,i} + p_{n,i}\tilde{\mathbf{A}}_{n,i} \ ,$$

where $\hat{\mathbf{A}}_{n,i} \equiv n^{-1}A_{n,i}(nt),, \ t \geq 0$. ■

If $c_n/\sqrt{n} \to \infty$, it will often happen that one of the limit processes $\tilde{\mathbf{A}}$ or $\mathbf{S}_i$ will be the zero function. If the burstiness in $\tilde{\mathbf{A}}$ dominates, so that $\mathbf{S}_i = 0\mathbf{e}$, then the limit in (5.4) becomes $p_i\tilde{\mathbf{A}}$, a deterministic-scalar multiple of the limit process $\tilde{\mathbf{A}}$. On the other hand, if the burstiness in $\mathbf{S}_i$ dominates, so that $\tilde{\mathbf{A}} = 0\mathbf{e}$, then the limit in (5.4) becomes $\mathbf{S}_i \circ \lambda \mathbf{e}$, a deterministic time change of the limit process $\mathbf{S}_i$.

It is instructive to contrast various routing methods. Variants of the round robin discipline approximate deterministic routing, in which every $(1/p_i)^{\text{th}}$ arrival from $\tilde{A}$ is assigned to A_i. With any reasonable approximation to round robin, we obtain $\mathbf{S}_i = 0\mathbf{e}$.

In contrast, with IID splittings, $\sum_{j=1}^{k} X_{n,i,j}$ has a binomial distribution for each n, i and k, so that $\mathbf{S}_{n,i} \Rightarrow \mathbf{S}_i$, where $c_n = \sqrt{n}$, $\mathbf{S}_i \stackrel{\mathrm{d}}{=} \sigma_i \mathbf{B}$ with $\mathbf{B}$ standard Brownian motion and $\sigma_i^2 = p_i(1-p_i)$. Then $\mathbf{S}$ is a zero-drift Brownian motion with covariance matrix $\Sigma \equiv (\sigma_{S,i,j}^2)$, where $\sigma_{S,i,i}^2 = p_i(1-p_i)$ and $\sigma_{S,i,j}^2 = -p_i p_j$ for $i \neq j$. We thus see that IID splitting produces greater variability in the split arrival processes than round robin, and thus produces greater congestion in subsequent queues. Moreover, with the heavy-traffic stochastic-process limits, we can quantify the difference.

9.6. Brownian Approximations

In this section we continue the discussion begun in Section 5.7 of Brownian limits that occur in the light-tailed weak-dependent case and the associated Brownian (or RBM) approximations that stem from them.

In the standard light-tailed weak-dependent case, the conditions of Theorems 9.3.1–9.3.4 hold with space scaling by $c_n = \sqrt{n}$ and limits

$$\mathbf{S} \stackrel{d}{=} \sigma_S \mathbf{B} + \eta \mathbf{e} \;, \tag{6.1}$$

where $\mathbf{B}$ is standard Brownian motion and η is obtained from the limit (3.11). Just as in Section 5.7, we can obtain such limits by considering essentially a single model. Here the single model is based on a single sequence of interarrival times and service times $\{(U_k, V_k) : k \geq 1\}$. Let the associated partial sums be $S_k^u \equiv U_1 + \cdots U_k$ and $S_k^v \equiv V_1 + \cdots + V_k$, $k \geq 1$. We then construct the sequences $\{(U_{n,k}, V_{n,k}) : k \geq 1\}$ for a sequence of models indexed by n by scaling the service times, i.e., by letting

$$U_{n,k} = U_k \quad \text{and} \quad V_{n,k} \equiv \rho_n V_k \;. \tag{6.2}$$

Then, in the setting of Section 9.3, $\lambda_n = \lambda$ and $\mu_n^{-1} = \lambda^{-1} \rho_n$ for all n. Then condition (3.11) becomes

$$\sqrt{n}(1 - \rho_n) \to \zeta \equiv -\eta \lambda > 0 \quad \text{as} \quad n \to \infty \;. \tag{6.3}$$

The required FCLT for (S_n^u, S_n^v) in condition (3.9) then follows from Donsker's theorem in Section 4.3 or one of its generalizations for dependent sequences in Section 4.4, applied to the partial sums of the single sequences $\{(U_k, V_k)\}$, under the assumptions there.

As in Section 5.5, it is natural to index the family of queueing systems by the traffic intensity ρ, where $\rho \uparrow 1$. Then, focusing on the waiting-time and queue-length processes and replacing n by $\zeta^2(1-\rho)^{-2}$ for ζ in (3.38) and (6.3), we have the Brownian approximations

$$W_{\rho,k} \approx \lambda \sigma_S^2 (1-\rho)^{-1} \mathbf{R}(\lambda^{-2} \sigma_S^{-2}(1-\rho)^2; -1, 1, 0) \tag{6.4}$$

and

$$Q_\rho(t) \approx \lambda^2 \sigma_S^2 (1-\rho)^{-1} \mathbf{R}(\lambda^{-1} \sigma_S^{-2}(1-\rho)^2; -1, 1, 0) \;, \tag{6.5}$$

where $\{\mathbf{R}(t; -1, 1, 0) : t \geq 0\}$ is canonical RBM. The Brownian approximation in (6.4) is the same as the Brownian approximation in (7.8) in Section 5.7 with λ^{-1} replacing μ. Approximation (6.5) follows from (6.4) because $\mathbf{Q} = \lambda \mathbf{W} \circ \lambda \mathbf{e}$; see (3.20).

9.6.1. Variability Parameters

For the GI/GI/1 queue, where the basic sequences $\{U_k\}$ and $\{V_k\}$ are independent sequences of IID random variables, the heavy-traffic variance constant is

$$\sigma_S^2 = \sigma_u^2 + \sigma_v^2 \;, \tag{6.6}$$

where

$$\sigma_u^2 \equiv Var\, U_1 \quad \text{and} \quad \sigma_v^2 \equiv Var\, V_1 \;.$$

For better understanding, it is helpful to replace the variances by dimensionless variability parameters: It is convenient to use the *squared coefficients of variation*

(SCV's), defined by
$$c_u^2 \equiv \frac{Var\, U_1}{(EU_1)^2} \quad \text{and} \quad c_v^2 \equiv \frac{Var\, V_1}{(EV_1)^2}\,. \tag{6.7}$$

Combining (6.5)–(6.7), we have
$$\sigma_S^2 = \frac{c_u^2 + c_v^2}{\lambda^2} \equiv \frac{c_{HT}^2}{\lambda^2}\,, \tag{6.8}$$

where c_{HT}^2 is the dimensionless overall variability parameter.

For the more general G/G/1 queue, in which $\{(U_k, V_k) : k \geq 1\}$ is a stationary sequence, we must include covariances. In particular,
$$\sigma_S^2 = c_{HT}^2/\lambda^2, \tag{6.9}$$

where
$$c_{HT}^2 = c_U^2 + c_V^2 - 2c_{U,V}^2\,, \tag{6.10}$$

with
$$c_U^2 \equiv \lim_{k\to\infty} k^{-1} \frac{Var\, S_k^u}{(EU_1)^2} \equiv \lim_{k\to\infty} k^{-1} \sum_{j=1}^{k} (k-j) \frac{Cov(U_1, U_j)}{(EU_1)^2}\,,$$

$$c_V^2 \equiv \lim_{k\to\infty} k^{-1} \frac{Var\, kS_k^v}{(EV_1)^2} \equiv \lim_{k\to\infty} k^{-1} \sum_{j=1}^{k} (k-j) \frac{Cov(V_1, V_0)}{(EV_1)^2}\,,$$

$$c_{U,V}^2 \equiv \lim_{k\to\infty} k^{-1} \frac{Cov(S_k^u, S_k^v)}{(EU_1)(EV_1)} \equiv \lim_{k\to\infty} k^{-1} \sum_{j=1}^{k} (k-j) \frac{Cov(U_1, V_j)}{(EU_1)(EV_1)}\,. \tag{6.11}$$

We call c_U^2, c_V^2 and $c_{U,V}^2$ in (6.11) the *asymptotic variability parameters* for the arrival and service processes.

We can combine (6.4), (6.5) and (6.9) to obtain general Brownian approximations in terms of the dimensionless variability parameter c_{HT}^2:
$$\begin{aligned} W_{\rho,k} &\approx \lambda^{-1} c_{HT}^2 (1-\rho)^{-1} \mathbf{R}(c_{HT}^{-2}(1-\rho)^2 k; -1, 1, 0) \\ Q_\rho(t) &\approx c_{HT}^2 (1-\rho)^{-1} \mathbf{R}(c_{HT}^{-2} \lambda (1-\rho)^2 t; -1, 1, 0)\,. \end{aligned} \tag{6.12}$$

For example, as a consequence, the approximation for the mean steady-state waiting time is
$$EW_{\rho,\infty} \approx \lambda^{-1} c_{HT}^2 / 2(1-\rho)\,. \tag{6.13}$$

(Recall that the mean service time in model ρ is $\lambda^{-1}\rho$ here.)

The dimensionless variability parameter c_{HT}^2 helps to understand the heavy-traffic limits for queues with superposition arrival processes. If the arrival process is the superposition of m IID component arrival processes, then c_{HT}^2 is independent of the number m of processes. (See Remark 9.4.1.)

For the GI/GI/1 model, $c_U^2 = c_u^2$, $c_V^2 = c_v^2$ and $c_{U,V}^2 = 0$. However, in many more general G/G/1 applications, these relations do not nearly hold. For example, that usually is the case with superposition arrival processes arising in models of statistical multiplexing in communication networks.

Example 9.6.1. *A packet network example.* In a detailed simulation of a packet network link (specifically, an X.25 link) with 25 independent sources, Fendick, Saksena and Whitt (1989) found that

$$c_u^2 \approx 1.89, \quad c_v^2 \approx 1.06 \quad \text{and} \quad c_{u,v}^2 \approx 0.03 , \tag{6.14}$$

where c_u^2 and c_v^2 are in (6.7) and

$$c_{u,v}^2 \equiv \frac{Cov(U_1, V_1)}{(EU_1)(EV_1)} . \tag{6.15}$$

In contrast, they found that

$$c_U^2 \approx 17.6, \quad c_V^2 \approx 35.1 \quad \text{and} \quad c_{U,V}^2 \approx -6.7 . \tag{6.16}$$

The differences between (6.16) and (6.14) show that there are significant correlations: (i) among successive interarrival times, (ii) among successive service times and (iii) between interarrival times and service times. The dependence among service times and between service times and interarrival times occur because of bursty arrivals from multiple sources with different mean service times (due to different packet lengths).

Note that the variability parameter c_{HT}^2 based on (6.10) and (6.16) is very different from the one based on (6.7), (6.8), (6.14) and (6.16). The variability parameter based on (6.10) and (6.16) is

$$c_{HT}^2 \approx 17.6 + 35.1 - 2(6.7) = 66.1 . \tag{6.17}$$

If, instead, we acted as if we had a GI/GI/1 queue and used (6.7), (6.8) and (6.14), we would obtain $c_{HT}^2 = 2.79$.

Moreover, under moderate to heavy loads, the average steady-state queue lengths in the simulation experiments were consistent with formulas (6.10) and (6.16) using the variability parameter c_{HT}^2 in (6.17). However, under lighter loads there were significant differences between the observed average queue lengths and the heavy-traffic approximations, which motivate alternative parametric approximations that we discuss in Section 9.9 below.

This simulation experiment illustrates that correlations can be, not only an important part of the relevant variability, but the dominant part; in this example,

$$c_U^2 + c_V^2 - 2c_{U,V}^2 \gg c_u^2 + c_v^2 . \tag{6.18}$$

Moreover, in this example the lag-k correlations, defined by

$$\begin{aligned} c_{u,k}^2 &= \frac{Cov(U_1, U_{1+k})}{(EU_1)^2}, \quad c_{v,k}^2 \equiv \frac{Cov(V_1, V_{1+k})}{(EV_1)^2} \\ c_{u,v,k}^2 &= \frac{Cov(U_1, V_{1+k})}{(EU_1, EV_1)} , \end{aligned} \tag{6.19}$$

were individually small for all k. The values in (6.16) were substantially larger than those in (6.14) because of the cumulative effect of many small correlations (over all k). See Albin (1982) for similar experiments. ∎

9.6.2. Models with More Structure

The heavy-traffic Brownian approximation is appealing because it is often not difficult to compute the variability parameter c_{HT}^2 in (6.10) for models. Indeed, in Section 4.4 we indicated that it is often possible to compute the normalization constant in a CLT involving dependent summands. There the specific formulas and algorithms depended on Markov structure. Now we illustrate by considering a model from Fendick, Saksena and Whitt (1989, 1991) that has more structure.

We consider a multiclass batch-renewal-process model that might serve as a model for a packet arrival process in a communication network. In that context, a customer class can be thought of as a particular kind of traffic such as data, video or fax. As an approximation, we assume that all packets (customers) in the same batch (message, burst or flow) arrive at the same instant. We discuss generalizations afterwards in Remark 9.6.1. For this model, we show how to determine the variability parameters c_U^2, c_V^2 and $c_{U,V}^2$.

We assume that the arrival process of k customer classes come as mutually independent batch-renewal processes. For class i, batches arrive according to a renewal process with rate λp_i where the interrenewal-time cdf has SCV $c_{u,i}^2$; the successive batch sizes are IID with mean m_i and SCV $c_{b,i}^2$; the packet service times are IID with mean τ_i and SCV $c_{v,i}^2$. (We assume that $p_1 + \cdots p_k = 1$, so that the total arrival rate of batches is λ. The total arrival rate of customers (packets) is thus $\bar{\lambda} \equiv \lambda m_B$ where

$$m_B = \sum_{i=1}^{k} p_i m_i \ . \tag{6.20}$$

Let q_i be the probability that an arbitrary packet belongs to class i, i.e.,

$$q_i \equiv p_i m_i / \sum_{j=1}^{k} p_j m_j \ . \tag{6.21}$$

Let

$$\tau \equiv \frac{\sum_{i=1}^{k} p_i m_i \tau_i}{\sum_{i=1}^{k} p_i m_i} \quad \text{and} \quad r_i \equiv \frac{\tau_i}{\tau} \ . \tag{6.22}$$

We do not describe the full distributions of intervals between batches, batch sizes and service times, because the heavy-traffic limit does not depend on that extra detail. The model can be denoted by $\sum (GI^{B_i}/GI_i)/1$, since the service times are associated with the arrivals.

Let $c_{U,i}^2$ be the heavy-traffic variability parameter for the class-i arrival process alone.

Theorem 9.6.1. (Heavy-traffic limit for the $\sum (GI^{B_i}/GI)/1$ model) *For the single-server queue with multiclass batch-renewal input above, the conditions of Theorems 9.3.1 and 9.3.3 hold with $c_n = n^{1/2}$ and $(\mathbf{S}^u, \mathbf{S}^v)$ being two-dimensional zero-drift Brownian motion, supporting the approximations in (6.12) with*

$$\begin{aligned}
c_U^2 &= \sum_{i=1}^{k} q_i c_{U,i}^2, \\
c_V^2 &= \sum_{i=1}^{k} q_i [r_i^2 c_{v,i}^2 + (r_i - 1)^2 c_{U,i}^2], \\
c_{U,V}^2 &= \sum_{i=1}^{k} q_i (1 - r_i) c_{U,i}^2, \\
c_{U,i}^2 &= m_i (c_{b,i}^2 + c_{u,i}^2).
\end{aligned} \qquad (6.23)$$

Proof. We only give a quick sketch. The independence assumptions allow us to obtain FCLTs for the partial sums of the batch interarrival times, the batch sizes and the service times. Given that initial FCLT, we can apply the Skorohod representation theorem to replace the convergence in distributions by convergence w.p.1 for special versions. Then note that the packet arrival process can be represented as a random sum: the number of packet arrivals in $[0, t]$ is the sum of the IID batches up to the number of batches to arrive in $[0, t]$. Hence we can apply Corollary 13.3.2 for random sums. The overall packet counting process is the sum of the k independent class packet counting processes. The partial sums of the interarrival times can be treated as the inverses of the counting processes. We thus obtain the limits for all arrival processes and $\mathbf{S}_n^u$, and the variability parameters $c_{U,i}^2$ and c_U^2 in (6.23). We treat the total input of work by adding over the classes, with the total input of work for each class being a random sum of the IID service times up to the number of packet arrivals. Hence we can apply Corollary 13.3.2 for random sums again. From the total input of work, we can directly obtain the limit for the workload by applying the reflection map. From the limit for the total input of work, we can also obtain a limit for the service times presented in order of their arrival to the queue. (This is perhaps the only tricky step.) In general, we have

$$C_n(S_{n,k}^u -) \leq S_{n,k}^v \leq C_n(S_{n,k}^u) \quad \text{for all} \quad n, k, \qquad (6.24)$$

where $S_{n,k}^v$ is the kth partial sum of the service times associated with successive arrivals in model n. We first obtain the limit for $C_n(S_{n,k}^u)$ by applying the random sum result in Corollary 13.3.2 once more. Since the limit process has continuous sample paths, from (6.24) we can conclude that $\mathbf{S}_n^v$ has the same limit; see Corollary 12.11.6. Given the limit for $(\mathbf{S}_n^u, \mathbf{S}_n^v)$, we can apply Theorems 9.3.3 and 9.3.4. ∎

It is helpful to further interpret the asymptotic variability parameters in (6.23). Note that c_U^2 is a convex combination of $c_{U,i}^2$ weighted by q_i in (6.21), where $q_1 + \cdots + q_k = 1$. Note that $c_{U,i}^2$ is directly proportional to the mean batch size m_i. Note that c_V^2 and $c_{U,V}^2$ also can be represented as weighted sums of $c_{V,i}^2$ and $c_{U,V,i}^2$, where

$$c_{V,i}^2 \equiv r_i^2 c_{v,i}^2 + (r_i - 1)^2 c_{U,i}^2$$
$$c_{U,V,i}^2 \equiv (1 - r_i) c_{U,i}^2 \qquad (6.25)$$

The class-i asymptotic service variability parameter $c_{U,i}^2$ and the class-i covariance asymptotic parameter $c_{U,V,i}^2$ depend upon the non-class-i processes only via the parameter $r_i \equiv \tau_i/\tau \equiv \tau_i$ in (6.22). Note that r_i is large (small) when class-i service times are larger (smaller) than usual. Note that $c_{V,i}^2$ has the component $r_i^2 c_{v,i}^2$ that is directly proportional to r_i^2 and $c_{v,i}^2$.

Remark 9.6.1. *Extra dependence.* In Theorem 9.6.1 we assumed that the basic model sequences are independent sequences of IID random variables. Using Section 4.4 that can be greatly relaxed. In the spirit of Section 9.3, we could have started with a joint FCLT.

For the model in Theorem 9.6.1, we let all arrivals in a batch arrive at the same instant. We could instead allow the arrivals from each batch to arrive in some arbitrary manner in the interval between that batch arrival and the next. It is significant that Theorem 9.6.1 is unchanged under that modification. However, both the original model and the generalization above implicitly assume that each batch size is independent of the interval between batch arrivals. That clearly is not realistic in many scenarios, e.g., for packet queues, where larger batch sizes usually entail longer intervals between batch arrivals. It is not difficult to create models that represent this feature. In particular, let $\{(B_n^i, L_n^i) : n \geq 1\}$ be the sequence of successive pairs of successive batch sizes and interval length between successive batch arrivals for class i. Assume that successive pairs are IID, but allow B_n^i and L_n^i to be dependent for each n. As above, let m_i and $c_{b,i}^2$ be the two parameters for B_n^i and let $(\lambda p_i)^{-1}$ and $c_{r,i}^2$ be the two parameters for L_n^i. Let $\gamma_{b,r,i}$ be the correlation between B_n^i and L_n^i. Then Theorem 9.6.1 remains valid with $c_{U,i}^2$ in (6.23) replaced by

$$c_{U,i}^2 = m_i(c_{b,i}^2 + c_{r,i}^2 - 2\gamma_{r,b,i} c_{b,i} c_{r,i}) . \quad \blacksquare \qquad (6.26)$$

The multiclass batch-renewal-process model above illustrates that the asymptotic variability parameters c_U^2, c_V^2 and $c_{U,V}^2$ appearing in the expression for c_{HT}^2 in (6.10) can often be computed for quite rich and complex models. We conclude this section by illustrating this feature again for point processes in a random environment, such as the Markov-modulated Poisson process (MMPP).

Example 9.6.2. *Point processes in random environments.* In this example we suppose that the arrival process can be represented as a counting process in a random environment, such as

$$A(t) = X(Y(t)), \quad t \geq 0 , \qquad (6.27)$$

where

$$(\mathbf{X}_n, \mathbf{Y}_n) \Rightarrow (\mathbf{B}_1, \mathbf{B}_2) \quad \text{in} \quad D^2 \qquad (6.28)$$

with $(\mathbf{B}_1, \mathbf{B}_2)$ being two-dimensional Brownian motion and

$$(\mathbf{X}_n, \mathbf{Y}_n)(t) \equiv n^{-1/2}(X(nt) - xnt, Y(nt) - ynt), \quad t \geq 0. \tag{6.29}$$

For example, an MMPP satisfies (6.27)–(6.29) where X is a homogeneous Poisson process and Y is a function of an irreducible finite-state continuous-time Markov chain (CTMC). Indeed, the representation (6.27) was already exploited for the cumulative-input processes of the fluid queue in (2.6) in Chapter 8.

Given (6.27), we can obtain the required FCLT for A from an established FCLT for (X, Y) in (6.28) by applying Corollary 13.3.2. Without loss of generality (by deterministically scaling X and Y in (6.27)), we can obtain (6.27) with X, Y and A all being rate-1 processes, i.e., for $x = y = 1$ in (6.29). Then the FCLT for $\mathbf{A}_n$ yields the dimensionless asymptotic variability parameter

$$c_A^2 = c_X^2 + c_Y^2 .$$

For example, if A is a rate-1 MMPP and X is a rate-1 Poisson process, then

$$c_A^2 = 1 + c_X^2 ,$$

where c_X^2 is the asymptotic variability parameter of a function of a stationary CTMC having mean 1, which is given in Theorem 2.3.4 in the Internet Supplement. ∎

9.7. Very Heavy Tails

When the interarrival times and service times come from independent sequences of IID random variables with heavy-tailed distributions, we obtain heavy-traffic stochastic-process limits with reflected-stable-Lévy-motion (RSLM) limit processes from Sections 4.5 and 9.3, the same way we obtained heavy-traffic stochastic-process limits with RSLM limit processes for fluid queues in Section 8.5 from Sections 4.5, 5.4 and 8.3. For the most part, the story has already been told in Section 8.5. Hence, now we will only discuss the case of very heavy tails, arising when the random variables have infinite mean. See Resnick and Rootzén (2000) for related results.

Specifically, as in (5.26) in Section 4.5, we assume that the service-time distribution has a power tail, satisfying

$$P(V_1 > x) \sim Ax^{-\alpha} \quad \text{as} \quad x \to \infty \tag{7.1}$$

for positive constants α and A with $0 < \alpha < 1$.

We note that $\alpha = 1$ is a critical boundary point, because if (7.1) holds for $\alpha > 1$, then the service-time distribution has a finite mean, which implies that the waiting-time process has a proper steady-state distribution. However, if (7.1) holds for $\alpha < 1$, then the service time distribution has infinite mean, which implies that the waiting time process fails to have a proper steady-state distribution, in particular,

$$W_k \to \infty \quad \text{as} \quad k \to \infty \quad \text{w.p.1.}$$

9.7.1. Heavy-Traffic Limits

We can use the heavy-traffic stochastic-process limits to show how the waiting times should grow over finite time intervals. Similar limits will hold for the queue-length processes. For these limits, it suffices to consider a single queueing system. Since the mean service time is infinite, the traffic intensity is infinite here, and thus plays no role. Let the random elements of D be defined by

$$\mathbf{S}_n(t) \equiv n^{-1/\alpha} S_{\lfloor nt \rfloor},$$
$$\mathbf{W}_n(t) \equiv n^{-1/\alpha} W_{\lfloor nt \rfloor}, \quad t \geq 0 . \tag{7.2}$$

Theorem 9.7.1. (service times with very heavy tails) *Consider the standard single-server queue with interarrival times and service times coming from a sequence of IID random vectors $\{(U_k, V_k)\}$. Suppose that $EU_1 < \infty$ and (7.1) holds with $0 < \alpha < 1$. Then*

$$(\mathbf{S}_n, \mathbf{W}_n) \Rightarrow (\mathbf{S}, \mathbf{S})) \quad in \quad D([0, \infty), \mathbb{R}^2, SJ_1) ,$$

for $\mathbf{S}_n$ and $\mathbf{W}_n$ in (7.2), where $\mathbf{S}$ is the α-stable Lévy motion with $\beta = 1$ characterized by Theorems 4.5.2 and 4.5.3 that arises as the stochastic-process limit for partial sums of the service times alone.

Proof. Since $EU_1 < \infty$, $\{U_k : k \geq 1\}$ obeys the strong law of large numbers, which in turn implies a functional strong law; see Corollary 3.2.1 in the Internet Supplement. Hence the FCLT for $\mathbf{S}_n$ follows for the FCLT for the partial sums of the service times alone (without translation term), by virtue of Theorem 11.4.5. The FCLT for the service times alone follows from Theorems 4.5.2 and 4.5.3. We obtain the limit theorem for the scaled waiting times by applying the continuous-mapping approach with the reflection map; in particular, we can apply Theorem 9.3.1 (a). Finally, the limit process $\mathbf{W}$ has the indicated form because $\mathbf{S}$ has nondecreasing sample paths since $\beta = 1$. ∎

As a consequence, of Theorem 9.7.1, we can approximate the transient waiting times by

$$W_k \approx n^{1/\alpha} \mathbf{S}(k/n) , \quad k \geq 0 ,$$

for any k.

We now show that we can calculate the pdf and cdf of the $S_\alpha(\sigma, 1, 0)$ stable distribution for $0 < \alpha < 1$. Paralleling the case $\alpha = 3/2$ described in Theorem 8.5.4, the case $\alpha = 1/2$ is especially tractable. As noted in Section 4.5, for $\alpha = 1/2$, we obtain the Lévy distribution; i.e., the $S_\alpha(1, 1, 0)$ distribution has cdf

$$G_{1/2}(x) = 2\Phi^c(1/\sqrt{x}), \quad x \geq 0$$

where $\Phi^c(x) \equiv P(N(0,1) > x)$ and pdf

$$g_{1/2}(x) = \frac{1}{\sqrt{2\pi x^3}} e^{-1/2x}, \quad x \geq 0 ;$$

see p. 52 of Feller (1971).

More generally, we can apply numerical inversion of Laplace transforms again to calculate the pdf and ccdf of the stable subordinator $\mathbf{S}(t)$. We exploit the fact that the distribution $S_\alpha(\sigma, 1, 0)$ of $S^\alpha(1)$ has support on the positive halfline. That makes the bilateral Laplace transform in (5.17) in Section 4.5 a bonafide Laplace transform. We exploit self-similarity to relate the distribution at any time t to the distribution at time 1, i.e.,

$$\mathbf{S}(t) \stackrel{\mathrm{d}}{=} t^{1/\alpha} \mathbf{S}(1) . \qquad (7.3)$$

Hence it suffices to consider the single-parameter family of distributions $S_\alpha(1, 1, 0)$.

By (5.12) in Section 4.5, we know that the ccdf of $S_\alpha(1, 1, 0)$ decays as $x^{-\alpha}$. Hence, for $0 < \alpha < 1$, the random variable $\mathbf{S}(t)$ has infinite mean. By (7.3), we expect $\mathbf{S}(t)$ to grow like $t^{1/\alpha}$ as t increases. However, we should expect much of the growth to be in large jumps. To illustrate the form of the ccdf's, we give the ccdf values of $S_\alpha(1, 1, 0)$ for three values of α in Table 9.1, again computed by numerical transform inversion, exploiting the Euler algorithm in Abate and Whitt (1995a).

	G_α^c ccdf of $S_\alpha(1,1,0)$		
x	$\alpha = 0.2$	$\alpha = 0.5$	$\alpha = 0.8$
$(0.01)2^0 = 0.01$	0.9037	1.0000	1.0000
$(0.01)2^1 = 0.02$	0.8672	1.0000	1.0000
$(0.01)2^2 = 0.04$	0.8251	0.9996	1.0000
$(0.01)2^4 = 0.16$	0.7282	0.9229	1.0000
$(0.01)2^6 = 0.64$	0.6233	0.6232	0.7371
$(0.01)2^8 = 2.56$	0.5197	0.3415	0.1402
$(0.01)2^{10} = 10.24$	0.4242	0.1749	$0.3739\ e{-}1$
$(0.01)2^{12} = 40.96$	0.3404	$0.8798\ e{-}1$	$0.1154\ e{-}1$
$(0.01)2^{16} = 655.36$	0.2112	$0.2204\ e{-}1$	$0.1220\ e{-}2$
$(0.01)2^{20} = 10486$	0.1269	$0.5510\ e{-}2$	$0.1324\ e{-}3$
$(0.01)2^{24} = 167,772$	$0.7477\ e{-}1$	$0.1377\ e{-}2$	$0.1440\ e{-}4$
$(0.01)2^{28} = 2,684,000$	$0.4359\ e{-}1$	$0.3444\ e{-}3$	$0.1567\ e{-}5$
$(0.01)2^{32} = 42,949,000$	$0.2525\ e{-}1$	$0.8609\ e{-}4$	$0.1705\ e{-}6$

Table 9.1. Tail probabilities of the stable law $S_\alpha(1, 1, 0)$ for $\alpha = 0.2, 0.5$ and 0.8 computed by numerical transform inversion.

The cdf of the stable law $S_\alpha(1, 1, 0)$ reveals the consequences of the heavy tail, but it does not directly show the jumps in the stable Lévy motion. We see the jumps more directly when we consider the first passage times to high levels. We can exploit the convergence to a stable Lévy motion to show, asymptotically, how the waiting-time process reaches new levels when the service-time distribution has such a heavy tail (with $0 < \alpha < 1$).

9.7.2. First Passage to High Levels

As observed in Section 4.5, a stable Lévy motion with $0 < \alpha < 2$ is a pure-jump stochastic process. Thus, the stable Lévy motion passes any specified level by making a jump. (See Bertoin (1996).) Hence the process is below the level just before the jump and above the level immediately after the jump. It is significant that we can obtain useful characterizations of the distributions of the values immediately before and after first passing any level for the limiting stable Lévy motion. We describe the asymptotic distribution of the last value before the jump as the level increases.

Stochastic-process limits for these quantities follow from the continuous-mapping approach with Theorem 13.6.5. Explicit expressions for the distributions associated with the limiting stable Lévy notion follow from the generalized arc sine laws; see Sections III and VIII of Bertoin (1996).

For $z > 0$, let τ_z be the *first passage time* to a level beyond z; i.e., for $x \in D$,

$$\tau_z(x) \equiv x^{-1}(z) \equiv \inf\{t \geq 0 : x(t) > z\} \tag{7.4}$$

with $\tau_z(x) = \infty$ if $x(t) \leq z$ for all t. Let γ_z be the associated *overshoot*; i.e.,

$$\gamma_z(x) = x(\tau_z(x)) - z . \tag{7.5}$$

Let λ_z be the *last value* before the jump; i.e.,

$$\lambda_z(x) \equiv x(\tau_z(x)-) . \tag{7.6}$$

Let these functions also be defined for the discrete-time process $W \equiv \{W_k\}$ (without scaling) in the same way.

Note that the scale parameter σ enters in simply to the first passage time, i.e., for $y > 0$

$$\tau_z(\mathbf{S}(y \cdot)) = y^{-1}\tau_z(\mathbf{S}) ,$$

and does not appear at all in the overshoot or the last value before passage (because σ corresponds to a simple time scaling).

Also note that we can determine the distribution of the overshoot and the jump size for the waiting times in a GI/GI/1 model if we know the distribution of the last value $\lambda_z(W)$: Because of the IID assumption for $\{(U_k, V_k)\}$,

$$P(\gamma_z(W) > x | \lambda_z(W) = y) = P(V_1 - U_1 > x + z - y | V_1 - U_1 > z - y) . \tag{7.7}$$

When x and $z - y$ are both large, we can exploit the service-time tail asymptotics in (7.1) to obtain the useful approximation

$$\begin{aligned} P(V_1 - U_1 > x + z - y | V_1 - U_1 > z - y) \\ \approx P(V_1 > x + z - y | V_1 > z - y) \\ \approx ((z-y)/(x+z-y))^\alpha . \end{aligned} \tag{7.8}$$

hence, much interest centers on determining the distribution of the last value before the jump for the waiting times. That exact distribution is hard to come by directly, so that the heavy-traffic limit is helpful.

Theorem 9.7.2. (limits for the first-passage time, overshoot and last value) *Under the conditions of Theorem 9.7.1, for $z > 0$,*

$$n^{-1}\tau_{zn^{1/\alpha}}(W) \Rightarrow \tau_z(\mathbf{S}) \quad in \quad \mathbb{R} \quad as \quad n \to \infty ,$$

so that

$$\lim_{n\to\infty} P(\tau_{zn^{1/\alpha}}(W) > nx) = P(\mathbf{S}(x) \leq z) ;$$

$$n^{-1/\alpha}\gamma_{zn^{1/\alpha}}(W) \Rightarrow \gamma_z(\mathbf{S}) \quad in \quad \mathbb{R} \quad as \quad n \to \infty ,$$

so that, for $b > z$,

$$\lim_{n\to\infty} P(\gamma_{zn^{1/\alpha}}(W) > (b-z)n^{1/\alpha})$$
$$= \frac{1}{\Gamma(\alpha)\Gamma(1-\alpha)} \int_0^z x^{\alpha-1}(b-x)^{-\alpha}dx ; \qquad (7.9)$$

$$n^{-1/\alpha}W(\lfloor \tau_{zn^{1/\alpha}}(W) - \rfloor) \Rightarrow \mathbf{S}(\tau_z(\mathbf{S})-)$$

and, for $0 < b < 1$,

$$\lim_{z\to\infty} P(\mathbf{S}(\tau_z(\mathbf{S})-) > zb) = \int_b^1 \frac{\sin(\alpha\pi)dt}{\pi t^{1-\alpha}(1-t)^\alpha} . \qquad (7.10)$$

Proof. By Theorem 13.6.5, the first-passage time, overshoot and last-value functions are almost surely continuous functions on D with respect to the limiting stable Lévy motion. Hence we can apply the continuous mapping theorem, Theorem 3.4.3. Note that

$$n^{-1}\tau_{zn^{1/\alpha}}(W) = \tau_z(n^{-1/\alpha}W_{\lfloor n\cdot \rfloor}) ,$$

$$n^{-1/\alpha}\gamma_{zn^{1/\alpha}}(W) = n^{-1/\alpha}W_{\lfloor \tau_{zn^{1/\alpha}}(W)\rfloor} - z = \gamma_z(n^{-1/\alpha}W_{\lfloor n\cdot \rfloor})$$

and

$$n^{-1/\alpha}W(\lfloor \tau_{zn^{1/\alpha}}(W) - \rfloor) = n^{-1/\alpha}W_{\lfloor n\tau_z(n^{-1/\alpha}W(n\cdot)-)\rfloor} .$$

For (7.9), see Exercise 3, p. 238, and p. 241 of Bertoin (1996). For (7.10), see Theorem 6, p. 81, of Bertoin (1996). ■

The limiting distribution in (7.10) is called the generalized arc sine law. Its density is in general an asymmetric U-shaped function. The case $\alpha = 1/2$ produces the standard arc sine density in Corollary 4.3.1. Consistent with intuition, as α decreases, the chance of the scaled last value being relatively small (making the final jump large for a large level z) increases.

9.8. An Increasing Number of Arrival Processes

In this section we establish heavy-traffic limits for queues with superposition arrival processes, where the number of arrival processes being superposed increases in the limit. Related results for fluid queues were established in Section 8.7.

9.8.1. Iterated and Double Limits

From Theorem 9.4.1, we see that the FCLT for a superposition of m IID counting processes is the same as the FCLT for a single counting process, except for the obvious deterministic scaling. Indeed, in Section 9.6 we observed that the dimensionless variability parameter c_{HT}^2 defined in (6.9) and (6.10) is independent of m.

However, we obtain a different picture from the fundamental limit for superpositions of point processes, where the number of superposed processes gets large with the total rate held fixed. (That limit is sometimes called the law of small numbers.) Then the superposition process converges to a Poisson process; e.g., see Çinlar (1972) or Daley and Vere Jones (1988). For this limit, convergence in (D, J_1) is equivalent to convergence of the finite-dimensional distributions.

Theorem 9.8.1. (Poisson limit for superposition processes) *Suppose that A^i are IID counting processes with stationary increments and without multiple points (all jumps in A^i are of size 1). Then*

$$A_m \Rightarrow A \quad in \quad (D, J_1) \quad as \quad m \to \infty, \tag{8.1}$$

where

$$A_m(t) \equiv \sum_{i=1}^{m} A^i(t/m), \quad t \geq 0, \quad m \geq 1, \tag{8.2}$$

and A is a homogeneous Poisson process with intensity

$$\lambda \equiv E[A(t+1) - A(t)] = E[A^1(t+1) - A^1(t)]. \tag{8.3}$$

In view of Theorem 9.8.1, we might well expect the superposition arrival process for large m to behave like a Poisson process in the heavy-traffic limit. However, if A^1 is a Poisson process, then $\mathbf{S}_{n,1}^u \Rightarrow \mathbf{S}_1^u$ with $c_n = \sqrt{n}$ and $\mathbf{S}_1^u$ a standard Brownian motion; i.e., the dimensionless variability parameter is $c_U^2 = 1$. Clearly, Theorem 9.4.1 does not capture this Poisson tendency associated with large m. The two iterated limits $\lim_{\rho \to 1} \lim_{m \to \infty}$ and $\lim_{m \to \infty} \lim_{\rho \to 1}$ are not equal. The reason that these iterated limits do not coincide is that the superposition process looks different in different time scales. The iterated limits do not agree because the Poisson superposition limit focuses on the short-time behavior, while the heavy-traffic limit focuses on long-time behavior.

Remark 9.8.1. *Different variability at different time scales.* A Poisson process is relatively simple in that it tends to have the same level of variability at all time scales. For example, if A is a Poisson counting process with rate λ, then both the mean and the variance of $A(t)$ are λt for all $t > 0$. In contrast, a superposition of a large number m of IID stationary point processes (without multiple points), where each component process is not nearly Poisson, tends to have different levels of variability at different time scales. Consistent with Theorem 9.8.1, for large m, the superposition process tends to look like a Poisson process in a short time scale, but it looks like a single component point process in a long time scale.

For example, for a superposition of IID point processes with large m and small t, the variance of $A(t)$ tends to be approximately λmt, where λ is the rate of one component process, just as if A were a Poisson process. However, consistent with Theorem 9.4.1, under regularity conditions, for any given m, the variance of $A(t)$ approaches $\lambda c_a^2 mt$, where

$$c_a^2 = \lim_{t \to \infty} Var(A_1(t)/\lambda t ,$$

with A_1 being the counting process of one source. If A_1 is a Poisson process, then $c_a^2 = 1$, but more generally c_a^2 can be very different from 1.

The heavy-traffic limits in Section 9.4 for queues with a superposition of a fixed number of component processes capture only the large-time-scale variability of the superposition process. That is appropriate for the queue for any number m of component processes provided that the traffic intensity ρ is large enough. However, in practice ρ may not be large enough.

The problem, then, is to understand how variability in the input, with levels varying over different time scales, affects queueing performance. Consistent with intuition, it can be shown that the large-time-scale variability tends to determine queue performance at very high traffic intensities, while the short-time-scale variability tends to determine queue performance at very low traffic intensities. More generally, we conclude that variability at longer times scales become more important for queue performance as the traffic intensity increases. See Section 9.9, Sriram and Whitt (1986), Fendick, Saksena and Whitt (1989, 1991) and Fendick and Whitt (1989) for more discussion. As shown by Whitt (1985a), for superposition processes, we gain insight into the effect of different variability at different time scales upon queueing performance by considering the double limit as $\rho \uparrow 1$ and $m \to \infty$. ∎

In order to have a limit that captures some of the structure of the superposition process not seen in either a single component process or the Poisson process, we consider a double limit, letting the number of component processes be n, and then letting $\rho_n \uparrow 1$ as $n \to \infty$. As in Theorem 9.8.1, we rescale time in the superposition process so that the total arrival rate is fixed, say at 1. Thus the superposition arrival process alone approaches a rate-1 Poisson process as the number n of components increases. In the nth queueing model with n component arrival processes, we let the service times have mean ρ_n^{-1}, so that the traffic intensity in model n is ρ_n. We achieve heavy traffic by letting $\rho_n \uparrow 1$ as $n \to \infty$.

The double limit considered in this section has advantages and disadvantages. Its first advantage is that it may more faithfully describe queues with superpositions of a large number of component arrival processes. Its second advantage is that, even if the interarrival times have heavy-tailed distributions, the limit process is likely to have continuous sample paths. However, a disadvantage is that the limit process is more complicated, because it does not have independent increments.

We start by considering the superposition arrival process alone. Treating the arrival process alone, we first scale time by n^{-1} to keep the rate fixed. Then we scale time again by n to establish the FCLT. These two time scalings cancel out, so that there is no time scaling inside the arrival process. In particular, the scaled arrival process is

$$
\begin{aligned}
\mathbf{A}_n(t) &\equiv c_n^{-1}(A_n(nt/n) - \lambda nt) \\
&= c_n^{-1}(A_n(t) - \lambda nt) \\
&= c_n^{-1}\left(\sum_{i=1}^{n} A^i(t) - \lambda nt\right) \\
&= c_n^{-1}\sum_{i=1}^{n}(A^i(t) - \lambda t), \quad t \geq 0.
\end{aligned}
\qquad (8.4)
$$

From the final line of (8.4), we see that the final scaled process $\mathbf{A}_n$ can be represented as the scaled sum of the IID processes $\{A^i(t) - \lambda t : t \geq 0\}$. Thus limits for $\mathbf{A}_n$ follow from the CLT for processes in Section 7.2.

Now, following and extending Whitt (1985a), we establish a general heavy-traffic stochastic-process limit for a queue with a superposition arrival process, where the number of component arrival processes increases in the limit. (See Knessl and Morrison (1991), Kushner and Martins (1993, 1994), Kushner, Yang and Jarvis (1995), Brichet et al. (1996, 2000) and Kushner (2001) for related limits.) We consider the general space scaling by c_n, where $c_n \to \infty$ and $n/c_n \to \infty$.

Theorem 9.8.2. (general heavy-traffic limit for a queue with a superposition arrival process having an increasing number of components) *Consider a sequence of single-server queueing models indexed by n, where the service times are independent of the arrival times and the arrivals come from the superposition of n IID component arrival processes A^i. Suppose that*

$$\mathbf{S}_n^v \Rightarrow \mathbf{S}^v \quad in \quad (D, M_1) \qquad (8.5)$$

for $\mathbf{S}_n^v$ in (3.1), $P(\mathbf{S}^v(0) = 0) = 1$, and

$$P(t \in Disc(\mathbf{S}^v)) = 0 \quad for\ all\ \ t. \qquad (8.6)$$

Suppose that

$$\mathbf{A}_n \Rightarrow \mathbf{A} \quad in \quad (D, M_1) \qquad (8.7)$$

for $\mathbf{A}_n$ in (8.4), $P(\mathbf{A}(0) = 0) = 1$ and

$$P(t \in Disc(\mathbf{A})) = 0 \quad for\ all\ \ t. \qquad (8.8)$$

If $c_n \to \infty$, $n/c_n \to \infty$ and

$$\eta_n \equiv n(\mu_n^{-1} - \lambda^{-1})/c_n \to \eta \quad as \quad n \to \infty, \qquad (8.9)$$

then the conditions and conclusions of Theorems 9.3.3 and 9.3.4 hold with $\lambda_n \to \lambda$, $\mathbf{S}^u = \lambda^{-1}\mathbf{A} \circ \lambda^{-1}\mathbf{c}$ and the WM_1 topology on the product space D^k.

Proof. It is easy to verify that the conditions here imply the conditions in Theorems 9.3.3 and 9.3.4: First, we can apply Theorem 7.3.2 to get

$$\mathbf{S}_n^u \Rightarrow \mathbf{S}^u = -\lambda^{-1}\mathbf{A} \circ \lambda^{-1}\mathbf{e} \ .$$

Then we can apply Theorem 11.4.4 to get

$$(\mathbf{S}_n^u, \mathbf{S}_n^v) \Rightarrow (\mathbf{S}^u, \mathbf{S}^v) \quad \text{in} \quad (D^2, WM1) \ .$$

Conditions (8.6) and (8.8) imply condition (3.10). The conditions also imply (3.19). ∎

Remark 9.8.2. *The case of a Lévy counting process.* If A is a Lévy process, then the scaled superposition process in (8.4) satisfies

$$\mathbf{A}_n(t) \stackrel{\mathrm{d}}{=} c_n^{-1}[A^1(nt) - \lambda nt], \quad t \geq 0 \ ,$$

as in (3.4) with constant λ, using the reasoning in Remark 9.4.1. In that case, Theorem 9.8.2 adds nothing new. The counting process A is a Lévy process if it is a Poisson process or, more generally, a batch Poisson process, with the batches coming from a sequence of IID integer-valued random variables. The scaled batch-Poisson process can converge to a nonBrownian stable Lévy motion. ∎

We now focus on the way the number of sources, n, and the traffic intensity, ρ, should change as $n \to \infty$ and $\rho \uparrow 1$. For that purpose, suppose that $c_n = n^H$ for $0 < H \leq 1$, then (8.9) implies that

$$n^{1-H}(1 - \rho_n) \to |\lambda \eta| \quad \text{as} \quad n \to \infty \ . \tag{8.10}$$

As the component arrival processes get more bursty, H increases. As H increases, n^{1-H} increases more slowly as a function of n. Thus, with greater burstiness, ρ can approach 1 more slowly to have the heavy-traffic limit in Theorem 9.8.2.

We now have criteria to determine when the two iterated limits tell the correct story: If

$$n >> (|\lambda \eta|/(1-\rho))^{1/(1-H)} \ ,$$

then the arrival process should behave like a Poisson process in the heavy-traffic limit; if

$$n << (|\lambda \eta|/(1-\rho))^{1/(1-H)} \ ,$$

then the arrival process should behave like a single component arrival process in the heavy-traffic limit. The intermediate case covered by (8.10) is more complicated.

In Section 7.2 we have given sufficient conditions for the condition $\mathbf{A}_n \Rightarrow \mathbf{A}$ in (8.7). We illustrate by giving a result from Whitt (1985a) for superpositions of renewal processes, drawing on Theorem 7.2.3.

Theorem 9.8.3. (reflected Gaussian heavy-traffic limit for a queue with a superposition arrival process having an increasing number of renewal components)

Consider a sequence of single-server queueing models indexed by n, where the service times are independent of the arrival times and the arrivals come from the superposition of n IID component stationary renewal processes A^i with interrenewal cdf F having mean λ^{-1}. Suppose that

$$\mathbf{S}_n^v \Rightarrow \mathbf{S}^v \quad in \quad (D, J_1) \tag{8.11}$$

for $\mathbf{S}_n^v$ in (3.1) with $c_n = \sqrt{n}$ and $\mathbf{S}^v$ a zero-mean Brownian motion. Suppose that

$$\lim_{t \to 0} t^{-1}(F(t) - F(0)) < \infty . \tag{8.12}$$

Suppose that

$$\eta_n = \sqrt{n}(\mu_n^{-1} - \lambda^{-1}) \to \eta \quad as \quad n \to \infty . \tag{8.13}$$

Then

$$\mathbf{A}_n \Rightarrow \mathbf{A} \quad in \quad (D, J_1) \tag{8.14}$$

where $\mathbf{A}$ is a zero-mean Gaussian process with stationary increments and continuous sample paths. The limit process $\mathbf{A}$ has the covariance function of A^1, which is characterized in Theorem 7.2.4. Then the conditions and conclusions of Theorems 9.3.3 and 9.3.4 hold with $c_n = \sqrt{n}$, $\lambda_n = \lambda$ and

$$\mathbf{S}^u = -\lambda^{-1}\mathbf{A} \circ \lambda^{-1}\mathbf{e} . \tag{8.15}$$

Consequently, the limit processes $\mathbf{S}$ and $\mathbf{X}$ are Gaussian processes with stationary increments and continuous sample paths.

Proof. Apply Theorems 9.3.3, 9.3.4 and 7.2.3. ■

Unfortunately the limit processes for the waiting time, queue-length and workload processes stemming from Theorem 9.8.3 are relatively intractable, because the limit processes $\mathbf{S}$ and $\mathbf{X}$ here do not have independent increments. However, since $\mathbf{S}$ and $\mathbf{X}$ are Gaussian processes, we can obtain approximations for the steady-state distributions of the queueing-content limit processes $\mathbf{W}$, $\mathbf{L}$, $\mathbf{Q}$ and $\mathbf{Q}^A$ by applying Section 8.8. We can also establish a second limit to RFBM as in Section 8.7.

9.8.2. Separation of Time Scales

When we let the number of sources become large in a single-server queue, we change the relevant time scales of the sources relative to the server. With n IID sources, the interarrival times for each source become of order $O(n)$, while the service times remain of order $O(1)$. When we scale time by n for the heavy-traffic limit, the interarrival times for each source become of order $O(1)$, while the service times become of order $O(n^{-1})$. From either perspective, the interarrival times for each source are of order $O(n)$ times longer than the service times. Thus, as n increases, the relevant time scales for the sources and the server separate. Consequently, for large n, the small-time-scale behavior of the source arrival processes (from their own perspective) can significantly affect the large-time-scale or heavy-traffic behavior of the queue.

Consistent with that observation, the limit process **A** in Theorem 9.8.3 providing the contribution of the arrival process to the heavy-traffic limit depends on the component process A_1 through its correlation function. Thus, unlike the case of a single source, locally smoothing the input for each source with many sources can dramatically reduce the congestion in heavy traffic. In contrast, for a single source, the heavy-traffic behavior of the queue depends on the source arrival process only through its CLT behavior, which of course depends on the large-time-scale behavior of that source.

As noted by Whitt (1988), there is a separation of time scales for flows in multiclass queueing networks: When one source at a queue has an arrival rate much smaller than the service rate (usually because the server is shared by many sources), the departure process for that class tends to be very similar to the arrival process for that class, because the service and delay experienced at that queue tend to be in a shorter time scale. Thus, in a flow through a network from source to destination, where the arrival rate of that flow is much smaller than the service rate at all queues on its path, the arrival process at the destination will be very similar to the original flow emitted from the source. This property has been further exposed by Wischik (1999) using moderate-deviation limits.

As indicated before, time scales are important for understanding the performance of communication networks. In packet networks it is useful to identify three separate time scales: the packet (or cell) scale, the burst (or flow) scale and the call (or connection) scale; see Hui (1988) and Roberts (1992). For further discussion about time scales in queues associated with communication networks, see Sriram and Whitt (1986), Fendick and Whitt (1989), Tse, Gallager and Tsitsiklis (1995), Jelenković, Lazar and Semret (1997), Grossglauser and Tse (1999), Greenberg, Srikant and Whitt (1999) and Kunniyur and Srikant (2001).

Example 9.8.1. *Token-bank rate-control throttles.* The separation of time scales has implication for the effectiveness of devices to regulate traffic. One such device is a token-bank rate-control throttle; see Berger (1991), Berger and Whitt (1992a, b, 1994, 1995b) and references cited there. The operation of such a throttle is depicted in Figure 9.8.1. The throttle contains two finite buffers, one for jobs and one for tokens. The jobs may be packets in a high-speed packet network or call-setup requests in a telecommunications switching system. The buffer for tokens, called a token bank, is typically a fictitious buffer, because the token bank is usually implemented by a counter with a cap, but it is convenient to think of physical tokens. These tokens arrive deterministically and evenly spaced from an infinite source.

Tokens that arrive to a full token bank are blocked and lost. If the bank contains a token when a job arrives to the throttle, then the job is allowed to pass through, and one token is removed from the token bank. If the token bank does not contain any tokens when a job arrives, then the job queues in the job buffer if the job buffer is not full. If a job arrives to find a full job buffer, then the job is not admitted and is said to have "overflowed". In packet networks, the overflowed packet may be discarded or may be marked and later treated as a lower priority class.

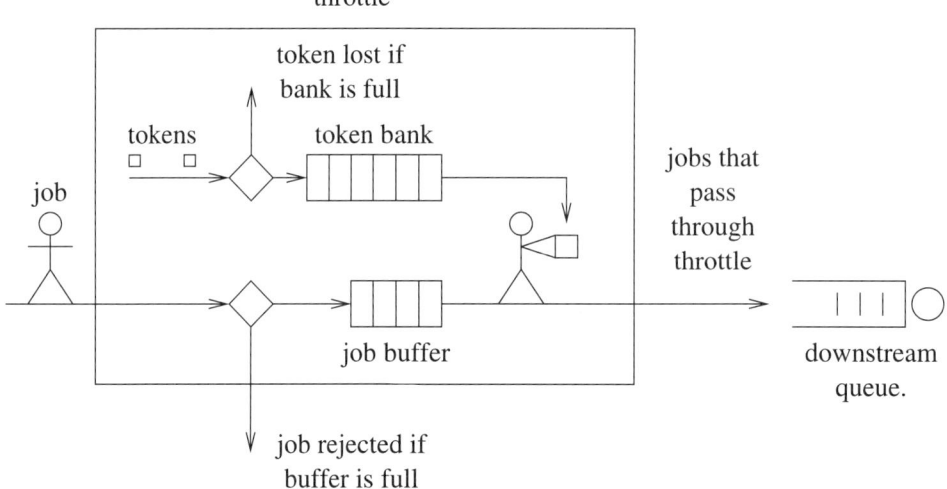

Figure 9.1. Diagram of a token-bank rate-control throttle with a job-buffer regulating traffic to a downstream queue.

The token-bank rate-control throttle is closely related to the *leaky-bucket regulator*. With the conventional definition, the leaky bucket has a constant *drain rate r* and a capacity C. At a job arrival, if the bucket content is below $C-1$, then the job is admitted and the bucket content is increased by 1. Otherwise, the job overflows. The bucket drains out at a deterministic rate r. When the bucket is empty, the draining stops. The draining process starts again upon the next job arrival. The arrival brings the bucket content to 1, and a new busy period of the bucket begins. Thus, the time epochs at which a unit of content drains out of the bucket do not remain synchronous in time, but instead experience a phase shift each time the bucket empties.

In contrast, with a token-bank rate-control throttle, the token arrival process continues to run independent of the state of the bank, so that the token arrival epochs do remain synchronous for all time. The leaky bucket is equivalent to a modified rate-control throttle, without job buffer, in which the deterministic token arrival process stops whenever the token bank becomes full, and starts again at the next job arrival epoch. Just like the rate-control throttle, the leaky bucket can be supplemented by adding a job buffer. Hence, our remarks here about token-bank rate-control throttles also apply to leaky-bucket regulators.

An important initial observation for understanding the performance of the throttle is the *overflow invariance property* established by Berger (1991) and Berger and Whitt (1992a): Except for a finite initial period to count for initial conditions, the job overflow process depends on the (finite) capacity of the token bank, C_T, and the (finite) capacity of the job buffer, C_J, only via their sum $C = C_T + C_J$. The overflow invariance property implies that we can decompose the question about the performance of the throttle into two separate parts: First, there is the traffic shaping caused by job rejections, which depends only on the total capacity C. Second, there is the additional traffic shaping provided by a job buffer given a fixed total capacity C.

The traffic shaping caused by job rejections can be studied by establishing heavy-traffic limits for the throttle; that was done by Berger and Whitt (1992b). Following Berger and Whitt (1992a, 1994), here we will focus on the second question: What is the traffic-shaping benefit provided by the job buffer, given fixed total capacity C?

For given total capacity C, we should prefer no job buffer ($C_J = 0$) if there were no performance differences, because a job buffer is an actual buffer requiring resources to implement. The reason for having a job buffer is that it can provide additional traffic shaping. The potential advantages of a job buffer are easy to see when we contrast an all-token-bank throttle (with $C_J = 0$) to an all-job-buffer throttle (with $C_T = 0$). With an all-token-bank throttle, the throttle can admit batches of jobs of size C. In contrast, with an all-job-buffer throttle, the successive admission epochs of jobs are always separated by at least the deterministic interval between successive token arrivals.

Early proponents of rate-control throttles with job buffers noted the smoothing properties of the throttle. For example, they showed that the throttle reduced the variability (e.g., as measured by the squared coefficient of variation) of the stationary interval between successive job admission epochs. However, through stochastic analysis and simulation, Berger and Whitt (1992a, 1994) showed that, while the traffic smoothing benefit of the throttle was dramatic in a short time scale, it was much less so in a long time scale. Indeed, they showed that the heavy-traffic limiting behavior at a downstream queue fed by a source with a rate-control throttle is independent of the job buffer, given fixed total capacity. More generally, simulations showed that the job buffer tends to provide only a relatively minor reduction of congestion in a downstream queue.

However, most systems actually have traffic from many sources entering the downstream system. As noted above, when the number of sources increases, the short-time behavior of the individual sources begins to have impact upon the large-time-scale behavior of the queue. In the limit, there is a separation of time scales. Consistent with that observation, simulations show that, in marked contrast to the case of a queue fed by a single source, the job buffer provides a dramatic smoothing benefit when 100 identical sources regulated by throttles feed a downstream queue. The separation of time scales provided by many sources makes the short-time-scale behavior of the individual sources relevant to the large-time-scale behavior of the queue.

Consistent with that observation, the simulations also show that the synchronization of many token arrival streams can be a major source of congestion: If there are many sources, and the token arrival epochs of these sources are synchronous, then there can be bursts of arrivals at each token arrival epoch. This effect tends not to appear, however, if all the throttles are not synchronized, i.e., if the phase is random for each throttle.

For recent work focusing on the impact of rate control throttles on long-range dependent input, see Vamvakos and Anantharam (1998) and Gonzáles-Arévalo and Samorodnitsky (2001). ∎

9.9. Approximations for Queueing Networks

Most systems experiencing congestion are not single queues, but networks of queues. However, a cardinal principle of performance analysis is: *Look for the bottleneck!* Often there is a critical resource that primarily determines system performance. When viewed correctly, the complex queueing network often reduces to a smaller system that is easier to analyze. Indeed, it often suffices to consider a single queue.

Hence, from a practical perspective, there is much justification for emphasizing single queues. However, it is also helpful to be able to analyze queueing networks.

9.9.1. Parametric-Decomposition Approximations

In this section we discuss heuristic approximations for queueing networks. These approximation are called *parametric-decomposition approximations* because the queues are analyzed separately after approximately characterizing the arrival process at each queue by two parameters; see Whitt (1983a,b, 1995) and Buzacott and Shanthikumar (1993). (The first work in this direction was done by Reiser and Kobayashi (1974), Sevcik et al. (1977) and Kuehn (1979).)

We discuss parametric-decomposition approximations here because heavy-traffic limits can play an important role in choosing appropriate variability parameters. Indeed, important insight is provided by the heavy-traffic limit for a queue with a superposition arrival process, where the number of component arrival processes increases in the heavy traffic limit, just considered in the previous section.

With parametric-decomposition approximations, the goal is to obtain improved performance predictions compared to more elementary one-parameter models such as the $M/M/1$ queue and the single-class open Jackson queueing network (a network of $M/M/1$ queues with Markovian routing); see Jackson (1957, 1963). Variability has an impact on the performance of these one-parameter models, but they provide no parameters to quantify the degree of variability.

In this section we are primarily interested in exploiting heavy-traffic limits to improve the quality of parametric-decomposition approximations. Along the way, we point out significant difficulties, where initial simple approaches break down. When considering approximation errors, it is good to keep in mind that in engineering applications the error in model fit is usually larger than the error in approximating the solution of the model.

We start by considering how to approximately characterize the distribution of a nonnegative real-valued random variable. It is natural to partially characterize the distribution by its mean and squared coefficient of variation (SCV, variance divided

9.9. Approximations for Queueing Networks

by the square of the mean). Thus it is natural to partially characterize a renewal process by the mean and SCV of the interrenewal time.

However, it is difficult to adequately characterize a general stationary arrival process by only two parameters, because in addition to the general interarrival-time distribution, there may be complicated dependence among the interarrival times. For models, the arrival rate can be determined exactly. The difficulty is in finding an appropriate second parameter to characterize the variability.

The variability of a general stationary arrival process often looks different in different time scales. Consequently, the variability impact on the congestion in a following queue often depends on the traffic intensity of that queue. Hence, following Whitt (1995), in order to partially characterize a general stationary arrival process, we propose using the arrival rate and a *variability function* that gives a variability parameter as a function of the traffic intensity in a following queue. When that arrival process appears in a queue, we obtain a variability parameter by evaluating the variability function at the traffic intensity of the queue. (Fendick and Whitt (1989) investigate how variability as a function of *time* in an arrival process can be converted into variability as a function of the *traffic intensity* in a following queue.)

In a typical queueing-network application, there are multiple classes of customers, each with their own arrival, service and routing pattern. It is often realistic to assume that the routing is primarily deterministic for each class, so we will consider the case of deterministic routing. With deterministic routing, there is an exogenous arrival process to some queue for each class, which we will regard as a renewal process partially characterized by the mean and the SCV of an interrenewal time. (We will later discuss extensions to nonrenewal arrival processes partially characterized by variability functions.) Each customer visits a sequence of queues in the network, possibly returning to the same queue more than once, and then leaves the network. At each queue on the customer's route, there is a service-time distribution, which is partially characterized by its mean and SCV. It is assumed that the arrival process and the service times are mutually independent. The service-time distributions may differ at different queues. The service-time distributions also may differ at the same queue for different customers or even for the same customer upon different visits to that queue.

The model data for one customer class might be the vector

$$(1, 2, 4; 2, 1, 0; 3, 1, 1; 2, 5, 1) \ . \tag{9.1}$$

The first triple describes the exogenous arrival process: Customers from that class enter the network at queue 1 with an exogenous renewal arrival process having arrival rate 2 and interarrival-time SCV 4. Afterwards, these customers visit queues 2, 3 and 2, in that order, and then leave the network. On the first visit to queue 2, the service time has mean 1 and SCV 0, while on the second visit to queue 2 the service time has mean 5 and SCV 1.

We must also specify the queues. For simplicity, we will consider only single-server queues with unlimited waiting room and the FCFS service discipline, but clearly the general approach can accomodate a wide variety of queues.

Given the specified model data partially characterizing the arrival and service processes of each customer class in an open queueing network of single-server FCFS queues, the goal is to describe the performance. We want to determine approximate queue-length distributions at each queue and approximate sojourn-time (time-in-system) distributions for each customer class.

Here we will only discuss the mean (steady-state) sojourn time for one class. The mean sojourn time is the sum of the mean waiting times (before beginning service) and the mean service times at all the queues on the customer's route. The mean service times are directly specified for each class as part of the model data, so we use them. (To do otherwise could introduce large errors unnecessarily.) In general, the waiting-time distribution and its mean can depend on the customer class, but we will use an approximation for the mean waiting time for an arbitrary customer at that queue. Hence, here our goal reduces to developing an approximation for the mean waiting time for an arbitrary customer at each queue in the network.

To determine the approximate mean waiting time at any single queue, we act as if we have a $GI/GI/1$ queue partially characterized by the mean λ^{-1} and SCV c_a^2 of an interarrival time and the mean μ^{-1} and SCV c_s^2 of a service time. We will use the heavy-traffic approximation, refined by the exact $M/GI/1$ formula, namely,

$$EW \equiv EW(\lambda, c_a^2, \mu, c_s^2) \approx \frac{\mu^{-1}\rho(c_a^2 + c_s^2)}{2(1-\rho)}, \qquad (9.2)$$

where $\rho \equiv \lambda/\mu$ is the traffic intensity. (We obtain (9.2) by multiplying (6.13) by ρ^2, which provides an asymptotically exact formula as $\rho \uparrow 1$ for any $GI/GI/1$ queue.

Part of the overall approximation error is due to using formula (9.2) for a $GI/GI/1$ queue. It is natural to ask about the range of possible mean-waiting-time values consistent with the four specified parameters; that is investigated for the $GI/M/1$ special case in Whitt (1984b,c) and Klincewicz and Whitt (1984). The range of possible values given the partial specification is quite large, e.g., the relative error could well be 100%, but for "typical" distributions, the range is not great, so that the relative error might be only 10%. However, touting much better accuracy, such as 1% relative error, for specific interarrival-time and service-time distributions is pointless because we can find different distributions that yield larger errors.

It is possible to improve (9.2), but we cannot escape the inevitable error caused by the partial characterization. A good refinement to approsimation (9.2), which makes EW smaller in some cases, was developed by Kraemer and Langenbach-Belz (1976). Possible refinements are discussed in Whitt (1983a, 1989b, 1993a) and references cited there.

Given approximation (9.2), the problem is to approximate the arrival and service processes at each queue in the network by the arrival and service processes in a $GI/GI/1$ queue partially characterized by the parameter four-tuple $(\lambda, c_a^2, \mu, c_s^2)$.

We start by treating the aggregate exogenous arrival process at each queue as the superposition of the single-class exogenous arrival processes at that queue. The exogenous arrival rate clearly should be the sum of the component single-class exogenous arrival rates at that queue. The variability function for the aggregate exogenous arrival process is more complicated and will be discussed later. The routing of customers within the network is treated as Markovian: The probability $P_{i,j}$ of a customer going next to queue j after completing service at queue i is made equal to the long-run proportion of departing customers from queue i that are routed next to queue j. At each queue, the first two moments of the aggregate service-time distribution is just the weighted (by the arrival rates) average of the moments of the individual service-time distributions. The service-time SCV is defined in terms of the first two moments in the usual way: $c_s^2 + 1 \equiv E[V^2]/(E[V])^2$.

We act as if the service times do not need great care, and that often is the case. However, Example 9.6.1 illustrates how there can be significant dependence among successive service times and significant dependence between interarrival times and service times. In any specific application setting, it is good to have verification by simulation and measurement. We can verify both the final performance predictions and the variability characterizations of arrival and service processes.

It is straightforward accounting to produce aggregate exogenous arrival rates, Markovian routing probabilities at each queue and service-time distributions partially characterized by their first two moments. Indeed, the first-order deterministic rate parameters are exact in the specified procedure.

In engineering applications of queueing network analyzers, e.g., in the design of a manufacturing facility, usually most of the benefit is gained from the initial phase of the analysis. In the initial planning stages, the model formulation and accounting identify queues with unacceptably high traffic intensities (e.g., $\rho_i > 1$).

A second benefit that occurs before solving the model comes from having a model with an explicit quantification of variability. The form of the required model data focuses attention on variability. It indicates what should be measured. To build the queueing-network model, the engineers must look at process variability. When engineers attempt to measure and quantify the variability of arrival and service processes, they often discover opportunities to reduce that variability and make dramatic improvements in system performance.

Returning to the parametric-decomposition approximation, it remains to determine the SCV of the renewal arrival process approximating the arrival process at each queue. We can decompose the final approximation of c_a^2 for one such queue into two steps: (i) approximating the exogenous arrival process at each queue by a renewal process partially characterized by its rate and SCV, and (ii) approximating the net arrival process at the queues in the network by renewal arrival processes partially characterized by their rates and SCV's. (The exact rates of both the exogenous and aggregate arrival processes have already been determined.)

The second step involves developing an approximation for a *generalized Jackson network*, which is a single-class queueing network with Markovian routing, mutually independent renewal exogenous arrival processes and IID service times at the queues. For the generalized Jackson network considered here, the interarrival-time and service-time distributions are only partially characterized by their first two moments or, equivalently, by their means and SCV's. Dividing the overall approximation into two steps allows us to focus on the accuracy of each step separately.

9.9.2. *Approximately Characterizing Arrival Processes*

We now discuss ways to approximate a general arrival process with stationary interarrival time sequence $\{U_k : k \geq 1\}$ by a renewal process partially characterized by the mean λ^{-1} and SCV c_a^2 of an interarrival time. Following Whitt (1982a), we observe that there are two natural ways: In both ways, we let the arrival rate be specified exactly by letting $\lambda^{-1} = EU_1$. The *stationary-interval method* lets the SCV c_a^2 be the SCV of one interval U_1, i.e., we let

$$c_a^2 \approx c_{SI}^2 \equiv c_u^2 \equiv Var(U_1)/(EU_1)^2 , \qquad (9.3)$$

as in (6.7). The *asymptotic method* lets c_a^2 be the scaled asymptotic variance

$$c_a^2 \approx c_{AM}^2 \equiv c_U^2 \equiv \lim_{n \to \infty} \frac{Var\, S_n^u}{n(EU_1)^2} , \qquad (9.4)$$

where $S_n^u \equiv U_1 + \cdots + U_n$, $n \geq 1$, as in (6.11).

Under the regularity condition of uniform integrability, the asymptotic method in (9.4) is equivalent to c_U^2 being the dimensionless space-scaling constant in the CLT for S_n^u or the associated arrival counting process $A(t)$, i.e.,

$$(c_U^2 \lambda^{-2} n)^{-1/2}(S_n^u - \lambda^{-1} n) \Rightarrow N(0,1) \qquad (9.5)$$

or, equivalently,

$$(\lambda c_U^2 t)^{-1/2}(A(t) - \lambda t) \Rightarrow N(0,1) ; \qquad (9.6)$$

see Sections 7.3 and 13.8.

The stationary-interval method in (9.3) ignores any correlations among successive interarrival times. At first glance, it might appear that the stationary-interval method is *implied* by a renewal-process approximation, because there are *no correlations* in a renewal process, but that is not so. Even though the approximating process is to be viewed as a renewal process, it is important *not* to ignore the correlations in the arrival process being approximated if significant correlations are there.

In contrast, the asymptotic method includes *all the correlations* in the arrival process being approximated. From the Brownian heavy-traffic limits for general $G/G/1$ queues discussed in Section 9.6, we know that the asymptotic method is asymptotically correct in heavy traffic, using the mean waiting-time formula in (9.2). Thus, the heavy-traffic limit provides a very important reference point for these heuristic approximations.

However, in light traffic the long-run correlations among interarrival times obviously are not relevant, so that the stationary-interval method seems intuitively better in light traffic. Indeed, the stationary-interval method usually performs well in light traffic. To appreciate this discussion, it is important to realize that the two approximation procedures can both perform well in their preferred regimes, and yet c_{AM}^2 can be very very different from c_{SI}^2. (We will give examples below.) Thus neither procedure alone can always work well.

Thus, an effective approximation procedure needs to involve a compromise between the two basic approaches. As mentioned in Section 5.7, one possible approach is to interpolate between light-traffic and heavy-traffic limits, but we do not discuss that approach.

9.9.3. A Network Calculus

A parametric-decomposition algorithm for open queueing networks provides an algorithm for calculating the approximate arrival-process variability parameter c_a^2 at each queue in the network. That variability parameter will subsequently be used, together with the exact arrival rate, to approximately characterize an approximating renewal arrival process at that queue. The overall algorithm for calculating the arrival-process variability parameters can be based on a *network calculus* that transforms arrival-process variability parameters for each of the basic network operations: superposition, splitting and departure (flow through a queue).

When the network is acyclic, the basic transformations can be applied sequentially, one at a time, but in general it is necessary to solve a system of equations in order to calculate the final variability parameters. Solving the equations becomes elementary if all the transformations are linear. Then the final algorithm involves solving a system of linear equations, with one equation for each queue. Hence there is motivation for developing linear approximations to characterize each transformation. The synthesis into a final system of linear equations is relatively straightforward, see Whitt (1983a, 1995); we will not discuss it here.

Here we will only discuss the basic transformations and the initial choice of variability parameters. As mentioned earlier, we will focus on variability functions instead of variability parameters. Given a variability function $\{c_a^2(\rho) : 0 \leq \rho \leq 1\}$, we obtain a specific variability parameter when we specify the traffic intensity at the queue.

Superposition. Superposition applies first to the exogenous arrival process at each queue and then to the aggregate or net arrival process at each queue, including departures routed from other queues. Suppose that we have the superposition of m independent renewal counting processes $A^i(t)$ with rates λ_i and SCV's $c_{a,i}^2$. As indicated above, these parameters can be determined from the first two moments of an interarrival time. Alternatively, the parameters can be determined from a CLT for A^i of the form

$$[A^i(t) - \lambda_i t]/\sqrt{\lambda_i c_{a,i}^2} \Rightarrow N(0,1) . \qquad (9.7)$$

Clearly the rate of the superposition process $A \equiv A^1 + \cdots A^m$ is $\lambda \equiv \lambda_1 + \cdots + \lambda_m$. It follows from Theorem 9.4.1 and Corollary 9.4.1 that the appropriate asymptotic-method approximation for c_a^2 is the weighted average of the component SCV's, i.e.,

$$c_{AM}^2 = \sum_{i=1}^{m} (\lambda_i/\lambda) c_{a,i}^2 \qquad (9.8)$$

The stationary-interval method for superposition processes is more complicated, as can be seen from exact formulas in Section 4.1 of Whitt (1982a). However, by

Theorem 9.8.1, for large m the superposition process behaves locally like a Poisson process, so that a large-m stationary-interval approximation is

$$c_{SI}^2 \approx 1 \ . \tag{9.9}$$

Notice that we have a demonstration of the inconsistency of the two basic approximation methods: For a superposition of m IID renewal processes, no matter how large is the interarrival-time SCV in a component arrival process, the superposition process approaches a Poisson process as $m \to \infty$. If the traffic intensity in a following queue is not too large, then the congestion at the queue is essentially the same as if the superposition arrival process were a Poisson process. On the other hand, for any fixed m, if the traffic intensity is high enough, the heavy-traffic limit is approximately correct. Since c_{AM}^2 can be arbitrarily large, the error from making the wrong choice can be arbitrarily large. (For further discussion, see Section 9.8.1 above.)

On the other hand, if $c_{AM}^2 \approx 1$, then the two basic methods are consistent and a Poisson-process approximation for the arrival process, which has $c_a^2 = 1$, is likely to perform well in many applications. However, if c_{AM}^2 is not near c_{SI}^2, then we can consider that a demonstration that the actual arrival process is not nearly a renewal process. Nevertheless, it may be possible to choose a variability parameter c_a^2 so that (9.2) is a reasonably good approximation for the mean waiting time.

The problem then is to find a compromise between the asymptotic method and the stationary-interval method that is appropriate for the queue. In general, that should depend upon the traffic intensity in the following queue. From the heavy-traffic limits in Section 9.3, it follows that the asymptotic method is asymptotically correct for the queue as $\rho \uparrow 1$, so that we should have $c_a^2(\rho) \to c_{AM}^2$ as $\rho \uparrow 1$. On the other hand, for very small ρ it is apparent that the stationary-interval method should be much better, so that we should have $c_a^2(\rho) \to c_{SI}^2$ as $\rho \downarrow 0$.

We can use Theorem 9.8.3 as a theoretical basis for a refined approximation. From Theorem 9.8.3, we know that, for superposition arrival processes with m component arrival processes, where $m \to \infty$, the asymptotic method is asymptotically correct for the queue as $\rho \to 1$ only if $m(1-\rho)^2 \to 0$. Thus, with superposition arrival processes, the weight on the asymptotic method should be approximately inversely proportional to $m(1-\rho)^2$.

In general, we want to treat superposition arrival processes where the component arrival processes have different rates. The number m has precise meaning in the expression $m(1-\rho)^2$ above only for identically distributed component processes. If one component process has a rate much larger than the sum of the rates of all other component processes, then the effective number should only be slightly larger than 1, regardless of m. However, it is not difficult to identify appropriate "equivalent numbers" of component processes that allow for unequal rates.

The considerations above lead to generalizations of the approximation used in the queueing network analyzer (QNA) software tool; see Whitt (1983a, 1995), Albin (1984) and Segal and Whitt (1989).

Specifically, an approximating variability function $c_a^2(\rho)$ for the superposition arrival process is

$$c_a^2(\rho) \approx wc_{AM}^2 + (1-w)c_{SI}^2$$
$$\approx w\left(\sum_{i=1}^m (\lambda_i/\lambda)c_{a,i}^2(\rho)\right) + (1-w) ,\qquad(9.10)$$

where

$$w \equiv w(\rho,\nu) \equiv [1 + 4(1-\rho)^2(\nu-1)]^{-1} \qquad(9.11)$$

with

$$\nu \equiv \left[\sum_{i=1}^m (\lambda_i/\lambda)^2\right]^{-1}. \qquad(9.12)$$

The parameter ν in (9.12) is the "equivalent number" of component arrival streams, taking account of unequal rates. When $m = 1$, $\nu = 1$ and $c_a^2(\rho) = c_{a,1}^2(\rho)$. In (9.10) we use the approximation $c_{SI}^2 \equiv c_{a,i}^2(0) \approx 1$ motivated by Theorem 9.8.1. Notice that w as a function of ρ and ν is roughly consistent with the scaling in (8.9) in Theorem 9.8.3: The complex limit occurs as $\nu(1-\rho)^2$ converges to a nondegenerate limit.

Splitting. When the routing is Markovian and we start with a renewal process, the split processes are also renewal processes, so that $c_{AM}^2 = c_{SI}^2$. If a renewal arrival process with interarrival times having mean λ^{-1} and SCV c_a^2 is split into m streams, with the probability being p_i of each point being assigned to the ith split stream, then the mean and SCV of the interarrival time in the ith split stream are

$$\lambda_i^{-1} = (\lambda p_i)^{-1} \quad \text{and} \quad c_{a,i}^2 = p_i c_a^2 + 1 - p_i ,\qquad(9.13)$$

as can be deduced from Theorem 9.5.1.

We now want to extend the splitting formula to independent splitting from more general nonrenewal processes. Now the original arrival process is partially characterized by its arrival state λ and its variability function $\{c_a^2(\rho) : 0 \leq \rho \leq 1\}$, where ρ is the traffic intensity at the following queue. A natural generalization of (9.13) is

$$\lambda_i = \lambda p_i \quad \text{and} \quad c_{a,i}^2(\rho) = p_i c_a^2(\rho) + 1 - p_i \qquad(9.14)$$

for $0 \leq \rho \leq 1$.

However, formulas (9.13) and (9.14) can perform poorly when the routing is not actually Markovian. Discussions of alternative approximations associated with nonMarkovian routing appear in Bitran and Tirupati (1988) and Whitt (1988, 1994, 1995). In particular, as noted in Section 9.8.2 above, when there are multiple classes with each class having its own deterministic routing, we can use the separation of time scales to deduce that the single class departure process is closely related to the single class arrival process: With many classes, the queue operates in a shorter time scale than the flow for one customer class. Then the customer sojourn times, being relatively short compared to the single-class interarrival times, tend to make the single-class departure process differ little from the single-class arrival process.

Suppose that there are m single-class arrival processes with variability functions $\{c_{a,i}^2(\rho) : 0 \leq \rho \leq 1\}$ for $1 \leq i \leq m$. Let $\{c_{d,i}^2(\rho) : 0 \leq \rho \leq 1\}$ be the associated variability functions for the single-class departure processes from that queue. The separation of time scales suggests that, when $\lambda_i \ll \lambda$ (which usually occurs because of large m), we should have

$$c_{d,i}^2(\rho) \approx c_{a,i}^2(\rho), \quad 0 \leq \rho \leq 1 , \tag{9.15}$$

The approximation (9.15) treats departure and splitting together in one step.

Departures. The stationary interval between departures in a $GI/GI/1$ queue partially characterized by the parameter four-tuple $(\lambda, c_a^2, \mu, c_s^2)$ has mean λ^{-1} and SCV

$$c_d^2 = c_a^2 + 2\rho^2 c_s^2 - 2\rho(1-\rho)\mu EW . \tag{9.16}$$

Hence we can use (9.2) to produce an approximation for the stationary-interval SCV of a departure process,

$$c_{SI}^2 \approx \rho^2 c_s^2 + (1 - \rho^2) c_a^2 ; \tag{9.17}$$

see Whitt (1984d). However, except for the $M/M/1$ queue, the departure process is not a renewal process. Hence there are correlations among successive interdeparture times that are not captured by approximation (9.17). Nevertheless, simulation experiments indicate that approximation (9.17) often performs remarkably well. For example, simulations indicate that approximations (9.17) and (9.2) together work well to determine the best order for queues in series (to minimize the mean steady-state sojourn time, given a fixed arrival process); see Whitt (1985b) and Suresh and Whitt (1990b).

As noted in Remark 5.3.1, for $0 < \rho < 1$, the departure process obeys the same CLT as the arrival process. Thus the asymptotic-method approximation for the departure process is

$$c_{AM}^2 = c_a^2 . \tag{9.18}$$

To highlight the difference between (9.17) and (9.18), consider two queues in series – the $GI/GI/1 \to /GI/1$ model. Let ρ_i be the traffic intensity, $c_{a,i}^2$ the interarrival-time SCV and $c_{s,i}^2$ the service-time SCV at queue i for $i = 1, 2$. First, it is evident that the departure process from queue 1 approaches the service process there as $\rho_1 \uparrow 1$. Consistent with that property, $c_{SI}^2 \to c_{s,1}^2$ as $\rho_1 \uparrow 1$ by (9.17). On the other hand, the asymptotic-method approximation is asymptotically correct for the arrival process at the second queue as $\rho_2 \uparrow 1$. Hence $c_{AM}^2 = c_{a,1}^2$ is asymptotically correct for $c_{a,2}^2$ as $\rho_2 \uparrow 1$ for fixed ρ_1.

Now, turning to the variability functions, a candidate approximation consistent with the reference point above is

$$c_{a,2}^2(\rho_2) = c_{d,1}^2(\rho_1, \rho_2) = \alpha(\rho_1, \rho_2) c_{s,1}^2 + (1 - \alpha(\rho_1, \rho_2)) c_{a,1}^2(\rho_2) , \tag{9.19}$$

where $\alpha(\rho_1, \rho_2) \uparrow 1$ as $\rho_1 \uparrow 1$ and $\alpha(\rho_1, \rho_2) \downarrow 0$ as $\rho_2 \uparrow 1$. A specific candidate that agrees with (9.17) unless $\rho_2 > \rho_1$ is

$$\alpha(\rho_1, \rho_2) = \rho_1^2 \min\{1, (1-\rho_2)^2/(1-\rho_1)^2\}, \qquad (9.20)$$

but further study is needed.

Example 9.9.1. *The heavy-traffic bottleneck phenomenon.* The purpose of this example is to demonstrate the need for variability functions instead of variability parameters to partially characterize arrival processes in parametric-decomposition approximations. We consider a large number n of queue in series, all with relatively low traffic intensity ρ_1, followed by a $(n+1)^{\text{st}}$ queue with high traffic intensity ρ_{n+1}.

To be concrete, we consider a $GI/M/1 \to /M/1 \to \cdots \to /M/1$ model with a rate-1 renewal arrival process partially characterized by its SCV $c_{a,1}^2$. The service-time distributions are all exponential, so that $c_{s,i}^2 = 1$ for all i. The mean service time and traffic intensity at each of the first n queues is ρ_1, while the traffic intensity at the final $(n+1)^{\text{st}}$ queue is ρ_{n+1}.

It is known that as n increases, the stationary departure process from the nth queue approaches a Poisson process; see Mountford and Prabhakar (1995), Mairesse and Prabhakar (2000) and references cited there. Consistent with that limit, the stationary-interval approximation in (9.17) for the SCV $c_{a,n+1}^2$ satisfies

$$c_{SI,n+1}^2 = (1-\rho_1^2)^n c_a^2 + (1-(1-\rho_1^2)^n) \to 1 \qquad (9.21)$$

as $n \to \infty$. On the other hand, for any fixed ρ_1, the final $(n+1)^{\text{st}}$ queue has a heavy-traffic limit that depends on the first n queues only through the exogenous arrival rate 1 and the SCV $c_{a,1}^2$.

We now describe a simulation experiment conducted by Suresh and Whitt (1990a) to show that this heavy-traffic bottleneck phenomenon is of practical significance. To consider "typical" values, they let $n = 8$, $\rho_1 = 0.6$ and $\rho_9 = 0.9$. (The initial traffic intensity is not too low, while the final traffic intensity is not too high.) Two renewal arrival processes are considered: hyperexponential interarrival times (mixtures of two exponential distributions) with $c_{a,1}^2 = 8.0$ and deterministic interarrival times with $c_{a,1}^2 = 0.0$, representing high and low variability.

We compare simulation estimates of the mean steady-state waiting times with three approximations. In all three approximations, the approximation formula is

$$EW \approx \frac{\rho^2(c_a^2+1)}{2(1-\rho)}, \qquad (9.22)$$

which is obtained from (9.2) by letting $\mu^{-1} = \rho$ and $c_s^2 = 1$. The three approximations differ in their choice of the arrival-process variability parameter c_a^2: The asymptotic-method (or heavy-traffic) approximation lets $c_a^2 = c_{a,1}^2$; the stationary-interval approximation lets $c_a^2 = c_{SI,n+1}^2$; the $M/M/1$ approximation lets $c_a^2 = 1$. The SI approximation yields $c_{a,9}^2 = 1.20$ and $c_{a,9}^2 = 0.97$ in the two cases.

Table 9.2 shows the results of the simulation experiment. From Table 9.2, we see that the asymptotic-method approximation is far more accurate than the other two

336   9. Single-Server Queues

		High variability $c_{a,1}^2 = 8.0$	Low variability $c_{a,1}^2 = 0.0$
Queue 9 $\rho_9 = 0.9$	Simulation estimate	30.1 ± 5.1	5.03 ± 0.22
	asymptotic-method approximation	36.5	4.05
	stationary-interval approximation	8.9	8.0
	M/M/1 approximation	8.1	8.1
Queue 8 $\rho_8 = 0.6$	Simulation estimate	1.42 ± 0.07	0.775 ± 0.013
	asymptotic-method approximation	4.05	0.45
	stationary-interval approximation	1.04	0.88
	M/M/1 approximation	0.90	0.90

Table 9.2. A comparison of approximations with simulation estimates of the mean steady-state waiting times at queue 9 and 8 in the network of nine queues in series.

approximations at the final queue 9, while the other two approximations are far more accurate at the previous queue 8 with lower traffic intensity. The appropriate variability parameter for the arrival process clearly depends on the traffic intensity at the final queue.

Consistent with the different approximations at the queues, the measured variability parameters differ. The stationary interarrival time at queue 9 has an SCV close to 1, while the estimated asymptotic variability parameter $c_{U,9}^2$ is close to $c_{a,1}^2$. Just as with superposition arrival processes (see Albin (1982)), the individual lag-k correlations are small; $c_{U,9}^2$ differs from $c_{u,9}^2$ because of the cumulative effect of many small correlations.

Just as in examples with superposition arrival processes, the heavy-traffic bottleneck phenomenon illustrates the need for variability functions. The heavy-traffic bottleneck phenomenon also illustrates that there can be long-range variability effects in networks. High or low variability in an exogenous arrival process can be unseen (can have little congestion impact) in some queues and then suddenly appear at a later queue with a much higher traffic intensity. The reason is that different levels of variability can exist at different time scales. The arrival process to the final queue in this example looks like a Poisson process in a small time scale, but looks like the exogenous arrival process in a long time scale.

9.9.4. *Exogenous Arrival Processes*

In applications of any method for analyzing the performance of queueing networks, it is necessary to specify the exogenous arrival processes. With the parametric-decomposition approximation, it is necessary to obtain initial variability functions chacterizing the exogenous arrival processes. If the exogenous arrival processes are actually renewal processes, then there is no difficulty: then we can simply let the variability function $c_a^2(\rho)$ be the SCV of an interarrival time for all traffic intensities ρ.

However, experience indicates that, in practice (as opposed to in models), an arrival process that fails to be nearly a Poisson process also fails to be nearly a renewal process. Indeed, exogenous arrival processes often fail to be renewal processes, so that it is necessary to take care in characterizing the variability of these exogenous arrival processes. Hence, instead of the route vector in (9.1), the model data for that customer class should be of the form

$$(1, 2, \{c_{a,0}^2(\rho) : 0 \leq \rho \leq 1\}; 2, 1, 0; 3, 1, 1; 2, 5, 1) \ . \tag{9.23}$$

With variability functions, then, we should be prepared to specify the variability functions of the exogenous arrival processes. Following Whitt (1981, 1983c) and Section 3 of Whitt (1995), we suggest fitting variability parameters indirectly by observing the congestion produced by this arrival process in a test queue. This can be done either through analytical formulas (if the arrival process is specified as a tractable mathematical model) or through simulation (if the arrival process is specified either as a mathematical model or via direct system measurements).

For example, we can use approximation formula (9.2). We might consider an exponential service-time distribution, which makes $c_s^2 = 1$. We then think of the queue as a $GI/M/1$ queue, but since the arrival process may not actually be a renewal process, we allow the variability parameter to depend on the traffic intensity. We estimate the mean waiting time as a function of ρ using the arrival process to be characterized. For each value of ρ, we let the variability function $c_a^2(\rho)$ assume the value c_a^2 that makes formula (9.2) match the observed mean waiting time.

This indirect procedure is illustrated by applying it to *irregular periodic deterministic arrival processes* in Section 4 of Whitt (1995). A simple example has successive interarrival times $3/2, 1/2, 3/2, 1/2, \ldots$. Consistent with intuition, for irregular periodic deterministic arrival processes, $c_a^2(\rho) = 0$ for all sufficiently small ρ and $c_a^2(\rho) \to 0$ as $\rho \uparrow 1$, but $c_a^2(\rho)$ can be arbitrarily large for intermediate values of ρ.

This indirect estimation procedure can also be used to refine parametric-decomposition approximations for the variability functions partially characterizing the internal flows in the network. Through simulations and measurements, we can appropriate variability functions that lead to accurate performance predictions for the internal flows, just as for the exogenous arrival processes. See Fendick and Whitt (1989) and Whitt (1995) for further discussion.

338   9. Single-Server Queues

9.9.5. Concluding Remarks

We conclude this section with two remarks.

Remark 9.9.1. *Heavy-traffic limits for queueing networks.* An alternative to the parametric-decomposition approximation is an approximation based directly on a heavy-traffic limit for the queueing network. The heavy-traffic limit ideally would be for the original multiclass queueing network, but it could be for the single-class Jackson network constructed in the first phase of the procedure described above. Heavy-traffic limits for the single-class generalized Jackson network are developed in Chapter 14. A specific algorithm based on the heavy-traffic limit is the QNET algorithm of Dai (1990), Dai and Harrison (1991, 1992), Harrison and Nguyen (1990) and Dai, Yeh and Zhou (1997). A direct heavy-traffic algorithm is an attractive alternative, but the limit process is usually complicated. The computational complexity of the QNET algorithm grows rapidly as the number of nodes increases.

When considering heavy-traffic limits for queueing networks, it is important to recognize that there is more than one way to take the heavy-traffic limit. With a queueing network, there is more than one traffic intensity: There is a traffic intensity at each queue. The standard limiting procedure involves *balanced loading*, in which all the traffic intensities approach 1 together; i.e., if ρ_i is the traffic intensity at queue i, then $\rho_i \uparrow 1$ for all i with $(1-\rho_i)/(1-\rho_1) \to c_i$, $0 < c_i < \infty$.

However, the bottleneck view, stemming from consideration of a fixed network with one traffic intensity larger than the others, has one traffic intensity approach 1 faster than the others. If the traffic intensity at one queue approaches 1 faster than the traffic intensities at the other queues, then we see a nondegenerate limit for the scaled queue-length process only at the bottleneck queue. With this form of heavy-traffic limit, one queue dominates. Just as in the heavy-traffic bottleneck phenomenon, the heavy-traffic approximation is equivalent to the heavy-traffic limit in which all the service times at the other queues are reduced to zero, and the other queues act as instantaneous switches.

A more general heavy-traffic approximation for a network of queues is the sequential-bottleneck decomposition method proposed by Reiman (1990a) and Dai, Nguyen and Reiman (1994). It is a heirarchical procedure similar to the one proposed for the priority queue in Section 5.10. The sequential-bottleneck procedure decomposes the network into groups of one or more queues with similar traffic intensities. Then heavy-traffic approximations are developed for the groups separately, starting with the group with highest traffic intensities. When analyzing a subnetwork associated with a group of queues, the remaining queues are divided into two sets, those with larger traffic intensities and those with smaller traffic intensities. Queues with smaller traffic intensities are treated as if their service times are zero, so they act as instantaneous switches. Queues with larger traffic intensities are treated as if they are overloaded, which turns them into sinks for flows into them and exogenous sources for flows out of them. Then the QNET approximation is applied to each subgroup. If the subgroups only contain a single queue, then we can apply the simple single-queue heavy-traffic approximation. The single-queue case was proposed by Reiman (1990a).

The sequential-bottleneck approximation is appealing, but note that it offers no way to achieve the needed non-heavy-traffic approximation at a queue with a superposition arrival process having many components. At first glance, the single-queue sequential-bottleneck approximation seems to perform well on Example 9.9.1: It produces the heavy-traffic approximation at the final queue with high traffic intensity, which is pretty good. However, the heavy-traffic approximation at the final queue would not be good if we lowered the traffic intensity of the final queue from 0.9 to 0.61, where it still is greater than all other traffic intensities. It still remains to develop an approximation for the special $GI/M/1 \to \cdots \to /M/1$ model with nine queues in series that can be effective for all possible traffic-intensity vectors. ∎

Remark 9.9.2. *Closed queueing networks.* For many applications it is natural to use closed queueing network models, which have fixed customer populations, instead of open queueing network models. There are convenient algorithms for a large class of Markovian closed queueing network models, but nonMarkovian closed queueing network models tend to be intractable.

Approximations for nonMarkovian open queueing networks can be applied via the *fixed-population-mean* (FPM) method: The steady-state performance of the closed queueing network is approximated by the steady-state performance of an associated open network in which the mean population in the open network is set equal to the specified population in the closed network; see Whitt (1984c). A search algorithm identifies the exogenous arrival rate in the open model producing the target mean. (A more complicated search algorithm is required if there are multiple customer classes with specified populations.) The FPM method provides good approximations when the population is not too small. A stochastic-process limit provides insight: The FPM method is asymptotically correct as the network grows (in a controlled way).

The FPM method can explain seemingly anomalous behavior in nonMarkovian closed queueing networks. If the variability of the service-time distribution increases at one queue, then it is possible for the mean queue length at that queue to decrease. Indeed that phenomenon routinely occurs at a bottleneck queue; see Bondi and Whitt (1986). That occurs because the bottleneck queue tends to act as an exogenous source for the rest of the network. Thus increased variability at the bottleneck queue is likely to cause greater congestion in the rest of the network. Since the total population is fixed, the mean queue length at the bottleneck queue is likely to go down. ∎

To summarize, parametric-decomposition approximations for queueing networks can be great aids in performance analysis. And heavy-traffic limits can help improve the performance of these algorithms. However, at the present time there is no one algorithm that works well on all examples. Nevertheless, there is sufficient understanding and there are sufficient tools to make effective algorithms for many specific classes of applications.

10
Multiserver Queues

10.1. Introduction

In this chapter we establish heavy-traffic stochastic-process limits for standard multiserver queueing models (with unlimited waiting space and the first-come first-served service discipline). There are two principal cases: first, when there is a moderate number of servers and, second, when there is a large number of servers. The first case commonly occurs in manufacturing systems, which may have workstations containing several machines. The second case commonly occurs in call centers, which may have agent groups containing hundreds of agents; e.g., see Borst, Mandelbaum and Reiman (2001), Garnett, Mandelbaum and Reiman (2000) and Whitt (1999). These two cases are sufficiently different to warrant different methods.

We start in Section 10.2 by considering the first case of a fixed finite number of servers. We show that the heavy-traffic behavior for a fixed finite number of servers is essentially the same as a single-server system (with an obvious scale adjustment to account for the multiple servers).

We consider the second case of a large number of servers in the remaining sections. The natural approximation for a large number of servers is an infinite number, provided that we find appropriate measures of congestion. In Section 10.3 we consider the case of infinitely many servers.

In Section 10.4 we consider the case in which the number of servers increases along with the traffic intensity in the heavy-traffic limit, so that the probability of delay converges to a nondegenerate limit. In Section 10.4 we also consider multiserver loss systems, without any extra waiting room. Paralleling Section 5.8, we show how heavy-traffic limits can be used to determine the simulation run lengths required to estimate, with prescribed precision, blocking probabilities in loss models with a large number of servers. Interestingly, the story is quite different from the single-server case in Section 5.8.

10.2. Queues with Multiple Servers

In this section we consider a queue with a fixed finite number, m, of servers. We should anticipate that the heavy-traffic behavior of a multiserver queue is essentially the same as the heavy-traffic behavior of a single-server queue with a superposition arrival process, treated in Section 9.4: Now we have a multichannel output process instead of a multichannel input process. Indeed, that is the case, but providing a proper demonstration is somewhat more difficult.

10.2.1. A Queue with Autonomous Service

One approach to this problem, used by Borovkov (1965) and Iglehart and Whitt (1970a, b), is to relate the standard multiserver queue to another model that is easier to analyze. Another model that is easier to analyze is a *queue with autonomous service*, in which servers are not shut off when they become idle. Associated with each of the m servers is a sequence of potential service times. If a server faces continued demand, then the actual service times coincide with these potential service times, but if there is no demand for service, then these potential service times are ignored and there is no actual service and no departure. New arrivals are assigned to the server that can complete their service first. After a server has been working in the absence of demand, the next demand will in general fall in the middle of a service time. The service time of that customer arriving after the server has been idle is the remaining portion of the potential service time in process at that time.

The queue with autonomous service is of some interest in its own right, but it was introduced primarily as a device to treat the standard multiserver queueing model. In the standard model, customers are assigned in order of arrival to the first available server, with some unspecified procedure to break ties. The queue with autonomous service is easy to analyze because the reflection map can be applied directly.

The service process in the queue with autonomous service is just like a superposition arrival process. Let $\{V_{i,k} : k \geq 1\}$ be the sequence of potential service times for server i for $1 \leq i \leq m$. Let the associated partial sums be

$$S_{i,k}^v \equiv V_{i,1} + \cdots + V_{i,m}, \quad k \geq 1, \tag{2.1}$$

and let $S_{i,0}^v \equiv 0$. Let N_i be the associated counting process, defined as in (2.9) by

$$N_i(t) \equiv max\{k \geq 0 : S_{i,k}^v \leq t\}, \quad t \geq 0. \tag{2.2}$$

Let N be the superposition process defined by

$$N(t) \equiv N_1(t) + \cdots + N_m(t), \quad t \geq 0 \tag{2.3}$$

10.2. Queues with Multiple Servers

and let S^v be the inverse partial-sum process associated with N, defined by

$$S_k^v \equiv inf\{t \geq 0 : N(t) \geq k\}, \quad k \geq 0, \tag{2.4}$$

with $S_0^v \equiv 0$. Let

$$V_k \equiv S_k^v - S_{k-1}^v, \quad k \geq 1 . \tag{2.5}$$

Note that the potential-service processes S_i^v, N_i, N, S^v and V are related the same way that superposition-arrival-process processes S_i^u, A_i, A, S^u and U are related in (4.1)–(4.4).

For simplicity, let the queue start out empty. Let $\{A(t) : t \geq 0\}$ be the arrival counting process and let $Q^a(t)$ be the queue length, with the superscript "a" denoting autonomous service. The queue with autonomous service is defined so that the queue-length process is directly the reflection of the net-input process $A - N$, i.e.,

$$Q^a(t) = \phi(A - N)(t), \quad t \geq 0 , \tag{2.6}$$

where ϕ is the one-sided reflection map in (2.5) of Section 9.2. By focusing on the m-server queue with autonomous service, the queue-length process becomes a special case of the fluid-queue model in Chapter 8 in which the cumulative-input and available-processing processes are integer-valued. The arrival process A is the cumulative-input process, while the service-time counting process N is the available-processing process.

Remark 10.2.1. *The case of IID exponential service times.* To better understand the queue with autonomous service, consider the special case in which the potential service times come from m independent sequences of IID exponential random variables with mean 1, independent of the arrival process. Then the queue with autonomous service is not equivalent to the $G/M/m$ queue, but is instead equivalent to the $G/M/1$ queue with the service rate m. In general, the $G/M/1$ queue with service rate m is quite different from the associated $G/M/m$ queue, having identical arrival process and m servers each with rate 1. However, for fixed m, in heavy traffic they behave essentially the same. Indeed, that is established as a special case of Theorem 10.2.2 below. ∎

Now let us consider a sequence of these queueing models with autonomous service indexed by n. Paralleling (4.5), define associated random elements of $D \equiv D([0, \infty), \mathbb{R})$ by letting

$$\begin{aligned}
\mathbf{S}_{n,i}^v(t) &\equiv c_n^{-1}[S_{n,i,\lfloor nt \rfloor}^v - \mu_{n,i}^{-1}nt], \\
\mathbf{N}_{n,i}(t) &\equiv c_n^{-1}[N_{n,i}(nt) - \mu_{n,i}nt], \\
\mathbf{N}_n(t) &\equiv c_n^{-1}[N_n(nt) - \mu_n nt], \\
\mathbf{S}_n^v(t) &\equiv c_n^{-1}[S_{n,\lfloor nt \rfloor}^v - \mu_n^{-1}nt] ,
\end{aligned} \tag{2.7}$$

where $\mu_n \equiv \mu_{n,1} + \cdots + \mu_{n,m}$. A limit for $(\mathbf{S}_{n,1}^v, \ldots, \mathbf{S}_{n,m}^v)$ implies an associated limit for $(\mathbf{S}_{n,1}^v, \ldots, \mathbf{S}_{n,m}^v, \mathbf{N}_{n,1}, \ldots \mathbf{N}_{n,m}, \mathbf{N}_n, \mathbf{S}_n^v)$ by Theorem 9.4.1.

Now define additional random elements of D associated with the arrival and queue-length processes A and Q^a by

$$\mathbf{A}_n(t) \equiv c_n^{-1}[A_n(nt) - \lambda_n nt],$$
$$\mathbf{Q}_n^a(t) \equiv c_n^{-1} Q_n^a(nt), \quad t \geq 0 \,. \tag{2.8}$$

The following heavy-traffic limit parallels Theorem 9.4.2, and is proved in the same way.

Theorem 10.2.1. (heavy-traffic limit for the m-server queue with autonomous service) *Suppose that*

$$(\mathbf{A}_n, \mathbf{S}_{n,1}^v, \ldots \mathbf{S}_{n,m}^v) \Rightarrow (\mathbf{A}, \mathbf{S}_1^v, \ldots, \mathbf{S}_m^v) \quad \text{in} \quad (D^{1+m}, WM_1) \tag{2.9}$$

for $\mathbf{A}_n$ in (2.8) and $\mathbf{S}_{n,i}^v$ in (2.7). Suppose that, for $1 \leq i \leq m$,

$$P(\mathbf{A}(0) = 0) = P(\mathbf{S}_i^v = 0) = 1 \,, \tag{2.10}$$

$c_n \to \infty$, $n/c_n \to \infty$, $\lambda_{n,i} \to \lambda_i$, $0 < \lambda_i < \infty$, *and*

$$\eta_n \equiv n(\lambda_n - \mu_n)/c_n \to \eta \quad \text{as} \quad n \to \infty \tag{2.11}$$

for λ_n in (2.8) and $\mu_n \equiv \mu_{n,1} + \cdots + \mu_{n,m}$. Suppose that

$$P(Disc(\mathbf{S}_i^v \circ \mu_i \mathbf{e}) \cap Disc(\mathbf{S}_j^v \circ \mu_j \mathbf{e}) = \phi) = 1 \tag{2.12}$$

and

$$P(Disc(\mathbf{S}_i^v \circ \mu_i \mathbf{e}) \cap Disc(\mathbf{A}) = \phi) = 1 \tag{2.13}$$

for all i, j with $1 \leq i, j \leq m$, $i \neq j$. Then

$$(\mathbf{A}_n, \mathbf{N}_n, \mathbf{Q}_n^a) \Rightarrow (\mathbf{A}, \mathbf{N}, \mathbf{Q}) \quad \text{in} \quad (D^3, WM_1) \,,$$

where

$$\mathbf{N} = \mathbf{N}_1 + \cdots + \mathbf{N}_m = -\sum_{i=1}^m \mu_i \mathbf{S}_i^v \circ \mu_i \mathbf{e} \tag{2.14}$$

and

$$\mathbf{Q} = \phi(\mathbf{A} - \mathbf{N} + \eta \mathbf{e}) \,. \tag{2.15}$$

Remark 10.2.2. *Resource Pooling with IID Lévy processes.* Just as in Remark 9.4.1, if the m limit processes $\mathbf{S}_1^v, \ldots, \mathbf{S}_m^v$ are IID Lévy processes, then the limit processes associated with the superposition process are deterministic time-scalings of the limit process associated with a single server, i.e.,

$$\mathbf{N} \stackrel{d}{=} \mathbf{N}_1 \circ m\mathbf{e} \quad \text{and} \quad \mathbf{S}^v \stackrel{d}{=} m^{-1} \mathbf{S}_1^v \,.$$

Then the m servers act as a "single super server" and we say that there is *resource pooling*. We discuss resource pooling with other queue disciplines in Remark 10.2.4 below. ∎

10.2.2. The Standard m-Server Model

We now want to consider the standard m-server queue, in which customers wait in a single queue and are assigned in order of arrival to the first available server, with some unspecified procedure to break ties. We now want to show that the queue-length process in the standard m-server queue has the same heavy-traffic limit.

Iglehart and Whitt (1970a, b) showed that the scaled queue-length processes in the two systems are asymptotically equivalent under regularity conditions. As before, let $\mathbf{Q}_n^a$ denote the scaled queue-length process in the m-server queue with autononous service, just considered, and let $\mathbf{Q}_n$ denote the scaled queue-length process in the standard system, with the same scaling as in (2.8). The next result follows from Theorem 10.2.1 and the reasoning on pages 159–162 in Iglehart and Whitt (1970a).

Theorem 10.2.2. (asymptotic equivalence with the standard m-server queue) *In addition to the assumptions of Theorem 10.2.1, suppose that* $\{V_{i,k}^v : k \geq 1\}$, $1 \leq i \leq m$, *are m independent sequences of IID random variables, independent of the arrival process. If, in addition,*

$$P(\mathbf{S}_i^v \in C) = 1 \quad (2.16)$$

for each i, $1 \leq i \leq m$, then there exist versions $\mathbf{Q}_n$ of the scaled standard queue-length process so that

$$\|\mathbf{Q}_n - \mathbf{Q}_n^a\|_t \to 0 \quad w.p.1 \quad as \quad n \to \infty, \quad (2.17)$$

for all $t > 0$, so that

$$\mathbf{Q}_n \Rightarrow \mathbf{Q} \quad in \quad (D, M_1),$$

where $\mathbf{Q}$ is as in (2.15).

Theorem 10.2.2 treats only one process in the standard multiserver queue. Under the assumptions of Theorem 10.2.2, heavy-traffic limits can also be obtained for other processes besides the queue-length process, as shown by Iglehart and Whitt (1970a, b).

Condition (2.16) requires the service-time limit processes $\mathbf{S}_i^v$ to have continuous paths. The asymptotic-equivalence argument in Iglehart and Whitt (1970a) does *not* apply if these processes can have jumps, because the difference is bounded above by the largest individual service time encountered, appropriately scaled. When the limit process has continuous sample paths, we can apply the maximum-jump function to conclude that this scaled maximum serice time is asymptotically negligible, but when the limit process has discontinuous sample paths, we cannot draw that conclusion.

Under the conditions of Theorem 10.2.2, the arrival-process limit process $\mathbf{A}$ can have discontinuous sample paths, so that in general we still need the M_1 topology to express the limit. The arrival process can force nonstandard scaling, but that may make the limit processes $\mathbf{S}_i^v$ be degenerate (zero processes), because the possibilities for the limit processes $\mathbf{S}_i^v$ are limited. Since the potential service times must

346   10. Multiserver Queues

come from sequences of IID random variables, to have the limit processes $\mathbf{S}_i^v$ be nondegenerate with continuous sample paths in Theorem 10.2.2, we are effectively restricted to the case in which $c_n = n^{1/2}$ and $\mathbf{S}_i^v$ is Brownian motion.

We can actually eliminate all the extra regularity conditions in Theorem 10.2.2 by using a different argument. Specifically, we can exploit bounds established by Chen and Shanthikumar (1994) to obtain convergence of the scaled standard queue-length processes under the assumptions of Theorem 10.2.1. Their argument actually applies to networks of multiserver queues.

Theorem 10.2.3. (heavy-traffic limit for the standard m-server queue) *Under the assumptions of Theorem 10.2.1,*

$$\mathbf{Q}_n \Rightarrow \mathbf{Q} \quad in \quad (D, M_1) \; ,$$

where $\mathbf{Q}_n$ is the standard queue-length process scaled as in (2.8) and $\mathbf{Q}$ is in (2.15).

Proof. We use extremal properties of the regulator map ψ_L defined in equations (2.6)–(2.10) of Section 5.2: Suppose that x, y and z are three functions in D satisfying $z = x + y \geq 0$. If y is nondecreasing, $y(0) = 0$ and y increases only when $z(t) \leq b$ for some positive constant b, then

$$\psi_L(x)(t) \leq y(t) \leq \psi_L(x-b)(t) \quad \text{for all} \quad t \geq 0 \; . \tag{2.18}$$

The lower bound is proved in Section 14.2 – see Theorem 14.2.1 – and the upper bound is proved in the same way (also see Theorem 14.2.3), as shown by Chen and Shanthikumar (1994).

We now proceed to treat the queue-length process just as we did for the single-server queue in Theorem 9.3.4, allowing for the fact that we now have m servers: Paralleling (3.23), we obtain

$$\mathbf{Q}_n = \mathbf{X}_n + \mathbf{Y}_n \; , \tag{2.19}$$

where

$$\mathbf{X}_n = \mathbf{A}_n - (\sum_{i=1}^m \mathbf{N}_n^i \circ \hat{\mathbf{B}}_n^i) + \eta_n \mathbf{e} \; , \tag{2.20}$$

$$\hat{\mathbf{B}}_n^i(t) \equiv n^{-1} B_n^i(nt), \quad t \geq 0, \tag{2.21}$$

$$\mathbf{Y}_n \equiv \sum_{i=1}^m \mu_n^i \mathbf{Y}_n^i \; , \tag{2.22}$$

$$\mathbf{Y}_n^i(t) \equiv c_n^{-1}(nt - B_n^i(nt)), \quad t \geq 0 \; , \tag{2.23}$$

and $B_n^i(t)$ is the cumulative busy time of server i in the interval $[0, t]$ in model n. Note that $\mathbf{Y}_n$ increases only when $\mathbf{Q}_n(t)$ is less than or equal to m/c_n. Thus, by (2.18),

$$\psi_L(\mathbf{X}_n) \leq \mathbf{Y}_n \leq \psi_L(\mathbf{X}_n - m/c_n) \; . \tag{2.24}$$

Since ψ_L is a Lipschitz map in the uniform norm, by Lemma 13.4.1,
$$\|\mathbf{Y}_n - \psi_L(\mathbf{X}_n)\|_t \leq m/c_n \to 0 \quad \text{as} \quad n \to \infty$$
for each $t \geq 0$. Hence,
$$\|\mathbf{Q}_n - \phi(\mathbf{X}_n)\|_t \leq 2m/c_n \to 0 \quad \text{as} \quad n \to \infty$$
for each $t \geq 0$. Since ϕ is continuous as a map from (D, M_1) to itself, it suffices to show that $\mathbf{X}_n \Rightarrow \mathbf{X}$ in (D, M_1). (We use the convergence-together theorem, Theorem 11.4.7.)

The key to establishing the limit $\mathbf{X}_n \Rightarrow \mathbf{X}$ is to show that
$$(\hat{\mathbf{B}}_n^1, \ldots, \hat{\mathbf{B}}_n^m) \Rightarrow (\mathbf{e}, \ldots, \mathbf{e}) \tag{2.25}$$
for $\hat{\mathbf{B}}_n^i$ in (2.21). To establish the limit in (2.25), we exploit the compactness approach in Section 11.6: Specifically, we can apply Theorems 11.6.2, 11.6.3 and 11.6.7 after observing that $\hat{\mathbf{B}}_n^i$ is uniformly Lipschitz: For $0 < t_1 < t_2$,
$$|\hat{\mathbf{B}}_n^i(t_2) - \hat{\mathbf{B}}_n^i(t_1)| = |n^{-1} B_n^i(nt_2) - n^{-1} B_n^i(nt_2)| \leq |t_2 - t_1|.$$
Hence, the sequence $\{(\hat{\mathbf{B}}_n^1, \ldots, \hat{\mathbf{B}}_n^m)\}$ has a convergent subsequence $\{(\hat{\mathbf{B}}_{n_k}^1, \ldots, \hat{\mathbf{B}}_{n_k}^m)\}$ in $C([0, T], \mathbb{R}^k, U)^m$ for every T. Suppose that
$$(\hat{\mathbf{B}}_{n_k}^1, \ldots, \hat{\mathbf{B}}_{n_k}^m) \to (\hat{\mathbf{B}}^1, \ldots, \hat{\mathbf{B}}^m) \tag{2.26}$$
as $n_k \to \infty$ in (D^m, U).

We now use the assumed FCLT in (2.9) and this convergence to establish a FWLLN along the subsequence $\{n_k\}$. In particular, it follows that
$$\hat{\mathbf{X}}_{n_k} \Rightarrow 0\mathbf{e},$$
where $\hat{\mathbf{X}}_n \equiv (c_n/n)\mathbf{X}_n$ for $\mathbf{X}_n$ in (2.20). By the continuous-mapping approach,
$$\hat{\mathbf{Y}}_{n_k} \Rightarrow \psi_L(0\mathbf{e}) = 0\mathbf{e},$$
where $\hat{\mathbf{Y}}_n \equiv (c_n/n)\mathbf{Y}_n$ for $\mathbf{Y}_n$ in (2.22). Consequently, we must have $\hat{\mathbf{B}}^i = \mathbf{e}$ for all i for $\hat{\mathbf{B}}^i$ in (2.26). Since the same limit holds for all subsequences, we actually have the desired convergence in (2.25). We can then apply the continuous-mapping approach to establish that $\mathbf{X}_n \Rightarrow \mathbf{X}$ for $\mathbf{X}_n$ in (2.20), $\mathbf{X} = \mathbf{A} - \mathbf{N} + \eta\mathbf{e}$ and $\mathbf{N}$ in (2.14). ■

In the general setting of Theorem 10.2.3, allowing limit processes with jumps, it remains to establish related stochastic-process limits for other queueing processes such as the workload and the waiting time.

Remark 10.2.3. *Bounds using other disciplines.* As shown by Wolff (1977), for the case in which all service times come from a single sequence of IID random variables, we can bound the queue length in the m-server FCFS model above stochastically by considering alternative service disciplines. For example, if the servers are allowed to have their own queues and arrivals are routed randomly or cyclically (in a round robin manner) to the servers, then the sum of the queue lengths stochastically dominates the queue length with the FCFS discipline. With the random or cyclic routing, we obtain heavy-traffic FCLT's for the single servers under the assumptions of Theorem 10.2.1, with the same scaling as in Theorem 10.2.1. We apply Theorem 9.5.1 to treat the arrival processes to the separate servers. We note that, even though the scaling is the same, the limit processes are different. Thus the differences in the service disciplines can be seen in the heavy-traffic limit.

Following Loulou (1973), we can also bound the workload process in the m-server FCFS model below by the workload process in a single-server queue with the same total input and constant output rate m. Thus we can obtain a nondegenerate limiting lower bound for the scaled workload process using the scaling in Theorem 10.2.1. ∎

Remark 10.2.4. *Resource pooling with other disciplines.* From Theorem 10.2.3 and Remark 10.2.2, we see that, under minor regularity conditions, there is resource pooling in heavy traffic for the standard multiserver model with homogeneous servers and the FCFS discipline. As noted in Remark 10.2.3 above, there is *not* resource pooling in heavy traffic when each server has its own queue and random or cyclic routing is used. However, there is resource pooling in heavy traffic with the join-the-shortest-queue rule and many related service disciplines that pay only a little attention to the system state; see Foschini and Salz (1978), Reiman (1984b), Laws (1992), Kelly and Laws (1993), Turner (1996, 2000) Harrison and Lopez (1999) and Bell and Williams (2001). The heavy-traffic behavior of random routing with periodic load balancing is analyzed by Hjálmtýsson and Whitt (1998). Once again, the time scaling in the heavy-traffic limit provides useful insight: The time scaling shows how the reconfiguration or balancing times should grow with the traffic intensity ρ in order to achieve consistent performance. For more on the impact of heavy-tailed distributions on load balancing, see Harchol-Balter and Downey (1997). For more on the great gains from only a little choice, see Azar et al. (1994), Vvedenskaya et al. (1996), Mitzenmacher (1996), Turner (1998) and Mitzenmacher and Vöcking (1999). ∎

10.3. Infinitely Many Servers

It may happen that there is a very large number of servers. An extreme case of a large number of servers is the case of infinitely many servers. With infinitely many servers, heavy-traffic is achieved by letting the arrival rate approach infinity. With infinitely many servers, we assume that the system starts out empty and that the service times are IID and independent of the arrival process.

Remark 10.3.1. *The power of infinite-server approximations.* More generally, when the load in a multiserver queue is light or there are a very large number of servers, it may be helpful to consider infinite-server models as approximations. The infinite-server assumption is attractive because the infinite-server model is remarkably tractable. For infinite-server models, it is even possible to obtain useful expressions for performance measures with time-varying arrival rates; e.g., see Eick, Massey and Whitt (1993), Massey and Whitt (1993) and Nelson and Taaffe (2000).

Infinite-server models are also called *offered-load models*, because they describe the load (number of busy servers) that would results if there were no capacity constraints, so that no customers are delayed or lost. At first glance, offered-load models may seem unrealistic as direct system models, but they offer great potential for engineering because they are tractable and because they actually do not differ greatly from associated, more complicated, delay and loss models when the capacity and offered load are large. The idea is to engineer so that the probability that the offered load exceeds capacity is sufficiently small; see Jennings, Mandelbaum, Massey and Whitt (1996), Leung, Massey and Whitt (1994), Massey and Whitt (1993, 1994a, b), Duffield and Whitt (1997, 1998, 2000) and Duffield, Massey and Whitt (2001). ∎

10.3.1. *Heavy-Traffic Limits*

Following Glynn and Whitt (1991), we show how to apply the continuous-mapping approach to establish heavy-traffic stochastic process limits for infinite-server queues when the service-time distribution takes values in a finite set. As shown by Borovkov (1967), heavy-traffic limits can be established for general $G/GI/\infty$ queues with general service-time distributions, but the argument is more elementary when the service-time distributions take values in a finite set. For other heavy-traffic limits and approximations for infinite-server queues, see Fleming and Simon (1999) and Krichagina and Puhalskii (1997).

The key observation from Glynn and Whitt (1991) is that, when the system is initially empty, the queue length (again number in system) is simply related to the arrival process when the service time is deterministic. Let $\{A(t) : t \geq 0\}$ and $\{Q(t) : t \geq 0\}$ be the arrival and queue-length processes, respectively. When the service time is x for all customers, the queue length at time t is simply

$$Q(t) = A(t) - A(t - x) , \qquad (3.1)$$

where we adopt the convention throughout this section that stochastic processes evaluated at negative arguments are identically 0; i.e., here $A(u) = 0$ for $u < 0$.

To exploit this simple representation more generally, we assume that all customers have service times in the finite set $\{x_1, \ldots, x_m\}$. We say that a customer with service time x_i is of type i and we let $A_i(t)$ count the number of type-i arrivals in the time interval $[0, t]$. We then let $D_i(t)$ count the number of type-i departures in $[0, t]$ and $Q_i(t)$ be the type-i queue length at time t. As in (3.1), we have the simple relations

$$\begin{aligned} D_i(t) &= A_i(t - x_i) \\ Q_i(t) &= A_i(t) - A_i(t - x_i), \quad t \geq 0 .\end{aligned}$$

We can also treat the workload process, representing the sum of the remaining service times of all customers in the system. To do so, we let $R_i(t, y)$ be the number of type-i customers with remaining service time greater than y in the system at time t and let $L_i(t)$ be the type-i workload at time t. Then we clearly have

$$R_i(t, y) = A_i(t) - A_i(t - [x_i - y]^+), \quad t \geq 0 ,$$

and
$$L_i(t) = \int_0^{x_i} R_i(t,y)\,dy.$$

We introduce the remaining-service-time processes as a means to treat the workload process, but the remaining-service-time process is of interest to describe the relevant state of the queue. To characterize the state of the customers in service, we need to consider the remaining-service-time $R_i(t,y)$ as a function of y for $y \geq 0$. In the $GI/GI/\infty$ model, when we append the elapsed interarrival time to the remaining-service-time process $\{R_i(t,y) : y \geq 0\}$, we obtain a Markov process as a function of t. In the heavy-traffic limit with a renewal arrival process, the interarrival time in process becomes negligible, so that the heavy-traffic limit for the remaining-service-time process is a Markov process as a function of t.

To apply the continuous mapping approach to treat the workload processes L_i, we need to regard $R_i(t,y)$ as a function mapping t into functions of y. We let the range be the subset of nonincreasing nonnegative functions in D with finite L_1 norm

$$\|x\|_1 \equiv \int_0^\infty |x(t)|\,dt. \tag{3.2}$$

Let $R_i : [0,\infty) \to (D, \|\cdot\|_1)$ be defined by $R_i(t) = \{R_i(t,y) : y \geq 0\}$. Since $(D, \|\cdot\|_1)$ is a Banach space, we can use the M_1 topology on the space $D([0,\infty),(D,\|\cdot\|_1))$, as noted in Section 11.5.

Let $A(t), D(t), Q(t), R(t,y), R(t)$ and $L(t)$ denote the associated m-dimensional vectors, e.g., $A(t) \equiv (A_1(t), \ldots, A_m(t))$, and let $\tilde{A}(t)$, etc. denote the associated partial sums, e.g., $\tilde{A}(t) \equiv A_1(t) + \cdots + A_m(t)$.

We form a sequence of infinite-server systems by scaling time in the original arrival process. Specifically, let

$$\begin{aligned}
A_{n,i}(t) &\equiv A_i(nt), \\
D_{n,i}(t) &\equiv A_{n,i}(t - x_i), \\
Q_{n,i}(t) &\equiv A_{n,i}(t) - A_{n,i}(t - x_i), \\
R_{n,i}(t,y) &\equiv A_{n,i}(t) - A_{n,i}(t - [x_i - y]^+), \\
L_{n,i}(t) &\equiv \int_0^{x_i} R_{n,i}(t,y)\,dy, \quad t \geq 0.
\end{aligned} \tag{3.3}$$

Now we define associated random elements of D by letting

$$\begin{aligned}
\mathbf{A}_n^i(t) &\equiv c_n^{-1}[A_{n,i}(t) - \lambda_i nt], \\
\mathbf{D}_n^i(t) &\equiv c_n^{-1}[D_{n,i} - nd_i(t)], \\
\mathbf{Q}_n^i(t) &\equiv c_n^{-1}[Q_{n,i}(t) - nq_i(t))], \\
\mathbf{L}_n^i(t) &\equiv c_n^{-1}[L_{n,i}(t) - nl_i(t)], \quad t \geq 0, \\
\mathbf{R}_n^i(t) &\equiv \{c_n^{-1}[Q_{n,i}(t,y) - nr_i(t,y)] : y \geq 0\},
\end{aligned} \tag{3.4}$$

where the translation functions are

$$d_i(t) \equiv \lambda_i([t - x_i]^+),$$

$$q_i(t) \equiv \lambda_i(t - [t-x_i]^+),$$
$$l_i(t) \equiv \int_0^{x_i} q_i(t,y)\,dy,$$
$$r_i(t,y) \equiv \lambda_i(t - [t - [x_i - y]^+]^+)\,. \tag{3.5}$$

Let the associated vector-valued random elements of D^m and partial sums in D be

$$\begin{aligned}
\mathbf{A}_n &\equiv (\mathbf{A}_n^1, \ldots, \mathbf{A}_n^m), \\
\mathbf{D}_n &\equiv (\mathbf{D}_n^1, \ldots, \mathbf{D}_n^m), \\
\mathbf{Q}_n &\equiv (\mathbf{Q}_n^1, \ldots, \mathbf{Q}_n^m), \\
\mathbf{L}_n &\equiv (\mathbf{L}_n^1, \ldots, \mathbf{L}_n^m), \\
\mathbf{R}_n &\equiv (\mathbf{R}_n^1, \ldots, \mathbf{R}_n^m)
\end{aligned} \tag{3.6}$$

and

$$\begin{aligned}
\tilde{\mathbf{A}}_n &\equiv \mathbf{A}_n^1 + \cdots + \mathbf{A}_n^m, \\
\tilde{\mathbf{D}}_n &\equiv \mathbf{D}_n^1 + \cdots + \mathbf{D}_n^m, \\
\tilde{\mathbf{Q}}_n &\equiv \mathbf{Q}_n^1 + \cdots + \mathbf{Q}_n^m, \\
\tilde{\mathbf{L}}_n &\equiv \mathbf{L}_n^1 + \cdots + \mathbf{L}_n^m, \\
\tilde{\mathbf{R}}_n &\equiv \mathbf{R}_n^1 + \cdots + \mathbf{R}_n^m \,.
\end{aligned} \tag{3.7}$$

We now can establish the basic heavy-traffic stochastic-process limit. For that purpose, let $\theta_s : D^m \to D^m$ denote the shift operator, defined by

$$\theta_s(x)(t) \equiv x(t+s), \quad t+s \geq 0\,,$$

with $\theta_s(x)(t) = 0$ for $t + s < 0$, all for $t \geq 0$. We exploit the fact that the shift operator θ_s is continuous for $s \leq 0$. However, note that the shift operator θ_s is *not* continuous for $s > 0$. To see this, let $s = 1$, $x = I_{[1,\infty)}$ and $x_n = I_{[1+n^{-1},\infty)}$.

Let the components of the vector-valued limit processes also be indexed by superscripts.

Theorem 10.3.1. (heavy-traffic limit for infinite-server queues) *If*

$$\mathbf{A}_n \Rightarrow \mathbf{A} \quad \text{in} \quad (D^m, WM_1)\,,$$

where

$$P(Disc(\mathbf{A}^i) \cap Disc(\theta_s(\mathbf{A}^j)) = \phi) = 1 \tag{3.8}$$

for all i, j and $s \leq 0$ for which $i \neq j$ or $i = j$ and $s \neq 0$, then

$$(\mathbf{A}_n, \tilde{\mathbf{A}}_n, \mathbf{D}_n, \tilde{\mathbf{D}}_n, \mathbf{Q}_n, \tilde{\mathbf{Q}}_n, \mathbf{L}_n, \tilde{\mathbf{L}}_n, \mathbf{R}_n, \tilde{\mathbf{R}}_n)$$
$$\Rightarrow (\mathbf{A}, \tilde{\mathbf{A}}, \mathbf{D}, \tilde{\mathbf{D}}, \mathbf{Q}, \tilde{\mathbf{Q}}, \mathbf{L}, \tilde{\mathbf{L}}, \mathbf{R}, \tilde{\mathbf{R}})$$

in $D([0,\infty), \mathbb{R})^{4(m+1)} \times D([0,\infty), \mathbb{R}, \|\cdot\|_1))^{m+1}$ with the WM_1 topology, where

$$\mathbf{D}^i(t) \equiv \mathbf{A}^i(t - x_i),$$

$$\begin{aligned}
\mathbf{Q}^i(t) &\equiv \mathbf{A}^i(t) - \mathbf{A}^i(t - x_i), \\
\mathbf{L}^i(t) &\equiv \int_0^{x_i} \mathbf{R}^i(t)(y)\, dy, \\
\mathbf{R}^i(t)(y) &\equiv \mathbf{A}^i(t) - \mathbf{A}^i(t - [x_i - y]^+) \\
\tilde{\mathbf{A}}(t) &\equiv \mathbf{A}^1(t) + \cdots + \mathbf{A}^m(t), \\
\tilde{\mathbf{D}}(t) &\equiv \mathbf{D}^1(t) + \cdots + \mathbf{D}^m(t), \\
\tilde{\mathbf{Q}}(t) &\equiv \mathbf{Q}^1(t) + \cdots + \mathbf{Q}^m(t), \\
\tilde{\mathbf{L}}(t) &\equiv \mathbf{L}^1(t) + \cdots + \mathbf{L}^m(t), \\
\tilde{\mathbf{R}}(t) &\equiv \mathbf{R}^1(t) + \cdots + \mathbf{R}^m(t)\,.
\end{aligned} \qquad (3.9)$$

Proof. We apply the continuous mapping theorem, Theorem 3.4.3, with a succession of maps that are measurable and almost surely continuous with respect to the limit process, by virtue of condition (3.8). The linearity of the model means that the scaled processes are related by the same maps the are used to construct the original processes. The maps are shown applied to the limit process $\mathbf{A}$ in (3.9). Note in particular that the map taking A_i into R_i, and thus $\mathbf{A}_n^i$ into $\mathbf{R}_n^i$, is continuous. ∎

Remark 10.3.2. *Stationarity after finite time.* If the limit process $\mathbf{A}$ has stationary increments, then the process $\theta_s(\mathbf{X}) \equiv \{\mathbf{X}(t+s) : t \geq 0\}$ is a stationary process when $s \geq max\{x_1, \ldots, x_m\}$ and $\mathbf{X}$ is any of the following limit processes: $\mathbf{A}(t+u) - \mathbf{A}(u)$, $\mathbf{Q}(t)$, $\mathbf{D}(t+u) - \mathbf{D}(u)$, $\mathbf{R}(t)(y)$ and $\mathbf{L}(t)$ for $u \geq 0$ and $y > 0$. Hence these limit processes have the property that they reach steady state in finite time (as the original processes do with a Poisson arrival process). If, in addition, $\mathbf{A}$ is a Gaussian process, such as a fractional Brownian motion, then these processes are stationary Gaussian processes. ∎

10.3.2. Gaussian Approximations

We will now focus on the common case in which the conditions of Theorem 10.3.1 hold with $c_n = \sqrt{n}$ and $\mathbf{A}$ an m-dimensional Brownian motion. As noted in Remark 10.3.2 above, when $\mathbf{A}$ is an m-dimensional Brownian motion, the limit processes $\mathbf{D}$, $\mathbf{Q}$, $\mathbf{L}$ and $\mathbf{R}(\cdot)(y)$ are all Gaussian processes. Assuming appropriate uniform integrability in addition to the condition of Theorem 10.3.1, so that the variances converge (see p. 32 of Billingsley (1968)), we can relate the covariance matrix Σ_A of the Brownian motion $\mathbf{A}$ to the original arrival processes. In particular, under that regularity condition,

$$\Sigma_{A,i,j} = \lim_{t\to\infty} t^{-1} cov(A_i(t), A_j(t))\,. \qquad (3.10)$$

We now describe the relatively simple stationary Gaussian approximations that hold for the aggregate departure, queue-length and workload processes when the limit process **A** is an m-dimensional Brownian motion and $t \geq \max\{x_1, \ldots, x_m\}$:

$$\tilde{D}(t+s) - \tilde{D}(t) \approx N(\sum_{i=1}^{m} \lambda_i s, \sum_{i=1}^{m} \sum_{j=1}^{m} \Sigma_{D,i,j}),$$

$$\tilde{Q}(t) \approx N(\sum_{i=1}^{m} \lambda_i x_i, \sum_{i=1}^{m} \sum_{j=1}^{m} \Sigma_{Q,i,j}),$$

$$\tilde{L}(t) \approx N(\sum_{i=1}^{m} \lambda_i x_i^2/2, \sum_{i=1}^{m} \sum_{j=1}^{m} \Sigma_{L,i,j}), \quad (3.11)$$

where

$$\Sigma_{D,i,j} \equiv \Sigma_{A,i,j}([s - |x_i - x_j|]^+),$$
$$\Sigma_{Q,i,j} \equiv \Sigma_{A,i,j}(x_i \wedge x_j),$$
$$\Sigma_{L,i,j} \equiv \Sigma_{A,i,j}[(x_i \vee x_j)(x_i \wedge x_j)/2 - (x_i \wedge x_j)^3/6]. \quad (3.12)$$

To obtain the covariance terms $\Sigma_{L,i,j}$ for the workload, we use the representation

$$\Sigma_{L,i,j} = \int_0^{x_i} \int_0^{x_j} cov[\mathbf{A}^i(t) - \mathbf{A}^i(t - x_i + y), \mathbf{A}^j(t) - \mathbf{A}^j(t - x_j + z)] \, dz \, dy.$$

The assumption of Theorem 10.3.1 starting out with a limit for $\mathbf{A}_n \equiv (\mathbf{A}_n^1, \ldots, \mathbf{A}_n^m)$ is natural when there are m classes of jobs each with their characteristic deterministic service time. However, we are often interested in a single arrival process, with each successive arrival being randomly assigned a service time, which we here take to be from the finite set $\{x_1, \ldots, x_m\}$. Sufficient conditions for the condition in Theorem 10.3.1 in that setting follow from Theorem 9.5.1 on split streams. Suppose that the limit process $(\tilde{\mathbf{A}}, \mathbf{S})$ there is a centered $(m+1)$-dimensional Brownian motion with covariance matrix Σ. Then the limit process **A** here is a centered m-dimensional Brownian motion with covariance matrix

$$\Sigma_{A,i,j} = \lambda \Sigma_{i,j} + p_i \lambda^{1/2} \Sigma_{i,m+1} + p_j \lambda^{1/2} \Sigma_{j,m+1} + p_i p_j \Sigma_{m+1,m+1}. \quad (3.13)$$

An important application of the setting above occurs when the service times are IID with distribution $P(V = x_i) = p_i$, and independent of the arrival process. If we further assume that the limit process $\tilde{\mathbf{A}}$ in Theorem 9.5.1 is $\lambda^{3/2}\sigma B$, where **B** is standard Brownian motion, as occurs for a renewal process when the interrenewal times have mean λ^{-1} and variance σ^2, then the covariance terms become $\Sigma_{i,i} = p_i(1-p_i)$ for $i \leq m$, $\Sigma_{m+1,m+1} = \lambda^3 \sigma^2$, $\Sigma_{i,j} = -p_i p_j$ for $i,j \leq m$ and $i \neq j$ and $\Sigma_{i,m+1} = 0$ for $i \leq m$. Then the covariance function of the limiting Gaussian process **Q** in Theorem 10.3.1 can be represented as

$$cov(\mathbf{Q}(s), \mathbf{Q}(s+t)) = \lambda \int_0^s H(u) H^c(t+u) \, du$$
$$+ \sigma^2 \lambda^3 \int_0^s H^c(t+u) H^c(u) \, du, \quad (3.14)$$

where $H(t) \equiv P(V \leq t)$ is the service-time cdf and $H^c(t) \equiv 1 - H(t)$ is the associated ccdf.

Borovkov (1967) established a heavy-traffic limit justifying (3.14) for general service-time distributions. Borovkov assumes that the scaled arrival process converges to $\lambda^{3/2}\sigma \mathbf{B}$, where $\mathbf{B}$ is standard Brownian motion. For general service-time cdf H and Brownian arrival-process limit, the heavy-traffic limit shows that the stationary queue length in the $G/GI/\infty$ model is approximately distributed according to

$$Q(\infty) \approx N(\gamma, \gamma z_Q) , \qquad (3.15)$$

where $\gamma \equiv \lambda/\mu$ is the *total offered load* and

$$\begin{aligned} z_Q &= \mu \int_0^\infty H(u) H^c(u)\, du + \lambda^2 \mu \sigma^2 \int_0^\infty H^c(u)^2\, du \\ &= 1 + (c_U^2 - 1)\mu \int_0^\infty H^c(u)^2\, du , \end{aligned} \qquad (3.16)$$

with $c_U^2 \equiv \lambda^2 \sigma^2$.

To put the normal approximation in (3.15) in perspective, the steady-state mean is exactly $EQ(\infty) = \gamma$ by Little's law, $L = \lambda W$. With a Poisson arrival process, $Q(\infty)$ has a Poisson distribution with mean γ, which is asymptotically normal as λ goes to infinity with fixed service rate μ. With a Poisson arrival process, $c_U^2 = c_u^2 = 1$; when $c_U^2 = 1$, $z_Q = 1$ and $\sigma_Q^2 = \gamma$.

More generally, we see that the asymptotic variance is γz_Q, so that $Q(\infty)$ tends to differ from the mean γ by amounts of order $O(\sqrt{\gamma z_Q})$ as γ gets large. The variance scale factor z_Q is called the *asymptotic peakedness*. More generally, the *peakedness* is the ratio

$$Var(Q(\infty))/EQ(\infty) = Var(Q(\infty))/\gamma .$$

The peakedness and the asymptotic peakedness are often used in approximations of loss and delay systems with finitely many servers; see Eckberg (1983, 1985), Whitt (1984a, 1992b) and Srikant and Whitt (1996).

A key quantity in the asymptotic peakedness is the integral $\mu \int_0^\infty H^c(u)^2\, du$. It varies from 1 when H is the cdf of the unit point mass on μ^{-1} to 0. That integral tends to *decrease* as the service-time distribution gets more variable with the mean held fixed. Note that the effect of sevice-time variability on the asymptotic peakedness z_Q depends on the sign of $(c_U^2 - 1)$.

Remark 10.3.3. *Exponential service times.* When the space scaling is by $c_n = \sqrt{n}$, the limit process for the arrival process is Brownian motion and the service times are exponential with ccdf $H^c(t) = e^{-\mu t}$, the limiting Gaussian process $\mathbf{Q}$ becomes an Ornstein-Uhlenbeck diffusion process with infinitesimal mean $-\mu x$ and diffusion coefficient $\gamma\mu(c_u^2+1)$ for $\gamma = \sum_{i=1}^m \lambda_i x_i$. Direct heavy-traffic limits for the $M/M/\infty$ and $GI/M/\infty$ models were established by Iglehart (1965) and Whitt (1982c). When H is exponential, $\mu \int_0^\infty H^c(u)\, du = 1/2$, so that $z_Q = (c_U^2 + 1)/2$. ∎

Remark 10.3.4. *Prediction with nonexponential service times.* The fact that the limit process is a Markov process when the limit process for the arrival process is Brownian motion and the service times are exponential implies that, for exponential service times, in the heavy-traffic limit the future evolution of the process depends on the past only through the present state. In the heavy-traffic limit there is no benefit from incorporating additional information about the past.

However, that is not the case for nonLévy limit processes for the arrival process and nonexponential service-time distributions. Then the elapsed service times of customers in service give information about the remaining service times of these customers.

As can be seen from the covariance formula in (3.14), the transient behavior depends on the full service-time cdf.

Example 10.3.1. *Transient behavior in the $M/GI/\infty$ queue.* Suppose that at some instant in steady state there happen to be n busy servers in an $M/GI/\infty$ model. In the case of a Poisson arrival process, we can decribe the future transient behavior, because, conditional on $Q(0) = n$, the n elapsed service times and the n residual service times are each distributed as n IID random variables with the stationary-excess cdf associated with the service-time cdf H, i.e.,

$$H_e(t) = \mu \int_0^t H^c(u) \, du, \quad t \geq 0 \,. \tag{3.17}$$

Moreover, the number of new arrivals after time 0 that are still in service at a later time t is independent of the customers initially in service and has a Poisson distribution with mean $\lambda \mu^{-1} H_e(t)$; see Duffield and Whitt (1997).

As a consequence, it is elementary to compute the mean and variance of the conditional number of customers in the system at any future time, given the initial number n. If the arrival rate and offered load are large, then the number of customers in the system is likely to be large. From the properties above, we obtain a normal approximation refining the conditional mean. We can exploit this structure to study various control schemes to recover from rare congestion events. ■

10.4. An Increasing Number of Servers

We now want to consider a third heavy-traffic limiting regime for multiserver queues. Just as we treated queues with superpositions of an increasing number of arrival processes in Section 9.8, we now want to treat queues with an increasing number of servers. Now we let the number m of servers go to infinity as the traffic intensity ρ approaches 1, the critical value for stability.

The m-server model we consider now is the standard m-server queue with unlimited waiting room and the FCFS service discipline. Customers are assigned in order of arrival to the first available server, with some unspecified procedure to break ties. the service times are independent of the arrival process and come from a sequence of IID random variables with mean μ^{-1}. The arrival rate is λ and the traffic intensity is $\rho \equiv \lambda/\mu m$. Heavy traffic will be obtained by increasing the arrival rate, using simple time scaling, just as done for the infinite-server model. But now we increase m as well as λ.

Just as in Remark 9.8.2, when the number of servers becomes large, the relevant time scale for an individual customer becomes different from the time scale of the system. The times between successive departures from the queue become much shorter than the individual service times. If there are many input streams as well as many servers, the time scale for individual customers is consistently much longer than the time scale for the queue. Thus, as in Section 9.8, the short-time behavior of individual customers will affect the large-time behavior of the queue.

We can use the infinite-server model to determine what the appropriate limiting regime for the m-server model should be. We will apply the infinite-server model to show that, when there is space scaling by $c_n = n^H$ for $0 < H < 1$, we should have $m \to \infty$ and $\rho \to 1$ with

$$m^{1-H}(1-\rho) \to \beta \tag{4.1}$$

for $0 < \beta < \infty$. (As before, the common case is $H = 1/2$.) Note that the growth rate here is the same as for superposition arrival processes in (8.10).

10.4.1. Infinite-Server Approximations

More generally, we can use an infinite-server model to estimate how many servers are needed in a finite-server system in order to achieve desired quality of service. To do so, we let the infinite-server model have the same arrival and service processes as the m-server model. We can use the probability that m or more servers are busy in the infinite-server model as a rough approximation for the same quantity in an m-server model.

For that purpose, we let $m = m_p$, where

$$P(Q(\infty) \geq m_p) = p \tag{4.2}$$

for a target tail probability p. We can use Theorem 10.3.1 to generate an approximation for the distribution of $Q(\infty)$ and determine how the threshold m_p should grow as the arrival rate increases. We assume that Theorem 10.3.1 holds with $c_n = n^H$ for $0 \leq H \leq 1$. Then

$$Q_n(t) \approx nq(t) + n^H \tilde{\mathbf{Q}}(t) . \tag{4.3}$$

If we focus on the steady-state behavior, which occurs for $t \geq max\{x_1, \ldots, x_m\}$ when the limit process $\mathbf{A}$ has stationary increments, then

$$Q_n(\infty) \approx n \sum_{i=1}^{m} \lambda_i x_i + n^H \tilde{\mathbf{Q}}(\infty) . \tag{4.4}$$

Let γ denote the total offered load, which here is

$$\gamma = EQ_n(\infty) = n \sum_{i=1}^{m} \lambda_i x_i . \tag{4.5}$$

Since the total offered load can be expressed in terms of the traffic intensity ρ and the number of servers m by $\gamma = m\rho$, we can replace (4.4) by

$$Q(\infty) \approx \rho m + (c\rho m)^H \tilde{\mathbf{Q}}(\infty) . \tag{4.6}$$

for a constant c. (For $\gamma = n \sum_{i=1}^m \lambda_i x_i$, $c = n$.)

Combining equations (4.2) and (4.6), we obtain the approximation

$$P(Q(\infty) \geq m) \approx P(\tilde{\mathbf{Q}}(\infty) \geq m(1-\rho)/(c\rho m)^H) . \tag{4.7}$$

Let x_p be the $(1-p)^{\text{th}}$ quantile of the distribution of $\tilde{\mathbf{Q}}(\infty)$, i.e.,

$$P(\tilde{\mathbf{Q}}(\infty) \geq x_p) = p . \tag{4.8}$$

Combining (4.7) and (4.8), we obtain $m_p(1-\rho) = (c\rho m_p)^H x_p$ or

$$m_p \approx (x_p(c\rho)^H/(1-\rho))^{1/(1-H)} . \tag{4.9}$$

Approximation (4.9) specifies the required number of servers, using the infinite-server constraint (4.2) and the heavy-traffic limit established in Theorem 10.3.1.

In the common case in which $c_n = n^{1/2}$ and $Q(\infty)$ has the normal approximation in (3.15), we have the approximation

$$P(Q(\infty) \geq m) \approx P(N(\gamma, \gamma z_Q) > m) = \Phi^c((m-\gamma)/\sqrt{\gamma z_Q}) , \tag{4.10}$$

where $\gamma \equiv \rho m$ is again the offered load, z_Q is the asymptotic peakedness and Φ^c is the standard normal ccdf. Then we obtain the special case of (4.9)

$$m_p \approx ((x_p\sqrt{\rho z_Q})/(1-\rho))^2 . \tag{4.11}$$

Thus, we have developed an approximation for the required number of servers in an m-server system based on an infinite-server-model approximation. This same approach applies to multiserver queues with time-varying arrival rates; see Jennings, Mandelbaum, Massey and Whitt (1996).

Note that equations (4.9) and (4.11) show that there is increased service efficiency as the number m of servers increases. The traffic intensity at which the system can satisfy the performance constraint (4.2) increases as m increases; see Smith and Whitt (1981) and Whitt (1992) for further discussion.

From (4.9) and (4.11), we also can determine the rate at which m should grow as $\rho \to 1$ in m-server systems so that the probability of delay approaches a nondegenerate limit (a limit p with $0 < p < 1$). In the infinite server model, we meet the tail probability constraint (4.2) as the arrival rate increases (by simple time scaling as in the previous subsection) if $m \to \infty$ and $\rho \equiv \gamma/m \to 1$ with m and ρ related by (4.9). In other words, we obtain (4.1) as an estimate of the way in which m should be related to ρ as $m \to \infty$ and $\rho \to 1$ in the m-server model.

10.4.2. Heavy-Traffic Limits for Delay Models

Of course, the infinite-server model is just an approximation. It remains to establish heavy-traffic limits as $\rho \to 1$ and $m \to \infty$ with (4.1) holding in an m-server model. However, this third limiting regime is more complicated, evidently requiring methods beyond the continuous-mapping approach. We will briefly summarize heavy-traffic limits established in this regime by Halfin and Whitt (1981) for the $GI/M/m$ model, having exponential service times and $H = 1/2$. Puhalskii and Reiman (2000) established heavy-traffic limits, with more complicated limit processes, for the more general $GI/PH/m$ model with phase-type service-time distributions.

Let $Q_m(t)$ be the queue length at time t and let $Q_m(\infty)$ be the steady-state queue length in a standard $GI/M/m$ model with m servers. Let the interarrival times be constructed from a sequence $\{U_k : k \geq 1\}$ of IID random variables with mean 1 and SCV c_u^2. When the number of servers is m, let the interarrival times be

$$U_{m,k} \equiv U_k/\lambda_m , \quad (4.12)$$

so that the arrival rate in model m is λ_m. Let the individual service rate be μ for all m, so that the traffic intensity as a function of m is $\rho_m = \lambda_m/\mu m$.

We now state the two main results from Halfin and Whitt (1981) without proof. The first concerns the steady-state distribution. The second is a FCLT.

Theorem 10.4.1. (necessary and sufficient conditions for asymptotically nondegenerate delay probability) *For the family of $GI/M/m$ models specified above,*

$$\lim_{m \to \infty} P(Q_m(\infty) \geq m) = p, \quad 0 < p < 1 , \quad (4.13)$$

if and only if the arrival rate λ_m increases with m so that

$$\lim_{m \to \infty} (1 - \rho_m)m^{1/2} = \beta, \quad 0 < \beta < \infty , \quad (4.14)$$

in which case

$$p = [1 + \xi\sqrt{2\pi}\Phi(\xi)exp(\xi^2/2)]^{-1} , \quad (4.15)$$

where

$$\xi = 2\beta/(1 + c_u^2) . \quad (4.16)$$

Moreover, if (4.14) holds, then

$$m^{1/2}(Q_m(\infty) - m) \Rightarrow Z \quad in \quad \mathbb{R} , \quad (4.17)$$

where

$$P(Z \geq 0) = p, \quad P(Z > x|Z \geq 0) = e^{-x\xi} \quad (4.18)$$

and

$$P(Z \leq x|Z \leq 0) = \Phi(x + \xi)/\Phi(\xi) \quad (4.19)$$

for ξ in (4.16).

To state the FCLT, we construct random elements of $D \equiv D([0, \infty), \mathbb{R})$ by letting

$$\mathbf{Q}_m(t) \equiv m^{-1/2}(Q_m(t) - m), \quad t \geq 0 . \quad (4.20)$$

There is no time scaling in (4.20) because the arrival rate λ_m is allowed to grow directly.

Theorem 10.4.2. (Heavy-traffic FCLT with an increasing number of servers) *If (4.14) holds and $\mathbf{Q}_m(0) \Rightarrow \mathbf{Q}(0)$ in $\mathbb{R}$, then*

$$\mathbf{Q}_m \Rightarrow \mathbf{Q} \quad in \quad (D, J_1) \quad as \quad m \to \infty , \qquad (4.21)$$

where $\mathbf{Q}$ is a diffusion process starting at $\mathbf{Q}(0)$ with infinitesimal mean

$$m(x) = \begin{cases} -\mu\beta, & x \geq 0 \\ -\mu(x+\beta), & x < 0 , \end{cases}$$

diffusion coefficient

$$\sigma^2(x) = \mu(1 + c_u^2) .$$

and the steady-state distribution of Z in Theorem 10.4.1.

As indicated earlier, generalizations of Theorem 10.4.2 to $GI/PH/m$ queues have been established by Puhalskii and Reiman (2000). Generalizations to Markovian service networks with time-varying arrivals have been established by Mandelbaum, Massey and Reiman (1998). Approximations for the steady-state queue-length distribution and other steady-state distributions in $GI/GI/m$ models based partly on Theorem 10.4.1 are discussed in Whitt (1992, 1993a).

A large number of servers also affects the departure process from a queue. As shown by Whitt (1984f), under regularity conditions, even with general service-time distributions the departure process can be approximated by a Poisson process. As with the superposition arrival process, the Poisson property applies in a relatively short time scale.

Remark 10.4.1. *State dependence.* Even when restricting attention to simple Markovian $M/M/m$ queues, the limit process in Theorem 10.4.2 differs significantly from the limit processes when $m = 1$ and $m = \infty$. As shown in Section 10.2, the heavy-traffic limit for any fixed m is the same as for $m = 1$, but if we let $m \to \infty$ so that (4.1) holds, then we obtain a limiting diffusion process with a *nonlinear drift function*, which shows that there is significant state-dependence.

In contrast, the limit processes for single-server and infinite-server queues have essentially linear drift. For the single-server queues, there is constant drift, modified only by the reflection at the barriers. As a consequence of the scaling, the barrier disappears in the heavy-traffic limit for infinite-server models. Then the limit process has linear drift.

State dependent behavior may be important to capture in queueing models. In addition to the state-dependent consequence of multiple servers, state-dependence occurs when there is balking (customers refusing to join a queue when it is congested) or reneging (customers abandoning the queue after waiting a long time). See Garnett, Mandelbaum and Reiman (2000) and Ward and Glynn (2001) for heavy-traffic limits for queues with reneging.

It is natural to model state-dependent behavior directly by birth-and-death processes. We can then establish heavy-traffic limits in which the birth-and-death processes converge to diffusion processes. However, diffusion processes are continuous analogs of birth-and-death processes, so we may not gain much from such a limit, if we cannot show that the same limit holds for more general processes. However, state-dependence is sufficiently complicated that we may have to be content doing all the analysis in a Markovian framework. For discussions of heavy-traffic limits for Markovian queues with state-dependence, see Browne and Whitt (1995), Pats (1994) and Mandelbaum and Pats (1995, 1998).

State dependence also arises when the service times and arrival times depend on the level of congestion; see Whitt (1990) for heavy-traffic limits in that setting.

10.4.3. Heavy-Traffic Limits for Loss Models

There is a heavy-traffic story for m-server loss models that closely parallels the heavy-traffic limits for m-server delay models just considered. We will consider $G/M/m$ loss models, in which all arrivals finding the m servers busy are blocked and lost, without affecting future arrivals. We will assume that the scaled arrival process satisfies a FCLT. Let A be a rate-1 counting process and let

$$\mathbf{A}_\lambda(t) \equiv \lambda^{-1/2}(A(\lambda t) - \lambda t), \quad t \geq 0 \ . \tag{4.22}$$

The first heavy-traffic limit is for the blocking probability. Let B_m be the blocking probability (the long-run proportion of arrivals that are blocked) as a function of the number of servers, m. The first heavy-traffic theorem is a *local limit theorem* due to Borovkov (1976). Related work is discussed in Whitt (1984a). A simple proof for the classical $M/M/m$ Erlang loss model is given in the appendix there. More detailed asymptotics for the Erlang model is contained in Jagerman (1974).

Theorem 10.4.3. (heavy-traffic limit for the blocking probability) *Consider a sequence of $GI/M/m$ loss models indexed by m with fixed individual service rate μ and arrival process $\{A(\lambda t) : t \geq 0\}$. Suppose that $\mathbf{A}_\lambda \Rightarrow \sigma_A \mathbf{B}$ as $\lambda \to \infty$ for $\mathbf{A}_\lambda$ in (4.22) and $\mathbf{B}$ standard Brownian motion, which for the renewal arrival process here is equivalent to the interarrival time having finite variance σ_A^2. If $\lambda \to \infty$ and $m \to \infty$ so that (4.14) holds or, equivalently, so that*

$$(m - \gamma)/\gamma \to \beta \tag{4.23}$$

for $\gamma \equiv \lambda/\mu$, then

$$\lim_{m \to \infty} \sqrt{\gamma} B_m = \sqrt{z}\, \phi(\beta/\sqrt{z})/\Phi(-\beta/\sqrt{z}) \ ,$$

where $z = (\sigma_A^2 + 1)/2$ is the asymptotic peakedness in (3.16).

There are significant differences between Theorems 10.4.3 and 10.4.1. Under the same conditions, the probability of delay in the $GI/M/m$ delay model approaches a nondegenerate limit, while the blocking probability in the $GI/M/m$ loss model is order $1/\sqrt{m}$ as $m \to \infty$. However, in both cases, the heavy-traffic limits show that the normal or critical operating regime when the offered load γ is large is $m = \gamma + c\sqrt{\gamma}$ for some constant c. The critical-loading regime is important for establishing asymptotic results for more general loss systems; e.g., see Hunt and Kelly (1989) and Reiman (1989, 1990b).

We now turn to the FCLT, in which we allow a general stationary arrival process. As above, we start with a rate-1 process. The FCLT is from Srikant and Whitt (1996), but it is closely related to Theorem 2 on p. 177 of Borovkov (1984).

Theorem 10.4.4. (FCLT for $G/M/m$ loss models) *Consider a sequence of $G/M/m$ loss models indexed by the number of servers, m, where the individual service rate is fixed at μ. Suppose that the arrival process is $\{A(\lambda t) : t \geq 0\}$, where A is a general stationary rate-1 process satisfying $\mathbf{A}_\lambda \Rightarrow \sigma_A \mathbf{B}$ as $\lambda \to \infty$ for $\mathbf{A}_\lambda$ in (4.22) and $\mathbf{B}$ standard Brownian motion. Suppose that $m \to \infty$ and $\gamma \equiv \lambda/\mu \to \infty$ with (4.23) holding and $\gamma^{-1/2}(Q_m(0) - m) \Rightarrow y$, where $\{Q_m(t) : t \geq 0\}$ is the queue-length process in model m. Then*

$$\mathbf{Q}_m \Rightarrow \mathbf{Q} \quad in \quad (D, J_1) \, ,$$

where

$$\mathbf{Q}_m(t) \equiv \gamma^{-1/2}(Q_m(t) - m), \quad t \geq 0 \, ,$$

and $\mathbf{Q}$ is a reflected Ornstein-Uhlenbeck process with infinitesimal mean $m(x) = -\mu(x + \beta)$ for $x \leq 0$, infinitesimal variance $\sigma^2(x) = \mu(1 + \sigma_A^2)$, initial position $\mathbf{Q}(0) = y$ and instantaneous reflecting barrier above at 0.

Remark 10.4.2. *Exponential service times.* In the case of exponential service times, the heavy-traffic limits for the infinite-server, delay and loss models involve essentially the same Ornstein-Uhlenbeck (OU) diffusion process. Of course, for the delay model, the diffusion acts like the OU diffusion only on part of its state space, i.e., when the servers are not all busy. The diffusion coefficient in the case of the infinite-server model in Remark 10.3.3 appears different only because the space scaling there is by $\sqrt{n}$ instead of by the square root of the offered load; the offered load there is $n \sum_{i=1}^{m} \lambda_i x_i$. ∎

10.4.4. Planning Simulations of Loss Models

Just as in Section 5.8, the heavy-traffic stochastic-process limits here can be used to help plan simulations. Assuming that our goal is to estimate the long-run blocking probability, we can use the heavy-traffic limits to produce approximations for the blocking probability and the asymptotic variance of the estimator appearing in the formulas for the required simulation run length to achieve desired statistical precision. Estimators of the blocking probability were investigated by Srikant and Whitt (1996, 1999).

The *natural estimator* for the steady-state blocking probability B based on observations of the system over the time interval $[0, t]$ is

$$\hat{B}_N(t) \equiv L(t)/A(t) \, , \tag{4.24}$$

where $L(t)$ is the number of lost arrivals and $A(t)$ is the number of arrivals in $[0, t]$. Since we may know the arrival rate in a stationary arrival process with rate λ, an alternative *simple estimator* is

$$\hat{B}_S(t) \equiv L(t)/\lambda t \ . \tag{4.25}$$

Another alternative estimator can be based on Little's law, $L = \lambda W$. From Little's law, we have the relation

$$EQ(\infty) = \lambda(1 - B)/\mu \ ; \tag{4.26}$$

here the effective arrival rate for admitted customers is $\lambda(1 - B)$. Hence an alternative *indirect estimator* is

$$\hat{B}_I(t) \equiv 1 - \hat{m}(t)/\gamma \ , \tag{4.27}$$

where $\gamma \equiv \lambda/\mu$ is the offered load and $\hat{m}(t)$ is an estimator of the steady-state mean $EQ(\infty)$. (Since we are considering a simulation, we assume that λ and μ are known.) It is natural for $\hat{m}(t)$ to be the sample mean

$$\hat{m}(t) \equiv t^{-1} \int_0^t Q(u) \, du \ , \tag{4.28}$$

As in Section 5.8, the required simulation run length t is proportional to the asymptotic variance of the estimator, denoted again by σ^2. (Let us use the criterion of absolute error.) However, the computational effort to simulate for time t is approximately proportional to λt, because that is the expected number of arrivals in the interval $[0, t]$. Hence it is natural to focus on the *workload factor* $\lambda\sigma^2$.

Srikant and Whitt (1996) apply heavy-traffic limits to develop approximations for the workload factors associated with the different estimators of the blocking probability. As a function of the basic parameters, they obtain the following approximation for the workload factor of the indirect estimator:

$$w_I \equiv w_I(m, \beta, c_a^2, c_s^2, z) \approx \frac{(c_a^2 + c_s^2)}{2} \psi_I(\beta/\sqrt{z}) \ , \tag{4.29}$$

where

$$\psi_I(x) \equiv w_I(\infty, x, 1, 1, 1) \tag{4.30}$$

is the *canonical workload factor* associated with the $M/M/m$ loss model asymptotically as $m \to \infty$, $\beta \equiv (\gamma - m)/\sqrt{\gamma}$ is the *scaled offered load* and z is the peakedness. The approximations for the workload factors of the natural and simple estimators have the same form in (4.29) except the canonical workload factors are different. Since $\psi_S \approx \psi_N$, we henceforth restrict attention to the simple estimator and the indirect estimator.

To partially demonstrate that there is indeed such statistical regularity, in Figures 10.1 and 10.2 we plot the exact workload factors for the simple and indirect estimators as functions of β for several different values of m, ranging from $m = 25$ to $m = 800$.

The fact that the curves tend to fall on top of each other in the figures (except in one tail) shows that there is indeed remarkable statistical regularity. The different shape shows that the simple estimator is much more efficient in light loading,

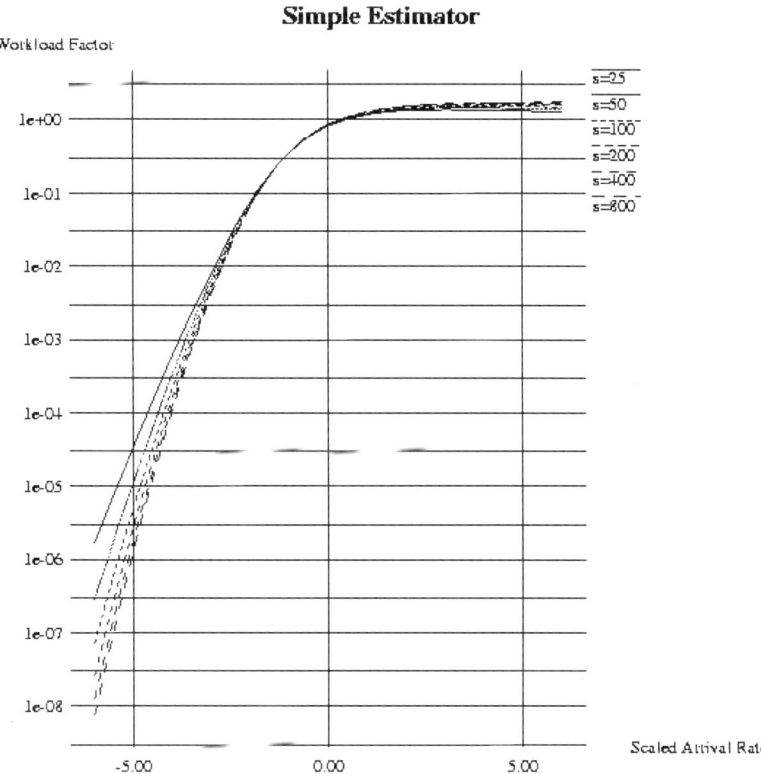

Figure 10.1. Workload factors $w_I \equiv \lambda \sigma_S^2$ for the simple estimator $\hat{B}_S(t)$ in the $M/M/m$ loss model with $\mu = 1$ as a function of the scaled arrival rate β for several numbers of servers, here denoted by s.

while the indirect estimator is much more efficient in heavy loading. Srikant and Whitt (1999) show that an empirically determined convex combination of these two estimators has the desirable qualities of both estimators, and is even more efficient.

364   10. Multiserver Queues

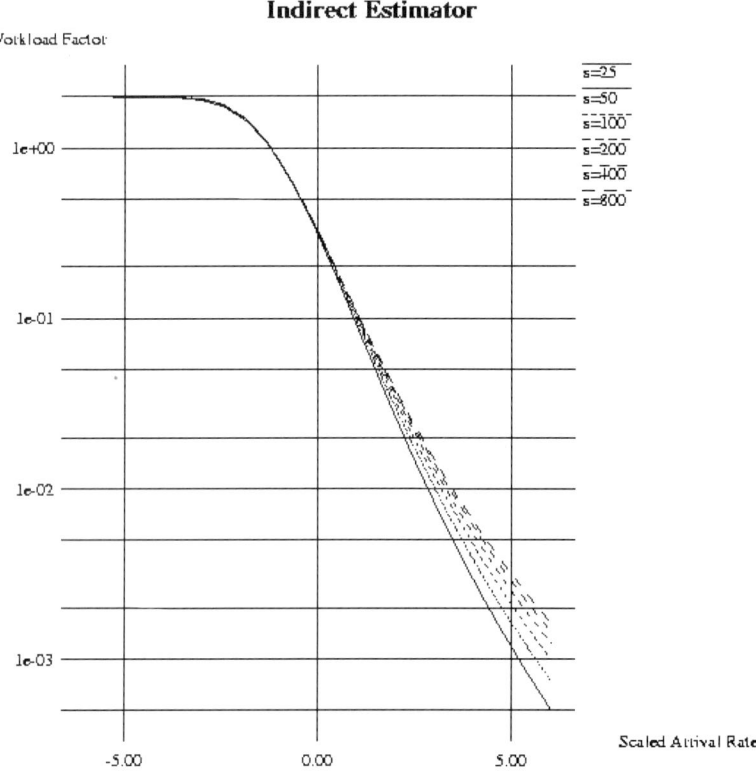

Figure 10.2. Workload factors $w_I \equiv \lambda \sigma_I^2$ for the as a function of the scaled arrival rate β for several numbers of servers, here denoted by s.

Srikant and Whitt (1996) present theoretical arguments supporting the approximations, some of which involve heavy-traffic stochastic-process limits. They also present numerical comparisons based on simulation experiments that support the approximations for $GI/GI/m$ models.

Paralleling the heavy-traffic analysis for the single-server queue in Section 5.8, it is natural to ask how the required simulation run length changes as the number of servers, m, increases with the blocking probability held fixed. Interestingly, the story here is very different from the single-server queue. If we fix the blocking probability, then the required computational effort starting in steady state actually decreases to 0 as $m \to \infty$. On the other hand, the required run length to eliminate the initialization bias, starting empty, is approximately independent of system size. Hence, when the system size grows, a greater proportion of the computational effort must be devoted to eliminating the initialization bias. Alternatively, different initial conditions must be used in order to reduce initialization bias.

11
More on the Mathematical Framework

11.1. Introduction

In this chapter we discuss the mathematical framework for stochastic-process limits, expanding upon the introduction in Chapter 3. For more details and discussion, see Billingsley (1968, 1999) and Parthasarathy (1967).

Throughout, we formalize the notion of convergence using topologies. Hence we start in Section 11.2 by reviewing basic topological concepts. In Section 11.3 we discuss the topology on the space $\mathcal{P}$ of all probability measures on a general metric space, expanding upon the introduction in Section 3.2.

In Section 11.4 we review basic properties of product spaces. We describe simple criteria for the joint convergence of random elements and we state the important (even if elementary) convergence-together theorem, which is used in many proofs.

In Section 11.5 we discuss the function space D containing the stochastic-process sample paths, expanding upon the introduction in Section 3.3. We introduce the other two Skorohod (1956) topologies – J_2 and M_2 – and provide additional details.

In Section 11.6 we briefly describe the standard approach to establish stochastic-process limits based on compactness and the convergence of the finite-dimensional distributions. The compactness approach complements the continuous-mapping approach described in Section 3.4.

11.2. Topologies

In this book we focus on stochastic-process limits, i.e., the convergence of a sequence of stochastic processes to a limiting stochastic process. We use *topology* to formalize that notion. Indeed, to a large extent, this is a book about topology. Of course, we use "topology" in the mathematical sense rather than the networking sense: Here we characterize the convergence of sequences of abstract objects (stochastic processes), rather than evaluate alternative configurations of nodes and links in a communication network.

11.2.1. Definitions

In particular, we define a topology on a set of stochastic processes. To explain, we briefly review basic topological concepts. Most of the concepts can be found in any introductory book on topology; e.g., see Simmons (1963) and Dugundji (1967).

A common way to define a topology on a set is via a metric, as defined in Section 3.2. In a metric space, we can regard the topology as a specification of which sequences converge. A direct definition of a topology involves subsets of the set S. We assume familiarity with elementary set theory. It is important to distinguish between *elements* and *subsets* of a set: When x is an element of a set S, we write $x \in S$; then $\{x\}$ is a subset of S and we write $\{x\} \subseteq S$. We write A^c, $A \cap B$, $A \cup B$ and $A - B \equiv A \cap B^c$ for complement, intersection, union and difference, respectively.

So here is the direct definition: A *topological space* is a nonempty set S together with a *topology* $\mathcal{T}$, with the topology $\mathcal{T}$ being a collection (set) of subsets of S called *open sets* satisfying certain axioms. In particular, a topology is any collection of subsets, including the whole set S and the empty set ϕ, that is closed under arbitrary unions and finite intersections. (Thus $S \in \mathcal{T}$.) By *closed under arbitrary unions*, we mean that arbitrary unions of sets in the topology are also in the topology.

Every metric determines a topology generated by (the smallest topology containing) the *open balls*

$$B_m(x, r) \equiv \{y \in S : m(x, y) < r\},$$

for $x \in S$ and $r > 0$, but not every topology can be induced by a metric. A topology that can be induced by a metric is called *metrizable*. We will primarily be concerned with metrizable topologies.

Given a topology $\mathcal{T}$ on a set S, we identify other sets (subsets of S) of interest. A set is *closed* if its complement is open. Thus the special subsets S and ϕ in every topology are both open and closed. We often use G to designate an open set and F to designate a closed set. For any subset A, its *closure* A^- is the intersection of all closed sets containing A, which is closed; its *interior* A° is the union of all open sets contained in A, which is open; and its *boundary* $\partial A \equiv A^- - A^\circ$ is the difference between the closure and the interior, which is closed.

The canonical example is the real line $\mathbb{R}$ with the usual distance $m(a, b) = |a - b|$. Let $(a, b] \equiv \{t \in \mathbb{R} : a < t \leq b\}$ and let other intervals be defined similarly. The intervals (a, b), $(-\infty, b)$ and (a, ∞) are all open sets (called open intervals), while the intervals $[a, b]$, $(-\infty, b]$ and $[a, \infty)$ are all closed sets (called closed intervals). The intervals (a, b), $(a, b]$ and $[a, b]$ all have boundary the two-point set $\{a, b\}$. The open intervals (a, b) for $-\infty < a < b < \infty$ are the open balls inducing the topology.

A second example is the k-dimensional product space $\mathbb{R}^k$. For any p, $0 < p < \infty$,

$$\|a\|_p \equiv (\sum_{i=1}^{k} |a^i|^p)^{1/p}$$

for $a \equiv (a^1, \ldots, a^k) \in \mathbb{R}^k$ is the L_p norm. The associated metric

$$m_p(a, b) \equiv \|a - b\|_p$$

induces the *Euclidean topology* on $\mathbb{R}^k$. As $p \to \infty$, the L_p norm approaches the L_∞ norm

$$\|a\|_\infty \equiv \max_{1 \leq i \leq k} |a^i|,$$

which also induces the Euclidean topology on $\mathbb{R}^k$.

Finite unions and finite intersections of open (closed) sets are again open (closed). We obtain new kinds of sets when we consider infinite unions or intersections. Because of our interest in probability measures on topological spaces, we are especially interested in countably infinite unions and intersections. A set is a G_δ if it is a countable intersection of open sets, an F_σ if it is a countable union of closed sets, a $G_{\delta\sigma}$ if it is a countable union of G_δ sets, and so forth.

A specific topology on a set is determined by specifying which subsets are open. That can be done by identifying a subbasis or a basis. A *subbasis* for the topology is any family of sets such that the given topology is the smallest topology containing that family. The topology generated by a subbasis contains the whole set S, the empty set ϕ, all finite intersections from the subbasis and all unions from these finite intersections. A *basis* is a family of open sets such that each open set is a union of basis sets. Thus, the family of all finite intersections from a subbasis forms a basis. For example, the collection of all open intervals with rational endpoints is a basis for the real line with the usual topology.

A topology on a set is also determined by specifying which functions from the given set to other topological spaces are continuous. A function f from one metric space (S, m) to another metric space (S', m') is *continuous* if $m'(f(x_n), f(x)) \to 0$ as $n \to \infty$ whenever $m(x_n, x) \to 0$ for a sequence $\{x_n : n \geq 1\}$ in S. A function f from one topological space S to another topological space S' is *continuous* if the inverse image of the open set G, $f^{-1}(G) \equiv \{s \in S : f(s) \in G\}$, is an open set in S for each open set G in S'.

One way to define a topology in terms of functions is to specify a class of functions from the given set to another topological space, and then stipulate that the topology is the smallest topology such that all functions in the designated class are continuous. (The inverse images of open sets for the functions form a subbasis.)

A one-to-one function $f : S \to S'$ mapping one topological space S onto another topological space S' such that both f and its inverse, mapping S' onto S (which we also denote by f^{-1}) are continuous is called a *homeomorphism*. (In the paragraph above, f^{-1} maps subsets of S' into subsets of S; here f^{-1} maps elements of S' into elements of S.) Two homeomorphic spaces are *topologically equivalent*. If we are only concerned about topological concepts, then two homeomorphic spaces can be regarded as two representations of the same space.

An important property held by some topological spaces is compactness. An *open cover* is a collection of open sets whose union is the entire space. A topological space is *compact* if each open cover has a finite subcover. In a metric space, a subset A is compact if every sequence in A has a convergent subsequence with limit in A.

Every subset B of a topological space S becomes a topological space in its own right with the *relative topology*, which contains all intersections of open subsets with B, i.e., all sets of the form $B \cap G$ where G is open in S. A subset of a compact topological space is itself compact if and only if the subset is closed. We often use K to denote a compact subset.

For example, in the real line $\mathbb{R}$ with the usual metric $m(a,b) = |a - b|$, the closed bounded interval $[a,b]$ is compact, but the intervals (a,b), $(a,b]$, $(-\infty, b)$ and $(-\infty, b]$ are not compact.

Every (Cartesian) product of topological spaces $\prod_{i \in I} S_i$ becomes a topological space with the *product topology*, which is defined by letting the subbasis contain all sets of elements $\{x_i : i \in I\}$ such that $x_{i_0} \in G_{i_0}$, where G_{i_0} is an open set in S_{i_0} for any single index i_0. By Tychonoff's theorem, arbitrary products of compact topological spaces are compact.

Under regularity conditions, implied by the topology being metrizable, the topology is determined by specifying which sequences of elements from the set converge to limits in the set. A sequence $\{x_n : n \geq 1\}$ in a topological space S *converges* to a limit x in S if, for each open subset G containing x, there is an integer n_0 such that $x_n \in G$ for all $n \geq n_0$. In a metric space (S, m), the sequence converges if, for all ϵ, there exists an integer n_0 such that $x_n \in B_m(x, \epsilon)$ for all $n \geq n_0$.

When sequences are not adequate, we can use nets. A sequence in S can be regarded as a map from the positive integers into S; a net in S is a map from a directed set into S. A *directed set*, say Δ, is a set with an order relation $\prec$ defined on it, so that for any $a, b \in \Delta$ there exists $c \in \Delta$ such that $a \prec c$ and $b \prec c$. Of course the positive integers is a directed set; another directed set that is not totally ordered is the set of all subsets of a given set ordered by set inclusion. A net $\{x_\delta : \delta \in \Delta\}$ in S in a topological space S *converges* to a limit x in S if, for each open set G containing x, there is $\delta_0 \in \Delta$ such that $x_\delta \in G$ for all δ with $\delta_0 \prec \delta$. However, as indicated above, in metric spaces it suffices to consider only sequences. We only mention nets when it has not yet been established that the topological space is metrizable.

Thus, in a metrizable topological space the topology can be specified in any of the following ways:

(i) specifying the open subsets, e.g., by specifying a subbasis or a basis,

(ii) specifying a class of functions that must be continuous,

(iii) defining a metric,

(iv) specifying which sequences converge.

11.2.2. Separability and Completeness

In addition to having the topological space be metrizable, we often want to impose two additional regularity properties, separability and completeness. A topological space is *separable* if it has a countable dense subset; a subset A is *dense* in a topological space S if $A^- = S$, i.e., if the closure of A is the whole space S. In metric spaces, separability is equivalent to *second countability*, i.e., the topology having a *countable basis*, which in turn is equivalent to every open cover having a countable subcover. In a separable metric space, the balls $B_m(x, r)$ of rational radius r centered at points x in a countable dense set form a countable basis. Separable metric spaces are quite general, but by the Urysohn embedding theorem, any separable metric space is homeomorphic to a subset (with the relative topology) of the space

$$[0, 1]^\infty \equiv [0, 1] \times [0, 1] \times \ldots$$

(with the product topology), which is metrizable as a compact metric space. The separable metric space itself is in general not compact, however.

A sequence $\{x_n : n \geq 1\}$ in a metric space (S, m) is *fundamental* (or satisfies the *Cauchy property*) if, for all $\epsilon > 0$, there exists $n_0 \equiv n_0(\epsilon)$ such that

$$m(x_n, x_m) < \epsilon \quad \text{for all} \quad n \geq n_0 \quad \text{and} \quad m \geq n_0 \ .$$

A metric space (S, m) is *complete* if each fundamental sequence converges to a limit in S. Completeness is useful for characterizing compactness, because a closed subset of a complete metric space is compact if and only if it is *totally bounded*, i.e., any cover by open balls has a finite subcover. (That is, in verifying that any open cover has a finite subcover, we may restrict attention to covers containing open balls.)

We are primarily concerned about the topology induced by a metric rather than the metric itself. We will often work with metrics that are not complete, but it will usually be possible to construct a topologically equivalent metric that is complete. When a topological space is metrizable as a complete metric space, we call the topological space *topologically complete*. When we are interested in the topology rather than the metric, the important property is topological completeness, not completeness.

A topological space that is metrizable as a complete separable metric space is said to be *Polish*. Closely related to Polish spaces are Lusin spaces. One topology $\mathcal{T}_1$ is a *stronger topology* (or finer topology) than another $\mathcal{T}_2$ if it contains the other as a proper subset, i.e., if $\mathcal{T}_2 \subseteq \mathcal{T}_1$. A set with a metrizable topology (or, more generally, a Hausdorff topology) on which there is a stronger topology that is Polish is called a *Lusin space*. A subset of a complete metric space is itself a complete metric space (with the same metric) if and only if it is closed; a subset of a Polish space is Polish if and only if it is a G_δ; a subset of a Lusin space is Lusin if and only if it is Borel measurable (see the next section). Countable products of Polish (Lusin) spaces are again Polish (Lusin). A nice account of Polish and Lusin spaces, and probability measures on them, is contained in Schwartz (1973). These spaces provide a natural setting for stochastic-process limits; only rarely is greater generality needed.

11.3. The Space $\mathcal{P}$

In this section we supplement the discussion in Section 3.2, in which we described the set $\mathcal{P}(S)$ of probability measures on a separable metric space (S, m), endowed with the topology of weak convergence.

11.3.1. Probability Spaces

In order to define probability measures on S, We make S a *measurable space* by endowing S with a σ-field of measurable sets (subsets of S). We let $\mathcal{S}$ denote a σ-field on S. Like a topology, a σ-*field* (on a set) is a collection of subsets of the designated set satisfying certain axioms. In particular, a σ-field contains the whole set and is closed under complements and countable unions. As usual, the sets in the σ-field are the sets to which we can assign probability. A σ-field generated by a collection of sets is the smallest σ-field containing those sets.

When we define a probability measure on a measurable space, we obtain a probability space. A probability measure on $(S, \mathcal{S})$ is a real-valued function on $\mathcal{S}$ satisfying $0 \leq P(A) \leq 1$ for all $A \in \mathcal{S}$, $P(S) = 1$ and $P(\cup_{n=1}^{\infty} A_n) = \sum_{n=1}^{\infty} P(A_n)$ whenever $\{A_n : n \geq 1\}$ is a sequence of mutually disjoint subsets in $\mathcal{S}$ (No two of the subsets have any points in common.).

We will want to consider functions mapping one measurable space $(S, \mathcal{S})$ into another $(S', \mathcal{S}')$. A function $h : (S, \mathcal{S}) \to (S', \mathcal{S}')$ is said to be *measurable* if $h^{-1}(A') \in \mathcal{S}$ for each $A' \in \mathcal{S}'$. A measurable map $h : (S, \mathcal{S}) \to (S, \mathcal{S}')$ induces an *image (probability) measure* Ph^{-1} on $(S', \mathcal{S}')$ associated with each probability measure P on $(S, \mathcal{S})$, defined by

$$Ph^{-1}(A') \equiv P(h^{-1}(A')) \equiv P(\{s \in S : h(s) \in A'\}).$$

forall $A' \in \mathcal{S}'$.

So far, we have not exploited the topology on the space S. For a topological space $(S, \mathcal{T})$, we always use the *Borel σ-field* $\mathcal{B}(S)$ generated by the open subsets of S. In a metric space (S, m), the topology is generated by the metric m, i.e., by the open sets determined by m. Assuming that the σ-fields are Borel σ-fields, all continuous functions are measurable. Indeed, the Borel σ-field can be characterized as the smallest σ-field such that all bounded continuous real-valued functions are measurable.

Since a topology is closed under arbitrary unions, while a σ-field is closed under countable unions, Borel σ-fields tend to be well behaved when the topological space is second countable, i.e., has a countable basis. Thus, as an important regularity condition, we require that the metric space (S, m) be *separable*. Separability plays an important role in product spaces; see Section 11.4 below.

11.3.2. *Characterizing Weak Convergence*

We are primarily interested in criteria for the convergence of a sequence of probability measures. As defined in Section 3.2, a sequence of probability measures $\{P_n : n \geq 1\}$ on (S, m) *converges weakly* or just *converges* to a probability measure P on (S, m), and we write $P_n \Rightarrow P$, if

$$\lim_{n \to \infty} \int_S f \, dP_n = \int_S f \, dP \tag{3.1}$$

for all functions f in $C(S)$, the space of all continuous bounded real-valued functions on S. (In Section 1.4 of the Internet Supplement we give a "Banach-space" explanation for the adjective "weak" in "weak convergence.")

Note that we could require more. We could require that (3.1) hold for all bounded measurable real-valued functions or for all indicator functions. That would be equivalent to requiring that

$$P_n(A) \to P(A) \quad \text{for all} \quad A \in \mathcal{B}(S),$$

but we do not. Indeed, that mode of convergence is often too strong. To see why, let

$$P_n(\{x_n\}) = 1 \quad \text{for} \quad n \geq 1 \quad \text{and} \quad P(\{x\}) = 1, \tag{3.2}$$

where

$$x_n \to x \quad \text{as} \quad n \to \infty. \tag{3.3}$$

If $x_n \neq x$ for infinitely many n, then $P_n(A) \not\to P(A)$ for $A = \{x\}$.

We can give a related equivalent characterization of weak convergence $P_n \Rightarrow P$. A measurable subset A for which $P(\partial A) = 0$, where ∂A is the boundary of A, is said to be a *P-continuity set*. Weak convergence $P_n \Rightarrow P$ is equivalent to "pointwise convergence" $P_n(A) \to P(A)$ for all P-continuity sets A in $\mathcal{B}(S)$. The following "Portmanteau theorem" gives several alternative characterizations of weak convergence.

Theorem 11.3.1. (alternative characterizations of weak convergence) *The following are equivalent characterizations of weak convergence $P_n \Rightarrow P$ on a metric space:*

(i) $\lim_{n \to \infty} \int_S f \, dP_n = \int_S f \, dP$ *for all* $f \in C(S)$;

(ii) $\lim_{n \to \infty} \int_S f \, dP_n = \int_S f \, dP$ *for all uniformly continuous f in $C(S)$;*

(iii) $\limsup_{n \to \infty} P_n(F) \leq P(F)$ *for all closed* F;

(iv) $\liminf_{n \to \infty} P_n(G) \geq P(G)$ *for all open* G;

(v) $\lim_{n \to \infty} P_n(A) = P(A)$ *for all P-continuity sets A;*

(vi) $P_n f^{-1} \Rightarrow P f^{-1}$ *on $\mathbb{R}$ for all $f \in C(S)$;*

(vii) $P_n f^{-1} \Rightarrow P f^{-1}$ *on $\mathbb{R}$ for all uniformly continuous f in $C(S)$.*

The different equivalent criteria in Theorem 11.3.1 can be understood by looking further at the deterministic example in (3.2)–(3.3). A key property used in the proof of Theorem 11.3.1 is the ability to approximate probabilities of measurable sets by the probabilities of open and closed sets.

Theorem 11.3.2. (approximation by closed and open sets) *Let P be an arbitary probability measure on a metric space (S, m) with the Borel σ-field $\mathcal{B}(S)$. For all $A \in \mathcal{B}(S)$ and all $\epsilon > 0$, there exists a closed set F and an open set G with*

$$F \subseteq A \subseteq G$$

such that

$$P(G - F) < \epsilon \ .$$

It is also useful to be able to approximate probabilities by the probability of compact subsets. A probability measure on a topological space S is said to be *tight* if, for all $\epsilon > 0$, there exists a compact supset K such that

$$P(K) > 1 - \epsilon \ .$$

The following is an important property of Lusin spaces; again see Schwartz (1973).

Theorem 11.3.3. (approximation by compact sets) *In a Lusin space S all probability measures are tight. Let P be an arbitrary probability measure on a Lusin space S with its Borel σ-field $\mathcal{B}(S)$. For all $A \in \mathcal{B}(S)$ and all $\epsilon > 0$, there exists a compact set K with $K \subseteq A$ such that*

$$P(A - K) < \epsilon \ .$$

We put the notion of weak convergence just given in a standard topological framework by observing that it can be characterized by a metric. That can be done in several ways; one is with the Prohorov metric defined in (2.2) in Section 3.2. We remark that in the definition of the Prohorov metric it suffices to restrict attention to A being a closed subset of S. The space $\mathcal{P}(S)$ tends to inherit properties from the underlying space S. Given that (S, m) is a separable metric space, the space $(\mathcal{P}(S), \pi)$ is topologically complete or compact if and only if (S, m) is. Moreover, the subset of probability measures in $\mathcal{P}(S)$ assigning unit mass to individual points in S with the relative topology is homeomorphic to S itself. See Chapter II of Parthasarathy (1967).

On the real line $\mathbb{R}$, we often use metrics applied to cumulative distribution functions (cdf's). The *Lévy metric*, say λ, is defined by (2.2) in Section 3.2 but only considering sets of the form $A = (-\infty, x]$. Clearly, $\lambda \leq \pi$, but both λ and π induce the topology of weak convergence in $\mathcal{P}(\mathbb{R})$.

11.3.3. Random Elements

As indicated in Section 3.2, instead of directly referring to probability measures, we often use random elements. We now restate Theorem 11.3.1 in terms of random elements. We say that a subset A in $\mathcal{B}(S)$ is an X-continuity set if $P(X \in \partial A) = 0$.

Theorem 11.3.4. (alternative characterizations of convergence in distribution) *The following are equivalent characterizations of convergence in distribution $X_n \Rightarrow X$ for random elements of a metric space:*

(i) $\lim_{n \to \infty} Ef(X_n) = Ef(X)$ *for all* $f \in C(S)$;

(ii) $\lim_{n \to \infty} Ef(X_n) = Ef(X)$ *for all uniformly continuous f in* $C(S)$;

(iii) $\limsup_{n \to \infty} P(X_n \in F) \leq P(X \in F)$ *for all closed* F;

(iv) $\liminf_{n \to \infty} P(X_n \in G) \geq P(X \in G)$ *for all open* G;

(v) $\lim_{n \to \infty} P(X_n \in A)$ *for all X-continuity sets* A;

(vi) $f(X_n) \to f(X)$ *in $\mathbb{R}$ for all* $f \in C(S)$;

(vii) $f(X_n) \Rightarrow f(X)$ *in $\mathbb{R}$ for all uniformly continuous f in* $C(S)$.

The adjective "weak" in "weak convergence" distinguishes convergence in distribution $X_n \Rightarrow X$ from the stronger convergence $X_n \to X$ with probability one (w.p.1), which is called a *strong limit*. (The strong limit can hold only when X_n and X are defined on a common probability space.) We will give an alternative explanation for using the adjective "weak" below.

We now elaborate further on the meaning of weak convergence $P_n \Rightarrow P$. First, notice that the definition of the Prohorov metric allows the probability measure P_2 to assign a mass $\pi(P_1, P_2)$ arbitrarily far from where P_1 assigns its mass, allowing for a small chance of a big error. We can better understand the Prohorov metric π by considering a special representation, originally due to Strassen (1965); also see Billingsley (1999) and Pollard (1984).

The Strassen representation theorem relates the Prohorov distance between two probability measures to the distance in probability between two specially constructed random elements with those probability laws. For two random elements X_1 and X_2 of a separable metric space (S, m) defined on the same underlying probability space $(\Omega, \mathcal{F}, P)$, the *in-probability distance* between X_1 and X_2 is defined by

$$p(X_1, X_2) \equiv \inf\{\epsilon > 0 : P(m(X_1, X_2) > \epsilon) < \epsilon\}. \quad (3.4)$$

(We need separability for $m(X_1, X_2)$ to be a legitimate random variable; see the next section. The distance p is only a pseudometric because $p(X_1, X_2) = 0$ does not imply that $X_1 = X_2$.)

It is easy to see that

$$\pi(PX_1^{-1}, PX_2^{-1}) \leq p(X_1, X_2)$$

for any random elements mapping an underlying probability space $(\Omega, \mathcal{F}, P)$ into a separable metric space (S, m). The Strassen representation theorem allows us to go the other way for specially constructed random elements.

Theorem 11.3.5. (Strassen representation theorem) *For any $\epsilon > 0$ and any two probability measures P_1 and P_2 on a separable metric space (S, m), there exist special S-valued random elements X_1 and X_2 on some common underlying probability space such that*

$$PX_i^{-1} = P_i \quad \text{for} \quad i = 1, 2$$

and

$$p(X_1, X_2) < \pi(P_1, P_2) + \epsilon , \tag{3.5}$$

where p is the in-probability distance in (3.4). If the two probability measures P_1 and P_2 are tight, which always holds if (S, m) is also a Lusin space, then the random elements X_1 and X_2 can be constructed so that

$$p(X_1, X_2) = \pi(P_1, P_2) . \tag{3.6}$$

The Strassen representation theorem says that the Prohorov distance between probability measures can be realized (possibly only via an infimum) as the distance in probability between two specially constructed random elements on a common probability space that have the given probability measures as their probability laws. It suffices to let the underlying probability space be the product space $(S, m) \times (S, m)$ and the random elements be the coordinate projections. The problem then is to construct the probability measure on $(S, m) \times (S, m)$ with the specified marginal probability laws satisfying (3.5) or (3.6).

Example 11.3.1. *A simple example.* To fix ideas it is useful to consider an example. In Table 11.1 we specify three different random variables defined on a simple probability space $(\Omega, \mathcal{F}, P)$. The sample space Ω contains only four elements; the σ-field $\mathcal{F}$ contains all subsets; and the probability measure P assigns equal probabilities to each set containing a single point.

Ω	$P(\{\omega\})$	$X_1(\omega)$	$X_2(\omega)$	$X_3(\omega)$
ω_1	1/4	1/4	3/4	1/4
ω_2	1/4	2/4	4/4	2/4
ω_3	1/4	3/4	1/4	3/4
ω_4	1/4	4/4	2/4	100

Table 11.1. Three possible random variables

In Table 11.2 we display the distances $\pi(PX_1^{-1}, PX_2^{-1})$ and $p(X_1, X_2)$, along with the uniform distance between the random variables $\| X_1 - X_2 \|$, where

$$\| X \| \equiv \sup_{\omega \in \Omega} |X(\omega)| .$$

distance	$i=1, j=2$	$i=1, j=3$	$i=2, j=3$
$\pi(PX_i^{-1}, PX_j^{-1})$	0	1/4	1/4
$p(X_i, X_j)$	1/2	1/4	1/2
$\|X_i - X_j\|$	1/2	99	99 1/4

Table 11.2. Distances between the random variables

Note that the random variables X_1 and X_2 are different, but they have the same distribution. The distances always increase as we go down in Table 11.2, but generalizations going sideways are hard to make. ∎

The Skorohod representation theorem, Theorem 3.2.2, also helps to understand the topology of weak convergence.

11.4. Product Spaces

We are often interested in joint convergence of random elements: We want to go beyond $X_n \Rightarrow X$ and $Y_n \Rightarrow Y$ to obtain $(X_n, Y_n) \Rightarrow (X, Y)$. We consider such joint limits because we want to understand the joint distribution of X_n and Y_n. We also often require the joint convergence in order to apply the continuous-mapping approach.

First, to have a vector random element (X, Y) well defined, we need X and Y to be defined on a common underlying probability space. Then, given random elements X of a separable metric space (S', m') and Y of a separable metric space (S'', m''), we can regard (X, Y) as a random element of the *product space* associated with two metric spaces (S', m') and (S'', m''), i.e.,

$$S \equiv S' \times S'' \equiv \{(x, y) : x \in S', y \in S''\}.$$

The *open rectangles* $G' \times G''$ with G' open in S' and G'' open in S'' are a basis for the *product topology* on $S' \times S''$. The product topology is characterized by having convergence $(x'_n, x''_n) \to (x', x'')$ if and only if $x'_n \to x'$ and $x''_n \to x''$. The product topology can be induced by several different metrics, one being the *maximum metric*

$$m((x_1, y_1), (x_2, y_2)) \equiv \max\{m'(x_1, x_2), m''(y_1, y_2)\}.$$

Similarly, the *measurable rectangles* $A' \times A''$ with A' measurable in S' and A'' measurable in S'' generate the *product σ-field* on the product space $S' \times S''$; i.e., the product σ-field is the smallest σ-field on $S' \times S''$ containing the measurable rectangles.

The following basic theorems explain why we need our metric spaces to be separable, i.e., to have countable dense subsets.

Theorem 11.4.1. (separability of product spaces) *The product space $S' \times S''$ with the product topology is separable if and only if the component spaces S' and S'' are separable.*

Theorem 11.4.2. (the Borel σ-field in product spaces) *The Borel σ-field $\mathcal{B}(S)$ associated with the product space $S = S' \times S''$ with product topology is the product σ-field $\mathcal{B}(S') \times \mathcal{B}(S'')$ if and only if the metric spaces (S', m') and (S'', m'') are separable.*

For a probability measure P on $S \equiv S' \times S''$, marginal probability measures P' and P'' are defined on $(S', \mathcal{S}')$ and $(S'', \mathcal{S}'')$ by setting

$$P'(A') \equiv P(A' \times S'') \quad \text{and} \quad P''(A'') \equiv P(S' \times A'')$$

for every $A' \in \mathcal{S}'$ and $A'' \in \mathcal{S}''$. Thus, if (X, Y) is a random element of $S = S' \times S''$ with probability law P, then P' and P'' are the probability laws of X and Y, respectively.

Theorem 11.4.3. (criteria for joint convergence) *Suppose that the product space $S \equiv S' \times S''$ with the product topology is separable. Then the following are each necessary and sufficient conditions for convergence in distribution $(X_n, Y_n) \Rightarrow (X, Y)$ in S:*

$$(i) \quad P(X_n \in A', Y_n \in A'') \to P(X \in A', Y \in A'')$$

for every X-continuity set A' and every Y-continuity set A''.

$$(ii) \quad \overline{\lim_{n \to \infty}} \, P(X_n \in F', Y_n \in F'') \leq P(X \in F', Y \in F'')$$

for every closed set F' in S' and every closed set F'' in S''.

$$(iii) \quad \underline{\lim_{n \to \infty}} \, P(X_n \in G', Y_n \in G'') \geq P(X \in G', Y \in G'')$$

for every open set G' in S' and every open set G'' in S''.

In general, we must verify one of the limits in Theorem 11.4.3 (i) – (iii) in order to establish convergence in distribution for vector random elements, but there are two special cases in which we easily get convergence in distribution for vector random elements. One case involves independence and the other involves a deterministic limit.

For given probability measures P' on $(S', \mathcal{S}')$ and P'' on $(S'', \mathcal{S}'')$, the *product probability measure* $P' \times P''$ on the product space $S' \times S''$ with the product σ-field $\mathcal{S}' \times \mathcal{S}''$ is defined by

$$(P' \times P'')(A' \times A'') = P'(A')P''(A'')$$

for every $A' \in \mathcal{S}'$ and $A'' \in \mathcal{S}''$. By Theorem 11.4.2, the product measure is defined on the Borel field of $S' \times S''$ when (S', m') and (S'', m'') are separable. When X and Y are independent random elements of S' and S'', the probability law of (X, Y) is the product probability law $P_1 X^{-1} \times P_2 Y^{-1}$, where P_i are the probability measures in the underlying probability spaces.

Theorem 11.4.4. (joint convergence for independent random elements) *Let X_n and Y_n be independent random elements of separable metric spaces (S', m') and*

(S'', m'') for each $n \geq 1$. Then there is joint convergence in distribution

$$(X_n, Y_n) \Rightarrow (X, Y) \quad in \quad S' \times S''$$

if and only if $X_n \Rightarrow X$ in S' and $Y_n \Rightarrow Y$ in S''.

Theorem 11.4.5. (joint convergence when one limit is deterministic) *Suppose that $X_n \Rightarrow X$ in a separable metric space (S', m') and $Y_n \Rightarrow y$ in a separable metric space (S'', m''), where y is deterministic. Then*

$$(X_n, Y_n) \Rightarrow (X, y) \quad in \quad S' \times S'' \ .$$

Proof. By Theorem 11.4.3, it suffices to show that

$$P(X_n \in A, Y_n \in B) \to P(X \in A, y \in B) \tag{4.1}$$

for each X-continuity set A and y-continuity set B (i.e., where $y \notin \partial B$). First suppose that $y \in B$, which implies that $P(Y_n \notin B) \to 0$. Then (4.1) holds because

$$P(X_n \in A) - P(Y_n \notin B) \leq P(X_n \in A, Y_n \in B) \leq P(X_n \in A) \ .$$

Now suppose that $y \notin B$. Then (4.1) again holds because

$$P(X_n \in A, Y_n \in B) \leq P(Y_n \in B) \to 0. \quad \blacksquare$$

Given two random elements X and Y of a common metric space (S, m), we can speak of the random distance $m(X, Y)$. Such a random distance is a legitimate real-valued random variable when (S, m) is a separable metric space, but not otherwise; see p. 225 of Billingsley (1968).

Theorem 11.4.6. (measurability of the distance between random elements) *If (S, m) is a separable metric space and X and Y are random elements of S defined on a common domain, then $m(X, Y)$ is a legitimate measurable real-valued random variable*

Note that convergence in distribution $Y_n \Rightarrow y$ to a deterministic limit in Theorem 11.4.5 (where (S'', m'') is a separable metric space) is equivalent to *convergence in probability*; i.e., $Y_n \Rightarrow y$ above if and only if

$$P(m''(Y_n, y) > \epsilon) \to 0 \quad \text{as} \quad n \to \infty$$

for all $\epsilon > 0$.

We now give a useful way to establish new weak convergence limits from given ones. We already used this result in our treatment of the Kolmogorov-Smirnov statistic in Section 2.2.

Theorem 11.4.7. (convergence-together theorem) *Suppose that X_n and Y_n are random elements of a separable metric space (S, m) defined on a common domain. If $X_n \to X$ in S and $m(X_n, Y_n) \Rightarrow 0$ in $\mathbb{R}$, then*

$$(X_n, Y_n) \Rightarrow (X, X) \quad in \quad (S, m) \times (S, m) \ .$$

Proof. For any closed subset F of S, let $F^{\bar{\epsilon}}$ be its closed ϵ-neighborhood, defined by

$$F^{\bar{\epsilon}} \equiv \{y \in S : m(x,y) \leq \epsilon \quad \text{for some} \quad x \in F\} \,.$$

For any two closed subsets F_1 and F_2 of S,

$$P(X_n \in F_1, Y_n \in F_2) \leq P(X_n \in (F_1 \cap F_2)^{\bar{\epsilon}}) + P(m(X_n, Y_n) \geq \epsilon) \,,$$

so that

$$\varlimsup_{n \to \infty} P(X_n \in F_1, Y_n \in F_2) \leq \varlimsup_{n \to \infty} P(X_n \in (F_1 \cap F_2)^{\bar{\epsilon}}) \leq P(X \in (F_1 \cap F_2)^{\bar{\epsilon}})$$

by Theorem 11.3.4 (iii), since $X_n \Rightarrow X$. Letting $\epsilon \downarrow 0$, we have $P(X \in (F_1 \cap F_2)^{\bar{\epsilon}}) \downarrow P(X \in (F_1 \cap F_2))$. Hence,

$$\varlimsup_{n \to \infty} P(X_n \in F_1, Y_n \in F_2) \leq P(X \in F_1, X \in F_2) \,,$$

which implies the conclusion by Theorem 11.4.3. ∎

We usually use Theorem 11.4.7 in proofs to obtain a desired limit $Y_n \Rightarrow X$ in S (one coordinate only) by treating a closely related sequence $\{X_n\}$ that is easier to analyze. There is a converse to Theorem 11.4.7 that adds insight into the significance of joint convergence to a common limit. We not only get the two marginal limits $X_n \Rightarrow X$ and $Y_n \Rightarrow X$, but we also get asymptotic equivalence of X_n and Y_n.

Theorem 11.4.8. (asymptotic equivalence from joint convergence) *Suppose that X_n and Y_n are random elements of a separable metric space (S, m) with a common domain. If $(X_n, Y_n) \Rightarrow (X, X)$ in $S \times S$, then*

$$m(X_n, Y_n) \Rightarrow 0 \quad \text{in} \quad \mathbb{R} \,.$$

Proof. Apply the Skorohod representation theorem to replace the convergence in distribution $(X_n, Y_n) \Rightarrow (X, X)$ in $S \times S$ by convergence w.p.1 for the special versions $\tilde{X}_n$ and $\tilde{Y}_n$. Then apply the triangle inequality in (S, m) to deduce that $m(\tilde{X}_n, \tilde{Y}_n) \to 0$ w.p.1. Finally, note that $m(X_n, Y_n)$ has the same distribution as $m(\tilde{X}_n, \tilde{Y}_n)$, so that we obtain the desired conclusion. ∎

11.5. The Space D

In this section we supplement the discussion of the functions space D in Section 3.3, primarily by introducing the Skorohod (1956) J_2 and M_2 topologies. For the discussion here, we assume that the functions are real-valued and that the function domain is $[0,1]$.

We start by making some observations about the J_1 metric defined in (3.2) of Section 3.3. The metric d_{J_1} is incomplete, but the topology is topologically complete; there exists a topologically equivalent metric that is complete; see Billingsley (1968).

The space (D, J_1) is separable; a countable dense set is made up of the rational-valued piecewise-constant functions with only finitely many discontinuities, all at rational time points in the domain $[0, 1]$. Thus the space (D, J_1) is Polish. The J_1 topology can also be defined on D spaces with more general ranges. We can let the range be $\mathbb{R}^k$ or any Polish space.

11.5.1. J_2 and M_2 Metrics

A metric inducing the J_2 topology on $D([0, 1], \mathbb{R})$ is defined by replacing the set of functions Λ by the larger set Λ' of all one-to-one maps of $[0, 1]$ onto $[0, 1]$, without requiring any continuity, i.e.,

$$d_{J_2}(x_1, x_2) \equiv \inf_{\lambda \in \Lambda'} \{\|x_1 \circ \lambda - x_2\| \vee \|\lambda - e\|\} . \tag{5.1}$$

Since $\Lambda \subseteq \Lambda'$, we obviously have

$$d_{J_2}(x_1, x_2) \leq d_{J_1}(x_1, x_2) \leq \|x_1 - x_2\| .$$

We will have little to say about the J_2 topology; see Bass and Pyke (1987) for an application.

As noted in Section 3.3, we need a different topology on D if we want the jump in a limit function to be unmatched in the converging functions. In order to establish limits with unmatched jumps in the limit function (or process), we use the Skorohod (1956) M topologies. We define the M topologies using the completed graphs of the functions, defined in (3.3) in Section 3.3.

The completed graph It is thus natural to consider established metrics defined on the set of all compact subsets of $\mathbb{R}^k$. Perhaps the best known such metric is the Hausdorff metric. Given compact subsets K_1 and K_2 of $\mathbb{R}^k$, the *Hausdorff metric* m_H is defined by

$$m_H(K_1, K_2) \equiv \sup_{x_1 \in K_1} m(x_1, K_2) \vee \sup_{x_2 \in K_2} m(x_2, K_1) , \tag{5.2}$$

where $m(x, A)$ is the *distance between the point x and the set A*, defined by

$$m(x, A) \equiv m(A, x) \equiv \inf_{y \in A} m(x, y) \tag{5.3}$$

and m is the metric used on $\mathbb{R}^k$; e.g., see Section 1.4 of Matheron (1975).

To illustrate, the Hausdorff distance between two compact subsets in $\mathbb{R}^k$ is depicted in Figure 11.1. The compact sets are the set of points inside the oval and the set of points inside the rectangle, including the boundaries. Note that the two sets overlap. The dashed lines identify the point in the oval furthest away from the rectangle and the point in the rectangle furthest away from the oval. The Hausdorff distance is the greater of these two distances. Thus, even if one compact subset is a proper subset of another, they are a positive distance apart.

Thus, for any $x_1, x_2 \in D$, the M_2 metric on D is defined by

$$d_{M_2}(x_1, x_2) \equiv m_H(\Gamma_{x_1}, \Gamma_{x_2}) , \tag{5.4}$$

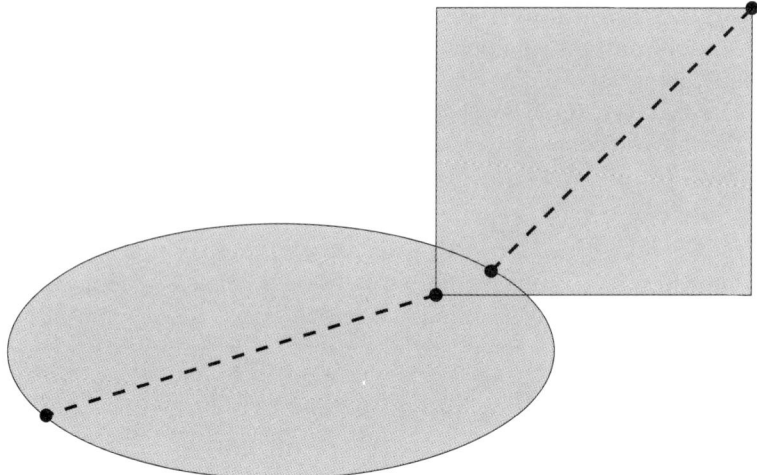

Figure 11.1. The Hausdorff distance between two compact subsets of the plane. The Hausdorff distance is the length of the longer dashed line.

where m_H is the Hausdorff metric in (5.2) and Γ_x is the completed graph of x, defined in (3.3) of Section 3.3. The topology on the space of completed graphs induced by the Hausdorff metric is the Skorohod M_2 topology (although that is not the way it was originally defined). The M_1 topology is stronger than the M_2 topology. It pays closer attention to order.

From the definitions above, it is not obvious how the M_1 and M_2 topologies are related. For understanding the relation beween the two topologies, it is significant that the M_2 topology can also be expressed via parametric representations. Indeed, an alternative M_2 metric (inducing the M_2 topology) can be defined by (3.4) in Section 3.3, after changing the definition of a parametric representation: Instead of requiring that (u,r) be nondecreasing, using the order on the completed graphs, we only require that the time component function r be nondecreasing. With that definition, it is evident that the M_1 topology is stronger than the M_2 topology, i.e., M_1 convergence implies M_2 convergence.

Unlike the J topologies, it is not possible to extend the M topologies by allowing the range to be an arbitrary Polish space, because the completed graphs require linear structure. However, the range can be a separable Banach space with the M topologies. That generalization is used to obtain heavy-traffic stochastic-process limits for the workload process in an infinite-server queue in Section 10.3.

11.5.2. The Four Skorohod Topologies

A unified approach to the four Skorohod topologies via graphs was provided in the thesis by Pomarede (1976). In that approach, the M_2 and J_2 topologies are generated by the Hausdorff metric applied to the completed and uncompleted graphs, respectively. Similarly, the M_1 and J_1 topologies are defined in terms of parametric representations of the completed and uncompleted graphs. That approach to the J_1 topology draws upon Kolmogorov (1956).

For applications, it is significant that previous limits for stochastic processes with the familiar J_1 topology on D will also hold when we use one of the other Skorohod (1956) nonuniform topologies instead, because the J_1 topology is stronger (or finer) than the other topologies. The four nonuniform Skorohod topologies are ordered by

$$J_1 > J_2 > M_2 \quad \text{and} \quad J_1 > M_1 > M_2 \,, \tag{5.5}$$

where $>$ means stronger than, with M_1 and J_2 not being comparable. Examples of functions x_n converging to the indicator function $x \equiv I_{[2^{-1},1]}$ in $D([0,1],\mathbb{R})$ in the different topologies are given in Figure 11.2. We contend that the M_1 topology is often the most appropriate one; we discuss this point further in Chapter 6.

We have indicated that all four nonuniform Skorohod topologies reduce to uniform convergence over $[0,1]$ when the limit function is continuous. More generally, convergence in all four of these topologies implies local uniform convergence at any continuity function of a limit function. Thus all four Skorohod topologies are stronger than the L_p topologies on D induced by the norms

$$\| x \|_p \equiv \left(\int_0^1 |x(t)|^p dt \right)^{1/p} \,. \tag{5.6}$$

More generally, we have the following continuity result.

Theorem 11.5.1. (continuity of integrals) *Suppose that* $g : \mathbb{R}^k \to \mathbb{R}$ *is a continuous function and let* $f : D^k \to (C, U)$ *be defined by*

$$f(x)(t) \equiv \int_0^t g(x)(s)ds, \quad t \geq 0 \,.$$

If D^k is endowed with any of the Skorohod nonuniform topologies, then f is continuous.

Proof. First note that

$$\int_0^t |x(s)|ds \leq \int_0^T |x(s)|ds$$

for all real-valued x and t with $0 \leq t \leq T$. Then use the bounded convergence theorem: Convergence $x_n \to x$ imples that $\sup_n \|x_n\| < \infty$ and that $x_n(t) \to x(t)$ at almost all t. ∎

Since the J_1 topology on D is Polish, the J_2, M_1, M_2 and L_p topologies on D are automatically Lusin. In Chapter 12 we will show that the M_1 topology is Polish.

As with the J_1 and M_1 topologies on D, the domain of the functions can be changed to $[0,\infty)$ for the J_2 and M_2 topologies. In all cases $x_n \to x$ is understood to mean that there is convergence of the restrictions of x_n to the restriction of x in $D([0,t],\mathbb{R}^k)$ for all t that are continuity points of the limit function x. However, it sometimes is desirable to obtain stronger control of the convergence at the end of

384   11. Mathematical Framework

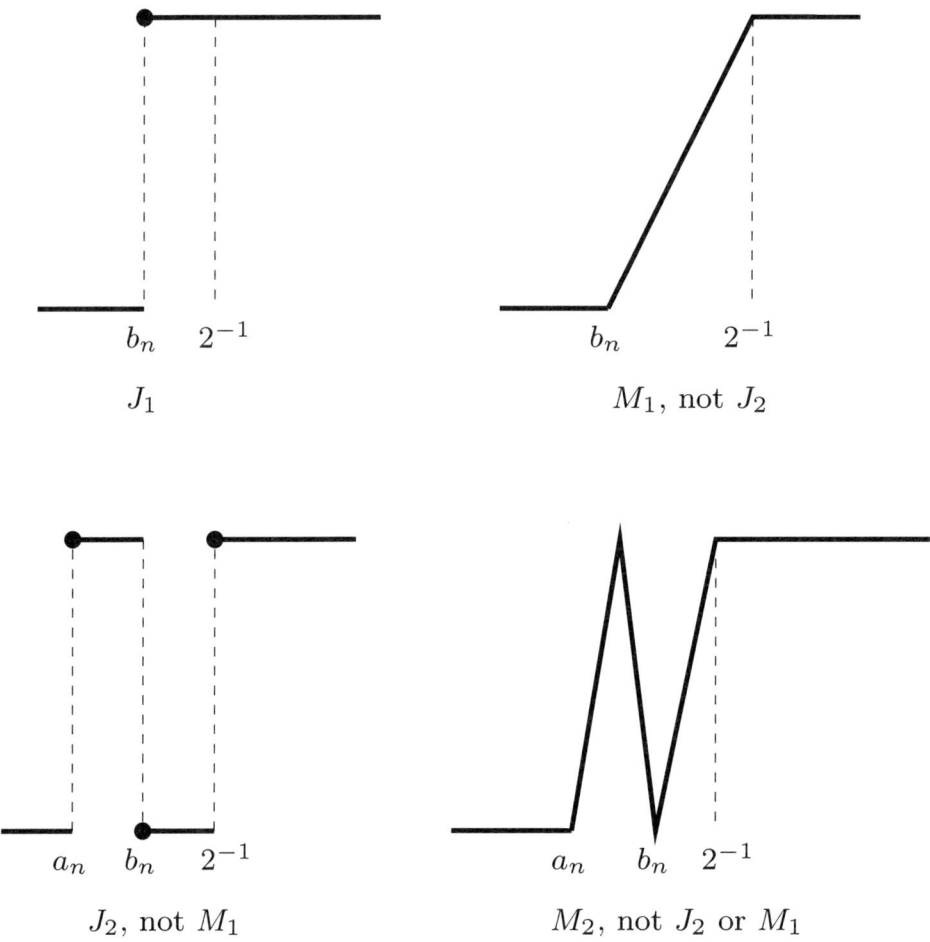

Figure 11.2. Four candidate sequences of functions $\{x_n : n \geq 1\}$ that might converge to $x \equiv I_{[1/2,1]}$ in $D([0,1], \mathbb{R})$, where $a_n = 2^{-1} - 2n^{-1}$ and $b_n = 2^{-1} - n^{-1}$.

the domain. By exploiting the SLLN, we can often assume that the sample paths x satisfy $\|x\|_w < \infty$ w.p.1, where $\|x\|_w$ is the *weighted-supremum norm*, i.e.,

$$\|x\|_w \equiv \sup_{0 \leq t < \infty} \{|x(t)|/(1+t)\} ,$$

so that we can use the metric $\|x_1 - x_2\|_w$ on the subset of functions in $D([0, \infty), \mathbb{R})$ with finite weighted-supremum norm. Weak convergence with the weighted-supremum norm was established by Müller (1968) and applied by Whitt (1972). Related weighted distances have been used extensively in the study of empirical processes, as can be seen from Shorack and Wellner (1986) and Csörgő and Horváth (1993).

In Section 3.3 we noted that addition is not continuous everywhere on $D \times D$, but that it is continuous at all pairs (x, y) in $D \times D$ that have no common discontinuity points (and, for the M_1 topology, at all pairs (x, y) that have no common discontinuity points with jumps of opposite sign). In many applications, we are able to show that the two-dimensional limiting stochastic process has sample paths in one of those subsets w.p.1., so that we can apply the continuous-mapping approach with addition.

11.5.3. Measurability Issues

Since addition is not continuous everywhere, we must not only show that it is continuous almost everywhere, but we must show that it is measurable. It is thus important to know more about the Borel σ-fields associated with the nonuniform Skorohod topologies. Fortunately, in each case, the Borel σ-field coincides with the usual σ-field, namely, the *Kolmorogov σ-field*, which is generated by the projection maps $\pi_{t_1,\ldots,t_k} : D \to \mathbb{R}^k$, defined by

$$\pi_{t_1,\ldots,t_k}(x) = [x(t_1),\ldots,x(t_k)] ,$$

so that measurability in $(D, \mathcal{B}(D))$ with any of the nonuniform topologies is consistent with the standard notion. The Kolmogorov σ-field generated by the coordinate projections (or "cylinder sets") is usually associated with the product space $\mathbb{R}^\infty$ and the Kolmogorov (1950) extension theorem; see Neveu (1965).

Theorem 11.5.2. (the Borel σ-fields on D) *The Borel σ-fields on D with any of the nonuniform Skorohod topologies coincides with the Kolmogorov σ-field generated by the coordinate projections.*

Theorem 11.5.2 can be proved by direct verification in each case, as done for the J_1 topology in Billingsley (1968). For the non-J_1 topologies, we can also exploit the established J_1 result and properties of Lusin spaces, see p. 101 of Schwartz (1973).

Theorem 11.5.3. (Borel σ-fields for comparable Lusin spaces) *Any two comparable Lusin topologies on a set have identical Borel σ-fields.*

It often happens that we have a limit for a sequence of stochastic processes with sample paths in D, where the limit process has continuous sample paths. Then there is considerable flexibility on the choice of the topology. In that case, the four nonuniform topologies on D reduce to uniform convergence over all bounded intervals, and all four topologies have the same Borel σ-field. Clearly, then it does not matter which of these topologies is used.

We might naturally try to simplify matters even further in such a situation. We might choose to work directly with the space (D, U), where U denotes the topology of uniform convergence (over closed bounded subintervals) on D, induced by the uniform metric in Section 3.3 when the function domain is $[0,1]$. There is a complication, however. Even though convergence $x_n \to x$ in D with the various nonuniform Skorohod topologies is equivalent to uniform convergence over all bounded intervals when the limit function x is continuous, in general we cannot simply work with the space (D, U), because there are measurability problems. The Borel σ-fields on D (generated by the open subsets) of all the nonuniform Skorohod topologies on D coincide with the usual Kolmogorov σ-field on D generated by the coordinate projections, but that is not true for (D, U). The Borel σ-field on (D, U) is much larger than the Kolmogorov σ-field. As a consequence, familiar stochastic processes such as the Poisson process cannot be regarded as random elements of (D, U) with the Borel σ-field; see Section 18 of Billingsley (1968).

The measurability problems arise because the space (D, U) is nonseparable. However, everything works well if we use D with a nonuniform Skorohod topology. The Borel σ-field then is the familiar one and, if a limit function is continuous, convergence is equivalent to uniform convergence.

If we want to consider only convergence of stochastic processes where the limiting stochastic process has continuous sample paths, then there is an alternative approach. We can then use the space (D, U) with the uniform topology, but use a smaller σ-field than the Borel σ-field, in particular, the σ-field generated by the open balls, called the em ball σ-field $B_m(x, r)$. Such a theory was developed by Dudley (1966, 1967) and is explained and used by Pollard (1984).

In contrast, in this book we are interested in the convergence of stochastic processes where the limiting stochastic process has discontinuous sample paths. Thus, for both measurability and convergence, we want to use a nonuniform topology on D. The particular nonuniform topology on D becomes important when the limit functions become discontinuous. The J_1 topology is useful because it allows some flexibility in the location of jumps, but it requires that the converging functions have jumps corresponding to each jump in the limit function. Thus, here we focus on the M_1 topology.

11.6. The Compactness Approach

In this book we focus on the continuous-mapping approach to establish stochastic-process limits. To put the continuous-mapping approach in perspective, we now describe the standard approach to establish stochastic-process limits based on compactness. Our accounts of both the compactness approach and the continuous-mapping approach are abridged versions of the excellent accounts in Billingsley (1968).

The compactness-approach applies to probability measures on a general metric space (S, m). In Section 3.2 we indicated that the space $(\mathcal{P}(S), \pi)$ of probability measures on (S, m) with the Prohorov metric π is a metric space. As in any metric space, we have convergence $\pi(P_n, P) \to 0$ as $n \to \infty$ for a sequence $\{P_n : n \geq 1\}$ in $(\mathcal{P}(S), \pi)$ if and only if every subsequence $\{P_{n'} : n' \geq 1\}$ contains a further subsequence $\{P_{n''} : n'' \geq 1\}$ with $P_{n''} \Rightarrow P$. We exploit a version of sequential compactness to provide conditions under which that characterization of convergence is satisfied. The compactness approach has been used to establish most of the initial stochastic-process limits we will use in the continuous-mapping approach.

11.6. The Compactness Approach

As in any metric space, a subset A of $(\mathcal{P}(S), \pi)$ has compact closure A^- if and only if the set A is *relatively compact*, i.e., if every sequence $\{P_n : n \geq 1\}$ in A has a subsequence $\{P_{n'} : n' \geq 1\}$ with $P_{n'} \Rightarrow P'$, where the limit P' is necessarily in the closure A^-. Thus, given a sequence $\{P_n : n \geq 1\}$, we can establish convergence $P_n \Rightarrow P$ in $\mathcal{P}(S)$ by showing, first, that the sequence $\{P_n : n \geq 1\}$ is relatively compact and, second, by showing that the limit of any convergent subsequence must be P. The second step can be established by establishing a weaker form of convergence, which is not strong enough to imply weak convergence (i.e., $\pi(P_n, P) \to 0$), but which is strong enough to uniquely determine the limit P. The two steps together imply that $P_n \Rightarrow P$.

A key step in the compactness-approach to limits for sequences of probability measures on a metric space is to relate compact subsets in the space $(\mathcal{P}(S), \pi)$ to compact subsets of the underlying space (S, m). That can be done by applying *Prohorov's theorem*, from Prohorov (1956). The key concept is tightness, which we now extend from a single probability measure to a set of probability measures. A subset A of probability measures in $\mathcal{P}(S)$ is said to be *tight* if, for all ϵ, there exists a compact subset K of (S, m) such that

$$P(K) > 1 - \epsilon \quad \text{for all} \quad P \in A \, .$$

The compact set K depends upon ϵ, but it must do the job for all P in A. Theorem 11.3.3 implies that every single probability measure on a Lusin space is tight.

Theorem 11.6.1. (Prohorov's theorem) *Let (S, m) be a metric space. If a subset A in $\mathcal{P}(S)$ is tight, then it is relatively compact. On the other hand, if the subset A is relatively compact and the topological space S is Polish, then A is tight.*

Thus, in Polish spaces tightness is necessary and sufficient for relative compactness. That implies that nothing is lost by focusing on tightness when we want to establish relative compactness. (That is the primary basis for interest in knowing whether a topological space is Polish when we are concerned about weak convergence of probability measures.)

From Prohorov's theorem, we obtain a useful way to establish convergence $P_n \Rightarrow P$.

Corollary 11.6.1. (tightness criterion for weak convergence) *Let $\{P_n : n \geq 1\}$ be a sequence of probability measures on a metric space (S, m). If the sequence $\{P_n\}$ is tight and the limit of any convergent subsequence from $\{P_n\}$ must be P, then $P_n \Rightarrow P$.*

These two conditions apply very naturally to establish criteria for convergence $X_n \Rightarrow X$ for stochastic processes $\{X_n(t) : 0 \leq t \leq 1\}$ in the function space $C \equiv C([0, 1], \mathbb{R})$ of continuous real-valued functions on the interval $[0, 1]$ (or any other closed bounded interval) with the uniform metric. First, it is natural to require convergence of all the finite-dimensional distributions, i.e., to show that

$$(X_n(t_1), \ldots, X_n(t_k)) \Rightarrow (X(t_1), \ldots, X(t_k)) \quad \text{in} \quad \mathbb{R}^k \qquad (6.1)$$

for all positive integers k and all k time points $t_1, \ldots, t_k$ with $0 \le t_1 < \cdots < t_k \le 1$. By the Kolmogorov extension theorem, it is known that the finite-dimensional distributions uniquely determine a probability distribution on the larger product space $\mathbb{R}^{[0,1]}$, endowed with the Kolmogorov σ-field generated by the coordinate projections. Since the Borel σ-field on C with the uniform norm coincides with the Kolmogorov σ-field generated by the coordinate projections, the finite-dimensional distributions also determine the distribution of a stochastic process X with sample paths in C.

However, convergence of the finite-dimensional distributions is *not* strong enough to imply convergence in distribution $X_n \Rightarrow X$ of the random elements of C.

Example 11.6.1. *Convergence of finite-dimensional distributions is not enough.* To see that convergence of the finite-dimensional distributions does not imply convergence in distribution $X_n \Rightarrow X$ for random elements of C, it suffices to consider a deterministic example. Let $P(X = x) = 1$ and $P(X_n = x_n) = 1$ for all $n \ge 1$, where $x(t) = 0, 0 \le t \le 1$, and

$$x_n(0) = 0, \quad x_n(n^{-1}) = 1 \quad \text{and} \quad x_n(2n^{-1}) = x_n(1) = 0 ,$$

with x_n defined by linear interpolation elsewhere. Clearly, $x_n(t) \to x(t)$ pointwise as $n \to \infty$, but $\| x_n - x \| = 1$ for all n. Hence, (6.1) holds for all positive integers k and all k-tuples $(t_1, \ldots, t_k)$ with $0 \le t_1 < \cdots < t_k \le 1$, but $P(\| X_n - X \|) = 1) = 1$. ∎

The observations above yield a simple criterion for convergence in distribution in C. We say that a set of random elements is tight if the associated set of image measures is tight,

Corollary 11.6.2. (criteria for convergence in distribution in C) *There is convergence $X_n \Rightarrow X$ in C if and only if the sequence $\{X_n : n \ge 1\}$ is tight and there is convergence of all the finite-dimensional distributions of X_n to those of X.*

Hence, in addition to the convergence of the finite-dimensional distributions in (6.1), we need to establish tightness of the sequence $\{\{X_n(t) : t \ge 0\} : n \ge 1\}$.

Fortunately compact subsets of the space C can be conveniently characterized by the *Arzelà-Ascoli theorem*. To state it, let $v(x; \delta)$ be a *modulus of continuity*, defined for any function x in C by

$$v(x, \delta) \equiv \sup\{|x(t_1) - x(t_2)| : \quad 0 \le t_1 < t_2 \le 1, \quad |t_1 - t_2| < \delta\} . \qquad (6.2)$$

Theorem 11.6.2. (Arzelà-Ascoli theorem) *A subset A of C has compact closure if and only if*

$$\sup_{x \in A} x(0) < \infty$$

and

$$\lim_{\delta \to 0} \sup_{x \in A} v(x, \delta) = 0 .$$

From the Arzelà-Ascoli theorem we easily obtain criteria for a sequence of probability measures on C to be tight.

Theorem 11.6.3. (tightness criterion for random elements of C) *A sequence $\{X_n : n \geq 1\}$ of random elements of C is tight if and only if, for every $\epsilon > 0$, there exists a constant c such that*

$$P(|X_n(0)| > c) < \epsilon \quad \text{for all} \quad n \geq 1, \tag{6.3}$$

and, for every $\epsilon > 0$ and $\eta > 0$, there exists $\delta > 0$ and n_0 such that

$$P(v(X_n, \delta) \geq \epsilon) \leq \eta \quad \text{for all} \quad n \geq n_0. \tag{6.4}$$

The tightness criterion in Theorem 11.6.3 in turn give us convenient necessary and sufficient conditions for convergence in distribution for random elements of C.

Theorem 11.6.4. (criteria for convergence in distribution in C) *There is convergence in distribution $X_n \Rightarrow X$ in C if and only if (6.1), (6.3) and (6.4) all hold.*

The modulus inequality in (6.4) can in turn be translated into various probability and moment inequalities. The following is a consequence of Theorem 12.3 of Billingsley (1968).

Theorem 11.6.5. (moment criterion for tightness in C) *A sequence $\{X_n : n \geq 1\}$ of random elements of $C \equiv C([0,1], \mathbb{R})$ is tight if $\{X_n(0)\}$ is tight in $\mathbb{R}$ and there exist constants $\gamma \geq 0$ and $\alpha > 1$ and a nondecreasing continuous function g on $[0,1]$ such that*

$$E[|X_n(t) - X_n(s)|^\gamma] \leq |g(t) - g(s)|^\alpha \tag{6.5}$$

for $0 \leq s \leq t \leq 1$.

The compactness approach to establish stochastic-process limits in (C, U) and (D, J_1) is developed in detail in Billingsley (1968). As illustrated by the treatment of the J_1 topology in Billingsley (1968), there are related criteria for convergence $X_n \Rightarrow X$ in $D \equiv D([0,1], \mathbb{R})$ with the nonuniform Skorohod topologies. Because of the discontinuities, we want to require convergence of the finite-dimensional distributions only for time points t that are almost surely continuity points of the limit process X, i.e., for which $P(t \in Disc(X)) = 0$, where $Disc(x)$ is the set of discontinuity points of x, and that suffices. Let

$$T_X \equiv \{t > 0 : P(t \in Disc(X)) = 0\} \cup \{1\} .$$

Theorem 11.6.6. (criteria for convergence in distribution in D) *There is convergence in distribution $X_n \Rightarrow X$ in D with one of the Skorohod nonuniform topologies if (6.1) holds for all $t_i \subset T_X$ and $\{X_n : n \geq 1\}$ is tight with respect to the topology. The conditions are necessary for the J_1 and M_1 topologies.*

From Theorem 11.0.1, we know that necessity in Theorem 11.6.6 depends on the space being Polish. Since we have separability, it remains to establish topological completeness. Topological completeness for J_1 was demonstrated by Kolmogorov

(1956) and Prohorov (1956); see Billingsley (1968) for a different approach. For the M_1 topology, we establish topological completeness in Section 12.8 by using Prohorov's argument.

We exploit analogs of the Arzelà-Ascoli theorem characterizing compact subsets of D with the relevant nonuniform Skorohod topology. For that purpose, we exploit generalizations of the modulus of continuity $v(x, \delta)$ in (6.2). The compactness-approach to stochastic-process limits via Prohorov's theorem explains our interest in characterizing compact subsets of D in Section 12.12.

We conclude this section by establishing an elementary result about tightness on product spaces. The result applies to finite or countably infinite products.

Theorem 11.6.7. (tightness on product spaces) *Let $S \equiv \prod_{i=1}^{\infty} S_i$ be a product of separable metric spaces with the product topology. A set A of probability measures on S is tight if and only if the sets $A_i \equiv \{P\pi_i^{-1} : P \in A\}$ of marginal probability measures on S_i, where π_i is the ith coordinate projection map, are tight for all i.*

Proof. First suppose that $A \in \mathcal{P}(S)$ is tight. Let ϵ be given. Thus there is a compact subst K in S with $P(K) > 1 - \epsilon$ for all $P \in A$. Then, for each i, $\pi_i(K)$ is compact in S_i and $K \subseteq \pi_i^{-1}(\pi_i(K))$, so that

$$P\pi_i^{-1}(\pi_i(K)) \geq P(K) > 1 - \epsilon \ .$$

Second, suppose that $P\pi_i^{-1}$ is tight for each i. For $\epsilon > 0$ given, choose compact K_i in S_i such that $P\pi_i^{-1}(K_i) > 1 - \epsilon 2^{-i}$ for all i. By Tychonoff's theorem, $K \equiv \prod_{i=1}^{\infty} K_i$ is compact in S. Moreover,

$$P(K^c) \leq \sum_{i=1}^{\infty} P\pi_i^{-1}(K_i^c) \leq \epsilon \sum_{i=1}^{\infty} 2^{-i} = \epsilon \ . \quad \blacksquare$$

12
The Space D

12.1. Introduction

This chapter is devoted to the function space $D \equiv D([0,T], \mathbb{R}^k)$ with the Skorohod M_1 and M_2 topologies, expanding upon the introduction in Sections 3.3 and 11.5 and the classic paper by Skorohod (1956). We omit most proofs here. Many are provided in Chapter 6 of the Internet Supplement.

Here is how the present chapter is organized: We start in Section 12.2 by discussing regularity properties of the function space D. A key property, which we frequently use, is the fact that any function in D can be approximated uniformly closely by piecewise-constant functions with only finitely many discontinuities.

In Section 12.3 we introduce the strong and weak versions of the M_1 topology on $D([0,T], \mathbb{R}^k)$, referred to as SM_1 and WM_1, and establish basic properties. We also discuss the relation among the nonuniform Skorohod topologies on D. In Section 12.4 we discuss local uniform convergence at continuity points and relate it to oscillation functions used to characterize different forms of convergence.

In Section 12.5 we provide several different alternative characterizations of SM_1 and WM_1 convergence. Some involve parametric representations of the completed graphs and others involve oscillation functions. It is significant that there are forms of the oscillation-function characterizations that involve considering one function argument t at a time. Consequently, the examples in Figure 11.2 tend to be more than illustrative: The topologies are characterized by the local behavior in the neighborhood of single discontinuities.

In Section 12.6 we discuss conditions that allow us to strengthen the mode of convergence from WM_1 to SM_1. The key condition is to have the coordinate limit functions have no common discontinuities. In Section 12.7 we study how SM_1 convergence in $D([0,T],\mathbb{R}^k)$ can be characterized by associated limits of mappings.

In Section 12.8 we exhibit a complete metric topologically equivalent to the incomplete metric inducing the SM_1 topology introduced earlier. As with the J_1 metric d_{J_1} in (3.2) of Section 3.3, the natural M_1 metric is incomplete, but there exists a topologically equivalent complete metric, so that D with the SM_1 topology is Polish (metrizable as a complete separable metric space).

In Section 12.9 we discuss extensions of the SM_1 and WM_1 topologies on $D([0,T],\mathbb{R}^K)$ to corresponding spaces of functions with noncompact domains. The principal example of such a noncompact domain is the interval $[0,\infty)$, but $(0,\infty)$ and $(-\infty,\infty)$ also arise.

In Section 12.10 we introduce the strong and weak versions of the M_2 topology, denoted by SM_2 and WM_2. In Section 12.11 we provide alternative characterizations of these topologies and discuss additional properties.

Finally, in Section 12.12 we discuss characterizations of compact subsets of D using oscillation functions. These characterizations are useful because they lead to characterizations of tightness for sequences of probability measures on D, which is a principal way to establish weak convergence of the probability measures; see Section 11.6.

12.2. Regularity Properties of D

Let $D \equiv D^k \equiv D([0,T],\mathbb{R}^k)$ be the set of all $\mathbb{R}^k$-valued functions $x \equiv (x^1,\ldots,x^k)$ on $[0,T]$ that are right continuous at all $t \in [0,T)$ and have left limits at all $t \in (0,T]$: If $x \in D$, then

$$\text{for } 0 \leq t < T, \quad x(t+) \equiv \lim_{s \downarrow t} x(s) \quad \text{exists with} \quad x(t+) = x(t)$$

and

$$\text{for } 0 < t \leq T, \quad x(t-) \equiv \lim_{s \uparrow t} x(s) \quad \text{exists}.$$

However, with the M_1 topology, we will be working with the completed graphs of the functions, which are obtained by adding segments joining the left and right limits to the graph at each discontinuity point. Thus the actual value of the function at discontinuity points does not matter, provided that the function value falls appropriately between the left and right limits. Such functions are said to have *discontinuities of the first kind*. In Chapter 15 we consider more general functions.

We use superscripts to designate coordinate functions, so that subscripts can index different functions in D. For example, x_3^2 denotes the second coordinate function in $D([0,T],\mathbb{R}^1)$ of $x_3 \equiv (x_3^1,\ldots,x_3^k)$ in $D([0,T],\mathbb{R}^k)$, where x_3 is the third element of the sequence $\{x_n : n \geq 1\}$. Let C be the subset of continuous functions in D.

12.2. Regularity Properties of D

Let $\|\cdot\|$ be the maximum (or l_∞) norm on $\mathbb{R}^k$ and the *uniform norm* on D; i.e., for each $b \equiv (b^1, \ldots, b^k) \in \mathbb{R}^k$, let

$$\|b\| \equiv \max_{1 \leq i \leq k} |b^i| \qquad (2.1)$$

and, for each $x \equiv (x^1, \ldots, x^k) \in D([0,T], \mathbb{R}^k)$, let

$$\|x\| \equiv \sup_{0 \leq t \leq T} \|x(t)\| = \sup_{0 \leq t \leq T} \max_{1 \leq i \leq k} |x^i(t)| \;. \qquad (2.2)$$

The maximum norm on $\mathbb{R}^k$ in (2.1) is topologically equivalent to the l_p norm

$$\|b\|_p \equiv \left(\sum_{i=1}^k (b^i)^p \right)^{1/p} .$$

For $p = 2$, the l_p norm is the Euclidean (or l_2) norm. For $p = 1$, the l_p norm is the sum (or l_1) norm. The uniform norm on D induces the uniform metric on D.

We first discuss regularity properties of D due to the existence of limits. Let $Disc(x)$ be the set of discontinuities of x, i.e.,

$$Disc(x) \equiv \{ t \in (0, T] : x(t-) \neq x(t) \} \qquad (2.3)$$

and let $Disc(x, \epsilon)$ be the set of discontinuities of magnitude at least ϵ, i.e.,

$$Disc(x, \epsilon) \equiv \{ t \in (0, T] : \|x(t-) - x(t)\| \geq \epsilon \} \;. \qquad (2.4)$$

The following is a key regularity property of D.

Theorem 12.2.1. (the number of discontinuities of a given size) *For each $x \in D$ and $\epsilon > 0$, $Disc(x, \epsilon)$ is a finite subset of $[0, T]$.*

Corollary 12.2.1. (the number of discontinuities) *For each $x \in D$, $Disc(x)$ is either finite or countably infinite.*

We say that a function x in D is *piecewise-constant* if there are finitely many time points t_i such that $0 \equiv t_0 < t_1 < \cdots < t_{m-1} \leq t_m \equiv T$ and x is constant on the intervals $[t_{i-1}, t_i)$, $1 \leq i \leq m-1$, and $[t_{m-1}, T]$. Let D_c be the subset of piecewise-constant functions in D. Let $v(x; A)$ be the *modulus of continuity* of the function x over the set A, defined by

$$v(x; A) \equiv \sup_{t_1, t_2 \in A} \{ \|x(t_1) - x(t_2)\| \} \qquad (2.5)$$

for $A \subseteq [0, T]$. The following is a second important regularity property of D.

Theorem 12.2.2. (approximation by piecewise-constant functions) *For each $x \in D$ and $\epsilon > 0$, there exists $x_c \in D_c$ such that $\|x - x_c\| < \epsilon$.*

We can deduce other useful consequences from Theorem 12.2.2.

Corollary 12.2.2. (oscillation function property) *For each $x \in D$ and $\epsilon > 0$, there exist finitely many points t_i with $0 \equiv t_0 < t_1 < \cdots < t_{m-1} \leq t_m \equiv T$ such that $v(x, [t_{i-1}, t_i)) < \epsilon$, $1 \leq i \leq m-1$, and $v(x, [t_{m-1}, T]) < \epsilon$.*

Corollary 12.2.3. (boundedness) *Each x in D is bounded, i.e., $\|x\| < \infty$.*

Corollary 12.2.4. (measurability) *Each x in D is a Borel measurable real-valued function on $[0, T]$.*

12.3. Strong and Weak M_1 Topologies

In this section we define strong and weak versions of the M_1 topology on the function space $D([0, T], \mathbb{R}^k)$, denoted by SM_1 and WM_1. The strong topology agrees with the standard topology introduced by Skorohod (1956). The strong and weak topologies coincide when $k = 1$ but differ for $k > 1$. We will show that the weak topology coincides with the product topology.

We consider functions with domain $[0, T]$, but our results can be applied to non-compact domains such as $[0, \infty)$, if as is customary we understand $x_n \to x$ as $n \to \infty$ in $D([0, \infty), \mathbb{R}^k)$ to mean that the restrictions of x_n to $[0, T]$ converge to the restriction of x to $[0, T]$ for all T that are continuity points of x. We discuss $D([0, \infty), \mathbb{R}^k)$ further in Section 12.9.

12.3.1. Definitions

The strong and weak topologies will be based on different notions of a segment in $\mathbb{R}^k$. For $a \equiv (a^1, \ldots, a^k)$, $b \equiv (b^1, \ldots, b^k) \in \mathbb{R}^k$, let $[a, b]$ be the *standard segment*, i.e.,

$$[a, b] \equiv \{\alpha a + (1 - \alpha)b : 0 \leq \alpha \leq 1\} \tag{3.1}$$

and let $[[a, b]]$ be the *product segment*, i.e.,

$$[[a, b]] \equiv \bigtimes_{i=1}^{k} [a^i, b^i] \equiv [a^1, b^1] \times \cdots \times [a^k, b^k], \tag{3.2}$$

where the one-dimensional segment $[a^i, b^i]$ coincides with the closed interval $[a^i \wedge b^i, a^i \vee b^i]$, with $c \wedge d = \min\{c, d\}$ and $c \vee d = \max\{c, d\}$ for $c, d \in \mathbb{R}$. Note that $[a, b]$ and $[[a, b]]$ are both subsets of $\mathbb{R}^k$. If $a = b$, then $[a, b] = [[a, b]] = \{a\} = \{b\}$; if $a^i \neq b^i$ for one and only one i, then $[a, b] = [[a, b]]$. If $a \neq b$, then $[a, b]$ is always a one-dimensional line in $\mathbb{R}^k$, while $[[a, b]]$ is a j-dimensional subset, where j is the number of coordinates i for which $a^i \neq b^i$. Always, $[a, b] \subseteq [[a, b]]$.

Remark 12.3.1. *More general range spaces.* We may want to consider the space D with a more general range space than $\mathbb{R}^k$. Generalizations of the M topologies are restricted by the linear structure in the definition of segments in (3.1) and (3.2). However, we can extend the M topologies to Banach-space valued functions. We use that extension to treat the workload process in the infinite-server queue in Section 10.3. ∎

12.3. Strong and Weak M_1 Topologies

We now define completed graphs of the functions: For $x \in D$, let the (standard) *thin graph* of x be

$$\Gamma_x \equiv \{(z,t) \in \mathbb{R}^k \times [0,T] : z \in [x(t-), x(t)]\}, \tag{3.3}$$

where $x(0-) \equiv x(0)$ and let the *thick graph* of x be

$$\begin{aligned} G_x &\equiv \{(z,t) \in \mathbb{R}^k \times [0,T] : z \in [[x(t-), x(t)]]\} \\ &= \{(z,t) \in \mathbb{R}^k \times [0,T] : z^i \in [x^i(t-), x^i(t)] \text{ for each } i\} \end{aligned} \tag{3.4}$$

for $1 \le i \le k$. Since $[a,b] \subseteq [[a,b]]$ for all $a, b \in \mathbb{R}^k$, $\Gamma_x \subseteq G_x$ for each x.

We now define *order relations* on the graphs Γ_x and G_x. We say that $(z_1, t_1) \le (z_2, t_2)$ if either (i) $t_1 < t_2$ or (ii) $t_1 = t_2$ and $|x^i(t_1-) - z_1^i| \le |x^i(t_1-) - z_2^i|$ for all i. The relation $\le$ induces a total order on Γ_x and a partial order on G_x.

It is also convenient to look at the ranges of the functions. Let the *thin range* of x be the projection of Γ_x onto $\mathbb{R}^k$, i.e.,

$$\rho(\Gamma_x) \equiv \{z \in \mathbb{R}^k : (z,t) \in \Gamma_x \text{ for some } t \in [0,T]\} \tag{3.5}$$

and let the *thick range* of x be the projection of G_x onto $\mathbb{R}^k$, i.e.,

$$\rho(G_x) \equiv \{z \in \mathbb{R}^k : (z,t) \in G_x \text{ for some } t \in [0,T]\}. \tag{3.6}$$

Note that $(z,t) \in \Gamma_x$ (G_x) for some t if and only if $z \in \rho(\Gamma_x)$ ($\rho(G_x)$). Thus a pair (z,t) cannot be in a graph of x if z is not in the corresponding range.

We now define strong (standard) and weak parametric representations based on these two kinds of graphs. A *strong parametric representation* of x is a continuous nondecreasing function (u, r) mapping $[0,1]$ onto Γ_x. A *weak parametric representation* of x is a continuous nondecreasing function (u, r) mapping $[0,1]$ into G_x such that $r(0) = 0$, $r(1) = T$ and $u(1) = x(T)$. (For the parametric representation, "nondecreasing" is with respect to the usual order on the domain $[0,1]$ and the order on the graphs defined above.) Here it is understood that $u \equiv (u^1, \ldots, u^k) \in C([0,1], \mathbb{R}^k)$ is the spatial part of the parametric representation, while $r \in C([0,1], [0,T])$ is the time (domain) part. Let $\Pi_s(x)$ and $\Pi_w(x)$ be the sets of strong and weak parametric representations of x, respectively. For real-valued functions x, let $\Pi(x) \equiv \Pi_s(x) = \Pi_w(x)$. Note that $(u, r) \in \Pi_w(x)$ if and only if $(u^i, r) \in \Pi(x^i)$ for $1 \le i \le k$.

We use the parametric representations to characterize the strong and weak M_1 topologies. As in (2.1) and (2.2), let $\|\cdot\|$ denote the supremum norms in $\mathbb{R}^k$ and D. We use the definition $\|\cdot\|$ in (2.2) also for the $\mathbb{R}^k$-valued functions u and r on $[0,1]$.

Now, for any $x_1, x_2 \in D$, let

$$d_s(x_1, x_2) \equiv \inf_{\substack{(u_j, r_j) \in \Pi_s(x_j) \\ j=1,2}} \{\|u_1 - u_2\| \vee \|r_1 - r_2\|\} \tag{3.7}$$

and

$$d_w(x_1, x_2) \equiv \inf_{\substack{(u_j, r_j) \in \Pi_w(x_j) \\ j=1,2}} \{\|u_1 - u_2\| \vee \|r_1 - r_2\|\}. \tag{3.8}$$

Note that $\|u_1 - u_2\| \vee \|r_1 - r_2\|$ can also be written as $\|(u_1, r_1) - (u_2, r_2)\|$, due to definitions (2.1) and (2.2). Of course, when the range is $\mathbb{R}$, $d_s = d_w = d_{M_1}$ for d_{M_1} defined in (3.4) in Section 3.3.

We say that $x_n \to x$ in D for a sequence or net $\{x_n\}$ in the SM_1 (WM_1) topology if $d_s(x_n, x) \to 0$ ($d_w(x_n, x) \to 0$) as $n \to \infty$. We start with the following basic result.

12.3.2. Metric Properties

Theorem 12.3.1. (metric inducing SM_1) d_s *is a metric on D.*

Proof. Only the triangle inequality is difficult. By Lemma 12.3.2 below, for any $\epsilon > 0$, a common parametric representation $(u_3, r_3) \in \Pi_s(x_3)$ can be used to obtain

$$\|u_1 - u_3\| \vee \|r_1 - r_3\| < d_s(x_1, x_3) + \epsilon$$

and

$$\|u_2 - u_3\| \vee \|r_2 - r_3\| < d_s(x_1, x_3) + \epsilon$$

for some $(u_1, r_1) \in \Pi_s(x_1)$ and $(u_2, r_2) \in \Pi_s(x_2)$. Hence

$$d_s(x_1, x_2) \leq \|u_1 - u_2\| \vee \|r_1 - r_2\| \leq d_s(x_1, x_3) + d_s(x_3, x_2) + 2\epsilon \ .$$

Since ϵ was arbitrary, the proof is complete. ∎

To prove Theorem 12.3.1, we use finite approximations to the graphs Γ_x. We first define an order-consistent distance between a graph and a finite subset. We use the notion of a finite ordered subset.

Definition 12.3.1. (order-consistent distance) *For $x \in D$, let A be a finite ordered subset of the ordered graph $(\Gamma_x, \leq)$, i.e., for some $m \geq 1$, A contains $m + 1$ points (z_i, t_i) from Γ_x such that*

$$(x(0), 0) \equiv (z_0, t_0) \leq (z_1, t_1) \leq \cdots \leq (z_m, t_m) \equiv (x(T), T) \ . \tag{3.9}$$

The order-consistent distance between A and Γ_x is

$$\hat{d}(A, \Gamma_x) \equiv \sup\{\|(z, t) - (z_i, t_i)\| \vee \|(z, t) - (z_{i+1}, t_{i+1})\|\} \ , \tag{3.10}$$

where the supremum is over all $(z_i, t_i) \in A$, $0 \leq i \leq m - 1$, and all $(z, t) \in \Gamma_x$ such that

$$(z_i, t_i) \leq (z, t) < (z_{i+1}, t_{i+1}) \ ,$$

using the order on the graph. ∎

We now observe that finite ordered subsets A can be chosen to make $\hat{d}(A, \Gamma_x)$ arbitrarily small. The missing proofs are in the Internet Supplement.

Lemma 12.3.1. (finite approximations to graphs) *For any $x \in D$ and $\epsilon > 0$, there exists a finite ordered subset A of Γ_x such that $\hat{d}(A, \Gamma_x) < \epsilon$ for $\hat{d}$ in (3.10).*

To complete the proof of Theorem 12.3.1, we need the following result, which we prove by applying Lemma 12.3.1.

Lemma 12.3.2. (flexibility in choice of parametric representations) *For any $x_1, x_2 \in D$, $(u_1, r_1) \in \Pi_s(x_1)$ and $\epsilon > 0$, it is possible to find $(u_2, r_2) \in \Pi_s(x_2)$ such that*

$$\|u_1 - u_2\| \vee \|r_1 - r_2\| \leq d_s(x_1, x_2) + \epsilon .$$

We will show that the metric d_s induces the standard M_1 topology defined by Skorohod (1956); see Theorem 12.5.1. Since $\Pi_s(x) \subseteq \Pi_w(x)$ for all x, we have $d_w(x_1, x_2) \leq d_s(x_1, x_2)$ for all x_1, x_2, so that the WM_1 topology is indeed weaker than the SM_1 topology. However, we show below in Example 12.3.2 that d_w in (3.8) is *not* a metric when $k > 1$.

For $x_1, x_2 \in D([0,T], \mathbb{R}^k)$, let d_p be a metric inducing the product topology, defined by

$$d_p(x_1, x_2) \equiv \max_{1 \leq i \leq k} d(x_1^i, x_2^i) \qquad (3.11)$$

for $x_j \equiv (x_j^1, \ldots, x_j^k)$ and $j = 1, 2$. (Note that $d_s = d_w = d_p$ when the functions are real valued, in which case we use the notation d.) It is an easy consequence of (3.8), (3.11) and the second representation in (3.4) that the WM_1 topology is stronger than the product topology, i.e., $d_p(x_1, x_2) \leq d_w(x_1, x_2)$ for all $x_1, x_2 \in D$. In Section 12.5 we will show that actually the WM_1 and product topologies coincide.

We now show that SM_1 is strictly stronger than WM_1. Let I_A denote the indicator function of a set A; i.e., $I_A(t) = 1$ if $t \in A$ and $I_A(t) = 0$ otherwise.

Example 12.3.1. *WM_1 convergence without SM_1 convergence.* To show that we can have $d_w(x_n, x) \to 0$ as $n \to \infty$ without $d_s(x_n, x) \to 0$ as $n \to \infty$, let $x \equiv (x^1, x^2) \in D([0,2], \mathbb{R}^2)$ be defined by $x^1 = x^2 = 2I_{[1,2]}$ and let $x_n^1 = 2I_{[1-n^{-1},2]}$ and $x_n^2 = I_{[1-n^{-1},1)} + 2I_{[1,2]}$. The thin range of x is the set $\{(0,0),(2,2)\}$ plus the line segment $[(0,0),(2,2)]$ connecting those two points, while the thin range of x_n is the set $\{(0,0),(2,1),(2,2)\}$ plus the line segments $[(0,0),(2,1)]$ and $[(2,1),(2,2)]$. Since $(2,1) \in \Gamma_{x_n}$ for all n but $(2,1) \notin \Gamma_x$, we must have $d_s(x_n, x) \not\to 0$ as $n \to \infty$. On the other hand, the thick ranges of x and x_n, $n \geq 1$ all are $[0,2] \times [0,2]$. To demonstrate that $d_w(x_n, x) \to 0$ as $n \to \infty$, we construct suitable parametric representations. Let

$$r(0) = 0, \; r(1/3) = 1 = r(2/3), \; r(1) = 2$$

$$r_n(0) = 0, \; r_n(1/3) = 1 - n^{-1} = r_n((1 - n^{-1})/2),$$

$$r_n((1 + n^{-1})/2) = 1 = r_n(2/3), \; r_n(1) = 2$$

$$u^1(0) = 0 = u^1(1/3), \; u^1(1/2) = 2 = u^1(1)$$

$$u_n^1(0) = 0 = u_n^1(1/3), \; u_n^1((1 - n^{-1})/2) = 2 = u_n^1(1)$$

$$u^2(0) = 0 = u^2(1/3), \ u^2(1/2) = 1, \ u^2(2/3) = 2 = u^2(1)$$

$$u_n^2(0) = 0 = u_n^2(1/3), \ u_n^2((1-n^{-1})/2) = 1 = u_n^2((1+n^{-1})/2),$$

$$u_n^2(2/3) = 2 = u_n^2(1)$$

with r, r_n, u^1, u_n^1, u^2, u_n^2 defined by linear interpolation in the gaps. With this construction, $(u_n, r_n) \in \Pi_w(x_n)$ and $(u, r) \in \Pi_w(x)$, $\|r_n - r\| = n^{-1}$, $\|u_n^1 - u^1\| = 6n^{-1}$ and $\|u_n^2 - u^2\| = 3n^{-1}$. Hence,

$$d_w(x_n, x) \leq \|u_n - u\| \vee \|r_n - r\| = 6n^{-1} \to 0 \quad \text{as} \quad n \to \infty \ . \ \blacksquare$$

Example 12.3.2. d_w *is not a metric.* We now show that d_w in (3.8) is not a metric. For this purpose, we use a minor modification of Example 12.3.1. Let $x^1 = x^2 = 2I_{[1,2]}$ as before. For even n, let $x_n^1 = 2I_{[1-n^{-1},2]}$ and $x_n^2 = I_{[1-n^{-1},1)} + 2I_{[1,2]}$ as before. Then let $x_{2n+1}^1 = x_{2n}^2$ and $x_{2n+1}^2 = x_{2n}^1$. We show that $d_w(x_{2n}, x_{2n+1}) \not\to 0$ as $n \to \infty$ even though $d_w(x_n, x) \to 0$ as $n \to \infty$, contradicting the triangle inequality property of a metric. The thick range of x_n is $([0, 2] \times [0, 1]) \cup (\{2\} \times [1, 2])$ for n even and $([0, 1] \times [0, 2]) \cup ([1, 2] \times \{2\})$ for n odd. The points $(2, 1)$ and $(1, 2)$ appear for n even and odd, respectively, but are distance 1 from the other thick range. Any parametric representation must pass through $(2, 1, 1-n^{-1})$ in $\mathbb{R}^2 \times [0, 2]$ for n even and $(1, 2, 1-n^{-1})$ for n odd. However, for n odd (n even) all points on G_{x_n} are at least a distance 1 from $(2, 1, 1-n^{-1})$ $((1, 2, 1-n^{-1}))$. This example shows that we cannot find a constant K such that $d_w(x_1, x_2) \leq K d_p(x_1, x_2)$ for all $x_1, x_2 \in D$. $\blacksquare$

We now relate the metrics $d_{M_1} \equiv d_s$ and d_{J_1} for d_{J_1} in (3.2) of Section 3.3.

Theorem 12.3.2. (*comparison of J_1 and M_1 metrics*) *For each $x_1, x_2 \in D$,*

$$d_s(x_1, x_2) \leq d_{J_1}(x_1, x_2) \ .$$

Remark 12.3.2. *Uses of the M_1 topology.* The M_1 topology has not been used extensively. It was used by Whitt (1971b, 1980, 2000b), Wichura (1974), Avram and Taqqu (1989, 1992), Kella and Whitt (1990), Chen and Whitt (1993), Mandelbaum and Massey (1995), Harrison and Williams (1996), Puhalskii and Whitt (1997, 1998), Resnick and van der Berg (2000), O'Brien (2000) and no doubt a few others. $\blacksquare$

12.3.3. Properties of Parametric Representations

We conclude this section by further discussing strong parametric representations. We first indicate how to construct a parametric representation (u, r) of Γ_x for any $x \in D$.

Remark 12.3.3. *How to construct a parametric representation.* Let t_j, $j \geq 1$, be a list of the discontinuity points of x (of which there are finitely or countably infinite many). For each j, select a subinterval $[a_j, b_j] \subseteq [0, 1]$ and let $r(s) = t_j$ for $a_j \leq s \leq b_j$, $u(a_j) = x(t_j-)$, $u(b_j) = x(t_j)$ and $u(\alpha a_j + (1-\alpha)b_j) = \alpha u(a_j) + (1-\alpha)u(b_j)$, $0 < \alpha < 1$. For successive discontinuities, do this in an order-preserving way; i.e., if $t_i < t_j < t_k$, then we require that $b_i < a_j < b_j < a_k$. Let this be done for all j. Next, suppose that t is not a discontinuity point but is the limit of discontinuity points. If $t_j \downarrow t$ as $j \to \infty$ where $t_j \in Disc(x)$, then let $r(a) = t$ and $u(a) = \lim_{j\to\infty} x(t_j-)$, where $a = \lim_{j\to\infty} a_j$ with $r(a_j) = t_j$. Similarly, if $t_j \uparrow t$ as $j \to \infty$ where $t_j \in Disc(x)$, then let $r(b) = t$ and $u(b) = \lim_{j\to\infty} x(t_j)$, where $b = \lim_{j\to\infty} b_j$ with $r(b_j) = t_j$. Finally, there may remain open intervals (a,b) over which (u,r) is undefined. Since (u,r) is already defined at the endpoints a and b, let $r(\alpha a + (1-\alpha)b) = \alpha r(a) + (1-\alpha)r(b)$ and $u(\alpha a + (1-\alpha)b) = x(r(\alpha a + (1-\alpha)b))$ for $0 < \alpha < 1$. This construction makes (u,r) a one-to-one function. This construction also makes r a generalization of piecewise linear; i.e., there are finite or countably many subintervals $[a_j, b_j]$ over which r is constant and there are finite or countably many intervals (b_k, a_k) over which r is linear. The union of all those points (where r is constant or linear) is dense in $[0,1]$. The function r is extended to all other points by continuity. ∎

Remark 12.3.4. *Parametric representations need not be one-to-one.* We do not require that a parametric representation be a one-to-one function. For example, even if x is continuous at t, we could have $r(s) = t$ for $a \leq s \leq b$. Then, necessarily, $u(s) = x(t)$, $a \leq s \leq b$. However, we get the same metric if the parametric representations (u,r) are required to be one-to-one with r nondecreasing, e.g., as done by Wichura (1974); see Remark 12.5.2 in Section 5. Skorohod (1956) only originally required that r be nondecreasing instead of (u,r), without the one-to-one property, in his Definitions 2.2.4 and 2.2.5. However, from his remarks after 2.2.5, it is evident that he meant to require that (u,r) be nondecreasing as we have defined it. As stated, Skorohod's version of the M_1 topology with only r nondecreasing is actually the M_2 topology. ∎

Example 12.3.3. *Need for monotonicity.* To see the importance of requiring that the parametric representation be nondecreasing, using the order on the graphs, let $x = I_{[1,2]}$, $x_n(1) = x_n(1 - 2n^{-1}) = x_n(2) = 1$ and $x_n(0) = x_n(1 - 3n^{-1}) = x_n(1 - n^{-1}) = 0$, with x_n defined by linear interpolation elsewhere. For these functions, $x_n \to x$ as $n \to \infty$ in the M_2 topology but not in the M_1 topology. If we did not require that parametric representations of x be nondecreasing in our M_1 definitions, then we would have $x_n \to x$ as $n \to \infty$ in the M_1 topology. To see this, we exhibit parametric representations. Let $u_n = u$, $n \geq 4$, and let

$$r(0) = 0, \; r(1/5) = r(4/5) = 1, \; r(1) = 2$$
$$u(0) = u(1/5) = u(3/5) = 0, \; u(2/5) = u(4/5) = u(2) = 1$$
$$r_n(0) = 0, \; r_n(1/5) = 1 - 3n^{-1}, \; r_n(2/5) = 1 - 2n^{-1},$$
$$r_n(3/5) = 1 - n^{-1}, \; r_n(4/5) = 1, \; r_n(1) = 2$$

with r, u, r_n and u_n defined by linear interpolation elsewhere. It is easy to see that $(u_n, r_n) \in \Pi_s(x_n)$, and $\|(u_n, r_n) - (u, r)\| = \|r_n - r\| = 3n^{-1}$, but $(u, r) \notin \Pi_s(x)$ because (u, r) fails to be nondecreasing, since it backtracks on the graph at $t = 1$. If r were only required to be nondecreasing, then we would have $(u, r) \in \Pi_s(x)$. ∎

We now continue characterizing parametric representations. For $x \in D$, $t \in Disc(x)$ and $(u,r) \in \Pi_s(x)$, there exists a unique pair of points $s_- \equiv s_-(t) \equiv s_-(t,x)$ and $s_+ \equiv s_+(t) \equiv s_+(t,x)$ such that $s_- < s_+$ and $r^{-1}(t) \equiv r^{-1}(\{t\}) = [s_-, s_+]$, i.e.,

(i) $r(s) < t$ for $s < s_-$ (3.12)

(ii) $r(s) = t$ for $s_- \leq s \leq s_+$

(iii) $r(s) > t$ for $s > s_+$.

We will exploit the fact that a parametric representation (u,r) in $\Pi_s(x)$ is *jump consistent*: for each $t \in Disc(x)$ and pair $s_- \equiv s_-(t,x) < s_+ \equiv s_+(t,x)$ such that (3.12) holds, there is a continuous nondecreasing function β_t mapping $[0,1]$ onto $[0,1]$ such that

$$u(s) = \beta_t\left(\frac{s - s_-}{s_+ - s_-}\right)u(s_+) + \left[1 - \beta_t\left(\frac{s - s_-}{s_+ - s_-}\right)\right]u(s_-) \quad \text{for} \quad s_- \leq s \leq s_+ . \tag{3.13}$$

Condition (3.13) means that u is defined within jumps by interpolation from the definition at the endpoints s_- and s_+, consistently over all coordinates. In particular, suppose that $t \in Disc(x^i)$. (Since $t \in Disc(x)$, we must have $t \in Disc(x^i)$ for some coordinate i.) Suppose that $x^i(t-) < x^i(t)$. Then we can let

$$\beta_t(s) = \frac{u^i(s) - u^i(s_-)}{u^i(s_+) - u^i(s_-)} . \tag{3.14}$$

We see that (3.13) and (3.14) are consistent in that

$$u^i(s) = \beta_t\left(\frac{s - s_-}{s_+ - s_-}\right)u^i(s_+) + \left[1 - \beta_t\left(\frac{s - s_-}{s_+ - s_-}\right)\right]u^i(s_-) \tag{3.15}$$

for β_t in (3.14). For another coordinate j, (3.13) and (3.14) imply that

$$u^j(s) = \left(\frac{u^i(s) - u^i(s_-)}{u^i(s_+) - u^i(s_-)}\right)u^j(s_+) + \left(\frac{u^i(s_+) - u^i(s)}{u^i(s_+) - u^i(s_-)}\right)u^j(s_-) . \tag{3.16}$$

It is possible that $t \notin Disc(x^j)$, in which case $u^j(s) = u^j(s_-) = u^j(s_+)$ for all s, $s_- \leq s \leq s_+$.

We can further characterize the behavior of a strong parametric representation at a discontinuity point. For $x \in D$, $t \in Disc(x)$ and $(u,r) \in \Pi_s(x)$, there exists a unique set of four points $s_- \equiv s_-(t,x) \leq s'_- \equiv s'_-(t,x) < s'_+ \equiv s'_+(t,x) \leq s_+ \equiv s_+(t,x)$ such that (3.12) holds and

(i) $u(s) = u(s_-)$ for $s_- \leq s \leq s'_-$,

(ii) for each i, either $u^i(s_-) < u^i(s) < u^i(s_+)$,

    or $u^i(s_-) > u^i(s) > u^i(s_+)$ for $s'_- < s < s'_+$,

(iii) $u(s) = u(s_+)$ for $s'_+ \leq s \leq s_+$.   (3.17)

Let D_1 be the subset of D containing functions all of whose jumps occur in only one coordinate, i.e., the set of x such that, for each $t \in Disc(x)$ there exists one and only one $i \equiv i(t)$ such that $t \in Disc(x^i)$. (The coordinate i may depend on t.)

Lemma 12.3.3. (strong and weak parametric representations coincide on D_1) *For each $x \in D_1$, $\Pi_s(x) = \Pi_w(x)$.*

We now show that parametric representations are preserved under linear functions of the coordinates when $x \in \Pi_s(x)$. That is *not* true in $\Pi_w(x)$.

Lemma 12.3.4. (linear functions of parametric representations) *If $(u, r) \in \Pi_s(x)$, then $(\eta u, r) \in \Pi_s(\eta x)$ for any $\eta \in \mathbb{R}^k$.*

12.4. Local Uniform Convergence at Continuity Points

In this section we provide alternative characterizations of local uniform convergence at continuity points of a limit function. The nonuniform Skorohod topologies on D all imply local uniform convergence at continuity points of a limit function. They differ by their behavior at discontinuity points.

We first observe that pointwise convergence is weaker than local uniform convergence.

Example 12.4.1. *Pointwise convergence is weaker than local uniform convergence.* To see that pointwise convergence in D at all continuity points of the limit is strictly weaker than local uniform convergence at continuity points of the limit, let $x(t) = 0$, $0 \leq t \leq 2$, and $x_n = I_{[1+n^{-1}, 1+2n^{-1})}$, $n \geq 1$. Then $x_n(t) \to x(t) = 0$ as $n \to \infty$ for all t, but $x_n(1 + n^{-1}) = 1 \not\to 0$ as $n \to \infty$, so we do not have local uniform convergence at $t = 1$. We also do not have $x_n \to x$ as $n \to \infty$ in D in any of the Skorohod topologies. ∎

We start by defining two basic *uniform-distance functions*.
For $x_1, x_2 \in D$, $t \in [0, T]$ and $\delta > 0$, let

$$u(x_1, x_2, t, \delta) \equiv \sup_{0 \vee (t-\delta) \leq t_1 \leq (t+\delta) \wedge T} \{\|x_1(t_1) - x_2(t_1)\|\}, \quad (4.1)$$

$$v(x_1, x_2, t, \delta) \equiv \sup_{0 \vee (t-\delta) \leq t_1, t_2 \leq (t+\delta) \wedge T} \{\|x_1(t_1) - x_2(t_2)\|\}, \quad (4.2)$$

We also define an *oscillation function*. For $x \in D$, $t \in [0, T]$ and $\delta > 0$, let

$$\bar{v}(x, t, \delta) \equiv \sup_{0 \vee (t-\delta) \leq t_1 \leq t_2 \leq (t+\delta) \wedge T} \{\|x(t_1) - x(t_2)\|\} \quad (4.3)$$

We next define oscillation functions that we will use with the M_1 topologies. They use the distance $\|z - A\|$ between a point z and a subset A in $\mathbb{R}^k$ defined in (5.3) in Section 11.5. The SM_1 and WM_1 topologies use the standard and product segments in (3.1) and (3.2). For each $x \in D$, $t \in [0,T]$ and $\delta > 0$, let

$$w_s(x, t, \delta) \equiv \sup_{0 \vee (t-\delta) \leq t_1 < t_2 < t_3 \leq (t+\delta) \wedge T} \{\|x(t_2) - [x(t_1), x(t_3)]\|\} \quad (4.4)$$

and

$$w_w(x, t, \delta) \equiv \sup_{0 \vee (t-\delta) \leq t_1 < t_2 < t_3 \leq (t+\delta) \wedge T} \{\|x(t_2) - [[x(t_1), x(t_3)]]\|\} \quad (4.5)$$

We now turn to the M_2 topology, which we will be studying in Sections 12.10 and 12.11. We define two uniform-distance functions. We use $\bar{w}$ as opposed to w to denote an M_2 uniform-distance function. Just as with the M_1 topologies, the SM_2 and WM_2 topologies use the standard and product segments in (3.1) and (3.2). For $x_1, x_2 \in D$, let

$$\bar{w}_s(x_1, x_2, t, \delta) \equiv \sup_{0 \vee (t-\delta) \leq t_1 \leq (t+\delta) \wedge T} \{\|x_1(t_1) - [x_2(t-), x_2(t)]\|\} \quad (4.6)$$

$$\bar{w}_w(x_1, x_2, t, \delta) \equiv \sup_{0 \vee (t-\delta) \leq t_1 \leq (t+\delta) \wedge T} \{\|x_1(t_1) - [[x_2(t-), x_2(t)]]\|\} \quad (4.7)$$

It is easy to establish the following relations among the uniform-distance and oscillation functions.

Lemma 12.4.1. (inequalities for uniform-distance and oscillation functions) *For all $x, x_n \in D$, $t \in [0,T]$ and $\delta > 0$,*

$$u(x_n, x, t, \delta) \leq v(x_n, x, t, \delta) \leq u(x_n, x, t, \delta) + \bar{v}(x, t, \delta) ,$$

$$w_w(x_n, t, \delta) \leq w_s(x_n, t, \delta) \leq \bar{v}(x_n, t, \delta) \leq 2v(x_n, x, t, \delta) + \bar{v}(x, t, \delta) ,$$

$$\bar{w}_w(x_n, x, t, \delta) \leq \bar{w}_s(x_n, x, t, \delta) \leq v(x_n, x, t, \delta) \leq 2\bar{w}_w(x_n, x, t, \delta) + \bar{v}(x, t, \delta) .$$

Since the M_1-oscillation functions $w_s(x_n, t, \delta)$ and $w_w(x_n, t, \delta)$ do not contain the limit x, their convergence to 0 as $n \to \infty$ and then $\delta \downarrow 0$ does not directly imply local uniform convergence at a continuity point of a prospective limit function x.

Example 12.4.2. *Characterizations of local uniform convergence at continuity points.* We show that it is possible to have $t \notin Disc(x)$, $x_n(t) \to x(t)$ as $n \to \infty$ and

$$\lim_{\delta \downarrow 0} \overline{\lim_{n \to \infty}} \ w_s(x_n, t, \delta) = 0$$

without having

$$\lim_{\delta \downarrow 0} \overline{\lim_{n \to \infty}} \ v(x_n, x, t, \delta) = 0 .$$

That occurs for $t = 1$ when $x(t) = 0$, $0 \leq t \leq 2$, and $x_n = I_{[1+n^{-1},2]}$, $n \geq 1$. In this example, we have $\bar{v}(x, t, \delta) = 0$ and $w_s(x_n, t, \delta) = 0$ for all n, t and $\delta > 0$, but $v(x_n, x, 1, \delta) = 1$ for $n > 1/\delta$. ∎

We relate convergence of $w_s(x_n,t,\delta)$ and $w_w(x_n,t,\delta)$ to 0 as $n \to \infty$ and $\delta \downarrow 0$ to local uniform convergence by requiring pointwise convergence in a neighborhood of t; see (vi) in Theorem 12.4.1 below.

Theorem 12.4.1. (characterizations of local uniform convergence at continuity points) *If $t \notin Disc(x)$, then the following are equivalent:*

$$(i) \quad \lim_{\delta \downarrow 0} \overline{\lim_{n \to \infty}} \; u(x_n,x,t,\delta) = 0 \;, \tag{4.8}$$

$$(ii) \quad \lim_{\delta \downarrow 0} \overline{\lim_{n \to \infty}} \; v(x_n,x,t,\delta) = 0 \;, \tag{4.9}$$

$$(iii) \quad \lim_{\delta \downarrow 0} \overline{\lim_{n \to \infty}} \; \bar{w}_s(x_n,x,t,\delta) = 0 \;, \tag{4.10}$$

$$(iv) \quad \lim_{\delta \downarrow 0} \overline{\lim_{n \to \infty}} \; \bar{w}_w(x_n,x,t,\delta) = 0 \;, \tag{4.11}$$

(v) $x_n(t_1) \to x(t_1)$ for all t_1 in a dense subset of a neighborhood of t (including 0 if $t=0$ or T if $t=T$) and

$$\lim_{\delta \downarrow 0} \overline{\lim_{n \to \infty}} \; w_s(x_n,t,\delta) = 0 \;,$$

(vi) $x_n(t_1) \to x(t_1)$ for all t_1 in a dense subset of a neighborhood of t (including 0 if $t=0$ or T if $t=T$) and

$$\lim_{\delta \downarrow 0} \overline{\lim_{n \to \infty}} \; w_w(x_n,t,\delta) = 0 \;. \tag{4.12}$$

We now show that local uniform convergence at all points in a compact interval implies uniform convergence over the compact interval.

Lemma 12.4.2. (local uniform convergence everywhere in a compact interval) *If (4.8) holds for all $t \in [a,b]$, then*

$$\lim_{\delta \downarrow 0} \overline{\lim_{n \to \infty}} \sup_{0 \vee (a-\delta) \leq t \leq (b+\delta) \wedge T} \{\|x_n(t) - x(t)\|\} = 0 \;.$$

12.5. Alternative Characterizations of M_1 Convergence

We now give alternative characterizations of SM_1 and WM_1 convergence.

12.5.1. SM_1 Convergence

We first give several alternative characterizations of SM_1-convergence (or, equivalently, d_s-convergence) in D, one being a minor variant of the original one involving an oscillation function established by Skorohod (1956). Another one – (v) below – involves only the local behavior of the functions. It helps us establish sufficient conditions to have $d_s((x_n, y_n), (x, y)) \to 0$ in $D([0,T], \mathbb{R}^{k+l})$ when $d_s(x_n, x) \to 0$ in $D([0,T], \mathbb{R}^k)$ and $d_s(y_n, y) \to 0$ in $D([0,T], \mathbb{R}^l)$; see Section 12.6. For the SM_1 topology, we define another oscillation function. For any $x_1, x_2 \in D$ and $\delta > 0$, let

$$w_s(x, \delta) \equiv \sup_{0 \le t \le T} w_s(x, t, \delta) , \qquad (5.1)$$

for $w_s(x, t, \delta)$ in (4.4). We include the proof here, except for the supporting lemmas, which are proved in the Internet Supplement.

Theorem 12.5.1. (characterizations of SM_1 convergence) *The following are equivalent characterizations of convergence $x_n \to x$ as $n \to \infty$ in (D, SM_1):*

(i) *For any $(u, r) \in \Pi_s(x)$, there exists $(u_n, r_n) \in \Pi_s(x_n)$, $n \ge 1$, such that*

$$\|u_n - u\| \vee \|r_n - r\| \to 0 \quad as \quad n \to \infty . \qquad (5.2)$$

(ii) *There exist $(u, r) \in \Pi_s(x)$ and $(u_n, r_n) \in \Pi_s(x_n)$ for $n \ge 1$ such that (5.2) holds.*

(iii) *$d_s(x_n, x) \to 0$ as $n \to \infty$; i.e., for all $\epsilon > 0$ and all sufficiently large n, there exist $(u, r) \in \Pi_s(x)$ and $(u_n, r_n) \in \Pi_s(x_n)$ such that*

$$\|u_n - u\| \vee \|r_n - r\| < \epsilon .$$

(iv) *$x_n(t) \to x(t)$ as $n \to \infty$ for each t in a dense subset of $[0,T]$ including 0 and T, and*

$$\lim_{\delta \downarrow 0} \overline{\lim_{n \to \infty}} \, w_s(x_n, \delta) = 0 \qquad (5.3)$$

for $w_s(x, \delta)$ in (5.1) and $w_s(x, t, \delta)$ in (4.4).

(v) *$x_n(T) \to x(T)$ as $n \to \infty$; for each $t \notin Disc(x)$,*

$$\lim_{\delta \downarrow 0} \overline{\lim_{n \to \infty}} \, v(x_n, x, t, \delta) = 0 \qquad (5.4)$$

for $v(x_1, x_2, t, \delta)$ in (4.2); and, for each $t \in Disc(x)$,

$$\lim_{\delta \downarrow 0} \overline{\lim_{n \to \infty}} \, w_s(x_n, t, \delta) = 0 \qquad (5.5)$$

for $w_s(x, t, \delta)$ in (4.4).

(vi) *For all $\epsilon > 0$, , there exist integers m and n_1, a finite ordered subset A of Γ_x of cardinality m as in (3.9) and, for all $n \ge n_1$, finite ordered subsets A_n of Γ_{x_n} of cardinality m such that, for all $n \ge n_1$, $\hat{d}(A, \Gamma_x) < \epsilon$, $\hat{d}(A_n, \Gamma_{x_n}) < \epsilon$ for $\hat{d}$ in (3.10) and $d^*(A, A_n) < \epsilon$, where*

$$d^*(A, A_n) \equiv \max_{1 \le i \le m} \{\|(z_i, t_i) - (z_{n,i}, t_{n,i})\| : (z_i, t_i) \in A, (z_{n,i}, t_{n,i}) \in A_n\} . \qquad (5.6)$$

In preparation for the proof of Theorem 12.5.1, we establish some preliminary results. We first show that SM_1 convergence implies local uniform convergence at all continuity points.

Lemma 12.5.1. (local uniform convergence) *If $d_s(x_n, x) \to 0$ as $n \to \infty$, then (4.9) holds for each $t \notin Disc(x)$.*

We next relate the modulus w_s applied to x and the modulus applied to corresponding points on the graph Γ_x. The following lemma is established in the proof of Skorohod's (1956) 2.4.1.

Lemma 12.5.2. (extending the modulus from a function to its graph) *If (z_1,t_1), (z_2,t_2), $(z_3,t_3) \in \Gamma_x$ with $0 \vee (t-\delta) \leq t_1 < t_2 < t_3 \leq (t+\delta) \wedge T$, then $\|z_2 - [z_1, z_3]\| \leq w_s(x,\delta)$.*

Lemma 12.5.3. (asymptotic negligibility of the modulus) *For any $x \in D$, $w_s(x,\delta) \downarrow 0$ as $\delta \downarrow 0$.*

Proof of Theorem 12.5.1. The implications (i)→(ii)→(iii) are trivial. We establish the others exploiting transitivity.

(iii)→(iv). First, the convergence $x_n(T) \to x(T)$ is assumed directly. Next, by Lemma 12.5.1, if $d_s(x_n, x) \to 0$ as $n \to \infty$, then $x_n(t) \to x(t)$ for all $t \in Disc(x)^c$, which is a dense subset of $[0,T]$. We now want to show that, for any $\epsilon > 0$, there exists n_0 and δ such that $w_s(x_n, \delta) < \epsilon$ for all $n \geq n_0$. For $x \in D$ and $\epsilon > 0$ given, start by choosing η so that $w_s(x,\eta) < \epsilon/2$, which we can do by Lemma 12.5.3. Then apply (iii) to choose n_0 so that $(u_n, r_n) \in \Pi_s(x_n)$, $(u,r) \in \Pi_s(x)$ and

$$\|u_n - u\| \vee \|r_n - r\| < (\epsilon \wedge \eta)/4 \quad \text{for} \quad n \geq n_0 \ .$$

Suppose that $(t-\delta) \vee 0 \leq t_1 < t_2 \leq t_3 < (t+\delta) \wedge T$. Let $s_{n,i}$ be such that $r_n(s_{n,i}) = t_i$ and $u_n(s_{n,i}) = x_n(t_i)$ for $i = 1,2,3$ and all n. Then, apply Lemma 12.5.2 to obtain, for $n \geq n_0$,

$$\begin{aligned}
\|x_n(t_2) - [x_n(t_1), x_n(t_3)]\| &= \|u_n(s_{n,2}) - [u_n(s_{n,1}), u_n(s_{n,3})]\| \\
&\leq \|u(s_{n,2}) - [u(s_{n,1}), u(s_{n,3})]\| + 2\|u_n - u\| \\
&\leq w_s(x, \delta + 2(\eta \wedge \epsilon)/4)) + 2((\eta \wedge \epsilon)/4) \\
&\leq w_s(x, \delta + (\eta/2)) + \epsilon/2 \ ,
\end{aligned}$$

so that, for $\delta < \eta/2$ and $n \geq n_0$, $w_s(x_n, \delta) < \epsilon$.

(iv)→(vi). First, for $\epsilon > 0$ given, apply (iv) to find $\eta < \epsilon/16$ and n_0 such that $w_s(x_n, \eta) < \epsilon/32$ for $n \geq n_0$. Next find a finite set A of points (z_i, t_i) in Γ_x with

$$(x(0), 0) = (z_1, t_1) < (z_2, t_2) < \cdots < (z_m, t_m) = (x(T), T) \ ,$$

using the order defined on Γ_x below (3.3), where for each i, either $t_i \in Disc(x, \epsilon/2)$ or $t_i \in S$, with $Disc(x, \epsilon/2)$ being as in (2.4) and S being a subset of $[0,T]$ including 0, T and the points in $Disc(x, \epsilon/2)^c$ at which x_n converges pointwise to x. Use the left and right limits of x to include in A for each $t \in Disc(x, \epsilon/2)$ points $t' \equiv t'(t)$ and $t'' = t''(t)$ in S such that $t' < t < t''$, t' is greater than all elements of $Disc(x, \epsilon/2)$

less than t, t'' is less than all elements of $Disc(x, \epsilon/2)$ greater than t, $|t' - t| < \eta$, $|t'' - t| < \eta$, $\|x(t') - x(t-)\| < \epsilon/32$ and $\|x(t'') - x(t)\| < \epsilon/32$. In addition, assume that $|t_{i+1} - t_i| < \eta$ for all i and $\hat{d}(A, \Gamma_x) < \epsilon/2$, for which we apply Lemma 12.3.1. Moreover, if $t \in Disc(x, \epsilon/2)$ and

$$t_r < t_{r+1} = t = t_{r+2} = \cdots = t_{r+j} < t_{r+j+1}, \tag{5.7}$$

then we require that $\|z_{r+1} - x(t-)\| > \epsilon/4$, $\|z_{r+j} - x(t)\| > \epsilon/4$ and $\|z_{r+i+1} - z_{r+i}\| > \epsilon/4$ for all i, $1 \le i \le j-1$. Since $\hat{d}(A, \Gamma_x) < \epsilon/2$, we also have the upper bound $\|z_{r+i+1} - z_{r+i}\| < \epsilon/2$. For $t_i \in S \cap A$, let $z_i = x(t_i)$. Now, for all $t_i \in S \cap A$, let $n_1 \ge n_0$ be such that $\|x_n(t_i) - x(t_i)\| < \epsilon/32$ for all i and $n \ge n_1$, using (iv). We now want to construct the subset A_n of Γ_{x_n}. First for all $t_i \in S \cap A$, let $(z_{n,i}, t_{n,i}) = (x_n(t_i), t_i)$. Now we consider time points in $Disc(x, \epsilon/2)$. By the construction above, given (5.7),

$$\|[x(t_r), x(t_{r+j+1})] - [x_n(t_r), x_n(t_{r+j+1})]\| < \epsilon/32$$

and

$$\|[x(t_r), x(t_{r_j+1})] - [x(t-), x(t)]\| < \epsilon/32 . \tag{5.8}$$

Since $w_s(x_n, \eta) < \epsilon/32$, for each (r, i) there is a point $(z_{n,r+i}, t_{n,r+i}) \in \Gamma_{x_n}$ such that

$$\|z_{n,r+i} - z_{r+i}\| < 3\epsilon/32 \quad \text{and} \quad |t_{n,r+i} - t| < \eta < \epsilon/16 . \tag{5.9}$$

Moreover, we must have $(z_{n,r+i+1}, t_{n,r+i+1}) > (z_{n,r+i}, t_{n,r+i})$ for $0 \le i \le j$. For $i = 0$ and $i = j$, we can conclude that $t_r < t < t_{r+j+1}$. For other i, a reversal of order can occur only if $w_s(x_n, t, \eta) > \epsilon/16$ because the construction implies that $\|z_{n,r+i+1} - z_{n,r+i}\| > \epsilon/16$, but that is prohibited by the condition that $w_s(x_n, t, \eta) < \epsilon/32$. Hence, the set of points A_n is ordered properly. Moreover, the construction yields $\hat{d}^*(A, A_n) < \epsilon/16$. Finally, it remains to bound $\hat{d}(A_n, \Gamma_{x_n})$ for $n \ge n_1$. Consider (z_n, t_n) such that $(z_{n,i}, t_{n,i}) < (z_n, t_n) < (z_{n,i+1}, t_{n,i+1})$. Since $\|z_{n,i} - z_i\| < 3\epsilon/32$ for all i and $\|z_{i+1} - z_i\| < \hat{d}(A, \Gamma_x) < \epsilon/2$, $\|z_{n,i+1} - z_{n,i}\| < 5\epsilon/8$ by the triangle inequality. Since $w_s(x_n, \eta) < \epsilon/32$, invoking Lemma 12.5.2, we have

$$\|(z_n, t_n) - [(z_{n,i}, t_{n,i}), (z_{n,i+1}, t_{n,i+1})]\| < \epsilon/32 ,$$

so that

$$\|z_n - z_{n,i}\| \vee \|z_n - z_{n,i+1}\| < 21\epsilon/32 < \epsilon$$

and $|t_{n,i} - t_{n,i+1}| < 2\eta < \epsilon/8$. Hence $\hat{d}(A_n, \Gamma_{x_n}) < \epsilon$ for $n \ge n_1$, so that the proof is complete.

(vi)→(i). Suppose that the conditions in (vi) hold and $\epsilon > 0$ is given. Let $(u, r) \in \Pi_s(x)$ and $(u_n, r_n) \in \Pi_s(x_n)$, $n \ge 1$, be arbitrary parametric representations. Let $s_1 = 0 < s_2 < \cdots < s_m = 1$ and $s_{n,1} = 0 < s_{n,2} < \cdots < s_{n,m} = 1$ be points such that $(u(s_i), r(s_i)) = (z_i, t_i) \in A$ and $(u_n(s_{n,i}), r_n(s_{n,i})) = (z_{n,i}, t_{n,i}) \in A_n$ for $1 \le i \le m$. Let $\lambda_n : [0, 1] \to [0, 1]$ be a continuous nondecreasing function such that $\lambda_n(s_i) = s_{n,i}$ for each i and n. We will show that $(u_n \circ \lambda_n, r_n \circ \lambda_n)$ is a parametric

representation of Γ_{x_n} for each n such that
$$\|u_n \circ \lambda_n - u\| \vee \|r_n \circ \lambda_n - r\| < 3\epsilon \quad \text{for} \quad n \geq n_1 . \tag{5.10}$$
Property (5.10) holds because, for $s_i \leq s \leq s_{i+1}$, $\lambda_n(s_i) = s_{n,i} \leq \lambda_n(s) \leq s_{n,i+1} = \lambda_n(s_{i+1})$ and

$$\begin{aligned}
&\|u_n \circ \lambda_n(s) - u(s)\| \vee \|r_n \circ \lambda_n(s) - r(s)\| \\
<\ & \|(u_n \circ \lambda_n(s), r_n \circ \lambda_n(s)) - (u_n(s_{n,i}), r_n(s_{n,i}))\| \vee \|(u_n \circ \lambda_n(s), r_n \circ \lambda_n(s)) \\
&- (u_n(s_{n,i+1}), r_n(s_{n,i+1}))\| \\
+\ & \|(u(s), r(s)) - (u(s_i), r(s_i))\| \vee \|(u(s), r(s)) - (u(s_{i+1}), r(s_{i+1}))\| \\
+\ & \|(u_n(s_{n,i}), r_n(s_{n,i})) - (u(s_i), r(s_i))\| \\
\leq\ & \hat{d}(A_n, \Gamma_{x_n}) + \hat{d}(A, \Gamma_x) + d^*(A, A_n) \leq 3\epsilon .
\end{aligned}$$

(v)→(iv). First, the convergence $x_n(T) \to x(T)$ is assumed directly. Next, (5.4) implies that $x_n(t) \to x(t)$ as $n \to \infty$ for each $t \notin Disc(x)$. Since $[0,T] - Disc(x)$ is a dense subset of $[0,T]$, the first part of (iv) is established. Condition (5.4) also implies that (5.5) holds for each $t \notin Disc(x)$ by Theorem 12.4.1. Finally, we show that (5.5) for each $t \in [0,T]$ implies (5.3). Condition (5.5) for each t implies that for each $\epsilon > 0$ and t, there is $\delta \equiv \delta(t)$ and $n(t, \epsilon, \delta)$ such that $w_s(x_n, t, \delta) < \epsilon$ for all $n \geq n(t, \epsilon, \delta)$. Now suppose that (5.3) does not hold. Then there must exist $\epsilon > 0$ such that for all $\delta > 0$ there is a sequence $\{t_k\}$ of points in $[0,T]$ and a sequence of integers n_k such that $n_k \to \infty$ and $w_s(x_{n_k}, t_k, \delta/2) > \epsilon$ for all k. However, the sequence $\{t_k\}$ has a subsequence $\{t_{k_j}\}$ with $t_{k_j} \to t \in [0,T]$ as $k_j \to \infty$. Thus, for all k_j suitably large,
$$w_s(x_{n_{k_j}}, t, \delta) > w_s(x_{n_{k_j}}, t_{n_{k_j}}, \delta/2) > \epsilon ,$$
which is a contradiction, so that (5.3) must in fact hold.

(iii)+(iv)→(v) By Lemma 12.5.1, (iii) implies (5.4) for each $t \in Disc(x)^c$. Trivially, (iv) implies (5.3), which in turn implies (5.5). ■

Remark 12.5.1. *Connection to Skorohod (1956).* Part (iv) of Theorem 12.5.1 is essentially Skorohod's (1956) original characterization, established in his 2.4.1. Instead of (5.1) with (4.4), Skorohod (1956) actually considered (5.1) with $w_s(x,t,\delta)$ replaced by
$$w'_s(x,t,\delta) \equiv \sup_{0 \vee (t-\delta) \leq t_1 \leq t \leq t_3 \leq (t+\delta) \wedge T} \{\|x(t) - [x(t_1), x(t_3)]\|\} , \tag{5.11}$$
but when the supremum over $t \in [0,T]$ is applied, w_s and w'_s are equivalent. In particular, clearly $w'_s(x,t,\delta) \leq w_s(x,t,\delta)$ for each t. On the other hand, if $w_s(x,t,\delta) > \epsilon$ for all t, then $w'_s(x,t,2\delta) > \epsilon$ for all t. Hence (iv) is equivalent to Skorohod's original characterization. We have introduced $w_s(x,t,\delta)$ in (4.4) in order to get characterization (v) in Theorem 12.5.1. We cannot use Skorohod's (5.11) instead of (4.4) in characterization (v) in Theorem 12.5.1, because it does not rule out multiple large oscillations on the same side of t. ■

Remark 12.5.2. *Possibility of using one-to-one parametric representations.* The proof of the implication (vi)→(i) shows that the SM_1 topology is unaltered if all the parametric representations are required to be one-to-one functions from $[0, 1]$ onto the graph. In the proof we would then let the transformations λ_n be homeomorphisms of $[0, 1]$, so that $(u_n \circ \lambda_n, r_n \circ \lambda_n)$ become one-to-one functions. ∎

We can apply Theorem 12.5.1 to develop a simple criterion for M_1 convergence for monotone functions.

Corollary 12.5.1. (the case of monotone functions) *If x_n is monotone for each n, then $d_s(x_n, x) \to 0$ for $x \in D$ if and only if $x_n(t) \to x(t)$ for all t in a dense subset of $[0, T]$ including 0 and T.*

Proof. Apply Theorem 12.5.1 (iv). Note that condition (5.3) always holds for monotone functions. ∎

12.5.2. WM_1 Convergence

We now establish an analog of Theorem 12.5.1 for the WM_1 topology. Several alternative characterizations of WM_1 convergence will follow directly from Theorem 12.5.1 because we will show that convergence $x_n \to x$ as $n \to \infty$ in WM_1 is equivalent to $d_p(x_n, x) \to 0$. To treat the WM_1 topology, we define another oscillation function. Let

$$w_w(x, \delta) \equiv \sup_{0 \leq t \leq T} w_w(x, t, \delta) \tag{5.12}$$

for $w_w(x, t, \delta)$ in (4.5). Recall that $w_w(x, t, \delta)$ in (4.5) is the same as $w_s(x, t, \delta)$ in (4.4) except it has the product segment $[[x(t_1), x(t_3)]]$ in (3.2) instead of the standard segment $[x(t_1), x(t_3)]$ in (3.1).

Paralleling Definition 12.3.1, let an ordered subset A of G_x of cardinality m be such that (3.9) holds, but now with the order being the order on G_x. Paralleling (3.10), let the *order-consistent distance* between A and G_x be

$$\hat{d}(A, G_x) \equiv \sup\{\|(z, t) - (z_i, t_i)\| \vee \|(z, t) - (z_{i+1}, t_{i+1})\| : (z, t) \in G_x\} \tag{5.13}$$

with the supremum being over all $(z, t) \in G_x$ such that $(z_i, t_i) \leq (z, t) \leq (z_{i+1}, t_{i+1})$ for all i, $1 \leq i \leq m - 1$.

Theorem 12.5.2. (characterizations of WM_1 convergence) *The following are equivalent characterizations of $x_n \to x$ as $n \to \infty$ in (D, WM_1):*

(i) $d_w(x_n, x) \to 0$ as $n \to \infty$.

(ii) $d_p(x_n, x) \to 0$ as $n \to \infty$.

(iii) $x_n(t) \to x(t)$ as $n \to \infty$ for each t in a dense subset of $[0, T]$ including 0 and T, and

$$\lim_{\delta \downarrow 0} \varlimsup_{n \to \infty} w_w(x_n, \delta) = 0 . \tag{5.14}$$

(iv) $x_n(T) \to x(T)$ as $n \to \infty$; for each $t \notin Disc(x)$,
$$\lim_{\delta \downarrow 0} \overline{\lim_{n \to \infty}} \; v(x_n, x, t, \delta) = 0 \qquad (5.15)$$

for $v(x_n, x, t, \delta)$ in (4.2); and, for each $t \in Disc(x)$,
$$\lim_{\delta \downarrow 0} \overline{\lim_{n \to \infty}} \; w_w(x_n, t, \delta) = 0 \qquad (5.16)$$

for $w_w(x_n, t, \delta)$ in (4.5).

(v) for all $\epsilon > 0$ and all n sufficiently large, there exist finite ordered subsets A of G_x (in general depending on n) and A_n of G_{x_n} of common cardinality such that $\hat{d}(A, G_x) < \epsilon$, $\hat{d}(A_n, G_{x_n}) < \epsilon$ and $d^*(A, A_n) < \epsilon$ for $\hat{d}$ in (5.13) and d^* in (5.6).

Example 12.5.1. *Need for changing parametric representations.* In general, there is no analog of characterizations (i) and (ii) in Theorem 12.5.1 for the parametric representations in $\Pi_w(x)$ and $\Pi_w(x_n)$; i.e., if $d_w(x_n, x) \to 0$ as $n \to \infty$, there need not exist $(u, r) \in \Pi_w(x)$ and $(u_n, r_n) \in \Pi_w(x_n)$ such that (5.2) holds. To see this, let $x^1 = x^2 = I_{[1,2]}$, $x^1_{2n+1} = x^2_{2n} = I_{[1-n^{-1},2]}$ and $x^2_{2n+1} = x^1_{2n} = I_{[1+n^{-1},2]}$ for $n \geq 2$. Property (i) of Theorem 12.5.2 holds, but different parametric representations of x are needed for even and odd n. ∎

12.6. Strengthening the Mode of Convergence

In this section we apply the characterizations of M_1 convergence in Sections 12.3 and 12.5 to establish conditions under which the mode of convergence can be strengthened: We seek conditions under which WM_1 convergence can be replaced by SM_1 convergence. We use the following Lemma.

Lemma 12.6.1. (modulus bound for (x_n, y_n)) *For* $x_n \in D([0,T], \mathbb{R}^k)$, $y_n, y \in D([0,T], \mathbb{R}^l)$, $t \in [0, T]$ *and* $\delta > 0$,
$$w_s((x_n, y_n), t, \delta) \leq w_s(x_n, t, \delta) + 2v(y_n, y, t, \delta).$$

Theorem 12.6.1. (extending SM_1 convergence to product spaces) *Suppose that* $d_s(x_n, x) \to 0$ *in* $D([0,T], \mathbb{R}^k)$ *and* $d_s(y_n, y) \to 0$ *in* $D([0,T], \mathbb{R}^l)$ *as* $n \to \infty$. *If*
$$Disc(x) \cap Disc(y) = \phi.$$
then
$$d_s((x_n, y_n), (x, y)) \to 0 \quad \text{in} \quad D([0,T], \mathbb{R}^{k+l}) \quad \text{as} \quad n \to \infty.$$

Proof. We use characterization (v) in Theorem 12.5.1. First, for each $t \notin Disc((x,y))$, $t \notin Disc(x) \cup Disc(y)$, (5.4) holds and
$$\lim_{\delta \downarrow 0} \overline{\lim_{n \to \infty}} \; v(y_n, y, \delta, t) = 0, \qquad (6.1)$$

which implies that

$$\lim_{\delta \downarrow 0} \overline{\lim_{n \to \infty}} \ v((x_n, y_n), (x, y), \delta, t) = 0 \ .$$

Now, for each $t \in Disc(x)$, (5.5) and (6.1) hold (because $Disc(x) \cap Disc(y) = \phi$). Thus, for those t, by Lemma 12.6.1,

$$\lim_{\delta \downarrow 0} \overline{\lim_{n \to \infty}} \ w_s((x_n, y_n), t, \delta) = 0 \ . \tag{6.2}$$

By the same reasoning (6.2) also holds for each $t \in Disc(y)$, so that (6.2) holds for all $t \in Disc((x, y)) = Disc(x) \cup Disc(y)$. ∎

Remark 12.6.1. *The discontinuity condition is not necessary.* The discontinuity condition $Disc(x) \cap Disc(y) = \phi$ in Theorem 12.6.1 is not necessary. To see that, note that if $x_n \to x$ as $n \to \infty$ in $D([0,T], \mathbb{R}^k)$, then $(x_n, x_n) \to (x, x)$ as $n \to \infty$ in $D([0,T], \mathbb{R}^{2k})$. However, some condition is needed, as can be seen from the fact that the WM_1 topology is strictly weaker than the SM_1 topology on $D([0,T], \mathbb{R}^k)$ for $k > 1$, as shown by Example 12.3.1. ∎

Remark 12.6.2. *The J_1 and M_2 analogs.* Analogs of Theorem 12.6.1 hold in the J_1 and M_2 topologies. For J_1, see Propositions 2.1 (a) and 2.2 (b) on p. 301 of Jacod and Shiryaev (1987). For M_2, see Theorem 12.11.3 below. ∎

As in Lemma 12.3.3, let $D_1 \equiv D_1([0,T], \mathbb{R}^k)$ be the subset of x in D with discontinuities in only one coordinate at a time; i.e., $x \in D_1$ if $x^i(t-) \neq x^i(t)$ for at most one coordinate i for each t. (The coordinate $i \equiv i(t)$ may depend upon t.)

Corollary 12.6.1. (from WM_1 convergence to SM_1 convergence when the limit is in D_1) *If $d_p(x_n, x) \to 0$ as $n \to \infty$ and $x \in D_1$, then $d_s(x_n, x) \to 0$.*

Example 12.3.3 shows that it is not enough to have $x \in D_s$ in Corollary 12.6.1.

12.7. Characterizing Convergence with Mappings

The strong topology SM_1 differs from the weak topology WM_1 by the behavior of linear functions of the coordinates. Example 12.3.1 shows that linear functions of the coordinates are not continuous in the product topology (there $(x_n^1 - x_n^2) \not\to (x^1 - x^2)$ as $n \to \infty$), but they are in the strong topology, as we now show. Note that there is no subscript on d on the left in (7.1) below because ηx is real valued.

Theorem 12.7.1. (Lipschitz property of linear functions of the coordinate functions) *For any $x_1, x_2 \in D([0,T], \mathbb{R}^k)$ and $\eta \in \mathbb{R}^k$,*

$$d(\eta x_1, \eta x_2) \leq (\|\eta\| \vee 1) d_s(x_1, x_2) \ . \tag{7.1}$$

Example 12.7.1. *Difficulties with the weak topology.* To see that $(\eta z, t)$ need not be an element of $\Gamma_{\eta x}$ when $(z, t) \in G_x$ and that $(\eta u, r)$ need not be an element of

$\Pi(\eta x)$ when $(u, r) \in \Pi_w(x)$, let $x^1 = x^2 = I_{[1,2]}$ and consider $\eta x = x^1 - x^2$. The flexibility allowed by G_x allows $(z, t) \in G_x$ with $\eta z \neq 0$ and $(u, r) \in \Pi_w(x)$ with $u(s) \neq 0$. ∎

We now obtain a sufficient condition for addition to be continuous on $(D, d_s) \times (D, d_s)$, which is analogous to the J_1 result in Theorem 4.1 of Whitt (1980).

Corollary 12.7.1. (SM_1-continuity of addition) *If $d_s(x_n, x) \to 0$ and $d_s(y_n, y) \to 0$ in $D([0, T], \mathbb{R}^k)$ and*

$$Disc(x) \cap Disc(y) = \phi,$$

then

$$d_s(x_n + y_n, x + y) \to 0 \quad in \quad D([0, T], \mathbb{R}^k).$$

Proof. First apply Theorem 12.6.1 to get $d_s((x_n, y_n), (x, y)) \to 0$ in $D([0, T], \mathbb{R}^{2k})$. Then apply Theorem 12.7.1. ∎

Remark 12.7.1. *Measurability of addition.* The measurability of addition on $(D, d_s) \times (D, d_s)$ holds because the Borel σ-field coincides with the Kolmogorov σ-field. It also follows from part of the proof of Theorem 4.1 of Whitt (1980). ∎

In Theorem 12.7.1 we showed that linear functions of the coordinates are Lipschitz in the SM_1 metric. We now apply Theorem 12.5.1 to show that convergence in the SM_1 topology is characterized by convergence of all such linear functions of the coordinates.

Theorem 12.7.2. (characterization of SM_1 convergence by convergence of all linear functions) *There is convergence $x_n \to x$ in $D([0, T], \mathbb{R}^k)$ as $n \to \infty$ in the SM_1 topology if and only if $\eta x_n \to \eta x$ in $D([0, T], \mathbb{R}^1)$ as $n \to \infty$ in the M_1 topology for all $\eta \in \mathbb{R}^k$.*

We can get convergence of sums under more general conditions than in Corollary 12.7.1. It suffices to have the jumps of x^i and y^i have common sign for all i. We can express this property by the condition

$$(x^i(t) - x^i(t-))(y^i(t) - y^i(t-)) \geq 0 \tag{7.2}$$

for all t, $0 \leq t \leq T$, and all i, $1 \leq i \leq k$.

Theorem 12.7.3. (continuity of addition at limits with jumps of common sign) *If $x_n \to x$ and $y_n \to y$ in $D([0, T], \mathbb{R}^k, SM_1)$ and if condition (7.2) above holds, then*

$$x_n + y_n \to x + y \quad in \quad D([0, T], \mathbb{R}^k, SM_1).$$

Proof. Apply the characterization of SM_1 convergence in Theorem 12.5.1 (v). At points t in $Disc(x)^c \cup Disc(y)^c$, use the local uniform convergence in Lemma 12.5.1 and Corollary 12.11.1. For other t not in $Disc(x) \cap Disc(y)$, use Theorem 12.6.1. For $t \in Disc(x) \cap Disc(y)$, exploit condition (7.2) to deduce that, for all $\epsilon > 0$, there exists δ and n_0 such that

$$w_s(x_n + y_n, t, \delta) \leq w_s(x_n, t, \delta) + w_s(y_n, t, \delta) + \epsilon$$

for all $n \geq n_0$. ∎

In Sections (2.2.7)–(2.2.13) of Skorohod (1956), convenient characterizations of convergence in each topology are given for real-valued functions. We can apply Theorem 12.7.2 to develop associated characterizations for $\mathbb{R}^k$-valued functions. For each $x \in D([0,T], \mathbb{R}^1)$, $0 \leq t_1 < t_2 \leq T$ and, for each $a < b$ in $\mathbb{R}$, let $v_{t_1,t_2}^{a,b}(x)$ be the number of visits to the strip $[a,b]$ on the interval $[t_1, t_2]$; i.e., $v_{t_1,t_2}^{a,b}(x) = k$ if it is possible to find k (but not $k+1$) points t'_i such that $t_1 < t'_1 < \cdots < t'_k \leq t_2$ such that either

$$x(t_1) \in [a,b],\ x(t'_1) \notin [a,b],\ x(t'_2) \in [a,b], \ldots,$$

or

$$x(t_1) \notin [a,b],\ x(t'_1) \in [a,b],\ x(t'_2) \notin [a,b], \ldots$$

We say that $x \in D([0,T], \mathbb{R})$ has a *local maximum (minimum) value at t relative to (t_1, t_2)* in $(0,T)$ if $t_1 < t < t_2$ and either

(i) $\sup\{x(s) : t_1 \leq s \leq t_2\} \leq x(t)$ $\quad (\inf\{x(s) : t_1 \leq s \leq t_2\} \geq x(t))$

or

(ii) $\sup\{x(s) : t_1 \leq s \leq t_2\} \leq x(t-)$ $\quad (\inf\{x(s) : t_1 \leq s \leq t_2\}) \geq x(t-))$.

We say that x has a *local maximum (minimum) value* at t if it has a local maximum (minimum) value at t relative to some interval (t_1, t_2) with $t_1 < t < t_2$. We call local maximum and minimum values *local extreme values*.

Lemma 12.7.1. (local extreme values) *Any $x \in D([0,T], \mathbb{R})$ has at most countably many local extreme values.*

If b is not a local extreme value of x, then x crosses level b whenever x hits b; i.e., if b is not a local extreme value and if $x(t) = b$ or $x(t-) = b$, then for every t_1, t_2 with $t_1 < t < t_2$ there exist t'_1, t'_2 with $t_1 < t'_1, t'_2 < t_2$ such that $x(t'_1) < b$ and $x(t'_2) > b$. This property implies the following lemma.

Lemma 12.7.2. *Consider an interval $[t_1, t_2]$ with $0 < t_1 < t_2 < T$. If $x(t_i) \notin \{a,b\}$ for $i = 1, 2$ and a, b are not local extreme values of x, then x crosses one of the levels a and b at each of the $v_{t_1,t_2}^{a,b}(x)$ visits to the strip $[a,b]$ in $[t_1, t_2]$.*

Theorem 12.7.4. (characterization of SM_1 convergence in terms of convergence of number of visits to strips) *There is convergence $d_s(x_n, x) \to 0$ as $n \to \infty$ in $D([0,T], \mathbb{R}^k)$ if and only if*

$$v_{t_1,t_2}^{a,b}(\eta x_n) \to v_{t_1,t_2}^{a,b}(\eta x) \quad \text{as} \quad n \to \infty$$

for all $\eta \in \mathbb{R}^k$, all points $t_1, t_2 \in \{T\} \cup Disc(x)^c$ with $t_1 < t_2$ and almost all a, b with respect to Lebesgue measure.

12.8. Topological Completeness

In this section we exhibit a complete metric topologically equivalent to the incomplete metric d_s in (3.7) inducing the SM_1 topology. Since a product metric defined as in (3.11) inherits the completeness of the component metrics, we also succeed in constructing complete metrics inducing the associated product topology. We make no use of the complete metrics beyond showing that the topology is topologically complete. Another approach to topological completeness would be to show that D is homeomorphic to a G_δ subset of a complete metric space, as noted in Section 11.2.

In our construction of complete metrics, we follow the argument used by Prohorov (1956, Appendix 1) to show that the J_1 topology is topologically complete; we incorporate an oscillation function into the metric. For M_1, we use $w_s(x, \delta)$ in (5.1). Since $w_s(x, \delta) \to 0$ as $\delta \to 0$ for each $x \in D$, we need to appropriately "inflate" differences for small δ. For this purpose, let

$$\hat{w}_s(x, z) \equiv \begin{cases} w_s(x, e^z), & z < 0 \\ w_s(x, 1), & z \geq 1. \end{cases} \quad (8.1)$$

Since $w_s(x, \delta)$ is nondecreasing in δ, $\hat{w}_s(x, z)$ is nondecreasing in z. Note that $\hat{w}_s(x, z)$ as a function of z has the form of a cumulative distribution function (cdf) of a finite measure. On such cdf's, the Lévy metric λ is known to be a complete metric inducing the topology of pointwise convergence at all continuity points of the limit; i.e.,

$$\lambda(F_1, F_2) \equiv \inf\{\epsilon > 0 : F_2(x - \epsilon) - \epsilon \leq F_1(x) \leq F_2(x + \epsilon) + \epsilon\}. \quad (8.2)$$

The Helly selection theorem, p. 267 of Feller (1971), can be used to show that the metric λ is complete.

Thus, our new metric is

$$\hat{d}_s(x_1, x_2) \equiv d_s(x_1, x_2) + \lambda(\hat{w}_s(x_1, \cdot), \hat{w}_s(x_2, \cdot)). \quad (8.3)$$

Theorem 12.8.1. (a complete SM_1 metric) *The metric $\hat{d}_s$ on D in (8.3) is complete and topologically equivalent to d_s.*

Example 12.8.1. *The counterexample for d_s is not fundamental under $\hat{d}_s$.* Recall that Example 12.10.1 was used to show that the metric d_s is not complete. That example has $x_n = I_{[1,1+1/n)}$, so that $d_s(x_m, x_n) \to 0$ as $m, n \to \infty$, i.e., the sequence $\{x_n\}$ is fundamental for d_s even though it does not converge. Note that $w_s(x_n, \delta) = 1$ for $\delta > 1/2n$ and $w_s(x_n, \delta) = 0$ otherwise. Hence, $\hat{w}_s(x_n, z) = 1$ for $z > \log(1/2n) = -\log(2n)$ and $\hat{w}_s(x_n, z) = 0$ otherwise. Note that $\hat{w}_s(x_n, \cdot)$ corresponds to the cdf of a unit point mass at $-\log(2n)$. Consequently, $\hat{d}_s(x_m, x_n) \not\to 0$ as $m, n \to \infty$.

Remark 12.8.1. *An alternative complete metric.* An alternative complete metric topologically equivalent to d_s is

$$d_s^\dagger(x_1, x_2) = m_s(x_1, x_2) + \lambda(\hat{w}_s(x_1, \cdot), \hat{w}_s(x_2, \cdot)) , \qquad (8.4)$$

where $m_s \equiv d_{M_2}$ is the M_2 metric in (5.4) of Section 11.5. That is actually what Prohorov did for J_1 (with $\hat{w}_s$ in (8.4) replaced by the J_1 oscillation function). ∎

12.9. Noncompact Domains

It is often convenient to consider the function space $D([0, \infty), \mathbb{R}^k)$ with domain $[0, \infty)$ instead of $[0, T]$. More generally, we may consider the function space $D(I, \mathbb{R}^k)$, where I is a subinterval of the real line. Common cases besides $[0, \infty)$ are $(0, \infty)$ and $(-\infty, \infty) \equiv \mathbb{R}$.

Given the function space $D(I, \mathbb{R}^k)$ for any subinterval I, we define convergence $x_n \to x$ with some topology to be convergence in $D([a, b], \mathbb{R}^k)$ with that same topology for the restrictions of x_n and x to the compact interval $[a, b]$ for all points a and b that are elements of I and either boundary points of I or are continuity points of the limit function x. For example, for $I = [c, d)$ with $-\infty < c < d < \infty$, we include $a = c$ but exclude $b = d$; for $I = [c, d]$, we include both c and d.

For simplicity, we henceforth consider only the special case in which $I = [0, \infty)$. In that setting, we can equivalently define convergence $x_n \to x$ as $n \to \infty$ in $D([0, \infty), \mathbb{R}^k)$ with some topology to be convergence $x_n \to x$ as $n \to \infty$ in $D([0, t], \mathbb{R}^k)$ with that topology for the restrictions of x_n and x to $[0, t]$ for $t = t_k$ for each t_k in some sequence $\{t_k\}$ with $t_k \to \infty$ as $k \to \infty$, where $\{t_k\}$ can depend on x. It suffices to let t_k be continuity points of the limit function x; for the J_1 topology, see Stone (1963), Lindvall (1973), Whitt (1980) and Jacod and Shiryaev (1987). We will discuss only the SM_1 topology here, but the discussion applies to the other nonuniform topologies as well. We also will omit most proofs.

As a first step, we consider the case of closed bounded intervals $[t_1, t_2]$. The space $D([t_1, t_2], \mathbb{R}^k)$ is essentially the same as (homeomorphic to) the space $D([0, T], \mathbb{R}^k)$ already studied, but we want to look at the behavior as we change the interval $[t_1, t_2]$. For $[t_3, t_4] \subseteq [t_1, t_2]$, we consider the restriction of x in $D([t_1, t_2], \mathbb{R}^k)$ to $[t_3, t_4]$, defined by

$$r_{t_3, t_4} : D([t_1, t_2], \mathbb{R}^k) \to D([t_3, t_4], \mathbb{R}^k)$$

with $r_{t_3, t_4}(x)(t) = x(t)$ for $t_3 \leq t \leq t_4$. Let d_{t_1, t_2} be the metric d_s on $D([t_1, t_2], \mathbb{R}^k)$. We want to relate the distance $d_{t_1, t_2}(x_1, x_2)$ and convergence $d_{t_1, t_2}(x_n, x) \to 0$ as $n \to \infty$ for different domains. We first state a result enabling us to go from the domains $[t_1, t_2]$ and $[t_2, t_3]$ to $[t_1, t_3]$ when $t_1 < t_2 < t_3$.

Lemma 12.9.1. (metric bounds) *For $0 \leq t_1 < t_2 < t_3$ and $x_1, x_2 \in D([t_1, t_3], \mathbb{R}^k)$,*

$$d_{t_1, t_3}(x_1, x_2) \leq d_{t_1, t_2}(x_1, x_2) \vee d_{t_2, t_3}(x_1, x_2) .$$

We now observe that there is an equivalence of convergence provided that the internal boundary point is a continuity point of the limit function.

Lemma 12.9.2. *For $0 \leq t_1 < t_2 < t_3$ and $x, x_n \in D([t_1, t_3], \mathbb{R}^k)$, with $t_2 \in Disc(x)^c$, $d_{t_1,t_3}(x_n, x) \to 0$ as $n \to \infty$ if and only if $d_{t_1,t_2}(x_n, x) \to 0$ and $d_{t_2,t_3}(x_n, x) \to 0$ as $n \to \infty$.*

For $x \in D([0,T], \mathbb{R}^k)$ and $0 \leq t_1 < t_2 \leq T$, let $r_{t_1,t_2} : D([0,T], \mathbb{R}^k) \to D([t_1, t_2], \mathbb{R}^k)$ be the restriction map, defined by $r_{t_1,t_2}(x)(s) = x(s)$, $t_1 \leq s \leq t_2$.

Corollary 12.9.1. (continuity of restriction maps) *If $x_n \to x$ as $n \to \infty$ in $D([0,T], \mathbb{R}^k, SM_1)$ and if $t_1, t_2 \in Disc(x)^c$, then*

$$r_{t_1,t_2}(x_n) \to r_{t_1,t_2}(x) \quad \text{as} \quad n \to \infty \quad \text{in} \quad D([t_1, t_2], \mathbb{R}^k, SM_1).$$

Let $r_t : D([0, \infty), \mathbb{R}^k) \to D([0, t], \mathbb{R}^k)$ be the *restriction map* with $r_t(x)(s) = x(s)$, $0 \leq s \leq t$. Suppose that $f : D([0, \infty), \mathbb{R}^k) \to D([0, \infty), \mathbb{R}^k)$ and $f_t : D([0, t], \mathbb{R}^k) \to D([0, t], \mathbb{R}^k)$ for $t > 0$ are functions with

$$f_t(r_t(x)) = r_t(f(x))$$

for all $x \in D([0, \infty), \mathbb{R}^k)$ and all $t > 0$. We then call the functions f_t *restrictions of the function f*.

Theorem 12.9.1. (continuity from continuous restrictions) *Suppose that $f : D([0, \infty), \mathbb{R}^k) \to D([0, \infty), \mathbb{R}^l)$ has continuous restrictions f_t with some topology for all $t > 0$. Then f itself is continuous in that topology.*

We now consider the extension of Lipschitz properties to subsets of $D([0, \infty), \mathbb{R}^k)$. For this purpose, suppose that μ_t is one of the M_1 metrics on $D([0, t], \mathbb{R}^k)$ for $t > 0$. As in Section 2 of Whitt(1980), an associated metric μ can be defined on $D([0, \infty), \mathbb{R}^k)$ by

$$\mu(x_1, x_2) = \int_0^\infty e^{-t}[\mu_t(r_t(x_1), r_t(x_2)) \wedge 1]dt. \tag{9.1}$$

The following result implies that the integral in (9.1) is well defined.

Theorem 12.9.2. (regularity of the metric $\mu_t(x_1, x_2)$ as a function of t) *Let μ_t be one of the M_1 metrics on $D([0, t], \mathbb{R}^k)$. For all $x_1, x_2 \in D([0, \infty), \mathbb{R}^k)$, $\mu_t(x_1, x_2)$ as a function of t is right-continuous with left limits in $(0, \infty)$ and has a right limit at 0. Moreover, $\mu_t(x_1, x_2)$ is continuous at $t > 0$ whenever x_1 and x_2 are both continuous at t.*

We also have the following result, paralleling Lemma 2.2 and Theorem 2.5 of Whitt (1980). For (iii), we exploit Theorem 12.5.1 (i).

Theorem 12.9.3. (characterizations of SM_1 convergence with domain $[0, \infty)$) *Suppose that μ and μ_t, $t > 0$ are the SM_1 (or WM_1) metrics on $D([0, \infty), \mathbb{R}^k)$ and $D([0, t], \mathbb{R}^k)$. Then the following are equivalent for x and x_n, $n \geq 1$, in $D([0, \infty), \mathbb{R}^k)$.*

(i) $\mu(x_n, x) \to 0$ as $n \to \infty$;

(ii) $\mu_t(r_t(x_n), r_t(x)) \to 0$ as $n \to \infty$ for all $t \notin Disc(x)$;

(iii) there exist parametric representations (u, r) and (u_n, r_n) of x and x_n mapping $[0, \infty)$ into the graphs such that

$$\|u_n - u\|_t \vee \|r_n - r\|_t \to 0 \quad as \quad n \to \infty$$

for each $t > 0$.

We now show that the Lipschitz property extends from $D([0,t], \mathbb{R}^k)$ to $D([0,\infty), \mathbb{R}^k)$.

Theorem 12.9.4. (functions with Lipschitz restrictions are Lipschitz) *If a function*

$$f : D([0,\infty), \mathbb{R}^k) \to D([0,\infty), \mathbb{R}^k)$$

has restrictions

$$f_t : D([0,T], \mathbb{R}^k) \to D([0,T], \mathbb{R}^k)$$

satisfying

$$\mu_t^2(f_t(r_t(x_1)), f_t(r_t(x_2))) \leq K\mu_t^1(r_t(x_1), r_t(x_2)) \quad for\ all \quad t > 0\ ,$$

where K is independent of t, then

$$\mu^2(f(x_1), f(x_2)) \leq (K \vee 1)\mu^1(x_1, x_2).$$

Proof. By (9.1) and the conditions,

$$\begin{aligned}
\mu^2(f(x_1), f(x_2)) &= \int_0^\infty e^{-t}[\mu_t^2(r_t(f(x_1)), r_t(f(x_2))) \wedge 1]dt \\
&= \int_0^\infty e^{-t}[\mu_t^2(f_t(r_t(x_1)), f_t(r_t(x_2))) \wedge 1]dt \\
&\leq \int_0^\infty e^{-t}[K\mu_t^1(r_t(x_1), r_t(x_2)) \wedge 1]dt \\
&\leq (K \vee 1)\int_0^\infty e^{-t}[\mu_t^1(r_t(x_1), r_t(x_2)) \wedge 1]dt \\
&\leq (K \vee 1)\mu^1(x_1, x_2)\ . \quad \blacksquare
\end{aligned}$$

12.10. Strong and Weak M_2 Topologies

We now define strong and weak versions of Skorohod's M_2 topology. In Section 12.11 we will show that it is possible to define the M_2 topologies by a minor modification of the definitions in Section 12.3, in particular, by simply using parametric representations in which only r is nondecreasing instead of (u,r), but now we will use Skorohod's (1956) original approach, and relate it to the Hausdorff metric on the space of graphs.

12.10. Strong and Weak M_2 Topologies

The weak topology will be defined just like the strong, except it will use the thick graphs G_x instead of the thin graphs Γ_x. In particular, let

$$\mu_s(x_1, x_2) \equiv \sup_{(z_1,t_1) \in \Gamma_{x_1}} \inf_{(z_2,t_2) \in \Gamma_{x_2}} \{\|(z_1, t_1) - (z_2, t_2)\|\} \tag{10.1}$$

and

$$\mu_w(x_1, x_2) \equiv \sup_{(z_1,t_1) \in G_{x_1}} \inf_{(z_2,t_2) \in G_{x_2}} \{\|(z_1, t_1) - (z_2, t_2)\|\} . \tag{10.2}$$

Following Skorohod (1956), we say that $x_n \to x$ as $n \to \infty$ for a sequence or net $\{x_n\}$ in the strong M_2 topology, denoted by SM_2 if $\mu_s(x_n, x) \to 0$ as $n \to \infty$. Paralleling that, we say that $x_n \to x$ as $n \to \infty$ in the weak M_2 topology, denoted by WM_2, if $\mu_w(x_n, x) \to 0$ as $n \to \infty$. We say that $x_n \to x$ as $n \to \infty$ in the product topology if $\mu_s(x_n^i, x^i) \to 0$ (or equivalently $\mu_w(x_n^i, x^i) \to 0$) as $n \to \infty$ for each i, $1 \le i \le k$.

We can also generate the SM_2 and WM_2 topologies using the Hausdorff metric in (5.2) of Section 11.5. As in (5.4) in Section 11.5, for $x_1, x_2 \in D$,

$$m_s(x_1, x_2) \equiv m_H(\Gamma_{x_1}, \Gamma_{x_2}) = \mu_s(x_1, x_2) \vee \mu_s(x_2, x_1) , \tag{10.3}$$

$$m_w(x_1, x_2) \equiv m_H(G_{x_1}, G_{x_2}) = \mu_w(x_1, x_2) \vee \mu_w(x_2, x_1) \tag{10.4}$$

and

$$m_p(x_1, x_2) \equiv \max_{1 \le i \le k} m_s(x_1^i, x_2^i) . \tag{10.5}$$

We will show that the metric m_s induces the SM_2 topology.

That will imply that the metric m_p induces the associated product topology. However, it turns out that the metric m_w does *not* induce the WM_2 topology. We will show that the WM_2 topology coincides with the product topology, so that the Hausdorff metric can be used to define the WM_2 topology via m_p in (10.5).

Closely paralleling the d or M_1 metrics, we have $m_p \le m_s$ on $D([0,T], \mathbb{R}^k)$ and $m_p = m_w = m_s$ on $D([0,T], \mathbb{R}^1)$. Just as with d, we use m without subscript when the functions are real valued. Example 12.3.1, which showed that WM_1 is strictly weaker than SM_1 also shows that WM_2 is strictly weaker than SM_2. Example 12.3.3 shows that the SM_2 topology is strictly weaker than the SM_1 topology.

Note that μ_s in (10.1) is *not* symmetric in its two arguments. We first show that if $\mu_s(x, x_n) \to 0$ as $n \to \infty$, we need not have $\mu_s(x_n, x) \to 0$ as $n \to \infty$.

Example 12.10.1. *Lack of symmetry of μ_s in its arguments.* To see that we can have $\mu_s(x, x_n) \to 0$ as $n \to \infty$ without $\mu_w(x_n, x) \to 0$ or $\mu_s(x_n, x) \to 0$ as $n \to \infty$, let $x(t) = 0$, $0 \le t \le 2$, and let $x_n = I_{[1,1+n^{-1})}$ in $D([0,2], \mathbb{R}^1)$. Clearly $m_w(x_n, x) \not\to 0$, but for any $(0, t) \in \Gamma_x = G_x$, we can find $(0, t_n) \in \Gamma_{x_n} = G_{x_n}$ such that $|t_n - t| \to 0$. ∎

We now observe that m_s induces the SM_2 topology.

Theorem 12.10.1. (the Hausdorff metric m_s induces the SM_2 topology) If $\mu_s(x_n, x) \to 0$ as $n \to \infty$, then $\mu_s(x, x_n) \to 0$ as $n \to \infty$. Hence, $\mu_s(x_n, x) \to 0$ as $n \to \infty$ if and only if $m_s(x_n, x) \to 0$ as $n \to \infty$.

It may seem natural to consider a weak M_2 topology defined by the metric $m_w(x_1, x_2)$ in (10.4), but this does not yield a desirable topology.

Example 12.10.2. *Deficiency of the m_w metric.* To see a deficiency of the m_w metric in (10.4), we show that convergence $d_s(x_n, x) \to 0$ as $n \to \infty$, which implies $m_s(x_n, x) \to 0$, does not imply $\mu_w(x, x_n) \to 0$ or $m_w(x_n, x) \to 0$ as $n \to \infty$. For this purpose, consider x and x_n, $n \geq 1$, in $D([0,2], \mathbb{R}^2)$ defined by $x^1 = x^2 = I_{[1,2]}$ and $x_n^1(t) = x_n^2(t) = n(t-1)I_{[1,1+n^{-1})}(t) + I_{[1+n^{-1},2]}(t)$ for $n \geq 1$. Then $d_s(x_n, x) \to 0$ as $n \to \infty$ and thus $m_s(x_n, x) \to 0$ as $n \to \infty$, but the thick ranges of the graphs of x and x_n are $\rho(G_x) = [0,2] \times [0,2]$ and $\rho(G_{x_n}) = \{\alpha(0,0) + (1-\alpha)(2,2) : 0 \leq \alpha \leq 1\}$, so that $\mu_w(x, x_n) \not\to 0$ and $m_w(x_n, x) \not\to 0$ as $n \to \infty$. in this case, $\mu_w(x_n, x) \to 0$ as $n \to \infty$.

We now observe that m_p induces the WM_2 topology.

Theorem 12.10.2. (WM_2 is the product topology) $\mu_w(x_n, x) \to 0$ as $n \to \infty$ for μ_w in (10.2) if and only if $m_p(x_n, x) \to 0$ as $n \to \infty$ for m_p in (10.5), so that the WM_2 topology on $D([0,T], \mathbb{R}^k)$ coincides with the product topology.

We conclude this section by summarizing the relations among the primary distances under consideration in the following theorem.

Theorem 12.10.3. (comparison of distances) *For each $x_1, x_2 \in D$,*

$$d_p \leq d_w \leq d_s \leq d_{J_1} \leq \|\cdot\|,$$

$$m_p \leq d_p \quad \text{and} \quad m_p \leq m_s \leq d_s.$$

Remark 12.10.1. *Relating the J and M topologies.* The J_i topologies were related to the M_i topologies in a revealing way in Pomarede (1976). The J_2 topology is induced by the Hausdorff metric on the space of incomplete graphs; that shows that J_2 is stronger than M_2. Similarly, the J_1 topology can be defined in terms of a metric applied to parametric representations of the incomplete graphs; that shows that J_1 is stronger than M_1. ∎

12.11. Alternative Characterizations of M_2 Convergence

We now give alternative characterizations of the SM_2 and WM_2 topologies.

12.11.1. M_2 Parametric Representations

We first observe that the SM_2 and WM_2 topologies can be defined just like the SM_1 and WM_1 topologies in Section 12.3. For this purpose, we say that a *strong M_2*

12.11. Alternative Characterizations for M_2

(SM_2) *parametric representation* of x is a continuous function (u,r) mapping $[0,1]$ onto Γ_x such that r is nondecreasing. A *weak M_2* (WM_2) *parametric representation* of x is a continuous function mapping $[0,1]$ into G_x such that r is nondecreasing with $r(0) = 0$, $r(1) = T$ and $u(1) = x(T)$. The corresponding M_1 parametric representations are nondecreasing using the order defined on the graphs Γ_x and G_x in Section 2. In contrast, only the component function r is nondecreasing in the M_2 parametric representations. Let $\Pi_{s,2}(x)$ and $\Pi_{w,2}(x)$ be the sets of all SM_2 and WM_2 parametric representations of x.

Paralleling (3.7) and (3.8), define the distance functions

$$d_{s,2}(x_1, x_2) \equiv \inf_{\substack{(u_j,r_j) \in \Pi_{s,2}(x_j) \\ j=1,2}} \{\|u_1 - u_2\| \vee \|r_1 - r_2\|\} \quad (11.1)$$

and

$$d_{w,2}(x_1, x_2) \equiv \inf_{\substack{(u_j,r_j) \in \Pi_{w,2}(x_j) \\ j=1,2}} \{\|u_1 - u_2\| \vee \|r_1 - r_2\|\} \;. \quad (11.2)$$

We then can say that $x_n \to x$ as $n \to \infty$ for a sequence or net $\{x_n\}$ if $d_{s,2}(x_n, x) \to 0$ or $d_{w,2}(x_n, x) \to 0$ as $n \to \infty$. A difficulty with this approach, just as for the WM_1 topology, is that neither $d_{s,2}$ nor $d_{w,2}$ is a metric.

Example 12.11.1. *Neither $d_{s,2}$ nor $d_{w,2}$ is a metric.* To see that neither $d_{s,2}$ nor $d_{w,2}$ is a metric, consider real-valued functions, so that $d_{s,2} = d_{w,2} = d_2$. Let $x = 2I_{[1,2]}$, $x_{2n+1} = 2I_{[1-2n^{-1},1-n^{-1}]} + 2I_{[1,2]}$ and $x_{2n} = I_{[1-n^{-1},1)} + 2I_{[1,2]}$ in $D([0,2],\mathbb{R})$ for $n \geq 3$. For each n, it is possible to choose parametric representations of x_n and x such that $d_2(x_{2n+1}, x) \leq 2n^{-1}$ and $d_2(x_{2n}, x) \leq n^{-1}$. However, $d_2(x_{2n}, x_{2n+1}) \geq 1$ for all n. We cannot simultaneously match the points in $\{2\} \times [1-2n^{-1}, 1-n^{-1}] \subseteq \Gamma_{2x_{n+1}}$ to $\{2\} \times [1,2] \subseteq \Gamma_{x_{2n}}$ and the points in $\{0\} \times [(1-n^{-1}),1) \subseteq \Gamma_{x_{2n+1}}$ to $\{0\} \times [0, 1-n^{-1}) \subseteq \Gamma_{x_{2n}}$ because the times are inconsistently ordered. ∎

12.11.2. SM_2 Convergence

We now establish the equivalence of several alternative characterizations of convergence in the SM_2 topology. To have a characterization involving the local behavior of the functions, we use the uniform-distance function $\bar{w}_s(x, x_2, t, \delta)$ in (4.6). We also use the related uniform-distance functions

$$\bar{w}_s(x_1, x_2, \delta) \equiv \sup_{0 \leq t \leq T} \bar{w}(x_1, x_2, t, \delta) \;. \quad (11.3)$$

$$\bar{w}_s^*(x_1, x_2, t, \delta) = \|r_1(t) \quad [x_2((t-\delta) \vee 0), x_2((t+\delta) \wedge T)]\| \quad (11.4)$$

$$\bar{w}_s^*(x_1, x_2, \delta) \equiv \sup_{0 \leq t \leq T} \bar{w}_s^*(x_1, x_2, t, \delta) \;. \quad (11.5)$$

We now define new oscillation functions. The first is

$$\bar{w}_s^*(x, t, \delta) \equiv \sup\{\|x(t) - [x(t_1), x(t_2)]\|\} \;, \quad (11.6)$$

where the supremum is over

$$0 \vee (t - \delta) \leq t_1 \leq [0 \vee (t - \delta)] + \delta/2 \text{ and } [T \wedge (t + \delta)] - \delta/2 \leq t_2 \leq (t + \delta) \wedge T.$$

The second is

$$\bar{w}_s^*(x, \delta) \equiv \sup_{0 \leq t \leq T} \bar{w}_s^*(x, t, \delta) . \tag{11.7}$$

The uniform-distance function $\bar{w}_s^*(x_1, x_2, \delta)$ in (11.5) and the oscillation function $\bar{w}_s^*(x, \delta)$ in (11.7) were originally used by Skorohod (1956).

As before, T need not be a continuity point of x in $D([0,T], \mathbb{R}^k)$. Unlike for the M_1 topology, we can have $x_n \to x$ in (D, M_2) without having $x_n(T) \to x(T)$.

Example 12.11.2. *M_2 convergence does not imply pointwise convergence at the right endpoint.* To see that M_2 convergence does not imply that $x_n(T) \to x(T)$, let $x(0 = x(T-) = 0$, $x(T) = 1$,

$$x_n(0) = x_n(T - 2n^{-1}) = x_n(T) = 0$$

and $x_n(T - n^{-1}) = 1$ for $n \geq 1$ with x and x_n defined by linear interpolation elsewhere. It is easy to see that $x_n \to x$ (M_2), but $x_n(T) \not\to x(T)$.

Let $v(x, A)$ represent the oscillation of x over the set A as in (2.5).

Theorem 12.11.1. *(characterizations of SM_2 convergence)* The following are equivalent characterizations of $x_n \to x$ as $n \to \infty$ in (D, SM_2):

(i) $d_{s,2}(x_n, x) \to 0$ as $n \to \infty$ for $d_{s,2}$ in (11.1); i.e., for any $\epsilon > 0$ and n sufficiently large, there exist $(u, r) \in \Pi_{s,2}(x)$ and $(u_n, r_n) \in \Pi_{s,2}(x_n)$ such that $\|u_n - u\| \vee \|r_n - r\| < \epsilon$.

(ii) $m_s(x_n, x) \to 0$ as $n \to \infty$ for the metric m_s in (10.3).

(iii) $\mu_s(x_n, x) \to 0$ as $n \to \infty$ for μ_s in (10.1).

(iv) Given $\bar{w}_s(x_1, x_2, \delta)$ defined in (11.3),

$$\lim_{\delta \downarrow 0} \varlimsup_{n \to \infty} \bar{w}_s(x_n, x, \delta) = 0 .$$

(v) For each t, $0 \leq t \leq T$,

$$\lim_{\delta \downarrow 0} \varlimsup_{n \to \infty} \bar{w}_s(x_n, x, t, \delta) = 0$$

for $\bar{w}_s(x_1, x_2, t, \delta)$ in (4.6).

(vi) For all $\epsilon > 0$ and all n sufficiently large, there exist finite ordered subsets A of Γ_x and A_n of Γ_{x_n}, as in (3.9) where $(z_1, t_1) \leq (z_2, t_2)$ if $t_1 \leq t_2$, of the same cardinality such that $\hat{d}(A, \Gamma_x) < \epsilon$, $\hat{d}(A_n, \Gamma_{x_n}) < \epsilon$ and $d^*(A, A_n) < \epsilon$ for $\hat{d}$ in (3.10) and d^* in (5.6).

(vii) Given $\bar{w}_s^*(x_1, x_2, \delta)$ defined in (11.5),
$$\lim_{\delta \downarrow 0} \overline{\lim_{n \to \infty}} \bar{w}_s^*(x_n, x, \delta) = 0 .$$

(viii) $x_n(t) \to x(t)$ as $n \to \infty$ for each t in a dense subset of $[0, T]$ including 0 and
$$\lim_{\delta \downarrow 0} \overline{\lim_{n \to \infty}} \bar{w}_s^*(x_n, \delta) = 0$$

for $\bar{w}_s^*(x, \delta)$ in (11.7).

Remark 12.11.1. The equivalence (iii)↔(vii)↔(viii) was established by Skorohod (1956). ∎

Remark 12.11.2. There is no analog to characterization (v) involving $\bar{w}_s^*(x_n, x, t, \delta)$ in (11.4) instead of $\bar{w}_s(x_n, x, t, \delta)$. For $t \in Disc(x)^c$,
$$\lim_{\delta \downarrow 0} \overline{\lim_{n \to \infty}} \bar{w}_s^*(x_n, x, t, \delta) = 0$$

implies pointwise convergence $x_n(t) \to x(t)$, but not the local uniform convergence in Theorem 12.4.1. ∎

12.11.3. WM_2 Convergence

Corresponding characterizations of WM_2 convergence follow from Theorem 12.11.1 because the WM_2 topology is the same as the product topology, by Theorem 12.10.2. Let
$$\bar{w}_w(x_1, x_2, \delta) \equiv \sup_{0 \leq t \leq T} \bar{w}_w(x_1, x_2, t, \delta) \tag{11.8}$$

for $\bar{w}_w(x_1, x_2, t, \delta)$ in (4.7).

Theorem 12.11.2. (characterizations of WM_2 convergence) *The following are equivalent characterizations of $x_n \to x$ as $n \to \infty$ in (D, WM_2):*

(i) $d_{w,2}(x_n, x) \to 0$ as $n \to \infty$ for $d_{w,2}$ in (11.2); i.e., for any $\epsilon > 0$ and all n sufficiently large, there exist $(u, r) \in \Pi_{w,2}(x)$ and $(u_n, r_n) \in \Pi_{w,2}(x_n)$ such that $\|u_n - u\| \vee \|r_n - r\| < \epsilon$.

(ii) $m_p(x_n, x) \to 0$ as $n \to \infty$ for the metric m_p in (10.5).

(iii) Given $\bar{w}_w(x_1, x_2, \delta)$ defined in (11.8),
$$\lim_{\delta \downarrow 0} \overline{\lim_{n \to \infty}} \bar{w}_w(x_n, x, \delta) = 0 .$$

(iv) For each t, $0 \leq t \leq T$,
$$\lim_{\delta \downarrow 0} \overline{\lim_{n \to \infty}} \bar{w}_w(x_n, x, t, \delta) = 0 .$$

(v) For all $\epsilon > 0$ and all sufficiently large n, there exist finite ordered subsets A of G_x and A_n of Γ_{x_n}, of common cardinality m as in (3.9) with $(z_1, t_1) \leq (z_2, t_2)$ if

$t_1 \leq t_2$, such that $\hat{d}(A, G_x) < \epsilon$, $\hat{d}(A_n, \Gamma_{x_n}) < \epsilon$ and $d^*(A, A_n) < \epsilon$ for all $n \geq n_0$, for $\hat{d}$ in (5.13) and d^* in (5.6).

Theorem 12.11.2 and Section 12.4 show that all forms of M convergence imply uniform convergence to continuous limit functions.

Corollary 12.11.1. (from WM_2 convergence to uniform convergence) *Suppose that $m_p(x_n, x) \to 0$ as $n \to \infty$.*
  (i) If $t \in Disc(x)^c$, then

$$\lim_{\delta \downarrow 0} \overline{\lim_{n \to \infty}} \, v(x_n, x, t, \delta) = 0 \ .$$

(ii) If $x \in C$, then $\lim_{n \to \infty} \|x_n - x\| = 0$.

Proof. For (i) combine Theorems 12.4.1 and 12.11.2. For (ii) add Lemma 12.4.2. ∎

Convergence in WM_2 has the advantage that jumps in the converging functions must be inherited by the limit function.

Corollary 12.11.2. (inheritance of jumps) *If $x_n \to x$ in (D, WM_2), $t_n \to t$ in $[0, T]$ and $x_n^i(t_n) - x_n^i(t_n-) \geq c > 0$ for all n, then $x^i(t) - x^i(t-) \geq c$.*

Proof. Apply Theorem 12.11.2 (iv). ∎

Let $J(x)$ be the maximum magnitude (absolute value) of the jumps of the function x in D. We apply Corollary 12.11.2 to show that J is upper semicontinuous.

Corollary 12.11.3. (upper semicontinuity of J) *If $x_n \to x$ in (D, M_2), then*

$$\overline{\lim_{n \to \infty}} \, J(x_n) \leq J(x) \ .$$

Proof. Suppose that $x_n \to x$ in (D, WM_2) and there exists a subsequence $\{x_{n_k}\}$ such that $J(x_{n_k}) \to c$. Then there exist further subsubsequences $\{x_{n_{k_j}}\}$ and $\{t_{n_{k_j}}\}$, and a coordinate i, such that $t_{n_{k_j}} \to t$ for some $t \in [0, T]$ and $|x_{n_{k_j}}^i(t_{n_{k_j}}) - x_{n_{k_j}}^i(t_{n_{k_j}}-)| \to c$. Then Corollary 12.11.2 implies that $|x^i(t) - x^i(t-)| \geq c$. ∎

12.11.4. Additional Properties of M_2 Convergence

We conclude this section by discussing additional properties of the M_2 topologies. First we note that there are direct M_2 analogs of the M_1 results in Theorems 12.6.1, 12.7.1, 12.7.2 and 12.7.3.

Theorem 12.11.3. (extending SM_2 convergence to product spaces) *Suppose that $m_s(x_n, x) \to 0$ in $D([0, T], \mathbb{R}^k)$ and $m_s(y_n, y) \to 0$ in $D([0, T], \mathbb{R}^l)$ as $n \to \infty$. If*

$$Disc(x) \cap Disc(y) = \phi \, ,$$

then

$$m_s((x_n, y_n), (x, y)) \to 0 \quad in \quad D([0, T], \mathbb{R}^{k+l}) \quad as \quad n \to \infty \, .$$

12.11. Alternative Characterizations for M_2

Corollary 12.11.4. (from WM_2 convergence to SM_2 convergence when the limit is in D_1) *If $m_p(x_n, x) \to 0$ as $n \to \infty$ and $x \in D_1$, then $m_s(x_n, x) \to 0$ as $n \to \infty$.*

Theorem 12.11.4. (Lipschitz property of linear functions of the coordinate functions) *For any $x_1, x_2 \in D([0,T], \mathbb{R}^k)$ and $\eta \in \mathbb{R}^k$,*

$$m(\eta x_1, \eta x_2) \leq (\|\eta\| \vee 1) m_s(x_1, x_2) \; .$$

We have an analog of Corollary 12.7.1 for the M_2 topology.

Corollary 12.11.5. (SM_2-continuity of addition) *If $m_s(x_n, x) \to 0$ and $m_s(y_n, y) \to 0$ in $D([0,T], \mathbb{R}^k)$ and*

$$Disc(x) \cap Disc(y) = \phi \, ,$$

then

$$m_s(x_n + y_n, x + y) \to 0 \quad in \quad D([0,T], \mathbb{R}^k) \, .$$

Theorem 12.11.5. (characterization of SM_2 convergence by convergence of all linear functions of the coordinates) *There is convergence $x_n \to x$ in $D([0,T], \mathbb{R}^k)$ as $n \to \infty$ in the SM_2 topology if and only if $\eta x_n \to \eta x$ in $D([0,T], \mathbb{R}^1)$ as $n \to \infty$ in the M_2 topology for all $\eta \in \mathbb{R}^k$.*

Just as with the M_1 topology, we can get convergence of sums under more general conditions than in Corollary 12.11.5. It suffices to have the jumps of x^i and y^i have common sign for all i. We can express this property by the condition (7.2).

Theorem 12.11.6. (continuity of addition at limits with jumps of common sign) *If $x_n \to x$ and $y_n \to y$ in $D([0,T], \mathbb{R}^k, SM_2)$ and if condition (7.2) holds, then*

$$x_n + y_n \to x + y \quad in \quad D([0,T], \mathbb{R}^k, SM_2) \, .$$

We now apply Theorem 12.11.5 to extend a characterization of convergence due to Skorohod (1956) to $\mathbb{R}^k$-valued functions. For each $x \in D([0,T], \mathbb{R}^1)$ and $0 \leq t_1 < t_2 \leq T$, let

$$M_{t_1, t_2}(x) \equiv \sup_{t_1 \leq t \leq t_2} x(t) \; . \tag{11.9}$$

The proof exploits the SM_2 analog of Corollary 12.9.1.

Theorem 12.11.7. (characterization of SM_2 convergence in terms of convergence of local extrema) *There is convergence $m_s(x_n, x) \to 0$ as $n \to \infty$ in $D([0,T], \mathbb{R}^k)$ if and only if*

$$M_{t_1, t_2}(\eta x_n) \to M_{t_1, t_2}(\eta x) \quad as \quad n \to \infty$$

for all $\eta \in \mathbb{R}^k$ and all points $t_1, t_2 \in \{T\} \cup Disc(x)^c$ with $t_1 < t_2$.

We can apply the characterization of M_2 convergence in Theorem 12.11.7 to show the preservation of convergence under bounding functions in the M_2 topology.

Corollary 12.11.6. (*preservation of WM_2 convergence within bounding functions*) *Suppose that*
$$y_n^i(t) \le x_n^i(t) \le z_n^i(t)$$
for all $t \in [0,T]$, $1 \le i \le k$, and all n. If $m_p(y_n, x) \to 0$ and $m_p(z_n, x) \to 0$ as $n \to \infty$, then $m_p(x_n, x) \to 0$ as $n \to \infty$.

Example 12.11.3. *Failure with other topologies.* To see that there is no analog of Corollary 12.11.6 for the M_1 and J_1 topologies, for $n \ge 1$, let $x = I_{[1,2]}$, $y_n = I_{[1+n^{-1},2]}$, $z_n = I_{[1-n^{-1},2]}$,
$$x_n(0) = x_n(1 - n^{-1}) = x_n(1 - (3n)^{-1}) = x_n(1 - (5n)^{-1}) = 0$$
and
$$x_n(1 - (2n)^{-1}) = x_n(1 - (4n)^{-1}) = x_n(1) = x_n(2) = 1 ,$$
with x_n defined by linear interpolation elsewhere. Then $y_n(t) \le x_n(t) \le z_n(t)$ for all t and n, $y_n \to x$ and $z_n \to x$ as $n \to \infty$ in $D([0,2], \mathbb{R})$ with the J_1 topology, while $x_n \to x$ with the M_2 topology, but not with the M_1, J_2 and J_1 topologies.

12.12. Compactness

We now characterize compact subsets in $D \equiv D([0,T], \mathbb{R}^k)$ in the M topologies, closely following Section 2.7 of Skorohod (1956). To do so, we define new oscillation functions that include more control of the behavior of the functions at the interval endpoints 0 and T. First let
$$\bar{w}_w^*(x, \delta) \equiv \max_{1 \le i \le k} \bar{w}_s^*(x^i, \delta) \tag{12.1}$$
for $\bar{w}_s^*$ in (11.7). Given $w_s(x, \delta)$ in (5.1), $w_w(x, \delta)$ in (5.12), $\bar{w}_s^*(x, \delta)$ in (11.7), $\bar{w}_w^*(x, \delta)$ in (12.1) and $\bar{v}(x, t, \delta)$ in (4.3), let
$$w_s'(x, \delta) \equiv w_s(x, \delta) \vee \bar{v}(x, 0, \delta) \vee \bar{v}(x, T, \delta) , \tag{12.2}$$
$$w_w'(x, \delta) \equiv w_w(x, \delta) \vee \bar{v}(x, 0, \delta) \vee \bar{v}(x, T, \delta) , \tag{12.3}$$
$$\bar{w}_s'(x, \delta) \equiv \bar{w}_s^*(x, \delta) \vee \bar{v}(x, 0, \delta) \vee \bar{v}(x, T, \delta) , \tag{12.4}$$
$$\bar{w}_w'(x, \delta) \equiv \bar{w}_w^*(x, \delta) \vee \bar{v}(x, 0, \delta) \vee \bar{v}(x, T, \delta) . \tag{12.5}$$

Since
$$\bar{w}_w^*(x, \delta) \le \bar{w}_s^*(x, \delta) \quad \text{and} \quad \bar{w}_w^*(x, \delta) \le w_w(x, \delta) \le w_s(x, \delta)$$
for all $x \in D$ and $\delta > 0$,
$$\bar{w}_w'(x, \delta) \le \bar{w}_s'(x, \delta) \quad \text{and} \quad \bar{w}_w'(x, \delta) \le w_w'(x, \delta) \le w_s'(x, \delta)$$
for all $x \in D$ and $\delta > 0$.

We start by stating a characterization of WM_2 convergence. The proof draws on Theorem 12.11.1.

Theorem 12.12.1. (another characterization of WM_2 convergence) *If $\{x_n\}$ is a sequence in D such that $x_n(t)$ converges as $n \to \infty$ for all t in a dense subset of $[0,T]$ including 0 and T and*

$$\lim_{\delta \downarrow 0} \varlimsup_{n \to \infty} \bar{w}'_w(x_n, \delta) = 0 \tag{12.6}$$

for $\bar{w}'$ in (12.5), then there exists $x \in D$ such that $m_p(x_n, x) \to 0$.

Example 12.12.1. *Need for the $\bar{v}$ terms.* To see the need for the terms $\bar{v}(x, 0, \delta)$ and $\bar{v}(x, T, \delta)$ in $\bar{w}'_w(x, \delta)$, let $x_n(0) = 1$, $x_n(n^{-1}) = x_n(1) = 0$ with x_n defined by linear interpolation elsewhere on $[0, 1]$. Then $\bar{w}^*_s(x_n, \delta) = 0$ for all n and δ, but $\{x_n : n \geq 1\}$ does not converge and is not compact in $D([0,1], \mathbb{R}, M_2)$. Since $\sup_n \bar{v}(x_n, 0, \delta) = 1$ for all $\delta > 0$, (12.6) fails.

Corollary 12.12.1. (new characterizations of convergence in other topologies) *If the conditions of Theorem 12.12.1 hold with $\bar{w}'_w$ in (12.5) replaced by $\bar{w}'_s$ in (12.4), w'_w in (12.3) or w'_s in (12.2), then the convergence can be strengthened to SM_2, WM_1 or SM_1, respectively.*

Theorem 12.12.2. (characterizations of compactness) *A subset A of D has compact closure in the SM_1, WM_1, SM_2 or WM_2 topology if*

$$\sup_{x \in A}\{\|x\|\} < \infty \tag{12.7}$$

and

$$\lim_{\delta \downarrow 0} \sup_{x \in A}\{w'(x, \delta)\} < \infty , \tag{12.8}$$

where w' is w'_s in (12.2) for SM_1, w'_w in (12.3) for WM_1, $\bar{w}'_s$ in (12.4) for SM_2 and $\bar{w}'_w$ in (12.5) for SM_2. The conditions are necessary for SM_1 and WM_1.

Example 12.12.2. *The conditions are not necessary for M_2.* To see that the conditions in Theorem 12.12.2 are not necessary for the M_2 topologies, for $s \in [1/4, 1/2]$, let

$$x_s = I_{[s, 1/4+s/2)} + I_{[1/2, 1]}$$

in $D([0, 1], \mathbb{R})$. The set $\{x_s : 1/4 \leq s \leq 1/2\}$ is clearly M_2 compact, but

$$\sup_{1/4 \leq s \leq 1/2} \bar{w}_w(x_s, \delta) = 1$$

for all δ, $0 < \delta < 1/4$. ∎

Compactness characterizations on D translate into tightness characterizations for sets of probability measures on D. Recall from Chapter 11 that a set A of probability measures on a metric space (S, m) is said to be tight if for all $\epsilon > 0$ there exists a compact subset K of (S, m) such that

$$P(K) > 1 - \epsilon \quad \text{for all} \quad P \in A .$$

Theorem 12.12.3. (characterizations of tightness) *A sequence $\{P_n : n \geq 1\}$ of probability measures on D with the SM_1, WM_1, SM_2 or WM_2 topology is tight if the following two conditions hold:*

(i) For each $\epsilon > 0$, there exists c such that

$$P_n(\{x \in D : \|x\| > c\}) < \epsilon, \quad n \geq 1 \ .$$

(ii) For each $\epsilon > 0$ and $\eta > 0$, there exists $\delta > 0$ such that

$$P_n(\{x \in D : w'(x,\delta) \geq \eta\}) < \epsilon, \quad n \geq 1 \ ,$$

for w' being the appropriate oscillation function in (12.2)–(12.5). The conditions are also necessary for the SM_1 and WM_1 topologies.

Proof. Suppose that conditions (i) and (ii) hold, where w' is w'_s in (12.2) for SM_1 w'_w in (12.3) for WM_1, $\bar{w}'_s$ in (12.4) for SM_2 and $\bar{w}'_w$ in (12.5) for WM_2. For $\epsilon > 0$ given, choose c and δ_k such that $P_n(A_k^c) < \epsilon 2^{-(k+1)}$, $k \geq 0$, where

$$A_0 = \{x \in D : \|x\| \leq c\} \tag{12.9}$$

and

$$A_k = \{x \in D : w'(x,\delta_k) < k^{-1}\}, \quad k \geq 1 \ . \tag{12.10}$$

Then let $A = \cap_{k \geq 0} A_k$. By the construction,

$$P_n(A^c) = P_n\left(\cup_{k \geq 0} A_k^c\right) \leq \sum_{k=0}^{\infty} P_n(A_k^c) \leq \epsilon \ . \tag{12.11}$$

Since $A \subseteq A_0$ and

$$\limsup_{\delta \downarrow 0} {}_{x \in A} w'(x,\delta) = 0 \ , \tag{12.12}$$

the set A has compact closure by Theorem 12.12.1. Going the other way, assume that the topology is SM_1 or WM_1 and suppose that $\{P_n : n \geq 1\}$ is tight, so that for any $\epsilon > 0$ there exists a compact subset K of D such that $P_n(K) > 1 - \epsilon$. By Theorem 12.12.2, for any $\eta > 0$ given, $K \subseteq \{x : \|x\| \leq c\}$ for some c and $K \subseteq \{x : w'(x,\delta) \leq \eta\}$ for small enough δ; by the monotonicity of $w'(x,\delta)$ in δ for the SM_1 and WM_1 topologies. Hence conditions (i) and (ii) hold for all n. ∎

For an alternative characterization of M_1 tightness in $D([0,T], \mathbb{R})$, see Avram and Taqqu (1989).

13
Useful Functions

13.1. Introduction

In this chapter we consider several useful functions from D or $D \times D$ to D that can be exploited to establish new stochastic-process limits from given ones. We concentrate on four basic functions introduced in Section 3.5: composition, supremum, reflection and inverse. Another basic function is addition, but it has already been treated in Sections 12.6, 12.7 and 12.11. Our treatment of useful functions follows Whitt (1980), but the emphasis there was on the J_1 topology, even though the M_1 topology was used in places. In contrast, here the emphasis is on the M_1 and M_2 topologies, although we also give results for the J_1 topology. As in the last chapter, many proofs are omitted. Most of the missing proofs appear in Chapter 7 of the Internet Supplement.

Here is how this chapter is organized: We start in Section 13.2 by considering the composition map, which plays an important role in establishing FCLTs involving a random time change. We consider composition without centering in Section 13.2; then we consider composition with centering in Section 13.3.

In Section 13.4 we study the supremum function, both with and without centering. In Section 13.5 we apply the supremum results to treat the (one-sided one-dimensional) reflection map, which arises in queueing applications. We study the two-sided reflection map in Section 14.8.

We start studying the inverse function in Section 13.6. We study the inverse map without centering in Section 13.6 and with centering in Section 13.7. In Section 13.8 we apply the results for inverse functions to obtain corresponding results for closely related counting functions.

Application of these convergence-preservation results to stochastic-process limits are described in Sections 7.3 and 7.4. Section 7.3 contains FCLT's for counting processes, while Section 7.4 contains FCLT's for renewal-reward processes. When there are heavy-tailed distributions, the M_1 topology plays an important role.

In Chapter 3 of the Internet Supplement we discuss pointwise convergence and its preservation under mappings. The perservation of pointwise convergence focuses on relations for individual sample paths, as in the queueing book by El-Taha and Stidham (1999). From Chapter 3 of the Internet Supplement, we see that a function-space setting is not required for all convergence preservation.

13.2. Composition

This section is devoted to the composition function, mapping (x, y) into $x \circ y$, where

$$(x \circ y)(t) \equiv x(y(t)) \quad \text{for all} \quad t.$$

We have in mind a map from $D^k \times D$ into D^k, where $D^k \equiv D([0,\infty), \mathbb{R}^k)$. The situation is much easier when we consider single times and the map is from $D^k \times \mathbb{R}_+$ to $\mathbb{R}^k$. We can still take advantage of the Skorohod topology on D, though. The following is an elementary, but important, consequence of the local uniform convergence established in Section 12.4.

Proposition 13.2.1. (local uniform convergence) *If*

$$(x_n, t_n) \to (x, t) \quad in \quad (D^k, WM_2) \times \mathbb{R}_+ ,$$

where $t \in Disc(x)^c$, then

$$x_n(t_n) \to x(t) \quad in \quad \mathbb{R}^k .$$

We now consider the composition map as a map from $D^k \times D$ to D, where we allow the domains of x and y to be $\mathbb{R}_+ \equiv [0, \infty)$ and we restrict the range of y to be $\mathbb{R}_+$. However, that is not enough; we need additional regularity conditions to have $x \circ y \in D$.

Example 13.2.1. *The need for a condition on y.* To see that $x \circ y$ need not be in D without additional conditions on y, let $x = I_{[2^{-1},\infty)}$ and $y = 2^{-1} + \sum_{n=1}^{\infty}(-2)^{-n}I_{[2^{-1}-2^{-n},2^{-1}-2^{-(n+1)})}$. Then $x, y \in D$, but $x \circ y$ has no limit from the left at $t = 1/2$. ∎

Henceforth in this chapter, unless stipulated otherwise, when $D \equiv D^k$, so that the range of functions is $\mathbb{R}^k$, we let D be endowed with the strong version of the J_1, M_1 or M_2 topology, and simply write J_1, M_1 or M_2. It will be evident that most results also hold with the corresponding weaker product topology.

To ensure that $x \circ y \in D$, we will assume that y is also nondecreasing. We begin by defining subsets of $D \equiv D^k \equiv D([0,\infty), \mathbb{R}^k)$ that we will consider. Let D_0 be the subset of all $x \in D$ with $x^i(0) \geq 0$ for all i. Let $D_\uparrow$ and $D_{\uparrow\uparrow}$ be the subsets of functions in D_0 that are nondecreasing and strictly increasing in each coordinate. Let D_m be the subset of functions x in D_0 for which the coordinate functions x^i are monotone (either increasing or decreasing) for each i. Let C_0, $C_\uparrow$, $C_{\uparrow\uparrow}$ and C_m be the corresponding subsets of C; i.e., $C_0 \equiv C \cap D_0$, $C_\uparrow \equiv C \cap D_\uparrow$, $C_{\uparrow\uparrow} = C \cap D_{\uparrow\uparrow}$, and $C_m = C \cap D_m$.

It is important that all of these subsets are measurable subsets of D with the Borel σ-fields associated with the non-uniform Skorohod topologies, which all coincide with the Kolmogorov σ-field generated by the projection maps; see Theorems 11.5.2 and 11.5.3.

Lemma 13.2.1. (Measurability of C in D) *C is a closed subset of (D, J_1) and so a measurable (but not closed) subset of D with the M_1 and M_2 topologies.*

Recall that a subset of a topological space is a G_δ subset if it is a countably intersection of open subsets. Clearly, a G_δ subset belongs to the Borel σ-field.

Lemma 13.2.2. (measurability of subsets of C) *C_m is a closed subset of C, $C_\uparrow$ is a closed subset of C_m and $C_{\uparrow\uparrow}$ is a G_δ subset of $C_\uparrow$.*

Proof. For the third relation, note that

$$C_{\uparrow\uparrow} = \cap_{p \in Q} \cap_{\substack{q \in Q \\ q > p}} \cap_{i=1}^k \{x \in C : x^i(q) - x^i(p) > 0\}$$

where Q is the set of rationals in $\mathbb{R}_+$. ∎

Lemma 13.2.3. (measurability of subsets of D) *With any of the non-uniform Skorohod topologies, D_0 is a closed subset of D, D_m is a closed subset of D_0, $D_\uparrow$ is a closed subset of D_m and $D_{\uparrow\uparrow}$ is a G_δ subset of $D_\uparrow$.*

Proof. For the last relation, let $\{t_j\}$ be a countable dense subset of $\mathbb{R}_+$. For each (j, l), let

$$D_{i,j,l} = \{x \in D_\uparrow : x^i \text{ is constant over } [t_j \wedge t_l, t_j \vee t_l]\} .$$

Then $D_{i,j,l}$ is a closed subset of $D_\uparrow$ and

$$D_{\uparrow\uparrow} = \cap_{j=1}^\infty \cap_{l=1}^\infty \cap_{i=1}^k (D_\uparrow - D_{i,j,l}) ,$$

so that $D_{\uparrow\uparrow}$ is indeed a G_δ subset of $D_\uparrow$. ∎

We now return to the composition map in (12.2), stating the condition for $x \circ y \in D$ as a lemma.

Lemma 13.2.4. (criterion for $x \circ y$ to be in D) *For each $x \in D([0, \infty), \mathbb{R}^k)$ and $y \in D_\uparrow([0, \infty), \mathbb{R}_+)$, $x \circ y \in D([0, \infty), \mathbb{R}^k)$.*

A basic result, from pp. 145, 232 of Billingsley (1968), is the following. The continuity part involves the topology of uniform convergence on compact intervals.

Theorem 13.2.1. (continuity of composition at continuous limits) *The composition map from $D^k \times D_\uparrow^1$ to D^k is measurable and continuous at $(x, y) \in C^k \times C_\uparrow^1$.*

Example 13.2.2. *Composition is not continuous everywhere.* To see that the composition on $D^1 \times D^1_\uparrow$ is not continuous in any of the Skorohod topologies, let $x_n = x = I_{[1/2,1]}$, $n \geq 1$, $y(t) = 2^{-1}$ and $y_n(t) = 2^{-1} - n^{-1}$, $0 \leq t \leq 1$. Then $x_n = x$ and $\|y_n - y\| = n^{-1} \to 0$, but $(x_n \circ y_n)(t) = 0$ and $(x \circ y)(t) = 1$, $0 \leq t \leq 1$. ∎

Our goal now is to obtain additional positive continuity results under extra conditions. We use the following elementary lemma.

Lemma 13.2.5. *If $y(t) \in Disc(x)$ and y is strictly increasing and continuous at t, then $t \in Disc(x \circ y)$.*

Example 13.2.3. *The need for y to be strictly increasing.* To see the need for the condition that y be strictly increasing at t in Lemma 13.2.5, let $x = I_{[1,\infty)}$ and $y(t) = 1$, $t \geq 0$. Then $(x \circ y)(t) = 1$ for all t, so that $x \circ y$ is continuous. Moreover, if $x_n = x$ and $y_n(t) = 1 - n^{-1}$, $t \geq 0$, $n \geq 1$, then $(x_n \circ y_n)(t) = 0$ for all n and t, so that $x_n \circ y_n$ fails to converge to $x \circ y$ for any t. ∎

The following is the J_1 result, taken from Whitt (1980). As indicated before, the proof appears in the Internet Supplement. The first J_1 composition results were established by Silvestrov; see Silvestrov (2000) for an account. See Serfozo (1973, 1975) and Gut (1988) for stochastic-process limits involving composition.

Theorem 13.2.2. (J_1-continuity of composition) *The composition map from $D^k \times D^1_\uparrow$ to D^k taking (x,y) into $(x \circ y)$ is continuous at $(x,y) \in (C^k \times D^1_\uparrow) \cup (D^k \times C^1_{\uparrow\uparrow})$ using the J_1 topology throughout.*

We have a different result for the M topologies:

Theorem 13.2.3. (M-continuity of composition) *If $(x_n, y_n) \to (x,y)$ in $D^k \times D^1_\uparrow$ and $(x,y) \in (D^k \times C^1_{\uparrow\uparrow}) \cup (C^k_m \times D^1_\uparrow)$, then $x_n \circ y_n \to x \circ y$ in D^k, where the topology throughout is M_1 or M_2.*

In most applications we have $(x,y) \in D^k \times C^1_{\uparrow\uparrow}$, as is illustrated by the next section. That part of the M conditions is the same as for J_1. The mode of convergence in Theorem 13.2.3 for $y_n \to y$ does not matter, because on $D^1_\uparrow$, convergence in the M_1 and M_2 topologies coincides with pointwise convergence on a dense subset of $[0,\infty)$, including 0; see Corollary 12.5.1.

It is easy to see that composition cannot in general yield convergence in a stronger topology, because $x \circ y = x$ and $x_n \circ y_n = x_n$, $n \geq 1$, when $y_n = y = e$, where $e(t) = t$, $t \geq 0$. Unlike for the J_1 topology, the composition map is in general *not* continuous at $(x,y) \in C \times D^1_\uparrow$ in the M topologies.

Example 13.2.4. *Why the J_1 and M conditions differ.* To see that composition is not continuous at $(x,y) \in C \times D^1_\uparrow$ in the M topologies, let y, y_n, $x = x_n$ be elements of $D([0,\infty), \mathbb{R})$ defined by

$$y(0) = y(.5-) = 0, y(.5) = .25, y(1) = 1, y(t) = t, t > 1,$$
$$y_n(0) = y_n(.5 - n^{-1}) = 0, y_n(.5) = .25, y_n(1) = 1, y_n(t) = t, t > 1,$$

$$x(0) = x(.25) = x(t) = 0 \quad \text{for} \quad t > 0.25, x(.125) = 1,$$

with the functions defined by linear interpolation elsewhere. Note that y jumps from 0 to 0.25 at 0.5, while y_n increases from 0 to 0.25 linearly over the interval $[2^{-1} - n^{-1}, 2^{-1}]$ for each n. Hence $y_n \to y$ in the M topologies but not in the J topologies. Note that $x(y(t)) = 0, t \geq 0$, while $x_n(y_n(2^{-1}-(2n)^{-1})) = x_n(.125) = 1$. Hence $x_n \circ y_n \not\to x \circ y$ as $n \to \infty$ in any of the Skorohod topologies. ∎

We actually prove a more general continuity result, which covers Theorem 13.2.3 as a special case.

Theorem 13.2.4. (more general M-continuity of composition) *Suppose that $(x_n, y_n) \to (x, y)$ in $D^k \times D^1_\uparrow$. If (i) y is continuous and strictly increasing at t whenever $y(t) \in Disc(x)$ and (ii) x is monotone on $[y(t-), y(t)]$ and $y(t-), y(t) \notin Disc(x)$ whenever $t \in Disc(y)$, then $x_n \circ y_n \to x \circ y$ in D^k, where the topology throughout is M_1 or M_2.*

Theorem 13.2.3 follows easily from Theorem 13.2.4: First, on $D^k \times C^1_\uparrow$, y is continuous, so only condition (i) need be considered; it is satisfied because y is continuous and strictly increasing everywhere. Second on $C^k_m \times D^1_\uparrow$, x is continuous so only condition (ii) need be considered; it is satisfied because x is monotone everywhere. Hence it suffices to prove Theorem 13.2.4, which is done in the Internet Supplement. The general idea in our proof of Theorem 13.2.4 is to work with the characterization of convergence using oscillation functions evaluated at single arguments, exploiting Theorems 12.5.1 (v), 12.5.2 (iv), 12.11.1 (v) and 12.11.2 (iv).

13.3. Composition with Centering

We now consider the composition map with centering. To obtain results, we apply both composition and addition. The results yield sufficient conditions for random sums and other processes transformed by a random time change to satisfy FCLTs, as we show in Section 7.4.

We start by establishing convergence properties of composition plus addition. We state results for the J_1 topology as well as the M_1 and M_2 topologies. As before, let e be the identity map on $[0, \infty)$.

Theorem 13.3.1. (convergence preservation for composition plus addition) *Let x, z and $x_n, n \geq 1$ be elements of D^k; let y, y_n and $v_n, n \geq 1$ be elements of $D^1_\uparrow$; and let $c_n \in \mathbb{R}^k$ for $n \geq 1$. If*

$$(x_n - c_n e, y_n, c_n(y_n - v_n)) \to (x, y, z) \quad \text{in} \quad D^k \times D^1_\uparrow \times D^k, \quad (3.1)$$

$y \in C^1_{\uparrow\uparrow}$ *and*

$$Disc(x \circ y) \cap Disc(z) = \phi, \quad (3.2)$$

then

$$x_n \circ y_n - c_n v_n \to x \circ y + z \quad \text{in} \quad D^k, \quad (3.3)$$

where the topology throughout is J_1, M_1 or M_2. If the topology is M_1 or M_2, then instead of (3.2) it suffices for $x^i \circ y$ and z^i to have no common discontinuities with jumps of the opposite sign for $1 \leq i \leq k$.

Proof. Note that
$$x_n \circ y_n - c_n v_n = (x_n - c_n e) \circ y_n + c_n(y_n - v_n) \,.$$

For the M topologies, apply Theorem 13.2.3 for composition, using the condition $y \in C^1_{\uparrow\uparrow}$, and Corollaries 12.7.1 and 12.11.5 for addition with the M_1 and M_2 topologies, respectively. The J_1 result is proved similarly, using Theorem 13.2.2 instead of Theorem 13.2.3. For addition with J_1, use Remark 12.6.2. Use Theorems 12.7.3 and 12.11.6 for the weaker condition for addition to be continuous with the M topologies. ∎

The standard application of Theorem 13.3.1 has $c_n^i \to \infty$ as $n \to \infty$ for each i and $v_n = b_n e$, where $b_n \to b$. We describe that case below.

Corollary 13.3.1. (convergence preservation for composition with linear centering) Let x, z and x_n, $n \geq 1$, be elements of D^k; let y_n, $n \geq 1$, be elements of $D^1_{\uparrow}$; let $c_n \in R^k$ and $b_n \in \mathbb{R}^1$ satisfy $|c_n^i| \to \infty$ for each i and $b_n \to b$ as $n \to \infty$. If
$$(x_n - c_n e, c_n(y_n - b_n e)) \to (x, z) \quad in \quad D^k \times D^k \tag{3.4}$$

and
$$Disc(x \circ be) \cap Disc(z) = \phi \,, \tag{3.5}$$

then
$$(x_n \circ y_n - c_n b_n e) \to x \circ y + z \quad in \quad D^k \,, \tag{3.6}$$

where the topology throughout is J_1, M_1 or M_2. If the topology is M_1 or M_2, then instead of condition (3.5) it suffices for $x^i \circ be$ and z^i to have no common discontinuities with jumps of opposite sign, $1 \leq i \leq k$. ∎

Proof. Since $|c_n^i| \to \infty$ as $n \to \infty$ for each i, the limit in (3.4) implies that $\|y_n - b_n e\| \to 0$ as $n \to \infty$. Hence $\|y_n - be\| \to 0$ as $n \to \infty$ and
$$(x_n - c_n e, y_n, c_n(y_n - be)) \to (x, y, z) \quad in \quad D^k \times D^1_{\uparrow} \times D^k \,,$$

where $y = be$. Hence we can apply Theorem 13.3.1 to obtain the desired conclusion. ∎

We now consider an application of the convergence-preservation results above to obtain a FCLT involving a random time change. Specifically, we consider an application of Corollary 13.3.1. Let $(X_n(t), Y_n(t)) : t \geq 0\}$ be random elements of $D^k \times D^1_{\uparrow}$ for each $n \geq 1$, with one of the topologies under consideration. Let $\mathbf{X}_n$, $\mathbf{Y}_n$ and $\mathbf{Z}_n$ be normalized processes constructed by

$$\begin{aligned}
\mathbf{X}_n(t) &\equiv \delta_n^{-1}[X_n(nt) - \mu_n nt], \quad t \geq 0 \\
\mathbf{Y}_n(t) &\equiv \delta_n^{-1}[Y_n(nt) - \lambda_n nt], \quad t \geq 0 \\
\mathbf{Z}_n(t) &\equiv \delta_n^{-1}[(X_n(Y_n(nt)) - \lambda_n \mu_n nt], \quad t \geq 0 \,.
\end{aligned} \tag{3.7}$$

Corollary 13.3.2. (stochastic consequence with linear centering) *Suppose that (X_n, Y_n) is a random element of $D^k \times D^1_\uparrow$ for each n. If*

$$(\mathbf{X}_n, \mathbf{Y}_n) \Rightarrow (\mathbf{U}, \mathbf{V}) \quad in \quad D^k \times D^1 \tag{3.8}$$

with topology J_1, M_1 or M_2, for the scaled processes $\mathbf{X}_n$, $\mathbf{Y}_n$ in (3.7) with $\delta_n \to \infty$, $n\delta_n^{-1} \to \infty$, $\mu_n \to \mu$ with $\mu^i \neq 0$ for all i and $\lambda_n \to \lambda$, and if

$$P(Disc(\mathbf{U} \circ \lambda \mathbf{c}) \cap Disc(\mathbf{V}) = \phi) = 1 , \tag{3.9}$$

then

$$\mathbf{Z}_n \Rightarrow \mathbf{U} \circ \lambda \mathbf{e} + \mu \mathbf{V} \quad in \quad D^k \tag{3.10}$$

for $\mathbf{Z}_n$ in (3.7) and the same topology. If the topology is M_1 or M_2, then instead of condition (3.9) it suffices for $\mathbf{U}^i \circ \lambda \mathbf{e}$ and $\mathbf{V}^i$ to almost surely have no common discontinuities with jumps of opposite sign, $1 \leq i \leq k$.

Proof. First, since $\mu_n \to \mu$ as $n \to \infty$ in $\mathbb{R}^k$, from condition (3.8) we obtain

$$(\mathbf{X}_n, \mu_n \mathbf{Y}_n) \Rightarrow (\mathbf{U}, \mu \mathbf{V}) \quad in \quad D^k \times D^k \tag{3.11}$$

from the continuous mapping theorem. Now apply Corollary 13.3.1 with $c_n = n\delta_n^{-1}\mu_n$, $b_n = \lambda_n$,

$$x_n(t) = \delta_n^{-1} X_n(nt) \quad \text{and} \quad y_n(t) = n^{-1} Y_n(nt) .$$

By the Skorohod (1956) representation theorem, there exist versions of the processes such that almost surely

$$(x_n - c_n e, c_n(y_n - b_n e)) \to (x, z) \quad \text{as} \quad n \to \infty$$

where $x = \mathbf{U}$ and $z = \mu \mathbf{V}$. Corollary 13.3.1 then yields

$$(x_n \circ y_n - c_n b_n e) \to x \circ y + z \quad \text{as} \quad n \to \infty \tag{3.12}$$

almost surely in D^k, where $y = \lambda \mathbf{e}$, but the limit process in (3.12) is distributed the same as the limit process in (3.10). The almost sure convergence in (3.12) implies the convergence in distribution in (3.10). ∎

A standard application of Corollary 13.3.2 is to random sums. Then, for each $n \geq 1$, $\{X_n(nt) : t \geq 0\}$ corresponds to a sequence of partial sums; i.e.,

$$X_n(nt) = \sum_{j=1}^{\lfloor nt \rfloor} Z_{n,j}, \quad t \geq 0 ,$$

where $\lfloor x \rfloor$ is the greatest integer less than or equal to x and $\{Z_{n,j} : j \geq 1\}$ is a sequence of random vectors in $\mathbb{R}^k$ for each n. The composition then yields a random sum, i.e.,

$$(x_n \circ y_n)(t) = \delta_n^{-1} X_n(Y_n(nt)) = \delta_n^{-1} \sum_{j=1}^{Y_n(nt)} Z_{n,j} ,$$

so that the limit (3.10) becomes for a random sum. We consider the special case in which the summands $Z_{n,j}$ come from a single IID sequence and the random index $Y_n(t)$ is a renewal process in Section 7.4.

Another application of Corollary 13.3.2 is to establish stochastic-process limits that imply asymptotic validity of sequential stopping rules in stochastic simulations. The asymptotic validity occurs in the limit as the desired volume of the target confidence set decreases. See Chapter 4 of the Internet Supplement.

We now establish a variant of Theorem 13.3.1 with nonlinear centering terms. In the proof we apply continuity of multiplication, which we now establish. By multiplication of x and y in D, we mean $(xy)(t) \equiv x(t)y(t)$ for all t.

For the composition results in Theorem 13.3.3 below, one of the functions is continuous, so we do not require the multiplication result in full generality. The general results is of independent interest. For the M topologies, the condition on the behavior at common discontinuities is more stringent for multiplication than for addition because of the way signs multiply.

Example 13.3.1. *The need for stronger conditions.* To see the need for stronger conditions with multiplication, let $x_n \equiv -1 + 2I_{[2^{-1}-n^{-1},\infty)}$ and let $y_n \equiv y \equiv -1 + 2I_{[2^{-1},\infty)}$ for $n \geq 2$. Then $x_n \to y$ in (D, J_1) as $n \to \infty$, but $x_n y_n = 1 - 2I_{[2^{-1}-n^{-1}, 2^{-1}]}$, which does not converge to $y^2 = 1$ in any of the Skorohod topologies. ∎

Theorem 13.3.2. (continuity of multiplication) *Suppose that $x_n \to x$ and $y_n \to y$ in $D([0,\infty), \mathbb{R})$ with one of the Skorohod topologies J_1, M_1 or M_2. If the topology is J_1, then assume that $Disc(x) \cap Disc(y) = \phi$. If the topology is M_1 or M_2, then assume for each $t \in Disc(x) \cap Disc(y)$ that $x(t), x(t-), y(t)$ and $y(t-)$ all have common sign and $[x(t) - x(t-)][y(t) - y(t-)] \geq 0$. Then $x_n y_n \to xy$ in $D([0,\infty), \mathbb{R})$ with the same topology, where $(xy)(t) \equiv x(t)y(t)$ for $t \geq 0$.*

Proof. For J_1, we can conclude that $(x_n, y_n) \to (x, y)$ in D^2 with the SJ_1 topology by the J_1 analog of Theorem 12.6.1; see Remark 12.6.2. It is then easy to show that $x_n y_n \to xy$. Use the fact that $x_n \to x$ implies that $\sup_n \{\|x_n\|\} < \infty$. For M_1, apply the characterization in Theorem 12.5.1 (v). For M_2, apply the characterization in Theorem 12.11.7. ∎

Theorem 13.3.3. (convergence preservation for composition with nonlinear centering) *Let $x, x_n \in D^k$, $y, y_n \in D^1_\uparrow$, $y \in C_{\uparrow\uparrow}$, x have a continuous derivative $\dot{x}$ and $c_n \to \infty$. If*

$$c_n(x_n - x, y_n - y) \to (u, v) \quad in \quad D^k \times D^1 \qquad (3.13)$$

with one of the topologies J_1, M_1 or M_2, where

$$Disc(u \circ y) \cap Disc(v) = \phi , \qquad (3.14)$$

then

$$c_n(x_n \circ y_n - x \circ y) \to u \circ y + (\dot{x} \circ y)v \quad in \quad D^k \qquad (3.15)$$

with the same topology, where

$$[(\dot{x} \circ y)v](t) \equiv [\dot{x}^1(y(t))v(t), \ldots, \dot{x}^k(y(t))v(t)]. \tag{3.16}$$

If the topology is M_1 or M_2, then instead of condition (3.14) it suffices to have $\dot{x}(t) \geq (\leq)0$ for all t and the functions $u \circ y$ and v to have no common discontinuities with jumps of opposite (common) sign.

Proof. Note that

$$c_n(x_n \circ y_n - x \circ y) = c_n(x_n - x) \circ y_n + c_n(x \circ y_n - x \circ y),$$

Given condition (3.13), we obtain

$$[c_n(x_n - x), c_n(y_n - y), y_n] \to [u, v, y] \quad \text{in} \quad D^k \times D^1 \times D^1$$

and then, applying composition, multiplication and addition,

$$[c_n(x_n \circ y_n - x \circ y_n) + (\dot{x} \circ y)c_n(y_n - y)] \to u \circ y + (\dot{x} \circ y)v$$

by virtue of Theorems 13.2.2, 13.2.3 and 13.3.2 and condition (3.14) (or the alternative M-topology condition). Note that

$$\|c_n(x_n \circ y_n - x \circ y) - c_n(x_n \circ y_n - x \circ y_n) - c_n(\dot{x} \circ y)(y_n - y)\|$$
$$\leq \|c_n(x \circ y_n - x \circ y) - c_n(\dot{x} \circ y)(y_n - y)\|. \tag{3.17}$$

However, the term on the right in (3.17) is asymptotically negligible because

$$c_n(x \circ y_n - x \circ y)(t) = c_n \int_{y(t)}^{y_n(t)} \dot{x}(s) ds$$

and

$$\sup_{0 \leq s \leq t} \left| c_n \int_{y(s)}^{y_n(s)} \dot{x}(u) du - \dot{x}(y(s)) c_n(y_n(s) - y(s)) \right| \to 0 \text{ as } n \to \infty,$$

because $\dot{x}$ is uniformly continuous over bounded intervals and $\|y_n - y\|_t \to 0$ as a consequence of $d(c_n(y_n - y), v) \to 0$. ∎

13.4. Supremum

In this section we consider the supremum function, mapping $D \equiv D([0,T], \mathbb{R})$ into itself according to

$$x^\uparrow(t) = \sup_{0 \leq s \leq t} x(s), \quad 0 \leq t \leq T. \tag{4.1}$$

We are primarily interested in the supremum function because it is closely related to the reflection map, discussed in the next section. Another motivation is extreme-value theory; see Resnick (1987) and Embrechts et al. (1997).

We have already observed that the map from D to $\mathbb{R}$ taking x into $x^\uparrow(t)$ is continuous in the M_2 topology at all $t \in Disc(x)^c$; that is a consequence of Theorem

12.11.7. Now we consider the map from D to D taking x into the function $x^\uparrow$ in (4.1).

The supremum function can be thought of as the *nondecreasing majorant*: It is easy to see that
$$x^\uparrow = \inf\{y \in D : y \geq x, \ y \text{ nondecreasing}\},$$
where $y \geq x$ if $y(t) \geq x(t)$ for all t. If $x \in D_0$, then $x^\uparrow \in D_\uparrow$.

It is easy to see that the supremum function is Lipschitz in the uniform norm:

Lemma 13.4.1. (Lipschitz property of the supremum function with the uniform norm) *For any $x_1, x_2 \in D([0,T],\mathbb{R})$,*
$$\|x_1^\uparrow - x_2^\uparrow\| \leq \|x_1 - x_2\|.$$

As consequences of Lemma 13.4.1, we obtain corresponding Lipschitz properties with the J_1, M_1 and M_2 metrics d_{J_1}, d_s and m_s, here denoted by d_{J_1}, d_{M_1} and d_{M_2}. For the M_1 topology, we use the following result.

Lemma 13.4.2. (inheritance of parametric representations) *For any $x \in D$, if $(u, r) \in \Pi(x)$ ($\Pi_{s,2}(x)$), then $(u^\uparrow, r) \in \Pi(x^\uparrow)$ ($\Pi_{s,2}(x)$).*

Theorem 13.4.1. (Lipschitz property of the supremum function) *For any $x_1, x_2 \in D([0,T],\mathbb{R})$,*
$$\begin{aligned} d_{J_1}(x_1^\uparrow, x_2^\uparrow) &\leq d_{J_1}(x_1, x_2), \\ d_{M_1}(x_1^\uparrow, x_2^\uparrow) &\leq d_{M_1}(x_1, x_2), \\ d_{M_2}(x_1^\uparrow, x_2^\uparrow) &\leq d_{M_2}(x_1, x_2). \end{aligned}$$

Example 13.4.1. *Convergence preservation fails with pointwise convergence.* It is significant that analogs of Lemma 13.4.1 and Theorem 13.4.1 do not hold for pointwise convergence: Let $x_n = I_{[n^{-1}, 2n^{-1}]}$. Then $x_n(t) \to 0$ as $n \to \infty$ for all t, while $x_n^\uparrow(t) \to 1$ as $n \to \infty$ for all $t > 0$. ∎

On the other hand, there is a pointwise-convergence analog of Theorem 13.4.1 for a single function; see Section 3.3 of the Internet Supplement.

Moreover, the conclusion in Theorem 13.4.1 can be recast in terms of pointwise convergence: Since $x^\uparrow$ is nondecreasing, convergence $x_n^\uparrow \to x^\uparrow$ in the M topologies is equivalent to pointwise convergence at continuity points of $x^\uparrow$, because on $D_\uparrow$ the M_1 and M_2 topologies coincide with pointwise convergence on a dense subset of $\mathbb{R}_+$ including 0 and T; see Corollary 12.5.1. Thus the M topologies have not contributed much so far. We obtain more useful convergence-preservation results for the supremum map with the M topologies when we combine supremum with centering. As before, let e be the identity map, i.e., $e(t) = t$, $0 \leq t \leq T$. The proof is in the Internet Supplement.

Theorem 13.4.2. (convergence preservation with the supremum function and centering) *Suppose that $c_n(x_n - e) \to y$ as $n \to \infty$ in $D([0,T],\mathbb{R})$ with one of the topologies J_1, M_1 or M_2, where $c_n \to \infty$.*

(a) If the topology is M_1 or M_2, then $c_n(x_n^\uparrow - e) \to y$ in the same topology.

(b) If the topology is J_1, then $c_n(x_n^\uparrow - e) \to y$ if and only if y has no negative jumps.

Example 13.4.2. *Pointwise convergence is not enough.* To see that a pointwise convergent analog of Theorem 13.4.2 does not hold, let $x_n = c_n^{-1} I_{[n^{-1}, 2n^{-1}]} + e$ where $c_n \to \infty$. Then $c_n(x_n - e)(t) = I_{[n^{-1}, 2n^{-1}]}(t) \to 0$ as $n \to \infty$ for all $t > 0$, while $x_n^\uparrow(t) = c_n^{-1} + t$ and $c_n(x_n^\uparrow - e)(t) = 1$ for all n sufficiently large, for $t > 0$. ∎

A common case covered by Theorem 13.4.2 is $y \in C$. If $y \in C$, then all modes of convergence in Theorem 13.4.2 reduce to uniform convergence and we have $c_n(x_n^\uparrow - e) \to y$ whenever $c_n(x_n - e) \to y$. Since $c_n \to \infty$, under the conditions of Theorem 13.4.2, $\|x_n - e\| \to 0$ as $n \to \infty$. By Theorem 13.4.1, $\|x_n^\uparrow - e\| \to 0$ as well.

We use the following lemma in the proof of both Theorem 13.4.2 above and Theorem 13.4.3 below.

Lemma 13.4.3. *If $x \in D([0,T], \mathbb{R})$ and x has no negative jumps, then for any $\epsilon > 0$ there is a $\delta > 0$ such that*

$$v^-(x, \delta) \equiv \sup_{\substack{0 \vee (t-\delta) \leq t' \leq t \\ 0 \leq t \leq T}} \{x(t') - x(t)\} < \epsilon . \tag{4.2}$$

We can easily extend Theorem 13.4.2 to cover a case of nonlinear centering. Recall that $\Lambda \equiv \Lambda([0,T])$ is the set of increasing homeomorphisms of $[0,T]$. We use elements of Λ as the centering term.

Corollary 13.4.1. (convergence preservation with the supremum and nonlinear centering) *Suppose that $c_n(x_n - \lambda_n) \to y$ as in $D([0,T], \mathbb{R})$ with one of the topologies J_1, M_1 or M_2, where $\lambda_n \to \lambda$ with $\lambda, \lambda_n \in \Lambda([0,T])$ and $c_n \to \infty$.*

(a) If the topology is M_1 or M_2, then $c_n(x_n^\uparrow - \lambda_n) \to y$ in the same topology.

(b) If the topology is J_1, then $c_n(x_n^\uparrow - \lambda_n) \to y$ if and only if y has no negative jumps.

Proof. Given $c_n(x_n - \lambda_n) \to y$, we have $c_n(x_n \circ \lambda_n^{-1} - e) \to y \circ \lambda^{-1}$ by applying Theorems 13.2.2 and 13.2.3. Then Theorem 13.4.2 implies that $c_n(x_n^\uparrow \circ \lambda_n^{-1} - e) \to y \circ \lambda^{-1}$ with the limit holding J_1 if and only if $y \circ \lambda^{-1}$ has no negative jumps. Clearly, $y \circ \lambda^{-1}$ has no negative jumps if and only if y does. Finally, apply Theorems 13.2.2 and 13.2.3 again to get $c_n(x_n^\uparrow \circ \lambda_n^{-1} \circ \lambda_n - \lambda_n) \to y \circ \lambda^{-1} \circ \lambda$, which implies the conclusion because $\lambda_n^{-1} \circ \lambda_n = \lambda^{-1} \circ \lambda = e$. ∎

We now obtain joint convergence in the stronger topologies on $D([0,T], \mathbb{R}^2)$ under the condition that the limit function have no negative jumps.

Theorem 13.4.3. (criterion for joint convergence) *Suppose that $c_n(x_n - e) \to y$ as $n \to \infty$ in $D([0,T], \mathbb{R})$ with one of the J_1, M_1 or M_2 topologies, where $c_n \to \infty$. If, in addition, y has no negative jumps, then*

$$c_n(x_n - e, x_n^\uparrow - e) \to (y, y) \quad as \quad n \to \infty \tag{4.3}$$

in $D([0,T], \mathbb{R}^2)$ with the strong version of the same topology, i.e., with SJ_1, SM_1 or SM_2.

Since addition is continuous on D^2 with the strong topologies, we obtain the following corollary.

Corollary 13.4.2. *Under the conditions of Theorem 13.4.3,*
$$\|c_n(x_n^\uparrow - x_n)\| \to 0 \quad as \quad n \to \infty.$$

Example 13.4.3. *The problem with negative jumps.* To see that Corollary 13.4.2 does not hold and the simple direct argument with parametric representations in the proof of Theorem 13.4.3 does not work for Theorem 13.4.2 when there are negative jumps, let $y = -I_{[1/2,1]}$, $c_n = n$ and $c_n(x_n - e) = y$, i.e., $x_n = e + n^{-1}y$. First,
$$c_n(x_n^\uparrow - x_n)(1/2) = 1 \quad \text{for all} \quad n \ge 1.$$

We now show what goes wrong with the parametric representations. let $u_n = u$ and $r_n = r$ with
$$u(0) = u(1/3) = 0, \quad u(2/3) = u(1) = -1 \tag{4.4}$$
and
$$r(0) = 0, \ r(1/3) = 1/2 = r(2/3), \ r(1) = 1,$$
with u and r defined by linear interpolation elsewhere. Then $(u_n', r) \in \Pi(c_n(x_n^\uparrow - e))$ for $u_n' = (u + nr)^\uparrow - nr$, so that
$$u_n'(0) = u_n'(1/3) = u_n'((2/3)) = 0, \ u_n'((2/3) + n^{-1}) = u_n'(1) = -1 \tag{4.5}$$
with u_n' defined by linear interpolation elsewhere. From (4.4) and (4.5), we see that $|u_n'(2/3) - u(2/3)| = 1$ for all n. Thus, to get the positive result, different parametric representations are needed for $c_n(x_n^\uparrow - e)$. ■

We next give an elementary result about the supremum function when the centering is in the other direction, so that x_n must be rapidly decreasing. Convergence $x_n^\uparrow(t) \to x(0)$ as $n \to \infty$ is to be expected, but that conclusion can not be drawn if the M_2 convergence in the condition is replaced by pointwise convergence.

Theorem 13.4.4. (*convergence preservation with the supremum function when the centering is in the other direction*) *Suppose that $c_n \to \infty$ and $x_n + c_n e \to y$ in $D([0,T], \mathbb{R}, M_2)$. Then*
$$\|x_n^\uparrow - z(y)\| \to 0 \quad as \quad n \to \infty,$$
where $z(y)(t) \equiv y(0)$, $0 \le t \le T$.

Example 13.4.4. *M_2 convergence cannot be replaced by pointwise convergence.* To see that the M_2 convergence cannot be replaced by pointwise convergence in the condition in Theorem 13.4.4, even to get pointwise convergence in the conclusion, let $x(t) = 0$, $0 \le t \le 1$, and $x_n(t) = I_{[n^{-1}, 2n^{-1}]}(t) - t$, $0 \le t \le 1$, $n \ge 1$. Then $x_n + e \to x$ pointwise (and not M_2), but $x_n^\uparrow(t) \to 1$ as $n \to \infty$ for all $t > 0$.

13.5. One-Dimensional Reflection

Closely related to the supremum function is the one-dimensional (one-sided) reflection mapping, which we have used to construct queueing processes. Indeed, the reflection mapping can be defined in terms of the supremum mapping as

$$\phi(x) \equiv x + (-x \vee 0)^\uparrow \;;$$

i.e.,

$$\phi(x)(t) = x(t) - (\inf\{x(s) : 0 \leq s \leq t\} \wedge 0) \,, \quad 0 \leq t \leq T \,, \tag{5.1}$$

as in (2.5) in Section 5.2.

The Lipschitz property for the supremum function with the uniform topology in Lemma 13.4.1 immediately implies a corresponding result for the reflection map ϕ in (5.1).

Lemma 13.5.1. (Lipschitz property with the uniform metric) *For any $x_1, x_2 \in D([0,T], \mathbb{R})$,*

$$\|\phi(x_1) - \phi(x_2)\| \leq 2\|x_1 - x_2\| \,.$$

Proof. By (5.1),

$$\begin{aligned}\|\phi(x_1) - \phi(x_2)\| &\leq \|x_1 - x_2\| + \|(-x_1 \vee 0)^\uparrow - (-x_2 \vee 0)^\uparrow\| \\ &\leq \|x_1 - x_2\| + \|(-x_1 \vee 0) - (-x_2 \vee 0)\| \leq 2\|x_1 - x_2\| \,. \quad \blacksquare\end{aligned}$$

Example 13.5.1. *The bound is tight.* To see that the bound in Lemma 13.5.1 is tight, let $x_1(t) = 0$, $0 \leq t \leq 1$, and $x_2 = -I_{[1/3, 1/2)} + I_{[1/2, 1]}$ in $D([0,1], \mathbb{R})$. Then $\phi(x_1) = x_1$, while $\phi(x_2) = 2I_{[1/2,1]}$, so that $\|x_1 - x_2\| = 1$ and $\|\phi(x_1) - \phi(x_2)\| = 2$.

Unfortunately, however, the Lipschitz property for the reflection map ϕ with the uniform topology does not even imply continuity in all the Skorohod topologies. In particular, ϕ is not continuous in the M_2 topology.

Example 13.5.2. *Continuity fails in M_2.* To see that the reflection map ϕ in (5.1) is not continuous in the M_2 topology, let $x = -I_{[1,2]}$ and

$$x_n(0) = x_n(1 - 3n^{-1}) = x(1 - n^{-1}) = 0$$

and

$$x_n(1 - 2n^{-1}) = x_n(1) = x_n(2) = -1$$

with x_n defined by linear interpolation elsewhere. Then $x_n \to x$ in $D([0,2], \mathbb{R})$, but $\phi(x)(t) = 0$, $0 \leq t \leq 2$, and $\phi(x_n)(1 - n^{-1}) = 1$, so that $\phi(x_n) \not\to \phi(x)$. This example fails to be a counterexample for the M_1 topology because then $x_n \not\to x$ as $n \to \infty$. $\blacksquare$

We do obtain positive results with the J_1 and M_1 topologies. As before, let d_{J_1} and d_{M_1} be the metrics in equations (3.2) and (3.4) in Section 3.3.. For the J_1 result, we use the following elementary lemma.

Lemma 13.5.2. *For any $x \in D$ and $\lambda \in \Lambda$,*
$$\phi(x) \circ \lambda = \phi(x \circ \lambda) \ .$$

For the M_1 result, we use the following lemma. A fundamental difficulty for treating the more general multidimensional reflection map is that Lemma 13.5.3 below does not extend to the multidimensional reflection map; see Chapter 14.

Lemma 13.5.3. (preservation of parametric representations under reflections) *For any $x \in D$, if $(u, r) \in \Pi(x)$, then $(\phi(u), r) \in \Pi(\phi(x))$.*

Proof. First, $(\phi(u), r)$ is continuous since (u, r) is, by Lemma 13.5.1. It suffices to show that $(\phi(u)(s), r(s)) \in \Gamma_{\phi(x)}$ for all s and that $(\phi(u), r)$ is nondecreasing in the order on $\Gamma_{\phi(x)}$. If $t \in Disc(x^c)$, then by (5.1) $\phi(u)(s) = \phi(x)(t)$ for each s such that $r(s) = t$. It remains to consider $t \in Disc(x)$. There exists an interval $[a, b] \subseteq [0, 1]$ such that $r(s) = t$ for $s \in [a, b]$, $u(a) = x(t-)$ and $u(b) = x(t)$. Moreover, by (5.1), $\phi(u)(a) = \phi(x)(t-)$ and $\phi(u)(b) = \phi(x)(t)$, with $\phi(u)(s)$ moving continuously and monotonically from $\phi(u)(a)$ to $\phi(u)(b)$ as s increases over $[a, b]$. Hence $(\phi(u)(s), r(s)) \in \Gamma_{\phi(x)}$ for all $s \in [0, 1]$ and $(\phi(u), r)$ is nondecreasing in the order on $\Gamma_{\phi(x)}$. ∎

Theorem 13.5.1. (Lipschitz property with the J_1 and M_1 metrics) *For any $x_1, x_2 \in D([0, T], \mathbb{R})$,*
$$d_{J_1}(\phi(x_1), \phi(x_2)) \le 2 d_{J_1}(x_1, x_2)$$
and
$$d_{M_1}(\phi(x_1), \phi(x_2)) \le 2 d_{M_1}(x_1, x_2)) \ ,$$
where ϕ is the reflection map in (5.1).

Proof. First, for the J_1 metric, by Lemmas 13.5.2 and 13.5.1,
$$\begin{aligned}
d_{J_1}(\phi(x_1), \phi(x_2)) &= \inf_{\lambda \in \Lambda} \{\|\phi(x_1) \circ \lambda - \phi(x_2)\| \vee \|\lambda - e\|\} \\
&= \inf_{\lambda \in \Lambda} \{\|\phi(x_1 \circ \lambda) - \phi(x_2)\| \vee \|\lambda - e\|\} \\
&\le \inf_{\lambda \in \Lambda} \{2\|x_1 \circ \lambda - x_2\| \vee \|\lambda - e\|\} \le 2 d_{J_1}(x_1, x_2) \ .
\end{aligned}$$

Turning to M_1, we use Lemma 13.5.3 to conclude that $(\phi(u), r) \in \Pi(\phi(x))$ whenever $(u, r) \in \Pi(x)$. Then, by Lemma 13.5.1,
$$\begin{aligned}
d_{M_1}(\phi(x_1), \phi(x_2)) &= \inf_{\substack{(u_i, r_i) \in \Pi(\phi(x_i)) \\ i=1,2}} \{\|u_1 - u_2\| \vee \|r_1 - r_2\|\} \\
&\le \inf_{\substack{(u_i, r_i) \in \Pi(x_i) \\ i=1,2}} \{\|\phi(u_1) - \phi(u_2)\| \vee \|r_1 - r_2\|\} \\
&\le \inf_{\substack{(u_i, r_i) \in \Pi(x_i) \\ i=1,2}} \{2\|u_1 - u_2\| \vee \|r_1 - r_2\|\} \le 2 d_{M_1}(x_1, x_2) \ .
\end{aligned}$$

Remark 13.5.1. *The Lipschitz constant.* Example 13.5.1 shows that the bounds in Theorem 13.5.1 are tight; i.e., the Lipschitz constant is 2. ∎

Theorem 13.5.1 covers the standard heavy-traffic regime for one single-server queue when $\rho = 1$, where ρ is the traffic intensity. The next result covers the other cases: $\rho < 1$ and $\rho > 1$. We use the following elementary lemma in the easy case of the uniform metric.

Lemma 13.5.4. *Let d be the metric for the U, J_1, M_1 or M_2 topology. Let $x \vee a : D \to D$ be defined by*

$$(x \vee a)(t) \equiv x(t) \vee a, \quad 0 \leq t \leq T. \tag{5.2}$$

Then, for any $x_1, x_2 \in D$,

$$d(x_1 \vee a(x_1), x_2 \vee a(x_2)) \leq d(x_1, x_2).$$

Theorem 13.5.2. (convergence preservation with centering) *Suppose that $x_n - c_n e \to y$ in $D([0,T], \mathbb{R})$ with the U, J_1, M_1 or M_2 topology.*
(a) *If $c_n \to +\infty$, then*

$$\phi(x_n) - c_n e \to y + \gamma(y) \quad \text{as} \quad n \to \infty \quad \text{in} \quad D$$

with the same topology, where

$$\gamma(y)(t) \equiv (-y(0)) \vee 0 = -(y(0) \wedge 0), \quad 0 \leq t \leq T.$$

(b) *If $c_n \to -\infty$, $y(0) \leq 0$ and y has no positive jumps, then*

$$\|\phi(x_n) - 0e\| \to 0 \quad \text{as} \quad n \to \infty \quad \text{in} \quad D,$$

where $e(t) = t$, $0 \leq t \leq T$.

Example 13.5.3. *The necessity of the condition on $y(0)$.* To see the need for the condition $y(0) \leq 0$ in Theorem 13.5.2 (b), let $y(t) = 1$, $0 \leq t \leq T$, $c_n = -n$ and $x_n(t) = (c_n e + y)(t) = 1 - nt$ for all t. Then $x_n - c_n e = y$ for all n, but $\phi(x_n)(0) = 1$ and $\phi(x_n)(t) \to 0$ for all $t > 0$.

13.6. Inverse

We now consider the inverse map, which arises in the study of renewal processes, first passage times and extremal processes; see Billingsley (1968), Gut (1988) and Resnick (1987).

It is convenient to consider the inverse map on the subset D_u of x in $D \equiv D([0,\infty), \mathbb{R})$ that are unbounded above and satisfy $x(0) \geq 0$. For $x \in D_u$, let the inverse of x be

$$x^{-1}(t) \equiv \inf\{s \geq 0 : x(s) > t\}, \quad t \geq 0. \tag{6.1}$$

As before, let D_0 be the subset of x in D with $x(0) \geq 0$, and let $D_\uparrow$ and $D_{\uparrow\uparrow}$ be the subsets of nondecreasing and strictly increasing functions in D_0. Let $D_{u,\uparrow} \equiv D_u \cap D_\uparrow$ and $D_{u,\uparrow\uparrow} \equiv D_u \cap D_{\uparrow\uparrow}$. Clearly,

$$D_{u,\uparrow\uparrow} \subseteq D_{u,\uparrow} \subseteq D_u \subseteq D_0.$$

13.6.1. The Standard Topologies

Recall that on $D_\uparrow$ the M_1 and M_2 topologies reduce to pointwise convergence on a dense subset including 0. The following result supplements Lemmas 13.2.1–13.2.3.

Lemma 13.6.1. (measurability of D_u) *Let D have one of the topologies J_1, M_1 or M_2. The subset D_u is a G_δ subset of D_0.*

Proof. Note that
$$D_u = \cap_{n=1}^\infty (D_0 - \bar{D}_n) \;,$$
where $\bar{D}_n$ is the subset of functions in D_0 bounded above by n. In the non-uniform Skorohod topologies, $\bar{D}_n$ is a closed subset of D_0, so that D_u is a G_δ subset of D_0. ■

We begin our study of the inverse function by stating some basic results. Our first result shows that the inverse map is closely related to the supremum.

Lemma 13.6.2. (duality) *For any $x \in D_u$, $x^{-1} \in D_{u,\uparrow}$ and $(x^{-1})^{-1} = x^\uparrow$.*

Corollary 13.6.1. *For any $x \in D_{u,\uparrow}$, $(x^{-1})^{-1} = x$.*

Remark 13.6.1. *The left-continuous inverse.* As part of Lemma 13.6.2, x^{-1} is right-continuous. In some circumstances it is convenient to work instead with the left-continuous inverse

$$x^\leftarrow(t) \equiv \inf\{s \geq 0 : x(s) \geq t\}, \quad t \geq 0 \;. \tag{6.2}$$

For $x \in D_u$, $x^\leftarrow(t) = x^{-1}(t-)$, $t \geq 0$, with $x^{-1}(0-) \equiv 0$. Note that $x^\leftarrow$ need not be right-continuous at 0. Indeed, $x^\leftarrow(0) > 0 = x^\leftarrow(0)$ if and only if $x^{-1}(0) > 0$. If $x^{-1}(0) = 0$, then the completed graphs of x^{-1} in (6.1) and $x^\leftarrow$ in (6.2) are identical, which implies that many M_1 and M_2 results for x^{-1} apply directly to $x^\leftarrow$ as well under that condition. ■

The left-continuous inverse has an appealing inverse property not shared by the right-continuous inverse:

Lemma 13.6.3. (inverse relation) *For any $x \in D_{u,\uparrow}$ and $t_1, t_2 \geq 0$,*

$$x^\leftarrow(t_1) \leq t_2 \quad \text{if and only if} \quad x(t_2) \geq t_1 \;. \tag{6.3}$$

Lemma 13.6.4. *For any $x \in D_{u,\uparrow}$,*

$$0 \leq (x \circ x^{-1})(t) - t \leq x(x^{-1}(t)) - x(x^{-1}(t)-) \;, \tag{6.4}$$
$$0 \leq (x^{-1} \circ x)(t) - t \leq x^{-1}(x(t)) - x^{-1}(x(t)-)) \;, \tag{6.5}$$
$$0 \leq (x \circ x^\leftarrow)(t) - t \leq x(x^\leftarrow(t)) - x(x^\leftarrow(t)-) \;, \tag{6.6}$$
$$0 \leq t - (x^\leftarrow \circ x)(t) \leq x^{-1}(x(t)) - x^\leftarrow(x(t)) \;, \tag{6.7}$$

where $x(0-)$ is interpreted as 0.

Let $J_t(x)$ be the maximum jump of x over $[0, t]$, i.e.

$$J_t(x) \equiv \sup_{0 \leq s \leq t} \{x(t) - x(t-)\} \;. \tag{6.8}$$

where again $x(0-) \equiv 0$.

Corollary 13.6.2. *For any $x \in D_{u,\uparrow}$ and $t > 0$,*
$$\|x \circ x^{-1} - e\|_t \leq J_{x^{-1}(t)}(x) \tag{6.9}$$
and
$$\|x^{-1} \circ x - e\|_t \leq J_{x(t)}(x^{-1}) , \tag{6.10}$$
for $J_t(x)$ in (6.8).

Lemma 13.6.5. *Suppose that $x \in D_{u,\uparrow}$. Then $x \in D_{u,\uparrow\uparrow}$ if and only if $x^{-1} \in C_{u,\uparrow}$.*

We now consider the inverse together with composition applied to elements of $\Lambda \equiv \Lambda([0, \infty))$, i.e., to homeomorphisms of $\mathbb{R}_+ \equiv [0, \infty)$. For each $\lambda \in \Lambda$, $\lambda(0) = 0$ and there is an inverse λ^{-1} with $\lambda, \lambda^{-1} \in C_{\uparrow\uparrow}$ and $\lambda \circ \lambda^{-1} = \lambda^{-1} \circ \lambda = e$.

Lemma 13.6.6. *If $x \in D_{u,\uparrow}$ and $\lambda_1, \lambda_2 \in \Lambda([0, \infty))$, then*
$$(\lambda_1 \circ x \circ \lambda_2)^{-1} = \lambda_2^{-1} \circ x^{-1} \circ \lambda_1^{-1} .$$

Proof. Note that
$$\begin{aligned}
(\lambda_1 \circ x \circ \lambda_2)^{-1}(t) &= \inf\{s \geq 0 : (\lambda_1 \circ x \circ \lambda_2)(s) > t\} \\
&= \inf\{s \geq 0 : (x \circ \lambda_2)(s) > \lambda_1^{-1}(t)\} \\
&= \inf\{\lambda_2^{-1}(s) \geq 0 : x(s) > \lambda_1^{-1}(t)\} \\
&= (\lambda_2^{-1} \circ x^{-1} \circ \lambda_1^{-1})(t) . \quad \blacksquare
\end{aligned}$$

We now turn to continuity properties of the inverse map. First we note that the inverse map from (D_u, J_1) to (D_u, J_1) or even from (D_u, U) to (D_u, J_1) is in general not continuous.

Example 13.6.1. *The inverse is not continuous when the range has the J_1 topology.* To see that the inverse map from $(D_{u,\uparrow}, U)$ to $(D_{u,\uparrow}, J_1)$ is not continuous, let $x = 2I_{[0,2]} + eI_{[2,\infty)}$ and
$$x_n = (2 - n^{-1})I_{[0,1)} + (2 + n^{-1})I_{[1,2+n^{-1})} + eI_{[2+n^{-1},\infty)} .$$
Then $\|x_n - x\| = n^{-1} \to 0$ and $x_n^{-1} \to x^{-1}$ (M_1), but $x_n^{-1} \not\to x^{-1}$ (J_1). $\blacksquare$

Even for the M_1 topology, there are complications at the left endpoint of the domain $[0, \infty)$.

Example 13.6.2. *Complications at the left endpoint of the domain.* To see that the inverse map from $(D_{u,\uparrow}, U)$ to $(D_{u,\uparrow}, M_1)$ is in general not continuous, let $x(t) = 0$, $0 \leq t < 1$, and $x(t) = t$, $t \geq 1$; Let $x_n = t/n$, $0 \leq t < 1$ and $x_n(t) = t$, $t \geq 1$. Then $\|x_n - x\|_\infty = n^{-1} \to 0$, but $x_n^{-1}(0) = 0 \not\to 1 = x^{-1}(0)$, so that $x_n^{-1} \not\to x^{-1}$ (M_1). $\blacksquare$

To avoid the problem in Example 13.6.2, we can require that $x^{-1}(0) = 0$. To develop an equivalent condition, let $D_{u,\epsilon}^\uparrow$ be the subset of functions x in D_u such that $x(t) = 0$ for $0 \leq t \leq \epsilon$.

Then let
$$D_u^* \equiv \cap_{n=1}^{\infty}(D_{u,n^{-1}})^c . \tag{6.11}$$

Lemma 13.6.7. (measurability of D_u^*) *With the J_1, M_1 or M_2 topology, D_u^* in (6.11) is a G_δ subset of D_u and*
$$D_u^* = \{x \in D_u : x^{-1}(0) = 0\} . \tag{6.12}$$

Let $D_{u,\uparrow}^* \equiv D_\uparrow \cap D_u^*$. A key property of $D_{u,\uparrow}^*$, not shared by $D_{u,\uparrow}$ because of the complication at the origin, is that parametric representation (u,r) for x directly serve as parametric representations for x^{-1} when we switch the roles of the components u and r.

Lemma 13.6.8. (switching the roles of u and r) *For $x \in D_{u,\uparrow}^*$, the graph Γ_x serves as the graph of $\Gamma_{x^{-1}}$ with the axes switched. Thus, $(u,r) \in \Pi(x)$ if and only if $(r,u) \in \Pi(x^{-1})$, where $\Pi(x)$ is the set of M_1 parametric representations.*

Corollary 13.6.3. (continuity on (D_u^*, M_1)) *The inverse map from (D_u^*, M_1) to $(D_{u,\uparrow}, M_1)$ is continuous.*

Proof. First apply Theorem 13.4.1 for the supremum. Then apply Lemma 13.6.8. ■

We now generalize Corollary 13.6.3 by only requiring that the limit be in D_u^*. As before, the missing proof is in the Internet Supplement.

Theorem 13.6.1. (measurability and continuity at limits in D_u^*) *The inverse map in (6.1) from (D_u, M_2) to $(D_{u,\uparrow}, M_1)$ is measurable and continuous at $x \in D_u^*$, i.e., for which $x^{-1}(0) = 0$.*

Corollary 13.6.4. . (continuity at strictly increasing functions) *The inverse map from (D_u, M_2) to $(D_{u,\uparrow}, U)$ is continuous at $x \in D_{u,\uparrow\uparrow}$.*

Proof. First, $D_{u,\uparrow\uparrow} \subseteq D_{u,\uparrow}^*$, so that we can apply Theorem 13.6.1 to get $x_n^{-1} \to x^{-1}$ in $(D_{u,\uparrow}, M_1)$. However, by Lemma 13.6.4, $x^{-1} \in C$ when $x \in D_{u,\uparrow\uparrow}$. Hence the M_1 convergence $x_n^{-1} \to x^{-1}$ actually holds in the stronger topology of uniform convergence over compact subsets. ■

13.6.2. The M_1' Topology

For cases in which the condition $x^{-1}(0) = 0$ in Theorem 13.6.1 is not satisfied, we can modify the M_1 and M_2 topologies to obtain convergence, following Puhalskii and Whitt (1997). With these new weaker topologies, which we call M_1' and M_2', we do not require that $x_n(0) \to x(0)$ when $x_n \to x$. We construct the new topologies by extending the graph of each function x by appending the segment $[0, x(0)] \equiv \{\alpha 0 + (1-\alpha)x(0) : 0 \leq \alpha \leq 1\}$. Let the new graph of $x \in D$ be

$$\begin{aligned}\Gamma_x' &= \{(z,t) \in \mathbb{R}^k \times [0,\infty) : z = \alpha x(t) + (1-\alpha)x(t-) \\ &\quad \text{for } 0 \leq \alpha \leq 1 \text{ and } t \geq 0\},\end{aligned} \tag{6.13}$$

where $x(0-) \equiv 0$. Let $\Pi'(x)$ and $\Pi'_2(x)$ be the sets of all M_1 and M_2 parametric representations of Γ'_x, defined just as before. We say that $x_n \to x$ in (D, M'_1) if there exist parametric representations $(u_n, r_n) \in \Pi'(x_n)$ and $(u, r) \in \Pi'(x)$, where $\Pi'(x)$ is the set of M'_1 parametric representations of x, such that

$$\|u_n - u\|_t \vee \|r_n - r\|_t \to 0 \quad \text{as} \quad n \to \infty \quad \text{for each} \quad t > 0. \quad (6.14)$$

We have a corresponding definition of convergence in (D, M'_2) using the parametric representations in $\Pi'_2(x)$ instead of $\Pi'(x)$. With the M'_i topologies, we obtain a cleaner statement than Lemma 13.6.8.

Lemma 13.6.9. (graphs of the inverse with the M'_i topology) *For $x \in D_{u,\uparrow}$, the graph Γ'_x serves as the graph $\Gamma'_{x^{-1}}$ with the axes switched, so that $(u, r) \in \Pi'(x)$ ($\Pi'_2(x)$) if and only if $(r, u) \in \Pi'(x^{-1})$ ($\Pi'_2(x^{-1})$).*

Thus we get an alternative to Theorem 13.6.1.

Theorem 13.6.2. (continuity in the M'_1 topology) *The inverse map in (6.1) from (D_u, M'_2) to $(D_{u,\uparrow}, M'_1)$ is continuous.*

Proof. By the M'_2 analog of Theorem 13.4.1, if $x_n \to x$ in (D_u, M'_2), then $x_n^\uparrow \to x^\uparrow$ in $(D_{u,\uparrow}, M'_2)$. Since the M'_2 topology coincides with the M'_1 topology on $D_\uparrow$, we get $x_n^\uparrow \to x^\uparrow$ in $(D_{u,\uparrow}, M'_1)$. By Lemma 13.6.9, we get $(x_n^\uparrow)^{-1} \to (x^\uparrow)^{-1}$ in $(D_{u,\uparrow}, M'_1)$. That gives the desired result because $(x^\uparrow)^{-1} = x^{-1}$ for all $x \in D_u$. ∎

An alternative approach to the difficulty at the origin besides M'_i topology on $D_u([0, \infty), \mathbb{R})$ is the ordinary M_i topology on $D_u((0, \infty), \mathbb{R})$. The difficulty at the origin goes away if we ignore it entirely, which we can do by making the function domain $(0, \infty)$ for the image of the inverse functions.

In particular, Theorem 13.6.2 implies the following corollary.

Corollary 13.6.5. (continuity when the origin is removed from the domain) *The inverse map in (6.1) from $D_u([0, \infty), M_2)$ to $D_{u,\uparrow}((0, \infty), M_1)$ is continuous.*

Proof. Since the M'_2 topology is weaker than M_2, if $x_n \to x$ in $D_u([0, \infty), M_2)$, then $x_n \to x$ in $D_u([0, \infty), M'_2)$. Apply Theorem 13.6.2 to get $x_n^{-1} \to x^{-1}$ in $D_{u,\uparrow}([0, \infty), M'_1)$. That implies $x_n^{-1} \to x^{-1}$ for the restrictions in $D_\uparrow([t_1, t_2], M_1)$ for all $t_1, t_2 \in Disc(x^{-1})^c$, which in turn implies that $x_n^{-1} \to x^{-1}$ in $D_{u,\uparrow}((0, \infty), M_1)$. ∎

However, in general we cannot work with the inverse on $D_u((0, \infty), \mathbb{R})$.

Example 13.6.3. *Difficulty with the domain $(0, \infty)$.* To see the problem with having the function domain be $(0, \infty)$, let $x = e$ and $x_n(0) = x_n(2n^{-1}) = 0$, $x_n(n^{-1}) = 1$, $x_n(t) = t - 2n^{-1}$, $t \geq 2n^{-1}$, with x_n defined by linear interpolation elsewhere. Then $x_n \to x$ in $D((0, \infty), \mathbb{R}, U)$, but $x_n^{-1} \not\to x^{-1} \equiv e$, because $x_n^{-1}(t) \to 1$ as $n \to \infty$ for each t with $0 < t < 1$. ∎

We can obtain positive results if all the functions are required to be monotone. The following result is elementary.

Theorem 13.6.3. *(equivalent characterizations of convergence for monotone functions) For x_n, $n \geq 1$, $x \in D_{u,\uparrow}([0,\infty),\mathbb{R})$, the following are equivalent:*

$$x_n \to x \quad in \quad D_{u,\uparrow}((0,\infty),\mathbb{R},M_1) ; \tag{6.15}$$

$$x_n \to x \quad in \quad D_{u,\uparrow}([0,\infty),\mathbb{R},M_1') ; \tag{6.16}$$

$$x_n(t) \to x(t) \quad \text{for all } t \text{ in a dense subset of } (0,\infty) ; \tag{6.17}$$

$$x_n^{-1} \to x^{-1} \quad in \quad D((0,\infty),\mathbb{R},M_1) ; \tag{6.18}$$

$$x_n^{-1} \to x^{-1} \quad in \quad D([0,\infty),\mathbb{R},M_1') ; \tag{6.19}$$

$$x_n^{-1}(t) \to x^{-1}(t) \quad \text{for all } t \text{ in a dense subset of } (0,\infty). \tag{6.20}$$

Example 13.6.4. *The need for monotonicity.* To see the advantage of M_1' on $[0,\infty)$ over M_1 on $(0,\infty)$, let $x(t) = 1$, $t \geq 0$,

$$x_n^1(0) = 0, x_n^1(n^{-1}) = 1 = x_n^1(t), \quad t \geq n^{-1} , \tag{6.21}$$

and

$$x_n^2(0) = 0 = x_n^2(2n^{-1}), x_n^2(n^{-1}) = x_n^2(3n^{-1}) = 1 = x_n^2(t), \quad t \geq 3n^{-1} , \tag{6.22}$$

with x_n^1 and x_n^2 defined by linear interpolation elsewhere. Then $x_n^1 \to x$ in both $D((0,\infty),\mathbb{R},M_1)$ and in $D([0,\infty),\mathbb{R},M_1')$, but $x_n^2 \to x$ only in $D((0,\infty),\mathbb{R},M_1)$. The monotonicity condition provides the equivalence in Theorem 13.6.3.

13.6.3. First Passage Times

In this final subsection we consider some real-valued functions closely related to the inverse function. Sometimes we are interested in the first passage time to or beyond some specified level. Given any specified level $z \in \mathbb{R}$, the *first passage time* beyond z is the function $\tau_z : D_u \to \mathbb{R}$ defined in terms of the inverse function by

$$\tau_z(x) \equiv x^{-1}(z) . \tag{6.23}$$

It is elementary that τ_z has the following two scaling invariance properties: For any $c > 0$,

$$\tau_{cz}(cx) = \tau_z(x) \tag{6.24}$$

and

$$c\tau_z(x \circ ce) = \tau_z(x) , \tag{6.25}$$

where e is the identity map, i.e., $e(t) = t$ for $t \geq 0$.

Three functions closely related to the first-passage-time function τ_z are the *overshoot function* $\gamma_z : D_u \to \mathbb{R}$ defined by

$$\gamma_z(x) \equiv x(\tau_z(x)) - z , \tag{6.26}$$

the *last-value function* $\lambda_z : D_u \to \mathbb{R}$ defined by

$$\lambda_z(x) \equiv x(\tau_z(x)-) \tag{6.27}$$

and the *final-jump functions* $\delta_z : D_u \to \mathbb{R}$ defined by

$$\delta_z(x) \equiv x(\tau_z(x)) - x(\tau_z(x)-) . \tag{6.28}$$

The following continuity properties are elementary, but of course important. It clearly does not suffice to have pointwise convergence.

Theorem 13.6.4. (continuity of first-passage-time functions) *Let x be an element of D_u that is not equal to z throughout the interval $(\tau_z(x) - \epsilon, \tau_z(x))$ for any $\epsilon > 0$. If $x_n \to x$ in (D, M_2), then*

$$(\tau_z(x_n), \gamma_z(x_n), \lambda_z(x_n), \delta_z(x_n)) \to (\tau_z(x), \gamma_z(x), \lambda_z(x), \delta_z(x))$$

as $n \to \infty$ in $\mathbb{R}^4$.

The regularity condition holds almost surely for Lévy processes. Hence we have the following consequence of Theorem 13.6.4, which we apply to queues in Section 9.7.

Theorem 13.6.5. (convergence of first-passage-time functions for Lévy limit processes) *Let X be a Lévy process such that*

$$P(\varlimsup_{t\to\infty} X(t) = \infty) = 1 . \tag{6.29}$$

If $X_n \Rightarrow X$ in (D_u, M_2), then

$$(\tau_z(X_n), \gamma_z(X_n), \lambda_z(X_n), \delta_z(X_n)) \to (\tau_z(X), \gamma_z(X), \lambda_z(X), \delta_z(X))$$

in $\mathbb{R}^4$ for any $z > 0$.

13.7. Inverse with Centering

We continue considering the inverse map, but now with centering. We start by considering linear centering. In particular, we consider when a limit for $c_n(x_n - e)$ implies a limit for $c_n(x_n^{-1} - e)$ when $x_n \in D_u \equiv D_u([0,\infty), \mathbb{R})$ and $c_n \to \infty$. By considering the behavior at one t, it is natural to anticipate that we should have $c_n(x_n^{-1} - e) \to -y$ when $c_n(x_n - e) \to y$. A first step for the M topologies is to apply Theorem 13.4.2, which yields limits for $c_n(x_n^\uparrow - e)$ Thus for the M topologies, it suffices to assume that $x_n \in D_{u,\uparrow}$.

For the J_1 topology, however, a different argument is needed to get limits when $y \notin C$, as the following result shows.

Lemma 13.7.1. *Suppose that $x_n \in D_u$, $n \geq 1$, and $c_n \to \infty$. if $c_n(x_n^\uparrow - e) \to y$ and $c_n(x_n^{-1} - e) \to -y$ (J_1), then $y \in C$.*

Proof. Since $x_n^\uparrow \in D_{u,\uparrow}$, $c_n(x_n^\uparrow - e)$ has no negative jumps. Since the topology is J_1 and $c_n(x_n - e) \to y$, y has no negative jumps; e.g., see p. 301 of Jacod and Shiryaev (1987). Similarly, $c_n(x_n^{-1} - e)$ has no negative jumps. Since $c_n(x_n^{-1} - e) \to -y$ (J_1), $-y$ has no negative jumps. ∎

The following lemma establishes a necessary condition in any of the topologies.

Lemma 13.7.2. *If $x_n \in D_{u,\uparrow}$, $c_n(x_n - e)(0) \to y(0)$ and $c_n(x_n^{-1} - e)(0) \to -y(0)$, where $c_n \to \infty$, then $y(0) = 0$.*

Proof. Since $x_n \in D_{u,\uparrow}$, $x_n(0) \geq 0$ and $x_n^{-1}(0) \geq 0$. Since $e(0) = 0$, the convergence $c_n(x_n - e)(0) \to y(0)$ implies that $y(0) \geq 0$. Similarly, the convergence $c_n(x_n^{-1} - e)(0) \to -y(0)$ implies that $y(0) \leq 0$. ∎

Now we state the main limit theorem for inverse functions with centering.

Theorem 13.7.1. (inverse with linear centering) *Suppose that $c_n(x_n - e) \to y$ as $n \to \infty$ in $D([0,\infty), \mathbb{R})$ with one of the topologies M_2, M_1 or J_1, where $x_n \in D_u$, $c_n \to \infty$ and $y(0) = 0$.*

(a) If the topology is M_2 or M_1, then $c_n(x_n^{-1} - e) \to -y$ as $n \to \infty$ with the same topology.

(b) If the topology is J_1 and if y has no positive jumps, then $c_n(x_n^{-1} - e) \to -y$ as $n \to \infty$.

We can combine Lemma 13.6.6 and Theorem 13.7.1 to obtain the following corollary. Let Λ be the space of homeomorphisms of $\mathbb{R}_+$.

Corollary 13.7.1. *Suppose that $x_n \in D_{u,\uparrow}$ and $\lambda_{1,n}, \lambda_{2,n} \in \Lambda$, $n \geq 1$. Let $c_n \to \infty$ and $y(0) = 0$. Then*

$$c_n(\lambda_{2,n} \circ x_n \circ \lambda_{1,n} - e) \to y \quad in \quad D([0,\infty), \mathbb{R}, M_i) \tag{7.1}$$

if and only if

$$c_n(\lambda_{1,n}^{-1} \circ x_n^{-1} \circ \lambda_{2,n}^{-1} - e) \to -y \quad in \quad D([0,\infty), \mathbb{R}, M_i), \tag{7.2}$$

where the topology in both cases is either M_1 or M_2.

We can apply Corollary 13.7.1 to obtain generalizations of Theorem 13.7.1 with nonlinear centering terms. (We obtain a more general result at the end of the section.)

Corollary 13.7.2. (centering functions from Λ) *Suppose that, in addition to the conditions of Corollary 13.7.1, $\lambda_{i,n} \to \lambda_i$ as $n \to \infty$ for each i, where $\lambda_i \in \Lambda$. Then*

$$c_n(\lambda_{2,n} \circ x_n - \lambda_{1,n}^{-1}) \to y \circ \lambda_1^{-1} \quad in \quad (D, M_i) \tag{7.3}$$

if and only if

$$c_n(\lambda_{1,n}^{-1} \circ x_n^{-1} - \lambda_{2,n}^{-1}) \to -y \circ \lambda_2^{-1} \quad in \quad (D, M_i). \tag{7.4}$$

Proof. Apply Theorem 13.2.3 with the composition map to show that (7.3) is equivalent to (7.1) and (7.4) is equivalent to (7.2). ∎

We can use Corollary 13.7.1 to obtain the following consequence.

13.7. Inverse with Centering

Corollary 13.7.3. *Suppose that $x_n \in D_u$, $y(0) = 0$, $c_n \to \infty$ and $a_n \to a > 0$. If*
$$c_n(x_n - a_n e) \to y \quad in \quad D$$
with the M_1 or M_2 topology, then
$$c_n(x_n^{-1} - a_n^{-1} e) \to -a^{-1} y \circ a^{-1} e \quad in \quad D$$
with the same topology.

Proof. Under the condition, $(a_n c_n)(a_n^{-1} x_n - e) \to x$, so that by Corollary 13.7.1, $(a_n c_n)(x_n^{-1} \circ a_n e - e) \to -y$. Now applying the composition map with $a_n^{-1} e$, $a_n c_n(x_n^{-1} - a_n^{-1} e) \to x \circ a^{-1} e$. Dividing by a_n yields the conclusion. ∎

Stochastic limit theorems are not often expressed directly in the form of Corollaries 13.7.1 or 13.7.3. We now state consequences of Corollary 13.7.1 that have more direct applications.

Corollary 13.7.4. *Let $y_n \in D_{u,\uparrow}$ and $\phi_{1,n}, \phi_{2,n} \in \Lambda$, $n \geq 1$; let $u(0) = 0$ and $n/\psi(n) \to \infty$ as $n \to \infty$. Let*
$$w_n(t) \equiv \psi(n)^{-1}[(\phi_{2,n} \circ y_n \circ \phi_{1,n})(nt) - nt], \quad t \geq 0, \tag{7.5}$$
and
$$x_n(t) \equiv \psi(n)^{-1}[(\phi_{1,n}^{-1} \circ y_n^{-1} \circ \phi_{2,n}^{-1})(nt) - nt], \quad t \geq 0, \tag{7.6}$$
for all $n \geq 1$. Then
$$w_n \to u \quad in \quad D([0,\infty), \mathbb{R}) \tag{7.7}$$
if and only if
$$x_n \to -u \quad in \quad D([0,\infty), \mathbb{R}), \tag{7.8}$$
where the topology throughout is either M_1 or M_2.

Proof. Apply Corollary 13.7.1 with $x_n(t) = n^{-1} y_n(t)$, $\lambda_{i,n}(t) = n^{-1} \phi_{i,n}(nt)$ and $c_n = n/\psi(n)$. Then $w_n = c_n(\lambda_{2,n} \circ x_n \circ \lambda_{1,n} - e)$ and $x_n = c_n(\lambda_{1,n}^{-1} \circ x_n^{-1} \circ \lambda_{2,n}^{-1} - e)$. ∎

We now consider the special case of Corollary (13.7.4) in which the homeomorphisms $\phi_{i,n}$ are linear, i.e., $\phi_{i,n} = a_{i,n} e$, $n \geq 1$.

Corollary 13.7.5. *Suppose that $y_n \in D_{u,\uparrow}$, $w(0) = 0$, $a_n \to a > 0$ and $n/\psi(n) \to \infty$ as $n \to \infty$. Let*
$$\tilde{w}_n = \psi(n)^{-1}[y_n(nt) - a_n nt], \quad t \geq 0, \tag{7.9}$$
and
$$\tilde{x}_n = \psi(n)^{-1}[y_n^{-1}(nt) - a_n^{-1} nt], \quad t \geq 0. \tag{7.10}$$
Then
$$\tilde{w}_n \to w \quad in \quad D([0,\infty), \mathbb{R}) \tag{7.11}$$
if and only if
$$\tilde{x}_n \to a^{-1} w \circ a^{-1} e \quad in \quad D([0,\infty), \mathbb{R}), \tag{7.12}$$

where the topology throughout is M_1 or M_2.

Proof. Apply Corollary 13.7.1 with $x_n(t) = n^{-1}y_n(nt)$, $\lambda_{2,n} = a_n^{-1}e$, $\lambda_{1,n} = e$ and $c_n = na_n/\psi(n)$. Then $\tilde{w}_n = c_n(\lambda_{2,n} \circ x_n \circ \lambda_{1,n} - e)$, so that $\tilde{w}_n \to w$ if and only if $c_n(\lambda_{1,n}^{-1} \circ x_n^{-1} \circ \lambda_{2,n}^{-1} - e) \to -w$. However,

$$c_n(\lambda_{1,n}^{-1} \circ x_n^{-1} \circ \lambda_{2,n}^{-1} - e) = a_n \tilde{x}_n \circ a_n e \qquad (7.13)$$

and

$$-a_n \tilde{x}_n \circ a_n e \to -w \quad \text{if and only if} \quad \tilde{x}_n \to -a^{-1} w \circ a^{-1} e. \quad \blacksquare \qquad (7.14)$$

Following Puhalskii (1994), we can generalize Theorem 13.7.1 by allowing nonlinear centering terms. We present several results of this kind.

Theorem 13.7.2. *Suppose that*

$$c_n(x_n - \lambda) \to u \quad as \quad n \to \infty \quad in \quad D$$

with one of the topologies M_2, M_1 or J_1, where $x_n \in D_u$, $u(0) = 0$, u has no positive jumps if the topology is J_1, $\lambda \in \Lambda$ and $c_n \to \infty$. Then

$$c_n(\lambda \circ x_n^{-1} - e) \to -u \circ \lambda^{-1} \quad as \quad n \to \infty \qquad (7.15)$$

with the same topology. If, in addition, λ is absolutely continuous with continuous positive derivative $\dot{\lambda}$, then

$$c_n(x_n^{-1} - \lambda^{-1}) \to \frac{-u \circ \lambda^{-1}}{\dot{\lambda} \circ \lambda^{-1}} \quad as \quad n \to \infty, \qquad (7.16)$$

where $(u/v)(t) \equiv u(t)/v(t)$, $t \geq 0$.

Proof. Apply Theorems 13.2.2 and 13.2.3 with the composition map to get $c_n(x_n \circ \lambda^{-1} - \lambda \circ \lambda^{-1}) \to u \circ \lambda^{-1}$ as in the same topology. Since $\lambda \circ \lambda^{-1} = e$, we can apply Theorem 13.7.1 or Corollary 13.7.1 to get (7.15) with the same topology. Now suppose that λ is absolutely continuous with continuous positive derivative $\dot{\lambda}$. Then

$$c_n(\lambda \circ x_n^{-1} - e)(t) = c_n(\lambda \circ x_n^{-1} - \lambda \circ \lambda^{-1})(t)$$
$$= c_n \int_{\lambda^{-1}(t)}^{x_n^{-1}(t)} \dot{\lambda}(s) ds. \qquad (7.17)$$

Since $c_n(x_n - \lambda) \to u$, $\|x_n - \lambda\|_t \to 0$ and $\|x_n^{-1} - \lambda^{-1}\|_t \to 0$ as $n \to \infty$ for all t. Since $\dot{\lambda}$ is continuous, it is uniformly continuous over bounded intervals. Hence

$$\sup_{0 \leq s \leq t} \left| c_n \int_{\lambda^{-1}(s)}^{x_n^{-1}(s)} \dot{\lambda}(u) du - \dot{\lambda}(\lambda^{-1}(s)) c_n(x_n^{-1}(s) - \lambda^{-1}(s)) \right| \to 0. \qquad (7.18)$$

Then (7.15), (7.17) and (7.18) imply that

$$(\dot{\lambda} \circ \lambda^{-1}) c_n(x_n^{-1} - \lambda^{-1}) \to -u \circ \lambda^{-1} \quad as \quad n \to \infty \qquad (7.19)$$

in the same topology, where $(uv)(t) \equiv u(t)v(t)$ for $u, v \in D$. Finally (7.19) implies (7.16). $\blacksquare$

Corollary 13.7.6. *Suppose that $x_n \in D_{u,\uparrow}$, $u(0) = 0$, $\lambda \in \Lambda$, λ is absolutely continuous with continuous positive derivative $\dot{\lambda}$ and $c_n \to \infty$. Then*

$$c_n(x_n - \lambda) \to u \quad in \quad D \tag{7.20}$$

with one of the topologies M_1 or M_2 if and only if

$$c_n(x_n^{-1} - \lambda^{-1}) \to \frac{-u \circ \lambda^{-1}}{\dot{\lambda} \circ \lambda^{-1}} \quad in \quad D \tag{7.21}$$

with the same topology.

Proof. The implication (7.20)$\to$(7.21) is directly covered by Theorem 13.7.2. to go the other way, note that $\lambda^{-1} \in \Lambda$ and λ^{-1} is absolutely continuous with continuous positive derivative $1/\dot{\lambda}(\lambda^{-1}(t))$. Moreover, if $v = -(u \circ \lambda^{-1})/\dot{\lambda} \circ \lambda^{-1}$ in (7.21), then $v(0) = 0$ and $-(v \circ \lambda)/(\dot{\lambda}^{-1}) \circ \lambda = u$. ∎

We can often apply the basic convergence-preservation results in combination. We can combine Theorems 13.3.1 and 13.7.2 to obtain limits for functions $x_n \circ y_n^{-1}$ and $x_n^{-1} \circ y_n$ with nonlinear centering.

Theorem 13.7.3. *(composition plus inverse with centering) Suppose that $x_n \in D$, $y_n \in D_u$, $c_n \to \infty$,*

$$c_n(x_n - x, y_n - y) \to (u, v) \quad in \quad D \times D \tag{7.22}$$

with one of the J_1, M_1 or M_2 topologies, where $v(0) = 0$ and v has no positive jumps if the topology is J_1, $y \in \Lambda$, x and y are absolutely continuous with continuous derivative $\dot{x}$ and $\dot{y}$ with $\dot{y} > 0$ and

$$Disc(u) \cap Disc(v) = \phi . \tag{7.23}$$

Then

$$c_n(x_n \circ y_n^{-1} - x \circ y^{-1}) \to u \circ y^{-1} - \left(\frac{\dot{x} \circ y^{-1}}{\dot{y} \circ y^{-1}}\right)(v \circ y^{-1}) \quad in \quad D . \tag{7.24}$$

If the topology is M_1 or M_2, then instead of (7.23) it suffices for u and v to have no common discontinuities with jumps of common (opposite) sign with $\dot{x}(t) \geq (\leq)0$ for all t.

Proof. The conditions imply that the conditions of Theorem 13.7.2 hold for y_n, so that

$$c_n(y_n^{-1} - y^{-1}) \to -\frac{v \circ y^{-1}}{\dot{y} \circ y^{-1}} \quad in \quad D . \tag{7.25}$$

The conditions then imply that the conditions of Theorem 13.3.3 hold with y_n^{-1} here playing the role of y_n there. We need

$$Disc(u \circ y^{-1}) \cap Disc(v \circ y^{-1}) = \phi \tag{7.26}$$

but that is equivalent to (7.23). With the M topologies, we can apply Theorems 12.7.3 and 12.11.6 to treat addition and Theorem 13.3.2 to treat multiplication. ∎

We now turn to the general first passage times

$$(x_n^{-1} \circ y_n)(t) = \inf\{s \geq 0 : x_n(s) > y_n(t)\}, \quad t \geq 0, \qquad (7.27)$$

which are elements of D when $x_n \in D_u$ and $y_n \in D_\uparrow$. The following is Puhalskii's (1994) result extended to allow discontinuous limits. For an application to obtain heavy-traffic stochastic-process limits for waiting times directly from corresponding heavy-traffic stochastic-process limits for queue lengths, see Section 5.4 of the Internet Supplement.

Theorem 13.7.4. (Puhalskii's theorem) *Suppose that* $x_n \in D_u$, $y_n \in D_\uparrow$, $c_n \to \infty$,

$$c_n(x_n - x, y_n - y) \to (u, v) \quad in \quad D \times D \qquad (7.28)$$

with one of the J_1, M_1 *or* M_2 *topologies, where* $u(0) = 0$, u *has no positive jumps if the topology is* J_1,

$$Disc(u \circ x^{-1} \circ y) \cap Disc(v) = \phi, \qquad (7.29)$$

$y \in C_{\uparrow\uparrow}$ *and* x *is absolutely continuous with a continuous positive derivative* $\dot{x}$, *then*

$$c_n(x_n^{-1} \circ y_n - x^{-1} \circ y) \to \frac{v - u \circ x^{-1} \circ y}{\dot{x} \circ x^{-1} \circ y} \quad in \quad D \qquad (7.30)$$

with the same topology. If the topology is M_1 *or* M_2, *then instead of condition* (7.29) *it suffices for* $u \circ x^{-1} \circ y$ *and* v *to have no common discontinuities with jumps of common sign.*

Proof. Since x is absolutely continuous with continuous positive derivative $\dot{x}$, $x \in C_{\uparrow\uparrow}$. Hence the conditions of Theorem 13.7.2 hold, so that

$$c_n(x_n^{-1} - x^{-1}) \to \frac{-u \circ x^{-1}}{\dot{x} \circ x^{-1}} \quad in \quad D \qquad (7.31)$$

with the same topology. We now apply Theorem 13.3.3, noting that x^{-1} has a continuous derivative $1/\dot{x}(x^{-1}(t))$. Condition (7.29) implies condition (3.14) for u in (3.14) equal to $-(u \circ x^{-1})/\dot{x} \circ x^{-1}$. Then (3.15) becomes (7.30). With the M topologies, we can apply Theorems 12.7.3 and 12.11.6. ∎

Remark 13.7.1. *Relating the theorems under extra conditions.* Under extra regularity conditions, we can apply Theorem 13.7.2 to obtain limits for $y_n \circ x_n^{-1}$ from limits for $x_n \circ y_n^{-1}$ provided by Theorem 13.7.3. We need $u(0) = v(0) = 0$, $x, y \in \Lambda$, $x_n, y_n \in D_u$ and both $\dot{x}$ and $\dot{y}$ to be continuous and positive. Since $x_n, y_n \in D_u$, $x_n^{-1}, y_n^{-1} \in D_{u,\uparrow}$. Then $\lambda \equiv x \circ y^{-1} \in \Lambda$ and $(x_n \circ y_n^{-1})^{-1} = y_n \circ x_n^{-1}$. From (7.16) and (7.24), we obtain

$$c_n(y_n \circ x_n^{-1} - y \circ x^{-1}) \to z \qquad (7.32)$$

where

$$z = \frac{-1}{\dot{\lambda} \circ \lambda^{-1}} \left(u \circ y^{-1} - \left(\frac{\dot{x} \circ y^{-1}}{\dot{y} \circ y^{-1}} \right) (v \circ y^{-1}) \right) \circ \lambda^{-1} \qquad (7.33)$$

for $\lambda = x \circ y^{-1}$. Since $\lambda^{-1} = y \circ x^{-1}$,

$$\dot{\lambda} = \frac{\dot{x} \circ y^{-1}}{\dot{y} \circ y^{-1}}, \quad \dot{\lambda} \circ \lambda^{-1} = \frac{\dot{x} \circ x^{-1}}{\dot{y} \circ x^{-1}} \tag{7.34}$$

and

$$z = -\frac{(\dot{y} \circ x^{-1})}{\dot{x} \circ x^{-1}}(u \circ x^{-1}) + v \circ x^{-1}, \tag{7.35}$$

which coincides with (7.24) with the labels changed, i.e., with (x, y, u, v) replaced by (y, x, v, u).

Similarly, under extra regularity conditions, we can apply Theorem 13.7.2 to obtain limits for $y_n^{-1} \circ x_n$ from limits for $x_n^{-1} \circ y_n$ provided by Theorem 13.7.4. We now need $x_n, y_n \in D_{u,\uparrow}$. We obtain

$$c_n(y_n^{-1} \circ x_n - y^{-1} \circ x) \to z, \tag{7.36}$$

where

$$z = \frac{-1}{\dot{\lambda} \circ \lambda^{-1}} \left(\frac{v - u \circ x^{-1} \circ y}{\dot{x} \circ x^{-1} \circ y} \right) \circ \lambda^{-1} \tag{7.37}$$

for $\lambda = x^{-1} \circ y$. Since $\lambda^{-1} = y^{-1} \circ x$,

$$\dot{\lambda} = \frac{\dot{y}}{\dot{x} \circ x^{-1} \circ y}, \quad \dot{\lambda} \circ \lambda^{-1} = \frac{\dot{y} \circ y^{-1} \circ x}{\dot{x}}, \tag{7.38}$$

and

$$z = \frac{u - v \circ y^{-1} \circ x}{\dot{y} \circ y^{-1} \circ x}, \tag{7.39}$$

which agrees with (7.30) with the labels changed, i.e., with (x, y, u, v) replaced by (y, x, v, u). ∎

13.8. Counting Functions

Inverse functions or first-passage-time functions are closely related to counting functions. A counting function is defined in terms of a sequence $\{s_n : n \geq 0\}$ of nondecreasing nonnegative real numbers with $s_0 = 0$. We can think of s_n as the partial sum

$$s_n \equiv x_1 + \cdots + x_n, \quad n \geq 1, \tag{8.1}$$

by simply writing $x_i \equiv s_i - s_{i-1}$, $i \geq 1$. The associated *counting function* $\{c(t) : t \geq 0\}$ is defined by

$$c(t) \equiv \max\{k \geq 0 : s_k \leq t\}, \quad t > 0. \tag{8.2}$$

To have $c(t)$ finite for all $t > 0$, we assume that $s_n \to \infty$ as $n \to \infty$. We can reconstruct the sequence $\{s_n\}$ from $\{c(t) : t \geq 0\}$ by

$$s_n = \inf\{t \geq 0 : c(t) \geq n\}, \quad n \geq 0. \tag{8.3}$$

The sequence $\{s_n\}$ and the associated function $\{c(t) : t \geq 0\}$ can serve as sample paths for a stochastic point process on the nonnegative real line. Then there are (countably) infinitely many points with the nth point being located at s_n. The summands x_n are then the intervals between successive points. The most familiar case is when the sequence $\{x_n : n \geq 1\}$ constitutes the possible values from a sequence $\{X_n : n \geq 1\}$ of IID random variables with values in $\mathbb{R}_+$. Then the counting function $\{c(t) : t \geq 0\}$ constitutes a possible sample path of an associated renewal counting process $\{C(t) : t \geq 0\}$; see Section 7.3.

Paralleling Lemma 13.6.3, we have the following basic inverse relation for counting functions.

Lemma 13.8.1. (inverse relation) *For any nonnegative integer n and nonnegative real number t,*

$$s_n \leq t \quad \text{if and only if} \quad c(t) \geq n. \tag{8.4}$$

We can put counting functions in the setting of inverse functions on $D_\uparrow$ by letting

$$y(t) \equiv s_{\lfloor t \rfloor}, t \geq 0. \tag{8.5}$$

To have $y \in D_\uparrow$, we use the assumption that $s_n \to \infty$ as $n \to \infty$. if all the summands are strictly positive, then

$$y^{-1}(t) = c(t) + 1, \quad t \geq 0, \tag{8.6}$$

where y^{-1} is the image of the inverse map in (6.1) applied to y in (8.5). With (8.6), limits for counting functions can be obtained by applying results in the previous two sections.

The connection to the inverse map can also be made when the summands x_i are only nonnegative. To do so, we observe that the counting function c is a time-transformation of y^{-1}. both are right-continuous, but $c(t) < y^{-1}(t)$. In particular, c and y can be expressed in terms of each other.

Lemma 13.8.2. (relation between counting functions and inverse functions) *For y in (8.5) and c in (8.2),*

$$c(t) = y^{-1}(y(y^{-1}(t)-)-), \quad t \geq 0, \tag{8.7}$$

$$c(t) = y^{-1}(t-) \quad \text{for all} \quad t \in Disc(c) = Disc(y^{-1}), \tag{8.8}$$

$$y^{-1}(t) = c(c^{-1}(c(t))), \quad t \geq 0. \tag{8.9}$$

The three functions y, y^{-1} and c are depicted for a typical initial segment of a sequence $\{s_n : n \geq 0\}$ in Figure 13.1.

We can apply (8.7)–(8.9) in Lemma 13.8.1 to show that limits for scaled counting functions with centering, are equivalent to limits for scaled inverse functions. We use the fact that the M topologies are not altered by changing to the left limits, because the graph is unchanged. We first consider the case of no centering; afterwards we consider the case of centering. When there is no centering, the M_1 and M_2 topologies coincide and reduce to pointwise convergence on a dense subset of $\mathbb{R}_+$ including 0.

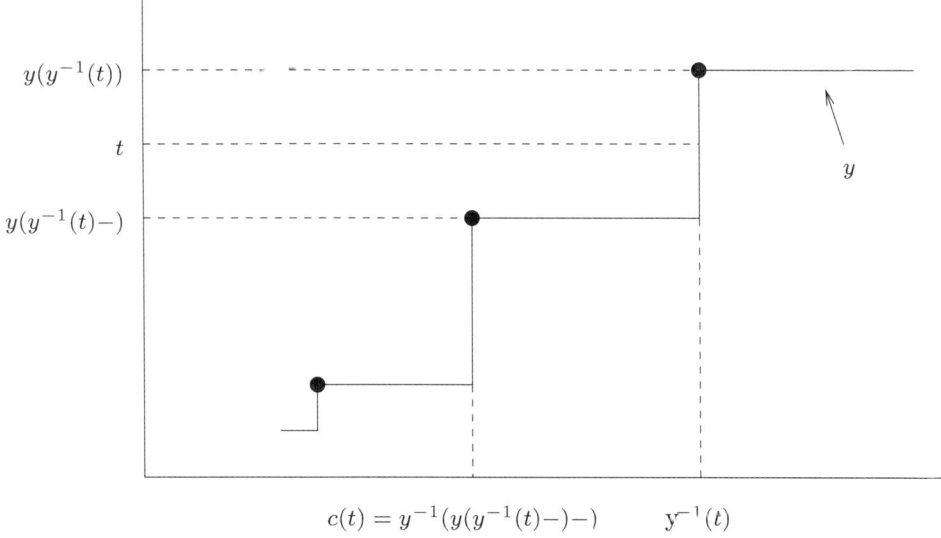

Figure 13.1. The relation between the counting function c and the inverse function y^{-1} for a typical function y.

Consider a sequence of counting functions $\{\{c_n(t) : t \geq 0\} : n \geq 1\}$ with associated processes

$$y_n^{-1}(t) \equiv c_n(c_n^{-1}(c_n(t))), \quad t \geq 0, \tag{8.10}$$

$y_n = (y_n^{-1})^{-1}$. Form scaled functions by setting

$$\hat{c}_n(t) = n^{-1}c_n(a_n t) \quad \text{and} \quad \hat{y}_n(t) = a_n^{-1}y_n(nt), \quad t \geq 0, \tag{8.11}$$

where a_n are positive real numbers with $a_n \to \infty$. Note that

$$\hat{c}_n^{-1}(t) = a_n^{-1}c_n^{-1}(nt) \quad \text{and} \quad \hat{y}_n^{-1}(t) = n^{-1}y_n(a_n t), \quad t \geq 0. \tag{8.12}$$

Theorem 13.8.1. (asymptotic equivalence of limits for scaled processes) *Suppose that $\hat{y}_n \in D_{u,\uparrow}$, $n \geq 1$, for $\hat{y}_n$ in (8.11). Then any one of the limits $\hat{y}_n \to \hat{y}$, $\hat{y}_n^{-1} \to \hat{y}^{-1}$, $\hat{c}_n \to \hat{y}^{-1}$ or $\hat{c}_n^{-1} \to \hat{y}$ in $D_\uparrow([0,\infty),\mathbb{R})$ with the M_2 $(= M_1)$ topology, for $\hat{y}_n^{-1}$, $\hat{c}_n$ and $\hat{c}_n^{-1}$ in (8.11) and (8.12), implies the others.*

We now apply the results for inverse maps with centering in Section 13.7 to obtain limits for counting functions with centering. Consider a sequence of counting functions $\{\{c_n(t) : t \geq 0\} : n \geq 1\}$ associated with a sequence of nondecreasing sequences of nonnegative numbers $\{\{s_{n,k} : k \geq 0\} : n \geq 1\}$ defined as in (8.2). Let the scaled functions $\hat{c}_n$, $\hat{y}_n$, c_n^{-1} and $\hat{y}_n^{-1}$ be defined as in (8.10)–(8.12).

Theorem 13.8.2. (asymptotic equivalence of counting and inverse functions with centering) *Consider $\hat{y}_n$, $\hat{c}_n$, and $\hat{y}_n^{-1}$ and $\hat{c}_n^{-1}$ as defined in (8.11) and (8.12). Suppose that $\hat{y}_n \in D_{u,\uparrow}$, $n \geq 1$, $b_n \to \infty$ and $\mathbf{z}(0) = 0$. Then any one of the limits*

$b_n(\hat{\mathbf{y}}_n - e) \to \mathbf{z}$, $b_n(\hat{\mathbf{c}}_n - e) \to -\mathbf{z}$, $b_n(\hat{\mathbf{y}}_n^{-1} - e) \to -\mathbf{z}$ or $b_n(\hat{\mathbf{c}}_n^{-1} - e) \to \mathbf{z}$ in $D([0,\infty), \mathbb{R})$ with the M_1 or M_2 topology implies the others with the same topology.

Corollary 13.8.1. *Consider a sequence of nondecreasing nonnegative sequences $\{\{s_{n,k} : k \geq 0\} : n \geq 1\}$ with $s_{n,0} = 0$ and $s_{n,k} \to \infty$ as $k \to \infty$ for all n. Let*

$$\mathbf{x}_n(t) = \delta_n^{-1}[s_{n,\lfloor nt \rfloor} - m_n nt], \quad t \geq 0,$$

and

$$\mathbf{y}_n(t) = \delta_n^{-1}[c_n(nt) - m_n^{-1} nt], \quad t \geq 0,$$

for $c_n(t)$ defined as in (8.2). Suppose that $\mathbf{u}(0) = 0$, $\delta_n \to \infty$, $n/\delta_n \to \infty$ and $m_n \to m > 0$ as $n \to \infty$. Then $\mathbf{x}_n \to \mathbf{u}$ in $D([0,\infty), \mathbb{R})$ with the M_1 or M_2 topology if and only if $\mathbf{y}_n \to -m^{-1}\mathbf{u} \circ m^{-1}\mathbf{e}$ in $D([0,\infty), \mathbb{R})$ with the same topology.

Proof. Apply Theorem 13.8.2, letting $\hat{\mathbf{c}}_n(t) = (a_n n)^{-1} c_n(nt)$ for $a_n = m_n^{-1}$ and, necessarily, $\hat{\mathbf{y}}_n(t) = n^{-1} s_{n, \lfloor a_n nt \rfloor}$. Then $b_n(\hat{\mathbf{y}}_n - e) \to \mathbf{z}$ if and only if $b_n(\hat{\mathbf{c}}_n - e) \to -\mathbf{z}$ for $b_n \to \infty$ and $\mathbf{z}(0) = 0$. However, $b_n(\hat{\mathbf{y}}_n - e) \to \mathbf{u} \circ m^{-1}\mathbf{e}$ if and only if $\mathbf{x}_n \to \mathbf{u}$, while $b_n(\hat{\mathbf{c}}_n - e) \to m^{-1}\mathbf{z}$ if and only if $\mathbf{y}_n \to \mathbf{z}$, for $b_n = n/\delta_n \to \infty$. ∎

14
Queueing Networks

14.1. Introduction

In this chapter we continue applying the continuous-mapping approach to establish heavy-traffic stochastic-process limits for queues, giving special attention to the possibility of having unmatched jumps in the limit process. Paralleling our application of the one-dimensional reflection map to obtain heavy-traffic limits for single queues in Chapters 5, 8 and 9, we now apply the multidimensional reflection map to obtain heavy-traffic limits for queueing networks. As before, we omit some proofs. These proofs plus additional supporting material appear in Chapter 8 of the Internet Supplement.

For background on queueing (or stochastic) networks, see Kelly (1979), Whittle (1986), Walrand (1988) and Serfozo (1999). For related discussions of heavy-traffic limits for queueing networks, see Chen and Mandelbaum (1994a,b), Harrison (1988, 2000, 2001a,b), Chen and Yao (2001) and Kushner (2001). The literature is discussed further at the end of this chapter.

The (standard) multidimensional reflection map was used with the continuous mapping theorem by Harrison and Reiman (1981a,b) and Reiman (1984a) to establish heavy-traffic limits with reflected Brownian motion limit processes for vector-valued queue-length, waiting-time and workload stochastic processes in single-class open queueing networks. Since Brownian motion and reflected Brownian motion have continuous sample paths, the topology of uniform convergence over bounded intervals could be used for those results. Variants of the M_1 topologies are needed to obtain alternative stochastic-process limits with discontinuous sample paths, such as reflected Lévy processes, when the discontinuities in the sample paths of the limit process are approached gradually in the sample paths of the converging processes.

However, unlike the one-dimensional reflection map, studied in Section 13.5 and applied in Chapters 5, 8 and 9, the multidimensional reflection map is not simply continuous, using either the SM_1 or WM_1 topology on D. Nevertheless, there is continuity with appropriate qualifications. In particular, the multidimensional reflection map is continuous at limits in the subset D_s of functions without simultaneous jumps of opposite sign in the coordinate functions, provided that the WM_1 topology is used on the range. As a consequence, the reflection map is continuous in the WM_1 topology at limits in the subset D_1 of functions that have discontinuities in only one coordinate at a time. That continuity property also holds for more general reflection maps and is sufficient to support limit theorems for stochastic processes in most applications.

We apply the continuity of the reflection map to obtain heavy-traffic limits for vector-valued buffer-content stochastic processes in single-class open stochastic fluid networks and for vector-valued queue-length stochastic processes in single-class open queueing networks. The M_1 topologies play a crucial role in these stochastic-process limits. With the stochastic fluid networks, just as with the single fluid queues in Chapters 5 and 8, the limit processes for the properly-scaled exogenous cumulative-input processes may have jumps even though the exogenous cumulative-input processes themselves have continuous sample paths. For example, this phenomenon occurs when the exogenous cumulative-input processes at the nodes are associated with the superpositions of independent on-off sources, where the busy (on or activity) periods and/or idle (off or inactivity) periods have heavy-tailed probability distributions (with infinite variances). Thus, in order to obtain heavy-traffic limits for properly scaled vector-valued buffer-content stochastic processes in the stochastic fluid networks, we need to invoke the continuous mapping theorem, using the continuity of the multidimensional reflection map on D with appropriate M_1 topologies.

Similarly, the M_1 topologies are needed to obtain heavy-traffic limits for open single-class queueing networks with single-server queues, where the servers are subject to rare long service interruptions. If the times between interruptions and the durations of the interruptions are allowed to increase appropriately as the traffic intensity increases in the heavy-traffic limit (with the durations of the interruptions being asymptotically negligible compared to the times between interruptions), then in the limit the interruptions occur instantaneously according to a stochastic point process, but nevertheless the interruptions have a spatial impact, causing jumps in the sample paths of the limiting queue-length process. Since these jumps in the limit process are approached gradually in the converging processes, again the M_1 topologies are needed to obtain heavy-traffic limits for the properly-scaled vector-valued queue-length stochastic processes.

Here is how this chapter is organized: We start in Section 14.2 by carefully defining the multidimensional reflection map and establishing its basic properties. Since the definition (Definition 14.2.1) is somewhat abstract, a key property is having the reflection map be well defined; i.e., we show that there exists a unique function

14.1. Introduction

satisfying the definition (Theorem 14.2.1). We also provide multiple characterizations of the reflection map, one alternative being as the unique fixed point of an appropriate operator (Theorem 14.2.2), while another is a basic complementarity property (Theorem 14.2.3).

A second key property of the multidimensional reflection map is Lipschitz continuity in the uniform norm on $D([0,T], \mathbb{R}^k)$ (Theorem 14.2.5). We also establish continuity of the multidimensional reflection map as a function of the reflection matrix, again in the uniform topology (Theorems 14.2.8 and 14.2.9). It is easy to see that the Lipschitz property is inherited when the metric on the domain and range is changed to d_{J_1} (Theorem 14.2.7). However, a corresponding direct extension for the SM_1 metric d_s does not hold. Much of the rest of the chapter is devoted to obtaining positive results for the M_1 topologies.

Section 14.3 provides yet another characterization of the multidimensional reflection map via an associated instantaneous reflection map on $\mathbb{R}^k$. The alternative characterization is as the limit of the multidimensional reflection map defined on the subset D_c of piecewise-constant functions in D (Theorem 14.3.4). On D_c, the reflection map can be defined in terms of the recursive application of the instantaneous reflection map on $\mathbb{R}^k$ (Theorem 14.3.1). The instantaneous reflection on $\mathbb{R}^k$ can be calculated by solving a linear program. Thus the instantaneous reflection map provides a useful way to simulate piecewise-constant approximations to reflected stochastic processes. We develop properties of the instantaneous reflection map that help us obtain the desired M_1 results (Theorem 14.3.2). In particular, we apply a monotonicity result (Corollary 14.3.2) to establish key properties of reflections of parametric representations (Lemma 14.3.4).

Sections 14.4 and 14.5 are devoted to obtaining the M_1 continuity results. In Section 14.4 we establish properties of reflection of parametric representations. We are able to extend Lipschitz and continuity results from the uniform norm to the M_1 metrics when we can show that the reflection of a parametric representation can serve as the parametric representation of the reflected function. The results are somewhat complicated, because this property holds only under certain conditions.

The basic M_1 Lipschitz and continuity results are established in Section 14.5. In addition to the positive results, we give counterexamples showing the necessity of the conditions in the theorems. Particularly interesting is Example 14.5.4, which shows that, without the regularity conditions, the fluctuations of the sequence of reflected processes can exceed the fluctuations in the reflection of the limit, thus exhibiting a kind of Gibbs phenomenon (see Remark 14.5.1). A proper limit in that example can be obtained, however, if we work in one of the more general spaces E or F introduced in Chapter 15.

In Sections 14.6 and 14.7, respectively, we apply the previous results to obtain heavy-traffic stochastic-process limits for stochastic fluid networks and conventional queueing networks. In the queueing networks we allow service interruptions. When there are heavy-tailed distributions or rare long service interruptions, the M_1 topologies play a critical role.

In Section 14.8 we consider the two-sided regulator and other reflection maps. The two-sided regulator is used to obtain heavy-traffic limits for single queues with finite waiting space, as considered in Section 2.3 and Chapter 8. With the scaling, the size of the waiting room is allowed to grow in the limit as the traffic intensity increases, but at a rate such that the limit process involves a two-sided regulator (reflection map) instead of the customary one-sided one. Like the one-sided reflection map, the two-sided regulator is continuous on (D^1, M_1). Moreover, the content portion of the two-sided regulator is Lipschitz, but the two regulator portions (corresponding to the two barriers) are only continuous; they are not Lipschitz.

We also give general conditions for other reflection maps to have M_1 continuity and Lipschitz properties. For these, we require that the limit function to be reflected belong to D_1, the subset of functions with discontinuities in only one coordinate at a time. We conclude the chapter with notes on the literature.

In Section 8.9 of the Internet Supplement we show that reflected stochastic processes have proper limiting stationary distributions and proper limiting stationary versions (stochastic-process limits for the entire time-shifted processes) under very general conditions. Our main result, Theorem 8.9.1 in the Internet Supplement, establishes such limits for stationary ergodic net-input stochastic processes satisfying a natural drift condition. It is noteworthy that a proper limit can exist even if there is positive drift in some (but not all) coordinates. Theorem 8.9.1 there is limited by having a special initial condition: starting out empty. Much of the rest of Section 8.9.1 in the Internet Supplement is devoted to obtaining corresponding results for other initial conditions. We establish convergence for all proper initial contents when the net input process is also a Lévy process with mutually independent coordinate processes (Theorem 8.9.6 there). Theorem 8.9.6 covers limit processes obtained in the heavy-traffic limits for the stochastic fluid networks in Section 14.6.

14.2. The Multidimensional Reflection Map

The definition of the multidimensional reflection map is somewhat indirect. So we begin by motivating the definition. Since the multidimensional reflection map arises naturally in the definition of the vector-valued buffer-content stochastic process in a stochastic fluid network, we first define the multidimensional reflection map in that special context.

14.2.1. A Special Case

We consider a single-class open stochastic fluid network with k nodes, each with a buffer of unlimited capacity. We let exogenous fluid input come to each of the k nodes. At each node the fluid is processed and released at a deterministic rate. The processed fluid from each node is then routed in a Markovian manner to other nodes or out of the network. The principal stochastic process of interest is the k-dimensional buffer-content process. The stochastic fluid network is a generalization of the fluid queue models in Chapters 5 and 8 in the case of unlimited waiting space (buffer capacity).

A single-class open stochastic fluid network with Markovian routing can be specified by a four-tuple $\{C, r, P, X(0)\}$, where $C \equiv (C^1, \ldots, C^k)$ is the vector of exogenous input stochastic processes at the k nodes, $r \equiv (r^1, \ldots, r^k)$ is the vector of deterministic output rates at the k nodes, $P \equiv (P_{i,j})$ is the $k \times k$ routing matrix and $X(0) \equiv (X^1(0), \ldots, X^k(0))$ is the nonnegative random vector of initial buffer contents at the k nodes. The stochastic process C is an element of $D_\uparrow^k \equiv D_\uparrow^1 \times \cdots \times D_\uparrow^1$, the subset of functions in $D \equiv D^k \equiv D([0,T], \mathbb{R}^k)$ that are nondecreasing and nonnegative in each coordinate. The random variable $C^i(t)$ represents the cumulative exogenous input to node i during the time interval $[0, t]$. It is natural to let the sample paths of C^i be continuous, but we do not require it.

When the buffer at node i is nonempty, there is fluid output from node i at constant rate r_i. When buffer i is empty, the output rate equals the minimum of the combined external (exogenous) plus internal input rate and the potential output rate r_i (formalized below). A proportion $P_{i,j}$ of all output from node i is immediately routed to node j, while a proportion $p_i \equiv 1 - \sum_{j=1}^k P_{ij}$ is routed out of the network. We assume that the routing matrix P is substochastic, so that $P_{i,j} \geq 0$ and $p_i \geq 0$ for all i, j. We also assume that $P^n \to 0$ as $n \to \infty$, where P^n is the nth power of P, so that all input eventually leaves the network. (We can think of P as the transition matrix of a transient k-state Markov chain.)

We now proceed to mathematically define the vector-valued buffer content process. We will work with column vectors. Hence it is convenient to use the transpose of the routing matrix P. Let Q be this transpose, i.e., $Q \equiv P^t$, so that Q is a column-substochastic matrix. We start by defining a potential buffer-content (or net-input) process, which represents the potential content at each node, ignoring the emptiness condition. The *potential buffer-content process* is

$$X(t) \equiv X(0) + C(t) - (I - Q)rt, \quad t \geq 0, \qquad (2.1)$$

where the vectors are regarded as column vectors and the transpose Q^t is the routing matrix P. The component $X^i(t)$ in (2.1) represents what the content of buffer i would be at time t if the output occurred continuously at rate r^j from node j for all j, regardless whether or not station j had fluid to emit. That is, the content vector $X(t)$ at time t, would be the initial value $X(0)$ plus the exogenous input $C(t)$ minus the output rt plus the internal input Qrt.

We obtain the actual buffer content by disallowing the potential output (and associated internal input) that cannot occur because of emptiness. We use the componentwise partial order on D and $\mathbb{R}^k$; i.e., $c_1 \equiv (c_1^1, \ldots, c_1^k) \leq c_2 \equiv (c_2^1, \ldots, c_2^k)$ in $\mathbb{R}^k$ if $c_2^i \leq c_2^i$ in $\mathbb{R}$ for all i, $1 \leq i \leq k$, and $x_1 \leq x_2$ in D if $x_1(t) \leq x_2(t)$ in $\mathbb{R}^k$ for all t, $0 \leq t \leq T$. We then let the buffer content be

$$Z(t) \equiv X(t) + (I - Q)Y(t), \qquad (2.2)$$

where X is defined in (2.1) and Y is the least possible element of $D_\uparrow^k$ (the subset of functions in D^k that are nonnegative and nondecreasing in each coordinate) such that $Z \geq 0$. We call the map from X to (Y, Z) the *reflection map*.

14.2.2. Definition and Characterization

We will prove that the vector-valued buffer-content process Z is properly defined by (2.1) and (2.2) by showing that such a process Y is well defined (exists and is unique). It turns out that the reflection map from X to Y and Z is well defined, even if X does not have the special structure in (2.1).

More generally, we call the transposed routing matrix Q the *reflection matrix*. Let $\mathcal{Q}$ be the set of all reflection matrices, i.e., the set of all column-stochastic matrices Q (with $Q_{i,j}^t \geq 0$ and $\sum_{j=1}^k Q_{i,j}^t \leq 1$) such that $Q^n \to 0$ as $n \to \infty$, where Q^n is the nth power of Q.

Definition 14.2.1. (reflection map) *For any $x \in D^k \equiv D([0,T], \mathbb{R}^k)$ and any reflection matrix $Q \in \mathcal{Q}$, let the feasible regulator set be*

$$\Psi(x) \equiv \{w \in D_\uparrow^k : x + (I - Q)w \geq 0\} \tag{2.3}$$

and let the reflection map be $R \equiv (\psi, \phi) : D^k \to D^{2k}$ with regulator component

$$y \equiv \psi(x) \equiv \inf \Psi(x) \equiv \inf\{w : w \in \Psi(x)\} , \tag{2.4}$$

i.e.,

$$y^i(t) \equiv \inf\{w^i(t) \in \mathbb{R} : w \in \Psi(x)\} \quad \text{for all} \quad i \quad \text{and} \quad t , \tag{2.5}$$

and content component

$$z \equiv \phi(x) \equiv x + (I - Q)y . \tag{2.6}$$

It remains to show that the reflection map is well defined by Definition 14.2.1; i.e., we need to know that the feasible regulator set $\Psi(x)$ is nonempty and that its infimum y (which necessarily is well defined and unique for nonempty $\Psi(x)$) is itself an element of $\Psi(x)$, so that $z \in D^k$ and $z \geq 0$.

To show that $\Psi(x)$ in (2.3) is nonempty, we exploit the well known fact that the matrix $I - Q$ has nonnegative inverse.

Lemma 14.2.1. (nonnegative inverse of reflection matrix) *For all $Q \in \mathcal{Q}$, $I - Q$ is nonsingular with nonnegative inverse*

$$(I - Q)^{-1} = \sum_{n=0}^{\infty} Q^n ,$$

where $Q^0 = I$.

The key to showing that the infimum belongs to the feasibility set is a basic result about semicontinuous functions. Recall that a real-valued function x on $[0, T]$ is

upper semicontinuous at a point t in its domain if
$$\limsup_{t_n \to t} x(t_n) \leq x(t)$$
for any sequence $\{t_n\}$ with $t_n \in [0, T]$ and $t_n \to t$ as $n \to \infty$. The function x is upper semicontinuous if it is upper semicontinuous at all arguments t in its domain.

Lemma 14.2.2. (preservation of upper semicontinuity) *Suppose that $\{x_s : s \in S\}$ is a set of upper semicontinuous real-valued function on a subinterval of $\mathbb{R}$. Then the infimum $\underline{x} \equiv \inf\{x_s : s \in S\}$ is also upper semicontinuous.*

Recall that $x^\uparrow \equiv \sup_{0 \leq s \leq t} x(s)$, $t \geq 0$, for $x \in D^1$. For $x \equiv (x^1, \ldots, x^k) \in D^k$, let $x^\uparrow \equiv ((x^1)^\uparrow, \ldots, (x^k)^\uparrow)$.

Theorem 14.2.1. (existence of the reflection map) *For any $x \in D^k$ and $Q \in \mathcal{Q}$,*
$$(I - Q)^{-1}[(-x)^\uparrow \vee 0] \in \Psi(x) , \qquad (2.7)$$
so that $\Psi(x) \neq \phi$.
$$y \equiv \psi(x) \in \Psi(x) \subseteq D_\uparrow^k \qquad (2.8)$$
for y in (2.4) and
$$z \equiv \phi(x) = x + (I - Q)y \geq 0 . \qquad (2.9)$$

Proof. Using Lemma 14.2.1, it is easy to see that (2.7) holds, because
$$x + (I - Q)(I - Q)^{-1}[(-x)^\uparrow \vee 0] \geq x + [(-x)^\uparrow \vee 0] \geq x + (-x)^\uparrow \geq x - x \geq 0$$
and $(-x)^\uparrow \vee 0 \in D_\uparrow^k$. Let y be the infimum in (2.4). Since $\Psi(x) \subseteq D_\uparrow^k$, necessarily y is nondecreasing and nonnegative. Since the elements of $D_\uparrow^k$ are nondecreasing, right-continuity coincides with upper semicontinuity, so we can apply Lemma 14.2.2 to conclude that $y \in D_\uparrow^k$. It now remains to show that $x + (I - Q)y \geq 0$. Fix ϵ, i and t. By the definition of the infimum, there exists $w \in \Psi(x)$ such that $w^i(t) \leq y^i(t) + \epsilon$ and $w^j(t) \geq y^j(t)$ for all j. Thus
$$x^i(t) + y^i(t) - \sum_{j=1}^k Q_{j,i} y^j(t) \geq x^i(t) + w^i(t) - \sum_{j=1}^k Q_{j,i} w^j(t) - \epsilon .$$
Since ϵ, i and t were arbitrary,
$$x + (I - Q)y \geq x + (I - Q)w \geq 0$$
so (2.8) holds and the proof is complete. ∎

We now characterize the regulator function $y = \psi(x)$ as the unique fixed point of a mapping $\pi = \pi_{x,Q} : D_\uparrow^k \to D_\uparrow^k$, defined by
$$\pi(w) = (Qw - x)^\uparrow \vee 0 \qquad (2.10)$$
for $w \in D_\uparrow^k$. For this purpose, we use two elementary lemmas.

Lemma 14.2.3. (feasible regulator set characterization) *The feasible regulator set $\Psi(x)$ in (2.3) can be characterized by*

$$\Psi(x) = \{w \in D_\uparrow^k : w \geq \pi(w)\}$$

for π in (2.10).

Proof. For each $w \in \Psi(x)$, $x + (I - Q)w \geq 0$ or, equivalently, $w \geq Qw - x$. Since $\Psi(x) \subseteq D_\uparrow^k$, we must also have $w \geq \pi(y)$. On the other hand, if $w \in D_\uparrow^k$ and $w \geq \pi(y)$, then we must have $w \geq Qw - x$, so that $w \in \Psi(x)$. ∎

Remark 14.2.1. *Semilattice structure.* It is also easy to see that if $y_1, y_2 \in \Psi(x)$, then $y_1 \wedge y_2 \in \Psi(x)$; that makes $\Psi(x)$ a meet semilattice.

Lemma 14.2.4. (closed subset of D) *With the uniform topology on D, The feasible regulator set $\Psi(x)$ is a closed subset of $D_\uparrow^k$, while $D_\uparrow^k$ is a closed subset of D.*

Theorem 14.2.2. (fixed-point characterization) *For each $Q \in \mathcal{Q}$, the regulator map $y \equiv \psi(x) \equiv \psi_Q(x) : D^k \to D_\uparrow^k$ can be characterized as the unique fixed point of the map $\pi \equiv \pi_{x,Q} : D_\uparrow^k \to D_\uparrow^k$ defined in (2.10).*

Proof. We use a standard argument to establish fixed points of monotone maps on ordered sets; e.g., see Section 3.8 of Edwards (1965). As before, we use the componentwise partial order on $\mathbb{R}^k$ and D^k. It is immediate that the map π in (2.10) is monotone. Note that

$$0 \leq \pi(0) = (-x)^\uparrow \vee 0 ,$$

where 0 is used as the vector and vector-valued function with zero values. Note that the functions 0 and $\pi(0)$ are both elements of $D_\uparrow^k$. Hence, the iterates $\pi^n(0) \equiv \pi(\pi^{n-1}(0))$ are elements of $D_\uparrow^k$ that are nondecreasing in n. On the other hand, $w \geq \pi(w)$ for any $w \in \Psi(x)$ by Lemma 14.2.3. Consequently, $\pi^n(w)$ is decreasing in n for any $w \in \Psi(x)$. Since $\Psi(x)$ is nonempty by Theorem 14.2.1, there exists $w \in \Psi(x)$. Since $0 \leq w$, $\pi^n(0) \leq \pi^n(w) \leq w$ for all n, so that $\pi^n(0)$ is bounded above. It is easy to see, using the addition and supremum maps, that $\pi : (D, U) \to (D, U)$ is also continuous. Hence $\pi^n(0) \uparrow w^*$ in $D_\uparrow^k$ as $n \to \infty$, where $\pi(w^*) = w^*$. Since $D_\uparrow^k$ is a closed subset of D by Lemma 14.2.4, $w^* \in D_\uparrow^k$. Since $w^* \geq \pi(w^*)$, $w^* \in \Psi(x)$ too. Since the regulator $y \equiv \psi(x)$ is the infimum by Definition 14.2.1, necessarily $y \leq w^*$. Since $0 \leq y \leq w^*$,

$$\pi^n(0) \leq \pi^n(y) \leq y \quad \text{for all} \quad n .$$

Letting $n \to \infty$, we see that $w^* \leq y$. Hence we must have $y = w^*$. ∎

Theorem 14.2.3. (complementarity characterization) *A function y in the feasible regulator set $\Psi(x)$ in (2.3) is the infimum $\psi(x)$ in (2.4) if and only if the pair (y, z) for $z \equiv x + (I - Q)y$ satisfies the complementarity property*

$$\int_0^\infty z^i dy^i = 0, \quad 1 \leq i \leq k . \tag{2.11}$$

14.2. The Multidimensional Reflection Map

Proof. We first prove that the infimum satisfies the complementarity property. We will show that failing to satisfy the complementarity implies failing the infimum property. Hence, suppose that (y, z) fails to satisfy the complementarity property (2.11). Thus there is t and j such that $z^j(t) > 0$ and y^j increases at t. We consider two cases:

Case 1. Suppose that $y^j(t) > y^j(t-)$. There must exist $\epsilon, \delta > 0$ such that $y^j(t) - y^j(t-) > \epsilon$ and $z^j(s) \geq \epsilon$, $t \leq s \leq t + \delta$. Let $\tilde{y}$ be defined by

$$\tilde{y}^j(s) = \begin{cases} y^j(s), & 0 \leq s < t, \\ y^j(s) - \epsilon, & t \leq s < t + \delta, \\ y^j(s), & t + \delta \leq s, \end{cases}$$

with $\tilde{y}^i = y^i$ for $i \neq j$. Then $\tilde{y} \in D_\uparrow^k$ and $x + (I - Q)\tilde{y} \geq 0$, so that $\tilde{y} \in \Psi(x)$, $\tilde{y} \leq y$ and $\tilde{y} \neq y$, which implies that y is not the infimum.

Case 2. Suppose that $y^j(t) = y^j(t-)$. Now there must exist $\epsilon, \delta > 0$ such that $z^j(s) \geq \epsilon$ and $0 \leq y^j(s) - y^j(t) \leq \epsilon$ for $t \leq s < t + \delta$ and $y^j(s) - y^j(t) > 0$ for $s > t$. Now let $\tilde{y}$ be defined by

$$\tilde{y}^j(s) = \begin{cases} y^j(s), & 0 \leq s < t, \\ y^j(t), & t \leq s < t + \delta, \\ y^j(s), & t + \delta \leq s, \end{cases}$$

with $\tilde{y}^i = y^i$ for $i \neq j$. Again $\tilde{y} \in D_\uparrow^k$ and $x + (I - Q)\tilde{y} \geq 0$. Since $\tilde{y} \leq y$ and $\tilde{y} \neq y$, y must not be the infimum. We now prove that the complementarity property implies the infimum property. Invoking Theorem 14.2.2, it suffices to show that if (y, z) satisfies the complementarity property, then necessarily y is the unique solution to the fixed point equation $y = \pi(y)$. Thus let $v = \pi(y)$. Since $y \in \Psi(x)$, $y \geq \pi(y)$ by Lemma 14.2.3, so that it suffices to show that $v = \pi(y) \geq y$. Suppose that $y(t) > v(t)$ for some t. Then necessarily $y(t_0) > v(t_0)$ for some t_0 that is a point of increase of y, but $y(t_0) > \pi(y)(t_0)$ implies that

$$z(t_0) = y(t_0) - (Qy - x)(t_0) \geq y(t_0) - \pi(y)(t_0) > 0,$$

which contradicts the complementarity condition. ∎

14.2.3. *Continuity and Lipschitz Properties*

We now establish continuity and Lipschitz properties of the reflection map as a function of the function x and the reflection matrix Q. We use the *matrix norm*, defined for any $k \times k$ real matrix A by

$$\|A\| \equiv \max_j \sum_{i=1}^{k} |A_{i,j}|. \tag{2.12}$$

We use the maximum column sum in (2.12) because we intend to work with the column-substochastic matrices in $\mathcal{Q}$. Note that

$$\|A_1 A_2\| \leq \|A_1\| \cdot \|A_2\|$$

for any two $k \times k$ real matrices A_1 and A_2. Also, using the sum (or l_1) norm

$$\|u\| \equiv \sum_{i=1}^{k} |u^i| \tag{2.13}$$

on $\mathbb{R}^k$, we have

$$\|Au\| \leq \|A\| \cdot \|u\| \tag{2.14}$$

for each $k \times k$ real matrix A and $u \in \mathbb{R}^k$. Indeed, we can also define the matrix norm by

$$\|A\| \equiv \max\{\|Au\| : u \in \mathbb{R}^n, \|u\| = 1\}, \tag{2.15}$$

using the sum norm in (2.13) in both places on the right. (Note that the sum norm in (2.13) is a change from the maximum norm in (2.1) of Section 12.2, but they are topologically equivalent.) Then (2.12) becomes a consequence. Consistent with (2.13), we let

$$\|x\| \equiv \sup_{0 \leq t \leq T} \|x(t)\| \equiv \sup_{0 \leq t \leq T} \sum_{i=1}^{k} \|x^i(t)\| \tag{2.16}$$

for $x \in D([0,T], \mathbb{R}^k)$. Combining (2.14) and (2.16), we have

$$\|Ax\| \leq \|A\| \cdot \|x\| \tag{2.17}$$

for each $k \times k$ real matrix A and $x \in D([0,T], \mathbb{R}^k)$.

We use the following basic lemma.

Lemma 14.2.5. (reflection matrix norms) *For any $k \times k$ matrix $Q \in \mathcal{Q}$,*

$$\|Q\| \leq 1, \quad \|Q^k\| = \gamma < 1 \tag{2.18}$$

and

$$\|(I-Q)^{-1}\| \leq \frac{k}{1-\gamma}. \tag{2.19}$$

Example 14.2.1. *Need for k-stage contraction.* The standard example in which $\|Q^j\| = 1$ for all j, $1 \leq j \leq k-1$, has $Q^t_{i,i+1} = 1$ for $1 \leq i \leq k-1$, $Q^t_{k,1} = \gamma$, $0 < \gamma < 1$, and $Q^t_{k,i} = 0$, $2 \leq i \leq k$. Then $Q^k = \gamma I$ and $\|Q^k\| = \gamma$. ∎

We now show that $\pi \equiv \pi_{x,Q}$ in (2.10) is a k-stage contraction map on $D_\uparrow^k$. Recall that for $x \in D$, $|x|$ denotes the function $\{|x(t)| : t \geq 0\}$ in D, where $|x(t)| = (|x^1(t)|, \ldots, |x^k(t)|) \in \mathbb{R}^k$. Thus, for $x \in D$, $|x|^\uparrow = (|x^1|^\uparrow, \ldots, |x^k|^\uparrow)$, where $|x^i|^\uparrow(t) = \sup_{0 \leq s \leq t} |x^i(s)|$, $0 \leq t \leq T$.

Lemma 14.2.6. (π *is a k-stage contraction*) *For any $Q \in \mathcal{Q}$ and $w_1, w_2 \in D_\uparrow^k$,*

$$|\pi^n(w_1) - \pi^n(w_2)|^\uparrow \leq |Q^n(|w_1 - w_2|^\uparrow)| \quad \text{for} \quad n \geq 1, \tag{2.20}$$

so that

$$\|\pi^n(w_1) - \pi^n(w_2)\| \leq \|Q^n\| \cdot \|w_1 - w_2\| \leq \|w_1 - w_2\| \tag{2.21}$$

for $n \geq 1$ and
$$\|\pi^n(w_1) - \pi^n(w_2)\| \leq \gamma \|w_1 - w_2\| \quad \text{for} \quad n \geq k,$$
where
$$\|Q^k\| \equiv \gamma < 1.$$
Hence
$$\|\pi^n(w) - \psi(x)\| \to 0 \quad \text{as} \quad n \to \infty.$$

Proof. First,
$$\begin{aligned}
|\pi(w_1) - \pi(w_2)|^\uparrow &= |(Qw_1 - x)^\uparrow \vee 0 - (Qw_2 - x)^\uparrow \vee 0|^\uparrow \\
&\leq |(Qw_1 - x)^\uparrow - (Qw_2 - x)^\uparrow|^\uparrow \\
&\leq |Qw_1 - x - Qw_2 + x|^\uparrow \\
&\leq Q|w_1 - w_2|^\uparrow,
\end{aligned}$$
which implies (2.20) for $n = 1$. We prove (2.20) for arbitrary n by induction. Suppose that it has been established up to n. Then
$$\begin{aligned}
|\pi^{n+1}(w_1) - \pi^{n+1}(w_2)|^\uparrow &= |(Q\pi^n(w_1) - x)^\uparrow \vee 0 - (Q\pi^n(w_2) - x)^\uparrow \vee 0|^\uparrow \\
&\leq Q|\pi^n(w_1) - \pi^n(w_2)|^\uparrow \leq Q^{n+1}|w_1 - w_2|^\uparrow,
\end{aligned}$$
using the induction hypothesis. Finally,
$$\|x\| = \sum_{i=1}^{k} |x^i|^\uparrow(T),$$
so that (2.21) follows directly from (2.20). By the Banach-Picard contraction fixed-point theorem, for any $w \in D_\uparrow^k$,
$$\|\pi^n(w) - w^*\| \to 0 \quad \text{as} \quad n \to \infty$$
geometrically fast, where w^* is the unique fixed point of π, but by Theorem 14.2.2 that fixed point is $\psi(x)$. ∎

We now establish inequalities that imply that the reflection map is a Lipschitz continuous map on $(D, \|\cdot\|)$. We use the stronger inequalities themselves to establish the existence of stationary versions of reflected processes in Section 8.9 of the Internet Supplement.

Theorem 14.2.4 (one-sided bounds) *For any $Q \in \mathcal{Q}$ and $x_1, x_2 \in D$,*
$$-(I - Q)^{-1}\eta_1(x_1 - x_2) \leq \psi(x_1) - \psi(x_2) \leq (I - Q)^{-1}\eta_1(x_2 - x_1) \quad (2.22)$$
where $\eta_1(x) \equiv (\hat{\eta}_1(x^1), \ldots, \hat{\eta}_1(x^k))$ with $\hat{\eta}_1 : D^1 \to D^1$ defined by
$$\hat{\eta}_1(x^i) \equiv (x^i)^\uparrow \vee 0.$$

Proof. Use the map π_x in (2.10). Assuming that

$$\psi(x_1) \leq \pi_{x_2}^n(\psi(x_1)) + \sum_{i=0}^{n-1} Q^i \eta_1(x_2 - x_1) ,$$

with $\pi_{x_2}^0$ being the identity (which is true for $n = 0$), we have

$$\begin{aligned}
\psi(x_1) &= \pi_{x_1}(\psi(x_1)) = \eta_1(Q\psi(x_1) - x_1) \\
&\leq \eta_1(Q\pi_{x_2}^n(\psi(x_1)) - x_2 + Q\sum_{i=0}^{n-1} Q^i \eta_1(x_2 - x_1) + (x_2 - x_1)) \\
&\leq \eta_1(Q\pi_{x_2}^n(\psi(x_1)) - x_2 + \sum_{i=0}^{n} Q^i \eta_1(x_2 - x_1)) \\
&\leq \eta_1(Q\pi_{x_2}^n(\psi(x_1) - x_2) + \sum_{i=0}^{n} Q^i \eta_1(x_2 - x_1) \\
&\leq \pi_{x_2}^{n+1}(\psi(x_1)) + \sum_{i=0}^{n} Q^i \eta_1(x_2 - x_1) .
\end{aligned}$$

Letting $n \to \infty$, we obtain the second inequality in (2.22). The first inequality in (2.22) follows by symmetry. ∎

We also have the following consequence.

Corollary 14.2.1. (more bounds) *For any $Q \in \mathcal{Q}$, $x \in D^k$ and $w \in \mathbb{R}_+^k$,*

$$0 \leq \psi(x + w) \leq \psi(x) \leq \psi(x + w) + (I - Q)^{-1} w .$$

Proof. Apply Theorem 14.2.4, letting $x_1 \equiv x$ and $x_2 = x + w$. ∎

As a direct consequence of Theorem 14.2.4, we obtain the desired Lipschitz property.

Theorem 14.2.5. (Lipschitz property with uniform norm) *For any $Q \in \mathcal{Q}$ and $x_1, x_2 \in D$,*

$$\begin{aligned}
\|\psi(x_1) - \psi(x_2)\| &\leq \|(I - Q)^{-1}\| \cdot \|x_1 - x_2\| \\
&\leq \sum_{n=0}^{\infty} \|Q^n\| \cdot \|x_1 - x_2\| \\
&\leq \frac{k}{1 - \gamma} \|x_1 - x_2\| ,
\end{aligned} \qquad (2.23)$$

where $\gamma \equiv \|Q^k\| < 1$, and

$$\begin{aligned}
\|\phi(x_1) - \phi(x_2)\| &\leq (1 + \|I - Q\| \cdot \|(I - Q)^{-1}\|)\|x_1 - x_2\| \\
&\leq \left(1 + \frac{2k}{1 - \gamma}\right) \|x_1 - x_2\| .
\end{aligned} \qquad (2.24)$$

Proof. Since
$$\left\|\sum_{n=0}^{\infty} Q^n\right\| \le \sum_{n=0}^{\infty} \|Q^n\| \le \sum_{n=0}^{k-1} \|Q^n\| + \gamma \sum_{n=0}^{\infty} \|Q^n\|,$$
we have
$$\sum_{n=0}^{\infty} \|Q^n\| \le \frac{\sum_{n=0}^{k-1} \|Q^n\|}{1-\gamma} \le \frac{k}{1-\gamma}.$$
By (2.6) and (2.23),
$$\begin{aligned} \|\phi(x_1) - \phi(x_2)\| &\le \|x_1 - x_2\| + \|I - Q\| \cdot \|\psi(x_1) - \psi(x_2)\| \\ &\le (1 + \|I - Q\| \cdot \|(I-Q)^{-1}\|)\|x_1 - x_2\|. \quad \blacksquare \end{aligned}$$

Remark 14.2.2. *Alternate proof.* Instead of applying Theorem 14.2.4, we could prove Theorem 14.2.5 by reasoning as in Lemma 14.2.6 to get
$$|\pi_{x_1}^n(0) - \pi_{x_2}^n(0)|^{\uparrow} \le (I + Q + \cdots + Q^{n-1})|x_1 - x_2|^{\uparrow},$$
which implies that
$$|\psi(x_1) - \psi(x_2)|^{\uparrow} \le (I - Q)^{-1}|x_1 - x_2|^{\uparrow}$$
and then (2.23). $\blacksquare$

Remark 14.2.3. *Lipschitz constant.* The upper bounds in Theorem 14.2.5 are minimized by making $Q_{i,j} = 0$ for all i, j. Let K^* be the infimum of K such that
$$\|R(x_1) - R(x_2)\| \le K\|x_1 - x_2\| \quad \text{for all} \quad x_1, x_2 \in D. \tag{2.25}$$
We call K^* the *Lipschitz constant*. The bounds yield $K^* \le 2$ when $Q_{i,j} = 0$ for all i, j, but the following example in shows that $K^* = 2$ in that case. Hence $K^* \ge 2$ in general.

Example 14.2.2. *Lower bound on the Lipschitz constant.* Let $k = 1$, $Q = 0$, $x_1(t) = 0$, $0 \le t \le 1$, and $x_2 = -I_{[1/3,1/2)} + I_{[1/2,1]}$ in $D([0,1], \mathbb{R})$. Then $y_1 = z_1 = x_1$, but $y_2 = I_{[1/3,1]}$ and $z_2 = 2I_{[1/2,1]}$, so that $\|z_1 - z_2\| = 2$. Hence $K^* \ge 2$ for all Q.

Example 14.2.3. *No upper bound on the Lipschitz constant.* To see that there is no upper bound on the Lipschitz constant K^* independent of Q, let $x_1(t) = 0$, $0 \le t \le 2$, and $x_2 = -I_{[1,2]}$ in $D([0,2], \mathbb{R})$, so that $\|x_1 - x_2\| = 1$. Let $Q = 1 - \epsilon$, so that (2.6) becomes $z = x + \epsilon y$. Then $z_2 = z_1 = y_1 = x_1$, but $y_2 = \epsilon^{-1} I_{[1,2]}$, so that $\|y_1 - y_2\| = \epsilon^{-1}$.

Example 14.2.4. *No upper bound on the Lipschitz constant for ϕ.* To see that the Lipschitz constant for the component map ϕ can be arbitrarily large as well, consider the two-dimensional example with $Q_{1,1} = 1 - \epsilon$, $Q_{2,1} = 1$ and $Q_{2,2} = Q_{1,2} = 1/2$, so that

$$(I - Q^t) = \begin{pmatrix} \epsilon & -1 \\ -1/2 & 1/2 \end{pmatrix}.$$

Let $x_1^1 = -I_{[1,2]}$, $x_2^1(t) = 0$, $0 \le t \le 2$, and $x_1^2 = x_2^2 = \epsilon^{-1}I_{[0,2]}$ in $D([0,2], \mathbb{R}^2)$. Then $\|x_1 - x_2\| = 1$, but $z_1^1(t) = z_2^1(t) = 0$, $0 \le t \le 2$, $z_1^2 = \epsilon^{-1}I_{[0,1)}$ and $z_2^2 = \epsilon^{-1}I_{[0,2]}$, so that $\|z_1 - z_2\| = \epsilon^{-1}$.

We now summarize some elementary but important properties of the reflection map.

Theorem 14.2.6. (reflection map properties) *The reflection map satisfies the following properties:*

(i) adaptedness: *For any $x \in D$ and $t \in [0, T]$, $R(x)(t)$ depends upon x only via $\{x(s) : 0 \le s \le t\}$.*

(ii) monotonicity: *If $x_1 \le x_2$ in D, then $\psi(x_1) \ge \psi(x_2)$.*

(iii) rescaling: *For each $x \in D([0,T], \mathbb{R}^k)$, $\eta \in \mathbb{R}^k$, $\beta > 0$ and γ nondecreasing right-continuous function mapping $[0, T_1]$ into $[0, T]$, $\eta + \beta(x \circ \gamma) \in D([0, T_1], \mathbb{R}^k)$ and*

$$R(\eta + \beta(x \circ \gamma)) = \beta R(\beta^{-1}\eta + x) \circ \gamma.$$

(iv) shift: *For all $x \in D$ and $0 < t_1 < t_2 < T$,*

$$\psi(x)(t_2) = \psi(x)(t_1) + \psi(\phi(x)(t_1) + x(t_1 + \cdot) - x(t_1))(t_2 - t_1)$$

and

$$\phi(x)(t_2) = \phi(\phi(x)(t_1) + x(t_1 + \cdot) - x(t_1))(t_2 - t_1)$$

(v) continuity preservation: *If $x \in C$, then $R(x) \in C$.*

We can apply Theorems 14.2.5 and 14.2.6 (iii) to deduce that the reflection map inherits the Lipschitz property on (D, J_1) from (D, U). Unfortunately, we will have to work harder to obtain related results for the M_1 topologies.

Theorem 14.2.7. (Lipschitz property with d_{J_1}) *For any $Q \in \mathcal{Q}$, there exist constants K_1 and K_2 (the same as in Theorem 14.2.5) such that*

$$d_{J_1}(\psi(x_1), \psi(x_2)) \le K_1 d_{J_1}(x_1, x_2) \qquad (2.26)$$

and

$$d_{J_1}(\phi(x_1), \phi(x_2)) \le K_2 d_{J_1}(x_1, x_2) \qquad (2.27)$$

for all $x_1, x_2 \in D$.

Proof. The argument is the same for (2.26) and (2.27), so we prove only (2.26). By the definition of d_{J_1} and Theorems 14.2.6(iii) and 14.2.5,

$$\begin{aligned} d_{J_1}(\psi(x_1), \psi(x_2)) &\equiv \inf_{\lambda \in \Lambda} \{\|\psi(x_1) \circ \lambda - \psi(x_2)\| \vee \|\lambda - e\|\} \\ &= \inf_{\lambda \in \Lambda} \{\|\psi(x_1 \circ \lambda) - \psi(x_2)\| \vee \|\lambda - e\|\} \end{aligned}$$

$$\leq \inf_{\lambda \in \Lambda} \{K_1 \|x_1 \circ \lambda - x_2\| \vee \|\lambda - e\|\}$$
$$\leq (K_1 \vee 1) d_{J_1}(x_1, x_2) ,$$

which implies (2.26) because $K_1 \geq 1$. ∎

Even when we establish convergence to stochastic-process limits with continuous sample paths, we need maps on D to be measurable as well as continuous at $x \in C$ in order to apply the continuous mapping theorem on D. From Theorem 14.2.5, it follows that the reflection map is measurable with respect to the Borel σ-field associated with the uniform topology (on both the domain and range), but that does not imply measurability with respect to the usual Kolmogorov σ-field generated by the coordinate projections, because the Borel σ-field on (D, U) is much larger than the Kolmogorov σ-field; see Section 11.5.3. Fortunately, Theorem 14.2.7 directly implies the desired measurability.

Corollary 14.2.2. (measurability) *The reflection map R mapping D^k to D^{2k} is measurable, using the Kolmogorov σ-field on the domain and range.*

Proof. The continuity established in Theorem 14.2.7 implies measurability with respect to the Borel σ-fields, but the Borel σ-field on (D, J_1) coincides with the Kolmogorov σ-field generated by the projection map; see Theorem 11.5.2. ∎

We now want to consider the reflection map R as a function of the reflection matrix Q as well as the net input function x. We first consider the maps $\pi \equiv \pi^n_{x,Q}(0)$ in (2.10) and $\psi \equiv \psi_Q$ in (2.4) as functions of Q when Q is a strict contraction in the matrix norm (2.12), i.e., when $\|Q\| < 1$.

Theorem 14.2.8. (stability bounds for different reflection matrices) *Let $Q_1, Q_2 \in \mathcal{Q}$ with $\|Q_1\| = \gamma_1 < 1$ and $\|Q_2\| = \gamma_2 < 1$. For all $n \geq 1$,*

$$\|\pi^n_{x,Q_j}(0)\| \leq (1 + \gamma_j + \cdots + \gamma_j^{n-1})\|x\| \qquad (2.28)$$

and

$$\|\pi^n_{x,Q_1}(0) - \pi^n_{x,Q_2}(0)\| \leq (1 + \gamma_2 + \cdots + \gamma_2^{n-1})\frac{\|x\| \cdot \|Q_1 - Q_2\|}{1 - \gamma_1} , \qquad (2.29)$$

so that

$$\|\psi_{Q_j}(x)\| \leq \frac{\|x\|}{1 - \gamma_j} \qquad (2.30)$$

and

$$\|\psi_{Q_1}(x) - \psi_{Q_2}(x)\| \leq \frac{\|x\| \cdot \|Q_1 - Q_2\|}{(1 - \gamma_1)(1 - \gamma_2)} . \qquad (2.31)$$

Corollary 14.2.3. (continuity as a function of Q) *If $Q_n \to Q$ in $\mathcal{Q}$,*

$$\|R_{Q_n}(x) - R_Q(x)\| \to 0 \quad as \quad n \to \infty$$

for each $x \in D$.

Proof. Apply Theorem 14.2.8, noting that $\|Q_n\| \to \|Q\|$ if $Q_n \to Q$. ∎

We now exploit the fact that any $Q \in \mathcal{Q}$ can be transformed into another matrix Q_* in $\mathcal{Q}$ with $\|Q_*\| < 1$ by letting

$$Q_* \equiv \Lambda^{-1} Q \Lambda \qquad (2.32)$$

for a suitable strictly positive diagonal matrix Λ. We note that a nonnegative matrix Q is in $\mathcal{Q}$ if and only if $|sp(Q)| < 1$, where $sp(Q)$ is its spectrum (set of eigenvalues) and $|sp(Q)|$ is its spectral radius (supremum of the eigenvalue norms): There is one real eigenvalue equal to $|sp(Q)|$ and all other eigenvalues λ (in general complex valued) satisfy $|\lambda| \leq |sp(Q)|$; e.g., see Seneta (1981). Next note that the transformation from Q to Q_* in (2.32) leaves the eigenvalues unchanged: If λ is an eigenvalue of Q with an associated left eigenvector u, i.e., if $uQ = \lambda u$, then

$$(u\Lambda) Q_* = u\Lambda (\Lambda^{-1} Q \Lambda) = (uQ\Lambda) = \lambda(u\Lambda) \; ,$$

so that λ is also an eigenvalue of Q_* with left eigenvector $u\Lambda$.

Lemma 14.2.7. (equivalence to a contractive reflection matrix) *For any $Q \in \mathcal{Q}$, there exists a positive diagonal matrix Λ such that $\|Q_*\| < 1$ for Q_* in (2.32).*

Proof. By Corollary 14.3.3 in Section 14.3 below, for any $u \in \mathbb{R}^k$, the map $\pi_0(v) \equiv (Qv - u)^+$ has a unique fixed point in $\mathbb{R}^k$. Let v be a vector in $\mathbb{R}^k$ such that $1 + Qv = v$. Necessarily $v \geq 1$. Let $\Lambda \equiv diag(1/v^j)$. Then

$$\Lambda v - \Lambda 1 = \Lambda(1 + Qv) - \Lambda 1 \; ,$$

so that

$$1 - \Lambda 1 = \Lambda Q v = \Lambda Q \Lambda^{-1} \equiv Q_* \; .$$

Since $0 \leq 1 - \Lambda 1 < 1$, $\|Q_*\| < 1$. ∎

By Theorem 14.2.6(iii), we can relate the reflection maps R_Q and R_{Q_*}.

Lemma 14.2.8. (reflections associated with equivalent reflection matrices) *Let Q and Q_* be reflection matrices in $\mathcal{Q}$ related by (2.32) for some strictly positive diagonal matrix Λ. Then*

$$\Psi_{Q_*}(\Lambda^{-1} x) = \Lambda^{-1} \Psi_Q(x) \equiv \{\Lambda^{-1} x : x \in \Psi_Q(x)\}$$

and

$$R_{Q_*}(\Lambda^{-1} x) = \Lambda^{-1} R_Q(x)$$

for all $x \in D$.

Proof. For any $w \in \Psi_Q(x)$,

$$\begin{aligned}
0 \leq x + (I - Q)w &= x + (I - \Lambda Q_* \Lambda^{-1})w \\
&= \Lambda \Lambda^{-1} x + (\Lambda \Lambda^{-1} - \Lambda Q_* \Lambda^{-1})w \\
&= \Lambda(\Lambda^{-1} x + (I - Q_*)(\Lambda^{-1} w)) \; ,
\end{aligned}$$

so that $\Lambda^{-1}w \in \Psi_{Q_*}(\Lambda^{-1}x)$, with $w_1 \le w_2$ if and only if $\Lambda^{-1}w_1 \le \Lambda^{-1}w_2$, so that $\Psi_{Q_*}(\Lambda^{-1}x) = \Lambda^{-1}\psi_Q(x)$ and

$$\Lambda^{-1}z = \Lambda^{-1}x + (I - Q_*)(\Lambda^{-1}y) ,$$

so that $\phi_{Q_*}(\Lambda^{-1}x) = \Lambda^{-1}\psi_Q(x)$. ∎

Theorem 14.2.9. (continuity as a function of x and Q) If $\|x_n - x\| \to 0$ in D^k and $Q_n \to Q$ in $\mathcal{Q}$, then

$$\|R_{Q_n}(x_n) - R_Q(x)\| \to 0 \quad \text{in} \quad D^{2k} .$$

Proof. By Lemma 14.2.7, we can find a positive diagonal matrix Λ so that $Q_* = \Lambda^{-1}Q\Lambda$ and $\|Q_*\| = \gamma < 1$. Since $Q_n \to Q$ as $n \to \infty$, $\|Q_{n*}\| \equiv \gamma_n \to \gamma$, where $Q_{n*} \equiv \Lambda^{-1}Q_n\Lambda$ with the same diagonal matrix used above. Consider n sufficiently large that $\gamma_n < 1$. Since $\psi_{Q_*}(\Lambda^{-1}x) = \Lambda^{-1}\psi_Q(x)$, for such n we have

$$\begin{aligned}
\|\psi_{Q_n}(x_n) - \psi_Q(x)\| &= \|\Lambda\Lambda^{-1}\psi_{Q_n}(x_n) - \Lambda\Lambda^{-1}\psi_Q(x)\| \\
&\le \|\Lambda\| \cdot \|\psi_{Q_{n*}}(\Lambda^{-1}x_n) - \psi_{Q_*}(\Lambda^{-1}x)\| \\
&\le \|\Lambda\|(\|\psi_{Q_{n*}}(\Lambda^{-1}x_n) - \psi_{Q_{n*}}(\Lambda^{-1}x)\| \\
&\quad + \|\psi_{Q_{n*}}(\Lambda^{-1}x) - \psi_{Q_*}(\Lambda^{-1}x)\|) .
\end{aligned}$$

Thus, by (2.30) and (2.31),

$$\begin{aligned}
\|\psi_{Q_n}(x_n) - \psi_Q(x)\| &\le \|\Lambda^{-1}\|\left(\frac{\|\Lambda x_n - \Lambda x\|}{1 - \gamma_n} + \frac{\|\Lambda x\| \cdot \|Q_{n*} - Q_*\|}{(1-\gamma_n)(1-\gamma)}\right) \\
&\le M_n\left(\|x_n - x\| + \frac{\|x\| \cdot M_n \cdot (1-\gamma_n) \cdot \|Q_n - Q\|}{1-\gamma}\right)
\end{aligned}$$

for

$$M_n \equiv \frac{\|\Lambda^{-1}\| \cdot \|\Lambda\|}{1 - \gamma_n} .$$

Hence,

$$\|\psi_{Q_n}(x_n) - \psi_Q(x)\| \to 0 \quad \text{as} \quad n \to \infty . \quad \blacksquare$$

14.3. The Instantaneous Reflection Map

In this section we introduce yet another characterization of the reflection map. We represent the reflection map on D as the limit of reflections of functions in D_e, the subset of piecewise constant functions in D. (Recall from Section 12.2 that any function in D can be approximated uniformly by functions in D_c.) On the subset D_c, the reflection map reduces to an iterative application of an instantaneous reflection map on $\mathbb{R}^k$, which is of interest itself.

The instantaneous reflection map describes how

$$(y(0), z(0)) \equiv (\psi(x)(0), \phi(x)(0))$$

depends upon $x(0)$ and characterizes the behavior of the full reflection map at each discontinuity point. We will apply the instantaneous reflection map to establish monotonicity results (in particular, Corollary 14.3.2 below), which we will in turn apply to establish the M_1 continuity results. We use the final Lemma 14.3.4 in the next section to study reflections of parametric representations.

14.3.1. Definition and Characterization

Let the instantaneous reflection map be $R_0 \equiv (\phi_0, \psi_0) : \mathbb{R}^k \to \mathbb{R}^{2k}$, where $\psi_0 : \mathbb{R}^k \to \mathbb{R}^k$ is defined by

$$\psi_0(u) \equiv \inf\{v \in \mathbb{R}_+^k : u + (I - Q)v \geq 0\}, \qquad (3.1)$$

where $u_1 \leq u_2$ in $\mathbb{R}^k$ if $u_1^i \leq u_2^i$ in $\mathbb{R}$, $1 \leq i \leq k$. It turns out that the infimum in (3.1) is attained (so that we can refer to it as the minimum) and there are useful expressions for it. Given the solution to (3.1), we can define the other component of the instantaneous reflection map $\phi_0 : \mathbb{R}^k \to \mathbb{R}^k$ by

$$\phi_0(u) = u + (I - Q)\psi_0(u). \qquad (3.2)$$

The instantaneous reflection map is a version of the *linear complementarity problem* (LCP), which has a long history; see Cottle, Pang and Stone (1992). Given a vector $u \in R^k$ and a $k \times k$ matrix M, the LCP is to find a vector $v \in \mathbb{R}_+^k$ such that

$$u + Mv \geq 0 \qquad (3.3)$$

and

$$v_i(u + Mv)_i = 0, \quad 1 \leq i \leq k. \qquad (3.4)$$

It turns out that the vector v satisfying (3.1) also satisfies (3.4) for $M = I - Q$ with $Q \in \mathcal{Q}$; see Corollary 14.3.1 below. Moreover, it turns out that the LCP based on (u, M) has solutions for all vectors u for more general matrices M than $I - Q$ for $Q \in \mathcal{Q}$. Such generalizations provide a basis for defining reflection maps in terms of other matrices, but we do not pursue that generalization.

Since the instantaneous reflection map is defined on $\mathbb{R}^k$ instead of D^k, it is much easier to calculate. For example, we can calculate v satisfying (3.1) by solving the *linear program*

$$\min c^t v$$

$$\text{subject to:} \quad u + (I - Q)v \geq 0 \qquad (3.5)$$

$$v \geq 0$$

for any strictly positive vector c in $\mathbb{R}^k$. We can thus simulate piecewise-constant approximations to reflected stochastic processes by generating finitely many linear programs. (We briefly discuss simulation some more at the end of the section.)

Paralleling the stochastic network example used to motivate the definition of the reflection map in the beginning of Section 14.2, we can introduce a discrete-time fluid-network model to motivate the instantaneous reflection map. At each transition epoch there is an input and a potential output. We might have an instantaneous nonnegative input vector u_1 and potential instantaneous nonnegative output vector u_2 which is routed to other queues by the stochastic matrix Q, so that the overall potential instantaneous net input is

$$u = u_1 - u_2 + u_2 Q . \tag{3.6}$$

However, if the potential output u_2 in (3.6) exceeds the available supply, then we may have to disallow some of the output u_2. That can be accomplished by adding a minimal $(I - Q^t)v$ to u in (3.6), which gives (3.1). In fact, as we will show, the instantaneous reflection map in (3.1) and (3.2) is well defined for any $u \in \mathbb{R}^k$, not just for u of the form (3.6).

Given the instantaneous reflection map R_0 in (3.1) and (3.2) (which we have yet to show is well defined), it is straightforward to define the associated reflection map R on D_c. For any $x \in D_c$, the set of discontinuities is $Disc(x) = \{t_1, \ldots, t_m\}$ for some integer m and some time points t_i satisfying and $t_0 \equiv 0 < t_1 < \cdots < t_m < T$. Clearly we should have

$$\psi(x)(t_i) \equiv y(t_i) = \psi_0(z(t_{i-1}) + x(t_i) - x(t_{i-1})) + y(t_{i-1}) \tag{3.7}$$

and

$$\phi(x)(t_i) \equiv z(t_i) = \phi_0(z(t_{i-1}) + x(t_i) - x(t_{i-1})) \tag{3.8}$$

for $0 \leq i \leq m$, where $z(t_{-1}) \equiv y(t_{-1}) \equiv x(t_{-1}) \equiv 0$, and we let (y, z) be piecewise constant with $Disc(y, z) = Disc(x)$.

Theorem 14.3.1. (reflection map for piecewise-constant functions) *For all $x \in D_c$, the reflection map defined by (3.1), (3.7) and (3.8) is equivalent to the reflection map in Definition* 14.2.1.

Proof. By induction, we can reexpress (3.7) as

$$\begin{aligned}
y(t_i) &= \psi_0(z(t_{i-1}) + x(t_i) - x(t_{i-1})) + y(t_{i-1}) \\
&= \min\{v \in \mathbb{R}_+^k : z(t_{i-1}) + x(t_i) - x(t_{i-1}) + (I - Q)v \geq 0\} + y(t_{i-1}) \\
&= \min\{v \in \mathbb{R}_+^k : x(t_i) + (I - Q)y(t_{i-1}) + (I - Q)v \geq 0\} + y(t_{i-1}) \\
&= \min\{v \geq y(t_{i-1}) : x(t_i) + (I - Q)v \geq 0\} ,
\end{aligned}$$

which corresponds to Definition 14.2.1. ∎

Since D_c is a subset of D, all the theorems in Section 14.2 apply to the reflection map on D_c in (3.7) and (3.8) by virtue of Theorem 14.3.1. The instantaneous reflection map itself corresponds to the reflection map in (3.7) and (3.8) applied to constant functions. Thus, from Section 14.2, we already know that the instantaneous reflection map is well defined. However, we can deduce additional structure of the reflection map by focusing directly on the instantaneous reflection map.

We first establish upper and lower bounds on $\psi_0(u)$. For $u \in \mathbb{R}^k$, let

$$u^+ \equiv u \vee 0 \equiv (u^1 \vee 0, \ldots, u^k \vee 0) \quad \text{and} \quad u^- \equiv u \wedge 0 \equiv (u^1 \wedge 0, \ldots, u^k \wedge 0) .$$

Lemma 14.3.1. (*bounds on $\psi_0(u)$*) *For any $u \in \mathbb{R}^k$,*
$$0 \leq -(u^-) \leq \psi_0(u) \leq -(I-Q)^{-1}u^- \ .$$

Proof. Let $v = -(I-Q^t)^{-1}u^-$ and note that
$$u + (I-Q^t)v = u - (I-Q^t)(I-Q^t)^{-1}u^- = u - u^- = u^+ \geq 0 \ .$$
Then, by the definition of ψ_0 in (3.1), $\psi_0(u) \leq v$, which establishes the upper bound. By (3.2),
$$\phi_0(u) = u + (I-Q^t)\psi_0(u) \geq 0.$$
Since $\psi_0(u) \geq 0$ and $Q \geq 0$,
$$\psi_0(u) \geq -u + Q^t\psi_0(u) \geq -u \ ,$$
which implies the lower bound. ∎

We now establish an additivity property of ψ_0.

Lemma 14.3.2. (*additivity of ψ_0*) *If $0 \leq v_0 \leq \psi_0(u)$ in $\mathbb{R}^k$, then*
$$\psi_0(u) = \psi_0(u + (I-Q)v_0) + v_0 \ .$$

Proof. By (3.1),
$$\begin{aligned}\psi_0(u) &= \min\{v \in \mathbb{R}^k_+ : u + (I-Q)v \geq 0\} \\ &= \min\{v \in \mathbb{R}^k_+ : u + (I-Q)v_0 + (I-Q)(v - v_0) \geq 0\} \\ &= v_0 + \min\{v' \in \mathbb{R}^k_+ : u + (I-Q)v_0 + (I-Q)v \geq 0\} \\ &= v_0 + \psi_0(u + (I-Q)v_0) \ ,\end{aligned}$$
using the condition in the penultimate step. ∎

We now characterize the instantaneous reflection map in terms of a linear map applied to the vector (u^+, u^-) in $\mathbb{R}^{2k}$ involving the positive and negative parts of the vector u. In particular, for any $u \in \mathbb{R}^k$, let
$$T(u) \equiv u^+ + Qu^- \tag{3.9}$$
and let T^k be the k-fold iterate of the map T, i.e., $T^k(u) = T(T^{k-1}(u))$ for $k \geq 1$ with $T^0(u) \equiv u$. Note that T is a nonlinear function from $\mathbb{R}^k$ to $\mathbb{R}^k$.

Theorem 14.3.2. (*characterization of the instantaneous reflection map*) *Let $u_n \equiv T(u_{n-1})$ for T in (3.9) and $u_0 \equiv u$. Then, for any $u \in \mathbb{R}^k$,*
$$u^+_{n-1} \geq u^+_n \geq 0 \tag{3.10}$$
and
$$0 \geq u^-_n \geq Q^n u^-_0 \quad \text{for all} \quad n \ , \tag{3.11}$$
so that
$$u^-_n \to 0, \quad u_n \to u_\infty \geq 0 \quad \text{and} \quad \psi_0(u_n) \to 0 \quad \text{as} \quad n \to \infty \ . \tag{3.12}$$

For each $n \geq 1$,

$$\psi_0(u) = -\sum_{k=0}^{n-1} u_k^- + \psi_0(u_n) \tag{3.13}$$

and

$$u_n = u - (I - Q)\sum_{k=0}^{n-1} u_k^-, \tag{3.14}$$

so that ψ_0 in (3.1) is well defined with

$$\psi_0(u) = -\sum_{k=0}^{\infty} u_k^- \equiv -\sum_{k=0}^{\infty} T^k(u)^- \tag{3.15}$$

and

$$\phi_0(u) = u_\infty \equiv \lim_{n \to \infty} T^n(u). \tag{3.16}$$

Proof. Since $u_n = u_{n-1}^+ + Qu_{n-1}^-$ by (3.9), $Qu_{n-1}^- \leq u_n \leq u_{n-1}^+$, which implies (3.10) and $Qu_{n-1}^- \leq u_n^- \leq 0$. By induction, these inequalities imply (3.11). Since $Q^n \to 0$ as $n \to \infty$, (3.10) and (3.11) imply the first two limits in (3.12). By Lemma 14.3.1,

$$\psi_0(u_n) \leq -(I-Q)^{-1} u_n^-. \tag{3.17}$$

Since $u_n^- \to 0$, (3.16) implies the last limit in (3.12). Formula (3.13) follows from Lemmas 14.3.1 and 14.3.2 by induction. From (3.9), $u_n - u_{n-1} = -(I-Q)u_{n-1}^-$, from which (3.14) follows by induction. Since $\psi_0(u_n) \to 0$, (3.13) implies (3.15), where the sum is finite. Moreover, (3.12)–(3.15) imply that $u_\infty = u + (I-Q)\psi_0(u)$, which in turn implies (3.16). ∎

We can apply Theorem 14.3.2 to deduce the complementarity property in (3.4). This corollary provides an alternative proof to half of Theorem 14.2.3.

Corollary 14.3.1. (complementarity) *For any* $u \in \mathbb{R}^k$,

$$\phi_0^i(u)\psi_0^i(u) = 0 \quad \text{for all} \quad i.$$

Proof. If $\phi_0^i(u) > 0$, then $u_n^i > 0$ for all n by (3.10), which implies that $(u_n^-)^i = 0$ for all n and $\psi_0^i(u) = 0$ by (3.15). On the other hand, if $\psi_0^i(u) > 0$, then $u_k^i < 0$ for some k by (3.15), which implies that $(u_k^+)^i = 0$ for some k, so that $u_\infty^i = 0$ by (3.10). ∎

Theorem 14.3.2 implies the following important monotonicity property.

Corollary 14.3.2. (monotonicity) *If* $u_1 \leq u_2$ *in* $\mathbb{R}^k$, *then*

$$\phi_0(u_1) \leq \phi_0(u_2) \quad \text{and} \quad \psi_0(u_1) \geq \psi_0(u_2).$$

We now apply Corollary 14.3.2 and Lemma 14.3.2 to show that there is regularity in the reflection map when increments have a common sign in all coordinates. First, it is elementary that, if $u \geq 0$ and $\Delta \geq 0$ in $\mathbb{R}^k$, then

$$\phi_0(u+\Delta) = u + \Delta \quad \text{and} \quad \psi(u+\Delta) = 0 \, .$$

The situation is more delicate when $u \geq 0$ but $\Delta \not\geq 0$, but we obtain regularity when $\Delta \leq 0$.

In particular, we now consider the instantaneous reflection map R_0 applied to $u - \sum_{i=1}^m \Delta_i$, where $u \geq 0$ and $\Delta_i \geq 0$ for all i. We show that the instantaneous reflection of $u - \sum_{i=1}^m \Delta_i$ is the same as the mth iteration of the iterative reflection introducing the increments Δ_i one at a time (in any order). Specifically, let

$$u_j \equiv \phi_0(u_{j-1} - \Delta_j), \quad 1 \leq j \leq m \, , \tag{3.18}$$

with $u_0 \equiv u$.

Theorem 14.3.3. (iterations of the instantaneous reflection mapping) *If $u \geq 0$ and $\Delta_i \geq 0$ for $i \geq 1$, then*

$$\phi_0(u - \sum_{i=1}^m \Delta_i) = u_m \tag{3.19}$$

for u_m in (3.18) and

$$\psi_0(u - \sum_{i=1}^m \Delta_i) = v_m \equiv \sum_{i=1}^m \psi_0(u_{i-1} - \Delta_i) \tag{3.20}$$

for all $m \geq 1$.

Proof. We use induction on m. For $m = 1$, relations (3.19) and (3.20) hold by definition. Suppose that (3.19) and (3.20) hold for all positive integers up to m; we will show that they must also hold for $m + 1$. By Corollary 14.3.2,

$$0 \leq \psi_0(u - \sum_{i=1}^m \Delta_i) \leq \psi_0(u - \sum_{i=1}^{m+1} \Delta_i) \, .$$

By Lemma 14.3.2,

$$\psi_0(u - \sum_{i=1}^{m+1} \Delta_i) = \psi_0(u - \sum_{i=1}^{m+1} \Delta_i + (I - Q^t)v_m) + v_m$$

for $v_j \equiv \psi_0(u - \sum_{i=1}^j \Delta_i)$ for $j \geq 1$. By the induction hypothesis,

$$\psi_0(u - \sum_{i=1}^{m+1} \Delta_i) = \psi_0(\phi_0(u - \sum_{i=1}^m \Delta_i) - \Delta_{m+1}) + v_m$$
$$= \psi_0(u_m - \Delta_{m+1}) + v_m = v_{m+1} \, .$$

Moreover, by (3.2),

$$\phi_0(u - \sum_{i=1}^{m+1} \Delta_i) = u - \sum_{i=1}^{m+1} \Delta_i + (I - Q^t)\psi_0(u - \sum_{i=1}^{m+1} \Delta_i)$$

$$\begin{aligned}
&= u - \sum_{i=1}^{m+1} \Delta_i + (I - Q^t)(\psi_0(u_m - \Delta_{m+1}) + v_m) \\
&= u_m - \Delta_{m+1} + (I - Q^t)(\psi_0(u_m - \Delta_{m+1})) \\
&= u_{m+1} \quad \blacksquare
\end{aligned}$$

It is easy to see that the conclusion of Theorem 14.3.3 does not hold even for $m = 2$ and $k = 1$ when $\Delta_1 > 0$ and $\Delta_2 < 0$.

Example 14.3.1. *Need for common signs.* To see the need for having $\Delta_i \geq 0$ in $\mathbb{R}^k$ for all i in Theorem 14.3.3, suppose that $k = 2$, Q is as in Example 14.5.4, $u = (1, 0)$, $\Delta_1 = (3, -1)$ and $\Delta_2 = (3, -10)$. Then $\phi_0(u - \Delta_1 - \Delta_2) = \phi_0(-5, 11) = (0, 6)$, but $\phi_0(\phi_0(u - \Delta_1) - \Delta_2)) = (0, 7)$. $\blacksquare$

We can also deduce the following alternative characterization of the instantaneous reflection map.

Corollary 14.3.3. (fixed-point characterization) *For any $u \in \mathbb{R}^k$, $\psi_0(u)$ can be characterized as the unique solution v in $\mathbb{R}_+^k$ to the equation*

$$v = \pi_0(v) \equiv (Qv - u)^+ . \tag{3.21}$$

Proof. We use the fact that

$$-Qu_k^- = u_k^+ - u_{k+1} ,$$

drawing on (3.9). Then

$$\begin{aligned}
(Q\psi(u) - u)^+ &= \lim_{n \to \infty} \left(Q \sum_{k=0}^{n-1} (-u_k^-) - u_0 \right)^+ \\
&= \lim_{n \to \infty} \left(\sum_{k=0}^{n-1} u_k^+ - \sum_{k=1}^{n} u_k - u_0 \right)^+ \\
&= \lim_{n \to \infty} \left(-u_n^+ - \sum_{k=0}^{n} u_k^- \right)^+ \\
&= (-\phi(u) + \psi(u))^+ .
\end{aligned}$$

By Corollary 14.3.1, the last expression equals $\psi(u)$: If $\psi^i(u) > 0$, then $\phi^i(u) = 0$ and $(-\phi^i(u) + \psi^i(u)) = \psi^i(u) > 0$; if $\phi^i(u) > 0$, then $\psi^i(u) = 0$ and $-\phi^i(u) + \psi^i(u) < 0$, so that $(-\phi^i(u) + \psi^i(u))^+ = 0 = \psi^i(u)$. It remains to establish uniqueness. Suppose that v_1 and v_2 are two solutions to equation (3.21). Then

$$\begin{aligned}
\|v_1 - v_2\| &= \|(Qv_1 - u)^+ - (Qv_2 - u)^+\| \\
&\leq \|(Qv_1 - u) - (Qv_2 - u)\| = \|Q(v_1 - v_2)\|
\end{aligned}$$

Suppose that $\|z\| = \|Qz\|$. Then, by induction, $\|z\| = \|Q^n z\|$ for all $n \geq 1$, but $Q^n \to 0$ as $n \to \infty$, so that we must have $z = 0$. Hence $v_1 = v_2$ and the solution to (3.21) is unique. ∎

14.3.2. Implications for the Reflection Map

We can now extend the reflection map defined on D_c to D. We can *define $R(x)$* for $x \in D$ by

$$R(x) \equiv \lim_{n \to \infty} R(x_n) \qquad (3.22)$$

for $x_n \in D_c$ with $\|x_n - x\| \to 0$.

Theorem 14.3.4. (extension of the reflection map from D_c to D) *For all $x \in D$, the limit in (3.22) exists and is unique. Moreover, R is Lipschitz as a map from $(D, \|\cdot\|)$ to $(D, \|\cdot\|)$ and satisfies properties (2.4)–(2.6).*

Proof. For $x \in D$ given, choose $x_n \in D_c$ with $\|x_n - x\| \to 0$. Since $\|x_n - x\| \to 0$ for $x_n \in D_c$, $\|x_n - x_m\| \to 0$ as $m, n \to \infty$. As noted above, we can deduce the Lipschitz property of R on D_c by applying Theorem 14.2.5. By that Lipschitz property on D_c, $\|R(x_n) - R(x_m)\| \leq K\|x_n - x_m\| \to 0$. Since $(D, \|\cdot\|)$ is a complete metric space, there exists $(y, z) \in D$ such that $\|R(x_n) - (y, z)\| \to 0$. To show uniqueness, suppose that $\|x_{jn} - x\| \to 0$ for $j = 1, 2$. Then $\|x_{1n} - x_{2n}\| \to 0$ and $\|R(x_{1n}) - R(x_{2n})\| \leq K\|x_{1n} - x_{2n}\| \to 0$, so that the limits necessarily coincide. Given that $x_n \in D_c$, so that (x_n, y_n, z_n) satisfy (2.4), (2.6) and (2.11) with $\|(x_n, y_n, z_n) - (x, y, z)\| \to 0$, it follows that (x, y, z) satisfies (2.4), (2.6) and (2.11) too. (If (2.11) were to be violated for (z^i, y^i) for some i, then it follows that (2.11) would necessarily be violated by (z_n^i, y_n^i) for some n, because there would exist an interval $[a, b]$ in $[0, T]$ such that $z^i(t) \geq \epsilon > 0$ for $a \leq t \leq b$ and $y^i(b) > y^i(a)$.) Alternatively, since there exists a unique solution to (2.4) and (2.6), it must coincide with the one obtained via the limit (3.22). To directly verify the Lipschitz property on D given the Lipschitz property on D_c, for any $x_1, x_2 \in D$, let $x_{1n}, x_{2n} \in D_c$ with $\|x_{1n} - x_1\| \to 0$ and $\|x_{2n} - x_2\| \to 0$. Then, for any $\epsilon > 0$, there is an n_0 such that

$$\begin{aligned}
\|R(x_1) - R(x_2)\| &\leq \|R(x_1) - R(x_{1n})\| + \|R(x_{1n}) - R(x_{2n})\| \\
&\quad + \|R(x_{2n}) - R(x_2)\| \\
&\leq K(\|x_1 - x_{1n}\| + \|x_{1n} - x_{2n}\| + \|x_{2n} - x_2\|) \\
&\leq K\|x_1 - x_2\| + 2K(\|x_1 - x_{1n}\| + \|x_{2n} - x_2\|) \\
&\leq K\|x_1 - x_2\| + \epsilon
\end{aligned}$$

for all $n \geq n_0$. Since ϵ was arbitrary, the Lipschitz property is established. ∎

From Theorems 14.2.6(iv), 14.3.1 and 14.3.2, we have the following result.

Lemma 14.3.3. (the reflection map at discontinuity points) *For any $x \in D$ and t, $0 < t < T$,*

$$\psi(x)(t) \equiv y(t) = \psi_0(z(t-) + x(t) - x(t-)) + y(t-)$$

and
$$\phi(x)(t) \equiv z(t) = \phi_0(z(t-) + x(t) - x(t-)) \ .$$

We can apply Lemma 14.3.3 to relate the set of discontinuity points of $R(x)$ to the set of discontinuity points of x, which we denote by $Disc(x)$.

Corollary 14.3.4. (the set of discontinuity points) *For any $x \in D$,*
$$Disc(R(x)) = Disc(x) \ .$$

Proof. By Lemma 14.3.3, we can write
$$z(t) - z(t-) = x(t) - x(t-) + (I - Q)(y(t) - y(t-)) \ ,$$
where $y^i(t) - y^i(t-)$ is minimal, $1 \leq i \leq k$. If $x(t) - x(t-) = 0$ (where here 0 is the zero vector), then necessarily $y(t) - y(t-) = 0$, which then forces $z(t) - z(t-) = 0$. On the other hand, if $x(t) - x(t-) \neq 0$, then we cannot have both $z(t) - z(t-) = 0$ and $y(t) - y(t-) = 0$, so we must have $t \in Disc(R(x))$. ∎

We obtain our strongest results for the case in which no coordinate of x has a negative jump. Let D_+ be the subset of functions x for which $x(t) - x(t-) \geq 0$ for all t.

Corollary 14.3.5. (stronger result in D_+) *For any $x \in D_+$, we have $\psi(x) \in C$, $\phi(x) \in D_+$ and*
$$\phi(x)(t) - \phi(x)(t-) = x(t) - x(t-) \ .$$

Finally, we can apply Lemma 14.3.3 and Corollary 14.3.2 to show how reflections of parametric representations perform. We consider the reflection map applied to α times the increment as a function of α for $0 \leq \alpha \leq 1$. We will apply the following lemma in the next section.

Lemma 14.3.4. (instantaneous reflection at discontinuity points) *Suppose that $x \in D$, $t \in Disc(x)$ and $0 \leq \alpha \leq 1$.*
(a) If $x(t) \geq x(t-)$, then
$$\hat{\psi}(x, t, \alpha) \equiv \psi_0(z(t-) + \alpha[x(t) - x(t-)]) + y(t-) = \hat{\psi}(x, t, 0) = y(t-) \quad (3.23)$$

and
$$\hat{\phi}(x, t, \alpha) \equiv \phi_0(z(t-) + \alpha[x(t) - x(t-)]) = \hat{\phi}(x, t, 0) + \alpha[x(t) - x(t-)] \quad (3.24)$$

for $0 \leq \alpha \leq 1$.
(b) If $x(t) \leq x(t-)$ and $0 \leq \alpha_1 < \alpha_2 \leq 1$, then
$$\hat{\psi}(x, t, \alpha_1) \leq \hat{\psi}(x, t, \alpha_2)$$

and
$$\hat{\phi}(x, t, \alpha_1) \geq \hat{\phi}(x, t, \alpha_2)$$

for $\hat{\psi}$ in (3.23) and $\hat{\phi}$ in (3.24).

We conclude this section by further discussing the possibility of using the instantaneous reflection map to simulate reflected stochastic processes in D^k. Given a stochastic process X, we can approximate it by the associated discrete-time stochastic process

$$X_n(t) \equiv X(\lfloor nt \rfloor / n), \quad t \geq 0 ,$$

for some suitably large n. For each n, X_n has sample paths in D_c. Given that X has sample paths in D, it is easy to see that $X_n \to X$ in (D, J_1) w.p.1 as $n \to \infty$. We simulate X_n for suitably large n by generating the random vectors $X(k/n) - X((k-1)/n)$ for $k \geq 1$. When X is a Lévy process, $X(k/n) - X((k-1)/n)$ for $k \geq 1$ will be IID random vectors with an infinitely divisible distribution. When X is a stable Lévy motion, the random vectors will have a stable law. For discussion about simulation of stable random vectors and processes, see Janicki and Weron (1993).

Given a sample path of $\{X(k/n) : k \geq 1\}$, we can calculate the sample path of the associated reflected process $\phi(X_n)$ by solving the linear complementarity problem (LCP) at each transition epoch k/n. Thus the substantial literature on LCP can be applied; see Cottle, Pang and Stone (1992). For example, as noted before, we can use linear programming to solve the LCP, recognizing that only the trivial calculation associated with $v = 0$ in (3.1) occurs whenever $u \geq 0$. (See (3.5) above.)

Instead of linear programming, we can also use Theorem 14.3.2 to do the calculation. We can approximate $\phi(u)$ by $T^n(u)$ for the map in (3.9). Even though the operator T on $\mathbb{R}^k$ in (3.9) must be applied many times to calculate each instantaneous reflection, the algorithm can be effective, because T itself is remarkably simple. By (3.10) and (3.16), u_n^+ is an upper bound for $\phi(u)$, where $u_n = T^n(u)$. Moreover, we can apply (3.17) to bound the error $\|u_n^+ - \phi(u)\|$. Since

$$\phi(u) = \phi(u_n) = u_n + (I - Q)\psi(u_n) ,$$

$$\begin{aligned} \|u_n^+ - \phi(u)\| &\leq \|u_n - \phi(u_n)\| \leq \|I - Q\| \cdot \|\psi(u_n)\| \\ &\leq \|I - Q\| \cdot \|(I - Q)^{-1} u_n^-\| . \end{aligned}$$

So $T^n(u) \equiv u_n$ provides an upper bound on $\phi(u)$ via u_n^+ and an upper bound on the error $\|u_n^+ - \phi(u)\|$ via u_n^-.

14.4. Reflections of Parametric Representations

In order to establish continuity and stronger Lipschitz properties of the reflection map R on D with the M_1 topologies, we would like to have $(R(u), r)$ be a parametric representation of $R(x)$ when (u, r) is a parametric representation of x. However, that property does not always hold, as we show in Example 14.4.1 below. Nevertheless, we can get quite close under reasonable regularity conditions. We now obtain positive results in that direction. The (rather involved) proofs appear in the

Internet Supplement. For the following result, we exploit picewise-constant approximations and the instantaneous reflection map. We also use properties of the SM_1 topology.

Theorem 14.4.1. (reflections of parametric representations) *Suppose that $x \in D$, $(u,r) \in \Pi_s(x)$ and $r^{-1}(t) = [s_-(t), s_+(t)]$.*
(a) *If $t \in Disc(x)^c$, then*
$$R(u)(s) = R(x)(t) \quad for \quad s_-(t) \leq s \leq s_+(t) \ .$$

(b) *If $t \in Disc(x)$, then*
$$R(u)(s_-(t)) = R(x)(t-) \quad and \quad R(u)(s_+(t)) = R(x)(t) \ .$$

(c) *If $t \in Disc(x)$ and $x(t) \geq x(t-)$, then*
$$\phi(u)(s) = \phi(x)(t-) + \left(\frac{u^j(s) - u^j(s_-(t))}{u^j(s_+(t)) - u^j(s_-(t))} \right) [x(t) - x(t-)]$$

for any j, $1 \leq j \leq k$, and
$$\psi(u)(s) = \psi(x)(t-) = \psi(x)(t) \quad for \quad s_-(t) \leq s \leq s_+(t) \ ,$$

so that
$$R(u)(s) \subset [R(x)(t-), R(x)(t)] \quad for \quad s_-(t) \leq s \leq s_+(t) \ .$$

(d) *If $t \in Disc(x)$ and $x(t) \leq x(t-)$, then $\phi^i(u)$ and $\psi^i(u)$ are monotone in $[s_-(t), s_+(t)]$ for each i, so that*
$$R(u)(s) \in [[R(x)(t-), R(x)(t)]] \quad for \quad s_-(t) \leq s \leq s_+(t) \ .$$

We can draw the desired conclusion that $(R(u), r)$ is a parametric representation of $R(x)$ if we can apply parts (c) and (d) of Theorem 14.4.1 to all jumps. Recall that D_+ (D_s) is the subset of D for which condition (c) (condition (c) or (d)) holds at all discontinuity points of x. For $x \in D_s$, the direction of the inequality is allowed to depend upon t.

Theorem 14.4.2. (preservation of parametric representations under reflection) *Suppose that $x \in D$ and $(u,r) \in \Pi_s(x)$.*
(a) *If $x \in D_+$, then $(R(u), r) \in \Pi_s(R(x))$.*
(b) *If $x \in D_s$, then $(R(u), r) \in \Pi_w(R(x))$.*

We can weaken the condition in Theorem 14.4.2 (b): If $x \in D_s$, then it suffices to have $(u,r) \in \Pi_w(x)$.

Theorem 14.4.3. (preservation of weak parametric representations) *If $x \in D_s$ and $(u,r) \in \Pi_w(x)$, then $(R(u), r) \in \Pi_w(R(x))$.*

As a basis for proving Theorem 14.4.1, we exploit piecewise-constant approximations. We use the following lemma, established in the Internet Supplement.

Lemma 14.4.1. (left and right limits) *For any $x \in D_c$, $(u,r) \in \Pi_s(x)$ and $r^{-1}(t) = [s_-(t), s_+(t)]$,*

$$R(u)(s_-(t)) = R(x)(t-) \quad \text{and} \quad R(u)(s_+(t)) = R(x)(t) . \tag{4.1}$$

We now show that it is essential in Lemma 14.4.1 to have $(u,r) \in \Pi_s(x)$ instead of just $(u,r) \in \Pi_w(x)$. We also show that we cannot improve upon Lemma 14.4.1 to conclude that $(R(u), r) \in \Pi_w(R(x))$ when $(u,r) \in \Pi_s(x)$.

Example 14.4.1. *Impossibility of improvements.* To demonstrate the points above, let $x \in D_c$ and R be defined by

$$x^1 = I_{[0,1)} - 3I_{[1,2]}, \quad x^2 = I_{[0,1)} + 2I_{[1,2]} ,$$

$$Q^t = \begin{pmatrix} 0 & 1 \\ 0 & .9 \end{pmatrix}, \quad \text{so that} \quad I - Q = \begin{pmatrix} 1 & 0 \\ -1 & .1 \end{pmatrix} .$$

Then $z^1 = z^2 = I_{[0,1)}$, $y^1 = 3I_{[1,2]}$ and $y^2 = 10I_{[1,2]}$. To see that the conclusion of Lemma 14.4.1 fails when we only have $(u,r) \in \Pi_w(x)$, let a parametric representation (u,r) in $\Pi_w(x)$ be defined by

$$r(0) = 0, \quad r(1/3) = r(2/3) = 1, \quad r(1) = 2$$

$$u^1(0) = u^1(1/3) = 1, \quad u^1(1/2) = u^1(1) = -3 \tag{4.2}$$

$$u^2(0) = u^2(1/2) = 1, \quad u^2(2/3) = u^2(1) = 2$$

with r, u^1 and u^2 defined by linear interpolation elsewhere. Notice that $[s_-(1), s_+(1)] = [1/3, 2/3]$, $\phi^2(u)(1/2) = 0$ and $\phi^2(u)(2/3) = 1 > 0 = z^2(1)$. Moreover, $\phi^2(u)(s) = 1$ on $[2/3, 1]$.

Next, to see that we need not have $(R(u), r) \in \Pi_w(R(x))$ when $(u,r) \in \Pi_s(x)$, let r be defined in (4.2) and let the parametric representation (u,r) in $\Pi_s(x)$ be defined by

$$u^1(0) = u^1(1/3) = 1, \quad u^1(2/3) = u^1(1) = -3$$
$$u^2(0) = u^2(1/3) = 1, \quad u^2(2/3) = u^2(1) = 2$$

with r, u^1, u^2 defined at other points by linear interpolation. Clearly $(u,r) \in \Pi_s(x)$. Note that $u^i(s) \geq 0$ for all $s \leq 5/12$. Then $r(5/12) = 1$, $u^1(5/12) = 0$ and $u^2(5/12) = 5/4$. Clearly $\phi(u)(5/12) = u(5/12) = (0, 5/4)$, which is not in $[[(0,0),(1,1)]]$, the weak range of $z = \phi(x)$. Further analysis shows that $\phi^1(u)(s) = 0$ for $s \geq 5/12$, while $\phi^2(u) = u^2$ on $[0, 1/3]$, $\phi^2(u)(5/12) = 5/4$, $\phi^2(u)(5/9) = \phi^2(u)(1) = 0$, with $\phi^2(u)$ defined elsewhere by linear interpolation. Similarly, ψ has slope $(12, 0)$ over $(5/12, 5/9)$ and slope $(12, 90)$ over $(5/9, 2/3)$, so that $\psi(u)(5/12) = (0,0)$, $\psi(u)(5/9) = (5/3, 0)$, $\psi(u)(2/3) = \psi(u)(1) = (3, 10)$ and ψ is defined by linear interpolation elsewhere. ■

14.5. M_1 Continuity Results and Counterexamples

In this section we establish positive results and give counterexamples showing that candidate stronger results do not hold.

14.5.1. M_1 Continuity Results

We first state continuity and Lipschitz properties of the reflection map on $D \equiv D^k \equiv D([0,T], \mathbb{R}^k)$ with the M_1 topologies. Our first result establishes continuity of the reflection map R (for an arbitrary reflection matrix Q) as a map from (D, SM_1) to (D, L_1), where L_1 is the topology on D induced by the L_1 norm

$$\|x\|_{L_1} \equiv \int_0^T \|x(t)\| dt \ . \tag{5.1}$$

Under a further restriction, the map from (D, WM_1) to (D, WM_1) will be continuous.

Recall that D_s is the subset of functions in D without simultaneous jumps of opposite sign in the coordinate functions; i.e., $x \in D_s$ if, for all $t \in (0,T)$, either $x(t) - x(t-) < 0$ or $x(t) - x(t-) \geq 0$, with the sign allowed to depend upon t. The subset D_s is a closed subset of D in the J_1 topology and thus a measurable subset of D with the SM_1 and WM_1 topologies (since the Borel σ-fields coincide). The (again somewhat involved) proofs of the main theorems here appear in Section 8.5 of the Internet Supplement.

Theorem 14.5.1. (continuity with the SM_1 topology on the domain) *Suppose that $x_n \to x$ in (D, SM_1).*
(a) Then

$$R(x_n)(t_n) \to R(x)(t) \quad in \quad \mathbb{R}^{2k} \tag{5.2}$$

for each $t \in Disc(x)^c$ and sequence $\{t_n : n \geq 1\}$ with $t_n \to t$,

$$\sup_{n \geq 1} \|R(x_n)\| < \infty \ , \tag{5.3}$$

$$R(x_n) \to R(x) \quad in \quad (D, L_1) \tag{5.4}$$

and

$$\psi(x_n) \to \psi(x) \quad in \quad (D, WM_1) \ . \tag{5.5}$$

(b) If in addition $x \in D_s$, then

$$\phi(x_n) \to \phi(x) \quad in \quad (D, WM_1) \ , \tag{5.6}$$

so that

$$R(x_n) \to R(x) \quad in \quad (D, WM_1) \ . \tag{5.7}$$

Under the extra condition in part (b), the mode of convergence on the domain actually can be weakened. However, little positive can be said if only $x_n \to x$ in (D, WM_1) without $x \in D_s$; see Example 14.5.3 below.

Theorem 14.5.2. (continuity with the WM_1 topology on the domain) *If $x_n \to x$ in (D, WM_1) and $x \in D_s$, then (5.7) holds.*

Remark 14.5.1. *Gibbs phenomenon.* Interestingly, the limit of $\phi(x_n)(t_n)$ for $t_n \to t \in Disc(x)$ can fall outside the product segment $[[\phi(x)(t-), \phi(x)(t)]]$; see Example 14.5.4 below. Thus the asymptotic fluctuations in $\phi(x_n)$ can be greater than the fluctuations in $\phi(x)$. The behavior here is analogous to the Gibbs phenomenon associated with Fourier series; see Chapter 9 of Carslaw (1930) and Remark 5.1 of Abate and Whitt (1992a). ∎

Example 12.3.1 shows that convergence $x_n \to x$ can hold in (D, WM_1) but not in (D, SM_1) even when $x \in D_s$. Thus Theorems 14.5.1 (a) and 14.5.2 cover distinct cases. An important special case of both occurs when $x \in D_1$, where D_1 is the subset of x in D with discontinuities in only one coordinate at a time; i.e., $x \in D_1$ if $t \in Disc(x^i)$ for at most one i when $t \in Disc(x)$, with the coordinate i allowed to depend upon t. In Section 12.7 it is shown that WM_1 convergence $x_n \to x$ is equivalent to SM_1 convergence when $x \in D_1$.

Just as with D_s above, D_1 is a closed subset of (D, J_1) and thus a Borel measurable subset of (D, SM_1). Since $D_1 \subseteq D_s$, the following corollary to Theorem 14.5.2 is immediate.

Corollary 14.5.1. (common case for applications) *If $x_n \to x$ in (D, WM_1) and $x \in D_1$, then $R(x_n) \to R(x)$ in (D, WM_1).*

We can obtain stronger Lipschitz properties on special subsets. Let D_+ be the subset of x in D with only nonnegative jumps, i.e., for which $x^i(t) - x^i(t-) \geq 0$ for all i and t. As with D_s and D_1 above, D_+ is a closed subset of (D, J_1) and thus a measurable subset of (D, SM_1).

Theorem 14.5.3. (*Lipschitz properties*) *There is a constant K (the same as associated with the uniform norm in (2.25)) such that*

$$d_s(R(x_1), R(x_2)) \leq K d_s(x_1, x_2) \tag{5.8}$$

for all $x_1, x_2 \in D_+$, and

$$d_p(R(x_1), R(x_2)) \leq d_w(R(x_1), R(x_2)) \leq K d_w(x_1, x_2) \leq K d_s(x_1, x_2) \tag{5.9}$$

for all $x_1, x_2 \in D_s$.

We can actually do somewhat better than in Theorem 14.5.1 when the limit is in D_+.

Theorem 14.5.4. (*strong continuity when the limits is in D_+*) *If*

$$x_n \to x \quad in \quad (D, SM_1), \tag{5.10}$$

where $x \in D_+$, then
$$R(x_n) \to R(x) \quad in \quad (D, SM_1) . \tag{5.11}$$

Our final result shows how the reflection map behaves as a function of the reflection matrix Q, as well as x, with the M_1 topologies.

Theorem 14.5.5. (continuity as a function of (x, Q)) *Suppose that $Q_n \to Q$ in $\mathcal{Q}$.*
(a) *If $x_n \to x$ in (D^k, WM_1) and $x \in D_s$, then*
$$R_{Q_n}(x_n) \to R_Q(x) \quad in \quad (D^{2k}, WM_1) . \tag{5.12}$$
(b) *If $x_n \to x$ in (D^k, SM_1) and $x \in D_+$, then*
$$R_{Q_n}(x_n) \to R_Q(x) \quad in \quad (D^{2k}, SM_1) . \tag{5.13}$$

We can apply Section 12.9 to extend the continuity and Lipschitz results to the space $D([0, \infty), \mathbb{R}^k)$.

Theorem 14.5.6. (extension of continuity results to D with domain $[0, \infty)$) *The convergence-preservation results in Theorems 14.5.1, 14.5.2 and 14.5.4 and Corollary 14.5.1 extend to $D([0, \infty), \mathbb{R}^k)$.*

Proof. Suppose that $x_n \to x$ in $D([0, \infty), \mathbb{R}^k)$ with the appropriate topology and that $\{t_j : j \geq 1\}$ is a sequence of positive numbers with $t_j \in Disc(x)^c$ and $t_j \to \infty$ as $j \to \infty$. Then, $r_{t_j}(x_n) \to r_{t_j}(x)$ in $D([0, \infty), \mathbb{R}^k)$ with the same topology as $n \to \infty$ for each j, where r_t is the restriction map to $D([0, t], \mathbb{R}^k)$. Under the specified assumptions,
$$r_{t_j}(R(x_n)) = R_{t_j}(r_{t_j}(x_n)) \to R_{t_j}(r_{t_j}(x)) = r_{t_j}(R(x)) \tag{5.14}$$
in $D([0, t_j], \mathbb{R}^{2k})$ with the specified topology as $n \to \infty$ for each j, which implies that
$$R(x_n) \to R(x) \quad in \quad D([0, \infty), \mathbb{R}^{2k}) \tag{5.15}$$
with the same topology as in (5.14). ■

Theorem 14.5.7. (extension of Lipschitz properties to $D([0, \infty), \mathbb{R}^k)$) *Let $R : D([0, \infty), \mathbb{R}^k) \to D([0, \infty), \mathbb{R}^{2k})$ be the reflection map with function domain $[0, \infty)$ defined by Definition 14.2.1. Let metrics associated with domain $[0, \infty)$ be defined in terms of restrictions by (9.1) in Section 12.9. Then the conclusions of Theorems 14.2.5, 14.2.7 and 14.5.3 also hold for domain $[0, \infty)$.*

Proof. Apply Theorem 12.9.4. ■

14.5.2. Counterexamples

We now return to the space $D([0, T], \mathbb{R}^k)$ and present several counterexamples. We first show that the reflection map is actually not continuous on $D([0, T], \mathbb{R}^1)$ with the SM_1 topology. (This would not be a counterexample if we restricted attention to the component ϕ mapping x into z in (2.6) or, more generally, the WM_1 topology were used on the range.)

Example 14.5.1. *Not continuous on* (D, SM_1). To show that
$$R \equiv (\psi, \phi) : (D([0,2], \mathbb{R}^1), SM_1) \to (D([0,2], \mathbb{R}^2), SM_1)$$
is *not* continuous, let
$$x_n(t) = 1 - 2n(t-1)I_{[1,1+n^{-1}]}(t) - 2I_{[1+n^{-1},2]}(t)$$
and
$$x(t) = 1 - 2I_{[1,2]}(t), \quad 0 \le t \le 2 .$$
It is easy to see that $d(x_n, x) \to 0$ as $n \to \infty$,
$$z_n(t) = 1 - 2n(t-1)I_{[1,1+(2n)^{-1}]}(t) ,$$
$$y_n(t) = 2n(t - (1+(2n)^{-1}))I_{[1+(2n)^{-1}, 1+n^{-1}]}(t) + I_{[1+n^{-1},2]}(t) ,$$
$$z(t) = I_{[0,1)}(t) \quad \text{and} \quad y(t) = I_{[1,2]}(t) .$$
We use the fact that any linear function of the coordinate functions, such as addition or subtraction, is continuous in the SM_1 topology; see Section 12.7. Note that $z(t) + y(t) = 1, 0 \le t \le 2$, while
$$z_n(t) + y_n(t) = 1 - 2n(t-1)I_{[1,1+(2n)^{-1}]}(t) + 2n(t - (1+(2n)^{-1}))I_{[1+(2n)^{-1}, 1+n^{-1}]}(t)$$
so that $d(z_n + y_n, z + y) \not\to 0$ as $n \to \infty$, which implies that $(x_n, y_n) \not\to (z, y)$ as $n \to \infty$ in $D([0,T], \mathbb{R}^2)$ with the SM_1 metric. However, we do have $d(z_n, z) \to 0$ and $d(y_n, y) \to 0$ as $n \to \infty$, so the maps from x to y and z separately are continuous. ∎

Example 14.5.1 suggests that the difficulty might only be in simultaneously considering both maps ψ and ϕ. We show that this is not the case by giving a counterexample with ϕ alone (but again in two dimensions).

Example 14.5.2. ϕ *is not continuous on* (D^2, SM_1). We now show that
$$\phi : (D([0,2], \mathbb{R}^2), SM_1) \to (D([0,2], \mathbb{R}^2), SM_1)$$
is not continuous. We use the trivial reflection map corresponding to two separate queues, for which Q is the 2×2 matrix of 0's. Let x_n^1 be as in Example 14.5.1, i.e.,
$$x_n^1(t) = 1 - 2n(t-1)I_{[1,1+n^{-1}]}(t) - 2I_{[1+n^{-1},2]}(t)$$
and let
$$x_n^2(t) = 2 - 3n(t-1)I_{[1,1+n^{-1}]}(t) - 3I_{[1+n^{-1},2]}(t) .$$
It is easy to see that $d_s((x_n^1, x_n^2), (x^1, x^2)) \to 0$ as $n \to \infty$, where
$$x^1(t) = 1 - 2I_{[1,2]}(t) \quad \text{and} \quad x^2(t) = 2 - 3I_{[1,2]}(t) .$$
(The same functions r_n and r can be used in the parametric representations of the two coordinates.) Clearly $\phi((x^1, x^2)) = (z^1, z^2)$, where
$$z^1(t) = I_{[0,1)}(t) \quad \text{and} \quad z^2(t) = 2I_{[0,1)}(t) ,$$

14.5. M_1 Continuity Results   489

while
$$z_n^1(t) = 1 - 2n(t-1)I_{[1,1+(1/2n))}(t)$$
$$z_n^2(t) = 2 - 3n(t-1)I_{[1,1+(2/3n))}(t) \ .$$

Note that $2z^1(t) - z^2(t) = 0$, $0 \leq t \leq 2$, while
$$2z_n^1(1+(2n)^{-1}) - z_n^2(1+(2n)^{-1}) = -z_n^2(1+(2n)^{-1}) = -1/2 \quad \text{for all} \quad n \ .$$

Hence $d_s(2z_n^1 - z_n^2, 2z^1 - z^2) \not\to 0$ so that $d_s((z_n^1, z_n^2),(z^1,z^2)) \not\to 0$ as $n \to \infty$. However, in this example, ϕ is continuous if we use the WM_1 topology on the range. ∎

We now show that the reflection map is not continuous if the WM_1 topology is used on the domain without imposing extra conditions.

Example 14.5.3. *The difficulty with the WM_1 topology on the domain.* We show that neither ψ nor ϕ need be continuous when the WM_1 topology is used on the domain, without extra conditions. Consider $D([0,2],\mathbb{R}^2)$ and let $x^1 = I_{[1,2]}$, $x^2 = -2I_{[1,2]}$ and

$$Q = \begin{pmatrix} 0 & 1/2 \\ 1/2 & 0 \end{pmatrix} \ .$$

Then the reflection map yields $y^1(t) = \psi^1(x)(t) = z^i(t) = \phi^i(x)(t) = 0$, $0 \leq t \leq 2$, for $i = 1, 2$ and $y^2 = \psi^2(x) = 2I_{[1,2]}$. Let the converging functions be $x_n^1 = I_{[1+n^{-1},2]}$ and $x_n^2 = -2I_{[1-n^{-1},2]}$ for $n \geq 1$. It is easy to see that $x_n \to x$ as $n \to \infty$ in WM_1 but that $x_n \not\to x$ as $n \to \infty$ in SM_1, because $(2x_n^1 + x_n^2)(1) = -2$, while $(2x^1 + x^2)(t) = 0$, $0 \leq t \leq 2$. The reflection map applied to x_n works on the jumps at times $1 - n^{-1}$ and $1 + n^{-1}$ separately, yielding $y_n^1 = (4/3)I_{[1-n^{-1},2]}$, $y_n^2 = (8/3)I_{[1-n^{-1},2]}$, $z_n^1 = I_{[1+n^{-1},2]}$ and $z_n^2(t) = 0$, $0 \leq t \leq 2$. Clearly $z_n^1 \not\to z^1$ and $y_n^i \not\to y^i$ as $n \to \infty$ for $i = 1, 2$ for any reasonable topology on the range. In particular, conclusions (5.2) and (5.4) – (5.7) all fail in this example.

Moreover, when we choose suitable parametric representations $(u_n, r_n) \in \Pi_w(x_n)$ and $(u,r) \in \Pi_w(x)$ to achieve $x_n \to x$ in WM_1, $(R(u),r)$ is not a parametric representation for $R(x)$. To be clear about this, we give an example: We let all the functions u_n, r_n, u and r be piecewise-linear. We define the functions at the discontinuity points of the derivative. We understand that the functions are extended to $[0,1]$ by linear interpolation. Let

$$r(0) = 0, \ r(0.2) = r(0.8) = 1, \ r(1) = 2 \ ,$$
$$u^1(0) = u^1(0.4) = 0, \ u^1(0.8) = u^1(1) = 1 \ ,$$
$$u^2(0) = u^2(0.2) = 0, \ u^2(0.4) = u^2(1) = -2 \ ,$$
$$r_n(0) = 0, \ r_n(0.2(1-n^{-1})) = r_n(0.2(2-n^{-1})) = 1 - n^{-1} \ ,$$
$$r_n(0.2(2+n^{-1})) = r_n(0.2(4+n^{-1})) = 1 + n^{-1}, \ r_n(1) = 2 \ ,$$
$$u_n^1(0) = u_n^1(0.2(2+n^{-1})) = 0, \ u_n^1(0.2(4+n^{-1})) = u_n^1(1) = 1 \ ,$$
$$u_n^2(0) = u_n^2(0.2(1-n^{-1})) = 0, \ u_n^2(0.2(2-n^{-1})) = u_n^2(1) = -2 \ .$$

This construction yields $(u_n, r_n) \in \Pi_w(x_n)$, $n \geq 1$, $(u,r) \in \Pi_w(x)$, but $(\phi^1(u), r) \notin \Pi(\phi^1(x))$, because $\phi^1(x)(t) = 0$, $0 \leq t \leq 2$, while $\phi^1(u)(1) = 1$. Note that $(u,r) \in \Pi_w(x)$, but $(u,r) \notin \Pi_s(x)$. ■

We now show that we need not have $R(x_n) \to R(x)$ in (D, WM_1) when $x_n \to x$ in (D, SM_1) without the extra regularity condition $x \in D_s$. A difficulty can occur when $x^i(t) - x^i(t-) > 0$ for some coordinate i, while $x^j(t) - x^j(t-) < 0$ for another coordinate j.

Example 14.5.4. *Need for the condition $x \in D_s$.* We now show that the condition $x \in D_s$ in Theorem 14.5.1 is necessary even when $x_n \to x$ in (D, SM_1). In our limit $x \equiv (x^1, x^2)$, x^1 has a jump down and x^2 has a jump up at $t = 1$. Our example is the simple network corresponding to two queues in series. Let $x \equiv (x^1, x^2)$ and $x_n \equiv (x_n^1, x_n^2)$, $n \geq 1$, be elements of $D([0,2], \mathbb{R}^2)$ defined by

$$x^1(0) = x^1(1-) = 1, \quad x^1(1) = x^1(2) = -3$$

$$x^2(0) = x^2(1-) = 1, \quad x^2(1) = x^2(2) = 2$$

$$x_n^1(0) = x_n^1(1) = 1, \quad x_n^1(1 + n^{-1}) = x_n^1(2) = -3$$

$$x_n^2(0) = x_n^2(1) = 1, \quad x_n^2(1 + n^{-1}) = x_n^2(2) = 2,$$

with the remaining values determined by linear interpolation. Let the substochastic matrix generating the reflection be

$$Q^t = \begin{pmatrix} 0 & 1 \\ 0 & 0 \end{pmatrix}, \quad \text{so that} \quad I - Q = \begin{pmatrix} 1 & 0 \\ -1 & 1 \end{pmatrix}.$$

Then $z^1 = z^2 = I_{[0,1)}$, $y^1 = 3I_{[1,2]}$, $y^2 = I_{[1,2]}$ and

$$z_n^1(0) = z_n^1(1) = 1, \quad z_n^1(1 + (4n)^{-1}) = z_n^1(2) = 0$$

$$z_n^2(0) = z_n^2(1) = 1, \quad z_n^2(1 + (4n)^{-1}) = 5/4, \quad z_n^2(1 + 2(3n)^{-1}) = z_n^2(2) = 0$$

with the remaining values determined by linear interpolation. Since $z_n^2(1+(4n)^{-1}) = 5/4$ for all n and $z^2(t) \leq 1$ for all t, z_n^2 fails to converge to z^2 in any of the Skorohod topologies. We remark that the graphs $G_{\phi(x_n)}$ of $\phi(x_n)$ do converge in the Hausdorff metric to the graph $G_{\phi(x)}$ of $\phi(x)$ augmented by the set $\{1\} \times [1, 5/4]$. This example motivates considering larger spaces of functions than D, which we discuss in Chapter 15. ■

14.6. Limits for Stochastic Fluid Networks

In this section we provide concrete stochastic applications of the convergence-preservation results for the multidimensional reflection map.

We consider the single class open stochastic fluid network with Markovian routing introduced in Section 14.2.

Recall that the stochastic fluid network is characterized by a four-tuple $\{C, r, Q, X(0)\}$, where $C \equiv (C^1, \ldots, C^k)$ is the vector of exogenous cumulative input stochastic processes at the k stations, $r = (r^1, \ldots, r^k)$ is the vector of potential output rates at the stations, $P \equiv (P_{i,j})$ is the routing matrix and $X(0) \equiv (X^1(0), \ldots, X^k(0))$ is the nonnegative random vector of initial buffer contents. The stochastic processes $C^j \equiv \{C^j(t) : t \geq 0\}$ have nondecreasing nonnegative sample paths; $C^j(t)$ represents the cumulative input at station j during the time interval $[0, t]$. A proportion $P_{i,j}$ of all output from station i is routed to station j, while a proportion $p_i \equiv 1 - \sum_{j=1}^{k} P_{i,j}$ is routed out of the network. We assume that P is substochastic so that $P_{i,j} \geq 0$, $1 \leq j \leq k$, and $p_i \geq 0$, $1 \leq i \leq k$. Moreover, we assume that $P^n \to 0$ as $n \to \infty$, where P^n is the nth power of P. The associated reflection matrix is the transpose $Q \equiv P^t$.

As a more concrete example, suppose that the exogenous input to station j is the sum of the inputs from m_j separate on-off sources. Let (j, i) index the ith on-off source at station j. When the (j, i) source is on, it sends fluid input at rate $\lambda_{j,i}$; when it is off, it sends no input. Let $B_{j,i}(t)$ be the cumulative busy (on) time for source (j, i) during the time interval $[0, t]$. Then the exogenous input process at station j is

$$C^j(t) = \sum_{i=1}^{m_j} \lambda_{j,i} B_{j,i}(t), \quad t \geq 0 .$$

Since $B_{j,i}$ necessarily has continuous sample paths, the associated exogenous cumulative input processes C^j and C also have continuous sample paths in this special case.

In general, given the defining four-tuple $(C, r, Q, X(0))$, the associated $\mathbb{R}^k$-valued potential buffer-content process (or net-input process) is

$$X(t) \equiv X(0) + C(t) - (I - Q)rt, \quad t \geq 0 . \tag{6.1}$$

where Q is the transpose P^t. Since C^j has nondecreasing sample paths for each j, the sample paths of X are of bounded variation. In many special cases, the sample paths of X will be continuous as well.

The buffer-content stochastic process $Z \equiv (Z^1, \ldots, Z^k)$ is simply obtained by applying the reflection map to the potential buffer-content process X in (6.1), in particular,

$$Z = \phi(X) , \tag{6.2}$$

where $R = (\psi, \phi)$ in (2.4)–(2.6). Again, we regard (6.2) as the definition. This stochastic fluid network model is more elementary than the queue-length processes in the queueing network in the following Section 14.7, because here the content process of interest Z is defined directly in terms of the reflection map, requiring only (6.1) and (6.2).

We now want to establish some limits for the stochastic processes. First, we obtain a model continuity or stability result.

14.6.1. Model Continuity

For this purpose, we consider a sequence of fluid network models indexed by n characterized by four-tuples $(C_n, r_n, Q_n, X_n(0))$. Let $\Rightarrow$ denote convergence in distribution.

Theorem 14.6.1. (stability for stochastic fluid networks) *If*

$$(C_n, X_n(0)) \Rightarrow (C, X(0))$$

in $D([0,\infty), \mathbb{R}^k) \times \mathbb{R}^k$, *where the topology is either* SM_1 *or* WM_1, $r_n \to r$ *and* $Q_n \to Q$ *in* $\mathcal{Q}$ *as* $n \to \infty$, *then*

$$(X_n, Y_n, Z_n) \Rightarrow (X, Y, Z) \quad as \quad n \to \infty \quad in \quad D([0,\infty), \mathbb{R}^{3k}),$$

with the same topology, where X_n *and* X *are the associated potential buffer-content processes defined by* (6.1), Y_n *and* Y *are the associated regulator processes, and* Z_n *and* Z *as the associated buffer-content processes, with*

$$R(X_n) \equiv (\psi(X_n), \phi(X_n)) \equiv (Y_n, Z_n), \quad n \geq 1.$$

Proof. Apply the continuous mapping theorem with the continuous functions in (6.1) and (2.4)–(2.6), invoking Theorem 14.5.5 and Corollary 14.2.2. Note that C_n, C, X_n and X have sample paths in D_+. First apply the linear function in (6.1) mapping $(C_n, r_n, Q_n, X_n(0))$ into X_n; then apply R mapping X_n into (Y_n, Z_n). For the special case of common Q, we can invoke Theorem 14.5.3 instead of Theorem 14.5.5. ■

Remark 14.6.1. *Sufficient conditions for* SM_1 *convergence.* If $P(C \in D_1) = 1$, i.e., if

$$P(Disc(C^i) \cap Disc(C^j) = \phi) = 1 \tag{6.3}$$

for all i, j with $1 \leq i, j \leq k$ and $i \neq j$, then the assumed SM_1 convergence $C_n \Rightarrow C$ is implied by WM_1 convergence. Since C_n and C have nondecreasing sample paths, the condition $C_n^i \Rightarrow C^i$ $D([0,\infty), \mathbb{R}, M_1)$ is equivalent to convergence of the finite-dimensional distributions at all time points t for which $P(t \in Disc(C^i)) = 0$, where $Disc(C^i)$ is the set of discontinuity points of C^i; see Corollary 12.5.1. ■

Remark 14.6.2. *The case of continuous sample paths.* As we have indicated, it is natural for the cumulative input processes C_n to have continuous sample paths, but that does not imply that the limit C necessarily must have continuous sample paths. If C does in fact have continuous sample paths, then so do X, Y and Z. Then, the SM_1 topology reduces to the topology of uniform convergence on compact subsets. ■

We can also obtain a bound on the distance between (X_n, Y_n, Z_n) and (X, Y, X) using the Prohorov metric π on the probability measures on (D_+, SM_1). For random elements X_1 and X_2, let $\pi(X_1, X_2)$ denote the Prohorov metric in (2.2) applied to the probability laws of X_1 and X_2. The conclusion for the case of common Q then is:

Corollary 14.6.1. (bounds on the Prohorov distance) *For common Q, there exists a constant K such that*

$$\pi((X_n, Y_n, Z_n), (X, Y, Z)) \leq K\pi((C_n, X_n(0)), (C, X(0))) .$$

Proof. Apply Theorems 3.4.2 and 14.5.3. ∎

14.6.2. Heavy-Traffic Limits

We also can obtain heavy-traffic FCLTs for stochastic fluid networks by considering a sequence of models with appropriate scaling. The scaling allows for on-off sources with heavy-tailed busy-period and idle-period distributions, as in Section 8.5. The scaling also allows for strong dependence in the input processes.

Theorem 14.6.2. (heavy-traffic limit) *Consider a sequence of stochastic fluid networks $\{(C_n, r_n, Q_n, X_n(0)) : n \geq 1\}$. If there exist a constant H with $0 < H < 1$, an $\mathbb{R}^k$-valued random vector $\mathbf{X}(0)$, vectors $\alpha_n \in \mathbb{R}^k$, $n \geq 1$, and a stochastic process $\mathbf{C}$ such that*

$$(\mathbf{C}_n, \mathbf{X}_n(0)) \Rightarrow (\mathbf{C}, \mathbf{X}(0)) \quad in \quad D([0,\infty), \mathbb{R}^k, WM_1) \times \mathbb{R}^k , \tag{6.4}$$

where

$$\mathbf{C}_n(t) \equiv n^{-H}(C_n(nt) - \alpha_n nt), \quad t \geq 0 ,$$

$$P(\mathbf{C} \in D_s) = 1 , \tag{6.5}$$

and

$$n^{1-H}[\alpha_n - (I - Q_n)r_n] \to c \quad in \quad \mathbb{R}^k ,$$

then

$$(\mathbf{X}_n, \mathbf{Y}_n, \mathbf{Z}_n) \Rightarrow (\mathbf{X}, \mathbf{Y}, \mathbf{Z})$$

in $D([0,\infty), \mathbb{R}^k, SM_1) \times D([0,\infty), \mathbb{R}^{2k}, WM_1)$, where

$$(\mathbf{X}_n, \mathbf{Y}_n, \mathbf{Z}_n)(t) \equiv n^{-H}(X_n(nt), Y_n(nt), Z_n(nt)), \quad t \geq 0$$

$$\mathbf{X}(t) = \mathbf{X}(0) + \mathbf{C}(t) + ct, \quad t \geq 0 ,$$

and $(\mathbf{Y}, \mathbf{Z}) = R(\mathbf{X})$ for R in (2.4)–(2.6).

Proof. Since

$$n^{-H}X_n(nt) = n^{-H}[X_n(0) + [C_n(nt) - c_n nt] + [\alpha_n nt - (I - Q_n^t)r_n nt], \quad t \geq 0 ,$$

$$\mathbf{X}_n \Rightarrow \mathbf{X} \quad in \quad (D^k, SM_1) .$$

The proof is completed by applying the continuous mapping theorem, using Theorem 14.5.5. For common Q, we could use Theorem 14.5.2. ∎

Remark 14.6.3. *Convenient sufficient conditions.* In order for conditions (6.4) and (6.5) to hold, it suffices to have $X_n(0)$ be independent of $\{C_n(t) : t \geq 0\}$ for each n,

$$\mathbf{X}_n(0) \Rightarrow \mathbf{X}(0) \quad \text{in} \quad \mathbb{R}^k ,$$

and $\{\mathbf{C}_n^i(t) : t \geq 0\}$, $1 \leq i \leq k$, be k mutually independent processes for each n, with

$$\mathbf{C}_n^i \Rightarrow \mathbf{C}^i \quad \text{in} \quad D([0,\infty), \mathbb{R}^1, M_1) \quad \text{for} \quad 1 \leq i \leq k ,$$

where $P(t \in Disc(\mathbf{C}^i)) = 0$ for all i and t (so that $\mathbf{C}^i$ has no fixed discontinuities). Then, almost surely, the limit process $\mathbf{C}$ has discontinuities in only one coordinate at a time. Then convergence in the WM_1 topology is actually equivalent to convergence in the SM_1 topology. ∎

Remark 14.6.4. *Convergence in the L_1 topology under weaker conditions.* If condition (6.5) does not hold, but the limit in condition (6.4) holds in the SM_1 topology, then we obtain the limit $(Y_n, Z_n) \Rightarrow (Y, Z)$ in $D([0,\infty), \mathbb{R}^{2k})$ with the L_1 topology instead of the WM_1 topology, by Theorem 14.5.1(a). ∎

Remark 14.6.5. *The special case of Lévy processes.* In many applications the limiting form of the initial conditions can be considered deterministic; i.e. $P(\mathbf{X}(0) = x) = 1$ for some $x \in \mathbb{R}^k$. Then $(\mathbf{Y}, \mathbf{Z})$ is simply a reflection of $\mathbf{C}$, modified by the deterministic initial condition x and the deterministic drift ct. In Chapter 8 conditions are determined to have the convergence $\mathbf{C}_n^i \Rightarrow \mathbf{C}^i$. Then $\mathbf{C}^i$ is often a Lévy process. When $\mathbf{C}$ is a Lévy process, $\mathbf{Z}$ and $(\mathbf{Y}, \mathbf{Z})$ are reflected Lévy processes. In some cases explicit expressions for non-product-form steady-state distributions have been derived; see Kella and Whitt (1992a) and Kella (1993, 1996). ∎

Remark 14.6.6. *Extensions.* Clearly, we can obtain similar results for more general models by similar methods. For example, the prevailing rates might be stochastic processes. The potential output rate from station j at time t can be the random variable $R_j(t)$. Then the net-input process in (6.1) should be changed to

$$X(t) = X(0) + C(t) - (I - Q^t)S(t), \quad t \geq 0 ,$$

where $S \equiv (S^1, \ldots, S^k)$ is the $\mathbb{R}^k$-valued potential output process, having

$$S^j(t) = \int_0^t R^j(u)du, \quad t \geq 0 .$$

Similarly, with the on-off sources, the input rates during the on periods might be stochastic processes instead of the constant rates $\lambda_{j,i}$ in (4.1). Extensions of Theorems 14.6.1 and 14.6.2 are straightforward with such generalizations, but we must be careful that the assumptions of Theorems 14.5.1–14.5.4 are satisfied. ∎

As in Corollary 14.6.1, we can extend the heavy-traffic limit theorem to obtain bounds on the Prohorov distance π beween the probability laws of the random elements of the function space D_s.

Corollary 14.6.2. (bounds on the Prohorov distance in the heavy-traffic limit) *Suppose that $Q_n = Q$, $X_n(0) = 0$ and*

$$\alpha_n = (I - Q)r_n + n^{-(1-q)}\alpha$$

for all n. If eqnF4a holds, then there exists a constant K such that

$$\pi((\mathbf{X}_n, \mathbf{Y}_n, \mathbf{Z}_n), (\mathbf{X}, \mathbf{Y}, \mathbf{Z})) \leq K\pi(\mathbf{C}_n, \mathbf{C}) ,$$

where the SM_1 metric d_s on D is used on the domain and the WM_1 product metric d_p on D is used on the range.

Proof. Apply Theorems 3.4.2 and 14.5.3. ∎

14.7. Queueing Networks with Service Interruptions

In this section we apply the continuous mapping theorem with the multidimensional reflection map to obtain heavy-traffic limits for single-class open queueing networks, where the queues are subject to service interruptions. With light-tailed distributions (having finite variance) and with ordinary (fixed) service interruptions, we obtain convergence to multidimensional reflected Brownian motion (RBM). However, with either heavy-tailed distributions or rare long service interruptions (or both), we obtain convergence to a limit process with jumps in the space (D, WM_1) under appropriate regularity conditions.

14.7.1. Model Definition

The model we consider has k single-server queues, each with unlimited waiting space and the first-come first-served service discipline. Customers arrive at each queue, receive service and then are routed to other queues or out of the network. The servers at the queues are subject to service interruptions, which occur exogenously. When an interruption occurs, service stops. When the interruption ends, service resumes on the customer that was in service when the interruption began. The customer's remaining service time is the same as it was when the interruption began.

We now specify the basic random elements of the model. Let $A^j(t)$ be the cumulative number of customers that arrive at queue j from outside the network in the interval $[0, t]$ and let $S^j(t)$ be the cumulative number of customers that are served at queue j during the first t units of busy time at that queue. (For the stochastic fluid network in Section 14.6, the exogenous input process A^j assumed arbitrary real-values and often had continuous sample paths. In contrast, here A^j and S^j are counting processes with values in the nonnegative integers.) Successive service times are thus associated with the queue instead of the customer. We call

$A \equiv \{A^j : 1 \leq j \leq k\}$, where $A^j \equiv \{A^j(t) : t \geq 0\}$, and $S \equiv \{S^j : 1 \leq j \leq k\}$, where $S^j \equiv \{S^j(t) : t \geq 0\}$, the *arrival process* and *service process*, respectively.

The routing of customers is determined by sequences of indicator variables $\{\chi_{i,j}(n) : n \geq 1\}$, $1 \leq i \leq k$ and $1 \leq j \leq k$. We have $\chi_{i,j}(n) = 1$ if the nth departure from queue i goes next to queue j. It is understood that $\chi_{i,j}(n) = 1$ for at most one j, and if $\chi_{i,j}(n) = 1$, then $\chi_{i,l}(n) = 0$ for all l with $l \neq j$. There is one other alternative: We can have $\chi_{i,j}(n) = 0$ for all j, $1 \leq j \leq k$, which indicates that the nth departure from queue i leaves the network. For each pair (i,j), let

$$R^{i,j}(n) \equiv \sum_{l=1}^{n} \chi_{i,j}(l), \quad n \geq 1 .$$

Clearly, $R^{i,j}(n)$ is the total number of customers immediately routed from i to j among the first n departures from queue i. We call $R \equiv \{R^{i,j} : 1 \leq i \leq k, 1 \leq j \leq k\}$ with $R^{i,j} \equiv \{R^{i,j}(n) : n \geq 1\}$ the *routing process*.

Let the service interruptions be specified by sequences $\{(u_n^j, d_n^j) : n \geq 1\}$ of ordered pairs of positive random variables, $1 \leq j \leq k$. The variable u_n^j specifies the duration of the nth up time (activity period) at queue j, while the variable d_n^j specifies the duration of the nth down time (inactivity period or interruption) at queue j. To be concrete, we assume that the queues all start at the beginning of the first up time. Then the epoch beginning the $(n+1)^{\text{st}}$ up period at queue j is

$$T_n^j = \sum_{l=1}^{n}(u_l^j + d_l^j), \quad n \geq 1, \quad T_0^j = 0 .$$

We assume that $T_n^j \to \infty$ w.p.1 as $n \to \infty$ for each j, so that there are only finitely many interruptions (up-down cycles) in any finite time interval.

Now define *server-availability indicator processes* $I^j \equiv \{I^j : 1 \leq j \leq k\}$, where $I^j \equiv \{I^j(t) : t \geq 0\}$ with $I^j(t) = 1$ if server j is up at time t and $I^j(t) = 0$ if server j is down. Then we have

$$I^j(t) = \begin{cases} 1 & \text{if} \quad T_n^j \leq t < T_n^j + u_{n+1}^j \\ 0 & \text{if} \quad T_n^j + u_{n+1}^j \leq t < T_{n+1}^j \end{cases}$$

for some n.

We focus on the *queue-length process* $Z \equiv \{Z^j : 1 \leq j < k\}$ with $Z^j \equiv \{Z^j(t) : t \geq 0\}$, where $Z^j(t)$ is the number of customers at queue j at time t (including the one in service if any). We define Z in terms of the model data, but the initial queue length must be included in the model data. Let $Z^j(0)$ be the initial queue length at queue j, $1 \leq j \leq k$, and let $Z(0) \equiv (Z^1(0), \ldots, Z^k(0))$.

Thus the *model data* are the arrival process A, service process S, routing process R, server-availability indicator process I and the initial queue-length vector $Z(0)$. We assume that the sample paths of A and S are right continuous (as well as nonnegative and nondecreasing), so that $(A, S, R, I, Z(0))$ is a random element of

$$D^k \times D^k \times D^{k^2} \times D^k \times \mathbb{R}^k \equiv D^{k^2+3k} \times \mathbb{R}^k ,$$

where $D^1 \equiv D([0, \infty), \mathbb{R})$.

We now construct associated stochastic processes that describe the model behavior. First let $U^j(t)$ and $D^j(t)$ represent the cumulative up time and down time, respectively, at station j during the interval $[0, t]$. These are defined by

$$U^j(t) \equiv \int_0^t 1_{\{I^j(s)=1\}} ds, \quad t \geq 0,$$

and

$$D^j(t) \equiv t - U^j(t), \quad t \geq 0,$$

where 1_A is the indicator function of the event A, i.e., $1_A(x) = 1$ if $x \in A$ and $1_A(x) = 0$ otherwise. Let $B^j(t)$ be the cumulative busy time of the server at queue j during the interval $[0, t]$, i.e., the total amount of time during $[0, t]$ that the server at queue j is serving customers. The busy-time process will be expressed in terms of the model data below. Then

$$Y^j(t) = U^j(t) - B^j(t), \quad t \geq 0, \tag{7.1}$$

is the cumulative idle time of the server at queue j. Thus,

$$B^j(t) + Y^j(t) + D^j(t) = t, \quad t \geq 0, \tag{7.2}$$

We can now define the queue-length process as

$$Z^j(t) \equiv Z^j(0) + A^j(t) + \sum_{i=1}^k R^{i,j}(S^i(B^i(t))) - S^j(B^j(t)), \quad t \geq 0, \tag{7.3}$$

for $1 \leq j \leq k$. Note that $S^j(B^j(t))$ gives the actual number of departures from queue j during $[0, t]$, so that (7.3) expresses the basic conservation of customers at each queue: The number of customers present at time t equals the initial number there plus the arrivals (external plus internal) minus the departures.

To complete the process definition, we still need to specify the busy-time process B. Since the FCFS discipline is used (any work-conserving discipline would suffice here), we must have

$$B^j(t) = \int_0^t 1_{\{Z^j(s)>0, I^j(s)=1\}} ds . \tag{7.4}$$

In (7.3) and (7.4) we have defined Z in terms of B and B in terms of Z.

Theorem 14.7.1. (existence and uniqueness) *There exists a unique solution (Z, B) to equations (7.3) and (7.4). Equations (7.3) and (7.4) determine a measurable mapping from $D^{k^2+3k} \times \mathbb{R}^k$ to D^{2k} taking $(A, S, R, I, Z(0))$ into (Z, B), using the Kolmogorov σ-field on all D spaces.*

Proof Existence and uniqueness follow by doing an induction on the transition epochs of (A, S, I). (There necessarily are only finitely many such transitions in any bounded interval.) From (7.4), it follows that the sample paths of B are Lipschitz and thus continuous. Hence the sample paths of (B, Z) do indeed belong to D^{2k}.

Finally, since (Z, B) over any interval $[0, t]$ depends only upon $Z(0)$ and finitely many transitions of (A, S, I) over $[0, t]$, the map is measurable using the Kolmogorov σ-field on all D spaces. ∎

We now want to show that the process pair (Y, Z) can be represented as the image of the reflection map applied to an appropriate potential net-input process X. For that purpose, let $\lambda \equiv (\lambda^1, \ldots, \lambda^k)$ and $\mu \equiv (\mu^1, \ldots, \mu^k)$ be nonnegative vectors in $\mathbb{R}^k$ and let $P \equiv (P_{i,j})$ be a $k \times k$ nonnegative matrix. We think of λ^j as the long-run arrival rate to queue j, $1/\mu^j$ as the long-run average service time at queue j, and $P_{i,j}$ as the long-run proportion of departures from queue i that are routed immediately to queue j, but these definitions are not yet required. For each $t > 0$, let

$$\xi^j(t) \equiv A^j(t) - \lambda^j t + \sum_{i=1}^{k}[R^{i,j}(S^i(B^i(t))) - P_{i,j}S^i(B^i(t))]$$

$$+ \sum_{i=1}^{k} P_{i,j}[S^i(B^i(t)) - \mu^i B^i(t)] - [S^j(B^j(t)) - \mu^j B^j(t)], \quad (7.5)$$

$$\eta^j(t) \equiv \left(\lambda^j - \mu^j + \sum_{i=1}^{k} \mu^i P_{i,j}\right) t + \mu^j D^j(t) - \sum_{i=1}^{k} \mu^i P_{i,j} D^i(t) \quad (7.6)$$

and

$$X^j(t) \equiv Z^j(0) + \xi^j(t) + \eta^j(t), \quad 1 \leq j \leq k. \quad (7.7)$$

Let $diag(\mu)$ be the $k \times k$ diagonal matrix with μ^i the (i, i) element.

Theorem 14.7.2. (reflection map representation) *For all nonnegative vectors $\lambda, \mu \in \mathbb{R}^k$ and all nonnegative $k \times k$ matrices P with $P^t \equiv Q \in \mathcal{Q}$,*

$$Z = \phi(X) \quad \text{and} \quad \psi(X) = diag(\mu)Y \quad (7.8)$$

for Z in (7.3), X in (7.7), Y in (7.1) and (ψ, ϕ) the reflection map in Definition 14.2.1 associated with the column-substochastic matrix $Q \equiv P^t$; i.e.,

$$Z^j(t) = X^j(t) + \mu^j Y^j(t) - \sum_{i=1}^{k} P_{i,j} \mu^i Y^i(t), \quad t \geq 0, \quad (7.9)$$

or, equivalently,

$$Z = X + (I - Q)diag(\mu)Y \quad (7.10)$$

and

$$\int_0^\infty Z^j(t) dY^j(t) = 0, \quad 1 \leq j \leq k, \quad w.p.1. \quad (7.11)$$

Proof. First, (7.9) follows directly from the definitions (7.1)–(7.3) and (7.5)–(7.7) by adding and subtracting. Since $Z(t) \geq 0$ and

$$\int_0^\infty Z^j(t) dY^j(t) = \int_0^\infty Z^j(t) 1_{\{Z^j(t)=0, I^j(t)=1\}} dt,$$

14.7.2. Heavy-Traffic Limits

To establish heavy-traffic limits, we consider a sequence of models indexed by n. As $n \to \infty$, we will let the traffic intensity at queue j in model n, ρ_n^j, approach the critical value 1 from below for each j. Since the interruptions are exogenous, we can start by establishing a limit for the cumulative-down-time process. Anticipating a limit for the scaled queue-length processes

$$\mathbf{Z}_n \equiv \mathbf{Z}_n(t) \equiv n^{-H} Z_n(nt), \quad t \geq 0, \quad (7.12)$$

with $1/2 \leq H < 1$, we assume that the sequences of up and down times $\{(u_{n,m}^j, d_{n,m}^j) : m \geq 1\}$ in model n satisfy

$$\{(n^{-1} u_{n,m}^j, n^{-H} d_{n,m}^j) : m \geq 0, 1 \leq j \leq k\} \Rightarrow \{(u_m^j, d_m^j) : m \geq 0, 1 \leq j \leq k\} \quad (7.13)$$

in $(\mathbb{R}^{2k})^\infty$, where $0 \leq H < 1$, $\mu_1^j > 0$ and $\sum_{m=1}^\infty u_m^j = \infty$ w.p.1 for each j. We require that $u_1^j > 0$ so that in the limit there is no interruption at time 0. We require that $\sum_{m=1}^\infty u_m^j = \infty$ so that in the limit there are only finitely many interruptions in any bounded interval. The idea behind the scaling in (7.13) is that the up times have proper limits in the scaling (7.12), while the down times are asymptotically negligible. Thus, in the limit, there is a proper point process describing the occurrence of interruptions, but these interruptions occur instantaneously. However, because of the down-time scaling in (7.13), they have a spatial impact, causing jumps in the process $\mathbf{Z}_n$ with the scaling in (7.12). The standard Brownian motion case has $H = 1/2$ in (7.12) and (7.13). For any value of H, $0 \leq H < 1$, the interruptions introduce extra jumps.

To treat the cumulative down-time process, let $N^j \equiv \{N^j(t) : t \geq 0\}$ be the counting process associated with the limiting up times, i.e.,

$$N^j(t) \equiv \max\left\{m \geq 0 : \sum_{l=1}^m u_l^j \leq t\right\}, \quad t \geq 0,$$

and let $\mathbf{D}^j$ be the random sum

$$\mathbf{D}^j(t) \equiv \sum_{l=1}^{N^j(t)} d_l^j, \quad t \geq 0. \quad (7.14)$$

Let $\mathbf{D}_n \equiv (\mathbf{D}_n^1, \ldots, \mathbf{D}_n^k)$ be the normalized cumulative-down-time processes associated with model n, defined by

$$\mathbf{D}_n^j(t) \equiv n^{-H} D_n^j(nt), \quad t \geq 0, \quad 1 \leq j \leq k, \quad (7.15)$$

where $D_n^j(t)$ is the cumulative down time in $[0, t]$ associated with the sequence $\{(u_{n,m}^j, d_{n,m}^j) : m \geq 0\}$. As before, let $Disc(x)$ be the set of discontinuities of x.

Here is the basic down-time result.

Theorem 14.7.3. (cumulative down-time limit) *If (7.13) holds with*

$$P(u_1^j > 0) = 1 \ ,$$

$$P(\sum_{m=1}^{\infty} u_m^j = \infty) = 1, \quad 1 \leq j \leq k$$

and

$$P\left(\bigcup_{i=1}^{k} \bigcup_{\substack{j=1 \\ j \neq i}}^{k} (Disc(\mathbf{D}^i) \cap Disc(\mathbf{D}^j)) = \phi\right) = 1 \ , \quad (7.16)$$

then

$$\mathbf{D}_n \Rightarrow \mathbf{D} \quad in \quad D([0,\infty), \mathbb{R}^k, SM_1) \ . \quad (7.17)$$

for $\mathbf{D}$ *in (7.14).*

Proof. First apply the Skorohod representation theorem to replace the convergence in distribution by convergence w.p.1 (without introducing new notation for the special versions). Then it is elementary that $\mathbf{D}_n^j(t) \to \mathbf{D}^j(t)$ for each $t \notin Disc(\mathbf{D}^j)$. Since $\mathbf{D}_n^j$ and $\mathbf{D}^j$ are nondecreasing, this implies convergence in $D([0,\infty), \mathbb{R}, M_1)$ by Corollary 12.5.1. By Corollary 12.6.1, condition (7.16) allows us to strengthen the resulting WM_1 convergence in D^k to SM_1 convergence. ∎

Henceforth we make the conclusion (7.17) in Theorem 14.7.3 part of the conditions; e.g., see (7.20) below. Thinking of the interruptions as exogenous, we can assume that the up-and-down-time processes and, thus, the cumulative-down-time processes $\mathbf{D}_n$ are independent of the rest of the model data. Thus the limit (7.17) will hold jointly with the assumed limit for the rest of the model data using the product topology by virtue of Theorem 11.4.4. Then strengthening the convergence to overall SM_1 convergence can be done by imposing conditions on the discontinuities, paralleling condition (7.16).

We now introduce scaled random elements of D associated with the sequence of models for the main limit theorem. Let

$$\begin{aligned}
\mathbf{A}_n(t) &\equiv n^{-H}(A_n(nt) - \lambda_n nt), \quad t \geq 0 \ , \\
\mathbf{S}_n(t) &\equiv n^{-H}(S_n(nt) - \mu_n nt), \quad t \geq 0 \ , \\
\mathbf{R}_n(t) &\equiv n^{-H}(R_n(nt) - P_n nt), \quad t \geq 0 \ ,
\end{aligned} \quad (7.18)$$

where λ_n and μ_n are nonnegative vectors in $\mathbb{R}^k$ and $P_n \equiv (P_n(i,j))$ is a nonnegative $k \times k$ matrix with transpose $P_n^t \equiv Q_n$ in $\mathcal{Q}$. Let the associated scaled processes for which we want to establish convergence be $\mathbf{Z}_n$ in (7.12) and

$$\begin{aligned}
\mathbf{Y}_n(t) &\equiv n^{-H} Y_n(nt) \\
\mathbf{B}_n(t) &\equiv n^{-H}(B^n(nt) - nt), \quad t \geq 0.
\end{aligned} \quad (7.19)$$

Theorem 14.7.4. (heavy-traffic limit with rare long interruptions) *Suppose that*

$$(\mathbf{A}_n, \mathbf{S}_n, \mathbf{R}_n, \mathbf{D}_n, \mathbf{Z}_n(0)) \Rightarrow (\mathbf{A}, \mathbf{S}, \mathbf{R}, \mathbf{D}, \mathbf{Z}(0)) \quad as \quad n \to \infty \qquad (7.20)$$

in $D([0,\infty), \mathbb{R}^{k^2+3k}, WM_1) \times \mathbb{R}^k$ *for* $(\mathbf{A}_n, \mathbf{S}_n, \mathbf{R}_n)$ *in* (7.18), $\mathbf{D}_n$ *in* (7.15) *and* $\mathbf{Z}_n$ *in* (7.12) *with* $0 \leq H < 1$, *where*

$$P((\mathbf{A}, \mathbf{S}, \mathbf{R}, \mathbf{D}) \in D_1) = 1 \ .$$

If, in addition, there exist vectors λ *and* μ *in* $\mathbb{R}^k$ *and a matrix* P *with* $P^t \in \mathcal{Q}$ *such that*

$$\lambda_n \to \lambda, \quad \mu_n \to \mu > 0 \quad and \quad P_n^t \to P^t \quad in \quad \mathcal{Q} \qquad (7.21)$$

and

$$c_n^j \equiv n^{1-H}\left(\lambda_n^j - \mu_n^j + \sum_{i=1}^k \mu_n^i P_n(i,j)\right) \to c^j \quad as \quad n \to \infty \qquad (7.22)$$

with $-\infty < c^j < \infty$, $1 \leq j \leq k$, *then*

$$(\mathbf{Z}_n, \mathbf{Y}_n, \mathbf{B}_n) \Rightarrow (\mathbf{Z}, \mathbf{Y}, \mathbf{B}) \quad in \quad D([0,\infty), \mathbb{R}^{3k}, WM_1) \qquad (7.23)$$

for

$$\begin{aligned}
\mathbf{Z} &\equiv \phi(\mathbf{X}), \quad \mathbf{Y} \equiv diag(\mu^{-1})\psi(\mathbf{X}) \ , \\
\mathbf{X} &\equiv \mathbf{Z}(0) + \boldsymbol{\xi} + \boldsymbol{\eta} \ , \\
\xi^j &\equiv \mathbf{A}^j + \sum_{i=1}^k [\mathbf{R}^{i,j} \circ \mu^i \mathbf{e} + P_{i,j}\mathbf{S}^i] - \mathbf{S}^j \\
\boldsymbol{\eta} &\equiv c\mathbf{e} + (I - P^t)diag(\mu)\mathbf{D} \\
\mathbf{B} &\equiv -\mathbf{D} - \mathbf{Y} \ .
\end{aligned} \qquad (7.24)$$

Proof. As usual, we start by applying the Skorohod representation theorem to replace the assumed convergence in distribution in (7.20) by convergence w.p.1 for alternative versions, without introducing special notation for the alternative versions. We first want to show that, asymptotically, the servers are busy all the time. For that purpose, we establish a FWLLN for the cumulative-busy-time process with spatial scaling by n^{-1}. To do so, we establish a FWLLN for all the processes with spatial scaling by n^{-1}. For that purpose, let

$$\begin{aligned}
&(\hat{\mathbf{A}}_n(t), \hat{\mathbf{S}}_n(t), \hat{\mathbf{R}}_n(t), \hat{\mathbf{D}}_n(t), \hat{\mathbf{Z}}_n(0), \hat{\mathbf{B}}_n(t), \hat{\mathbf{Y}}_n(t), \hat{\mathbf{Z}}_n(t)) \\
&\equiv n^{-1}(A_n(nt), S_n(nt), R_n(nt), D_n(nt), Z_n(0), B_n(nt), Y_n(nt), Z_n(nt)) \ .
\end{aligned}$$

Conditions (7.20) and (7.21) imply that

$$(\hat{\mathbf{A}}_n, \hat{\mathbf{S}}_n, \hat{\mathbf{R}}_n, \hat{\mathbf{D}}_n, \hat{\mathbf{Z}}_n(0)) \to (\lambda \mathbf{e}, \mu \mathbf{e}, P\mathbf{e}, 0\mathbf{e}, 0\mathbf{e}) \quad as \quad n \to \infty$$

in $(D^{k^2+3k}, U) \times \mathbb{R}^k$, i.e., with the topology of uniform convergence over bounded intervals. Then note that $\{\hat{\mathbf{B}}_n : n \geq 1\}$ is relatively compact by the Arzela-Ascoli theorem (Theorem 11.6.2) because $\mathbf{B}_n$ is uniformly Lipschitz: For $0 < t_1 < t_2$,

$$|\hat{\mathbf{B}}_n(t_2) - \hat{\mathbf{B}}_n(t_1)| = |n^{-1}B_n(nt_2) - n^{-1}B_n(nt_2)| \leq |t_2 - t_1|.$$

Hence, $\{\hat{\mathbf{B}}_n\}$ has a convergent subsequence $\{\hat{\mathbf{B}}_{n_k}\}$ in $C([0,T], \mathbb{R}^k, U)$ for every T. Suppose that

$$\hat{\mathbf{B}}_{n_k} \to \hat{\mathbf{B}} \quad \text{as} \quad n_k \to \infty \quad \text{in} \quad (D^k, U).$$

Then, from Theorems 14.2.5 and 14.7.2,

$$(\hat{\mathbf{Z}}_{n_k}, \hat{\mathbf{Y}}_{n_k}) \to (\mathbf{0}, \mathbf{0}) \quad \text{in} \quad (D^{2k}, U) \quad \text{as} \quad n_k \to \infty.$$

As a consequence of (7.2), we must have $\hat{\mathbf{B}}^j = \mathbf{e}$, i.e., $\hat{\mathbf{B}}^j(t) = t$, $t \geq 0$, $1 \leq j \leq k$. Since the limit is the same for all subsequences, we have

$$(\hat{\mathbf{Z}}_n, \hat{\mathbf{Y}}_n, \hat{\mathbf{B}}_n) \to (\mathbf{0}, \mathbf{0}, \mathbf{e}) \quad \text{as} \quad n \to \infty \quad \text{in} \quad (D^{3k}, U),$$

where $\mathbf{e}$ is the vector-valued function, equal to e in each coordinate. Now we return to the processes with spatial scaling by n^{-H}. Let

$$
\begin{aligned}
\boldsymbol{\xi}_n^j(t) &\equiv n^{-H} \xi_n^j(nt) \\
&= n^{-H}[A_n^j(nt) - \lambda_n^j nt] + \sum_{i=1}^k n^{-H}[R_n^{i,j}(n[n^{-1}S_n^i(n[n^{-1}B_n^i(nt)])]) \\
&\quad - P_n(i,j)n(n^{-1}S_n^i(n[n^{-1}B_n^i(nt)]))] \\
&\quad + \sum_{i=1}^k P_n(i,j)n^{-H}[S_n^i(n[n^{-1}B_n^i(nt)]) - \mu_n^i n[n^{-1}B_n^i(nt)]] \\
&\quad - n^{-H}[S_n^j(n[n^{-1}B_n^j(nt)]) - \mu_n^j n(n^{-1}B_n^j(nt))], \\
\boldsymbol{\eta}_n^j(t) &\equiv n^{-H} \eta_n^j(nt) \\
&= n^{1-H}\left(\lambda_n^j - \mu_n^j + \sum_{i=1}^k \mu_n^i P_n(i,j)\right) t \\
&\quad + \mu_n^j n^{-H} D^j(nt) - \sum_{i=1}^k \mu_n^i P_n(i,j) n^{-q} D_n^i(nt)
\end{aligned}
$$

and

$$\mathbf{X}_n^j(t) \equiv n^{-H} X_n^j(nt) = n^{-H}(Z_n^j(0) + \xi_n^j(nt) + \eta_n^j(nt)), \quad t \geq 0,$$

so that we have

$$
\begin{aligned}
\boldsymbol{\xi}_n^j &= \mathbf{A}_n^j + \sum_{i=1}^k \mathbf{R}_n^{i,j} \circ \hat{\mathbf{S}}_n^i \circ \hat{\mathbf{B}}_n^i \\
&\quad + \sum_{i=1}^k P_n(i,j) \mathbf{S}_n^i \circ \hat{\mathbf{B}}_n^i - \mathbf{S}_n^j \circ \hat{\mathbf{B}}_n^j,
\end{aligned}
$$

$$\boldsymbol{\eta}_n^j = c_n^j \mathbf{e} + \mu_n^j \mathbf{D}_n^j - \sum_{i=1}^{k} \mu_n^i P_n(i,j) \mathbf{D}_n^i ,$$

for c_n in (7.22),

$$\mathbf{X}_n = \mathbf{Z}_n(0) + \boldsymbol{\xi}_n + \boldsymbol{\eta}_n , \qquad (7.25)$$

$$\mathbf{Z}_n = \phi_{Q_n}(\mathbf{X}_n), \quad \mathbf{Y}_n = diag(\mu_n)\psi_{Q_n}(\mathbf{X}_n) \qquad (7.26)$$

and

$$\mathbf{B}_n = -\mathbf{Y}_n - \mathbf{D}_n . \qquad (7.27)$$

Since $\mu^j \mathbf{e}$, $\mathbf{e} \in C_{\uparrow\uparrow}$, the subset of functions in C that are nonnegative and strictly increasing, we can apply the composition map plus addition to get $\mathbf{X}_n \to \mathbf{X}$ in $D([0,\infty), \mathbb{R}^k, SM_1)$; see Theorem 13.2.3. Then we can apply Theorem 14.5.5 with (7.25) and (7.26) to get the desired limit (7.23) with the limit processes in (7.24).

Remark 14.7.1. *The natural sufficient condition.* The natural sufficient condition for the limit in condition (7.20) in Theorem 14.7.4 is to have the $k^2 + 3k$ processes A_n^j, S_n^j, $R_n^{i,j}$ and D_n^j, $1 \leq i \leq k$, $1 \leq j \leq k$, be mutually independent and for the limit processes to have no fixed discontinuities, i.e., for $P(t \in Disc(V)) = 0$ for all V and $t \geq 0$, where V is one of the coordinate limit processes above. ∎

Remark 14.7.2. *The heavy-traffic condition.* Note that (7.22) can be rewritten as

$$n^{1-H}(\lambda_n - (I - Q_n)\mu_n) \to c ,$$

so that (7.21) and (7.22) together imply that

$$\lambda = (I - Q)\mu , \qquad (7.28)$$

which is a version of the traffic rate equation associated with a Markovian network having external arrival rate vector λ and routing matrix P. Then μ is the net (external plus internal) arrival rate at each queue, and the asymptotic traffic intensity at each queue is 1. Thus, we see that indeed Theorem 14.7.4 provides a heavy-traffic limit.

Remark 14.7.3. *A simple sequence of models.* We have allowed the vectors λ_n, μ_n and the routing matrix P_n all to vary with n. For applications it should usually suffice to let only λ_n vary with n. More generally, it should suffice to let the service processes S_n and the routing processes R_n be independent of n. Then it is natural to have μ and P be independent of n. Moreover, the arrival processes could be made to depend on n only through a single sequence of scaling vectors $\{\alpha_n : n \geq 1\}$ in $\mathbb{R}^k$ with $\alpha_n \to 1 \equiv (1,\ldots,1)$ as $n \to \infty$. We could start with a single arrival process A', where

$$\mathbf{A}'_n(t) = n^{-H}[A'(nt) - \lambda nt], \quad t \geq 0 ,$$

and

$$\mathbf{A}'_n \Rightarrow \mathbf{A} \quad \text{in} \quad D^k . \qquad (7.29)$$

We can then let

$$A_n^j(t) \equiv A^j(\alpha_n^j t) \quad \text{and} \quad \lambda_n^j \equiv \alpha_n^j \lambda^j, \tag{7.30}$$

so that by (7.29) and (7.30),

$$\mathbf{A}_n \Rightarrow \mathbf{A}$$

for $\mathbf{A}_n$ in (7.18) and (7.30). With the single-vector parameterization, since

$$\begin{aligned}
c_n^j &= n^{1-H}\left(\alpha_n^j \lambda^j - \mu^j + \sum_{i=1}^k \mu^i P(i,j)\right) \\
&= n^{1-H}\left(\alpha_n^j \lambda^j - \lambda^j + \lambda^j - \mu^j + \sum_{i=1}^k \mu^i P(i,j)\right),
\end{aligned} \tag{7.31}$$

condition (7.21) holds if and only if (7.28) holds and

$$n^{1-H}(\alpha_n^j - 1) \to \alpha^j,$$

in which case

$$c_n^j = n^{1-H}(\alpha_n^j - 1)\lambda^j \to \alpha^j \lambda^j \equiv c^j \quad \text{as} \quad n \to \infty. \quad \blacksquare$$

We now establish the corresponding limit with fixed up and down times. We now assume that the up and down time sequence $\{(u_m^j, d_m^j) : m \geq 1, 1 \leq 0 \leq k\}$ is independent of n when we consider the family of models indexed by n. Then instead of the limit in Theorem 14.7.3, we assume that $\mathbf{D}_n \Rightarrow \mathbf{D}$ in $D([0,\infty), \mathbb{R}^k, SM_1)$, where

$$\mathbf{D}_n^j(t) \equiv n^{-H}(D^j(nt) - (1-\nu)nt), \quad t \geq 0, \tag{7.32}$$

where $0 < \nu^j \leq 1$, $1 \leq j \leq k$. Then ν^j is the proportion of up time at queue j.

Theorem 14.7.5. (heavy-traffic limit with fixed up and down times) *Suppose that the assumptions of Theorem 14.7.4 hold with $\mathbf{D}_n$ in (7.15) replaced by (7.32) and (7.22) replaced by*

$$c_n^j \equiv n^{1-H}\left(\lambda_n^j - \mu_n^j \nu^j + \sum_{i=1}^k \mu_n^i \nu_n^i P_n(i,j)\right) \to c^j \tag{7.33}$$

with $-\infty < c^j < \infty$, $1 \leq j \leq k$. Then the conclusions of Theorem 14.7.4 hold with $\mathbf{B}_n$ in (7.19) replaced by

$$\mathbf{B}_n^j(t) \equiv n^{-H}(B_n^j(nt) - \nu^j nt), \quad t \geq 0, \quad 1 \leq j \leq k,, \tag{7.34}$$

and $\boldsymbol{\xi}^j$ in (7.24) replaced by

$$\boldsymbol{\xi}^j = \mathbf{A}^j + \sum_{i=1}^k \mathbf{R}^{i,j} \circ \mu^i \nu^i \mathbf{e} + P_{i,j} \mathbf{S}^i \circ \nu^i \mathbf{e} - \mathbf{S}^j \circ \nu^j \mathbf{e}. \tag{7.35}$$

Proof. The proof is essentially the same as for Theorem 14.7.4, except now $(\hat{\mathbf{D}}_n, \hat{\mathbf{B}}_n) \to (\hat{\mathbf{D}}, \hat{\mathbf{B}})$, where $\hat{\mathbf{D}}^j(t) = (1 - \nu_j)t$ and $\hat{\mathbf{B}}^j(t) = \nu^j t$, $t \geq 0$. ■

Remark 14.7.4. *The common case.* The common case has scaling exponent $H = 1/2$ and $(\mathbf{A}, \mathbf{S}, \mathbf{R}, \mathbf{D})$ Brownian motion, in which case $(\mathbf{Z}, \mathbf{Y})$ is reflected Brownian motion. The special case without down times is the heavy-traffic limit for a single-class open queueing network in Reiman (1984a). Theorem 14.7.5 also includes convergence to reflected stable processes when the interarrival times and service times are IID in the normal domain of attraction of a stable law with index α, $1 < \alpha < 2$; then $H = 1/\alpha$. ■

14.8. The Two-Sided Regulator

In this section we establish continuity and Lipschitz properties for the two-sided regulator (or reflection) map, which arises in heavy-traffic limits for single queues with finite waiting room (or buffer space); see Section 2.3 and Chapter 5. We also show that the continuity and Lipschitz properties of the multidimensional reflection map with the M_1 topologies extend to other more general reflection maps, such as those considered by Dupuis and Ishii (1991), Williams (1987, 1995) and Dupuis and Ramanan (1999a,b).

14.8.1. Definition and Basic Properties

Anticipating the more general reflection maps to be introduced later in the section, we allow the two-sided regulator to depend on the initial position as well as the net-input function. The initial position will be a point s in the set $S \equiv [0, c]$. Specifically, we let the two-sided regulator $R : S \times D([0,T], \mathbb{R}) \to D([0,T], \mathbb{R}^{3k})$ be defined by

$$R(s, x) \equiv (\phi(s,x), \psi_1(s,x), \psi_2(s,x)) \equiv (z, y_1, y_2),$$

where $S \equiv [0, c]$,

$$z = s + x + y_1 - y_2,$$
$$0 \leq z(t) \leq c, \quad 0 \leq t \leq T,$$
$$y_1(0) = -((s + x(0)) \wedge 0)^- \quad \text{and} \quad y_2(0) = [c - s - x(0)]^+,$$
$$y_1 \text{ and } y_2 \text{ are nondecreasing},$$
$$\int_0^T z(t) dy_1(t) = 0 \quad \text{and} \quad \int_0^T [c - z(t)] dy_2(t) = 0. \tag{8.1}$$

The two-sided regulator can also be defined using with the elementary instantaneous reflection map

$$R_0 \equiv (\phi_0, \psi_{0,1}, \psi_{0,2}) : [0, c] \times \mathbb{R} \to \mathbb{R}^3$$

defined by

$$\phi_0(s,u) \equiv (s+u) \vee 0 \wedge c,$$
$$\psi_{0,1}(s,u) \equiv -(s+u)^-,$$
$$\psi_{0,2}(s,u) \equiv [s+u-c]^+, \quad (8.2)$$

where again s is the initial position in $S \equiv [0,c]$ and u is the increment. We now can define the reflection map $R \equiv (\phi, \psi_1, \psi_2) : S \times D_c \to D_c^3$ recursively by letting

$$z(t_i) \equiv \phi(z(0-), x)(t_i) \equiv \phi_0(z(t_{i-1}), x(t_i) - x(t_{i-1}))$$
$$y_1(t_i) \equiv \psi_1(z(0-), x)(t_i) \equiv \psi_{0,1}(z(t_{i-1}), x(t_i) - x(t_{i-1})) + y_1(t_{i-1})$$
$$y_2(t_i) \equiv \psi_2(z(0-), x)(t_i) \equiv \psi_{0,2}(z(t_{i-1}), x(t_i) - x(t_{i-1})) + y_2(t_{i-1}), (8.3)$$

where $t_1, \ldots, t_m$ are the discontinuity points of x with $t_0 \equiv 0 < t_1 < \cdots < t_m < T$, $x^i(t_{-1}) \equiv 0$ for all i and $z(t_{-1}) \equiv z(0-) \in S$ is the initial position. We let (z, y_1, y_2) be constant in between discontinuities. Finally, we define $R : S \times D \to D^3$ by letting

$$R(s,x) \equiv \lim_{n \to \infty} R(s, x_n) \quad (8.4)$$

for $x_n \in D_c$ with $\|x_n - x\| \to 0$.

Theorem 14.8.1. (two-sided regulator) *There exists a unique reflection map $R : S \times D \to D^3$ defined by (8.2), (8.3) and (8.4), which coincides with the reflection map defined by (8.1). For any $(s_1, x_1), (s_2, x_2) \in S \times D$,*

$$\|\phi(s_1, x_1) - \phi(s_2, x_2)\| \leq 2(\|s_1 - s_2\| + \|x_1 - x_2\|). \quad (8.5)$$

If $\|s_n - s\| \to 0$ in S and $\|x_n - x\| \to 0$ in D, then

$$\|\psi_j(s_n, x_n) - \psi_j(s,x)\| \to 0 \quad \text{for} \quad j = 1, 2, . \quad (8.6)$$

Proof. First existence and uniqueness of the reflection map on D_c defined by (8.2) and (8.3) is immediate from the recursive definition. It is then easy to see that (8.1) is equivalent to (8.2) and (8.3) on $S \times D_c$; apply induction on the successive discontinuity points. We will show that there exists a unique extension of (8.2) and (8.3) to $S \times D$ specified by the limit in (8.4). For that purpose, we establish the Lipschitz property (8.5) when $x_1, x_2 \in D_c$. We use induction over the points at which at least one of these functions has a discontinuity. Suppose that $\|s_1 - s_2\| + \|x_1 - x_2\| = \epsilon$.

$$\Delta_n \equiv s_1 + x_1(t_n) - s_2 - x_2(t_n)$$

and

$$\Gamma_n \equiv z_1(t_n) - z_2(t_n).$$

We are given that $|\Delta_n| \leq \epsilon$ for all n. By induction we show that

$$\Delta_n - \epsilon \leq \Gamma_n \leq \Delta_n + \epsilon \quad \text{for all} \quad n, \quad (8.7)$$

from which the desired conclusion follows. Since

$$z_i(0) \equiv (s + x_i(0)) \vee 0 \wedge c,$$

(8.7) holds for $n = 0$. Suppose that (8.7) holds for all nonnegative integers up to n. Then, by considering the possible jumps at t_{n+1}, we see that

$$\Gamma_{n+1} \leq \begin{cases} 0 \leq \Delta_{n+1} + \epsilon & \text{if } z_2(t_{n+1}) = c \text{ or } z_1(t_{n+1}) = 0 \\ \Gamma_n + \Delta_{n+1} - \Delta_n \leq \Delta_{n+1} + \epsilon & \text{otherwise}, \end{cases}$$

and

$$\Gamma_{n+1} \geq \begin{cases} 0 \geq \Delta_{n+1} - \epsilon & \text{if } z_2(t_{n+1}) = 0 \text{ or } z_1(t_{n+1}) = c \\ \Gamma_n + \Delta_{n+1} - \Delta_n \geq \Delta_{n+1} - \epsilon & \text{otherwise}. \end{cases}$$

Clearly these two inequalities imply (8.7). For general $x \in D$, we have defined $R(x)$ by the limit in (8.4). To show that the limit actually exists, given $x \in D$ choose $x_n \in D_c$ such that $\|x_n - x\| \to 0$. Hence $\|x_n - x_m\| \to 0$ as $n, m \to 0$, where $x_n, x_m \in D_c$. By the Lipschitz property on D_c above, $\|R(s, x_n) - R(s, x_m)\| \leq 2\|x_n - x_m\| \to 0$. Since $(D, \|\cdot\|)$ is complete, there exists $z \in D$ such that $\|R(s, x_n) - z\| \to 0$. Let $R(s, x) \equiv z$. To establish uniqueness of the limit in (8.4), suppose that $\|x_{j,n} - x\| \to 0$ where $x_{j,n} \in D_c$ for $j = 1, 2$. By the triangle inequality, $\|x_{1,n} - x_{2,n}\| \to 0$. Then by the Lipschitz property $\|R(s, x_{1,n}) - R(s, x_{2n})\| \leq 2\|x_{1,n} - x_{2,n}\| \to 0$, so that the two limits must agree. To establish the Lipschitz property of ϕ on $S \times D$ let s_1, s_2, x_1 and x_2 be given and choose $x_{1,n}, x_{2,n} \in D_c$ such that $\|x_{1,n} - x_1\| \to 0$ and $\|x_{2,n} - x_2\| \to 0$. By the Lipschitz property on $S \times D^c$,

$$\|\phi(s_1, x_{1,n}) - \phi(s_2, x_{2,n})\| \leq 2(|s_1 - s_2| + \|x_{1,n} - x_{2,n}\|).$$

Letting $n \to \infty$ yields (8.5). To establish the remaining results, we reduce the remaining results to previous established results for the one-sided reflection map over subintervals. Suppose that s and x are given. By above, $z \equiv \phi(s, x)$ is well defined. Let

$$t_1 \equiv \inf\{t \geq 0 : \text{either } z(t) \leq \epsilon \text{ or } z(t) \geq c - \epsilon\}$$

for some ϵ with $0 < \epsilon < c - \epsilon < c$, with $t_1 \equiv T + 1$ if the infimum is not attained. Clearly $z(t) = s + x(t)$ and $y_1(t) = y_2(t) = 0$ for $0 \leq t < t_1$ if $t_1 > 0$. For simplicity, suppose that $z(t_1) \leq \epsilon$. Then let

$$t_{2m} \equiv \inf\{t : t_{2m-1} < t \leq T, z(t) \geq c - \epsilon\}$$

with $t_{2m} \equiv T + 1$ if the infimum is not attained, and

$$t_{2m+1} \equiv \inf\{t : t_{2m} < t \leq T, z(t) \leq \epsilon\}$$

with $t_{2m+1} \equiv T + 1$ if the infimum is not attained. Since $z \in D$, there are finitely many points $0 \leq t_1 < \cdots < t_m \leq T$ such that the infima above are attained. It suffices to apply the one-sided reflection map over each of the subintervals $[0, t_1), [t_1, t_2), \ldots, [t_{m-1}, t_m)$ and $[t_m, T]$. Suppose that $|s_n - s| \to 0$ and $\|x_n - x\| \to 0$ as $n \to \infty$. By (8.5), $\|z_n - z\| \to 0$. Thus, for all n sufficiently large, only the one-sided reflection map need be applied over each of the subintervals $[0, t_1), [t_1, t_2), \ldots, [t_m, T]$. Hence, from (2.23) in Theorem 14.2.5, we can deduce that (8.6) holds. Moreover, from Theorem 14.2.3, we can deduce the complementarity in (8.1). Indeed, with the other conditions there, the complementarity property characterizes the reflection map. ■

We have already seen that the Lipschitz bound in (8.5) is tight for the one-sided reflection map in Example 14.2.2, so it is tight here as well. Unlike for the one-sided reflection map in Chapter 13 and the multidimensional reflection map in Section 14.2, the regulator maps ψ_1 and ψ_2 here need not be Lipschitz.

Example 14.8.1. *Counterexample to the Lipschitz property for ψ_1 and ψ_2.* To see that ψ_1 and ψ_2 need not be Lipschitz, suppose that $c > \epsilon$ and, for $i = 1, 2$, let $x_i \equiv x_i^\uparrow + x_i^\downarrow$, where

$$x_1^\uparrow(t) \equiv \sum_{i=0}^{n-1}[c1_{[2iT/2n,T]}(t) + (c+\epsilon)1_{[(2i+1)T/2n,T]}(t)]$$

$$x_2^\uparrow(t) \equiv \sum_{i=0}^{n-1}[(c+\epsilon)1_{[2iT/2n,T]}(t) + c1_{[(2i+1)T/2n,T]}(t)]$$

and

$$x_1^\downarrow(t) \equiv x_2^\downarrow \equiv -\sum_{i=0}^{n-1}(2c+\epsilon)c1_{[(2i+1)T/2n,T]}(t), \quad 0 \le t \le T.$$

Then

$$\|x_1^\uparrow - x_2^\uparrow\| = \|x_1 - x_2\| = \epsilon, \quad \|z_1 - z_2\| = 0,$$

$y_1(t) = 0$, $0 \le t \le T$, but

$$y_2(t) = \sum_{i=0}^{n-1}\epsilon I_{[(2i+1)T/2n,T]}(t),$$

so that

$$\|y_1 - y_2\| = \|y_2\| = n\epsilon = n\|x_1 - x_2\| \to \infty \quad \text{as} \quad n \to \infty. \quad ■$$

Reasoning just as for Theorem 14.2.7 and Corollary 14.2.2, we have the following consequences of Theorem 14.8.1.

Theorem 14.8.2. (Lipschitz and continuity with d_{J_1}) *For the two-sided regulator map $R \equiv (\phi, \psi_1, \psi_2)$ in (8.1),*

$$d_{J_1}(\phi(s_1, x_1), \phi(s_2, x_2)) \le 2(d_{J_1}(x_1, x_2) + |s_1 - s_2|)$$

for all $s_1, s_2 \in [0, c]$ and $x_1, x_2 \in D$. If $s_n \to s$ and $d_{J_1}(x_n, x) \to 0$, then

$$d_{J_1}(R(s_n, x_n), R(s, x)) \to 0 \quad \text{as} \quad n \to \infty.$$

Corollary 14.8.1. (measurability) *The two-sided regulator map $R : S \times D \to D^3$ is measurable, using the Kolmogorov σ-fields on both the domain and range.*

14.8.2. With the M_1 Topologies

We now establish results for the two-sided reflection map with the M_1 topologies. At the same time, we show how M_1 results can be obtained for other reflection maps, but we proceed abstractly without going into the details.

We assume that the general reflected process has values in a closed subset S of $\mathbb{R}^k$. We assume that we are given an instantaneous reflection map $\phi_0 : S \times \mathbb{R}^k \to S$. The idea is that an initial position s_0 in S and an instantaneous net input u are mapped by ϕ_0 into the new position $s_1 \equiv \phi_0(s_0, u_0)$ in S. In many cases $\phi_0(s_0, u_0)$ will depend upon (s_0, u_0) only through their sum $s_0 + u_0$, but we allow more general possibilities. It is also standard to have S be convex and $\phi_0(s, u) = s+u$ if $s+u \in S$, while $\phi_0(s, u) \in \partial S$ if $s + u \notin S$, where ∂S is the boundary of S, but again we do not directly require it. Under extra regularity conditions, ϕ_0 corresponds to the projection in Dupius and Ramanan (1999a).

As in Section 14.3, we use ϕ_0 to define a reflection map on $D_c \equiv D_c([0,T], \mathbb{R}^k)$. However, we also allow dependence upon the initial position in S. Thus, we define $\phi : S \times D_c \to D_c$ by letting

$$\phi(z(0-), x)(t_i) \equiv z(t_i) \equiv \phi_0(z(t_{i-1}), x(t_i) - x(t_{i-1})), \quad 0 \le i \le m , \qquad (8.8)$$

where $t_1, \ldots, t_m$ are the discontinuity points of x, with $t_0 = 0 < t_1 < \cdots < t_m < T$, $x^i(t_{-1}) = 0$ for all i and $z(t_{-1}) \equiv z(0-) \in S$ is the initial position. A standard case is $x^i(0) = 0$ for all i and $z(0) = z(0-)$. We let z be constant in between these discontinuity points.

We then make two general assumptions about the instantaneous reflection map ϕ_0 and the associated reflection map ϕ on $S \times D_c$ in (8.8). One is a Lipschitz assumption and the other is a monotonicity assumption.

Lipschitz Assumption. There is a constant K such that

$$\|\phi(s_1, x_1) - \phi(s_2, x_2)\| \le K(\|x_1 - x_2\| \vee \|s_1 - s_2\|)$$

for all $s_1, s_2 \in S$ and $x_1, x_2 \in D_c$, where ϕ is the reflection map in (8.8).

We now turn to the monotonicity. Let e_i be the vector in $\mathbb{R}^k$ with a 1 in the ith coordinate and 0's elsewhere. Let $\phi_0^j(s, u)$ be the jth coordinate of the reflection. We require monotonicity of all these coordinate maps, but we allow the monotonicity to be in different directions in different coordinates.

Monotonicity Assumption. For all $s_0 \in \mathbb{R}^k$, i, $1 \le i \le k$ and j, $1 \le j \le k$, $\phi_0^j(s_0, \alpha e_i)$ is monotone in the real variable α for $\alpha > 0$ and for $\alpha < 0$.

Just as in Theorems 14.3.4 and 14.8.1, we can use the Lipschitz assumption to extend the reflection map from D_c to D. The proof is essentially the same as before.

Theorem 14.8.3. (extension of general reflection maps) *If the reflection map $\phi : S \times D_c \to D_c$ in (8.8) satisfies the Lipschitz assumption, then there exists a unique extension $\phi : S \times D \to D$ of the reflection map in (8.8) satisfying $\|\phi(s, x_n) - \phi(s, x)\| \to 0$ if $s \in S$, $x_n \in D_c$ and $\|x_n - x\| \to 0$. Moreover, $\phi : S \times D \to D$ inherits the Lipschitz property.*

We now want to establish sufficient conditions for the reflection map to inherit the Lipschitz property when we use appropriate M_1 topologies on D. From our previous analysis, we know that we need to impose regularity conditions. With the monotonicity assumption above, it is no longer sufficient to work in D_s. We assume that the sample paths have discontinuities in only one coordinate at a time, i.e., we work in the space D_1. We exploit another approximation lemma.

Let $D_{c,1}$ be the subset of D_c in which all discontinuities occur in only one coordinate at a time, i.e.,

$$D_{c,1} \equiv D_c \cap D_1 \ .$$

The following is another variant of Theorem 12.2.2, which can be established using it.

Lemma 14.8.1. (approximation in D_1) *For all $x \in D_1$, there exist $x_n \in D_{c,1}$, $n \geq 1$, such that $\|x_n - x\| \to 0$.*

We are now ready to state our M_1 result.

Theorem 14.8.4. (Lipschitz and continuity properties of other reflection maps) *Suppose that the Lipschitz and monotonicity assumptions above are satisfied. Let $\phi : S \times D \to D$ be the reflection mapping obtained by extending (8.8) by applying Theorem 14.8.3. For any $s \in S$, $x \in D_1$ and $(u,r) \in \Pi_w(x)$, $(\phi(s,u),r) \in \Pi_w(\phi(s,x))$. Thus there exists a constant K such that*

$$\begin{aligned}
d_p(\phi(s_1,x_1), \phi(s_2,x_2)) &\leq d_w(\phi(s_1,x_1), \phi(s_2,x_2)) \\
&\leq K(d_w(x_1,x_2) \vee \|s_1 - s_2\|) \\
&\leq K(d_s(x_1,x_2) \vee \|s_1 - s_2\|)
\end{aligned} \quad (8.9)$$

for all $s_1, s_2 \in S$ and $x_1, x_2 \in D_1$. Moreover, if $s_n \to s$ in $\mathbb{R}^k$ and $x_n \to x$ in (D, WM_1) where $x \in D_1$, then

$$\phi(s_n, x_n) \to \phi(s, x) \quad in \quad (D, WM_1) \ . \quad (8.10)$$

Proof. By Theorem 14.8.3, the extended reflection map $\phi : S \times D \to D$ is well defined and Lipschitz in the uniform norm. For any $x \in D_1$, apply Lemma 14.8.1 to obtain $x_n \in D_{c,1}$ with $\|x_n - x\| \to 0$. Since $x \in D_1$, the strong and weak parametric representations coincide. Choose $(u,r) \in \Pi_s(x) = \Pi_w(x)$. Since $\|x_n - x\| \to 0$ and $x_n \in D_{c,1}$, we can find $(u_n, r_n) \in \Pi_s(x_n) = \Pi_w(x_n)$ such that $\|u_n - u\| \vee \|r_n - r\| \to 0$. Now, paralleling Theorem 14.4.2, we can apply the monotonicity condition on $D_{c,1}$ to deduce that $(\phi(s, u_n), r_n) \in \Pi_w(\phi(s, x_n))$ for all n. (Note that we need not have either $\phi(s, x) \in D_1$ or $\phi(s, x_n) \in D_{c,1}$, but we do have $\phi(s, x_n) \in D_c$. Note that the componentwise monotonicity implies that $(\phi(s, u_n), r_n)$ belongs to $\Pi_w(\phi(s, x_n))$, but not necessarily to $\Pi_s(\phi(s, x_n))$.) By the Lipschitz property of ϕ,

$$\|\phi(s, u_n) - \phi(s, u)\| \vee \|r_n - r\| \to 0 \ . \quad (8.11)$$

Hence, we can apply Lemma 8.4.5 of the Internet Supplement to deduce that $(\phi(s,u),r) \in \Pi_w(\phi(s,x))$. We thus obtain the Lipschitz property (8.9), just as in Theorem 14.5.3. Finally, to obtain (8.10), suppose that $s_n \to s$ in S and $x_n \to x$

in (D, SM_1) with $x \in D_1$. Under that condition, by Lemma 14.8.1, we can find $x'_n \in D_{1,l} \subseteq D_1$ such that $\|x_n - x'_n\| \to 0$. Since ϕ is Lipschitz on $S \times (D, U)$, there exists a constant K such that

$$\|\phi(s_n, x_n) - \phi(s_n, x'_n)\| \leq K\|x_n - x'_n\| \to 0 . \tag{8.12}$$

By part (a), there exists a constant K such that

$$d_w(\phi(s_n, x'_n), \phi(s, x)) < K(d_o(r'_n, x) \vee \|s_n - s\|) \to 0 . \tag{8.13}$$

By (8.12), (8.13) and the triangle inequality for d_p, we obtain (8.10). ■

We now combine Theorems 14.8.1 and 14.8.4 to obtain continuity and Lipschitz properties for the two-sided regulator with the M_1 topology. From (8.2) it is immediate that the monotonicity condition is satisfied.

Theorem 14.8.5. (Lipschitz and continuity of the two-sided regularity with the M_1 topology) *Let $\phi : [0, c] \times D^1 \to D^1$ be the content portion of the two-sided regulator map defined by (8.1). Then*

$$d(\phi(s_1, x_1), \phi(s_2, x_2)) \leq 2(d(x_1, x_2) \vee |s_1 - s_2|)$$

for all $x_1, x_2 \in D^1$, where d is the M_1 metric. Moreover, if $s_n \to s$ in $[0, c]$ and $d(x_n, x) \to 0$, then

$$R(s_n, x_n) \to R(s, x) \quad in \quad (D^3, WM_1) .$$

We can apply Theorem 14.8.5 to obtain heavy-traffic limits for the queueing examples in Section 2.3 and Chapter 8.

For other reflection maps, we need to verify the Lipschitz and monotonicity assumptions above. Evidently the Lipschitz assumption is the more difficult condition to verify. However,

Dupuis and Ishii (1991) and Dupuis and Ramanan (1999a,b) have established general conditions under which the Lipschitz assumption is satisfied.

14.9. Related Literature

There is now a substantial literature on heavy-traffic limits for queueing networks, as can be seen from the books by Chen and Yao (2001) and Kushner (2001). Much of their attention and much of the recent interest is focused on multiple customer classes and control (e.g., routing and sequencing in the networks). We do not discuss either of these important issues. Multiple customer classes and control in settings requiring the M_1 topologies remain important directions for research.

This chapter is primarily based on the papers by Harrison and Reiman (1981a), Reiman (1984a), Chen and Whitt (1993) and Whitt (2001). Heavy-traffic stochastic process limits for acyclic networks of queues were obtained by application of the one-dimensional reflection map by Iglehart and Whitt (1970a,b). For tandem networks in the Brownian case, the multidimensional reflection map was defined and the limit process was characterized as a diffusion process by Harrison (1978). The multidimensional reflection map and multidimensional reflected Brownian motion were defined in the general case by Harrison and Reiman (1981a,b). Related early work on multidimensional reflection was done by Tanaka (1979) and Lions and Sznitman (1984). Reiman (1984a) applied the multidimensional reflection map to establish heavy-traffic limits with multidimensional reflected Brownian motion limit processes for single-class open queueing networks. Corresponding limits using the M_1 topologies for single-class open queueing networks with rare long service-interruptions, as in Section 14.7, were stated by Chen and Whitt (1993), but that paper contains errors. In particular, it failed to identity the conditions needed in the theorems establishing continuity and Lipschitz properties with the M_1 topologies here in Section 14.5. The corrected continuity and Lipschitz results in Section 14.5 as well as their application to obtain heavy-traffic limits for stochastic fluid networks in Section 14.6 come from Whitt (2001). Heavy-traffic stochastic-process limits for a single queue with service interruptions were obtained previously by Kella and Whitt (1990); those limits produce very tractable approximations, exploiting stochastic decomposition properties; also see Kella and Whitt (1991, 1992c).

In addition to the seminal papers by Harrison and Reiman (1981a) and Reiman (1984a), our discussion of the multidimensional reflection map in Section 14.2 draws upon Chen and Mandelbaum (1991a,b,c). The indirect definition of a reflected process is originally due to Skorohod (1961, 1962). Hence the reflection map is sometimes called the solution of a Skorohod problem. (See also Beneš (1963) for early focus on one-dimensional reflection.) The Lipschitz bounds for the reflection map with the uniform norm in Theorem 14.2.5 come from Chen and Whitt (1993), but there are relatively obvious errors in the proof there that are corrected by Lemma 14.2.6 and Theorem 14.2.4 here. (Remark 14.2.2 indicates the intended argument.) Theorem 14.2.4 itself is Lemma 2 from Kella and Whitt (1996). The stability results in Theorems 14.2.8 and 14.2.9 are from Whitt (2001).

The instantaneous reflection map and its connection to the linear complementarity problem are discussed by Chen and Mandelbaum (1991c). The book by Cottle, Pang and Stone (1992) gives a thorough overview of the linear complementarity problem. Theorem 14.3.2 is essentially Lemma 1 in Kella and Whitt (1996). Most of Sections 14.3–14.6 come from Whitt (2001).

The results on the two-sided regulator and other reflection maps in Section 14.8 are from Berger and Whitt (1992b) and Whitt (2001). The two sided regulator and its application to Brownian motion are discussed in Chapter 2 of Harrison (1985). More general reflection maps have been studied by Williams (1987, 1995), Dupuis and Ishii (1991) and Dupuis and Ramanan (1999a,b).

As indicated at the outset, much of the recent research on heavy-traffic limits for queues is aimed at multiclass queueing networks. To see some of the complications which arise in this more general context, see Whitt (1993c), Bramson (1994a, b) and Harrison and Williams (1996). Some heavy-traffic stochastic-process limits for multiclass queueing networks have exploited methods different from the continuous-mapping approach; see Bramson (1998) and Williams (1998a,b).

For recent developments, see Bell and Williams (2001), Chen and Yao (2001), Harrison (2000, 2001a,b), Kumar (2000), Kushner (2001) and Markowitz and Wein (2001).

15
The Spaces E and F

15.1. Introduction

In this final chapter we introduce new spaces of functions larger than D. These new spaces of functions are intended to serve as spaces of sample paths for new stochastic processes that are limits for sequences of appropriately scaled stochastic processes that have significant fluctuations in two different time scales. We have in mind applications to queueing models of communication networks and manufacturing systems, but there are many other possible applications, e.g., to stochastic models of earthquake dynamics, cancer growth or stock prices.

Here is how this chapter is organized: To provide motivation for the new function spaces, we start in Section 15.2 by discussing three important time scales for the performance of queues: the performance time scale, the service time scale and the failure time scale. When the failure time scale falls between a shorter service time scale and a longer performance time scale, stochastic-process limits with scaling of space and time may provide insight into the impact of the failures upon performance. However, the failures in an intermediate time scale can lead to more complicated oscillations in buffer content, which require a new framework for the stochastic-process limits. We discuss these oscillations and their impact in Section 15.3.

In order to have stochastic-process limits with greater oscillations, we need functions spaces larger than D. In Section 15.4 we define the space E and specify conditions under which the Hausdorff metric inducing the analog of the M_2 topology on E is well defined. In Section 15.5 we develop alternative characterizations of convergence in (E, M_2). In Section 15.6 we give an example of convergence of stochastic processes in (E, M_2), where there is no convergence in D. The example is closely related to extreme-value limits to extremal processes. We show how previous limits in that context can be viewed in a new way.

In Section 15.7 we define the space F and introduce the analog of the M_1 metric. We conclude in Section 15.8 by discussing queueing applications. We obtain a heavy-traffic stochastic-process limit describing a queue that experiences long rare failures, where the failures are more complicated than the service interruptions considered in Sections 6.5 and 14.7.

The present chapter is a brief introduction, identifying a direction for future research. The goal is to establish analogs for the spaces E and F, to the extent possible, of all the results for the space D in earlier chapters. There could even be another book!

15.2. Three Time Scales

In this section we discuss three important time scales for the performance of queues. A source of motivation for the queueing models is the desire to understand and control the performance of evolving communication networks. In some communication networks, such as Internet Protocol (IP) networks, there is evidence that performance degradation occurs, not only because of periods of exceptionally high user demand, but also because of various kinds of system failures (or by the combination of the two phenomena). One possible model of this phenomenon is a complex queueing system (network of queues) subject to occasional failures. This system alternates between periods of being "in control," and "out of control," where only some portion of the network may be experiencing difficulties when the system is out of control. The net-input process of packets into the network nodes (with buffers or queues) changes when the system changes state from in control to out of control, and so on. In this context, our goal is to obtain a suitable framework for establishing heavy-traffic stochastic-process limits for buffer-content stochastic processes that reveal the performance degradation caused by the system failures.

Similar problems arise in manufacturing systems. As above, a key factor in system performance is often system failures. We anticipate a customary randomness in the production process, e.g., associated with setups when changing a machine from one function to another, but occasional larger disruptions may be caused by system failures. A factory also alternates between periods of being "in control" and "out of control." The factory may be out of control when a critical machine breaks down or when a shipment of essential parts fails to arrive when scheduled. The system failures may cause work-in-process inventories to temporarily build up to high levels at various work centers.

Both the communication network and the manufacturing system can be modeled as a network of queues, with the queues containing the bits or packets to be transmitted or the work-in-process that needs further processing. In that context, our goal is to capture the impact of system failures upon performance through appropriate stochastic-process limits. As an initial abstraction, consider a single queue. The system failures may cause significant modification in the input or the service capacity. For example, failure at that queue might cause an interruption of service. A substantial backlog can build up if input keeps arriving during the service interruption. Similarly, failures elsewhere might cause sudden decrease or increase of input. These failure modes can have a dramatic impact on system performance, if the failures cannot be corrected quickly. Obviously it is desirable to eliminate the failures whenever possible, but to effectively manage the system it is also important to understand and control the system when occasional failures do occur.

To understand the impact of system failures upon the performance of queues, it is useful to examine the relevant time scales. We draw attention to three different times scales for the performance of queues:

$$\begin{aligned}&\text{(i) the performance time scale}\\&\text{(ii) the service time scale}\\&\text{(iii) the failure time scale.}\end{aligned} \qquad (2.1)$$

The *performance time scale* is the time scale of concern when judging system performance; the *service time scale* is the time scale of individual service times; the *failure time scale* is the time scale of failure durations.

We are primarily concerned with the case in which the performance time scale is much longer than the service time scale. For example, in communication networks, the performance time scale is often rooted in human perception. We often want to ensure satisfactory performance in the "human" time scale of seconds. In contrast, the service time scale corresponds to packet transmission times, which may be in milliseconds or less, depending upon the transmission rate. Similarly, in a make-to-order production facility, the performance time scale may be associated with product delivery times, i.e., the interval between a customer order and the product delivery, which might be measured in days or weeks. In contrast, the service time scale, determined by the production times on individual machines, might be measured in minutes.

Earthquake dynamics would seem to be a very different phenomenon, because the performance time scale is usually shorter than the "service" time scale: As before, our concern might be rooted in human perception, and thus may be measured in the time scale of seconds, minutes or days, whereas key factors in the system dynamics (which serve as analogs of the service times in a queue) may occur in the much longer time scale of months or years, or even longer.

When the time scale of interest in performance evaluation of a queueing system is much longer than the time scale of individual service times, it is likely that asymptotic analysis can provide useful insight. We may be able to usefully describe the macroscopic behavior of the system in the longer time scale of interest for performance by establishing heavy-traffic limits for stochastic processes after appropriately scaling space and time. Hopefully, the macroscopic description will capture the essential features of the microscopic model, while discarding inessential detail.

We want to include system failures in this framework. Assuming that the performance time scale is much greater than the service time scale, there are five possibilities for the failure time scale:

(i) failure time scale > performance time scale

(ii) failure time scale ≈ performance time scale

(iii) performance time scale > failure time scale > service time scale

(iv) failure time scale ≈ service time scale

(v) failure time scale < service time scale   (2.2)

Here we are primarily concerned with the common case (iii), but we are also interested in cases (ii) and (iv). From a performance-analysis perspective, cases (i) and (v) are relatively trivial. In case (i), the failures totally dominate, in which case attention should be concentrated there. In case (v), failures tend to be have little consequence, so that they probably can be ignored.

In cases (ii), (iii) and (iv), it is worth carefully studying the impact of failures upon performance. In case (iv), the failures can be treated as random perturbations of the service times, so that the failures can usually be analyzed by making minor modifications of models that do not account for the failures.

We are particularly interested in case (iii) in (2.2). We then say that the failures occur in an *intermediate time scale*. When the failure time scale falls between a shorter service time scale and a longer performance time scale, it is useful to consider heavy-traffic stochastic-process limits with time scaling. In such a limit, the failure durations may be asymptotically negligible in the performance time scale, while the overall performance impact of the failures may still be significant, being much greater than can be captured by inflating the mean and variance of service times in case (iv).

In fact, we have already considered examples illustrating failures occuring in an intermediate time scale: The examples were the queues with rare long service interruptions, discussed in Sections 6.5 and 14.7. In those models we assumed that the times between successive failures are in the long performance time scale, while the failure durations occur in an intermediate time scale, shorter than the performance time scale but longer than the service time scale. With appropriate definitions and scaling, heavy-traffic limits can be established in which failures occasionally occur in the (performance) time scale of the limit process. Since the failure durations are in the intermediate time scale, the failure durations are asymptotically negligible in the limit, so that failures occur instantaneously at single time points in the limit. However, the failures cannot be disregarded altogether. When the failure occurs, service stops while the input keeps arriving, so that a large backlog can quickly build up. With appropriate scaling, the failure leads to a sudden jump up in the limit process representing the asymptotic queue length or workload. The jump up represents the stronger consequence of the failure when the failure time scale is longer than the service time scale.

To be more explicit, we review the scaling that was used for the heavy-traffic stochastic-process limits for queues with rare long service interruptions. (See Section

5.5 for a general discussion of heavy-traffic scaling.) For the standard heavy-traffic limit for the single-server queue with unlimited waiting room, time is scaled by multiplying by n while space is scaled by dividing by $\sqrt{n}$, where n typically corresponds to $(1-\rho)^{-2}$, with ρ being the traffic intensity. When the traffic intensity is allowed to increase with n in this way, the queue length or waiting time unscaled tends to grow without bound. Since the steady-state means are of order $(1-\rho)^{-1}$ as $\rho \to 1$, the appropriate space scaling to obtain a nondegenerate limit is to divide by $\sqrt{n} = (1-\rho)^{-1}$. It turns out that the associated time scaling is to multiply by $n = (1-\rho)^{-2}$.

To capture the impact of failures in the way described, we let the time between failures be of order n and the failure durations be of order $\sqrt{n}$. In particular, assuming that a system failure corresponds to a service interruption, the queue buildup will be by the number of arrivals during the service interruption. Assuming that the service times are of order 1, the arrival rate is also order 1. Thus the queue buildup in a service interruption whose duration is of order $\sqrt{n}$ will also be of order $\sqrt{n}$. Hence the failures are represented in a content limit process by occasional jumps up. In performance analysis, these jumps quantitatively describe the performance impact of the failures. The jumps reveal sudden performance degradation from the perspective of the long performance time scale.

To establish such limits with jumps for queues with service interruptions, we need to exploit the M_1 topology on D, because the limiting jump is approached gradually as a consequence of many arrivals during the service interruption. Thus the system failures serve as a major source of motivation for considering the function space D with the M_1 topology.

15.3. More Complicated Oscillations

Our purpose now is to go beyond the space (D, M_1). We want to be able to treat failures that cause more complicated oscillations in the stochastic process of interest. On D, the M_2 topology is useful because it allows the converging functions to have quite general fluctuations in the neighborhood of a limiting discontinuity.

Example 15.3.1. *The use of the M_2 topology on D.* Let the limit function be $x = I_{[1,2]}$ in $D([0,2], \mathbb{R})$. Then $x_n \to x$ in (D, M_2) as $n \to \infty$, but not for any of the other Skorohod topologies if

$$\begin{aligned} x_n(0) &= x_n(1 - 3n^{-1}) = 0, \\ x_n(1 - 2n^{-1}) &= 3/4, \ x_n(1 - n^{-1}) = 1/5, \\ x_n(1) &= 7/8, \ x_n(1 + n^{-1}) = 1/16, \\ x_n(1 + 2n^{-1}) &= x_n(2) = 1, \end{aligned} \qquad (3.1)$$

with x_n defined by linear interpolation elsewhere. ∎

Example 15.3.1 shows the advantage of the space (D, M_2) over the space (D, M_1), but Example 15.3.1 also shows two shortcomings of the space (D, M_2): First, the nature of the fluctuations in x_n in the neighborhood of $t = 1$ are not evident from the limit function x; we only know that the process x_n is in the neighborhood of the interval $[0, 1]$ for $t \in (1 - \epsilon, 1 + \epsilon)$ for all sufficiently small ϵ. We might want functions to be defined so that the limit shows that x_n goes up from 0 to 3/4, then down to 1/5, then up to 7/8, then down to 1/16, and finally up to 1 in the neighborhood of time 1.

A second shortcoming of the space (D, M_2) is the requirement that all functions assume values at discontinuity points between the left and right limits there. Of course, we actually have required more; we have required that the functions actually be right-continuous. However, the space D with the M_1 or M_2 topology is unchanged if we only require that the functions have left and right limits everywhere, provided that the function value at each discontinuity point falls between the left and right limit, because the completed graph is then the same as for the right-continuous version.

Now we want to allow for more general fluctuations. If functions x_n do have significant fluctuations in the neighborhood of some point t, then the functions x_n might well visit regions outside the interval $[x(t-) \wedge x(t+), x(t-) \vee x(t+)]$ in the neighborhood of t. And we might well want to say that x_n converges as $n \to \infty$ in that situation.

To illustrate, consider the following example.

Example 15.3.2. *The need for spaces larger than D.* With the same limit function $x = I_{[1,2]}$ in Example 15.3.1, the functions x_n might be defined, instead of by (3.1), by

$$\begin{aligned} x_n(0) &= x_n(1 - 3n^{-1}) = 0, \\ x_n(1 - 2n^{-1}) &= 2, \; x_n(1 - n^{-1}) = -1, \\ x_n(1) &= 3, \; x_n(1 + n^{-1}) = 1/16, \\ x_n(1 + 2n^{-1}) &= x_n(2) = 1, \end{aligned} \quad (3.2)$$

with x_n again defined by linear interpolation elsewhere. In this case, $x_n \not\to x$ in (D, M_2) as $n \to \infty$. However, we might want to say that x_n does actually converge to a limit as $n \to \infty$. It is natural that the graph of the limit be the graph Γ_x of x augmented by the vertical line segment $[-1, 3] \times \{1\}$. Then the graph shows the range of points visited by x_n. In addition, we might want the limit to reflect the fact that x_n goes up from 0 to 2, then down to -1, then up to 3, then down to 1/16, and finally up to 1 in the neighborhood of time 1. ∎

Example 15.3.3. *Another example requiring a larger space.* Another simple example is the case in which $x_n = I_{[1+n^{-1}, 1+2n^{-1})}$ in $D([0,2], \mathbb{R})$. This is the classic example in which x_n converges pointwise for all t to x with $x(t) = 0$, $0 \leq t \leq 2$, but x_n fails to converge in any of the Skorohod topologies. However, we might well want to say that x_n does in fact converge, and that it converges to a limit that captures the fluctuation experienced by x_n. Clearly, the graph of the limit should be the graph of x, i.e., $\{0\} \times [0, T]$, augmented by the vertical line segment $[0, 1] \times \{1\}$. ∎

The first new space we introduce, the space E, allows for extra excursions at individual time points. We construct an element of $E([0,T],\mathbb{R})$ by augmenting a function x in $D([0,T],\mathbb{R})$ by adding vertical line segments to the graph of x at these specially designated excursion times. We do this in such a way that the graphs remain compact subsets of $\mathbb{R}^{k+1}$. Then, paralleling (D, M_2), we induce a topology (which we call the M_2 topology) on E by using the Hausdorff metric on the resulting set of graphs.

We also construct another space of functions, F, in order to capture more information about the fluctuations of the converging functions in the limit. One important source of motivation for the space F is our desire to apply reflection maps to establish heavy-traffic stochastic-process limits for queueing processes with limits in one of these larger spaces. Thus it is helpful to reconsider Example 13.5.2, which shows that the standard one-dimensional reflection map is not continuous on D with the M_2 topology.

Example 15.3.4. *The reflection map is not continuous on* (D, M_2). Recall that Example 13.5.2 shows that the one-sided one-dimensional reflection map $\phi : D \to D$ defined in Section 13.5 is not continuous with the M_2 topology. That example has $x = -I_{[1,2]}$ in $D([0,2],\mathbb{R})$,

$$x_n(0) = x_n(1 - 3n^{-1}) = x(1 - n^{-1}) = 0$$

and

$$x_n(1 - 2n^{-1}) = x_n(1) = x_n(2) = -1$$

with x_n defined by linear interpolation elsewhere. Then $x_n \to x$ in (D, M_2), but $\phi(x_n) \not\to \phi(x)$ in (D, M_2), where $\phi(x)(t) = 0$ for all t. This example fails to provide a counterexample in (D, M_1) because then $x_n \not\to x$.

However, we might well want to have a topology allowing us to say that $x_n \to \hat{x}$ and $\phi(x_n) \to \hat{y}$, where $\hat{x}$ and $\hat{y}$ are different from x and $\phi(x)$ above, being elements of F instead of elements of D. Indeed, for this example, it is natural to say that $\phi(x_n)$ converges to a limit in (E, M_2), which coincides with the zero-function $\phi(x)$ augmented by the vertical line $[0,1] \times \{1\}$. Our new space allows us to reach that conclusion.

For that purpose, we want the initial limit $\hat{x}$ to reflect the parametric representations that can be used to justify M_2 convergence $x_n \to x$. In particular, when $r(s) = 1$, $u(s)$ goes from 1 to 0, to 1 and back to 0, in order to match the fluctuation in x_n. Assuming that we keep track of the fluctuation detail in $\hat{x}$, the associated limit $\phi(\hat{x})$ should have a graph equal to the zero-function $\phi(x)$ augmented by the vertical line $[0,1] \times \{1\}$. When we introduce another space with an appropriate topology, we will be able to say that $x_n \to \hat{x}$. Then the reflection map will be continuous, so that we will have $\phi(x_n) \to \phi(\hat{x})$. Moreover, since this new topology will be stronger than the M_2 topology, we will be able to deduce that $\phi(x_n)$ converges to the graph of $\phi(\hat{x})$ in E, as desired. ■

The need for a new space can also be explained by the fact that the reflection map is not even well defined on E. In order to define the reflection of an element of E, we need to know the order in which the points in a vertical segment are visited.

Example 15.3.5. *Maximal and minimal reflection maps on E.* To see that the reflection map is not well defined on E, consider the function $\tilde{x}$ in E obtained from $x = I_{[0,1)}$ in $D([0,2], \mathbb{R})$ with graph

$$\Gamma_x = (\{1\} \times [0,1)) \cup (\{0\} \times [1,2]) \tag{3.3}$$

augmented by the vertical line $[-2, 3] \times \{1\}$, i.e., with graph

$$\Gamma_{\tilde{x}} = (\{1\} \times [0,1)) \cup ([-2,3] \times \{1\}) \cup (\{0\} \times [1,2]) . \tag{3.4}$$

A *maximal reflection map* ϕ^+ can be defined by assuming that, at time 1, $\tilde{x}$ first goes from 1 down to -2, then goes up to 3 and finally goes down to the right limit 0. The graph of $\phi^+(\tilde{x})$ is

$$\Gamma_{\psi^+(\tilde{x})} = (\{1\} \times [0,1)) \cup ([0,5] \times \{1\}) \cup (\{2\} \times (1,2]) . \tag{3.5}$$

On the other hand, a *minimal reflection map* ϕ^- can be defined by assuming that $\tilde{x}$ first goes up to 3, then down to -2 and then up to 1. The graph of $\phi^-(\tilde{x})$ is

$$\Gamma_{\psi^-(\tilde{x})} = (\{1\} \times [0,1)) \cup ([0,3] \times \{1\}) \cup (\{2\} \times (1,2]) . \tag{3.6}$$

Note that the maximum value of $\phi^+(\tilde{x})$ is 5, while the maximum value of $\phi^-(\tilde{x})$ is 3. Thus, to be well defined, the reflection map needs to know the order of the points visited during the excursions. ∎

To faithfully represent the fluctuations at excursions we introduce a new space F based on parametric representations. In particular, we consider parametric representations of the graphs of the elements of E. We say that two parametric representations of $\Gamma_{\tilde{x}}$ for $\tilde{x} \in E$ are equivalent if they visit the same points in the same order. We let F be the space of equivalence classes with respect to that equivalence relation. The space F is larger than E because there are many different elements of F with the same graph. We give F an analog of the M_1 metric. The space (D, M_1) itself is homeomorphic to a subset of F, in which the graphs correspond to functions in D and the parametric representations are monotone in the order put on the graphs in Sections 3.3 and 12.3.

The spaces (E, M_2) and (F, M_1) allow us to obtain a satisfactory treatment of reflection. First, working with F instead of E, we avoid the ambiguity about the order points are visited in Example 15.3.5. In particular, the reflection map ϕ is well defined and continuous on (F, M_1). Thus, starting from convergence $\hat{x}_n \to \hat{x}$ in (F, M_1), we obtain $\phi(\hat{x}_n) \to \phi(\hat{x})$ in (F, M_1). Since convergence in (F, M_1) implies convergence in (E, M_2) for the graphs of the elements of F, we obtain as a consequence associated convergence in (E, M_2) for the graphs of $\phi(\hat{x}_n)$ and $\phi(\hat{x})$. In particular, we obtain the proposed limit in (E, M_2) in Example 15.3.4.

The example above is for the one-dimensional reflection map. We can also treat the multidimensional reflection map. Paralleling Chapter 14, we need to make assumptions about the domain and the range: First for the domain, it suffices to assume that the limit $\hat{x}$ belongs to the space F_1, the subset of F in which the excursions and discontinuities occur only in one coordinate at a time. Next for the range, we need the product topology: First, we can use the space (F, WM_1); then we can go to (E, WM_2). In other words, convergence of functions in the domain (F, WM_1), where the limit belongs to F_1 will imply convergence of the graphs of the reflections in the range (E, WM_2).

In summary, we introduce new function spaces E and F to supplement the familiar ones C and D considered before. Hence, we have four spaces of functions, each larger than the one before:

C – *Continuous* functions
D – *Discontinuous* functions
E – functions with extra *Excursions*
F – functions with extra excursions that faithfully model *Fluctuations*

We omit most proofs in this chapter, which are similar to proofs for (D, M_2) and (D, M_1). A more extensive discussion of the spaces E and F is planned for the Internet Supplement.

15.4. The Space E

Our general approach to defining the new function spaces E and F is to base the definitions on the M_2 and M_1 topologies used on D. Now we use the graphs and parametric representations, not only to define the topologies, but also to define the functions themselves.

The space E is larger than D because the functions are allowed to have extra excursions, which occur at single time points. We initially represent $E \equiv E([0,T], \mathbb{R}^k)$ as the space of *excursion triples*

$$(x, S, \{I(t) : t \in S\}), \tag{4.1}$$

where $x \in D \equiv D([0,T], \mathbb{R}^k)$, the space of all right-continuous $\mathbb{R}^k$-valued functions on $[0,T]$ with left limits everywhere, S is a countable set with

$$Disc(x) \subseteq S \subseteq [0,T], \tag{4.2}$$

and, for each $t \in S$, $I(t)$ is a compact subset of $\mathbb{R}^k$ with at least two points such that

$$x(t), x(t-) \in I(t) \quad \text{for all} \quad t \in S. \tag{4.3}$$

We call the function x in D the *base function*. We call S the set of *excursion times* or the set of *discontinuity points* of the new function. We want to allow the new function to make excursions where the base function x is continuous, so $Disc(x)$ may be a proper subset of S. We call the set $I(t)$ the *set of excursion values*; $I(t)$ is the set of values assumed by the new function at time t for $t \in S$. By requiring that $I(t)$ contain at least two points, we ensure that $\{x(t)\} \stackrel{\subseteq}{\neq} I(t)$ when $t \in Disc(x)^c$. Hence, each $t \in S$ corresponds to some genuine form of discontinuity.

Associated with the excursion triple $(x, S, \{I(t), t \in S\})$ is the set-valued function

$$\tilde{x}(t) \equiv \begin{cases} I(t), & t \in S \\ \{x(t)\}, & t \notin S \end{cases}. \tag{4.4}$$

Associated with the set-valued function $\tilde{x}$ is its graph

$$\Gamma_{\tilde{x}} \equiv \{(z,t) \in \mathbb{R}^k \times [0,T] : z \in \tilde{x}(t)\}. \tag{4.5}$$

Clearly, it is possible to construct the excursion triple $(x, S, \{I(t), t \in S\})$, the set-valued function $\tilde{x}$ and the graph $\Gamma_{\tilde{x}}$, starting from any one of the three. Hence these are three equivalent representations for the elements of the space E. For brevity, we will let elements of E be denoted by $\tilde{x}$.

Unlike the graphs Γ_x and G_x for $x \in D$ defined in Section 12.3, the set $I(t)$ appearing in the graph $\Gamma_{\tilde{x}}$ in (4.5) need not be a segment. Indeed it need not even be connected. We will impose further restrictions below.

Paralleling our treatment of (D, M_2), we propose making E a separable metric space by using the Hausdorff metric on the space of graphs $\Gamma_{\tilde{x}}$ for $\tilde{x} \in E$. For that purpose, we want to be sure that each graph is a compact subset of $\mathbb{R}^{k+1}$. Without additional assumptions, we need not have that property.

Example 15.4.1. *The graphs need not be compact.* To see the need for additional assumptions in order to guarantee that $\Gamma_{\tilde{x}}$ in (4.5) is compact, consider the triple $\{x, S, \{I(t) : t \in S\}\}$ with $x(t) = 0$, $0 \le t \le 1$, S the set of rational numbers strictly less than $1/2$, and $I(t) = [0,1]$ for all $t \in S$. Then $\Gamma_{\tilde{x}}$ is a dense subset of $[0,1] \times [0, 1/2]$, which is bounded but not closed, and thus not compact. ∎

A simple way to ensure that $\Gamma_{\tilde{x}}$ is compact is to require that the set S of excursion times be a finite subset of $[0,T]$, and we think of that being an important case for applications. However, that assumption is unappealing because then D would no longer be a proper subset of E. Thus we want to allow countable sets of excursion times. We can allow countable subsets if we impose a restriction. We give equivalent characterizations of the condition below.

When we talk about convergence of sets, the sets will always be compact subsets of $\mathbb{R}^k$, and we use the Hausdorff metric, as defined in (5.2) in Section 11.5; i.e., for compact sets A_1, A_2,

$$m(A_1, A_2) \equiv \mu(A_1, A_2) \vee \mu(A_2, A_1) \tag{4.6}$$

where

$$\mu(A_1, A_2) \equiv \sup_{x \in A_1} \{\|x - A_2\|\} \tag{4.7}$$

and

$$\|x - A\| = \|A - x\| \equiv \inf_{y \in A} \{\|x - y\|\}. \tag{4.8}$$

For $A \subseteq \mathbb{R}^k$, let $\delta(A)$ be the *diameter* of A, i.e.,

$$\delta(A) \equiv \sup_{x,y \in A} \{\|x - y\|\}. \tag{4.9}$$

Theorem 15.4.1. (equivalent conditions) *The following conditions for elements of E are equivalent:*
(a) *For each $\epsilon > 0$, there are only finitely many t for which $\delta(I(t)) > \epsilon$.*
(b) *For each $t \in [0, T)$, $\tilde{x}$ has a single-point right limit*

$$\tilde{x}(t+) \equiv \lim_{s \downarrow t} \tilde{x}(s) = \{x(t)\} \tag{4.10}$$

and, for each $(0, T]$, $\tilde{x}$ has a single-point left limit

$$\tilde{x}(t-) \equiv \lim_{s \uparrow t} \tilde{x}(s) = \{x(t-)\} \ . \tag{4.11}$$

Theorem 15.4.2. (conditions to make the graphs compact) *Under the conditions in Theorem 15.4.1, the graph $\Gamma_{\tilde{x}}$ in (4.5) is a compact subset of $\mathbb{R}^{k+1}$.*

Henceforth let E be the set of graphs $\Gamma_{\tilde{x}}$ for which the conditions of Theorem 15.4.1 hold. Then, paralleling (D, M_2), we endow E with the Hausdorff metric, i.e.,

$$m(\tilde{x}_1, \tilde{x}_2) \equiv m(\Gamma_{\tilde{x}_1}, \Gamma_{\tilde{x}_2}) \ , \tag{4.12}$$

where m is the Hausdorff metric on the space of compact subsets of $\mathbb{R}^{k+1}$, as in (4.6). We call the topology induced by m on E the M_2 topology. Since the graphs are compact subsets of $\mathbb{R}^{k+1}$, we have the following result.

Theorem 15.4.3. (metric property) *The space (E, m) is a separable metric space.*

So far, we have defined the metric m on the space $E([0, T], \mathbb{R}^k)$ with the compact domain $[0, T]$. We extend to non-compact domains just as was done for D. We say that $\tilde{x}_n \to \tilde{x}$ in $E(I, \mathbb{R}^k)$ for an interval I in $\mathbb{R}$ if $\tilde{x}_n \to \tilde{x}$ for the restrictions in $E([t_1, t_2], \mathbb{R}^k)$ for all $t_1, t_2 \in I$ with $t_1 < t_2$ and $t_1, t_2 \notin S$.

We also define a stronger metric than the Hausdorff metric m in (4.12) on $E \equiv E([0, T], \mathbb{R}^k)$, which we call the *uniform metric*, namely,

$$m^*(\tilde{x}_1, \tilde{x}_2) = \sup_{0 \le t \le T} m(\tilde{x}_1(t), \tilde{x}_2(t)) \ , \tag{4.13}$$

where m is again the Hausdorff metric, here applied to compact subsets of $\mathbb{R}^k$. The following comparison is not difficult.

Theorem 15.4.4. (comparison with the uniform metric) *For any $\tilde{x}_1, \tilde{x}_2 \in E$,*

$$m(\tilde{x}_1, \tilde{x}_2) \le m^*(\tilde{x}_1, \tilde{x}_2)$$

for m in (4.12) and m^* in (4.13).

As with the metrics inducing the U and M_2 topologies on D, m^* is complete but not separable, while m is separable but not complete.

Example 15.4.2. *The metric m on E is not complete.* To see that the Hausdorff metric m on E is not complete, let

$$x_n = \sum_{k=1}^{n-1} I_{[2k/2n,(2k+1)/2n)}, \quad n \ge 2 \ . \tag{4.14}$$

Then $x_n \in D([0,1], \mathbb{R})$,

$$m(\Gamma_{x_n}, \Gamma_{x_m}) \to 0 \quad \text{as} \quad n, m \to \infty \tag{4.15}$$

$$m(\Gamma_{x_n}, [0,1] \times [0,1]) \to 0 \quad \text{as} \quad n \to \infty, \tag{4.16}$$

but the limit is not in E. Hence x_n does not converge to a limit in E. ∎

We now consider approximations of functions in E by piecewise-constant functions. Let E_c be the subset of *piecewise-constant functions* in E, i.e., the set of functions for which S is finite, x is constant between successive points in S and $I(t)$ is finite valued for all $t \in S$. Rational-valued piecewise-constant functions with $S = \{jT/k, 1 \leq j \leq k\}$, which are elements of E_c, form a countable dense subset of (E, m). The following result parallels Theorem 12.2.2.

Theorem 15.4.5. (approximation by piecewise-constant functions) *If $\tilde{x} \in (E, m)$, then for all $\epsilon > 0$ there exists $\tilde{x}_c \in E_c$ such that $m^*(\tilde{x}, \tilde{x}_c) < \epsilon$ for m^* in (4.13).*

Even though the excursion sets $I(t)$ associated with the functions $\tilde{x}$ in E must be compact because of the conditions imposed in Theorem 15.4.1, so far they can be very general. In particular, the excursion sets $I(t)$ need not be connected. It may well be of interest to consider the space E with disconnected excursions, which the framework above allows, but for the applications we have in mind, we want to consider graphs that are connected sets. We will make the stronger assumption that the graphs are *path-connected*, i.e., each pair of points can be joined by a path – a continuous map from $[0, 1]$ into the graph; e.g., see Section V.5 of Dugundji (1966). In fact, we will assume that the graph can be represented as the image of parametric representations.

We say that $(u, r) : [0, 1] \to \mathbb{R}^{k+1}$ is a *strong parametric representation* of $\Gamma_{\tilde{x}}$ or $\tilde{x}$ in E if (u, r) is a continuous function from $[0, 1]$ onto $\Gamma_{\tilde{x}}$ such that r is nondecreasing. We say that $\tilde{x}$ and $\Gamma_{\tilde{x}}$ are *strongly connected* if there exists a strong parametric representation of $\Gamma_{\tilde{x}}$.

Similarly, we say that (u, r) is a *weak parametric representation* of $\Gamma_{\tilde{x}}$ or $\tilde{x}$ in E, (u, r) is a continuous function from $[0, 1]$ into $\Gamma_{\tilde{x}}$ such that r is nondecreasing, $r(0) \in \tilde{x}(0)$ and $r(1) \in \tilde{x}(T)$. We say that $\tilde{x}$ and Γ_x are *weakly connected* if there exists a weak parametric representation of Γ_x and if the union of $(u(s), r(s))$ over all s, $0 \leq s \leq 1$, and all weak parametric representations of $\Gamma_{\tilde{x}}$ is $\Gamma_{\tilde{x}}$ itself.

Let E_{st} and E_{wk} represent the subsets of strongly connected and weakly connected functions $\tilde{x}$ in E. When the range of the functions is $\mathbb{R}$, $E_{st} = E_{wk}$ and E_{st} is a subset of E in which $I(t)$ is a closed bounded interval for each $t \in S$.

As in (3.1) and (3.2) in Section 12.3, for $a, b \in \mathbb{R}^k$, let $[a, b]$ and $[[a, b]]$ be the standard and product segments in $\mathbb{R}^k$. It is easy to identify (D, SM_2) and (D, WM_2) as subsets of (E_{st}, m) and (E_{wk}, m), respectively.

Theorem 15.4.6. (when $\tilde{x}$ reduces to x) *Consider an element $\tilde{x}$ in E. Suppose that $S = Disc(x)$.*
(a) If $\tilde{x} \in E_{st}$ and

$$I(t) = [x(t-), x(t)] \quad \text{for all} \quad t \in S, \tag{4.17}$$

then
$$\Gamma_{\tilde{x}} = \Gamma_x \quad (4.18)$$
for the thin graph Γ_x in (3.3) of Section 12.3.

(b) If $\tilde{x} \in E_{wk}$ and
$$I(t) = [[x(t-), x(t)]] \quad \text{for all} \quad t \in S , \quad (4.19)$$
then
$$\Gamma_{\tilde{x}} = G_x \quad (4.20)$$
for the thick graph G_x in (3.4) Section 12.3.

Let D_{st} and D_{wk} be the subsets of all $\tilde{x}$ in E_{st} and E_{wk}, respectively, satisfying the conditions of Theorem 15.4.6 (a) and (b). Since the SM_2 and WM_2 topologies on D can be defined by the Hausdorff metric on the graphs Γ_x and G_x, the following corollary is immediate.

Corollary 15.4.1. (identifying the spaces (D, SM_2) and (D, WM_2)) *The space (D, SM_2) is homeomorphic to the subset D_{st} in (E_{st}, m), while the space (D, WM_2) is homeomorphic to the subset D_{wk} in (E_{wk}, m).*

15.5. Characterizations of M_2 Convergence in E

Paralleling Section 12.11, we now want to develop alternative characterizations of M_2 convergence on E. For simplicity, we consider only real-valued functions. Thus there is no need to distinguish between the SM_2 and WM_2 topologies.

We consider M_2 convergence on $E \equiv E_{st} \equiv E_{st}([0,T], \mathbb{R})$, assuming that the conditions of Theorem 15.4.1 hold. By considering E_{st}, we are assuming that the functions are strongly connected (i.e., there exist parametric representations onto the graphs.) Thus the excursion sets $I(t)$ for t in the set S of excursion times are all closed bounded intervals.

Given that the functions $\tilde{x}$ in E are all strongly connected, it is natural to consider characterizing M_2 convergence in terms of parametric representations. For $\tilde{x}_1, \tilde{x}_2 \in E$, let
$$d_{s,2}(\tilde{x}_1, \tilde{x}_2) = \inf_{\substack{(u_i, r_i) \in \Pi_{s,2}(\tilde{x}_i) \\ i=1,2}} \{\|u_1 - u_2\| \vee \|r_1 - r_2\|\} , \quad (5.1)$$

where $\Pi_{s,2}(\tilde{x}_i)$ is the set of all strong parametric representations (u_i, r_i) of $\tilde{x}_i \in E$. As in Section 12.11, it turns out that $d_{s,2}(\tilde{x}_n, \tilde{x}) \to 0$ if and only if $m(\tilde{x}_n, \tilde{x}) \to 0$. (We will state a general equivalence theorem below.) However, as before, $d_{s,2}$ in (5.1) is not a metric on E. That is shown by Example 12.11.1.

We next introduce a characterization corresponding to local uniform convergence of the set functions. For this purpose, let the δ-neighborhood of $\tilde{x}$ at t be
$$N_\delta(\tilde{x})(t) = \cup_{0 \vee (t-\delta) \le s \le (t+\delta) \wedge T} \tilde{x}(s) . \quad (5.2)$$

(Since N_δ includes only perturbations horizontally, it is not actually the δ neighborhood in the metric m.)

Lemma 15.5.1. (compactness) *For each $\tilde{x} \in E$, $t \in [0,T]$ and $\delta > 0$, $N_\delta(\tilde{x})(t)$ in (5.2) is a compact subset of $\mathbb{R}^k$.*

Let
$$v(\tilde{x}_1, \tilde{x}_2, t, \delta) \equiv m(N_\delta(\tilde{x}_1)(t), \tilde{x}_2(t)) \tag{5.3}$$
where m is the Hausdorff metric in (4.6), and
$$v(\tilde{x}_1, \tilde{x}_2, \delta) \equiv \sup_{0 \le t \le T} v(\tilde{x}_1, \tilde{x}_2, t, \delta) . \tag{5.4}$$

An attractive feature of the function space $D \equiv D([0,T], \mathbb{R}^k)$ not shared by the larger space E is that all function values $x(t)$ for $x \in D$ are determined by the function values $x(t_k)$ for any countable dense subset $\{t_k\}$ in $[0,T]$. For $t \in [0,T)$, by the right continuity of x, $x(t) = \lim x(t_k)$ for $t_k \downarrow t$. In contrast, in E since we do not necessarily have $Disc(x) = S$, we cannot even identify the set S of excursion times from function values $\tilde{x}(t)$ for t outside S. In some settings it may be reasonable to assume that $S = Disc(x)$, in which case S can be identified from the left and right limits. For instance, in stochastic settings we might have $P(X(t-) = X(t)) = 0$ when $t \in S$, so that it may be reasonable to assume that $S = Disc(x)$.

However, even when $S = Disc(x)$, we are unable to discover the set $\tilde{x}(t)$ for $t \in S$ by observing the set-valued function $\tilde{x}$ at other time points t. Hence, in general *stochastic processes with sample paths in E are not separabile*; e.g., see p. 65 of Billingsley (1968). Thus it is natural to look for alternative representations for the functions $\tilde{x}$ that are determined by function values on any countable dense subset.

Thus, for $\tilde{x} \in E$, let the local-maximum function be defined by
$$M_{t_1, t_2}(\tilde{x}) = \sup\{z : z \in \tilde{x}(t) : t_1 \le t \le t_2\} \tag{5.5}$$
for $0 \le t_1 < t_2 \le T$. It is significant that we can also go the other way. For $\tilde{x} \in E$, if we are given $M_{t_1,t_2}(\tilde{x})$ and $M_{t_1,t_2}(-\tilde{x})$ for all t_1, t_2 in a countable dense subset A of $[0,T]$, we can reconstruct $\tilde{x}$. In particular, for $0 < t < T$,
$$\tilde{x}(t) = [a(t), b(t)] , \tag{5.6}$$
with $[a,a] \equiv \{a\}$, where
$$b(t) = \lim_{\substack{t_{1,k} \uparrow t \\ t_{2,k} \downarrow t}} M_{t_{1,k}, t_{2,k}}(\tilde{x}) \tag{5.7}$$
and
$$-a(t) = \lim_{\substack{t_{1,k} \uparrow t \\ t_{2,k} \downarrow t}} M_{t_{1,k}, t_{2,k}}(-\tilde{x}) \tag{5.8}$$
with $t_{1,k}$ and $t_{2,k}$ taken from the countable dense subset A. For $t = 0$,
$$b(t) = \lim_{t_{2,k} \downarrow t} M_{0, t_{2,k}}(\tilde{x}) \tag{5.9}$$

and
$$-a(t) = \lim_{t_{2,k}\downarrow t} M_{0,t_{2,k}}(-\tilde{x}) \,, \qquad (5.10)$$

where $t_{2,k}$ is again taken from A. A similar construction holds for $t = T$. It is significant that M_2 convergence in E is also determined by the maximum function.

We are now ready to state our convergence-characterization theorem. It is an analog of Theorem 12.11.1 for (D, SM_2).

Theorem 15.5.1. (alternative characterizations of M_2 convergence in E) *The following are equivalent characterizations of convergence $\tilde{x}_n \to \tilde{x}$ in $E([0,T], \mathbb{R})$:*
(i) $m(\tilde{x}_n, \tilde{x}) \to 0$ for the metric m in (4.12) and (4.6).
(ii) $\mu(\Gamma_{\tilde{x}_n}, \Gamma_{\tilde{x}}) \to 0$ for μ in (4.7).
(iii) $d_{s,2}(\tilde{x}_n, \tilde{x}) \to 0$ for $d_{s,2}$ in (5.1); i.e. for any $\epsilon > 0$ and n sufficiently large, there exist $(u,r) \in \Pi_{s,2}(\tilde{x})$ and $(u_n, r_n) \in \Pi_{s,2}(\tilde{x}_n)$ such that $\|u_n - u\| \vee \|r_n - r\| < \epsilon$.
(iv) Given $v(\tilde{x}_1, \tilde{x}_2, \delta)$ in (5.4),
$$\lim_{\delta \downarrow 0} v(\tilde{x}_n, \tilde{x}, \delta) = 0 \,. \qquad (5.11)$$

(v) For each t, $0 \le t \le T$,
$$\lim_{\delta \downarrow 0} \overline{\lim_{n \to \infty}} v(x_n, \tilde{x}, t, \delta) = 0 \qquad (5.12)$$

for $v(\tilde{x}_1, \tilde{x}_2, t, \delta)$ in (5.3).
(vi) For all t_1, t_2 in a countable dense subset of $[0,T]$ with $t_1 < t_2$, including 0 and T,
$$M_{t_1,t_2}(\tilde{x}_n) \to M_{t_1,t_2}(\tilde{x}) \quad in \quad \mathbb{R} \qquad (5.13)$$

and
$$M_{t_1,t_2}(-\tilde{x}_n) \to M_{t_1,t_2}(-\tilde{x}) \quad in \quad \mathbb{R} \qquad (5.14)$$

for the local maximum function M_{t_1,t_2} in (5.5).

For $\tilde{x} \in E$, let $S'(\tilde{x})$ be the subset of non-jump discontinuities in $S(\tilde{x})$, i.e.,
$$S'(\tilde{x}) \equiv \{t \in S(\tilde{x}) : I(t) \neq [x(t-), x(t+)]\} \,. \qquad (5.15)$$

We envision that in many applications $S'(\tilde{x})$ will be a finite subset and x_n will belong to D for all n. A more elementary characterization of convergence holds in that special case.

Theorem 15.5.2. (when there are only finitely many non-jump discontinuities) *Suppose that $\tilde{x} \in E$, $S'(\tilde{x})$ in (5.15) is a finite subset $\{t_1, \ldots, t_k\}$ and $x_n \subset D$ for all n. Then $x_n \to \tilde{x}$ in (E, M_2) holds if and only if*
(i) $x_n \to x$ in (D, M_2) for the restrictions over each of the subintervals $[0, t_1)$, $(t_1, t_2), \ldots, (t_{k-1}, t_k)$ and $(t_k, T]$, where x is the base function of $\tilde{x}$, and
(ii) (5.12) holds for $t = t_i$ for each $t_i \in S'(\tilde{x})$.

15.6. Convergence to Extremal Processes

In this section we give a relatively simple example of a stochastic-process limit in which random elements of D converge in E to a limiting stochastic-process whose sample paths are in E but not in D.

Let $\{X_n : n \geq 1\}$ be a sequence of real-valued random variables and let $\{S_n : n \geq 0\}$ be the associated sequence of partial sums, i.e.,

$$S_n = X_1 + \cdots + X_n, \quad n \geq 1, \tag{6.1}$$

with $S_0 = 0$. Let $\mathbf{S}_n$ and $\mathbf{X}_n$ be associated scaled random elements of $D \equiv D([0,T], \mathbb{R})$ defined by

$$\begin{aligned}\mathbf{S}_n(t) &= c_n^{-1}(S_{\lfloor nt \rfloor} - \mu nt) \\ \mathbf{X}_n(t) &= c_n^{-1} X_{\lfloor nt \rfloor}, \quad 0 \leq t \leq T,\end{aligned} \tag{6.2}$$

where $c_n \to \infty$ as $n \to \infty$. Since the maximum jump functional is continuous at continuous functions, p. 301 of Jacod and Shiryaev (1987), we have $\mathbf{X}_n \Rightarrow 0\mathbf{e}$ in D, where $\mathbf{e}$ is the identity function, or equivalently $\|\mathbf{X}_n\| \Rightarrow 0$, whenever $\mathbf{S}_n \Rightarrow \mathbf{S}$ in D and $P(\mathbf{S} \in C) = 1$.

However, we cannot conclude that $\|\mathbf{X}_n\| \Rightarrow 0$ when the limit $\mathbf{S}$ fails to have continuous sample paths, as noted in Sections 4.5.4 and 5.3.2. In this section we show that we can have $\mathbf{X}_n \Rightarrow \mathbf{X}$ in $E \equiv E([0,T], \mathbb{R})$ in that situation, where $\mathbf{X}$ is a random element of E and not a random element of D.

We establish our result for the case in which $\{X_n : n \geq 1\}$ is a sequence of IID nonnegative random variables with cdf F, where the complementary cdf $F^c \equiv 1 - F$ is regularly varying with index $-\alpha$, i.e., $F^c \in \mathcal{R}(-\alpha)$, with $\alpha < 2$. That is known to be a necessary and sufficient condition for F to belong to both the domain of attraction of a non-normal stable law of index α and the maximum domain of attraction of the Frechet extreme value distribution with index α; see Theorem 4.5.4.

We will use extremal processes to characterize the limit of $\mathbf{X}_n$ in E. Hence we now briefly describe extremal processes; see Resnick (1987) for more details. We can construct an extremal process associated with any cdf F on $\mathbb{R}$. Given the cdf F, we define the finite-dimensional distributions of the extremal process by

$$F_{t_1, \ldots, t_k}(x_1, \ldots, x_k) \equiv F^{t_1}\left(\bigwedge_{i=1}^{k} x_i\right) F^{t_2 - t_1}\left(\bigwedge_{i=2}^{k} x_i\right) \cdots F^{t_k - t_{k-1}}(x_k), \tag{6.3}$$

where

$$\bigwedge_{i=j}^{k} x_i \equiv \min\{x_i : j \leq i \leq k\}. \tag{6.4}$$

We are motivated to consider definition (6.3) because the first n successive maxima

$$\{M_k : 1 \leq k \leq n\}$$

have cdf $F_{1,2,...,n}$; e.g.,

$$P(M_k \leq x_k, M_n \leq x_n) = F^k(x_k \wedge x_n) F^{n-k}(x_n) \ . \tag{6.5}$$

It is easy to see that the finite-dimensional distributions in (6.3) are consistent, so that there is a stochastic process $Y \equiv \{Y(t) : t > 0\}$, with those finite-dimensional distributions. Moreover, there is a version in D. We summarize the basic properties in the following theorem; see Resnick (1987) for a proof.

Theorem 15.6.1. (characterization of the extremal process asociated with the cdf F) *For any cdf F on $\mathbb{R}_+$, there is a stochastic process $Y \equiv \{Y(t) : t \geq 0\}$, called the extremal process associated with F, with sample paths in $D((0,\infty), \mathbb{R}, J_1)$ such that*

(i) $P(t \in Disc(Y)) = 0$ for all $t > 0$,

(ii) Y has nondecreasing sample paths,

(iii) Y is a jump Markov process with

$$P(Y(t+s) \leq x \mid Y(s) = y) = \begin{cases} F^t(\alpha), & x \geq y \\ 0, & x < y, \end{cases}$$

(iv) the parameter of the exponential holding time in state x is $Q(x) \equiv -\log F(x)$; given that a jump occurs in state x, the process jumps from x to $(-\infty, y]$ with probability $1 - Q(y)/Q(x)$ if $y > x$ and 0 otherwise.

Having defined and characterized extremal processes, we can now state our limit in E.

Theorem 15.6.2. (stochastic-process limit with limit process in E) *If $\{X_n : n \geq 1\}$ is a sequence of IID nonnegative real-valued random variables with cdf $F \in \mathcal{R}(-\alpha)$, then*

$$\mathbf{X}_n \Rightarrow \mathbf{X} \quad in \quad E((0,\infty), \mathbb{R})$$

for $\mathbf{X}_n$ in (6.2) and

$$c_n \equiv (1/F^c)^{\leftarrow}(n) \equiv \inf\{s : (1/F^c)(s) \geq n\} = \inf\{s : F^c(s) \leq n^{-1}\} \ , \tag{6.6}$$

where $\mathbf{X}$ is characterized by the maxima $M_{t_1,t_2}(\mathbf{X})$ for $0 < t_1 < t_2$. These maxima satisfy the properties:

(i) for each $k \geq 2$ and k disjoint intervals $(t_1, t_2), (t_3, t_4), \ldots, (t_{2k-1}, t_{2k})$, the random variables $M_{t_1,t_2}(\mathbf{X}), M_{t_3,t_4}(\mathbf{X}), \ldots, M_{t_{2k-1},t_{2k}}(\mathbf{X})$ are mutually independent, and

(ii) for each t_1, t_2 with $0 < t_1 < t_2$,

$$M_{t_1,t_2}(\mathbf{X}) \stackrel{d}{=} Y(t_2 - t_1) \ , \tag{6.7}$$

where Y is the extremal process associated with the Frechet extreme-value cdf in (5.34) in Section 4.5.

Proof. To carry out the proof we exploit convergence of random point measures to a Poisson random measure, as in Chapter 4 of Resnick (1987). We will briefly outline the construction. We use random point measures on $\mathbb{R}^2$. For $a \in \mathbb{R}^2$, let ϵ_a be the measure on $\mathbb{R}^2$ with

$$\epsilon_a(A) = \begin{cases} 1, & a \in A \\ 0, & a \notin A \end{cases}. \tag{6.8}$$

A *point measure* on $\mathbb{R}^2$ is a measure μ of the form

$$\mu \equiv \sum_{i=1}^{\infty} \epsilon_{a_i}, \tag{6.9}$$

where $\{a_i : i \geq 1\}$ is a sequence of points in $\mathbb{R}^2$ and

$$\mu(K) < \infty \tag{6.10}$$

for each compact subset K of $\mathbb{R}^2$. Let $M_p(\mathbb{R}^2)$ be the space of point measures on $\mathbb{R}^2$. We say that μ_n *converges vaguely* to μ in $M_p(\mathbb{R}^2)$ and write $\mu_n \to \mu$ if

$$\int f d\mu_n \to \int f d\mu \quad \text{as} \quad n \to \infty \tag{6.11}$$

for all nonnegative continuous real-valued functions with compact support. It turns out that $M_p(\mathbb{R}^2)$ with vague convergence is metrizable as a complete separable metric space; see p. 147 of Resnick (1987).

We will consider random point measures, i.e., probability measures on the space $M_p(\mathbb{R}^2)$. A Poisson random measure N with mean measure ν is a random point measure with the properties

(i) $N(A_1), N(A_2), \ldots, N(A_k)$ are independent random variables for all k and all disjoint measurable subsets $A_1, A_2, \ldots, A_k$ of $\mathbb{R}^2$

(ii) for each measurable subset A of $\mathbb{R}^k$,

$$P(N(A) = k) = e^{-\nu(A)} \frac{\nu(A)^k}{k!} \tag{6.12}$$

where ν is a measure on $\mathbb{R}^2$. In our context, the random point measure limit associated with the sequence $\{X_k : k \geq 1\}$ is

$$\sum_{k=1}^{\infty} \epsilon_{\{k/n, X_k/c_n\}} \Rightarrow N \tag{6.13}$$

where c_n is again as in (6.6) and N is a Poisson random measure with mean measure ν determined by

$$\nu(A \times [x, \infty)) = \lambda(A) x^{-\alpha}, \quad x > 0, \tag{6.14}$$

where λ is Lebesgue measure; the proof of convergence is shown in Resnick (1987).

We now use the Skorohod representation theorem to replace the convergence in distribution in (6.13) by convergence w.p.1 for special versions. It then follows (p. 211 of Resnick) that

$$M_{t_1,t_2}(X_n) \to Y(t_2 - t_1) \quad \text{in} \quad \mathbb{R} \tag{6.15}$$

for almost all t_1, t_2 with $0 < t_1 < t_2$, again for the special versions. We then can apply Theorem 15.5.1 to deduce that

$$\mathbf{X}_n \to \mathbf{X} \quad \text{in} \quad E((0, \infty, \mathbb{R}) , \tag{6.16}$$

again for the special versions. However, the w.p.1 convergence implies the desired convergence in distribution. Clearly the distributions of $\mathbf{X}_n$ and $\mathbf{X}$ are characterized by the independence and the distributions of the maxima $M_{t_1,t_2}(\mathbf{X}_n)$ and $M_{t_1,t_2}(\mathbf{X})$. ■

Thus we have established a limit for the scaled process $\mathbf{X}_n$ in E and connected it to the convergence of successive maxima to extremal processes.

15.7. The Space F

We now use parametric representations to define the space F of functions larger than E. Our purpose is to more faithfully model the fluctuations associated with the excursions. In particular, we now want to describe the order in which the points are visited in the excursions.

Suppose that the graphs $\Gamma_{\tilde{x}}$ of elements $\tilde{x}$ of $E \equiv E([0,T], \mathbb{R}^k)$ are defined as in Section 15.4 and that the conditions of Theorem 15.4.1 hold. We will focus on the subset E_{st} of strongly connected functions in E, and call the space E.

We say that two (strong) parametric representations of a graph $\Gamma_{\tilde{x}}$ are *equivalent* if there exist nondecreasing continuous function λ_1 and λ_2 mapping $[0,1]$ onto $[0,1]$ such that

$$(u_1, r_1) \circ \lambda_1 = (u_2, r_2) \circ \lambda_2 . \tag{7.1}$$

A consequence of (7.1) is that, for each $t \in S$, the functions u_1 and u_2 in the two equivalent parametric representations visit all the points in $I(t)$, the same number of times and in the same order. (Staying at a value is regarded as a single visit.)

We let F be the set of *equivalence classes* of these parametric representations. We thus regard any two parametric representations that are equivalent in the sense of (7.1) as two representations of the same function in F. Let $\hat{x}$ denote an element of F and let $\Pi_s(\hat{x})$ be the set of all parametric representations of $\hat{x}$, i.e., all members of the equivalence class.

Paralleling the metric d_s inducing the M_1 topology on D in equation (3.7) in Section 12.3, let d_s be defined on $F \equiv F([0,T], \mathbb{R}^k)$ by

$$d_s(\hat{x}_1, \hat{x}_2) = \inf_{\substack{(u_i, r_i) \in \Pi_s(\hat{x}_i) \\ i=1,2}} \{\|u_1 - u_2\| \vee \|r_1 - r_2\|\} . \tag{7.2}$$

We call the topology on F induced by the metric d_s in (7.2) the M_1 topology.

Paralleling Theorem 12.3.1, we can conclude that d_s in (7.2) is a bonafide metric.

Theorem 15.7.1. *The space (F, d_s) for d_s in (7.2) is a separable metric space.*

To prove Theorem 15.7.1, we use the following lemma, which closely parallels Lemma 12.3.2.

Lemma 15.7.1. *For any $\hat{x}_1, \hat{x}_2 \in F$, $(u_1, r_1) \in \Pi_s(\hat{x}_1)$ and $\epsilon > 0$, it is possible to find $(u_2, r_2) \in \Pi_s(\hat{x}_2)$ such that*

$$\|u_1 - u_2\| \vee \|r_1 - r_2\| \leq d_s(\hat{x}_1, \hat{x}_2) + \epsilon \ . \tag{7.3}$$

In turn, to prove Lemma 15.7.1, we use a modification of Lemma 12.3.1. For $\hat{x} \in F$, we say that A is an *F-ordered finite subset* of the graph $\Gamma_{\hat{x}}$ if for some parametric representation (u, r) of $\hat{x}$ we have

$$A \equiv \{(z_i, t_i) : 0 \leq i \leq m\}$$

with

$$(z_i, t_i) = (u(s_i), r(s_i)) \quad \text{for} \quad 0 = s_0 < s_1 < \cdots < s_m = 1 \ .$$

Paralleling the order-consistent distance in Definition 12.3.1, we say that the *F-order-consistent distance* between the *F*-ordered subset A and the graph $\Gamma_{\hat{x}}$ is

$$\hat{d}(A, \Gamma_{\hat{x}}) \equiv \sup\{\|(u(s), r(s)) - (u(s_i), r(s_i))\| \tag{7.4}$$
$$\vee \|(u(s), r(s)) - (u(s_{i+1}), r(s_{i+1}))\|\},$$

where the supremum is over all s in $[0, 1]$ such that $s_i < s < s_{i+1}$ with $0 \leq i \leq m$.

We use the following analog of Lemma 12.3.1.

Lemma 15.7.2. *(finite approximations to F graphs) For any $\hat{x} \in F$ and $\epsilon > 0$, there exists an F-ordered finite subset A of the graph $\Gamma_{\hat{x}}$ such that $\hat{d}(A, \Gamma_{\hat{x}}) < \epsilon$ for $\hat{d}$ in (7.4).*

Given an element $\hat{x}$ in F, let $\gamma(\hat{x})$ denote the associated element of E. (Note that $\gamma : F \to E$ is a many-to-one map.) Let $\Gamma_{\hat{x}} \equiv \Gamma_{\gamma(\hat{x})}$ be the graph of $\gamma(\hat{x})$ or just $\hat{x}$. It is easy to relate convergence $\hat{x}_n \to \hat{x}$ in (F, M_1) to convergence $\gamma(\hat{x}_n) \to \gamma(\hat{x})$ in (E, M_2) because both modes of convergence have been characterized by parametric representations. Clearly, M_1 convergence in F implies M_2 convergence in E, but not conversely.

Theorem 15.7.2. *(relating the metrics $d_{s,2}$ and d_s) For $\tilde{x}_1, \tilde{x}_2 \in F$,*

$$d_{s,2}(\gamma(\hat{x}_1), \gamma(\hat{x}_2)) \leq d_s(\hat{x}_1, \hat{x}_2) \ ,$$

where $d_{s,2}$ and d_s are defined in (5.1) and (7.2). Hence, if $\hat{x}_n \to \hat{x}$ in (F, M_1), then $\gamma(\hat{x}_n) \to \gamma(\hat{x})$ in (E_{st}, SM_2).

To put (F, d_s) into perspective, recall that for D with the metric d_s in (3.7) of Section 12.3 inducing the SM_1 topology, all parametric representations were required to be nondecreasing, using an order introduced on the graphs. Thus, all parametric representations of x in D are equivalent parametric representations. In contrast, here with the more general functions in F, there is in general no one natural order to consider on the graphs.

Paralleling Theorem 15.4.6 and Corollary 15.4.1, we can identify (D, SM_1) as a subset of (F, M_1).

Theorem 15.7.3. *Let D' be the subset of functions $\hat{x}$ in F for which $\Gamma_{\hat{x}}$ is the graph of a function in D, i.e., for which*

$$I(t) = [x(t-), x(t+)] \quad in \quad \mathbb{R}^k \tag{7.5}$$

for all $t \in S$, and for which the parametric representation is monotone in the order on the graphs defined in Section 12.3. Then (D', d_s) for d_s in (7.2) is homeomorphic to (D, SM_1). Indeed,

$$d_s(\hat{x}_1, \hat{x}_2) = d_s(\gamma(\hat{x}_1), \gamma(\hat{x}_2)) \tag{7.6}$$

where $d_s(\gamma(\hat{x}_1), \gamma(\hat{x}_2))$ is interpreted as the metric on D in (3.7) of Section 12.3.

We now give an example to show how F allows us to establish new limits. We will show that we can have $x_n \to x$ in (D, M_2), $x_n \not\to x$ in (D, M_1) and $x_n \to \hat{x}$ in (F, M_1). An important point is that the new M_1 limit $\hat{x}$ in F is not equivalent to the M_2 limit x in D.

Example 15.7.1. *Convergence in (F, M_1) where convergence in (D, M_1) fails.* Consider the functions $x = I_{[1,2]}$ and

$$x_n(0) = x_n(1) = x_n(1 + 2n^{-1}) = 0$$
$$x_n(1 + n^{-1}) = x_n(1 + 3n^{-1}) = x_n(2) = 1 ,$$

with x_n defined by linear interpolation elsewhere. Note that $x_n \to x$ in (D, M_2), but $x_n \not\to x$ in (D, M_1). However, $\hat{x}_n \to \hat{x}$ in (F, M_1), where $\hat{x}$ is different from x. The convergence $\hat{x}_n \to \hat{x}$ in F implies that $\gamma(\hat{x}_n) \to \gamma(\hat{x})$ in (E, M_2), which in this case is equivalent to $x_n \to x$ in (D, M_2). ∎

15.8. Queueing Applications

In this final section we discuss queueing applications of the spaces E and F. First, for queueing networks, the key result is the following analog of results in Chapter 14. Let (F, WM_1) denote the space F with the product topology, using the M_1 topology on each component.

Theorem 15.8.1. (the multidimensional reflection map on F) *The multidimensional reflection map R defined in Section 14.2 is well defined and measurable as a map from (F, M_1) to (F, M_1). Moreover, R is continuous at all $\hat{r}$ in F_1, the subset of functions in F with discontinuities or excursions in only one coordinate at a time, provided that the range is endowed with the product topology. Hence, if $\hat{x}_n \to \hat{x}$ in (F, M_1), where $\hat{x} \in F_1$, then $R(\hat{x}_n) \to R(\hat{x})$ in (F, WM_1) and $\gamma(R(\hat{x}_n)) \to \gamma(R(\hat{x}))$ in (E_{wk}, WM_2).*

A natural way to establish stochastic-process limits in F and E is to start from stronger stochastic-process limits in D and then abandon some of the detail. In particular, we can start with a stochastic-process limit in (D, SM_1) that simultaneously describes the asymptotic behavior in two different time scales. We can then obtain a limit in (F, M_1) when we focus on only the longer time scale. From the perspective of the longer time scale, what happens in the shorter time scale happens instantaneously. We keep some of the original detail when we go from a limit in (D, SM_1) to a limit in (F, M_1), because the order in which the points are visited in the shorter time scale at each one-time excursion is preserved.

To illustrate, we consider a single infinite-capacity fluid queue that alternates between periods of being "in control" and periods of being "out of control." When the system is in control, the net-input process is "normal;" when the system is out of control, the net-input process is "exceptional." The model is a generalization of the infinite-capacity version of the fluid queue in Chapter 5. The alternating periods in which the system is in and out of control generalize the up and down times for the queueing network with service interruptions in Section 14.7, because here we allow more complicated behavior during the down times.

As in Section 14.7, let $\{(U_k, D_k) : k \geq 1\}$ be the sequence of successive up and down times. We assume that these random variables are strictly positive and that $\sum_{i=1}^{\infty} U_k = \infty$ w.p.1. The system is up or in control during the intervals $[T_k, T_k + U_k)$ and down or out of control during the intervals $[T_k + U_k, T_{k+1})$, where $T_k \equiv \sum_{i=1}^{k}(U_i + D_i)$, $k \geq 1$, with $T_0 \equiv 0$.

Let X_k^u and X_k^u be potential net-input processes in effect during the kth up period and down period, respectively. We regard X_k^u and X_k^d as random elements of $D \equiv D([0, \infty), \mathbb{R})$ for each k. However, X_k^u is only realized over the interval $[0, U_k)$, and X_k^u is only realized over the interval $[0, D_k)$. Specifically, the overall net-input process is the process X in D defined by

$$X(t) = \begin{cases} X_{k+1}^u(t - T_k) + X(T_k), & T_k \leq t < T_k + U_k , \\ X_{k+1}^d(t - T_k - U_k) + X(T_k + U_k), & T_k + U_k \leq t < T_{k+1} , \end{cases} \quad (8.1)$$

for $t \geq 0$.

Just as in equations (2.5) – (2.7) in Section 5.2, the associated workload or buffer-content process, starting out empty, is defined by

$$W(t) \equiv \phi(X)(t) \equiv X(t) - \inf_{0 \leq s \leq t} X(s), \quad t \geq 0 . \quad (8.2)$$

We want to establish a heavy-traffic stochastic-process limit for a sequence of these fluid-queue models. In preparation for that heavy-traffic limit, we need to show how to identify the limit process in F: We need to show how the net-input random element of D approaches a random element of F when the down times decrease to 0, while keeping the toital net inputs over the down intervals unchanged. What happens during the down time is unchanged; it just happens more quickly as the down time decreases.

To formalize that limit, suppose that we are given a fluid queue model as specified above. We now create a sequence of models by altering the variables D_k and the

processes X_k^d for all k. Specifically, we let
$$D_{n,k} \equiv c_n^{-1} D_k \tag{8.3}$$
for all k, where $c_n \to \infty$ as $n \to \infty$. To keep the associated net-input processes during these down times unchanged except for time scaling, we let
$$X_{n,k}^d(t) \equiv X_k^d(c_n t), \quad t \geq 0, \tag{8.4}$$
which implies that
$$X_{n,k}^d(p D_{n,k}) = X_k^d(p D_k) \quad \text{for} \quad 0 \leq p \leq 1.$$

With this special construction, it is easy to see that the original element of D approaches an associated element of F as $n \to \infty$.

Lemma 15.8.1. (decreasing down times) *Consider a fluid-queue model as specified above. Suppose that we construct a sequence of fluid-queue models indexed by n, where the models change only by (8.3) and (8.4) with $c_n \to \infty$ as $n \to \infty$. Then $X_n \to \hat{X}$ in (F, M_1).*

We are now ready to state the heavy-traffic stochastic-process limit for the fluid-queue model. We start with a more-detailed limit in D, and obtain the limit in F by applying Lemma 15.8.1 to the limit process in D.

For the heavy-traffic limit, we start with a sequence of fluid-queue models indexed by n. In this initial sequence of models, we let the initial up and down times be independent of n. We then introduce the following scaled random elements:

$$\begin{aligned}
U_{n,k} &\equiv n U_k, \quad k \geq 1, \\
D_{n,k} &\equiv n^\beta D_k, \quad k \geq 1, \\
\mathbf{X}_{n,k}^u(t) &\equiv n^{-H} X_{n,k}^u(nt), \quad t \geq 0, \\
\mathbf{X}_{n,k}^d(t) &\equiv n^{-H} X_{n,k}^u(n^\beta t), \quad t \geq 0, \\
\mathbf{X}_n(t) &\equiv n^{-H} X_n(nt), \quad t \geq 0, \\
\mathbf{W}_n(t) &\equiv n^{-H} W_n(nt), \quad t \geq 0,
\end{aligned} \tag{8.5}$$

where
$$0 < \beta < 1 \quad \text{and} \quad H > 0. \tag{8.6}$$

Note that in (8.5) the time scaling of U_k, $X_{n,k}^u$, $X_n(t)$ and $W_n(t)$ in (8.5) is all by n, while the time scaling of D_k and $X_{n,k}^d$ is by only n^β, where $0 < \beta < 1$. Thus the durations of the down periods are asymptotically negligible compared to the durations of the uptimes in the limit as $n \to \infty$.

Here is the result:

Theorem 15.8.2. (heavy-traffic limit in F) *Consider a sequence of fluid-queue models with the scaling in (8.5). Suppose that*

$$\{(U_{n,k}, D_{n,k}, \mathbf{X}_{n,k}^u, \mathbf{X}_{n,k}^d) : k \geq 1\}$$
$$\Rightarrow \{(U_k', D_k', \mathbf{X}_k^u, \mathbf{X}_k^d) : k \geq 1\} \tag{8.7}$$

in $(\mathbb{R} \times \mathbb{R} \times D \times D)^\infty$ as $n \to \infty$, where $U'_k > 0$ for all k and $\sum_{i=1}^\infty U'_i = \infty$ w.p.1. Then

$$(\mathbf{X}_n, \mathbf{W}_n) \Rightarrow (\hat{\mathbf{X}}, \hat{\mathbf{W}}) \quad in \quad (F, M_1) \times (F, M_1)$$

with the product topology on the product space $(F, M_1) \times (F, M_1)$, where $\hat{\mathbf{X}}$ is the limit in F obtained in Lemma 15.8.1 applied to the limiting fluid queue model with up and down times (U'_k, D'_k) and potential net-input processes $\mathbf{X}^u_k$ and $\mathbf{X}^d_k$ in (8.7) and $\hat{\mathbf{W}} = \phi(\hat{\mathbf{X}})$. Consequently,

$$\mathbf{W}_n \Rightarrow \gamma(\hat{\mathbf{W}}) \quad in \quad (E, M_2) \ .$$

A standard sufficient condition for condition (8.7) in Theorem 15.8.2 is to have the random elements $(U_{n,k}, D_{n,k}, \mathbf{X}^u_{n,k}, \mathbf{X}^d_{n,k})$ of $\mathbb{R} \times \mathbb{R} \times D \times D$ be IID for $k \geq 1$ and to have

$$(U_{n,1}, D_{n,1}, \mathbf{X}^u_{n,1}, \mathbf{X}^d_{n,1}) \quad \Rightarrow \quad (U'_1, D'_1, \mathbf{X}^u_1, \mathbf{X}^d_1) \tag{8.8}$$

in $\mathbb{R} \times \mathbb{R} \times D \times D$. In the standard case, we have $H = 1/2$ and $\mathbf{X}^u_1$ Brownian motion with drift down, but we have seen that there are other possibilities. In this framework, the complexity of the limit process is largely determined by the limit process $\mathbf{X}^d_1$ describing the asymptotic behavior of the net-input process during down times. We conclude by giving an example in which the limiting stochastic process, regarded as an element of F or E, is tractable.

Example 15.8.1. *The Poisson-excursion limit process.* A special case of interest in the IID cycle framework above occurs when the limit $\mathbf{X}^u_k$ in (8.7) is deterministic, e.g., $\mathbf{X}^u_k(t) = ct$, $t \geq 0$. A further special case is to have $c = 0$. Then the process is identically zero except for the excursions, so all the structure is contained in the excursions.

Suppose that we have this special structure; i.e., suppose that $\mathbf{X}^u_k(t) = 0$, $t \geq 0$. If, in addition, U'_k is exponentially distributed, then in the limit the excursions occur according to a Poisson process. We call such a process a *Poisson-excursion process*.

To consider a simple special case of a Poisson-excursion process, suppose that, for each k, the process $\mathbf{X}^d_{n,k}$ corresponds to a partial sum of three random variables $Z_{n,k,1}, Z_{n,k,2}, Z_{n,k,3}$ evenly spaced in the interval $[0, n^\beta D_k]$, i.e., occurring at times $n^\beta D_k/4$, $2n^\beta D_k/4$ and $3n^\beta D_k/4$. For example, we could have deterministic down times of the form $D_k = 4$ for all k, which implies that $D_{n,k} = 4n^\beta$ for all k. Consistent with (8.7), we assume that

$$n^{-H}(Z_{n,k,1}, Z_{n,k,2}, Z_{n,k,3}) \Rightarrow (Z_{k,1}, Z_{k,2}, Z_{k,3}) \quad as \quad n \to \infty \ . \tag{8.9}$$

Then the base process $\mathbf{X}$ associated with the graph of the limit $\hat{\mathbf{X}}$ is

$$\mathbf{X}(t) = \sum_{i=1}^{N(t)} Y_i, \quad t \geq 0 \ , \tag{8.10}$$

where $\{N(t) : t \geq 0\}$ is a Poisson process with rate $1/EU_1'$ and $\{Y_i\}$ is an IID sequence with $Y_i \stackrel{d}{=} \sum_{j=1}^{3} Z_{i,j}$.

The excursions in more detail are described by the successive partial sums $\{\sum_{j=1}^{n} Z_{i,j} : 1 \leq n \leq 3\}$; i.e., the first excursion occuring at time U_1' goes from 0 to Z_1, then to $Z_1 + Z_2$, and finally to $Z_1 + Z_2 + Z_3$. Note that we have not yet made any restrictive assumptions on the joint distribution of $(Z_{n,k,1}, Z_{n,k,2}, Z_{n,k,3})$, so that the limit $(Z_{k,1}, Z_{k,2}, Z_{k,3})$ can have an arbitrary distribution. However, if in addition, $Z_{k,1}$, $Z_{k,2}$ and $Z_{k,3}$ are IID, then we can characterize the limiting distribution of $\gamma(\hat{\mathbf{W}})$, the random element of E representing the buffer content. The base process W can be represented as

$$W(t) = Q_{3N(t)}, \quad t \geq 0, \qquad (8.11)$$

where $\{N(t) : t \geq 0\}$ is the Poisson process counting limiting excursions and Q_k is the queue-content in period k of a discrete-time queue with IID net inputs distributed as Z_1, i.e., where $\{Q_k : k \geq 0\}$ satisfies the Lindley equation

$$Q_k = \max\{Q_{k-1} + Z_k, 0\}, \quad k \geq 1, \qquad (8.12)$$

with $Q_0 = 0$. Consequently, $W(t) \Rightarrow W$ as $t \to \infty$, where the limiting workload W is distributed as Q with $Q_k \Rightarrow Q$ as $k \to \infty$.

The additional excursions in $\hat{\mathbf{W}}$ occur according to the Poisson process N. Given that the kth excursion occurs at time t, we can easily describe it in terms of the four random variables $\mathbf{W}(t-)$, $Z_{k,1}$, $Z_{k,2}$ and $Z_{k,3}$: The process moves from $A_0 \equiv \mathbf{W}(t-)$ to $A_1 \equiv max\{0, A_0 + Z_{k,1}\}$, then to $A_2 \equiv max\{0, A_1 + Z_{k,2}\}$ and finally to $\mathbf{W}(t) = A_3 \equiv max\{0, A_2 + Z_{k,3}\}$. For the process in F, these three steps all occur at the time t. The parametric representation moves continuously from A_0 to A_1, then to A_2 and A_3. The limiting workload then remains constant until the next excursion. ∎

References

[1] Abate, J., Choudhury, G. L., Lucantoni, D. M. and Whitt, W. (1995) Asymptotic analysis of tail probabilities based on the computation of moments. *Ann. Appl. Prob.* 5, 983–1007.

[2] Abate, J., Choudhury, G. L. and Whitt, W. (1993) Calculation of the GI/G/1 steady-state waiting-time distribution and its cumulants from Pollaczek's formula. *Archiv für Elektronik und Ubertragungstechnik (Special Issue in memory of F. Pollaczeck)* 47, 311–321.

[3] Abate, J., Choudhury, G. L. and Whitt, W. (1994a) Waiting-time tail probabilities in queues with long-tail service-time distributions. *Queueing Systems* 16, 311–338.

[4] Abate, J., Choudhury, G. L. and Whitt, W. (1994b) Asymptotics for steady-state tail probabilities in structured Markovian queueing models. *Stochastic Models* 10, 99–143.

[5] Abate, J., Choudhury, G. L. and Whitt, W. (1995) Exponential approximations for tail probabilities in queues, I: waiting times. *Operations Res.* 43, 885–901.

[6] Abate, J., Choudhury, G. L. and Whitt, W. (1999) An introduction to numerical transform inversion and its application to probability models. *Computational Probability*, W. Grassman (ed.), Kluwer, Boston, 257–323.

[7] Abate, J. and Whitt, W. (1987a) Transient behavior of regulated Brownian motion, I: starting at the origin. *Adv. Appl. Prob.* 19, 560–598.

[8] Abate, J. and Whitt, W. (1987b) Transient behavior of regulated Brownian motion, II: non-zero initial conditions. *Adv. Appl. Prob.* 19, 599–631.

[9] Abate, J. and Whitt, W. (1988a) Transient behavior of the M/M/1 queue via Laplace transform. *Adv. Appl. Prob.* 20, 145–178.

[10] Abate, J. and Whitt, W. (1988b) Simple spectral representations for the M/M/1 queue. *Queueing Systems* 3, 321–346.

[11] Abate, J. and Whitt, W. (1988c) The correlation functions of RBM and M/M/1. *Stochastic Models* 4, 315–359.

[12] Abate, J. and Whitt, W. (1988d) Approximations for the M/M/1 busy-period distribution. *Queueing Theory and its Applications – Liber Americorum for J. W. Cohen*, North Holland, Amsterdam, 149–191.

[13] Abate, J. and Whitt, W. (1992a) The Fourier-series method for inverting transforms of probability distributions. *Queueing Systems* 10, 5–88.

[14] Abate, J. and Whitt, W. (1992b) Solving probability transform functional equations for numerical inversion. *Operations Res. Letters* 12, 275–281.

[15] Abate, J. and Whitt, W. (1994a) Transient behavior of the M/G/1 workload process. *Operations Res.* 42, 750–764.

[16] Abate, J. and Whitt, W. (1994b) A heavy-traffic expansion for the asymptotic decay rates of tail probabilities in multi-channel queues. *Operations Res. Letters* 15, 223–230.

[17] Abate, J. and Whitt, W. (1995a) Numerical inversion of Laplace transforms of probability distributions. *ORSA J. Computing* 7, 36–43.

[18] Abate, J. and Whitt, W. (1995b) Limits and approximations for the busy-period distribution in single-server queues. *Prob. Eng. Inf. Sci.* 9, 581–602.

[19] Abate, J. and Whitt, W. (1996) An operational calculus for probability distributions via Laplace transforms. *Adv. Appl. Prob.* 28, 75–113.

[20] Abate, J. and Whitt, W. (1997a) Limits and approximations for the $M/G/1$ LIFO waiting-time distribution. *Operations Res. Letters* 20, 199–206.

[21] Abate, J. and Whitt, W. (1997b) Asymptotics for $M/G/1$ low-priority waiting-time tail probabilities. *Queueing Systems* 25, 173–233.

[22] Abate, J. and Whitt, W. (1998) Explicit M/G/1 waiting-time distributions for a class of long-tail service-time distributions. *Operations Res. Letters* 25, 25–31.

[23] Abate, J. and Whitt, W. (1999a) Computing Laplace transforms for numerical inversion via continued fractions. *INFORMS J. Computing* 11, 394–405.

[24] Abate, J. and Whitt, W. (1999b) Modelling service-time distributions with non-exponential tails: Beta mixtures of exponentials. *Stochastic Models* 15, 517–546.

[25] Abate, J. and Whitt, W. (1999c) Infinite-series representations of Laplace transforms of probability density functions. *J. Operations Res. Society Japan* 42, 268–285.

[26] Abbott, E. A. (1952) *Flatland*, Dover, New York.

[27] Abramowitz, M. and Stegun, I. A. (1972) *Handbook of Mathematical Functions*, National Bureau of Standards, Washington, D.C.

[28] Addie, R. G. and Zukerman, M. (1994) An approximation for performance evaluation of stationary single server queues. *IEEE Trans. Commun.* 42, 3150–3160.

[29] Adler, R. J., Feldman, R. E. and Taqqu, M. S. (1998) *A Practical Guide to Heavy Tails: Statistical Techniques and Applications*, Birkhauser, Boston.

[30] Albin, S. L. (1982) On Poisson approximations for superposition arrival processes in queues. *Management Sci.* 28, 126–137.

[31] Albin, S. L. (1984) Approximating a point process by a renewal process, II: superposition arrival processes. *Operations Res.* 32, 1133–1162.

[32] Aldous, D. (1989) *Probability Approximations via the Poisson Clumping Heuristic*, Springer, New York.

[33] Amdahl, G. (1967) Validity of the single processor approach to achieving large scale computing capabilities. *AFIPS Conference Proceedings, Atlantic City* 30, 483–485.

[34] Anantharam, V. (1999) Scheduling strategies and long-range dependence. *Queueing Systems* 33, 73–89.

[35] Anick, D., Mitra, D. and Sondhi, M. M. (1982) Stochastic theory of a data handling system with multiple sources. *Bell System Tech. J.* 61, 1871–1894.

[36] Anisimov, V. V. (1993) The averaging principle for switching processes. *Theor. Prob. Math. Stat.* 46, 1–10.

[37] Araujo, A. and Giné, E. (1980) *The Central Limit Theorem for Real and Banach valued Random Variables*, Wiley, New York.

[38] Arvidsson, A. and Karlsson, P. (1999) On traffic models for TCP/IP. *Traffic Engineering in a Competitive World, Proceedings* 16th *Int. Teletraffic Congress*, P. Key and D. Smith (eds.), North-Holland, Amsterdam, 457–466.

[39] Asmussen, S. (1987) *Applied Probability and Queues*, Wiley, New York.

[40] Asmussen, S. (1992) Queueing simulation in heavy traffic. *Math. Oper. Res.* 17, 84–111.

[41] Asmussen, S. (2000) *Ruin Probabilities*, World Scientific, Singapore.

[42] Asmussen, S. and Teugels, J. (1996) Convergence rates for M/G/1 queues and ruin problems with heavy tails. *J. Appl. Prob.* 33, 1181–1190.

[43] Astrauskas, A. (1983) Limit theorems for sums of linearly generated random variables. *Lithuanian Math. J.* 23, 127–134.

[44] Avram, F. and Taqqu, M. S. (1987) Noncentral limit theorems and Appell polynomials. *Ann. Probab.* 15, 767–775.

[45] Avram, F. and Taqqu, M. S. (1989) Probability bounds for M-Skorohod oscillations. *Stoch. Proc. Appl.* 33, 63–72.

[46] Avram, F. and Taqqu, M. S. (1992) Weak convergence of sums of moving averages in the α-stable domain of attraction. *Ann. Probab.* 20, 483–503.

[47] Ayhan, W. and Olsen, T. L. (2000) Scheduling multi-class single-server queues under nontraditional performance measures. *Operations Res.* 48, 482–489.

[48] Azar, Y., Broder, A., Karlin, A. and Upfal, E. (1994) Balanced allocations. *Proceedings* 26th *ACM Symposium on the Theory of Computing*, 593–602.

[49] Baccelli, F. and Brémaud, P. (1994) *Elements of Queueing Theory*, Springer, New York.

[50] Baccelli, F. and Foss, S. (1994) Ergodicity of Jackson-type queueing networks. *Queueing Systems* 17, 5–72.

[51] Barford, P. and Crovella, M. (1998) Generating representative web workloads for network and server performance evaluation. *Proc. 1998 ACM Sigmetrics*, 151–160.

[52] Bass, F. R. and Pyke, R. (1987) A central limit theorem for $D(A)$-valued processes. *Stochastic Proc. Appl.* 24, 109–131.

[53] Becker, R. A., Chambers, J. M. and Wilks, A. R. (1988) *The New S Language*, Chapman and Hall, New York.

[54] Bell, S. L. and Williams, R. J. (2001) Dynamic scheduling of a system with two parallel servers in heavy traffic with complete resource pooling: asymptotic optimality of a threshold policy. *Ann. Appl. Prob.* 11.

[55] Bender, C. M. and Orszag, S. A. (1978) *Advanced Mathematical Methods for Scientists and Engineers*, McGraw-Hill, New York.

[56] Beneš, V. E. (1963) *General Stochastic Processes in the Theory of Queues*, Addison-Wesley, Reading, MA.

[57] Beran, J. (1994) *Statistics for Long-Memory Processes*, Chapman and Hall, New York.

[58] Berger, A. W. and Whitt, W. (1992a) The impact of a job buffer in a token-bank rate-control throttle. *Stochastic Models* 8, 685–717.

[59] Berger, A. W. and Whitt, W. (1992b) The Brownian approximation for rate-control throttles and the G/G/1/C queue. *J. Discrete Event Dynamic Systems* 2, 7–60.

[60] Berger, A. W. and Whitt, W. (1994) The pros and cons of a job buffer in a token-bank rate-control throttle. *IEEE Trans. Commun.* 42, 857–861.

[61] Berger, A. W. and Whitt, W. (1995a) Maximum values in queueing processes. *Prob. Eng. Inf. Sci.* 9, 375–409.

[62] Berger, A. W. and Whitt, W. (1995b) A comparison of the sliding window and the leaky bucket. *Queueing Systems* 20, 117–138.

[63] Berger, A. W. and Whitt, W. (1998a) Effective bandwidths with priorities. *IEEE/ACM Trans. Networking*, 6, 447–460.

[64] Berger, A. W. and Whitt, W. (1998b) Extending the effective bandwidth concept to networks with priority classes. *IEEE Communications Magazine* 36, August, 78–83.

[65] Bertoin, J. (1996) *Lévy Processes*, Cambridge University Press, Cambridge, UK.

[66] Bertsekas, D. P. and Gallager, R. G. (1992) *Data Networks*, second ed., Prentice-Hall, Enlewood Cliffs, NJ.

[67] Billingsley, P. (1968) *Convergence of Probability Measures*, Wiley, New York.

[68] Billingsley, P. (1999) *Convergence of Probability Measures*, second edition, Wiley, New York.

[69] Bingham, N. H., Goldie, C. M. and Teugels, J. L. (1989) *Regular Variation*, Cambridge University Press, Cambridge, UK.

[70] Bitran, G. and Tirupati, D. (1988) Multiproduct queueing networks with deterministic routing: decomposition approach and the notion of interference. *Management Sci.* 34, 75–100.

[71] Bleistein, N. and Handelsman, R. A. (1986) *Asymptotic Expansions of Integrals*, Dover, New York.

[72] Bloznelis, M. (1996) Central limit theorem for stochastically continuous processes. convergence to stable limit. *J. Theor. Probab.* 9, 541–560.

[73] Bloznelis, M. and Paulauskas, V. (2000) Central limit theorem in $D[0,1]$. *Skorohod's Ideas in Probability Theory*, V. Korolyuk, N. Portenko and H. Syta (eds.), institute of Mathematics of the National Acafdemy of Sciences of the Ukraine, Kyiv, Ukraine, 99–110.

[74] Bondi, A. B. and Whitt, W. (1986) The influence of service-time variability in a closed network of queues. *Performance Evaluation* 6, 219–234.

[75] Borodin, A. N. and Salminen, P. (1996) *Handbook of Brownian Motion – Facts and Formulae*, Birkhauser, Boston.

[76] Borovkov, A. A. (1965) Some limit theorems in the theory of mass service, II. *Theor. Prob. Appl.* 10, 375–400.

[77] Borovkov, A. A. (1967) On limit laws for service processes in multi-channel systems. *Siberian Math. J.* 8, 746–763.

[78] Borovkov, A. A. (1976) *Stochastic Processes in Queueing Theory*, Springer, New York.

[79] Borovkov, A. A. (1984) *Asymptotic Methods in Queueing Theory*, Wiley, New York.

[80] Borovkov, A. A. (1998) *Ergodicity and Stability of Stochastic Processes*, Wiley, New York.

[81] Borst, S. C., Boxma, O. J. and Jelenković, P. R. (2000) Asymptotic behavior of generalized processor sharing with long-tailed traffic sources. *Proceedings IEEE INFOCOM 2000*, 912–921.

[82] Borst, S. C., Mandelbaum, A. and Reiman, M. I. (2001) Dimensioning large call centers. Bell Labs, Murray Hill, NJ.

[83] Botvich, D. D. and Duffield, N. G. (1995) Large deviations, the shape of the loss curve, and economies of scale in large multiplexers. *Queueing Systems* 20, 293–320.

[84] Box, G. E. P., Jenkins, G. M. and Reinsel, G. C. (1994) *Time Series Analysis, Forecasting and Control*, third edition, Prentice Hall, NJ.

[85] Boxma, O. J. and Cohen, J. W. (1998) The M/G/1 queue with heavy-tailed service time distribution. *IEEE J. Sel. Areas Commun.* 16, 749–763.

[86] Boxma, O. J. and Cohen, J. W. (1999) Heavy-traffic analysis for the GI/G/1 queue with heavy-tailed distributions. *Queueing Systems* 33, 177–204.

[87] Boxma, O. J. and Cohen, J. W. (2000) The single server queue: heavy tails and heavy traffic. In *Self-Similar Network Traffic and Performance Evaluation*, K. Park and W. Willinger (eds.), Wiley, New York, 143–169.

[88] Boxma, O. J., Cohen, J. W. and Deng, Q. (1999) Heavy-traffic analysis of the M/G/1 queue with priority classes. *Traffic Engineering in a Competitive World, Proceedings of the* 16th *Int. Teletraffic Congress*, P. Key and D. Smith (eds.), North-Holland, Amsterdam, 1157–1167.

[89] Boxma, O. J. and Dumas, V. (1998) Fluid queues with long-tailed activity period distributions. *Comput. Commun.* 21, 1509–1529.

[90] Boxma, O. J. and Takagi, H. (1992) Editors: Special issue on polling systems. *Queueing Systems* 11, 1 and 2.

[91] Bramson, M. (1994a) Instability of FIFO queueing networks. *Ann. Appl. Prob.* 4, 414–431. (Correction p. 952)

[92] Bramson, M. (1994b) Instability of FIFO queueing networks with quick service times. *Ann. Appl. Prob.* 4, 693–718.

[93] Bramson, M. (1998) State space collapse with application to heavy traffic limits for multiclass queueing networks. *Queueing Systems* 30, 89–148.

[94] Bratley, P., Fox, B. L. and Schrage, L. E. (1987) *A Guide to Simulation*, second edition, Springer, New York.

[95] Breiman, L. (1968) *Probability*, Addison-Wesley, Reading, MA.

[96] Brichet, F., Roberts, J. W., Simonian, A. and Veitch, D. (1996) Heavy traffic analysis of a storage model with long-range dependent on/off sources. *Queueing Systems* 23, 197–215.

[97] Brichet, F., Simonian, A. Massoulié, L. and Veitch, D. (2000) Heavy load queueing analysis with LRD on/off sources. In *Self-Similar Network Traffic and Performance Evaluation*, K. Park and W. Willinger (eds.), Wiley, New York, 115–141.

[98] Browne, S. and Whitt, W. (1995) Piecewise-linear diffusion processes. *Advances in Queueing*, J. Dshalalow, ed., CRC Press, Boca raton, FL, 463–480.

[99] Bu, T. and Towsley, D. (2001) Fixed point approximations for TCP behavior in an AQN network. *Proceedings ACM SIGMETRICS '01*.

[100] Burman, D. Y. and Smith, D. R. (1983) Asymptotic analysis of a queueing model with bursty traffic. *Bell System Tech. J.* 62, 1433–1453.

[101] Burman, D. Y. and Smith, D. R. (1986) An asymptotic analysis of a queueing system with Markov-modulated arrivals. *Operations Res.* 34, 105–119.

[102] Buzacott, J. A. and Shanthikumar, J. G. (1993) *Stochastic Models of Manufacturing Systems*, Prentice-Hall, Englewood Cliffs, NJ.

[103] Carslaw, H. S. (1930) *Introduction to the Theory of Fourier's Series and integrals*, third ed., Dover, New York.

[104] Chang, K.-H. (1997) Extreme and high-level sojourns of the single server queue in heavy traffic. *Queueing Systems* 27, 17–35.

[105] Chang, C. S. and Thomas, J. A. (1995) Effective bandwidths in high-speed digitial networks. *IEEE J. Sel. Areas Commun.* 13, 1091-1100.

[106] Chen, H. and Mandelbaum, A. (1991a) Discrete flow networks: bottleneck analysis and fluid approximation. *Math. Oper. Res.* 16, 408–446.

[107] Chen, H. and Mandelbaum, A. (1991b) Discrete flow networks: diffusion approximations and bottlenecks. *Ann. Prob.* 19, 1463–1519.

[108] Chen, H. and Mandelbaum, A. (1991c) Leontief systems, RBV's and RBM's. In *Proc. Imperial College Workshop on Applied Stochastic Processes*, eds. M. H. A. Davis and R. J. Elliot, Gordon and Breach, London.

[109] Chen, H. and Mandelbaum, A. (1994a) Heirarchical modeling of stochastic networks, part I: fluid models. *Stochastic Modeling and Analysis of Manufacturing Systems*, D. D. Yao, ed., Springer-Verlag, New York, 47–105.

[110] Chen, H. and Mandelbaum, A. (1994b) Heirarchical modeling of stochastic networks, part II: strong approximations. *Stochastic Modeling and Analysis of Manufacturing Systems*, D. D. Yao, ed., Springer-Verlag, New York, 107–131.

[111] Chen, H. and Shanthikumar, J. G. (1994) Fluid limits and diffusion approximations for networks of multi-server queues in heavy traffic. *J. Disc. Event Dyn. Systems* 4, 269–291.

[112] Chen, H. and Shen, X. (2000) Strong approximations for multiclass feedforward queueing networks. *Ann. Appl. Prob.* 10, 828–876.

[113] Chen, H. and Whitt, W. (1993) Diffusion approximations for open queueing networks with service interruptions. *Queueing Systems* 13, 335–359.

[114] Chen, H. and Yao, D. D. (2001) *Fundamentals of Queueing Networks: Performance, Asymptotics and Optimization*, Springer, New York.

[115] Choe, J. and Shroff, N. B. (1998) A central limit theorem based approach for analyzing queue behavior in high-speed networks. *IEEE/ACM Trans. Networking* 6, 659–671.

[116] Choe, J. and Shroff, N. B. (1999) On the supremum distribution of integrated stationary Gaussian processes with negative linear drift. *Adv. Appl. Prob.* 31, 135–157.

[117] Choudhury, G. L., Lucantoni, D. M. and Whitt, W. (1994) Multi-dimensional transform inversion with application to the transient $M/G/1$ queue. *Ann. Appl. Prob.* 4, 719–740.

[118] Choudhury, G. L., Lucantoni, D. M. and Whitt, W. (1996) Squeezing the most out of ATM. *IEEE Trans. Commun.* 44, 203–217.

[119] Choudhury, G. L., Mandelbaum, A., Reiman, M. I. and Whitt, W. (1997) Fluid and diffusion limits for queues in slowly changing environments. *Stochastic Models* 13, 121–146.

[120] Choudhury, G. L. and Whitt, W. (1994) Heavy-traffic asymptotic expansions for the asymptotic decay rates in the BMAP/G/1 queue. *Stochastic Models* 10, 453–498.

[121] Choudhury, G. L. and Whitt, W. (1996) Computing distributions and moments in polling models by numerical transform inversion. *Performance Evaluation* 25, 267–292.

[122] Choudhury, G. L. and Whitt, W. (1997) Long-tail buffer-content distributions in broadband networks. *Performance Evaluation* 30, 177–190.

[123] Chung, K. L. (1974) *A Course in Probability Theory*, second ed., Academic Press, New York.

[124] Çinlar, E. (1972) Superposition of point processes. In *Stochastic Point Processes, Statistical Analysis, Theory and Applications*, P.A.W. Lewis, ed., Wiley New York, pp. 549–606.

[125] Coffman, E. G., Jr., Puhalskii, A. A. and Reiman, M. I. (1995) Polling systems with zero switchover times: a heavy-traffic averaging principle. *Ann. Appl. Prob* 5, 681–719.

[126] Coffman, E. G., Jr., Puhalskii, A. A. and Reiman, M. I. (1998) Polling systems in heavy-traffic: a Bessel process limit. *Math. Oper. Res.* 23, 257–304.

[127] Cohen, J. W. (1982) *The Single Server Queue*, revised edition, North-Holland, Amsterdam.

[128] Cohen, J. W. (1998) A heavy-traffic theorem for the GI/G/1 queue with a Pareto-type service time distribution. *J. Applied Math. Stoch. Analysis*, special issue dedicated to R. Syski, 11, 247–254.

[129] Cooper, R. B. (1982) *Introduction to Queueing Theory*, second ed., North-Holland, Amsterdam.

[130] Cottle, R. W., Pang, J.-S., and Stone, R. E. (1992) *The Linear Complementarity Problem*, Academic Press, New York.

[131] Courtois, P. (1077) *Decomposability*, Academic Press, New York.

[132] Cox, D. R. (1962) *Renewal Theory*, Methuen, London.

[133] Cox, J. T. and Grimmett, G. (1984) Central limit theorems for associated random variables and the precolation model. *Ann. Probab.* 12, 514–528.

[134] Crovella, M. and Bestavros, A. (1996) Self-similarity in world wide web traffic: evidence and possible causes. *Proceedings 1996 ACM Sigmetrics.*

[135] Crovella, M., Bestavros, A. and Taqqu, M. (1998) Heavy-tailed distributions in the world wide web. *A Practical Guide to Heavy Tails: Statistical Techniques and Applications*, R. J. Adler, R. E., Feldman and M. S. Taqqu (eds.), Birkhauser, Boston.

[136] Csörgő, M. and Horváth, L. (1993) *Weighted Approximations in Probability and Statistics*, Wiley, New York.

[137] Dabrowski, A. R. and Jakubowski, A. (1994) Stable limits for associated random variables. *Ann. Probab.* 22, 1–16.

[138] Dai, J. G. (1990) *Steady-State Analysis of Reflected Brownian Motions: Characterization, Numerical Methods and Queueing Applications*, Ph.D. dissertation, Department of Mathematics, Stanford University.

[139] Dai, J. G. (1994) On positive Harris recurrence of multiclass queueing networks: a unified approach via fluid limit models. *Ann. Appl. Prob.* 4, 49–77.

[140] Dai, J. G. (1998) *Diffusion Approximations of Queueing Networks*, special issue, *Queueing Systems* 30.

[141] Dai, J. G. and Harrison, J. M. (1991) Steady-state analysis of RBM in a rectangle: numerical methods and a queueing application. *Ann. Applied Prob.* 1, 16–35.

[142] Dai, J. G. and Harrison, J. M. (1992) Reflected Brownian motion in the orthant: numerical methods for steady-state analysis. *Ann. Applied Prob.* 2, 65–86.

[143] Dai, J. G., Nguyen, V. and Reiman, M. I. (1994) Sequential bottleneck decomposition: an approximation method for generalized Jackson networks. *Operations Res.* 42, 119–136.

[144] Dai, J. G., Yeh, D. H. and Zhou, C. (1997) The QNET method for reentrant queueing networks with priority disciplines. *Operations Res.* 45, 610–623.

[145] Daley, D. J. and Vere-Jones, D. (1988) *An Introduction to the Theory of Point Processes*, Springer-Verlag, New York.

[146] Damerdji, H. (1994) Strong consistency of the variance estimator in steady-state simulation output analysis. *Math. Oper. Res.* 19, 494–512.

[147] Damerdji, H. (1995) Mean-square consistency of the variance estimator in steady-state simulation output analysis. *Operations Res.* 43, 282–291.

[148] Das, A. and Srikant, R. (2000) Diffusion approximations for a single node accessed by congestion controlled sources. *IEEE Trans. Aut. Control* 45, 1783–1799.

[149] Davis, R. and Resnick, S. (1985) Limit theory for moving averages of random variables with regularly varying tails. *Ann. Probab.* 13, 179–185.

[150] Davydov, Yu. A. (1970) The invariance principle for stationary processes. *Theor. Probability Appl.* 15, 487–498.

[151] Demers, A., Keshav, S. and Shenker, S. (1989) Design and analysis of a fair queueing system. *Proceedings ACM SIGCOMM 89*.

[152] de Meyer, A. and Teugels, J. L. (1980) On the asymptotic behavior of the distribution of the busy period and the service time in M/G/1. *Adv. Appl. Prob.* 17, 802–813.

[153] de Veciana, G., Kesidis, G. and Walrand, J. (1995) Resource management in wide-area ATM networks using effective bandwidths. *IEEE J. Sel. Areas Commun.* 13, 1081-1090.

[154] Devroye, L. (1997) *A Course in Density Estimation*, Birkhauser, Boston.

[155] Dobrushin, R. L. and Major, P. (1979) Noncentral limit theorems for nonlinear functions of Gaussian fields. *Zeitschrift für Wahrscheinlichkeitstheorie verw. Gebiete* 50, 27–52.

[156] Doetsch, G. (1974) *Introduction to the Theory and Application of the Laplace Transformation*, Springer-Verlag, New York.

[157] Donsker, M. (1951) An invariance principle for certain probability limit theorems. *Mem. Amer. Math. Soc.* 6.

[158] Donsker, M. (1952) Justification and extension of Doob's heuristic approach to the Kolmogorov-Smirnov theorems. *Ann. Math. Statist.* 23, 277–281.

[159] Doob, J. L. (1949) Heuristic approach to the Kolmogorov-Smirnov theorems. *Ann. Math. Statist.* 20, 393–403.

[160] Doytchinov, B., Lehoczky, J. and Shreve, S. (2001) Real-time queues in heavy traffic with earliest-deadline-first queue discipline. *Ann. Appl. Prob.* 11, 332–378.

[161] Dudley, R. M. (1966) Weak convergence of probabilities on non-separable metric spaces and empirical measures on Euclidean spaces. *Ill. J. Math.* 10, 109–126.

[162] Dudley, R. M. (1967) Measures on non-separable metric spaces. *Ill. J. Math.* 11, 449–453.

[163] Dudley, R. M. (1968) Distances of probability measures and random variables. *Ann. Math. Statist.* 39, 1563–1572.

[164] Duffield, N. G. (1997) Exponents for the tails of distributions in some polling models. *Queueing Systems* 26, 105–119.

[165] Duffield, N. G., Massey, W. A. and Whitt, W. (2001) A nonstationary offered-load model for packet networks. *Telecommunication Systems* 16, 271–296.

[166] Duffield, N. G. and O'Connell (1995) Large deviations and overflow probabilities for the general single-server queue, with applications. *Math. Proc. Camb. Phil. Soc.* 118, 363–374.

[167] Duffield, N. G. and Whitt, W. (1997) Control and recovery from rare congestion events in a large multi-server system. *Queueing Systems* 26, 69–104.

[168] Duffield, N. G. and Whitt, W. (1998) A source traffic model and its transient analysis for network control. *Stochastic Models* 14, 51–78.

[169] Duffield, N. G. and Whitt, W. (2000) Network design and control using on-off and multi-level source traffic models with heavy-tailed distributions. *Self-Similar Network Traffic and Performance Evaluation*, K. Park and W. Willinger (eds.), Wiley, Boston, 421–445.

[170] Dugundji, J. (1966) *Topology*, Allyn and Bacon, Boston.

[171] Dupuis, P. and Ishii, H. (1991) On when the solution to the Skorohod problem is Lipschitz continuous with applications. *Stochastics* 35, 31–62.

[172] Dupuis, P. and Ramanan, K. (1999) Convex duality and the Skorohod problem - I. *Prob. Theor. Rel. Fields*, 115, 153–195.

[173] Dupuis, P. and Ramanan, K. (1999) Convex duality and the Skorohod problem - II. *Prob. Theor. Rel. Fields*, 115, 197–236.

[174] Eberlein, E. and Taqqu, M. S. (1986) *Dependence in Probability and Statistics*, Birkhauser, Boston.

[175] Eckberg, A. E. (1983) Generalized peakedness of teletraffic processes. *Proceedings of the Tenth Intern. Teletraffic Congress*, Montreal, Canada, June 1983, paper 4.4b.3.

[176] Eckberg, A. E. (1985) Approximations for bursty (and smooth) arrival queueing delays based on generalized peakedness. *Proceedings Eleventh Int. Teletraffic Congress*, Kyoto, Japan.

[177] Edwards, R. E. (1965) *Functional Analysis*, Holt, Rinehart and Winston, New York.

[178] Eick, S. G., Massey, W. A. and Whitt, W. (1993) The physics of the $M_t/G/\infty$ queue. *Operations Res.* 41, 731–742.

[179] El-Taha, M. and Stidham, S., Jr. (1999) *Sample-Path Analysis of Queueing Systems*, Kluwer, Boston.

[180] Embrechts, P., Klüppelberg, C. and Mikosch, T. (1997) *Modelling Extremal Events*, Springer, New York.

[181] Erdős, P. and Kac, M. (1946) On certain limit theorems in the theory of probability. *Bull. Amer. Math. Soc.* 52, 292–302.

[182] Erdős, P. and Kac, M. (1947) On the number of positive sums of independent random variables. *Bull. Amer. Math. Soc.* 53, 1011–1020.

[183] Ethier, S. N. and Kurtz, T. G. (1986) *Markov Processes, Characterization and Convergence*, Wiley, New York.

[184] Feldmann, A., Greenberg, A. G., Lund, C. Reingold, N. and Rexford, J. (2000) Netscope: traffic engineering for IP networks. *IEEE Network Magazine*, special issue on Internet traffic engineering, March-April, 11–19.

[185] Feldmann, A., Greenberg, A. G., Lund, C. Reingold, N., Rexford, J. and True, F. (2001) Deriving traffic demands for operational IP networks: methodology and experience. *IEEE/ACM Trans. Networking* 9, to appear.

[186] Feldmann, A. and Whitt, W. (1998) Fitting mixtures of exponentials to long-tail distributions to analyze network performance models. *Performance Evaluation* 31, 245–279.

[187] Feller, W. (1968) An Introduction to Probability Theory and its Applications, vol. I, third edition, Wiley, New York.

[188] Feller, W. (1971) *An Introduction to Probability Theory and its Applications*, vol. II, second edition, Wiley, New York.

[189] Fendick, K. W. and Rodrigues, M. A. (1991) A heavy traffic comparison of shared and segregated buffer schemes for queues with head-of-the-line processor sharing discipline. *Queueing Systems* 9, 163–190.

[190] Fendick, K. W., Saksena, V. R. and Whitt, W. (1989) Dependence in packet queues. *IEEE Trans. Commun.* 37, 1173–1183.

[191] Fendick, K. W., Saksena, V. R. and Whitt, W. (1991) Investigating dependence in packet queues with the index of dispersion for work. *IEEE Trans. Commun.* 39, 1231–1243.

[192] Fendick, K. W. and Whitt, W. (1989) Measurements and approximations to describe the offered traffic and predict the average workload in a single-server queue. *Proceedings IEEE* 77, 171–194.

[193] Fendick, K. W. and Whitt, W. (1998) Verifying cell loss requirements in high-speed communication netyworks. *J. Appl. Math. Stoch. Anal.* 11, 319–338.

[194] Flatto, L. (1997) The waiting time distribution for the random order of service $M/M/1$ queue. *Ann. Appl. Prob.* 7, 382–409.

[195] Fleming, P. J. and Simon, B. (1999) Heavy traffic approximations for a system of infinite servers with load balancing. *Prob. Eng. Inf. Sci.* 13, 251–273.

[196] Foschini, G. J. and Salz, J. (1978) A basic dynamic routing problem and diffusion. *IEEE Trans. Commun.* 26, 320–327.

[197] Freidlin, M. I. and Wentzell, A. D. (1993) Diffusion processes on graphs and the averaging principle. *Ann. Probab.* 21, 2215–2245.

[198] Furrer, H., Michna, Z. and Weron, A. (1997) Stable Lévy motion approximation in collective risk theory. *Insurance: Math and Econ.*, 20, 97–114.

[199] Garnett, O., Mandelbaum, A. and Reiman, M. I. (2000) Designing a call center with impatient customers. Bell Laboratories, Murray Hill, NJ.

[200] Garrett, M. W. and Willinger, W. (1994) Analysis, modeling and generation of self-similar v.b.r. video traffic. *Proceedings ACM SIGCOMM, London*, 269–280.

[201] Glasserman, P. and Yao, D. D. (1994) *Monotone Structure in Discrete-Event Systems*, Wiley, New York.

[202] Glynn, P. W. (1990) Diffusion Approximations, Chapter 4 in *Stochastic Models*, D. P. Heyman and M. J. Sobel (eds.), North-Holland, Amsterdam, 145–198.

[203] Glynn, P. W. and Whitt, W. (1988) Ordinary CLT and WLLN versions of $L = \lambda W$. *Math. Oper. Res.* 13, 674–692.

[204] Glynn, P. W. and Whitt, W. (1989) Indirect estimation via $L = \lambda W$. *Operations Res.* 37, 82–103.

[205] Glynn, P. W. and Whitt, W. (1991) A new view of the heavy-traffic limit theorem for the infinite-server queue. *Adv. Appl. Prob.* 23, 188–209.

[206] Glynn, P. W. and Whitt, W. (1994) Logarithmic asymptotics for steady-state tail probabilities in a single-server queue. *Studies in Applied Probability, Papers in honor of Lajos Takács*, J. Galambos and J. Gani (eds.), Applied Probability Trust, Sheffield, U.K., 131–156.

[207] Glynn, P. W. and Whitt, W. (1995) Heavy-traffic extreme-value limits for queues. *Operations Res. Letters* 18, 107–111.

[208] Gnedenko, B. V. and Kolmogorov, A. N. (1968) *Limit Distributions for Sums of Independent Random Variables*, revised edition, Addison Wesley, Reading, MA.

[209] Gonzáles-Arévalo, B. and Samorodnitsky, G. (2001) Buffer content of a leaky bucket system with long-range dependent input traffic. School of Operations Research and Industrial Engineering, Cornell University, Ithaca, New York.

[210] Greenberg, A. G. and Madras, N. (1992) How fair is fair queueing? *Journal of ACM* 39, 568–598.

[211] Greenberg, A. G., Srikant, R. and Whitt, W. (1999) Resource sharing for book-ahead and instantaneous-request calls. *IEEE/ACM Trans. Networking* 7, 10–22.

[212] Grishechkin, S. (1994) GI/G/1 processor sharing queue in heavy traffic. *Adv. Appl. Prob.* 26, 539–555.

[213] Grossglauser, M. and Bolot, J.-C. (1999) On the relevance of long-range dependence in network traffic. *IEEE/ACM Trans. Networking* 7, 629–640.

[214] Grossglauser, M. and Tse, D. N. C. (1999) A time-scale decomposition approach to measurement-based admission control. *Proceedings IEEE INFOCOM '99.*

[215] Guerin, C. A., Nyberg, H., Perrin, O., Resnick, S., Rootzen, H. and Stărică, C. (2000) Empirical testing of the infinite source Poisson data traffic model. Technical Report 1257, School of ORIE, Cornell University, Ithaca, NY.

[216] Gunther, N. J. (1998) *The Practical Performance Analyst*, McGraw-Hill, New York.

[217] Gut, A. (1988) *Stopped Random Walks*, Springer-Verlag, New York.

[218] Hahn, M. G. (1978) Central limit theorems in $D[0,1]$. *Zeitschrift für Wahrscheinlichkeitstheorie verw. Gebiete* 44, 89–101.

[219] Halfin, S. and Whitt, W. (1981) Heavy-traffic limits for queues with many exponential servers. *Operations Res.* 29, 567–588.

[220] Hall, R. W. (1991) *Queueing Methods for Services and Manufacturing*, Prentice Hall, Englewood Cliffs, New Jersey.

[221] Harchol-Balter, M. and Downey, A. (1997) Exploiting process lifetime distributions for dynamic load balancing. *ACM Transactions on Computer Systems* 15, 253–285.

[222] Harrison, J. M. (1978) The diffusion approximation for tandem queues in heavy traffic. *Adv. Appl. Prob.* 10, 886–905.

[223] Harrison, J. M. (1985) *Brownian Motion and Stochastic Flow Systems*, Wiley, New York.

[224] Harrison, J. M. (1988) Brownian models of queueing networks with heterogeneous populations. In *Stochastic Differential Systems, Stochastic Control Theory and Applications*, W. Fleming and P. L. Lions (eds.), Springer, New York, 147–186.

[225] Harrison, J. M. (2000) Brownian models of open processing networks: canonical representation of workload. *Ann. Appl. Prob.* 10, 75–103.

[226] Harrison, J. M. (2001a) A broader view of Brownian networks. *Ann. Appl. Prob.*, to appear.

[227] Harrison, J. M. (2001b) Stochastic networks and activity analysis. Graduate School of Business, Stanford University.

[228] Harrison, J. M. and Lopez, M. J. (1999) Heavy traffic resource pooling in parallel server systems. *Queueing Systems* 33, 339–368.

[229] Harrison, J. M. and Nguyen, V. (1990) The QNET method for two-moment analysis of open queueing networks. *Queueing Systems* 6, 1–32.

[230] Harrison, J. M. and Reiman, M. I. (1981a) Reflected Brownian motion in an orthant. *Ann. Probab.* 9, 302–308.

[231] Harrison, J. M. and Reiman, M. I. (1981b) On the distribution of multidimensional reflected Brownian motion. *SIAM J. Appl. Math.* 41, 345–361.

[232] Harrison, J. M. and van Mieghem, J. A. (1997) Dynamic control of Brownian networks: state space collapse and equivalent workload formulations. *Ann. Appl. Prob.* 7, 747–771.

[233] Harrison, J. M. and Williams, R. J. (1987) Brownian models of open queueing networks with homogeneous customer population. *Stochastics* 22, 77–115.

[234] Harrison, J. M. and Williams, R. J. (1996) A multiclass closed queueing network with unconventional heavy traffic behavior. *Ann. Appl. Prob.* 6, 1–47.

[235] Harrison, J. M., Williams, R. J. and Chen, H. (1990) Brownian models of closed queueing network with homogeneous customer populations. *Stochastics and Stochastic Reports* 29, 37–74.

[236] Heyman, D. P. and Lakshman, T. V. (1996) What are the implications of long-range dependence for traffic engineering? *IEEE/ACM Trans. Networking* 4, 310–317.

[237] Heyman, D. P. and Lakshman, T. V. (2000) Long-range dependence and queueing effects for VBR video. Chapter 12 in *Self-Similar Network Traffic and Performance Evaluation*, K. Park and W. Willinger (eds.), Wiley, New York, 285–318.

[238] Hida, T. and Hitsuda, M. (1976) *Gaussian Processes*, American Math Society, Providence, RI.

[239] Hjálmtýsson, G. and Whitt, W. (1998) Periodic load balancing. *Queueing Systems* 30, 203–250.

[240] Hopp, W. J. and Spearman, M. L. (1996) *Factory Physics*, Irwin, Chicago.

[241] Hsing, T. (1999) On the asymptotic distributions of partial sums of functionals of infinite-variance moving averages. *Ann. Probab.* 27, 1579–1599.

[242] Hui, J. T. (1988) Resource allocation for broadband networks. IEEE J. Sel. Areas Commun. SAC-6, 1598–1608.

[243] Hunt, P. J. and Kelly, F. P. (1989) On critically loaded loss networks. *Adv. Appl. Prob.* 21, 831–841.

[244] Hurst, H. E. (1951) Long-term storage capacity of reservoirs. *Trans. Am. Soc. Civil Engineers* 116, 770–799.

[245] Hurst, H. E. (1955) Methods of using long-term storage in reservoirs. *Proc. Inst. Civil Engr.*, Part I, 519–577.

[246] Hüsler, J. and Piterbarg, V. (1999) Extremes of a certain class of Gaussian processes. *Stoch. Proc. Appl.* 83, 257–271.

[247] Iglehart, D. L. (1965) Limit diffusion approximations for the many server queue and the repairman problem. *J. Appl. Prob.* 2, 429–441.

[248] Iglehart, D. L. and W. Whitt (1970a) Multiple channel queues in heavy traffic, I. *Adv. Appl. Prob.* 2, 150–177.

[249] Iglehart, D. L. and W. Whitt (1970b) Multiple channel queues in heavy traffic, II. *Adv. Appl. Prob.* 2, 355–369.

[250] Iglehart, D. L. and W. Whitt (1971) The equivalence of functional central limit theorems for counting processes and associated partial sums. *Ann. Math. Statist.* 42, 1372–1378.

[251] Ivanoff, B. G. and Merzbach, E. (2000) A Skorohod topology for a class of set-indexed functions. *Skorohod's Ideas in Probability Theory*, V. Korolyuk, N. Portenko and H. Syta (eds.), Institute of Mathematics of the National Academy of Sciences of the Ukraine, Kyiv, Ukraine, 172–178.

[252] Jackson, J. R. (1957) Networks of waiting lines. *Operations Res.* 5, 518–521.

[253] Jackson, J. R. (1963) Jobshop-like queueing systems. *Management Sci.* 10, 131–142.

[254] Jacod, J. and Shiryaev, A. N. (1987) *Limit Theorems for Stochastic Processes*, Springer-Verlag, New York.

[255] Jagerman, D. L. (1974) Some properties of the Erlang loss function. *Bell System tech. J.* 53, 525–551.

[256] Jain, R. (1991) *The Art of Computer System Performance Analysis: Techniques for Experimental Desugn, Measurement, Simulation and Modeling*, Wiley, New York.

[257] Jakubowski, A. (1996) Convergence in various topologies for stochastic integrals driven by semimartingales. *Ann. Probab.* 24, 2141–2153.

[258] Janicki, A. and Weron, A. (1993) *Simulation and Chaotic Behavior of α-Stable Stochastic Processes*, Marcel Dekker, New York.

[259] Jelenković, P. (1999) Subexponential loss rates in a $GI/GI/1$ queue with applications. *Queueing Systems* 33, 91–123.

[260] Jelenković, P. (2000) Asymptotic results for queues with subexponential arrivals. Chapter 6 in *Self-Similar Network Traffic and Performance Evaluation*, K. Park and W. Willinger (eds.), Wiley, New York.

[261] Jelenković, P. R., Lazar, A. A. and Semret, N. (1997) The effect of multiple time scales and subexponentiality in MPEG video streams on queue behavior. *IEEE J. Selected Areas Commun.* 15, 1052–1071.

[262] Jennings, O. B., Mandelbaum, A., Massey, W. A. and Whitt, W. (1996) Server staffing to meet time varying demand. *Management Sci.* 42, 1383–1394.

[263] Johnson, N. L. and Kotz, S. (1969) *Distributions in Statistics – Discrete Distributions*, Wiley, New York.

[264] Johnson, N. L. and Kotz, S. (1970) *Distributions in Statistics – Continuous Univariate Distributions – I*, Wiley, New York.

[265] Karatzas, I. and Shreve, S. E. (1988) *Brownian Motion and Stochastic Calculus*, Springer-Verlag, New York.

[266] Karatzas, I. and Shreve, S. E. (1998) *Methods of mathematical Finance*, Springer-Verlag, New York.

[267] Karlin, S. and Taylor, H. M. (1981) *A Second Course in Stochastic Processes*, Academic Press, New York.

[268] Karpelovich, F. I. and Kreinin, A. Ya. (1994) *Heavy Traffic Limits for Multiphase Queues*, American Math. Society, Providence, RI.

[269] Kasahara, Y. and Maejima, M. (1986) Functional limit theorems for weighted sums of i.i.d. random variables. *Prob. Th. Rel. Fields* 72, 161–183.

[270] Kella, O. (1993) Parallel and tandem fluid networks with dependent Lévy inputs. *Ann. Appl. Prob.* 3, 682–695.

[271] Kella, O. (1996) Stability and non-product form of stochastic fluid networks with Lévy inputs. *Ann. Appl. Prob.* 6, 186–199.

[272] Kella, O. and Whitt, W. (1990) Diffusion approximations for queues with server vacations. *Adv. Appl. Prob.* 22, 706–729.

[273] Kella, O. and Whitt, W. (1991) Queues with server vacations and Lévy processes with secondary jump input. *Ann. Appl. Prob.* 1, 104–117.

[274] Kella, O. and Whitt, W. (1992a) A tandem fluid network with Lévy input. In *Queues and Related Models*, I. Basawa and N. V. Bhat (eds.), Oxford University Press, Oxford, 112–128.

[275] Kella, O. and Whitt, W. (1992b) A storage model with a two-state random environment. *Operations Res.* 40 (Supplement 2), S257–S262.

[276] Kella, O. and Whitt, W. (1992c) Useful martingales for stochastic storage processes with Lévy input, *J. Appl. Prob.* 29, 396–403.

[277] Kella, O. and Whitt, W. (1996) Stability and structural properties of stochastic storage networks. *J. Appl. Prob.* 33, 1169–1180.

[278] Kelly, F. P. (1979) *Reversibility and Stochastic Networks*, Wiley, Chichester, U. K.

[279] Kelly, F. P. (1991) Loss networks. *Ann. Appl. Prob.* 1, 319–378.

[280] Kelly, F. P. (1996) Notes on effective bandwidths. *Stochastic Networks, Theory and Applications*, Clarendon Press, Oxford, U.K., 141–168.

[281] Kelly, F. P. and Laws, C. N. (1993) Dynamic routing in open queueing networks: Brownian models, cut constraints and resource pooling. *Queueing Systems* 13, 47–86.

[282] Kelly, F. P. and Williams, R. J. (1995) *Stochastic Networks*, IMA Volumes in Mathematics and its Applications 71, Springer-Verlag, New York.

[283] Kelly, F. P., Zachary, S. and Ziedins, I. (1996) *Stochastic Networks, Theory and Application*, Clarendon Press, Oxford, U.K.

[284] Kemeny, J. G. and Snell, J. L. (1960) *Finite Markov Chains*, Van Nostrand, Princeton.

[285] Kennedy, D. (1973) Limit theorems for finite dams. *Stoch. Process Appl.* 1, 269–278.

[286] Kingman, J. F. C. (1961) The single server queue in heavy traffic. *Proc. Camb. Phil. Soc.* 57, 902–904.

[287] Kingman, J. F. C. (1962) On queues in heavy traffic. *J. Roy. Statist. Soc.* Ser B, 24, 383–392.

[288] Kingman, J. F. C. (1965) The heavy traffic approximation in the theory of queues. In *Proceedings of the Symposium on Congestion Theory*, W. L. Smith and W. E. Wilkinson (eds.), University of North Carolina Press, Chapter 6, 137–159.

[289] Kingman, J. F. C. (1982) Queue disciplines in heavy traffic. *Math. Oper. Res.* 7, 262–271.

[290] Kleinrock, L. (1975) *Queueing Systems, I*, Wiley, new York.

[291] Kleinrock, L. (1976) *Queueing Systems, II: Computer Applications*, Wiley, new York.

[292] Klincewicz, J. G. and Whitt, W. (1984) On approximations for queues, II: shape constraints. *AT&T Bell Labs. Tech. J.* 63, 139–161.

[293] Knessl, C. (1999) A new heavy traffic limit for the asymmetric shortest queue problem. *Eur. J. Appl. Math.* 10, 497–509.

[294] Knessl, C. and Morrison, J. A. (1991) Heavy traffic analysis of data handling system with multiple sources. *SIAM J. Appl. Math.* 51, 187–213.

[295] Knessl, C. and Tier, C. (1995) Applications of singular perturbation methods in queueing. Chapter 12 in *Advances in Queueing*, J. H. Dshalalow (ed.), CRC Press, Boca Raton, FL, 311–336.

[296] Knessl, C. and Tier, C. (1998) Heavy traffic analysis of a Markov-modulated queue with finite capacity and general service times. *SIAM J. Appl. Math.* 58, 257–323.

[297] Kokoszka, P. and Taqqu, M. S. (1995) Fractional ARIMA with stable innovations. *Stoch. Proc. Appl.* 60, 19–47.

[298] Kokoszka, P. and Taqqu, M. S. (1996a) Parameter estimation for infinite variance fractional ARIMA. *Ann. Statist.* 24, 1880–1913.

[299] Kokoszka, P. and Taqqu, M. S. (1996b) Infinite variance stable moving averages with long memory. *J. Econometrics* 73, 79–99.

[300] Kolmogorov, A. N. (1950) *Foundations of the Theory of Probability*, second English edition of 1933 German original, Chelsea, New York.

[301] Kolmogorov, A. N. (1956) On Skorohod convergence. *Theor. Probability Appl.* 1, 213–222. (also republished in *Skorohod's Ideas in Probability Theory*, V. Korolyuk, N. Portenko and H. Syta (eds.), institute of Mathematics of the National Acafdemy of Sciences of the Ukraine, Kyiv, Ukraine, 99–110.

[302] Konstantopoulos, T. (1999) The Skorohod reflection problem for functions with discontinuities (contractive case). University of Texas at Austin.

[303] Konstantopoulos, T. and S.-J. Lin (1996) Fractional Brownian motion as limits of stochastic traffic models. *Proc. 34th Allerton Conference on Communication, Control and Computing*, University of Illinois, 913–922.

[304] Konstantopoulos, T. and S.-J. Lin (1998) Macroscopic models for long-range dependent network traffic. *Queueing Systems* 28, 215–243.

[305] Korolyuk, V., Portenko, N. and Syta, H. (2000) *Skorohod's Ideas in Probability Theory*, Proceedings of the Institute of Mathematics of the National Academy of Sciences of Ukraine, vol. 32, Kyiv, Ukraine.

[306] Kraemer, W. and Langenbach-Belz, M. (1976) Approximate formulae for the delay in the queueing system $GI/G/1$. *Proceedings Eighth Int. Teletraffic Congress, Melbourne*, 235, 1–8.

[307] Krichagina, E. V. and Puhalskii, A. A. (1997) A heavy-traffic analysis of a closed queueing system with a GI/∞ service center. *Queueing Systems* 25, 235–280.

[308] Krishnamurthy, B. and Rexford, J. (2001) *Web Protocols and Practice*, Addison-Wesley, Boston.

[309] Kruk, L., Lehoczky, J., Shreve, S. and Yeung, S.-N. (2000) Multiple-input heavy-traffic real-time queues. Carnegie Mellon University.

[310] Kuehn, P. J. (1979) Approximate analysis of general queueing networks by decomposition. *IEEE Trans. Commun.* 27, 113–126.

[311] Kumar, S. (2000) Two-server closed networks in heavy traffic: diffusion limits and asymptotic optimality. *Ann. Appl. Prob.* 10, 930–961.

[312] Kunniyur, S. and Srikant, S. (2001) A time-scale decomposition approach to adaptive ECN marking. Coordinated Science Laboratory, University of Illinois.

[313] Kurtz, T. G. (1996) Limit theorems for workload input models. In *Stochastic Networks: Theory and Applications*, F. P. Kelly, S. Zachary and I. Ziedins (eds.) Oxford University Press, Oxford, U.K., 119–139.

[314] Kurtz, T. G. and Protter, P. (1991) Weak limit theorems for stochastic integrals and stochastic difference equations. *Ann. Probab.* 19, 1035–1070.

[315] Kushner, H. J. (2001) *Heavy Traffic Analysis of Controlled Queueing and Communication Networks*, Springer, New York.

[316] Kushner, H. J. and Dupuis, P. G. (2000) *Numerical Methods for Stochastic Control Problems in Continuous Time*, second edition, Springer, New York.

[317] Kushner, H. J. and Martins, L. F. (1993) Heavy traffic analysis of a data transmission system with many independent sources. *SIAM J. Appl. Math.* 53, 1095–1122.

[318] Kushner, H. J. and Martins, L. F. (1994) Numerical methods for controlled and uncontrolled multiplexing and queueing systems. *Queueing Systems* 16, 241–285.

[319] Kushner, H. J., Yang, J. and Jarvis, D. (1995) Controlled and optimally controlled multiplexing systems: a numerical exploration. *Queueing Systems* 20, 255–291.

[320] Lamperti, J. W. (1962) Semi-stable stochastic processes. *Trans. Amer. Math. Soc.* 104, 62–78.

[321] Laws, C. N. (1992) Resource pooling in queueing networks with dynamic routing. *Adv. Appl. Prob.* 24, 699–726.

[322] L'Ecuyer, P. (1998a) Random number generation. Chapter 4 in The Handbook of Simulation, J. Banks, ed., Wiley, New York, Chapter 4, 93–137.

[323] L'Ecuyer, P. (1998b) Uniform random number generation. Encyclopedia of Operations Research and Management Science, S. I. Gass and C. M. Harris, eds., Kluwer, Boston, 323–339.

[324] Leland, W. E., Taqqu, M. S., Willinger, W. and Wilson, D. V. (1994) On the self-similar nature of Ethernet traffic. *IEEE/ACM Trans. Networking* 2, 1-15.

[325] Leung, K. K., Massey, W. A. and Whitt, W. (1994) Traffic models for wireless communication networks. *IEEE J. Sel. Areas Commun.* 12, 1353-1364.

[326] Lévy, P. (1939) Sur certaines processus stochastiques homogènes. *Compos. Math.* 7, 283–339.

[327] Levy, J. and Taqqu, M. S. (1987) On renewal processes having stable inter-renewal intervals and stable rewards. *Ann. Sci. Math. Québec* 11, 95–110.

[328] Levy, J. B. and Taqqu, M. S. (2000) Renewal reward processes with heavy-tailed inter-renewal times and heavy-tailed rewards. *Bernoulli* 6, 23–44.

[329] Likhanov, N. and Mazumdar, R. R. (2000) Loss asymptotics in large buffers fed by heterogeneous long-tailed sources. *Adv. Appl. Prob.* 32, 1168–1189.

[330] Limic, V. (1999) A LIFO queue in heavy traffic. Dept. Math., University of California at San Diego. *Ann. Appl. Prob.* 11, 301–331.

[331] Lindvall, T. (1973) Weak convergence in the function space $D[0,\infty)$. *J. Appl. Prob.* 10, 109–121.

[332] Lions, P.-L. and Sznitman, A.-S. (1984) Stochastic differential equations with reflecting boundary conditions. *Commun. Pure Appl. Math.* 37, 511–553.

[333] Loulou, R. (1973) Multi-channel queues in heavy traffic. *J. Appl. Prob.* 10, 769–777.

[334] Mairesse, J. and Prabhakar, B. (2000) On the existence of fixed points for the $\cdot/GI/1/\infty$ queue. Stanford University.

[335] Malkiel, B. G. (1996) *A Random Walk Down Wall Street*, W. W. Norton, New York.

[336] Majewski, K. (2000) Path-wise heavy traffic convergence of single class queueing networks and consequences. Siemens, Germany.

[337] Mandelbaum, A. and Massey, W. A. (1995) Strong approximations for time-dependent queues. *Math. Oper. Res.* 20, 33–64.

[338] Mandelbaum, A., Massey, W. A. and Reiman, M. I. (1998) Strong approximations for Markovian service networks. *Queueing Systems* 30, 149–201.

[339] Mandelbaum, A., Massey, W. A., Reiman, M. I. and Stolyar, A. (1999) Waiting time asymptotics for time varying multiserver queues with abandonments and retrials. Proceedings of 37th Allerton Conference on Communication, Control and Computing, September 1999, 1095–1104.

[340] Mandelbaum, A. and Pats, G. (1995) State-dependent queues: approximations and applications. *Stochastic Networks*, IMA Volumes in Mathematics and its Applications, F. P. Kelly and R. J. Williams, eds., Springer, Berlin, 239—282.

[341] Mandelbaum, A. and Pats, G. (1998) State-dependent stochastic networks, Part I: approximations and applications with continuous diffusion limits. *Ann. Appl. Prob.* 8, 569–646.

[342] Mandelbrot, B. B. (1977) *Fractals: Form, Chance and Dimension*, W. H. Freeman and Co., San Francisco.

[343] Mandelbrot, B. B. (1982) *The Fractal Geometry of Nature.* W. H. Freeman and Co., San Francisco.

[344] Mandelbrot, B. B. and Wallis, J. R. (1968) Noah, Joseph and operational hydrology. *Water Resources Research* 4, 909–918.

[345] Markowitz, D. M. and Wein, L. M. (2001) Heavy traffic analysis of dynamic cyclic policies: a unified treatment of the single machine scheduling problem. *Operations Res.* 49, 246–270.

[346] Markowitz, D. M., Reiman, M. I. and Wein, L. M. (2000) The stochastic economic lot scheduling problem: heavy traffic analysis of dynamic cyclic policies. *Operations Res.* 48, 136–154.

[347] Massey, W. A. and Whitt, W. (1993) Networks of infinite-server queues with nonstationary Poisson input. *Queueing Systems* 13, 183–250.

[348] Massey, W. A. and Whitt, W. (1994a) Unstable asymptotics for nonstationary queues. *Math. Oper. Res.* 19, 267–291.

[349] Massey, W. A. and Whitt, W. (1994b) A stochastic model to capture space and time dynamics in wireless communication systems. *Prob. Eng. Inf. Sci.* 8, 541–569.

[350] Massey, W. A. and Whitt, W. (1994c) An analysis of the modified offered load approximation for the nonstationary Erlang loss model. *Ann. Appl. Prob.* 4, 1145–1160.

[351] Massoulie, L. and Simonian, A. (1999) Large buffer asymptotics for the queue with fractional Brownian input. *J. Appl. Prob.* 36, 894–906.

[352] Matheron, G. (1975) *Random Sets and Integral Geometry*, Wiley, New York.

[353] McDonald, D. R. and Turner, S. R. E. (2000) *Analysis of Communication Networks: Call Centres, Traffic and Performance*, Fields Institute Communications 28, The American Math. Soc., Providence, RI.

[354] Meyn, S. P. and Down, D. (1994) Stability of generalized Jackson networks. *Ann. Appl. Prob.* 4, 124–148.

[355] Mikosch, T. and Nagaev, A. V. (2001) Rates in approximations to ruin probabilities for heavy-tailed distributions. *Extremes* 4, 67–78.

[356] Mikosch, T., Resnick, S. I., Rootzén, H. and Stegeman, A. (2001) Is network traffic approximated by stable Lévy motion or fractional Brownian motion? *Ann. Appl. Prob.* 11, to appear.

[357] Mikosch, T. and Samorodnitsky, G. (2000) The supremum of a negative drift random walk with dependent heavy-tailed steps. *Ann. Appl. Prob.* 10, 1025–1064.

[358] Mitzenmacher, M. (1996) *The Power of Two Choices in Randomized Load Balancing*, Ph.D. dissertation, University of California, Berkeley.

[359] Mitzenmacher, M. (1997) How useful is old information? *Proceedings of* 16th *ACM Symposium on Principles of Distributed Computing*, 83–91.

[360] Mitzenmacher, M. and Vöcking, B. (1999) The asymptotics of selecting the shortest of two, improved. *Proceedings of 1999 Allerton Conference on Communication, Control and Computing*, University of Illinois.

[361] Montanari, A., Rosso, R. and Taqqu, M. S. (1997) Fractionally differenced ARIMA models applied to hydrologic time series: identification, estimation and simulation. *Water Resources Research* 33, 1035–1044.

[362] Moran, P. A. P. (1959) *The Theory of Storage*, Methuen, London.

[363] Mountford, T. and Prabhakar, B. (1995) On the weak convergence of departures from an infinite sequence of $\cdot/M/1$ queues. *Ann. Appl. Prob.* 5, 121–127.

[364] Müller, D. W. (1968) Verteilungs-Invarianzprinzipien für das starke Gesetz der grossen Zahl. *Zeitschrift Wahrscheinlichkeitsth. verw. Gebiete* 10, 173–192.

[365] Narayan, O. (1998) Exact asymptotic queue length distribution from fractional Brownian motion traffic. *Adv. Performance Anal.* 1, 39–64.

[366] Nelson, B. L. and Taaffe, M. R. (2001) The $Ph_t/Ph_t/\infty$ queueing system. *INFORMS J. Computing*, to appear.

[367] Neuts, M. F. (1989) *Structured Stochastic Matrices of M/G/1 Type and Their Applications*, Marcel Dekker, New York.

[368] Neveu, J. (1965) *Mathematical Foundations of the Calculus of Probability*, Holden-Day, San Francisco.

[369] Newell, G. F. (1982) *Applications of Queueing Theory*, second ed., Chapman and Hall, London.

[370] Newman, C. M. and Wright, A. L. (1981) An invariance principle for certain dependent sequences. *Ann. Probab.* 9, 671–675.

[371] Norros, I. (1994) A storage model with self-similar input. *Queueing Systems* 16 (1994) 387–396.

[372] Norros, I. (2000) Queueing behavior under fractional Brownian motion. Chapter 4 in *Self-Similar Network Traffic and Performance Evaluation*, K. Park and W. Willinger (eds.), Wiley, New York.

[373] O'Brien, G. L. (2000) An application of Skorohod's M_1 topology. *Skorohod's Ideas in Probability Theory*, V. Korolyuk, N. Portenko and H. Syta (eds.), institute of Mathematics of the National Acafdemy of Sciences of the Ukraine, Kyiv, Ukraine, 111–118.

[374] O'Brien, G. L. and Vervaat, W. (1985) Self-similar processes with stationary increments generated by point processes. *Ann. Probab.* 13, 28–52.

[375] Oliver, R. M. and Samuel, A. H. (1962) Reducing letter delays in post offices. *Operations Res.* 10, 839–892.

[376] Olsen, T. L. (2001) Limit theorems for polling models with increasing setups. *Stochastic Models* 15, 35–56.

[377] Olver, F. W. J. (1974) *Asymptotics and Special Functions*, Academic, New York.

[378] Parekh, A. K. and Gallager, R. G. (1993) A generalized processor sharing approach to flow control in integrated services networks: the single node case. *IEEE/ACM Trans. Networking* 1, 344–357.

[379] Parekh, A. K. and Gallager, R. G. (1994) A generalized processor sharing approach to flow control in integrated services networks: the multiple node case. *IEEE/ACM Trans. Networking* 2, 137–150.

[380] Park, K. and Willinger, W. (2000) *Self-Similar Network Traffic and Performance Evaluation*, Wiley, New York.

[381] Parthasarathy, K. R. (1967) *Probability Measures on Metric Spaces*, Academic, New York.

[382] Pats, G. (1994) *State-Dependent Queueing Networks: Approximations and Applications*, Ph.D. thesis, Faculty of Industrial Engineering and Management, Technion, Haifa, Israel.

[383] Paxson, V. and Floyd, S. (1995) Wide area traffic: the failure of Poisson modeling. *IEEE/ACM Trans. Networking* 3, 226–244.

[384] Padhye, J., Firoiu, V., Towsley, D. and Kurose, J. (2000) Modeling TCP Reno throughput: a simple model and its empirical validation. *IEEE/ACM Trans. Networking* 8, 133–145.

[385] Philipp, W. and Stout, W. (1975) *Almost Sure Invariance Principles for Partial Sums of Weakly Dependent Random Variables*, Mem. Amer. Math. Soc. **161**, Providence, RI.

[386] Pollard, D. (1984) *Convergence of Stochastic Processes*, Springer-Verlag, New York.

[387] Pomarede, J. L. (1976) *A Unified Approach Via Graphs to Skorohod's Topologies on the Function Space D*, Ph.D. dissertation, Department of Statistics, Yale University.

[388] Prabhu, N. U. (1998) *Stochastic Storage Processes*, second ed. Springer, New York.

[389] Prohorov, Yu. V. (1956) Convergence of random processes and limit theorems in probability. *Theor. Probability Appl.* 1, 157–214.

[390] Protter, P. (1992) *Stochastic Integration and Differential Equations*, Springer, New York.

[391] Puhalskii, A. A. (1994) On the invariance principle for the first passage time. *Math. Oper. Res.* 19, 946–954.

[392] Puhalskii, A. A. (1999) Moderate deviations for queues in critical loading. *Queueing Systems* 31, 359–392.

[393] Puhalskii, A. A. and Reiman, M. I. (2000) The multiclass $GI/PH/N$ queue in the Halfin-Whitt regime. *Adv. Appl. Prob.* 32, 564–595.

[394] Puhalskii, A. A. and Whitt, W. (1997) Functional large deviation principles for first-passage-time processes. *Ann. Appl. Prob.* 7, 362–381.

[395] Puhalskii, A. A. and Whitt, W. (1998) Functional large deviation principles for waiting and departure processes. *Prob. Eng. Inf. Sci.* 12, 479–507.

[396] Puterman, M. L. (1994) *Markov Decsion Processes: Discrete Stochastic Dynamic Programming*, Wiley, New York.

[397] Rachev, S. and Mittnik, S. (2000) *Stable Paretian Models in Finance*, Wiley, New York.

[398] Reiman, M. I. (1982) The heavy traffic diffusion approximation for sojourn times in Jackson networks. In *Applied Probability – Computer Science, the Interface, II*, R. L. Disney and T. J. Ott (eds.), Birkhauser, Boston, 409–422.

[399] Reiman, M. I. (1984a) Open queueing networks in heavy traffic. *Math. Oper. Res.* 9, 441–458.

[400] Reiman, M. I. (1984b) Some diffusion approximations with state space collapse. In *Modeling and Performance Evaluation Methodology*, F. Baccelli and G. Fayolle (eds.), INRIA Lecture Notes in Control and Information Sciences, 60, Springer, New York, 209–240.

[401] Reiman, M. I. (1989) Asymptotically optimal trunk reservation for large trunk groups. *Proc. 28th IEEE Conf. Decision and Control*, 2536-2541.

[402] Reiman, M. I. (1990a) Asymptotically exact decomposition approximations for open queueing networks. *Operations Res. Letters* 9, 363–370.

[403] Reiman, M. I. (1990b) Some allocation problems for critically loaded loss systems with independent links. *Proc. Performance '90*, Edinburgh, Scotland, 145–158.

[404] Reiman, M. I. and Simon, B. (1988) An interpolation approximation for queue systems with Poisson input. *Operations Res.* 36, 454–469.

[405] Reiman, M. I. and Simon, B. (1989) Open queueing systems in light traffic. *Math. Oper. Res.* 14, 26–59.

[406] Reiman, M. I. and Wein, L. M. (1998) Dynamic scheduling of a two-class queue with setups. *Operations Res.* 46, 532–547.

[407] Reiman, M. I. and Weiss, A. (1989) Light traffic derivatives via likelihood ratios. *IEEE Trans. Inf. Thy.* 35, 648–654.

[408] Reiser, M. and Kobayashi, H. (1974) Accuracy of the diffusion approximation for some queueing systems. *IBM J. Res. Dev.* 18, 110–124.

[409] Resnick, S. I. (1986) Point processes, regular variation and weak convergence. *Adv. Appl. Prob.* 18, 66–138.

[410] Resnick, S. I. (1987) *Extreme Values, Regular Variation and Point Processes*, Springer-Verlag, New York.

[411] Resnick, S. I. (1997) Heavy tail modeling and teletraffic data. *Ann. Statist.* 25, 1805–1869.

[412] Resnick, S. I. and Rootzén, H. (2000) Self-similar communication models and very heavy tails, *Ann. Appl. Prob.* 10, 753–778.

[413] Resnick, S. and Samorodnitsky, G. (2000) A heavy traffic approximation for workload processes with heavy tailed service requirements. *Management Sci.* 46, 1236–1248.

[414] Resnick, S. and van der Berg, E. (2000) Weak convergence of high-speed network traffic models. *J. Appl. Prob.* 37, 575–597.

[415] Roberts, J. (1992) *Performance Evaluation and Design of Multiservice Networks*, COST 224 Final Report, Commission of the European Communities, Luxembourg.

[416] Rogers, L. C. G. (2000) Evaluating first-passage probabilities for spectrally one-sided Lévy processes. *J. Appl. Prob.* 37, 1173–1180.

[417] Ross, S. M. (1993) *Introduction to Probability Models*, fifth edition, Academic, New York.

[418] Ryu, B. and Elwalid, A. I. (1996) The importance of long-range dependence of VBR video traffic in ATM traffic engineering: myths and realities. *Computer Communications Review* 13, 1017–1027.

[419] Samorodnitsky, G. and Taqqu, M. S. (1994) *Stable Non-Gaussian Random Processes*, Chapman-Hall, New York.

[420] Schwartz, L. (1973) *Radon Measures on Arbitrary Topological Spaces and Cylindrical Measures*, Tata Institute of Fundamental Research, Bombay.

[421] Segal, M. and Whitt, W. (1989) A queueing network analyzer for manufacturing. *Traffic Science for New Cost-Effective Systems, Networks and Services, Proceedings of ITC 12* (ed. M. Bonatti), North-Holland, Amsterdam, 1146–1152.

[422] Seneta, E. (1981) *Non-negative Matrices and Markov Chains*, Springer, New York.

[423] Sengupta, B. (1992) An approximation for the sojourn-time distribution for the GI/G/1 processor-sharing queue. *Stochastic Models* 8, 35–57.

[424] Serfozo, R. F. (1973) Weak convergence of superpositions of randomly selected partial sums. *Ann. Probab.* 1, 1044–1056.

[425] Serfozo, R. F. (1975) Functional limit theorems for stochastic processes based on embedded processes. *Adv. Appl. Prob.* 7, 123–139.

[426] Serfozo, R. F. (1999) *Introduction to Stochastic Networks*, Springer, New York.

[427] Sethi, S. P. and Zhang, Q. (1994) *Hierarchical Decision Making in Stochastic Manufacturing Systems*, Birkhäuser, Boston.

[428] Sevcik, K. C., Levy, A. I., Tripathi, S. K. and Zahorjan, J. L. (1977) Improving approximations of aggregated queueing network subsystems. *Computer Performance*, K. M. Chandy and M. Reiser (eds.), North Holland, Amsterdam, 1–22.

[429] Shorack, G. R. and Wellner, J. A. (1986) *Empirical Processes with Applications to Statistics*, Wiley, New York.

[430] Shwartz, A. and Weiss, A. (1995) *Large Deviations for Performance Analysis*, Chapman and Hall, New York.

[431] Sigman, K. (1999) *Queues with Heavy-Tailed Distributions*, Queueing Systems 33.

[432] Silvestrov, D. S. (2000) Convergence in Skorohod topology for compositions of stochastic processes. *Skorohod's Ideas in Probability Theory*, V. Korolyuk, N. Portenko and H. Syta (eds.), institute of Mathematics of the National Acafdemy of Sciences of the Ukraine, Kyiv, Ukraine, 298–306.

[433] Simmons, G. F. (1963) *Topology and Modern Analysis*, McGraw-Hill, New York.

[434] Skorohod, A. V. (1956) Limit theorems for stochastic processes. *Theor. Probability Appl.* 1, 261–290. (also republished in *Skorohod's Ideas in Probability Theory*, V. Korolyuk, N. Portenko and H. Syta (eds.), institute of Mathematics of the National Acafdemy of Sciences of the Ukraine, Kyiv, Ukraine, 23–52.)

[435] Skorohod, A. V. (1957) Limit theorems for stochastic processes with independent increments. *Theor. Probability Appl.* 2, 138–171.

[436] Skorohod, A. V. (1961) Stochastic equations for a diffusion process in a bounded region. *Theor. Probability Appl.* 6, 264–274.

[437] Skorohod, A. V. (1962) Stochastic equations for a diffusion process in a bounded region. II. *Theor. Probability Appl.* 7, 3–23.

[438] Smith, D. R. and Whitt, W. (1981) Resource sharing for efficiency in traffic systems. *Bell System Tech. J.* 60, 39–55.

[439] Srikant, R. and Whitt, W. (1996) Simulation run lengths to estimate blocking probabilities. *ACM Trans. Modeling and Computer Simulation* 6, 7–52.

[440] Srikant, R. and Whitt, W. (1999) Variance reduction in simulations of loss models. *Operations Res.* 47, 509–523.

[441] Srikant, R. and Whitt, W. (2001) Resource sharing for book-ahead and instantaneous-request calls using a CLT approximation. *Telecommunication Systems* 16, 235–255.

[442] Sriram, K. and Whitt, W. (1986) Characterizing superposition arrival processes in packet multiplexers for voice and data. *IEEE J. Sel. Areas Commun.* SAC-4, 833–846.

[443] Stewart, W. J. (1994) *Introduction to the Numerical Solution of Markov Chains*, Princeton University Press, Princeton, NJ.

[444] Stone, C. (1963) Weak convergence of stochastic processes defined on semi-infinite time intervals. *Proc. Amer. Math. Soc.* 14, 694–696.

[445] Strassen, V. (1965) The existence of probability measures with given marginals. *Ann. Math. Statist.* 36, 423–439.

[446] Suresh, S. and Whitt, W. (1990a) The heavy-traffic bottleneck phenomenon in open queueing networks. *Operations Res. Letters* 9, 355–362.

[447] Suresh, S. and Whitt, W. (1990b) Arranging queues in series: a simulation experiment. *Management Sci.* 36, 1080–1091.

[448] Szczotka, W. (1986) Stationary representation of queues II. *Adv. Appl. Prob.* 18, 849–859.

[449] Szczotka, W. (1990) Exponential approximation of waiting time and queue size for queues in heavy traffic. *Adv. Appl. Prob.* 22, 230–240.

[450] Szczotka, W. (1999) Tightness of the stationary waiting time in heavy traffic. *Adv. Appl. Prob.* 31, 788–794.

[451] Takács, L. (1967) *Combinatorial Methods in the Theory of Stochastic Processes*, Wiley, New York.

[452] Takagi, H. (1986) *Analysis of Polling Systems*, MIT Press, Cambridge, MA.

[453] Tanaka, H. (1979) Stochastic differential equations with reflecting boundary conditions in convex regions. *Hiroshima Math. J.* 9, 163–177.

[454] Taqqu, M. S. (1975) Weak convergence to fractional Brownian motion and to the Rosenblatt process. *Zeitschrift für Wahrscheinlichkeitstheorie verw. Gebiete* 31, 287–302.

[455] Taqqu, M. S. (1979) Convergence of iterated processes of arbitrary Hermite rank. *Zeitschrift für Wahrscheinlichkeitstheorie verw. Gebiete* 50, 53–83.

[456] Taqqu, M. S. (1986) A bibliographic guide to self-similar processes and long-range dependence. *Dependence in Probability and Statistics*, E. Eberlein and M. S. Taqqu (eds.), Birkäuser, Boston.

[457] Taqqu, M. S. and Levy, J. (1986) Using renewal processes to generate long-range dependence and high variability. *Dependence in Probability and Statistics*, E. Eberlein and M. S. Taqqu (eds.), Birkhäuser, Boston, 73–89.

[458] Taqqu, M. S., Willinger, W. and Sherman, R. (1997) Proof of a fundamental result in self-similar traffic modeling. *Computer Communications Review* 27, 5–23.

[459] Tse, D. N. C., Gallager, R. G. and Tsitsiklis, J. N. (1995) Statistical multiplexing of multiple time-scale Markov streams. *IEEE J. Selected Areas Commun.* 13, 1028–1038.

[460] Tsoukatos, K. P. and Makowski, A. M. (1997) Heavy-traffic analysis of a multiplexer driven by $M/G/\infty$ input processes. *Teletraffic Contributions for the Information Age, Proceedings of ITC 15*, V. Ramaswami and P. E. Wirth (eds.), Elsevier, Amsterdam, 497–506.

[461] Tsoukatos, K. P. and Makowski, A. M. (2000) Heavy traffic limits associated with $M/G/\infty$ input processes. *Queueing Systems*, 34, 101–130.

[462] Turner, S. R. E. (1996) *Resource Pooling in Stochastic Networks*, Ph.D. dissertation, University of Cambridge.

[463] Turner, S. R. E. (1998) The effect of increasing routing choice on resource pooling. *Prob. Eng. Inf. Sci.* 12, 109–124.

[464] Turner, S. R. E. (2000) A join the shorter queue model in heavy traffic. *J. Appl. Prob.* 37, 212–223.

[465] Vamvakos, S. and Anantharam, V. (1998) On the departure process of a leaky bucket system with long-range dependent input traffic. *Queueing Systems* 28, 191–214.

[466] van der Mei, R. D. (2000) Polling systems with switch-over times under heavy load: moments of the delay. *Queueing Systems* 36, 381–404.

[467] van der Mei, R. D. (2001) Polling systems with periodic server routing in heavy traffic. *Stochastic Models* 15, 273–292.

[468] van der Mei, R. D. and Levy, H. (1997) Polling systems in heavy traffic: exhaustiveness of service policies. *Queueing Systems* 27, 227–250.

[469] van der Waart, A. W. and Wellner, J. A. (1996) *Weak Convergence and Empirical Processes*, Springer, New York.

[470] van Mieghem, J. A. (1995) Dynamic scheduling with convex delay costs: the generalized $c - \mu$ rule. *Ann. Appl. Prob.* 5, 809–833.

[471] Venables, W. N. and Ripley, B. D. (1994) *Modern Applied Statistics with S-Plus*, Springer, New York.

[472] Vvedenskaya, N. D., Dobrushin, R. L. and Karpelovich, F. I. (1996) Queueing systems with selection of the shortest of two queues: an asymptotic approach. *Problems of Information Transmission* 32, 15–27.

[473] Vervaat, W. (1972) FCLT's for processes with positive drift and their inverses. *Zeitschrift für Wahrscheinlichkeitstheorie verw. Gebiete* 23, 245–253.

[474] Vervaat, W. (1985) Sample path properties of self-similar processes with stationary increments. *Ann. Probab.* 13, 1–27.

[475] Walrand, J. (1988) *An Introduction to Queueing Networks*, Prentice-Hall, Englewood Cliffs, NJ.

[476] Ward, A. R. and Glynn, P. W. (2001) A diffusion approximation for a Markovian queue with reneging. Stanford University.

[477] Ward, A. R. and Whitt, W. (2000) Predicting response times in processor-sharing queues. In *Analysis of Communication Networks: Call Centres, Traffic and Performance*, D. R. McDonald and S. R. E. Turner (eds.), Fields Institute Communications 28, American Math. Society, Providence, R. I., 1–29.

[478] Whitt, W. (1971a) Weak convergence theorems for priority queues: preemptive-resume discipline. *J. Appl. Prob.* 8, 74–94.

[479] Whitt, W. (1971b) Weak convergence of first passage time processes. *J. Appl. Prob.* 8, 417–422.

[480] Whitt, W. (1972) Stochastic Abelian and Tauberian theorems. *Zeitschrift für Wahrscheinlichkeitstheorie verw. Gebiete* 22, 251–267.

[481] Whitt, W. (1974a) Preservation of rates of convergence under mappings. *Zeitschrift für Wahrscheinlichkeitstheorie verw. Gebiete* 29, 39–44.

[482] Whitt, W. (1974b) Heavy traffic limits for queues: a survey. *Mathematical Methods in Queueing Theory*, Proceedings of a Conference at Western Michigan University, ed. A. B. Clarke, Lecture Notes in Econ. and Math. Systems 98, Springer-Verlag, New York, 1974, 307–350.

[483] Whitt, W. (1980) Some useful functions for functional limit theorems. *Math. Oper. Res.* 5, 67–85.

[484] Whitt, W. (1981) Approximating a point process by a renewal process: the view through a queue, an indirect approach. *Management Sci.* 27, 619–636.

[485] Whitt, W. (1982a) Approximating a point process by a renewal process: two basic methods. *Operations Res.* 30, 125–147.

[486] Whitt, W. (1982b) Refining diffusion approximations for queues. *Operations Res. Letters* 1, 165–169.

[487] Whitt, W. (1982c) On the heavy-traffic limit theorem for $GI/G/\infty$ queues. *Adv. Appl. Prob.* 14, 171–190.

[488] Whitt, W. (1983a) The queueing network analyzer. *Bell System Tech. J.* 62, 2779–2815.

[489] Whitt, W. (1983b) Performance of the queueing network analyzer. *Bell System Tech. J.* 62, 2817–2843.

[490] Whitt, W. (1983c) Queue tests for renewal processes. *Operations Res. Letters* 2, 7–12.

[491] Whitt, W. (1984a) Heavy-traffic approximations for service systems with blocking. *AT&T Bell Lab. Tech. J.* 63, 689–708.

[492] Whitt, W. (1984b) On approximations for queues, I: extremal distributions. *AT&T Bell Lab. Tech. J.* 63, 115–138.

[493] Whitt, W. (1984c) On approximations for queues, III: mixtures of exponential distributions. *AT&T Bell Lab. Tech. J.* 63, 163–175.

[494] Whitt, W. (1984d) Approximations for departure processes and queues in series. *Naval Res. Logistics Qtrly.* 31, 499–521.

[495] Whitt, W. (1984e) Open and closed models for networks of queues. *AT&T Bell Lab. Tech. J.* 63, 1911–1979.

[496] Whitt, W. (1984f) Departures from a queue with many busy servers. *Math. Oper. Res.* 9, 534–544.

[497] Whitt, W. (1985a) Queues with superposition arrival processes in heavy traffic. *Stoch. Proc. Appl.* 21, 81–91.

[498] Whitt, W. (1985b) The best order for queues in series. *Management Sci.* 31, 475–487.

[499] Whitt, W. (1985c) Blocking when service is required from several facilities simultaneously. *AT&T Tech. J.* 64, 1807–1856.

[500] Whitt, W. (1988) A light-traffic approximation for single-class departure processes from multi-class queues. *Management Sci.* 34, 1333–1346.

[501] Whitt, W. (1989a) Planning queueing simulations. *Management Sci.* 35, 1341-1366.

[502] Whitt, W. (1989b) An interpolation approximation for the mean workload in a GI/G/1 queue. *Operations Res.* 37, 936–952.

[503] Whitt, W. (1990) Queues with service times and interarrival times depending linearly and randomly upon waiting times. *Queueing Systems* 6, 335–351.

[504] Whitt, W. (1992) Understanding the efficiency of multi-server service systems. *Management Sci.* 38, 708–723.

[505] Whitt, W. (1993a) Approximations for the GI/G/m queue. *Production and Operations Management* 2, 114–161.

[506] Whitt, W. (1993b) Tail probabilities with statistical multiplexing and effective bandwidths in multi-class queues. *Telecommunication Systems* 2, 71–107.

[507] Whitt, W. (1993c) Large fluctuations in a deterministic multiclass network of queues. *Management Sci.* 39, 1020–1028.

[508] Whitt, W. (1994) Towards better multi-class parametric-decomposition approximations for open queueing networks. *Annals Operations Res.* 48, 221–248.

[509] Whitt, W. (1995) Variability functions for parametric decomposition approximations for queueing networks. *Management Sci.* 41, 1704–1715.

[510] Whitt, W. (1999a) Predicting queueing delays. *Management Sci.* 45, 870–888.

[511] Whitt, W. (1999b) Dynamic staffing in a telephone call center aiming to immediately answer all calls. *Operations Res. Letters* 24, 205–212.

[512] Whitt, W. (2000a) An overview of Brownian and non-Brownian FCLTs for the single-server queue. *Queueing Systems* 36, 39–70.

[513] Whitt, W. (2000b) Limits for cumulative input processes to queues. *Prob. Eng. Inf. Sci.* 14, 123–150.

[514] Whitt, W. (2000c) The impact of a heavy-tailed service-time distribution upon the $M/GI/s$ waiting-time distribution. *Queueing Systems* 36, 71–88.

[515] Whitt, W. (2001) The reflection map with discontinuities. *Math. Oper. Res.* 26, 447–484.

[516] Whittle, P. (1986) *Systems in Stochastic Equilibrium*, Wiley, New York.

[517] Wichura, M. J. (1974) Functional laws of the iterated logarithm for the partial sums of i.i.d. random variables in the domain of attraction of a completely asymmetric stable law. *Ann. Probab.* 2, 1108–1138.

[518] Williams, R. J. (1987) Reflected Brownian motion with skew symmetric data in a polyhedral domain. *Prob. Rel. Fields* 75, 459–485.

[519] Williams, R. J. (1992) Asymptotic variance parameters for the boundary local times of reflected Brownian motion on a compact interval. *J. Appl. Prob.* 29, 996–1002.

[520] Williams, R. J. (1995) Semimartingale reflecting Brownian motions in the orthant. In *Stochastic Networks*, F. P. Kelly and R. J. Williams (eds.), Springer-Verlag, New York, 125–137.

[521] Williams, R. J. (1996) On the approximation of queueing networks in heavy traffic. *Stochastic Networks: Theory and Applications*, F. P. Kelly, S. Zachary and I. Ziedins (eds.), Clarendon Press, Oxford, U.K., 35–56.

[522] Williams, R. J. (1998a) An invariance principle for semimartingale reflecting Brownian motions in an orthant. *Queueing Systems* 30, 5–25.

[523] Williams, R. J. (1998b) Diffusion approximations for open multiclass queueing networks: sufficient conditions involving state space collapse. *Queueing Systems* 30, 27–88.

[524] Willinger, W., Taqqu, M. S. and Erramilli, A. (1996) A bibliographic guide to self-similar traffic and performance modeling for modern high-spped networks. In *Stochastic Networks, Theory and Application*, F. P. Kelly, S. Zachary and I. Ziedins (eds.), Clarendon press, Oxford, U.K., 339–366.

[525] Willinger, W., Taqqu, M. S., Leland, M. and Wilson, D. (1995) Self-similarity in high-speed packet traffic: analysis and modeling of ethernet traffic measurements. *Statistical Science* 10, 67–85.

[526] Willinger, W., Taqqu, M. S., Sherman, R. and Wilson, D. V. (1997) Self similarity through high-variability: statistical analysis of Ethernet LAN traffic at the source level. *IEEE/ACM Trans. Networking* 5, 71–96.

[527] Wischik, D. (1999) The output of a switch, or, effective bandwidths for networks. *Queueing Systems* 32, 383–396.

[528] Wischik, D. (2001a) Sample path large deviations for queues with many inputs. *Ann Appl. Prob.* 11, 379–404.

[529] Wischik, D. (2001b) Moderate deviations in queueing theory. Statistical Laboratory, Cambridge.

[530] Wolff, R. W. (1977) An upper bound for multi-channel queues. *J. Appl. Prob.* 14, 884–888.

[531] Wolff, R. W. (1989) *Stochastic Modeling and the Theory of Queues*, Prentice-Hall, Englewood Cliffs, NJ.

[532] Yao, D. D. (1994) *Stochastic Modeling and Analysis of Manufacturing Systems*, Springer, New York.

[533] Yashkov, S. F. (1993) On a heavy-traffic limit theorem for the $M/G/1$ processor-sharing queue. *Stochastic Models* 9, 467–471.

[534] Zeevi, A. J. and Glynn, P. W. (2000) On the maximum workload of a queue fed by fractional Brownian motion. *Ann. Appl. Prob.* 10, 1084–1099.

[535] Zolotarev, V. M. (1986) *One-dimensional Stable Distributions*, Amer. Math. Soc., 65, Providence, RI

[536] Zwart, A. P. (2000) A fluid queue with a finite buffer and subexponential input. *Adv. Appl. Prob.* 32, 221–243

[537] Zwart, A. P. (2001) *Queueing Systems with Heavy Tails*, Ph.D. dissertation, Department of Mathematics and Computer Science, Eindhoven University of Technology.

[538] Zwart, A. P. and Boxma, O. J. (2000) Sojourn time asymptotics in the M/G/1 processor sharing queue. *Queueing Systems* 35, 141–166.

Appendix A
Regular Variation

Since we use regular variation at several places in this book, we give a brief account here without proofs, collecting together the properties we need. Bingham, Goldie and Teugels (1989) is the definitive treatment; it contains everything here. Nice accounts also appear in Feller (1971) and Resnick (1987).

We say that a real-valued function f defined on the interval (c, ∞) for some $c > 0$ is *asymptotically equivalent* to another such function g (at infinity) and write

$$f(x) \sim g(x) \quad \text{as} \quad x \to \infty \tag{A.1}$$

if

$$f(x)/g(x) \to 1 \quad \text{as} \quad x \to \infty . \tag{A.2}$$

We say that the real-valued function f has a *power tail* of index α (at infinity) if

$$f(x) \sim Ax^\alpha \quad \text{as} \quad x \to \infty \tag{A.3}$$

for a non-zero constant A. Regular variation is a generalization of the power-tail property that captures just what is needed in many mathematical settings.

A positive, Lebesgue measurable real-valued function (on some interval (c, ∞)) L is said to be *slowly varying* (at infinity) if

$$L(\lambda x) \sim L(x) \quad \text{as} \quad x \to \infty \quad \text{for each} \quad \lambda > 0 . \tag{A.4}$$

Examples of slowly varying functions are positive constants, functions that converge to positive constants, logarithms and iterated logarithms. (Note that $\log x$ is positive on $(1, \infty)$, while $L_2 x \equiv \log \log x$ is positive on (e, ∞).) The positive functions $2 + \sin x$ and e^{-x} are not slowly varying. It can happen that a slowly varying function

experiences infinite oscillation in the sense that
$$\liminf_{x\to\infty} L(x) = 0 \quad \text{and} \quad \limsup_{x\to\infty} L(x) = \infty \; ;$$
an example is
$$L(x) \equiv \exp\{(ln(1+x))^{1/2} \cos((ln(1+x))^{1/2})\}. \tag{A.5}$$

It is significant that there is local uniform convergence in (A.4).

Theorem A.1. (local uniform convergence) *If L is slowly varying, then $L(\lambda x)/L(x) \to 1$ as $x \to \infty$ uniformly over compact λ sets.*

Theorem A.2. (representation theorem) *The function L is locally varying if and only if*
$$L(x) = c(x) \exp\left(\int_a^x (b(u)/u)du\right), \quad x \geq a, \tag{A.6}$$
for some $a > 0$, where c and b are measurable functions, $c(x) \to c(\infty) \in (0, \infty)$ and $b(x) \to 0$ as $x \to \infty$.

In general a slowly varying function need not be smooth, but it is always asymptotically equivalent to a smooth slowly varying function; see p. 15 of Bingham et al. (1989).

Theorem A.3. *If L is slowly varying function, then there exists a slowly varying function L_0 with continuous derivatives of all orders (in C^∞) such that*
$$L(x) \sim L_0(x) \quad \text{as} \quad x \to \infty \;.$$
If L is eventually monotone, then so is L_0.

We now turn to regular variation. A positive, Lebesgue measurable function (on some interval (c, ∞)) is said to be *regularly varying* of index α, and we write $h \in \mathcal{R}(\alpha)$, if
$$h(\lambda x) \sim \lambda^\alpha h(x) \quad \text{as} \quad x \to \infty \quad \text{for all} \quad \lambda > 0. \tag{A.7}$$
Of course, (A.7) holds whenever h has a power tail with index α, but it also holds more generally.

The connection to slowly varying function is provided by the characterization theorem.

Theorem A.4. (characterization theorem) *If*
$$h(\lambda x) \sim g(\lambda)h(x) \quad \text{as} \quad x \to \infty \tag{A.8}$$
for all λ in a set of positive measure, then

(i) (A.8) holds for all $\lambda > 0$,

(ii) there exists a number α such that $g(\lambda) = \lambda^\alpha$ for all λ, and

(iii) $h(x) = x^\alpha L(x)$ for some slowly varying function L.

From Theorem A.4 (iii) we see that we could have defined a regularly varying function in terms of a slowly varying function L. On the other hand, (A.8) is an appealing alternative starting point, implying (A.7) and the representation in terms of slowly varying functions. As a consequence of Theorem A.4 (iii), we write $h \in \mathcal{R}(0)$ when h is slowly varying.

We now indicate how the local-uniform-convergence property of $\mathcal{R}(0)$ in Theorem A.1 extends to $\mathcal{R}$.

Theorem A.5. (local uniform convergence) *If $h \in \mathcal{R}(\alpha)$ and h is bounded on each interval $(0, c]$, then*

$$h(\lambda x)/h(x) \to \lambda^\alpha \quad as \quad x \to \infty$$

uniformly in λ:

$$\begin{aligned}
&\text{on each } [a, b], &&0 < a < b < \infty, &&\text{if } \alpha = 0 \\
&\text{on each } (0, b], &&0 < b < \infty, &&\text{if } \alpha > 0 \\
&\text{on each } [a, \infty), &&0 < a < \infty, &&\text{if } \alpha < 0.
\end{aligned}$$

The representation theorem for slowly varying functions in Theorem A.2 also extends to regularly varying functions.

Theorem A.6. (representation theorem) *We have $h \in \mathcal{R}(\alpha)$ if and only if*

$$h(x) = c(x) \exp\left\{\int_a^x (b(u)/u) du\right\}, \quad x \geq a,$$

for some $a > 0$, where c and b are measurable functions, $c(x) \to c(\infty) \in (0, \infty)$ and $b(x) \to \alpha$ as $x \to \infty$.

We now present some closure properties.

Theorem A.7. (closure properties) *Suppose that $h_1 \in \mathcal{R}(\alpha_1)$ and $h_2 \in \mathcal{R}(\alpha_2)$. Then:*

(i) $h_1^\alpha \in \mathcal{R}(\alpha \alpha_1)$.

(ii) $h_1 + h_2 \in \mathcal{R}(\alpha)$ for $\alpha = \max\{\alpha_1, \alpha_2\}$.

(iii) $h_1 h_2 \in \mathcal{R}(\alpha_1 + \alpha_2)$.

(iv) If, in addition, $h_2(x) \to \infty$, then $(h_1 \circ h_2)(x) \equiv h_1(h_2(x)) \in \mathcal{R}(\alpha_1 \alpha_2)$.

Regular variation turns out to imply local regularity conditions; see p. 18 of Bingham et al. (1989).

Theorem A.8. (local integrability theorem) *If $h \in \mathcal{R}(\alpha)$ for some α, then h and $1/h$ are both bounded and integrable over compact subintervals of (c, ∞) for suitably large c.*

It is significant that integrals of regularly varying functions are again regularly varying with the same slowly varying function (i.e., the slowly varying function passes through the integral).

Theorem A.9. *(Karamata's integral theorem) Suppose that $L \in \mathcal{R}(0)$ and that L is bounded on every compact subset of $[c, \infty)$ for some $c \geq 0$. Then*

(a) for $\alpha > -1$,

$$\int_c^x t^\alpha L(t)dt \sim \frac{x^{\alpha+1}}{\alpha+1} L(x) \quad as \quad x \to \infty \;;$$

(b) for $\alpha < -1$,

$$\int_x^\infty t^\alpha L(t)dt \sim -\frac{x^{\alpha+1}}{\alpha+1} L(x) \quad as \quad x \to \infty \;.$$

It is also possible to deduce regular variation of functions from the regular-variation properties of the integrals, i.e., the converse half of the Karamata integral theorem; see p. 30 of Bingham et al. (1989).

Appendix B
Contents of the Internet Supplement

Preface	iii
1 Fundamentals	**1**
1.1 Introduction	1
1.2 The Prohorov Metric	1
1.3 The Skorohod Representation Theorem	6
1.3.1 Proof for the Real Line	6
1.3.2 Proof for Complete Separable Metric Sspaces	7
1.3.3 Proof for Separable Metric Spaces	10
1.4 The "Weak" in Weak Convergence	16
1.5 Continuous-Mapping Theorems	17
1.5.1 Proof of the Lipschitz Mapping Theorem	17
1.5.2 Proof of the Continuous-Mapping Theorems	19
2 Stochastic-Process Limits	**23**
2.1 Introduction	23
2.2 Strong Approximations and Rates of Convergence	23
2.2.1 Rates of Convergence in the CLT	24
2.2.2 Rates of Convergence in the FCLT	25
2.2.3 Strong Approximations	27
2.3 Weak Dependence from Regenerative Structure	30
2.3.1 Discrete-Time Markov Chains	31
2.3.2 Continuous-Time Markov Chains	33
2.3.3 Regenerative FCLT	36

		2.3.4	Martingale FCLT	40
	2.4	\multicolumn{2}{l}{Double Sequences and Lévy Limits . . . .}	41
	2.5	\multicolumn{2}{l}{Linear Models . . . .}	46

3 Preservation of Pointwise Convergence   51
    3.1   Introduction 51
    3.2   From Pointwise to Uniform Convergence 52
    3.3   Supremum 54
    3.4   Counting Functions 55
    3.5   Counting Functions with Centering 62
    3.6   Composition 68
    3.7   Chapter Notes 71

4 An Application to Simulation   73
    4.1   Introduction 73
    4.2   Sequential Stopping Rules for Simulations 73
           4.2.1   The Mathematical Framework 75
           4.2.2   The Absolute-Precision Sequential Estimator 79
           4.2.3   The Relative-Precision Sequential Estimator 82
           4.2.4   Analogs Based on a FWLLN 83
           4.2.5   Examples 86

5 Heavy-Traffic Limits for Queues   97
    5.1   Introduction 97
    5.2   General Lévy Approximations 97
    5.3   A Fluid Queue Fed by On-Off Sources 100
           5.3.1   Two False Starts 101
           5.3.2   The Proof 103
    5.4   From Queue Lengths to Waiting Times 105
           5.4.1   The Setting 105
           5.4.2   The Inverse Map with Nonlinear Centering 106
           5.4.3   An Application to Central-Server Models 110

6 The Space D   113
    6.1   Introduction 113
    6.2   Regularity Properties of D 114
    6.3   Strong and Weak M_1 Topologies 117
           6.3.1   Definitions 117
           6.3.2   Metric Properties 118
           6.3.3   Properties of Parametric Representations 122
    6.4   Local Uniform Convergence at Continuity Points 124
    6.5   Alternative Characterizations of M_1 Convergence 128
           6.5.1 SM_1 Convergence 128
           6.5.2 WM_1 Convergence 130
    6.6   Strengthening the Mode of Convergence 134

6.7		Characterizing Convergence with Mappings	134
	6.7.1	Linear Functions of the Coordinates	135
	6.7.2	Visits to Strips	137
6.8		Topological Completeness	139
6.9		Non-Compact Domains	142
6.10		Strong and Weak M_2 Topologies	144
	6.10.1	The Hausdorff Metric Induces the SM_2 Topology	145
	6.10.2	WM_2 is the Product Topology	147
6.11		Alternative Characterizations of M_2 Convergence	148
	6.11.1	M_2 Parametric Representations	148
	6.11.2	SM_2 Convergence	148
	6.11.3	WM_2 Convergence	155
	6.11.4	Additional Properties of M_2	158
6.12		Compactness	161

7 Useful Functions   163
7.1		Introduction	163
7.2		Composition	164
	7.2.1	Preliminary Results	164
	7.2.2	M-Topology Results	166
7.3		Composition with Centering	173
7.4		Supremum	173
	7.4.1	The Supremum without Centering	173
	7.4.2	The Supremum with Centering	174
7.5		One-Dimensional Reflection	181
7.6		Inverse	183
	7.6.1	The M_1 Topology	183
	7.6.2	The M_1' Topology	186
7.7		Inverse with Centering	188
7.8		Counting Functions	190
7.9		Renewal-Reward Processes	194

8 Queueing Networks   195
8.1		Introduction	195
8.2		The Multidimensional Reflection Map	197
	8.2.1	Definition and Characterization	197
	8.2.2	Continuity and Lipschitz Properties	200
8.3		The Instantaneous Reflection Map	204
8.4		Reflections of Parametric Representations	204
8.5		M_1 Continuity Results	210
8.6		Limits for Stochastic Fluid Networks	217
8.7		Queueing Networks with Service Interruptions	217
8.8		The Two-Sided Regulator	217
8.9		Existence of a Limiting Stationary Version	218
	8.9.1	The Main Results	218

		8.9.2 Proofs	224
9	**Nonlinear Centering and Derivatives**		**235**
	9.1	Introduction	235
	9.2	Nonlinear Centering and Derivatives	237
	9.3	Derivative of the Supremum Function	243
	9.4	Extending Pointwise Convergence to M_1 Convergence	254
	9.5	Derivative of the Reflection Map	258
	9.6	Heavy-Traffic Limits for Nonstationary Queues	262
	9.7	Derivative of the Inverse Map	267
	9.8	Chapter Notes	276
10	**Errors Discovered in the Book**		**281**
11	**Bibliography**		**283**

Notation Index

Presented in Order of First Use.

Chapter 1, 1
$\equiv$, equality by definition, 1
$\lfloor t \rfloor$, floor function, 17, 96
$\Rightarrow$, convergence in dist., 21, 78
$\Phi(t) \equiv P(N(0,1) \leq x)$, standard normal cdf, 21, 101
$\stackrel{d}{=}$, equality in distribution, 22, 78
$F^{\leftarrow}$, left-cont. inverse of cdf, 32
F^{-1}, right-cont. inverse of cdf, 33
$F^c(t) \equiv 1 - F(t)$, ccdf, 38
$\sim$, asym. equiv., 40, 113, 569

Chapter 2, 49
ρ, traffic intensity, 56, 148
ϕ_K, two-sided ref. map, 57, 143

Chapter 3, 75
$\mathcal{P} \equiv \mathcal{P}(S)$, 77
A^ϵ, open ϵ-nbhd., 77
$\pi(P_1, P_2)$, Prohorov metric, 77
$a \vee b \equiv \max\{a, b\}$, 79
$a \wedge b \equiv \min\{a, b\}$, 83
$x \circ y$, composition, 87, 428
$x^\uparrow$, supremum map, 87, 435
$\phi(x) = x + (-x \vee 0)^\uparrow$, one-dim. reflection map, 87, 140, 439
$(x \vee 0)(t) \equiv x(t) \vee 0$, 87

x^{-1}, inverse map, 87, 441

Chapter 4, 95
α, stable index, 100, 111
$\Sigma \equiv (\sigma^2_{i,j})$, cov. matrix, 104, 352
$\mathcal{F}_n \equiv \sigma[X_k : k \leq n]$, 108
$\mathcal{G}_n \equiv \sigma[X_k : k \geq n]$, 108
σ, stable scale param., 111
β, stable skewness param., 111
μ, stable shift param., 111
$\mathcal{R}(\alpha)$, reg. var., index α, 114, 570
$(1/G^c)^{\leftarrow}(n)$, space scaling, 115
$\Phi_\alpha(x)$, Fréchet cdf, 118

Chapter 5, 137
λ, arrival rate, 146
μ, service rate, 146
σ^2_L, bdry. reg. asymp. var., 172
σ^2_U, bdry. reg. asymp. var., 172

Chapter 6, 193
$\dot{x}$, derivative of x, 222

Chapter 7, 225
$\hat{f}(s)$, Lapl. transf. of f, 230
ψ, reg. var. scaling fct., 234

Chapter 8, 243
$\Lambda(t)$, source rate process, 246
ϕ, empty set, 254, 368
ξ_n, source on rate, 252
γ_n^{-1}, mean source cycle time, 252
λ_n, rate of rate process Λ_n, 254
$\lambda_n \equiv \xi_n \hat{\lambda}_n$, source input rate, 254
η_n, net input rate of scaled pr., 255
$\hat{\phi}(s) \equiv \log E e^{-s[\eta + S_\alpha(\sigma,1,0)]}$, 268

Chapter 9, 287
$\delta(x_1, x_2) \equiv x_2 + (x_1 - x_2)^\uparrow$, 296
$\hat{\delta}(x_1, x_2) \equiv x_2 + (x_1 - x_2)^\downarrow$, 298
$\tau_z(x) \equiv x^{-1}(z)$, first-passage, 316, 446
$\gamma_z(x)$, overshoot fct., 316, 446
$\lambda_z(x)$, last-val. fct., 316, 447

Chapter 10, 341

Chapter 11, 367
$\mathcal{T}$, topology, 368
A^-, closure of set A, 368
A°, interior of set A, 368
∂A, boundary of set A, 368
σ-field, 372

Chapter 12, 391
$x(t-)$, left limit, 392
$|b|$, max norm on $\mathbb{R}^k$, 393
$|x|$, uniform norm, 393
$|b|_p$, l_p norm, 393
$[a, b]$, standard segment, 394
$[[a, b]]$, product segment, 394
Γ_x, thin graph, 395
G_x, thick graph, 395
$\rho(\Gamma_x)$, thin range, 395

$\rho(G_x)$, thick range, 395
$\Pi_s(x)$, set of strong par. reps., 395
$\Pi_w(x)$, set of weak par. reps., 395
$\beta_t : [0, 1] \to [0, 1]$, 400
$\mu_s(x_1, x_2)$, SM_2 dist., 417
$\Pi_{s,2}(x)$, SM_2 par. reps., 419
$\Pi_{w,2}(x)$, WM_2 par. reps., 419

Chapter 13, 427
$x^\leftarrow$, left-contin. inverse, 442
$\delta_z(x)$, final-jump fct., 447

Chapter 14, 457
$\Psi(x)$, feas. regulator set, 462
$R \equiv (\psi, \phi)$, mult. ref. map, 462
$\pi \equiv \pi_{x,Q} : D_\uparrow^k \to D_\uparrow^k$, 463
$|A|$, matrix norm, 465
$R_0 \equiv (\phi_0, \psi_0)$, inst ref. map, 473
$u^+ \equiv u \vee 0$, comp. max, 475
$u^- \equiv u \wedge 0$, comp. min, 475
$\pi_0(v) \equiv (Qv - u)^+$, 479
$\hat{\psi}(x, t, \alpha)$, 482
$\hat{\phi}(x, t, \alpha)$, 482
$(C, r, Q, X(0))$, fluid net. data, 491
$(A, S, R, I, Z(0))$, queueing netwk. data, 497
$R \equiv (\phi, \psi_1, \psi_2)$, two-sided reg., 505

Chapter 15, 515
ϕ^+, maximal ref. map, 522
ϕ^-, minimal ref. map, 522
$(x, S, \{I(t) : t \in S\})$, excursion triple, 523
$\tilde{x}$, set-val. fct., 524
$\Gamma_{\tilde{x}}$, graph of $\tilde{x}$, 524
$\hat{x}$, elt. of F, 533
$\Pi_s(\hat{x})$, par. reps of $\hat{x}$ in F, 533

Author Index

Abate, 156, 157, 166, 169, 173, 175–177, 183, 191, 229, 269, 270, 276, 277, 315, 486
Abbott, viii
Abramowitz, 28, 124, 269, 277
Addie, 283
Adler, x, 65, 116, 136
Albin, 310, 332, 336
Aldous, xiv
Amdahl, 165
Anantharam, 68, 326
Anick, 243, 244
Anisimov, 71
Araujo, 227
Arvidsson, 67
Asmussen, x, xiv, 156, 160, 183, 231
Astrauskas, 131, 134
Avram, 123, 125, 130–132, 398, 426
Ayhan, 68
Azar, 348

Baccelli, 156
Barford, 65
Bass, 227, 381
Becker, 2
Bell, xv, 73, 348, 513
Bender, 156

Beneš, 512
Beran, 98, 120, 125
Berger, 156, 171, 172, 177, 323–325, 512
Bertoin, 110, 267, 316, 317
Bertsekas, xiv
Bestavros, 65
Billingsley, viii, ix, xii, 28, 54, 55, 75–136, 146, 352, 367–390, 429, 441, 528
Bingham, 120, 569–572
Bitran, 333
Bleistein, 156
Bloznelis, 227, 233
Bolot, 67
Bondi, 339
Borodin, 28, 55, 166
Borovkov, 142, 153, 156, 260, 342, 349, 354, 360, 361
Borst, 68, 341
Botvich, 284
Box, 123
Boxma, 68, 60, 157, 191, 243, 245
Bramson, xv, 72, 156, 512
Bratley, 108
Breiman, 54, 107, 109
Brémaud, 156
Brichet, 245, 320
Broder, 348

Browne, 360
Bu, 67
Burman, 169
Buzacott, xiv, 326

Carslaw, 486
Chambers, 2
Chang, C. S., 156
Chang, K.-H., 177
Chen, xv, 189, 191, 218, 346, 398, 457, 511–513
Choe, 283, 284
Choudhury, 69, 70, 156, 157, 169, 173, 243, 270, 277
Chung, 20, 21
Çinlar, 318
Coffman, 71, 72, 191
Cohen, 159, 191, 245, 273, 276
Cooper, xiv
Cottle, 474, 482, 512
Courtois, 70
Cox, D. R., 229, 230
Cox, J. T., 109
Crovella, 65
Csörgő, viii, 384

Dabrowski, 109
Dai, xv, 156, 338
Daley, 318
Damerdji, 108
Das, 67
Davis, 131
Davydov, 125
de Meyer, 276
de Veciana, 156
Demers, 67
Deng, 191
Devroye, 10
Dobrushin, 122, 157, 348
Doetsch, 269
Donsker, viii, 25, 29, 55, 95
Doob, viii, 55
Down, 156
Downey, 348
Doytchinov, 192
Dudley, 77, 386
Duffield, 67, 157, 228, 233, 283–285, 349, 355
Dugundji, 368, 526

Dumas, 157, 243
Dupuis, 72, 505, 509, 511, 512

Eberlein, 108, 120
Eckberg, 354
Edwards, 464
Eick, 348
El-Taha, 428
Elwalid, 67
Embrechts, x, 16, 51, 110, 115, 116, 118, 136, 435
Erdös, viii, 29, 103
Erramilli, 65, 98
Ethier, 84

Feldman, x, 65, 116, 136
Feldmann, 139, 277
Feller, 16, 20, 21, 28, 36, 40, 44, 103, 104, 110, 114, 124, 175, 314, 413, 569
Fendick, 169, 173, 191, 309, 310, 319, 323, 327, 337
Firoiu, 67
Flatto, 157
Fleming, 349
Floyd, 65
Foschini, 348
Foss, 156
Fox, 108
Freidlin, 71
Furrer, 245

Gallager, xiv, 68, 323
Garnett, 341, 359
Garrett, 65
Giné, 227
Glasserman, 108
Glynn, 108, 156, 166, 177, 234, 282, 349, 359
Gnedenko, 114
Goldie, 120, 569–572
Gonzáles-Arévalo, 326
Greenberg, 67, 139, 323
Grimmett, 109
Grishechkin, 191
Grossglauser, 67, 323
Guerin, 67
Gunther, xiv, 165
Gut, 430, 441

Hahn, 226, 228
Halfin, 358
Hall, xiv, 148, 222
Handelsman, 156
Harchol-Balter, 348
Harrison, xiv, xv, 57, 72, 73, 142, 158, 166, 167, 171, 174, 338, 348, 398, 457, 511–513
Heyman, 67
Hida, 123
Hitsuda, 123
Hjálmtýsson, 348
Hopp, xiv, 163
Horváth, viii, 384
Hsing, 134
Hui, 323
Hunt, 361
Hurst, 97
Hüsler, 285

Iglehart, 153, 235, 342, 345, 354, 511
Ishii, 505, 511, 512

Jackson, 326
Jacod, x, 84, 107, 117, 119, 194, 410, 414, 448, 530
Jagerman, 360
Jain, xiv
Jakubowski, 109, 126, 130
Janicki, 110, 482
Jarvis, 320
Jelenković, 68, 157, 323
Jenkins, 123
Jennings, 349, 357
Johnson, 36, 277

Kac, viii, 29, 103
Karatzas, 51, 125, 166
Karlin, A., 348
Karlin, S., 54, 166
Karlsson, 67
Karpelovich, 152, 157, 348
Kasahara, 126
Kella, 169, 217, 218, 268, 398, 494, 512
Kelly, xv, 156, 157, 348, 361, 457
Kemeny, 55, 277
Kennedy, 153
Keshav, xiv, 67
Kesidis, 156

Kingman, 153, 191
Kleinrock, xiv, 170
Klincewicz, 328
Klüppelberg, x, 16, 51, 110, 115, 116, 118, 136, 435
Knessl, 156, 320
Kobayashi, 326
Kokoszka, 136
Kolmogorov, 114, 383, 385, 389
Konstantopoulos, 245
Korolyuk, 79
Kotz, 36, 277
Kraemer, 328
Kreinin, 152
Krichagina, 349
Krishnamurthy, xiv, 65, 67, 139
Kruk, 192
Kuehn, 326
Kumar, xv, 73, 513
Kunniyur, 323
Kurose, xiv, 67
Kurtz, 84, 126, 233, 245
Kushner, xv, 72, 73, 84, 320, 457, 511, 513

Lakshman, 67
Lamperti, 97, 98
Langenbach-Belz, 328
Laws, 348
Lazar, 323
L'Ecuyer, 17
Lehoczky, 192
Leland, x, 65, 98, 231
Leung, 349
Lévy, P., 103
Levy, A. I., 326
Levy, H., 72, 191
Levy, J., 233
Likhanov, 157
Limic, 191
Lin, 245
Lindvall, 414
Lions, 512
Lopez, 73, 348
Loulou, 348
Lucantoni, 156, 157, 173
Lund, 139

Madras, 67
Maejima, 126

Mairesse, 335
Majewski, 156
Major, 122
Makowski, 245, 259
Malkiel, 49
Mandelbaum, 70, 152, 222, 301, 341, 349, 357, 359, 360, 398, 457, 512
Mandelbrot, x, 97–99
Markowitz, xv, 72, 513
Martins, 320
Massey, 67, 152, 222, 234, 301, 348, 349, 357, 359, 398
Massoulié, 245, 285, 320
Matheron, 381
Mazumdar, 157
McDonald, xv
Meyn, 156
Michna, 245
Mikosch, x, 16, 51, 67, 110, 115, 116, 118, 136, 160, 232, 280, 435
Mitra, 243, 244
Mittnik, 51
Mitzenmacher, 157, 187, 348
Montanari, 136
Moran, 138
Morrison, 320
Mountford, 335
Müller, 384

Nagaev, 160
Narayan, 285
Nelson, 348
Neuts, 276
Neveu, 385
Newell, xiv, 148, 222
Newman, 109
Nguyen, 338
Norros, 279, 282–285
Nyberg, 67

O'Brien, 100, 398
O'Connell, 157, 284, 285
Oliver, 148, 222
Olsen, 68, 72, 191
Olver, 124, 156
Orszag, 156

Padhye, 67
Pang, 474, 482, 512

Parekh, 68
Park, x, xiv, xv, 65, 67
Parthasarathy, 367, 374
Pats, 360
Paulauskas, 227
Paxson, 65
Perrin, 67
Philipp, 107
Piterbarg, 285
Pollard, viii, 55, 375, 386
Pomarede, 382, 418
Portenko, 79
Prabhakar, 335
Prabhu, 138, 191
Prohorov, viii, 29, 77, 387, 390, 413
Protter, 125, 126
Puhalskii, 71, 72, 156, 191, 301, 349, 358, 359, 398, 444, 450, 452
Puterman, 69
Pyke, 227, 381

Rachev, 51
Ramanan, 505, 509, 511, 512
Reiman, 70–72, 152, 169, 187, 191, 301, 338, 341, 348, 358, 359, 361, 457, 505, 511, 512
Reingold, 139
Reinsel, 123
Reiser, 326
Resnick, 65, 67, 115, 116, 118, 119, 131, 177, 194, 232, 245, 259, 280, 313, 398, 435, 441, 530–532, 569
Rexford, xiv, 65, 67, 139
Ripley, 2, 8, 17
Roberts, 243, 245, 320, 323
Rodrigues, 191
Rogers, 272
Rootzén, 67, 232, 245, 280, 313
Ross, xiv, 16, 51, 54
Rosso, 136
Ryu, 67

Saksena, 309, 310, 319
Salminen, 28, 55, 166
Salz, 348
Samorodnitsky, 97–136, 245, 267, 326
Samuel, 148, 222
Schrage, 108
Schwartz, 372, 374, 385

Segal, 332
Semret, 323
Seneta, 472
Sengupta, 191
Serfozo, 430, 457
Sethi, 71
Sevcik, 326
Shanthikumar, xiv, 326, 346
Shen, 189, 191
Shenker, 67
Sherman, 65, 231, 232, 243, 279
Shiryaev, x, 84, 107, 117, 119, 194, 410, 414, 448, 530
Shorack, viii, 55, 384
Shreve, 51, 125, 166, 192
Shroff, 283, 284
Shwartz, xiv, 156
Sigman, 157
Silvestrov, 430
Simmons, 368
Simon, 169, 349
Simonian, 245, 285, 320
Skorohod, viii, xi, xiii, 29, 75–93, 117, 194, 367–426, 433, 456, 512
Smith, 164, 169, 357
Snell, 55, 277
Sondhi, 243, 244
Spearman, xiv, 163
Srikant, 67, 171, 283, 323, 354, 361–363
Sriram, 67, 319, 323
Starica, 67
Stegeman, 67, 232, 280
Stegun, 28, 124, 269, 277
Stewart, 55, 277
Stidham, 428
Stolyar, 301
Stone, C., 414
Stone, R. E., 474, 482, 512
Stout, 107
Strassen, 375
Suresh, 334, 335
Syta, 79
Szczotka, 167
Sznitman, 512

Taaffe, 348
Takacs, 276
Takagi, 69
Tanaka, 512

Taqqu, x, 65, 97–136, 231–233, 243, 267, 279, 398, 426
Taylor, 54, 166
Teugels, 120, 160, 276, 569–572
Thomas, 156
Tier, 156
Tirupati, 333
Towsley, 67
Tripathi, 326
True, 139
Tse, 323
Tsitsiklis, 323
Tsoukatos, 245, 259
Turner, xv, 157, 348

Upfal, 348

Vamvakos, 326
van der Berg, 245, 259, 398
van der Mei, 72, 191
van der Waart, viii
van Mieghem, 68, 72
Veitch, 245, 320
Venables, 2, 8, 17
Vere Jones, 318
Vervaat, 98, 100, 235
Vöcking, 348
Vvedenskaya, 157, 348

Wallis, 99
Walrand, 68, 156, 457
Ward, 277, 283, 359
Wein, xv, 72, 513
Weiss, xiv, 156, 169
Wellner, viii, 55, 384
Wentzell, 71
Weron, 110, 245, 482
Whittle, 457
Wichura, 398, 399
Wilks, 2
Williams, xv, 72, 73, 172, 348, 398, 505, 512, 513
Willinger, x, xiv, xv, 65, 67, 98, 231, 232, 243, 279
Wilson, x, 65, 98, 231, 243
Wischik, 156, 284, 323
Wolff, xiv, 347
Wright, 109

Yang, 320
Yao, xv, 108, 457, 511, 513
Yashkov, 191
Yeh, 338
Yeung, 192

Zachary, xv
Zahorjan, 326

Zeevi, 282
Zhang, 71
Zhou, 338
Ziedins, xv
Zolotarev, 110
Zukerman, 283
Zwart, 157, 191, 243, 276

Subject Index

adaptedness of ref. map, 470
addition, 86
   continuity of
      for M_1, 411
      for M_2, 423
      not everywhere, 84
   measurability of, 411
α-stable, *see* stable
approximation
   of functions
      in D by fcts. in D_c, 393
      in E by fcts. in E_c, 526
   of graphs by finite sets, 396
   of pt. pr. by ren. pr.
      asymptotic method, 330
      stationary-int. method, 330
   of queueing networks
      fixed-pop.-mean method, 339
      param.-decomp. method, 326
      seq. bottle. decomp., 338
      param.-decomp. method, 331, 332
      QNET method, 338
      via the bottleneck, 326
   of queues
      by infinite server models, 348
      heuristic, 169
arc sine law, 103, 316

Arzelà-Ascoli theorem, 388, 390
associated random variables, 109
asymptotic
   constants, 64
   meth., approx. pt. pr., 330
   methods, classical, 156
   peakedness, 354, 360
   variability params., 308
   variance, 107, 180
asymptotic equivalence of
   m-server queues, 345
   contin.-time reps., 195
   cting. and inv. fcts., 455
   fcts., $\sim$, 40, 113, 156, 569
   queue lgths and wait times, 209
   ran. elts., 195, 380
   renewal pr. and sums, 204
   scaled svc. processes, 294
automatic scaling by plotter, 2, 4
averaging principle, ix, 71

$\mathbf{B}$, standard BM, 22, 102
$\mathbf{B}_0$, Brownian bridge, 51
$\hat{R}(t)$, estimator of blk. prob., 361
$B_m(x, r)$, open ball, 77
balanced loading, 338
ball σ-field, 386

balls
  definition of, 77, 368
  used to define
    a countable basis, 371
    ball σ-field, 386
    totally bounded, 371
Banach
  -Picard fixed-point thm., 467
  space
    D is not a, 84
    explan. for "weak", 373
    valued fcts., 350, 382, 394
base function for E, 523
basis for a topology, 377
batching and unbatching, 163
Black-Scholes formula, 51
blow up the picture, 4
BM, *see* Brownian motion
Borel σ-field
  comparable Lusin spaces, 385
  on D, 385
  on a metric space, 77
  on a product space, 377
  on a top. space, 372
bottleneck in queueing network, 326
boundary of a set, 368
boundary regulator pr., 143, 290
  rate of, 171
bounded convergence theorem, 383
Brownian
  approximation
    fl. queue, on-off sources, 261
    for gen. fluid queue, 165
    single-server queue, 306
    superposition arrival pr., 304
  bridge, 51
  motion, 22
    k-dim., 105
    fractional, *see* fractional
    geometric, 50
    range of, 28
    standard, 22, 102
buffer-sizing equation, 64
busy pd. of a queue, 177
  Brownian approx., 177

C, the space, 79
$C_0 \equiv C \cap D_0$, 429
$C_\uparrow \equiv C \cap D_\uparrow$, 429

$C_{\uparrow\uparrow} \equiv C \cap D_{\uparrow\uparrow}$, 429
$C_m \equiv C \cap D_m$, 429
$C(S)$, cont. bdd. real-val. fcts. on S, 77
c_L^2, normalized asym. var., 172
c_U^2, normalized asym. var., 172
$c_u^2 \equiv Var\, U_1/(EU_1)^2$, SCV, 308
$c_v^2 \equiv Var\, V_1/(EV_1)^2$, SCV, 308
c_U^2, asymp. var. par., 308
c_V^2, asymp. var. par., 308
$c_{U,V}^2$, asymp. var. par., 308
c_{HT}^2, variability par., 308
$c_{u,v}^2 \equiv Cov(U_1, V_1)/(EU_1)(EV_1)$, 309
c_{SI}^2, stat. int. approx. SCV, 330
c_{AM}^2, as. meth. approx. SCV, 330
cadlag or càdlàg function, 79
call centers, 341
canonical
  Brownian motion, 174
  reflected Br. mo., 174
  scaling fcts. as fct. of ρ, 161
Cauchy
  distribution, a stable law, 112
  property of a sequence, 371
ccdf, $F^c(t) \equiv 1 - F(t)$, 38
cdf, *see* distribution function
centering
  for convergence preservation
    basic idea, 87
    linear, 88, 432, 448
    nonlinear, 88, 434, 451
  in other direction, 438
  of a random walk, 3, 37
  of partial sums, 4
central limit theorem
  k-dim., 105
  classical, viii, 20, 101, 102
  equiv., cting. pr., 234
  for processes, 226
    applied to get RFBM, 279
    finite-state CTMC's, 228
    Lipschitz processes, 228
    Markov processes, 228
    renewal processes, 229
  general, 96
  heavy-tailed version, 39, 114
characterizations of
  SM_1 convergence in D
    main theorem, 404
    by linear maps, 411

Subject Index 587

by visits to strips, 412
with domain $[0, \infty)$, 415
SM_2 convergence in D
  main theorem, 420
  by linear maps, 423
  by local extrema, 423
SM_2 convergence in E, 527
WM_1 convergence in D, 408
WM_2 convergence in D, 421
compact subsets in D, 424
convergence in distribution, 375
extremal processes, 531
feasible regulator set, 463
instantaneous ref. map, 476
  by fixed-point property, 479
local uniform convergence, 403
multidimensional reflection map
  by complementarity, 464
  by fixed-point property, 464
parametric reps., 400
tightness in D, 425
weak convergence, 373
closure of a set, 368
CLT, see central limit theorem
coef. of variation, see squared coef.
communication network
  bursty traffic, ix
  economy of scale, 164
  effective bandwidths, 156
  engineering challenges, x, 243
  engr. impact of queueing, 64
  failures in, 516
  FCLT for bursty traffic, 231
  fl. queue, on-off sources, 91
  fluid rep. of packet flow, 139
  impact of flow controls, 258
  learning network status, 187
  loss measurements, 173
  lost packets, 170
  model of data in a buffer, 55
  multiclass input model, 310
  multiplexing gain, 164
  rate-control throttle, 323
  scaling with size, 163
  separation of time scales, 322
  TCP, 67
  topology of a, 368
  traffic measurements, ix, 65, 243
  traffic shaping, 325

compactness
  approach to s-p limits, 386
  characterization for D, 424
  definition of, 370
  of graphs in E, 525
comparison of
  SJ_1 and SM_1 metrics, 398
  SM_1 and WM_1 topologies, 397
  Borel and Kolm. σ-fields, 385
  distances on D, 418
  metrics on F and E, 534
complementarity and reflection, 464
complementary cdf, see ccdf
complete
  graph, 80, 381
    order on, 81
    param. rep. of, 81
  metric space, 371, 413
completely monotone function, 175
composition map, 87, 428–435
  continuity of, 430
  convergence preservation
    with linear centering, 432
    with nonlinear centering, 434
  not continuous everywhere, 429
confidence interval, 180
continuity
  definition of, 369
  modulus of, 388
  of addition
    for M_1, 411
    for M_2, 423
    not everywhere, 84
  of composition, 430
  of first-passage fcts., 447
  of genl. ref. maps, 510
  of integrals, 383
  of multidim. reflect.
    in F, 535
    in uniform top., 468
    on (D, SJ_1), 470
    with M_1 tops., 485
  of multiplication, 434
  of the inverse map, 444
  of two-sided regulator
    with J_1, 508
    with the uniform norm, 506
    with M_1, 511
point

of a cdf, 21
of a limit function, 383
right, 392
continuous-mapping
  approach, s-p limits, ix, 25, 84
  theorem, 84, 86, 87
continuous-time rep. of a discrete-time process
  M_1 dist. between versions, 195
  lin. interp. version, 17, 195
  scaling in a, 19
  simulations to compare, 196
  step-fct. version, 17, 195
convergence
  characterization of
    SM_1, 404, 412
    SM_2, 420, 423
    WM_1, 408
    WM_2, 421
  extending to product spaces
    for SM_1, 409
    for SM_2, 422
  in distribution, 21, 78, 86, 375
    criterion for, 388
  in probability, 379
  joint on product spaces, 378
  local uniform, 401
  of a net, 370
  of a sequence
    in a metric space, 76
    in a topological space, 370
  of all finite-dim. dists., 387
  of cdf's, 21
  of first-passage vars., 447
  of plots, 26
  of probability measures, 77, 373
  of random point measures, 532
  of restrictions, 83, 383, 414
  rates of, 85
  strengthening the mode
    for WM_1, 409
    for WM_2, 422
  -together theorem, 54, 379
  with probability one, w.p.1, 86
convergence preservation
  WM_2 within bnding fcts., 423
  with centering, 88
    composition map, 432, 434
    inverse map, 448, 451

reflection map, 441
supremum map, 436, 438
counting fcts., 453–456
  asym. equiv., inv. fcts., 455
counting process, 233, 454
  Brownian FCLT for, 235
  CLT for, 234
  definition, 201
  FBM FCLT, 236
  FCLT equiv., 202, 235
  LFSM FCLT for, 237
  stable Lévy FCLT for, 236, 237
covariance
  fct. of stat. RBM, 175, 181
  fct. of stat. renewal pr., 229
  matrix, 104
cover, 370, 371
Cramér-Wold device, 104
cumulative distribution function, see distribution function

D, the space, 78, 380, 391–426
  SM_1 and WM_1 tops., 394
  SM_2 and WM_2 tops., 416
  characterization of
    M_1 convergence, 404
    M_2 convergence, 419
    compact subsets, 424
    tightness, 425
  regularity properties, 392
$D([0,\infty), \mathbb{R}^k)$, 83, 394, 414, 487
D_c, piecewise-const. fcts., 393, 473
D_0, subset with $x(0) \geq 0$, 428
$D_\uparrow$, nondecreasing x in D_0, 428
$D_{\uparrow\uparrow}$, increasing x in $D_\uparrow$, 428
D_m, x^i monotone, all i, 428
D_u, x in D_0 unbded. above, 441
$D_{u,\uparrow} \equiv D_u \cap D_\uparrow$, 441
$D_{u,\uparrow\uparrow} \equiv D_u \cap D_{\uparrow\uparrow}$, 441
$D_{u,\epsilon}$, subset of D_u, 443
D_u^*, subset of D_u, 443
$D_{u\uparrow}^*$, subset of $D_{u\uparrow}$, 444
D_s, jumps same sign, 458, 483, 486
D_1, jumps in one coord., 458, 486
D_+, jumps up, 481, 486
$D_{c,1} \equiv D_c \cap D_1$, 510
$Disc(x)$, set of disc. pts., 86, 393
(D, M_1) in (F, M_1), 535
(D, SM_2) in (E_{st}, m), 526

Subject Index   589

(D, WM_2) in (E_{wk}, m), 526
$d_{J_1}(x_1, x_2)$, J_1 metric, 79
$d_{M_1}(x_1, x_2)$, M_1 metric, 82, 396
$d_\infty(x_1, x_2)$, on $D([0, \infty), \mathbb{R})$, 83
$d_s(x_1, x_2)$, SM_1 metric on D, 395
$d_w(x_1, x_2)$, WM_1 dist. on D, 396
$\hat{d}(A, \Gamma_x)$, order-consist. dist., 396
$d_p(x_1, x_2)$, product metric, 397
$d^*(A, A_n)$, graph subsets, 405
$\hat{d}(A, G_x)$, order-const. dist., 408
$d_{s,2}(x_1, x_2)$, SM_2 dist., 419
$d_{w,2}(x_1, x_2)$, WM_2 dist., 419
$d_{s,2}(\tilde{x}_1, \tilde{x}_2)$, par. rep. dist., 527
$d_s(\hat{x}_1, \hat{x}_2)$, metric on F, 533
dense subset, 76, 371
density
    est. of final pos., 8, 38, 43
    of range of Br. motion, 29
departure pr., 144, 152, 184, 291
    approximations for, 334
determining the buffer size, 64
deterministic analysis of queues, 148
diameter of a set, 524
discontinuities of the first kind, 392
discontinuity points, 86, 393
    jumps common sign, 411, 423
    of an element of E, 523
distance, 375
    between a point and a set, 381
    in-probability, 375
    order-consistent, 396, 408
distribution
    empirical, 8, 11
    function, 10, 374
domain
    of a function, 83
    of attract. of stable law, 114
        normal, 114
Donsker's theorem, viii, 24, 102
    k-dim., 105
    weak-dep. generalizations, 107
double limits, 280
double sequence of r. vars., 110
double stochastic-process limit, 231
DTMC, see Markov chain

E, the space, 92, 515–539
    compactness of graphs in, 525
    Hausdorff metric on, 525

E_c, piecewise-constant fcts., 526
E_{st}, strongly connected, 526
E_{wk}, weakly connected, 526
e, identity map, 79, 146
$ess\,sup\,(Y)$, essential sup., 228
economy of scale, 164, 357
effective bandwidths, 156
effective processing rate, 165
eigenvalues, 472
empirical
    process, 226
empirical distribution
    and statistical tests, 51
    compared to the cdf, 10
    definition, 11
    in a QQ plot, 8
    of final position
        of random walk, 8
        with heavy tails, 38, 43
empty set, ϕ, 368
equality
    by definition, $\equiv$, 1
    in distribution, $\stackrel{d}{=}$, 22, 78
equivalence
    classes of parametric reps., 533
    CLT for cting. and sum pr., 234
    FCLT for cting. pr., 237
    of par. reps. in F, 533
    to contractive ref. matrix, 472
ergodicity, 109
essential supremum, 228
estimation
    BM range dist., 29
    final pos. r. walk, 8, 38
    planning q. sim. for, 178, 361
estimator of blocking prob.
    indirect, 362
    natural, 361
    simple, 362
excursion
    times, set of, S, 523
    triples for F, 523
    values, set of, $I(t)$, 523
exogenous input, 139, 337, 491
exponential smoothing, 30
exponential tail, 64
extending
    conv. to product spaces

for SM_1, 409
for SM_2, 422
graphs for M' tops., 444
ref. map from D_c to D, 480
extremal processes, 530
extreme-value engineering, 177
extreme-value limit, 118, 177

F, the space, 92, 515, 533–539
(D, M_1) as a subspace, 535
M_1 metric on, 533
failures
hvy-traf. limit in F for, 536
in queueing systems, 516
service interruptions, 217
fair queueing discipline, 67
FBM, *see* fractional
FCLT
Donsker's, viii, 24, 102
equiv. for cting. pr., 202, 235
for associated process, 109
for heavy tails plus dep., 130
for Markov processes, 109
for MMPP, 312
for renewal-reward pr., 238–456
for stationary martingale, 109
for strong dependence
  and heavy tails, 130, 134
  and light tails, 125
  Gaussian processes, 129
general, definition, 96
pt. pr. in ran. envt., 312
to stable Lévy motion, 117
with regenerative structure, 109
with weak dependence, 108
feasible regulator
definition, 462
semilattice structure in, 464
final pos. of ran. walk, 7, 8, 38, 43
final-jump function, 447
finance, x, 49
finite dam, 273
finite-dim. distributions, 23, 387
first-passage times
continuity results, 447
limits for, 176
of Brownian motion, 172
of RBM, 176
with very hvy. tails, 316

fixed-point char. of ref. map, 464
fixed-pop.-mean approx., 339
Flatland, viii
fluid limit
for fluid queue, 146
stochastic refinements, 149
fluid queue
discrete-time, 55, 140
general model
  Brownian approximation, 165
  fluid limit, LLN, 146
  model definition, 139
  stable, hvy-tr. lmts, 153
  unstable, 145
  with priorities, 187, 189
with $M/G/\infty$ input, 259
with on-off sources
  simulation example, 249
  FCLT, cum. busy time, 254
  heavy-traffic limit, 255
  many sources, 259, 280
  model definition, 245
  stable-Lévy approx., 264
FPM, *see* fixed-pop.
Fréchet cdf, $\Phi_\alpha(x)$, 118
fractional Brownian motion
convergence to, 125
definition, 125
from iterated lim., 232, 322
Joseph effect, 100
limit for traffic processes, 231
reflected, 279, 322
simulating, 126
function
addition, *see* addition
cadlag or càdlàg, 79
completely monotone, 175
composition, *see* composition
continuous, 369
distribution, *see* distribution
domain of a, 83
final-jump, 447
indicator, $I_A(x)$, 52
inverse, *see* inverse
last-value, 447
Lipschitz, *see* Lipschitz
max.-jump, 119
measurable, 372
oscillation, 401

overshoot, 446
piecewise-constant, 393
projection, 385
random, 23, 75
range of a, 83
reflection, *see* reflection
regularly varying, *see* regular
renewal, 230
scale-invariant, 25
set-valued, 524
supremum, *see* supremum
useful, 86
functional
 central limit theorem, *see* FCLT
 strong law of large numbers, *see* FSLLN
fundamental sequence, 371
FWLLN, funct. weak LLN, 146

G_δ set, 369, 429
G_x, thick graph, 395
game of chance, 1
Gaussian distribution, *see* normal distribution
Gaussian process, 110, 120
 FCLT for with str. dep., 129
 limit in process CLT, 226, 281
 limits for inf-svr. queue, 352
 reflected, 283
geometric Brownian motion, 50
Gibbs phenom. for reflection, 486
graph
 completed, 80, 381
  param.rep. of, 81
 extended for M' topologies, 444
 of a set-valued fct. in E, 524
 path-connected in E, 526
 thick, 395
 thin, 395
greed, 1
Gumbel cdf, 177

H, Hurst scaling expon., 97, 121
H-sssi, H-self sim., stat. incs., 99
Hausdorff metric
 general definition, 381, 524
 on graphs for D, 382, 417
 on graphs in E, 525
heavy-tailed dist.
 in ran.-walk steps, 35

stat. regularity with, 38
cause of jumps, 35, 38, 90
cause strong dep., 280
FCLT for ct. pr., 236, 237
FCLT with, 42, 44, 99, 109–120, 130–136
in $GI/GI/1$ queue, 213
in busy and idle pds., 231, 248, 264, 493
in commun. netwks., x, 65, 231
in ran.-walk steps, 196
in ren-rwd. pr., 240
in renewal pr., 201
new statistical regularity, xi
Noah effect, 99
Pareto inputs to queue, 61
sgl-svr. queue, 313
SLM FCLT, 304
stable law, 111
unmatched jumps, xi
heavy-traffic
 averaging principle, 71
 bottleneck phenom., 335
 snapshot principle, 187
 state space collapse, 72
heavy-traffic limit
 engineering significance, 63
 gen. fluid queue
  discrete-time, 55
  for stable queue, 153
  RBM limit, 165
  scaling as fct. of ρ, 158
  scaling with size, 162
  to plan simulations, 178
  with failures in F, 536
  with priorities, 187, 189
  LLN for unstable, 146
 multiserver queue
  autonom. svc., 344
  increasing no. servers, 357
  inf. many servers, 260, 351
  loss models, 360
  standard model, 346
 on-off fluid sources
  m gen. sources, 255
  Brownian limit, 262
  from dep. to jumps, 257
  RFBM, many sources, 280
  RSLM limit, 265
  struc. avail-proc. pr., 258

queueing network
  fluid model, 493
    rare long interruptions, 500
    std. interruptions, 504
    to get approximations, 338
  single-server queue
    approx. q. netwks., 326
    continuous-time pr., 297
    directly from ref. map, 293
    discrete-time processes, 295
    first passage, hvy tails, 317
    from q lgth. to wait, 301
    heavy-tailed dists., 313
    mlti-cls btch-ren. input, 310
    MMPP arrival process, 312
    need for M_1 top., 300
    split process, 305
    superpos. arr. pr., 303, 320
    to ref. Gaussian pr., 321
  state dependence, 359
heavy-traffic scaling as fct. of ρ
  identify scaling fcts., 58, 160
  performance impact, 158, 162
  plan simulations, 90, 178, 361
  scaling exponents, 161
  sing-svr. queue, 301
Helly selection theorem, 413
homeomorphism, 370, 374
H-self-sim. with stat. incs., *see* H
human perception in perform., 517
Hurst param., H, *see* scaling exp.

$I_A(x)$, indicator function, 52
$I(t)$, excursion values, 523
identity map, e, 79
IID, indep. and ident. distrib., 17
image probability measure, 77, 372
in-probability distance, 375
independent
  and identically distrib., *see* IID
  random elements, 378
  sample, 8, 10
index
  of a stable law, 100, 111
  of reg. var., *see* regular
  of self-similarity, 97
indicator function, $I_A(x)$, 52
infinite variance, *see* heavy-tailed
infinite-server queue

heavy-traffic limit, 348
  as approximation, 356
  input to fl. queue, 259
  peakedness, 354
  transient behavior, 355
inheritance of jumps from WM_2
  convergence, 422
innovation process, 130, 134
instantaneous reflection, 473–482
  and linear complementarity, 474
  and linear programming, 474
  characterization of, 476
  fixed-point characterization, 479
  iterations of, 478
  monotonicity of, 477
  two-sided, 505
insurance, x
integral, contin. of, 383
interior of a set, 368
Internet Supplement, xiii, 20, 77, 78, 85,
  86, 93, 99, 106, 107, 109, 110, 116,
  126, 148, 152, 189, 222, 226, 234,
  235, 253, 257, 264, 268, 276, 278,
  301, 313, 314, 373, 391, 427, 428,
  430, 431, 434, 436, 452, 457, 460,
  467, 483, 485, 510, 523, 573
invariance principle, 27, 42, 103
inverse
  processes, 201
  relation
    for counting functions, 454
    for left-continuous inverse, 442
inverse map, 87, 441–453
  continuity of, 444
  conv. pres. with centering, 301, 448, 451

J_1 metric, 79
J_2 metric, 381
J, max-jump fct., 119, 422
J_t^+, max. pos. jump fct., 296
J_t^-, max. neg. jump fct., 296
Jackson queueing network, 326
  generalized, 329
joint conv. of ran. elts.
  criteria for, 378
  for sup with centering, 437
Joseph effect, *see* strong dep.
  biblical basis, 99
  FBM FCLT for cting. pr., 236

FCLT for, 120–136
  in cum. input process, 231
  quantification of, 100
jumps in the limit process, x
  from strong dep., 257
  in plots
    of ctred ran. walks, 38
    of disc.-time fl. queues, 62
    of unctred ran. walks, 35
  unmatched, *see* unmatched jumps

Kendall notation, 211, 292
Kolmogorov
  σ-field, 385
  -Smirnov statistic, viii, 52, 226
  extension theorem, 385, 388

L_p, 383
L_1 topology on D, 485
lag-k correlations, 309
Laplace transform, 230
  num. inv., *see* numerical trans.
  of RSLM steady-state dist., 268
  RBM bdry-reg. vars., 173
  RBM first-pass. times, 176
  stable law ccdf, 314
large deviations, 156
last-value function, 316, 447
law of large numbers
  explaining jumps, 35
  explaining linear plots, 20
  for unstable queue, 146
leaky bucket regulator, 323
left limits, 392
left-continuous inverse
  of a cdf, 32
  of a function, 442
Lévy
  distribution, a stable law, 112
  metric, 374, 413
  process, 102, 116
    first-passage for, 447
    superposition of IID, 303
LFSM, *see* linear fractional
Lindley recursion, 289, 539
linear
  -process rep., 122, 126, 130
  centering, 88
  complementarity problem, 474

fcts. of coord. fcts., 411, 423
filter, 123
frac. stable motion, 100, 133
interpolation, 12
program, 474
Lipschitz
  assumption for ref. maps, 509
  constant, 85, 469
  function, 85, 416, 439
  mapping theorem, 85
  property of
    multidim. ref. map, 468, 470, 486, 487
    general ref. maps, 510
    lin. fct. of coord. fcts., 410, 423
    supremum map, 436
    two-sided reg., 506, 508, 511
local uniform convergence, 401
local-maximum function
  for M_2 top. on D, 423
  in E, 528
log-fractional stable motion, 101
log-log scale, 38
loss measurements, 173
loss rate in fl. queue, 171
Lusin space, 371, 374, 383, 385

M_1 metric
  on D, xi, 82, 395
    uses of, 398
  on F, 533
M_1' top. on $D([0,\infty),\mathbb{R})$, 444
M_2 metric
  on D, 382, 417
  on E, 525
M_2' top. on $D([0,\infty),\mathbb{R})$, 444
$M/G/1$ queue, 169
$M/G/1/K$ queue, 273
$M/P_p/1$, Kendall notation, 211
M_n, successive maxima, 118
$M(t)$, renewal function, 230
M_{t_1,t_2}, local max fct., 423, 528
$m_s(x_1,x_2)$, SM_2 metric on D, 417
$m_p(x_1,x_2)$, WM_2 product metric on D, 417
$m(\tilde{x}_1,\tilde{x}_2)$, Hausd. metric on E, 525
$m^*(x_1,\tilde{x}_2)$, unif. metric on E, 525
macroscopic view of uncert., vii, 2
manufacturing systems
  failures in, 516

594   Subject Index

scaling with size, 163
marginal probability measure, 378
Markov chain
   as an environment process, 31
   FCLT for, 108
   nearly-completely-decomposable, 70
   process CLT for, 228
   routing in fluid network, 460
   workload in fl. queue as, 55
Markov seq. decision pr., 69
Markov-mod. Pois. pr., 312
martingale, 109, 172
math. models, simulation and the physical
      world, 17
matrix
   covariance, 104
   norm, 465
   reflection, 462
   routing, 461
   spectral radius of a, 472
   spectrum of a, 472
   substochastic, 461
maximum
   -jump fct., *see* J
   norm on $\mathbb{R}^k$, 393
   of n r. vars., 118
measurability
   of C in (D, M_1), 429
   of addition on $D \times D$, 411
   of dist. betw. ran. elts., 379
   of functions in D, 394
   of multidim. ref. map, 471
   of subsets of D, 429, 442, 444
   of the inverse map, 444
   of two-sided regulator, 508
   problems on (D, U), 386
measurable
   function, 77, 372
   rectangle, 377
   set, 77, 372
   space, 76, 372
metric
   J_1, 79
   J_2, 381
   M_1, 82, 396
      on F, 533
   M_2, 382
      on E, 525
   SJ_1, 83

   SM_1, 83, 396
   SM_2, 417
   WM_2, 417
   general definition, 76
   Hausdorff, 381, 382, 417
   Lévy, 374, 413
   maximum, 377
   on space of stoch. pr., 76
   product, 83, 417
   Prohorov, 77, 374, 375
   space, *see* space, metric
   uniform, 79, 393
mixing conditions for CLT's, 108
MMPP, *see* Markov-mod.
modulus of continuity, 388, 390
   over a set, 393
monotonicity
   assump. genl. ref. map, 509
   of fcts, M tops., 408
multidimensional reflection, 457–513
   application to
      fluid networks, 490–495
      queueing networks, 495–505
   as function of
      x and Q, 473
      the ref. matrix Q, 471
   at discontinuity points, 480
   behavior at disc. points, 481
   characterization of
      by complementarity, 464
      by fixed-point property, 464
   continuity of
      as fct of x and Q, 487
      in M_1, 485
   counterexamples for M_1, 487–490
   definition of, 462
   discontinuity points of, 481
   existence of, 463
   extension from D_c to D, 480
   instantaneous, *see* instantaneous
   Lipschitz property
      for $D([0,\infty), \mathbb{R}^k)$, 487
      for uniform norm, 468
      for M_1, 486
      for SJ_1, 470
   measurability of, 471
   on F, 535
   one-sided bounds, 467
   properties of, 470

multiplexing, 164, 245
multiplication, contin., 434
multiserver queue, 341
  autonomous service, 342
    asymp. equiv. std., 345
    heavy-traffic limit, 344
  growing number servers, 355
    hvy-trfic limits, 357
  inf-srver. approx., 356
  infinitely many servers, 348
  non-FCFS disciplines, 347
multivariate normal distribution, 104

$N(m, \sigma^2)$, normal variable, 8, 102
$N(m, \Sigma)$, normal r. vector, 105
$N \equiv \{N(t) : t \geq 0\}$, cting pr., 233
$n(x)$, normal density, 102
negative dependence, 101
net, 370
net-input process, 140
network calculus, 331
Noah effect, see heavy-tailed dist.
  biblical basis, 99
  FCLT for, 109–120, 130–136
  in on and off periods, 231
  in ran. walks, 196
  quantification of, 100
noncompact domains, 414
nonlinear centering, 88
nonparam. density estimator, 8
norm
  L_p, 369
  matrix, 465
  maximum on $\mathbb{R}^k$, 83, 393
  weighted-supremum, 384
normal distribution, viii, 8, 101
  departures from, 9
  in the CLT, 21, 102
  indep. samples from, 8
  multivariate, 104
  quantile of, 8
normal domain of attraction, 114
normalized partial sums, 96, 102
number of visits to a strip, 412
numerical transform inversion, 173, 270, 314

offered load model, 67
open

ϵ-neighborhood, 77
  ball, 77, 368
  cover, 370
  queueing model, 139
  rectangle, 377
  set, 368
optimal fee to play a game, 15
option pricing, 51
order
  -consistent distance, 396, 408
  on completed graph, 81, 395
  statistic, 12, 53
  total, 81, 395
ordinary differential eq., ODE, 221
organization of book, xii, 89
Ornstein-Uhl. diff. pr., 260, 354
oscillation function, 401
overflow process in fl. queue, 170
overshoot function, 316, 446

P-continuity set, 373
$P \equiv (P_{i,j})$, routing matrix, 461
parallel processing, 165
parametric representation
  M_1
    definitions, 81, 395
    how to construct, 398
    need not be one-to-one, 399
    one-to-one, 408
  M_2
    definition, 419
    equivalence in F, 533
    reflection of, 483
    relating M_1 and M_2 tops., 382
    strong in E, 526
    weak in E, 526
Pareto distribution, 36
  of busy and idle periods, 249
parsimonious approximations, 98
partial sum, 1, 96
  centered, 4
path-connected graphs in E, 526
peakedness, 354
performance analysis, xiv
performance of queues, 516
piecewise
  -constant fct., 393, 473, 526
planning qing. sim., 90, 178, 361
plot

as a map from D to D, 25
automatic, 2, 4
efficient, 6
function, 2
of a queue-length process, 210
of centered renewal process, 202
QQ, 8
random version of a, 7
random walk, 2
plotting routine, 2
point proc. in rand. envt., 312
Poisson
excursion process in F, 538
process, 53, 259
limit for superpos. proc., 318
random measure, 532
Polish space, 371, 381, 383, 389
polling, 69
potential
-workload process, $X(t)$, 140
buffer-content process, 461
power tail
def. for fct., 569
for packet loss, 65
in queues, 157
of stable law, 100, 113
ran-wk. fin-pos. dist., 39, 43
RSLM stdy-st. dist., 269
see heavy-tailed, 73
prediction, 283, 355
exploiting dependence for, 128
preserv. par. rep. by ref., 483
priority discipline in a queue, 187
probability
density, see density
distribution
hvy-tld., see heavy-tailed
of a random element, 23, 77
of a random path, 4
power tail, see power
measure, 372
image, 77, 372
marginal, 378
product, 378
tight, 374, 376
uniquely determined by, 388
space, 372
processing time
definition, 145

heavy-traffic limit for, 184
with priorities, 190
product
metric, 417
probability measure, 378
space, 377
topology, 83
Prohorov
metric, 77, 85, 374, 375, 493, 495
thm., rel. compactness, 387
projection map, 385
pruning data, 6

$Q \equiv P^t$, reflection matrix, 462
$Q^a(t)$, q. lth., auton. svc., 343
$Q_* \equiv \Lambda^{-1}Q\Lambda$, 472
QNA, qing. netwk. analyzer, 332
QNET algorithm, 338
QQ plot, 8
quantile, 8
queue
first example, 55
fluid, see fluid
in slowly chg. envt., 70
multiserver, see multiserver
single-server, see single-server
svc. interruptions, 217, 518
time-varying arr. rate, 220
shift in arr. rate, 222
queueing network, 457–513
closed, 339
heavy-traffic limit, 338, 535
svc. interruptions, 495
queues in series, 152

$R \equiv (\psi, \phi)$, mult. ref. map, 462
$R \equiv (\phi, \psi_1, \psi_2)$, two-sided reg., 505
$R_0 \equiv (\phi_0, \psi_0)$, inst ref. map, 473
$r_H(s,t) \equiv Cov(\mathbf{Z}_H(s), \mathbf{Z}_H(t))$, FBM
covar. fct., 125
r_{t_1,t_2}, restriction map, 414
random
element, 77, 375
function, 23, 75
measure, 532
multiplier, 50
path, 4
sum
in renewal-reward process, 238

use of composition, 433
version of original plot, 7
random number
  generator, 1, 16
  seed, 3
  uniform, 1
random walk, 2
  applications, 49
    Kolmogorov-Smirnov stat., 51
    queueing model, 55, 289
    stock prices, 49
  centered, 3
  final position of, 7
  in a ran. envmnt, 31
  plot, 2, 25
  relative final pos. of, 8, 27
  simple, 28, 103
  uncentered, 4
  with diff. step dist., 31
range
  more general spaces, 394
  of a function, 25, 83
  of Brownian motion, 28
  of data, 2
  thick, 395
  thin, 395
rate of convergence, 85
  in Donsker's FCLT, 106
  in limits for fluid networks
    heavy-traffic, 495
    stability, 493
rate-control throttle, 323
ratio of maximum to sum, 120
rectangle
  measurable, 377
  open, 377
reflected
  BM, 59, 165, 262, 304
    covariance function, 175, 181
    transition prob. dist., 174
  fractional Brownian motion, 279, 322
  Gaussian process
    heavy-traffic limit, 321
  Ornstein-Uhlenbeck pr., 361
  stable Lévy motion, 62, 264, 304, 313
    $M/G/1/K$ approx. for, 273
reflection
  matrix, 462
    equivalence to contractive, 472
    norm, 466
  of a parametric rep., 483
reflection map
  instantaneous, see instantaneous
  multidimensional, see multidim.
  one-sided, one-dim., 87, 290, 439–441
    M_2-cont. fails, 439, 521
    conv. pres. with ctring., 441
    maximal in E, 522
    Lipschitz property, 439
  two-sided, 57, see two-sided
regenerative structure
  CLT with weak dep., 109
  for boundary reg. pr., 172
regular variation
  appendix on, 569
  in CLT for counting pr., 234
  in extreme vals., 118
  in Lamperti thm., 98
  in stable CLT, 114
  in strong dep., 121
regularity, 3
  statistical, 1, 4
rel. final pos. of r. walk, 8, 27
relatively compact, 387
relaxation time, 159
renewal
  function, 230
  process, 201
    covariance fct. of, 229
    process CLT for, 229
renewal reward process
  definition of, 238
  FCLT for
    finite variances, 239
    heavy-tailed ren. pr., 240
    heavy-tailed summands, 240
    both heavy-tailed, 241
rescaling of mult. ref. map, 470
restriction of fct., 83, 383, 414
risk management, x
routing process, 496

S, S-Plus, 2, 8
S, excursion times, 523
S_n, partial sum, 2
$S_\alpha(\sigma, \beta, \mu)$, stable law, 111
SJ_1 metric, 83
SM_1

598   Subject Index

converg. charact. of
   main theorem, 404
   by linear maps, 411
   visits to strips, 412
   with domain $[0, \infty)$, 415
metric, 83, 396
SM_2
converg. charact. of
   main theorem, 420
   by extrema, 423
   by linear maps, 423
metric, 417
param. rep., 419
topology, 416
sample
   independent, 8, 10
   mean, 180
   paths, no negative jumps, 268
scale-invariance property, 446
scale-invariant function, 25
scaling
   exponent, 97, 121, 232, 280
   heirarch., priorities, 189
   in queue, *see* heavy-traffic
   relations for stable law, 112
   with system size, 162
scaling of time and space
   canonical RBM, 174
   as fct. of ρ in queue, *see* heavy-traffic, scaling
   automatic by plotter, 2, 6
   for fluid limits, 146
   in a cont.-time rep., 19
   in a covariance function, 181
scheduling service, 69
SCV, *see* squared coef.
second countable space, 371, 372
seed, random number, 3
segment
   product, 394
   standard, 394
self-similarity, x
   in ntwk. traffic, 65, 98
   in ran.-walk plots, 6, 38
   of a stochastic process, 97
   of Brownian motion, 24, 102
   of stable Lévy motion, 117
semilattice, of ref. map, 464
separable

space, 76, 371–381
stochastic process, 528
separation of time scales, 70, 322
sequence
convergence of
   in a metric space, 76
   in a top. space, 370
definition, 76
double of r. vars., 110
fundamental, 371
server-avail. indicator pr., 496
service discipline
   fair queueing, 67
   gen. processor sharing, 67
service interruptions, 217, 495
set
   G_δ, 369, 429
   P-continuity, 373
   X-continuity, 375
   boundary of a, 368
   closed, 368
   closure of, 368
   compact, 370
   diameter of, 524
   empty, 368
   feasible regulator, 462
   interior of a, 368
   measurable, 77, 372
   of discontinuity points, 86, 393
   of excursion values in E, 523
   of param. reps., *see* param.
   open, 368
   relatively compact, 387
set-valued function in E, $\tilde{x}$, 524
σ-field
   ball, 386
   Borel, *see* Borel
   general definition, 372
   Kolmogorov, 385
   product, 377
simple random walk, 28, 103
simulation
   and the physical world, 17
   of disc.-time fl. queue
      with expon. inputs, 59
      with Pareto inputs, 61
   of empirical cdf's, 11
   of FBM, 126
   of game of chance, 1, 13

of hvy-tailed ren. pr., 203
of LFSM, 134
planning a queueing, 178, 361
   bias of an estimate, 182
   initial portion to delete, 182
   required run length, 181
with heavy tails and dependence, 132
simulation of queues
   hvy-tld. arr., svc. times, 205
   shift in arrival rate, 222
   svc. interruptions, 218
   fl. queue, on-off sources, 248
   hvy-tr. bottleneck phen., 335
simulation of random walks
   BM range dist., 29
   centered, 34
   compare ct.-time reps., 18, 196
   uncentered, 2, 3, 34
single-server queue, 206, 287
   Brownian approximation, 306
   heavy-traffic limits, 292
   model definition, 289
   split process, 305
   superposition arr. proc., 301
   with hvy-tld. distribs., 313
size-increase factor, 163
Skorohod rep. thm., 78, 85
SLLN, strong law large nos., 20
slowly varying fct., see regular
snapshot principle, ix, 187
space
   L_p, 369
   Banach, 84, 373
   function, 78
   measurable, 76, 372
   metric, 76
     complete, 371, 380
   of all probability measures, 77
   probability, 77, 372
     underlying, 77
   range, 394
   separable, see separable
   topological, 368
     compact, 370
     Lusin, see Lusin
     metrizable, 368, 371
     Polish, see Polish
     product, 370, 377
     second countable, 371, 372

   separable, see separable
   topologically complete, 413
   vector, 84
     topological, 84
spectral radius of a matrix, 472
spectrum of a matrix, 472
split process, 305
   approximations for, 333
squared coeff. of var., 307, 326
stable
   FCLT, 117
   index, 100, 111
   Lévy motion, 41, 116
     limit for ren-rew. pr., 240
   process, 100, 110
   subordinator, 117
stable law, 40, 111
   central limit theorem, 114
   characteristic fct. of, 111
   domain of attraction, 114
   in stable process, 100
   index α, 100, 111
   normal dom. of attract., 114
   scale param. σ, 111
   scaling relations for, 112
   shift param. μ, 111
   skewness param. β, 111
   totally skewed, 111, 268
standard
   Brownian motion, 22, 102
   normal r. var., 21, 101
   single-server queue, 206
state-dependence in queues, 360
state-space collapse, 72
stationary
   -int. approx. pt. pr., 330
   increments, 99
   process
     FCLT with weak dep., 108
stationary versions, 467
statistical
   package, 2
   regularity
     for queues, 63
     goal to see, vii
     seen by playing a game, 1
     seen through plots, 4
     using prob. to explain, 16
     with jumps, 38, 43

test, 11, 51
steady-state distribution
    of RSLM, 268
    of RBM, 167
    of refl. Gaussian pr., 283
    of RFBM, 283
    of the workload, 142, 264
        refined hvy-traf. approx., 169
        Brownian approx., 167
        for $M/G/1$, 169
stochastic fluid network
    applic. of ref. map, 490, 495
    heavy-traffic limit for, 493
    queue-length process
        definition of, 497
        ref. map rep., 498
    stability of, 492
    to motivate ref. map, 460
stochastic integral
    convergence of seq. of, 126
    rep. of FBM, 125
    rep. of LSFM, 133
stochastic-process limit
    compactness approach, 386
    cont-map. approach, 84
    definition, 20, 24, 75
    for limit process, 277
    for plotted random walks, 25
    in E, 531
    in F, 537
    overview, 95–136, 225–242
    second, 272
    topic of book, vii
    with jumps, 43
stock prices, 49, 515
Strassen
    representation theorem, 375
strengthening mode conv., 409
strictly stable, 111
strong
    limit, w.p.1, 375
    parametric rep. in E, 526
strong dependence
    FBM FCLT for ct. pr., 236
    FCLT for, 90, 120–136
    frac. Br. motion, 100
    hvy-tld. on-off pds., 231, 280
    in comm. ntwks., x, 231, 243
    in commun. ntwks., 279

    in fl. queue input, 493
    Joseph effect, 100
    negative, 101
    positive, 100, 101, 120
    quantification of, 100
    RFBM limit
        in fl. queue, 279
        in sg-svr. queue, 322
stronger topology, 371
strongly connected elt. of E, 526
subbasis for a topology, 370
subordinator, 117
substochastic matrix, 461
superposition arrival process
    approximations for, 331
    CLT for processes, 226
    fixed number, 255, 301
    growing number, 259, 318
    in fluid queue, 245
    Poisson limit for, 259, 318
supremum map, 87, 435–438
    conv. pres., centering, 436
    criterion for joint conv., 437
    ctring in other direction, 438
    Lipschitz properties, 436
    used in limits of plots, 25
switching u and r in par. rep. of inverse
    fct., 444

$T_{a,b}$, first-passage time, 172, 176
$T(u) \equiv u^+ + Qu^- : \mathbb{R}^k \to \mathbb{R}^k$, 476
tail-prob. asymptotics, 156
TCP, see Transmission Control Protocol
theorem
    Arzelà-Ascoli, 388, 390
    Banach-Picard fixed-point, 467
    bounded convergence, 383
    central limit, 20, 102
        k-dim., 105
    continuous-mapping, 84, 86, 87
    convergence-together, 54, 379
    Donsker, 24, 102
        k-dim., 105
    functional central limit, see FCLT
    Helly selection, 413
    joint-convergence
        for indep. ran. elts., 378
        one limit is determ., 379
    Kolmogorov extension, 385, 388

Lamperti self-similarity, 98
Lipschitz-mapping, 85
Portmanteau, 373
Prohorov, rel. comp., 387
Skorohod representation, 78, 85
Strassen representation, 375
Tychonoff, 370
Urysohn embedding, 371
thick
   graph, 395
   range, 395
thin
   graph, 395
   range, 395
tightness
   characterizations for D, 425
   moment criterion for, 389
   of a set of prob. measures, 387
   of one prob. meas., 374, 376
   on product spaces, 390
time scales
   for a gambling house, 12
   for polling models, 70
   for queueing performance, 516
   in communication networks, 323, 325
   in on-off source models, 244
   in priority queues, 191
   intermediate, 70, 518
   need for time-dep. analysis, 159
   of fluctuations, 6, 244
   of traffic shaping, 325
   separation of, ix, 70, 191, 322, 325
   slowly chging. envt., 70
   snapshot principle, 187
   three important, 516
   variability at different, 319
time-varying arrival rates, 220
topological
   equivalence, 370
   space, *see* space, topological
   vector space, 84
topologically complete, 371, 374, 380, 389, 413
topology, 76, 368
   L_p space, 383
   M, J, S, W, *see* $M, \ldots$
   basis for a, 369, 377
   Euclidean, 369
   nonuniform, 385

   of a commun. net., 368
   of weak convergence, 372, 374
   product, 83, 370, 377
   relative, 370, 374
   strong, standard, 83
   stronger, finer, 83, 371, 383
   subbasis for a, 369, 370
   uniform, 385
   weak, 83
totally bounded, 371
traffic intensity, 56, 153
traffic shaping, 325
translation scaling vector, 96
Transmission Control Protocol, 67
triangle inequality, 76
two-sided ref. map, 57, 143, 505

U, uniform ran. var., 1, 31
$u(x_1, x_2, t, \delta)$, unif. dist. fct., 401
uncentered random walk, 4
uniform
   distance functions, 401
   distribution, 1
   metric, 79, 393
      on E, 525
   mixing conditions, 108
   order statistic, 12, 53
   random number, 1
   random variable, 1
unit square, 2
units on the axes, 2, 4, 5
unmatched jumps, xi
   for queue-length process, 205
   examples, 193–224
   hvy-tld. renewal pr., 203
   need for diff. top., xi, 80, 381
upper semicontinuity
   of max. abs. jump fct., 422
   preservation by infimum, 463
useful functions, 86, 427–456

Var, variance, 21
$v(x; A)$, modulus of cont., 393
$v(x_1, x_2, t, \delta)$, unif. dist. fct., 401
$v_{t_1,t_2}^{a,b}(x)$, visits to strip $[a, b]$, 412
variability function, 327
volatility of stocks, 51

WJ_1 topology, 83

WM_1
  conv. characteriz. of, 408
  topology, 83, 396
WM_2
  convergence
    characterizations of, 421
    inherit jumps from, 422
    pres. within bnding fcts., 423
  param. rep., 419
  topology, 416
$w_s(x,t,\delta)$, SM_1 oscil. fct., 402
$w_I \equiv w_I(m,\beta,c_a^2,c_s^2,z)$, 362
$w_w(x,t,\delta)$, WM_1 oscil. fct., 402
$\bar{w}_s(x_1,x_2,t,\delta)$, SM_2 osc. fct., 402
$\bar{w}_w(x_1,x_2,t,\delta)$, WM_2 osc. fct., 402
$\bar{w}_s(x,\delta)$, SM_2 osc. fct., 404
$w_w(x,\delta)$, osc. fct., 408
$w'(x,\delta)$, oscil. fcts. for compactness, 424
waiting time, with priorities, 190

weak
  convergence, 77, 373
  dependence, 107
  parametric rep. in E, 526
weakly connected elt. of E, 526
Weibull tail, 66, 285
weighted-supremum norm, 384
work-conserving service policy, 69, 188
workload
  factor, 362
  process in a queue, 55, 140

X-continuity set, 375
$\tilde{x}$, set-val. fct. in E, 524
$\hat{x}$, elt. of F, 533

$\mathbf{Z}_H \equiv \{\mathbf{Z}_H(t) : t \geq 0\}$, FBM, 125
$\mathbf{Z}_{H,\alpha}$, LFSM, 133
z_Q, asymp. peakedness, 354